T0390394

Silicon in Plants

Advances and Future Prospects

Silicon in Plants

Advances and Future Prospects

Edited by

Durgesh Kumar Tripathi
Center of Advance Study
Department of Botany
Banaras Hindu University
Varanasi, India

Vijay Pratap Singh
Government Ramanuj Pratap Singhdev Post Graduate College
Baikunthpur, Korea-497335
Chhattisgarh, India

Parvaiz Ahmad
University of Kashmir
S.P. College, India

Devendra Kumar Chauhan
University of Allahabad, India

Sheo Mohan Prasad
Ranjan Plant Physiology and Biochemistry Laboratory
Department of Botany
University of Allahabad, India

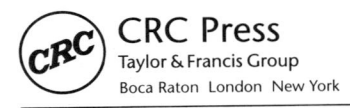

CRC Press
Taylor & Francis Group
Boca Raton London New York

CRC Press is an imprint of the
Taylor & Francis Group, an **informa** business

CRC Press
Taylor & Francis Group
6000 Broken Sound Parkway NW, Suite 300
Boca Raton, FL 33487-2742

© 2017 by Taylor & Francis Group, LLC
CRC Press is an imprint of Taylor & Francis Group, an Informa business

No claim to original U.S. Government works

Printed on acid-free paper
Version Date: 20160225

International Standard Book Number-13: 978-1-4987-3949-8 (Hardback)

Visit the Taylor & Francis Web site at
http://www.taylorandfrancis.com

and the CRC Press Web site at
http://www.crcpress.com

Contents

Preface

In the present era, rapid industrialization and urbanization have resulted in unwanted physiological, chemical, and biological changes in the environment that have had harmful effects on crop quality and productivity. This situation is further worsened by the increasing demands for food by the ever-increasing population. This has forced plant scientists and agronomists to look forward for alternative strategies to enhance crop production, as well as produce safer foods from a health point of view. Biotic and abiotic stresses are major constraints to crop productivity and have become important challenges to the scientists working on the agricultural and agronomical aspects because both considerably reduce agriculture production worldwide per year.

Silicon is the most available element in the earth crust (28.8%) after oxygen, and it is a sister element of carbon. It has several significant roles in agriculture, mainly in plant pathology, stress physiology, nanotechnology, biotechnology, and plant physiology. From the beginning of the nineteenth century, scientific groups have worked on silicon applicability and its various aspects in plant biology and agriculture sciences and have shown various benefits in plant life. However, there is a lack of availability of recently compiled works on silicon and its role in regulating many physiological and biochemical processes under stress, as well as nonstress, conditions. Therefore, in this book we have tried to compile the recent advances made worldwide in different leading laboratories regarding the role of silicon in plant biology in order to make these outcomes easily accessible to academicians, researchers, industrialists, and students in one place. *Silicon in Plants: Advances and Future Prospects* summarizes the diverse beneficial role of silicon against the different stresses on crop production. This book also relates how to obtain a maximum yield through the addition of silicon to limited resources.

The book is composed of 19 chapters. Chapter 1 deals with how Si mitigates the negative effects of selected abiotic stresses on plants. Chapter 2 provides full insight related to the role of Si in enhancing the availability of various nutrients, considering both the extrinsic and intrinsic factors, to mitigate their deficiencies in plants. Chapter 3 deals with the mechanism of nutrient regulation by Si in mobilization and utilization during salt stress. Particularly, deleterious effects of NaCl on the deficiency of K, Ca, and N, and either or both deficiency and toxicity on P, B, Fe, Cu, Zn, and Mn are covered in this chapter. Chapter 4 discusses the tolerance mechanism of plants, with special reference to the photosynthetic machinery subjected to different abiotic stresses. Chapter 5 provides a synthesis of the current knowledge regarding Si-mediated alleviation of metal stress in plants and aims to integrate the effects of Si in the soil and the physiological reactions mediated by Si in plants exposed to heavy metal stresses. Chapter 6 explains the application of nanosilicon in agriculture systems, which has provided new solutions to problems in plants and food science to enhance the quality of plant products. It may help plants absorb more nutrients, and it is also applied to mitigate stress conditions. The main objective of Chapter 7 is to summarize research on the effects of Si supplementation on plant growth and changes in stress resistance under abiotic and biotic stress conditions, as well as to evaluate Si uptake and deposition in floricultural crops. Chapter 8 discusses the agronomic, physiological and biochemical, and genetic and proteomic responses of different crop genotypes in the alleviation of salt stress that result by supplying Si. Further, advancements with a combination of molecular and genetic studies in Si transport and functions in different crop genotypes will help understand its complete role under abiotic stress conditions like salinity. Chapter 9 reveals the Si-mediated regulations of genes and secondary metabolites under abiotic and biotic stress. Chapter 10 demonstrates the role of Si in plant roots and their response to elevated concentrations of heavy metals, like Cd, Zn, Cr, and Cu, and toxic concentrations of other metals and metalloids, like Al, Mn, As, and Sb. Chapter 11 deals with silicon uptake and translocation mechanisms in plants, recent advances, and the future perspective. Chapter 12 briefly discusses the mechanisms involved in the alleviation of chromium toxicity by the application of silicon under chromium toxicity. Chapter 13 describes the role of silicon in plants, as well as its present scenario and future prospects. Chapter 14 discusses the beneficial impact of silicon, which could reduce Cd stress in plants by complexation and coprecipitation and deposition of Cd with Si in different plant parts. Chapter 15 is based on the positive roles of silicon in improving drought tolerance in plants, and a variety of genes are also overexpressed by silicon in many drought-stressed plants. Chapter 16 demonstrates that because of all its beneficial effects, silicon should be included as one of the important nutrients in modern crop fertilization in the near future. Chapter 17 reveals the beneficial impact of silicon in the regulation of the antioxidant defense system against abiotic stresses in plants. Chapter 18 discusses silicon nutrition and rice crop improvement in Iran, including recent advances and future perspectives. Chapter 19 demonstrates the efficacy of silicon against aluminum toxicity in plants and the mechanisms involved.

This volume has a wide range of information regarding the role of silicon in plants and their growth and development, physiological and molecular responses, and responses against various abiotic stresses. All the chapters in this book have been published keeping intact the authors' justifications; consequently, appropriate editorial changes were made wherever they were found compulsory. We tried our best to gather information on different aspects related to the silicon and plant intraction research for this volume; however, there is still a possibility of some errors creeping into the book, for which we seek readers' indulgence and feedback. We are very thankful to the authors for their valuable contribution. We are also very thankful to CRC Press, Taylor & Francis Group, particularly Randy Brehm, sectional editor of biological science; Amanda Parida, editorial assistant of chemical and life sciences; and Jill J. Jurgensen, senior project coordinator, editorial project development, all of whom were directly or indirectly associated with us in this project, for their constant help, valuable suggestions, and efforts in bringing out the timely publication of this volume.

Durgesh Kumar Tripathi
Vijay Pratap Singh
Parvaiz Ahmad
Devendra Kumar Chauhan
Sheo Mohan Prasad

About the Editors

Dr. Durgesh Kumar Tripathi works as a UGC Dr. D. S. Kothari Postdoctoral Fellow in the Centre of Advanced Study in Botany, Banaras Hindu University, Varanasi, India. Dr. Tripathi earned his D. Phil. degree from D. D. Pant Interdisciplinary Research Laboratory, Department of Botany, University of Allahabad, Allahabad, India. His main research area is Crop stress physiology, Plant biotechnology, Agro-nanotechnology, Plant molecular biology, Phytolith, and applications of Laser-spectroscopy for the study of several plant materials. His research interest is to obtain novel abiotic stress tolerance mechanisms to plants. He has published more than 50 national and international research papers, review articles and several book chapters in international books. He is a life member of several academic and professional societies and an editor and reviewer of several international journals of repute.

Dr. Vijay Pratap Singh is an assistant professor, Department of Botany, Government Ramanuj Pratap Singhdev Post Graduate College, Baikunthpur, Koriya, an affiliated degree college of Sarguja University, Chhattisgarh, India. Dr. Singh earned his D. Phil. degree from the University of Allahabad, Allahabad, India, in the study of oxidative stress and antioxidant systems in some cyanobacteria simultaneously exposed to ultraviolet B and heavy metals. He has authored 49 publications, as well as editorials, in international journals of repute. His research interest is abiotic stress tolerance in cyanobacteria and plants. Dr. Singh is also an editor and reviewer of several international journals of repute.

Dr. Parvaiz Ahmad is senior assistant professor in the Department of Botany at Sri Pratap College, Srinagar, Jammu and Kashmir, India. He completed his postgraduation in botany in 2000 from Jamia Hamdard, New Delhi, India. After earning a doctorate degree from the Indian Institute of Technology, Delhi, India, he joined the International Centre for Genetic Engineering and Biotechnology, New Delhi, in 2007. His main research area is stress physiology and molecular biology. He has published more than 50 research papers in peer-reviewed journals and 35 book chapters. He is also an editor of 14 volumes (1 with Studium Press Pvt. India Ltd., 9 with Springer, 3 with Elsevier, and 1 with John Wiley). He is a recipient of the Junior Research Fellowship and Senior Research Fellowship by the Council of Scientific and Industrial Research, New Delhi, India. Dr. Parvaiz has been awarded the Young Scientist Award under the fast-track scheme in 2007 by the Department of Science and Technology, India. Dr. Parvaiz is actively engaged in studying the molecular and physiobiochemical responses of different agricultural and horticultural plants under environmental stress.

Prof. Devendra Kumar Chauhan is a professor in the Department of Botany, University of Allahabad, Allahabad, India. Professor Chauhan earned his M. Sc. and D. Phil. degrees from the University of Allahabad. He has published more than 100 research papers, review articles, and book chapters in journals of national and international repute and has also published one edited book. His main research areas are agro-nanotechnology, anatomy, stress physiology, evolutionary botany, morphology, and biodiversity. Professor Chauhan is a member of the National Academy of Science, Allahabad, India, and a life member and fellow of several academic and professional societies. He is also an editor and reviewer for several international journals of repute.

x

Prof. Sheo Mohan Prasad is a professor in the Department of Botany, University of Allahabad, Allahabad, India. Professor Prasad earned his academic degrees from Banaras Hindu University, Varanasi, India. He has authored more than 133 scientific publications. His main area of research is the physiology and biochemistry of plants, as well as cyanobacteria under abiotic stresses, for example, ultraviolet B, heavy metals, pesticides, temperature, salinity, and high light intensity. Professor Prasad is an editor and reviewer of several international journals of repute.

Contributors

Muhammad Adrees
Department of Environmental Sciences
    and Engineering
Government College University
Faisalabad, Pakistan

Parvaiz Ahmad
Department of Botany
Sri Pratap College
Srinagar, India

Rehan Ahmad
Department of Environmental Sciences
    and Engineering
Government College University
Faisalabad, Pakistan

Zahoor Ahmad
Department of Agronomy/Crop Physiology
University of Agriculture
Faisalabad, Pakistan

Fatima Akmal
Institute of Soil and Environmental Sciences
University of Agriculture
Faisalabad, Pakistan

Shafaqat Ali
Department of Environmental Sciences
    and Engineering
Government College University
Faisalabad, Pakistan

Arkadiusz Artyszak
Department of Agronomy
Warsaw University of Life Sciences (SGGW)
Warsaw, Poland

Gausiya Bashri
Ranjan Plant Physiology and Biochemistry
    Laboratory
Department of Botany
University of Allahabad
Allahabad, India

Boris Bokor
Faculty of Natural Sciences
Comenius University in Bratislava
Bratislava, Slovakia

Devendra Kumar Chauhan
D. D. Pant Interdisciplinary Research
    Laboratory
Department of Botany
University of Allahabad
Allahabad, India

Karina Patrícia Vieira da Cunha
Federal University of Rio Grande do Norte
Department of Civil Engineering
Natal, Rio Grande do Norte, Brazil

Clístenes Williams Araújo do Nascimento
Federal Rural University of Pernambuco
Department of Agronomy
Recife, Pernambuco, Brazil

Nawal Kishore Dubey
Centre of Advanced Study in Botany
Department of Botany
Banaras Hindu University
Varanasi, India

Allahyar Fallah
Rice Research Institute of Iran
Amol, Iran

Fakhir Hannan
Department of Environmental Sciences
    and Engineering
Government College University
Faisalabad, Pakistan

Muhammad Ibrahim
Department of Environmental Sciences
    and Engineering
Government College University
Faisalabad, Pakistan

Muhammad Iqbal
Department of Environmental Sciences
    and Engineering
Government College University
Faisalabad, Pakistan

Byoung Ryong Jeong
Division of Applied Life Science (BK21 Plus)
and
Institute of Agriculture and Life Science
and
Research Institute of Life Science
Gyeongsang National University
Jinju, South Korea

Hinnan Khalid
Institute of Soil and Environmental Sciences
University of Agriculture
Faisalabad, Pakistan

Muhammad Daud Khan
Department of Biotechnology and Genetic
    Engineering
Kohat University of Science and Technology
Kohat, Pakistan

Hemmat Khattab
Botany Department
Faculty of Science
Ain Shams University
Cairo, Egypt

Chung Ho Ko
Division of Applied Life Science (BK21 Plus)
Gyeongsang National University
Jinju, South Korea

Katarzyna Kucińska
Department of Agronomy
Warsaw University of Life Sciences (SGGW)
Warsaw, Poland

Jitendra Kumar
Ranjan Plant Physiology and Biochemistry
    Laboratory
Department of Botany
University of Allahabad
Allahabad, India

Bishwajit Kumar Kushwaha
D. D. Pant Interdisciplinary Research Laboratory
Department of Botany
University of Allahabad
Allahabad, India

Denis Líška
Department of Plant Physiology
Faculty of Natural Sciences
Comenius University in Bratislava
Bratislava, Slovakia

Zuzana Lukačová
Department of Plant Physiology
Faculty of Natural Sciences
Comenius University in Bratislava
Bratislava, Slovakia

Abinaya Manivannan
Division of Applied Life Science (BK21 Plus)
Gyeongsang National University
Jinju, South Korea

Sajid Masood
State Key Laboratory of Soil and Sustainable
    Agriculture
Institute of Soil Science
Chinese Academy of Sciences
Nanjing, China

Noorani Mir
Institute of Soil and Environmental Sciences
University of Agriculture
Faisalabad, Pakistan

Sowbiya Muneer
Division of Applied Life Science (BK21 Plus)
Gyeongsang National University
Jinju, South Korea

Muhammad Nadeem
Boreal Ecosystem Research Initiative
Grenfell Campus Memorial University
    of Newfoundland
Corner Brook, Newfoundland, Canada

and

Department of Environmental Sciences
COMSATS Institute of Information Technology
Vehari, Pakistan

Asif Naeem
Institute of Soil and Environmental Sciences
University of Agriculture
and
Soil Science Division
Nuclear Institute for Agriculture and Biology
    (NIAB)
Faisalabad, Pakistan

Yoo Gyeong Park
Institute of Agriculture and Life Science
Gyeongsang National University
Jinju, South Korea

Anuradha Patel
Ranjan Plant Physiology and Biochemistry
    Laboratory
Department of Botany
University of Allahabad
Allahabad, India

Sheo Mohan Prasad
Ranjan Plant Physiology and Biochemistry
    Laboratory
Department of Botany
University of Allahabad
Allahabad, India

Abdul Qadir
COMSATS Institute of Information Technology
Department of Environmental Sciences
Abbottabad, Pakistan

Farooq Qayyum
Department of Soil Science
Bahaudin Zikria University
Multan, Pakistan

Hina Rizvi
Department of Environmental Sciences and
    Engineering
Government College University
Faisalabad, Pakistan

Muhammad Rizwan
Department of Environmental Sciences
    and Engineering
Government College University
Faisalabad, Pakistan

Muhammad Sabir
Institute of Soil and Environmental Sciences
University of Agriculture
Faisalabad, Pakistan

Muhammad Shahzad
COMSATS Institute of Information Technology
Department of Environmental Sciences
Abbottabad, Pakistan

Anita Singh
Ranjan Plant Physiology and Biochemistry
    Laboratory
Department of Botany
University of Allahabad
Allahabad, India

Madhulika Singh
Ranjan Plant Physiology and Biochemistry
    Laboratory
Department of Botany
University of Allahabad
Allahabad, India

Shikha Singh
Ranjan Plant Physiology and Biochemistry
    Laboratory
Department of Botany
University of Allahabad
Allahabad, India

Shweta
D. D. Pant Interdisciplinary Research
    Laboratory
Department of Botany
University of Allahabad
Allahabad, India

Swati Singh
D. D. Pant Interdisciplinary Research
    Laboratory
Department of Botany
University of Allahabad
Allahabad, India

Vijay Pratap Singh
Government Ramanuj Pratap Singhdev Post
    Graduate College
Chhattisgarh, India

Milan Soukup
Department of Plant Physiology
Faculty of Natural Sciences
Comenius University in Bratislava
Bratislava, Slovakia

Prabhakaran Soundararajan
Division of Applied Life Science (BK21 Plus)
Gyeongsang National University
Jinju, South Korea

Durgesh Kumar Tripathi
Centre of Advanced Study in Botany
Banaras Hindu University
Varanasi, India

Marek Vaculík
Department of Plant Physiology
Faculty of Natural Sciences
Comenius University in Bratislava
and
Institute of Botany
Slovak Academy of Sciences
Bratislava, Slovakia

Miroslava Vaculíková
Institute of Botany
Slovak Academy of Sciences
Bratislava, Slovakia

Maqsooda Waqar
Institute of Soil and Environmental Sciences
University of Agriculture
Faisalabad, Pakistan

Muhammad Zia-ur-Rehman
Institute of Soil and Environmental Sciences
University of Agriculture
Faisalabad, Pakistan

1 Mechanisms of Silicon-Mediated Alleviation of Abiotic Stress in Plants
Recent Advances and Future Perspective

Denis Líška, Milan Soukup, Zuzana Lukačová, Boris Bokor, and Marek Vaculík

CONTENTS

ABSTRACT

Abiotic stress factors negatively influence crop production all over the world. The human population is globally increasing every year. Therefore, to ensure sufficient food and feed, plants need to be grown on sites where they are exposed to negative abiotic stresses like heavy metals and other toxic elements, drought, and increased salinity. The mechanisms that help plants to overcome these stresses are intensively studied. Silicon (Si) was shown to be a promising agent to alleviate various forms of biotic as well as abiotic stresses on plants. Therefore, the prospective use of Si in agriculture is gaining attention. This chapter reveals how Si mitigates the negative effects of selected abiotic stresses on plants.

1.1 INTRODUCTION

Research focused on the impact of various abiotic stresses has gained importance in the field of plant science in recent years. An increasing demand for natural resources is joined with extensive mining activity and metal processing, which represents the greatest source of metal pollution of air, water, and soils. Similarly, industrial activities and intense traffic contribute to an increased level of heavy metals and toxic elements in the environment. Another stress factor for plants is high salinity. It has been estimated that more than 800 million ha of land on earth is salt affected, representing more than

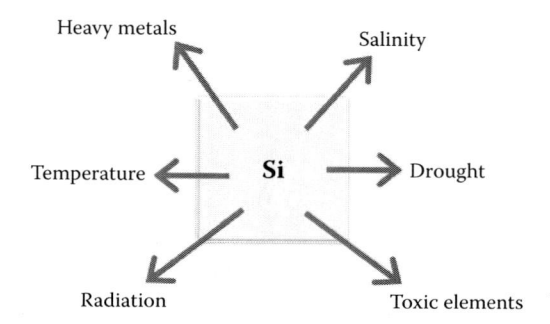

FIGURE 1.1 The most common types of abiotic stress factors that can be alleviated by Si application.

6% of the world's total land area. Of the current 230 million ha of irrigated land, about 45 million ha is salt affected (19.5%). Of the 1500 million ha of land farmed by dryland agriculture, 32 million ha (2.1%) is affected by secondary salinization (FAO 2005). The detrimental effects of high salinity on plants can be observed at the whole plant level, as decreases in yield productivity may lead to plant death. Together with salinity, a lack of water during the growth season is another limiting factor for agriculture. This is mainly a problem for feeding humans and grazing animals in many dryland regions of the world, like sub-Saharan Africa (Sahel zone); however, due to increasing disturbances between dry and rainy seasons caused by climatic change, these problems have also recently been observed in Europe. Therefore, there is an active search for possibilities on how to decrease all of the above-mentioned abiotic stresses that negatively affect agriculture and decrease crop production.

In recent decades, silicon was shown to decrease the harmful effect of various stresses on plant growth and development. Besides the whole spectra of biotic stressors, like pathogens or microorganisms (e.g., Ma 2004; Van Bockhaven et al. 2013), Si was shown to ameliorate the negative effect of the whole spectra of abiotic stresses, including heavy metals, salinity, and water imbalance in plants (e.g., Epstein 1999; Guntzer et al. 2012; Adrees et al. 2015). In this chapter, the role of Si in the mitigation of those abiotic stresses that mainly influence agriculture and crop productivity (Figure 1.1) will be discussed.

1.2 SILICON AND HEAVY METALS

Ninety-two chemical elements in the earth crust have been identified, and 17 of these elements are known to be essential to all plants (Williams and Salt 2009). The essential elements that are required in amounts greater than 0.1% of dry biomass are known as macronutrients (C, H, O, N, S, P, Ca, K, and Mg), and those elements required in amounts less than 0.01% of dry biomass are referred to as microelements or trace elements (Fe, B, Mn, Zn, Cu, Mo, Ni, and Cl). Moreover, those elements that improve the growth and development of plants and are essential to some taxa, but not in general, are considered beneficial elements and include Na, Co, Al, Se, and Si (Williams and Salt 2009). Mineral nutrients are essential components of every living cell and are used to regulate the electrochemical balance of cellular compartments. They may react as cofactors in various biochemical reactions and are structural components in biological molecules and complexes (Baxter 2009). For plants, it is necessary to have these elements in optimal concentrations for normal growth and development. Excesses and deficiencies of macronutrients and micronutrients can have serious effects on plant metabolism, affecting the type and amounts of metabolites. In crops, this has an impact on fruit and seed nutritional quality and taste, yield, processing and storage, and pathogen resistance (Williams and Salt 2009).

With the exception of carbon dioxide, which is absorbed from the air through the leaves, all other elements, including metals, are taken up from the soil solution by the roots. To cope with a shortage of elements in the rhizosphere, several mechanisms in plants have evolved to increase the availability of metals in the root environment, for example, by lowering the pH through exudation of protons and organic acids by roots (Kinraide et al. 2005), or by excretion of metal-complexing agents (Blaylock et al. 1997). On the other hand, plants also use several mechanisms to deal with an excess of metals. Deciduous plants store some elements in those parts that can later be shed, or toxic doses of metals can be deposited in special compartments of plant cells, like vacuoles or trichomes, and in such a manner plants render the metal metabolically inactive (Callahan et al. 2006). A second alternative may also provide a reserve of metals in times of low supply, when the plant is able to remobilize and transport the required metal from one tissue to another. Studies on the interaction of genes, elements, and the environment may be advanced by performing genetics and modeling that further determine the mechanism through which plants adapt to different environments and explain the whole elemental and ionic composition of the plant body. Plants are therefore well adapted to various environmental conditions and different element supplies, having evolved mechanisms to deal with conditions of deficiency or threshold (Singh et al. 2013).

Many studies of plant nutrition have shown that some mineral elements are toxic in higher concentrations. In modern high-production conditions, substantial nutrients are removed from the soil and then, to prevent plant deficiencies, various prosperous elements, including metals, can be added back to the soil in the form of fertilizers. Secondarily, some elements and metals are added back to the soil through different human activities, such as metallurgy, traffic, and wastewater (Vojtáš 2000; Clemens 2006). Soil pH has a large influence on the availability of these elements to plants (Broadley et al. 2007; Palmer and Guerinot 2009). Acid rains increase the acidity of the soils, which leads to higher metal mobility and solubility, presenting phytotoxic problems (Hutchinson and Whitby 1977). Metals are absorbed primarily as inorganic ions and are kept as a nontoxic form in the root system or through the vascular system are translocated to the shoots (Peer et al. 2006). If excess metals are present in the soil, plant growth may be diversely affected. Metal toxicity can create reactive oxygen species (ROS) that damage metabolism and cell compartments (Yadav 2010). Those changes are seen as chlorotic and necrotic spots on the leaves (Lukačová Kuliková and Lux 2010).

Müller (2007) and Housecroft and Sharpe (2008) characterized and distinguished metals from nonmetals by their physical properties—an electrical resistance and the ability to conduct heat that is directly proportional to temperature, malleability, ductibility, and even luster. The term *metals* refers to elements with very good electric conductance (property declines with decreasing temperature) and that exhibit an electrical resistance that is proportional to the absolute temperature (Lyubenova and Schröder 2010). Metals can be classified as hard and soft acids and bases, meaning that hard acids form strong bonds with hard bases, whereas soft acids form bonds with soft bases. Thus, some metals occur in the earth crust as ores of oxides and carbonates, whereas some elements can be found in the form of sulfides (Lyubenova and Schröder 2010). The reason is that positively charged ions are able to accept electrons—these chemical properties of metal ions determine their ability to form complexes (Pearson 1968). Nieboer and Richardson (1980), based on ionic and covalent indices, classified metals and metalloids as hard acceptors (group A), soft acceptors (group B) and borderline (e.g., Ni^{2+}, Mn^{2+}, and V^{2+}). Ions in group A (e.g., Mg^{2+}, Ca^{2+}, Al^{3+}, and As^{3+}) are expected to interact with oxygen-containing ligands, and ions in group B (e.g., Ag^+, Tl^+, Hg^{2+}, and Cd^{2+}) are often very toxic and form stable bonds with S- and N-containing ligands.

Element interactions can affect the absorption and bioavailability of other nutrients for plants. Mineral elements with chemical similarities (in size and charge) can compete for the same transport proteins or other uptake mechanisms (Singh et al. 2013). Singh et al. (2013) studied a transport protein that is usually highly specific for a certain ion, but its specificity is often not absolute. Because of this, the coordinated accumulation of chemically similar element species may occur when the shared-membrane transport proteins or chelating metabolites are up- or downregulated (Baxter and Dilkes 2012). For example, iron transporter IRT1 has a broad substrate range and is responsible for the transport of manganese, zinc, cobalt, and cadmium, in addition to iron (Vert et al. 2003; Kerkeb et al. 2008; Puig and Peñarrubia 2009). Also, the broad substrate range of the CAX2 transporter responsible for Ca transport allows plants to accumulate other metal ions—Cd and Mn—and, moreover, increases the tolerance of the plants to Mn^{2+} stress (Hirschi et al. 2000). In these cases, it has been observed that alteration in the movement of one ion is influenced by the other ion. Thus, the transport of all the elements in the cells may be connected and correlated with each other. This correlation may vary with spatial and environmental factors, while some correlations between elements may exist regardless of the tissue, species, or environment (Singh et al. 2013).

Marschner (1995) lists two important facts: (1) the effect of any substance on a living system is always dependent on the available concentration to the cells, and (2) several ions are crucial to the metabolism of cells at low concentrations, but are toxic at high concentrations. Micronutrients (Co, Cu, Fe, Mn, Mo, Ni, and Zn) have essential functions in plant cells. When the internal concentration of these ions in a biological system exceeds a certain threshold value, their toxic effects are clearly visible, and then they are commonly termed heavy metals (Appenroth 2010). Heavy metals are common environmental contaminants in industrialized societies. Lasat (2002) detailed that soil pollution by heavy metals varies from water or air pollution, because these substances persist in soil much longer than in other compartments of the biosphere. The major problem is in affecting the agricultural productivity of plants (Yadav 2010). Heavy metals are elements with metallic properties and an atomic number of >20. The most common soil pollutants are Pb, Hg, Cd, Zn, Cu, Cr, Co, Ni, and Mn (Yadav 2010).

Lead (Pb) is a bluish or silvery-gray metal and exists in many forms in natural sources throughout the world. Lead belongs to one of the most widely and evenly distributed trace metals. Contamination by Pb^{2+} is long term, because it is not biodegradable, and is caused by car exhaust, dust, and gases from various industrial sources. Most of the inorganic salts of Pb^{2+} have poor solubility in water (Tangahu et al. 2011). Contamination by Pb is a serious problem that damages human health, for example, brain damage and retardation (Cho-Ruk et al. 2006). A high level of Pb affects morphology, growth, and photosynthetic activities, as well as causes inhibition of enzyme activities, modifies water balance, and disturbs mineral nutrition (Sharma and Dubey 2005). Li et al. (2012) observed that addition of Si markedly decreased the proportion of exchangeable Pb, thus reducing Pb availability in the soil. As a result, Pb uptake by banana plants was decreased and the plant growth was improved. The addition of Si also increased the activities of peroxidase (POD), superoxide dismutase (SOD), and catalase (CAT) in banana roots (Li et al. 2012). Similarly, Yan et al. (2014) described the decreased Pb uptake into tobacco leaves due to the reduced Pb mobility and availability in the soil plant system caused by Si.

Cadmium (Cd) is a silver-white metal with toxic and carcinogenic influence to human and plant cells (Bertin and Averbeck 2006). Several human activities, such as traffic, metallurgy, or wastewater, cause a threshold of Cd in the soil (Clemens 2006). Cadmium is water soluble, and in acid soils and Cd^{2+} ions easily dissociated. These ions have a negative influence on the environment (Styk 2001) and also on human health, which affects the cardiovascular, endocrine, immune, reproductive, and motoric (itai-itai) systems (Kaji 2012; Al Bakheet et al. 2013). Plants influenced by Cd show a visible reduction in photosynthesis and water and nutrient uptake, and symptoms of injury reflected in terms of chlorosis, necrosis, growth inhibition, browning of root tips, and finally death (Wójcik and Tukiendorf 2004; Lukačová Kuliková and Lux 2010). Cadmium is toxic to most plants even at low concentrations (Prasad 1995; Lux et al. 2011). There are many records that Si can alleviate the negative effects of Cd on plant growth. A decrease in plant Cd uptake and translocation from root to shoot was suggested as one of the beneficial roles of Si in many species, including maize (Liang et al. 2005; Da Cunha et al. 2008), *Solanum nigrum* (Liu et al. 2013), and mangrove seedlings (Zhang et al. 2013). In maize, the alleviation of Cd toxicity was partially attributed to Si-enhanced cell wall elasticity and plasticity (Vaculík et al. 2009). Moreover, the addition of Si to a low Cd concentration does not change the distribution of the symplasmic and apoplasmic Cd concentrations in roots, but it decreases the symplasmic and increases the apoplasmic Cd concentration in shoots. This indicates that Si causes lower Cd availability for leaf cells, and that the cells are protected from the toxic effects of Cd (Vaculík et al. 2012). Also, the modification of apoplasmic barriers by Si takes place in the plant Cd uptake and root-to-shoot translocation (Vaculík et al. 2009, 2012; Vatehová et al. 2012). The mitigation effect of Si on Cd toxicity may also be caused by enhancement of the late metaxylem lignification and development of endodermal cell walls. This effect prevents Cd translocation into the shoot of plants (Lukačová et al. 2013). The alleviative effect of Si on Cd toxicity was also partially attributed to the changes in the activity of important antioxidative enzymes involved in scavenging free radicals formed by the presence of Cd in pakchoi (Song et al. 2009), maize (Lukačová et al. 2013), *Solanum nigrum* (Liu et al. 2013), and other species. A positive effect of Si on photosynthesis in plants treated by Cd was also reported (Mihaličová Malčovská et al. 2014), and recently, Vaculík et al. (2015) documented that improved thylakoid formation in bundle sheath's cell chloroplasts may contribute to Si-induced enhancement of photosynthesis and a related increase in the biomass production in C4 plants.

Zinc (Zn) is a bluish white lustrous metal that is an essential microelement involved in various physiological processes in plants. It is the 23rd most abundant element in the earth crust and the second most abundant transition metal in organisms after iron (Broadley et al. 2007). Because of the chemical similarity between Zn and Cd, these metals compete for the same or similar binding transporters. Thus, Zn deficiency is responsible for a higher accumulation of Cd in plants (Grant and Sheppard 2008). Increased soil Zn concentration due to industrial, mining, and agricultural activities causes toxicity symptoms in plants growing in such a polluted environment (Marschner 1995; Broadley et al. 2007). Higher levels of Zn in soil inhibit many plant metabolic functions, which results in retarded growth of roots and shoots, chlorosis in the younger leaves, and finally senescence (Ebbs and Kochian 1997). Lee et al. (1996) indicated the appearance of a purplish red color in leaves as a typical effect of Zn toxicity on plants. Da Cunha and do Nascimento (2009) studied the alleviation effect of Si on Zn phytotoxicity in maize plants grown on Cd- and Zn-enriched soil treated with doses of Si. The results indicated that the maize treated with Si presented not only a biomass increase, but also a higher metal accumulation. The authors concluded that the deposition of Si in the endodermis and pericycle of roots seems to play an important role in maize tolerance to Cd and Zn stress. Also, Kaya et al. (2009) reported a mitigation effect of silicon on maize plants grown at high zinc concentration. The authors concluded that the addition of Si to a high Zn concentration in solution may protect membrane permeability, thus mitigating Zn toxicity and improving the growth of experimental plants. Song et al. (2011) studied two rice cultivars (Zn sensitive and Zn tolerant) in growth parameters; the addition of Si to both cultivars alleviated the negative effect of Zn. Moreover, some antioxidant activities increase, whereas malondialdehyde (MDA) and hydrogen peroxide decreased in Zn + Si treatments of both cultivars in comparison to Zn treatment alone. The authors concluded that the Si-mediated alleviation of Zn toxicity is mainly attributed to the Si-mediated antioxidant defense capacity and membrane integrity, and that Si may play a role in the reduction of zinc root-to-shoot translocation. On the other hand, silicon does not always mitigate zinc toxicity in plants (Masarovič et al. 2012; Bokor et al. 2014). The effect of Si on high zinc supply was not significantly alleviating; the biomass production was not positively influenced in Zn + Si treatments. A combination of Si and Zn also negatively affected the activity of some antioxidant enzymes in sorghum (Masarovič et al. 2012), and in maize it resulted in decreased biomass and increased physiological stress, although the concentration of Zn was reduced due to Si in plant tissues (Bokor et al. 2014).

Copper (Cu) is a reddish orange color ductile chemical element with atomic number 29 and 29 known isotopes. This micronutrient plays an important role in carbon dioxide assimilation and ATP synthesis. Cu is also an essential component of various proteins of the photosynthetic system or respiratory electron transport chain (e.g., plastocyanin and cytochrome oxidase, respectively) (Demirevska-Kepova et al. 2004). Soil and environment contamination by Cu is caused by mining, smelting of Cu-containing ores, and enhanced industrial activities. The threshold of Cu in soil is cytotoxic to plant cells, induces oxidative stress, generates ROS, and causes injury that leads to plant growth retardation and leaf chlorosis (Lewis et al. 2001). Nowakowski and Nowakowska

(1977) observed that application of Si in Cu-contaminated conditions increased the biomass and water content of plants, and decreased the Cu concentrations in the shoots and roots of spring wheat compared to Cu stress alone. Also, Collin et al. (2014) examined Cu absorption, distribution, and toxicity and the role of Si in the bamboo. These authors found out that Si application did not significantly improve bamboo tolerance to Cu, but Si supplementation led to a visual improvement in the plants while modifying Cu speciation in aboveground parts by increasing the proportion of organic and inorganic CuS compounds. Also, Khandekar and Leisner (2011) studied the influence of Si to Cu toxicity and found that Si apparently allows plants to more efficiently respond to Cu toxicity by maintaining or upregulating Cu-binding molecules and increasing the expression of enzymes metabolizing free radicals. They suggested that these, as well as other Si-attributed mechanisms, help to reduce the Cu stress of leaves. They assumed that Si modulated Cu tolerance via more than one mechanism. In the recent study of Mateos-Naranjo et al. (2015), Si appears to play an important role in the Cu tolerance of *Spartina densiflora*, not by avoiding its uptake by roots, but via some mechanism that avoids Cu translocation from roots to leaves, resulting in a general reduction of the Cu-induced deleterious effect on the leaf photosynthetic apparatus.

Chromium (Cr) is steely gray, lustrous, hard, and brittle heavy metal with atomic number 24 (Brandes et al. 1956). Chromium is a serious environmental problem; it causes contamination of the soil and groundwater (Shanker et al. 2005). The major polluter of water and also the major producer of wastewater contaminated by Cr is the tanning industry. Chromium (IV) is a very toxic and carcinogenic element to human and plant cells. This toxicity has a deleterious effect on plant physiological processes such as photosynthesis—the inhibition of chlorophyll biosynthesis (Vajpayee et al. 2000), water relation, and mineral nutrition, and it also causes inhibition of plant growth, chlorosis, and nutrient and root injury (Sharma et al. 2003). Cr(VI) is a potentially hazardous phytotoxic environmental pollutant that is preferentially absorbed by plants through roots, and it induces toxicity symptoms like reduced growth, chlorosis, blackening of the root system, and some modifications in plant processes (Tripathi et al. 2015). Tripathi et al. (2015) studied the effect of Si on Cr(VI) toxicity. The authors detailed that Si addition to growth media significantly alleviated Cr toxicity and induced growth and photosynthetic activities in wheat. These findings correlated with reduction in Cr accumulation following Si addition. Zeng et al. (2011) studied the alleviative effect of Si on Cr toxicity in hydroponically cultivated rice plants. Silicon alleviates Cr toxicity mainly through inhibiting the uptake and translocation of Cr. Additionally, the role of Si in enhancing the capacity of defense against oxidative stress was also suggested. Ali et al. (2013) found a positive effect of Si on Cr toxicity in hydroponically cultivated barley. The results showed that Si application significantly enhanced plant growth and also increased photosynthetic parameters. Apparently, Si and Cr behaved antagonistically, indicating that Si could be a candidate for Cr detoxification in crops in Cr-contaminated soil.

Manganese (Mn) is a silvery-gray metal. The threshold dose of Mn in leaves causes a reduction of the photosynthetic rate (Kitao et al. 1997). It was observed that Mn toxicity caused chlorosis in younger leaves that was attributed to Mn-induced Fe deficiency. An excess of Mn inhibits the synthesis of chlorophyll by blocking a Fe-concerning process (Clarimont et al. 1986). As a common symptom of Mn toxicity, Wu (1994) reported necrotic brown spotting on leaves, petioles, and stems. Roots exhibiting Mn toxicity are commonly brown and sometimes crack (Foy et al. 1995). It is known that exogenous application of Si increased the biomass and decreased lipid peroxidation of cucumber under Mn toxicity. The application of Si also increased activities of enzymatic and nonenzymatic antioxidants (Shi et al. 2005). In Mn-stressed maize plants, Doncheva et al. (2009) found that callose may be involved in Si-mediated alleviation of Mn toxicity. Horst and Marschner (1978) reported that in bean plants, Si-enhanced Mn tolerance was associated with a diffuse distribution of Mn rather than a reduction of Mn uptake or translocation. Si pretreatment increased Mn concentration in the leaves compared with the nonpretreated plants under Mn excess (Doncheva et al. 2009).

Many of the toxic responses induced by heavy metals have been identified and classified as general stress responses (Appenroth 2010). Heavy metals cause oxidative damage to plants due to formation of ROS either directly through Haber–Weiss reactions (Wojtaszek 1997; Mithofer et al. 2004) or indirectly, including via interaction with the antioxidant system (Schützendübel et al. 2001; Srivastava et al. 2004). The alleviating effect of Si on metal toxicity is also shown in the increasing activities of antioxidant enzymes in some plant species (e.g., Song et al. 2011; Li et al. 2012; Lukačová et al. 2013); however, there are also reports on the negative effects of Si on the antioxidative system in plants (e.g., Masarovič et al. 2012; Bokor et al. 2014). In addition, the positive role of Si in the mitigation of heavy metal stress might be attributed to decreased availability, reduced uptake, and the translocation of metals from roots to shoots in relation to the changes in the root-and-shoot anatomy. Additional mechanisms responsible for heavy metal detoxification might be based on the reduced availability of heavy metals in cytosol and forced deposition of these harmful substances in cell walls, as well as changes in cell wall plasticity and elasticity.

1.3 SILICON AND OTHER TOXIC ELEMENTS

Besides heavy metals, there are other elements that can not be listed within this group due to their chemical properties. This group mainly represents elements like boron (B), selenium (Se), arsenic (As), and antimony (Sb), with metalloid character. The first two elements in this group are essential micronutrients important for the plant body in small doses; however, the latter ones are dangerous contaminants of the environment with no known positive function for plant growth. Below we discuss the role of Si in those plants exposed to elevated amounts of As and B.

Arsenic is a class 1 carcinogen and is toxic to all living organisms. Its toxicity has become a global concern owing to the increasing contamination of water, soil, and crops in many regions of the world. Contamination by As has occurred in some areas as a result of mining activities; the use of arsenical herbicides, insecticides, and wood preservatives; and irrigation with As-contaminated groundwater. Behind the anthropogenic sources, As is also widely distributed in the environment from natural sources. It is a component of about 200 minerals occurring in the lithosphere. A study by Nordstrom (2002) revealed that in recent decades, millions of people have suffered from As poisoning because of drinking As-contaminated water from shallow tube wells in South and Southeast Asia, but definitely it is a widespread problem all over the world. Its effective uptake and transport into plant parts used as a source of food presents a serious problem to humans. Food surveys have revealed that among cereals, paddy rice generally has the highest As content compared to others (Tao and Bolger 1998; Williams et al. 2007) because of the increased availability of soil As during anaerobic flooded conditions (Xu et al. 2008); the diet of many rice consumers is therefore especially under threat from As contamination (Meharg 2004; Williams et al. 2005).

Understanding how plants take up and metabolize As is important for developing mitigation measures to counter the problem of food chain contamination by As. The negative effects of As can be observed on the level of plant morphology and physiology. Arsenic inhibits root and shoot growth, decreases photosynthetic activity, alters carbohydrate and amino acid metabolism, and induces oxidative stress (Hoffmann and Schenk 2011; Finnegan and Chen 2012; Tripathi et al. 2012). Arsenic occurs naturally in inorganic and organic, trivalent, or pentavalent chemical species. The most abundant inorganic species are arsenite As(III) and arsenate As(V). Arsenite is considered to be more phytotoxic than arsenate (Zhao et al. 2010), but arsenic (V) is toxic because it is a nonfunctional phosphate analogue that disturbs phosphate assimilation (Meharg and Hartley-Whitaker 2002). The reduced form As(III) has a high affinity for the imidazolium nitrogen of histidine residues and for sulfhydryl groups in cysteine residues, both of which are essential for the structure and function of numerous proteins. Arsenite therefore interferes with redox signaling, induces oxidative stress, and affects protein function in general (Requejo and Tena 2006).

Physiological studies have shown different transport pathways for arsenite and arsenate; arsenate shares the transporters involving cotransport of phosphate or arsenate and protons (having a higher affinity for phosphate than for arsenate) (e.g., Asher and Reay 1979; Ullrich-Eberius et al. 1989;

Meharg et al. 1994). Arsenite uptake in rice shares the highly efficient silicon pathway of entry to root cells and efflux toward the xylem.

Two strategies can help counter the toxic effect of As: remove As from the environment or develop safe crops that can be grown in the presence of As contamination with minimized As uptake or translocation to edible plant parts. For both scenarios, understanding the details of As uptake and detoxification is necessary. Important research dealing with possibilities of how to decrease As uptake has recently been done. Evidence that some plant aquaporin channels can mediate arsenite influx has been obtained from various studies (Bienert et al. 2008; Isayenkov and Maathuis 2008; Ma et al. 2008). Ma et al. (2008) have identified the specific transporter OsNIP2;1, named Lsi1 because of its primary function as a silicon transporter (Ma et al. 2006). This represents a major pathway for the entry of As(III), but not As(V), into rice roots. It is probably due to two important similarities of arsenite and silicic acid: both have a high pKa (9.2 and 9.3 for arsenous acid and silicic acid, respectively), and both molecules are tetrahedral with similar sizes. Also, another Si carrier (efflux, Lsi2) affects the As(III) transport to the xylem and As accumulation in shoots (Ma et al. 2008).

The inhibition effect of As uptake by Si application based on their competition was reported in several studies (Bogdan and Schenk 2008; Seyfferth and Fendorf 2012; Fleck et al. 2013), mainly demonstrated in Si accumulator crops like rice and sugarcane. This implies that soils that contain high plant-available Si concentrations result in low plant As accumulation. However, the results of Lee et al. (2014) show that the concentration of As in soil solutions increased after Si applications due to competitive adsorption between As and Si on soil solids. The authors indicated that there was an initial aggravation in As toxicity before the beneficial effects of Si fertilization to rice were revealed. Lee et al. (2014) also did not observe any decrease of As concentrations in shoots and roots of rice seedlings by Si application. However, Bogdan and Schenk (2008) reported the negative relationships between Si concentrations in soil solutions and As concentrations in both polished grain and shoots of rice grown in six As uncontaminated and untreated soils. Similarly, Li et al. (2009) and Fleck and Schenk (2011) reported an increase in As concentration in soil solutions, but still a decrease in total As concentrations in straw, grains, and husk after the application of silica gel; they also showed that rice yields increased with Si supply. Ma et al. (2008), Guo et al. (2009), and Tripathi et al. (2013) suggested that external Si-mediated reduction in As(III) uptake by rice might be attributed to the direct competition between Si and As(III) during the uptake process. The role of Si in minimizing uptake and the root-to-shoot transport of metal ions has also been confirmed in seedlings of rice grown with arsenate (Guo et al. 2005; Tripathi et al. 2013).

The difference in the effect of Si application on As toxicity and uptake by rice seedlings between various reports can be explained by Si/As ratios in soil solutions, but it also depends on the different soils. According to Lee et al. (2014), in soils that have higher contents of iron oxides, much lower amounts of As were released into the solutions, and thus the As toxicity to rice seedlings was less obvious. The physiological mechanisms through which Si might alleviate As toxicity in plants remain poorly elucidated, but from other physiological studies, it is obvious that this alleviation is not only due to decreased As uptake or transportation. As shown by Sanglard et al. (2014), Si application is as an important target in an attempt not only to decrease As concentrations in edible plant parts, but also to improve the photosynthetic performance of rice leaves modified by As. The authors concluded that there is a genotype-dependent (wild-type [WT] and lsi1 mutant defective in Si uptake) pattern of As-Si interactions, and that the increase of Si levels was probably a result of growth inhibition associated with As (also see Yu et al. 2012), and thus concentrating Si levels in plant tissues. Marmiroli et al. (2014) measured tomato (*Solanum lycopersicum* L.) seed germination and seedling growth under arsenic or arsenate stress in plants treated with Si. Several cultivars were tested, and the authors concluded that As entry into the fruit varied from cultivar to cultivar. In the presence of Si supplementation, As uptake was reduced or increased, depending on the cultivar. Therefore, they supposed that intraspecific variations in As-Si applications might be also genetically determined.

Arsenic generates oxidative stress, and thus causes lipid degradation of various biomolecules. Application of Si (1 mM) significantly ameliorated the oxidative stress in a study by Tripathi et al. (2013), and effects were more significant in tolerant than sensitive rice in either the presence or absence of As(III). Addition of Si to As-exposed plants reduces the O_2^{-}, H_2O_2, electrical conductivity, and lipid peroxidation more in tolerant than in sensitive cultivars. The study demonstrated that addition of Si alleviates the As phytotoxicity by subsequent reduction of shoot As accumulation, and Si was responsible for an increase of antioxidant level in rice plants.

Boron (B) is an essential microelement for higher plants. The majority of the water-soluble B seems to be localized in the apoplasmic region as boric acid. The water-insoluble B is associated with rhamnogalacturonan II (RG-II), and the complex is ubiquitous in higher plants (Matoh 1997). To date, a primary function of B is undoubtedly its structural role in the cell wall; however, there is increasing evidence for a possible role of B in other processes, such as the maintenance of plasma membrane function and several metabolic pathways. There has been increasing knowledge about the molecular basis of B deficiency and toxicity responses in plants in recent years (Reid 2014). Boron in excess is toxic to plants and limits their growth (Alpaslan and Gunes 2001; Camacho-Cristobal et al. 2008). A high concentration of B may occur naturally, or it can be added to the soils from fertilizers or mining (Nable et al. 1990). Boron is also often found in high concentrations in association with saline soils (Alpaslan and Gunes 2001; Ben-Gal and Shani 2002) and in soils with high pH in semiarid regions.

There are several studies dealing with the ameliorative effect of Si on B toxicity. The formation of B-Si (boron–silicate) complexes in soil that reduces B availability is probably the first reason how Si alleviates B toxicity. Supplemental Si has been shown to reduce B root-to-shoot translocation in spinach plants (Gunes et al. 2007a), to increase shoot and root dry matter, and to increase the net photosynthetic rate under B deficiency (Liang and Shen 1994; Hanafy Ahmed et al. 2008).

Under stress caused by B ROS such as superoxide radicals (O_2^{-}), hydroxyl radicals (OH^{-}) and hydrogen peroxide (H_2O_2) are generated and accumulated in plants (Gunes et al. 2007b; Kaya et al. 2011); hence, several studies deal with possibilities of how to alleviate oxidative stress in the plant facing B. Soylemezoglu et al. (2009) studied the effect of Si on antioxidant and stomatal responses of two grapevine (*Vitis vinifera* L.) rootstocks grown in B toxic conditions and concluded that Si alleviates the adverse effects of B toxicity on grapevine rootstocks by preventing both oxidative membrane damage and translocation of B from roots to shoots or soil to plant, and also lowers the phytotoxic effects of B within plant tissues. Silicon treatment significantly affected the enzyme activities of rootstocks by lowering SOD and CAT but increasing ascorbate peroxidase (APX). Similarly, Kaya et al. (2011) achieved reduced SOD activity but unchanged POD activity in B + Si treatment in tomato plants. Behind this, in grapevine Si lowered MDA, H_2O_2, and proline content in the combined treatment, in comparison with plants facing B toxicity. Growth reduction caused by B toxicity was also partly alleviated by Si supply in the study by Soylemezoglu et al. (2009) in the grapevine. Interactive effects of Si and high B on growth and yield of tomato (*Lycopercison esculentum*) plants were studied by Kaya et al. (2011). They found that high B reduced dry matter, fruit yield, and chlorophyll (*Chl*) in tomato plants compared to the control treatment, but increased the proline accumulation. By addition of Si, the deleterious effects of high B on plant biomass, fruit yield, and chlorophyll concentrations were alleviated.

Two independent studies by Gunes et al. (2007b) and Inal et al. (2009) were done to investigate interactions between these two elements in typical Si-accumulating plants—wheat and barley. Both of them concluded that the contents of proline, H_2O_2, MDA, and lipoxygenase (LOX) activity of wheat grown in B toxic soil were significantly reduced by Si treatments. Also in this research, a decrease in the major antioxidant enzyme activities was noticed in the combined treatment, and the authors concluded that Si alleviates B toxicity of wheat by preventing oxidative membrane damage and also translocation of B from root to shoot or soil to plant Si.

Like Si, B is also taken up in the form of an undissociated molecule. In the 1990s, no competitive interaction was observed in the uptake of boron and silicon (Nable et al. 1990). In 2006, when Ma et al. identified an influx Si transporter (Lsi1), the authors showed its permeability to water, urea, and boric acid (Takano et al. 2006). However, physiological evidence has shown that transporters for B and Si are different, at least in rice (Tamai and Ma 2003). This indicates that Lsi1 is a highly specific transporter for silicic acid (Ma and Yamaji 2006, 2008).

1.4   SILICON AND WATER DEFICIENCY

Water deficiency and high salinity are the key factors intrinsic to water and osmotic stresses. They cause serious limitations in agricultural productivity and can lead to remarkable losses in crop yields (Ashraf 2010; Deinlein et al. 2014; Landridge and Reynolds 2015). Improving plant tolerance to drought and high salinity by breeding and genetic modifications brought several advances, but their applicability also possesses biological, legislative, and other limitations. The worldwide problem of arable land loss thus still requires more steps to be taken.

Plant water deficit occurs when the rate of transpiration exceeds the rate of water uptake. It is a component of several stresses, including drought, salinity, and low temperature (Bray 1997). Water deficit and drought stress reduce seed germination rates and plant growth and development (Farooq et al. 2009). Nonetheless, some aspects of a water deficit represent normal components of plant ontogenesis, such as the desiccation of seeds (Bray 1997). The response of plant organisms to drought exposure mostly depends on the inherent life strategy of the species and the duration and severity of the stress (Cruz de Carvalho 2008). At the cellular level, a water deficit results in an increase of solute concentrations, changes in cell volume and turgor, disruption of membrane shape and integrity, and denaturation of proteins (Bray 1997). During a water deficit, the generation of ROS is enhanced and may cause significant oxidative damages to plant cell membranes and organelles. In this manner, drought may disrupt crucial metabolic pathways and result in irreversible or even fatal consequences (Cruz de Carvalho 2008). The control of oxidation agent levels in cells and tissues is provided either enzymatically or by nonenzymatic antioxidant molecules, and their efficiency is highly correlated with plant stress tolerance (Kar 2011). The strategies of ROS scavenging vary among the species, and they are dependent on the age of the plant and both the duration and intensity of stress. Many times, this has led to contradictory observations of plant antioxidant stress responses, and therefore they should be interpreted specifically for species and environmental conditions (Cruz de Carvalho 2008).

Reduced water availability also generally results in limitations of total nutrient uptake, and thus also in several nutrient deficiencies. The aforementioned limitations occur either because of less ion mobility within a water-deficient environment or because of the limited amounts of energy available for active ion uptake. Moreover, further conversion of nutrients into utilizable forms may also be affected. Overall, probably the major negative effect of drought stress is the reduction of photosynthetic activity, which may cause remarkable developmental defects with subsequent yield losses, and thus may have a serious impact on agricultural production (Farooq et al. 2009).

The adaptations on drought stress basically include mechanisms of maximizing the water uptake and minimizing the water loss. Osmotic adjustment provided by organic solutes, such as proline and glycine betaine, is one of the key mechanisms to overcome the water deficit by maintaining water absorption and cell turgor (Cattivelli et al. 2008). However, energy costs for their biosynthesis are relatively high, and the drought-tolerant species mostly favor different strategies of drought stress avoidance (Munns 2002). Water loss from the aboveground organs is mostly prevented by the deposition of cuticle and waxes on the epidermal surface (Kerstiens 1996). Transpiration is then predominantly restricted to stomata and should be adjusted via regulation of stomatal conductance (Munns 2002). Moreover, in order to reduce water leakage to the soil, several anatomical barriers are also present in the roots. The deposition of hydrophobic substances such as suberin and lignin occurs in root exodermal and endodermal cell layers and restricts water movement from the plant to the surrounding environment (Steudle and Peterson 1998). Advances in crop breeding and genetic engineering seem to be relatively efficient means of crop modification to enable their growth in water-limiting conditions. However, both possess many intrinsic limitations (Ashraf 2010). Because there are currently no economically viable technological means to facilitate crop production under drought, new techniques and procedures are still demanded (Farooq et al. 2009).

Water deficiency and drought stress alleviation provided by silicon supplementation include several aspects, but not all of the mechanisms are clearly understood. The most reported effects are the decrease in stress-induced growth reduction, the improvement of water leakage barriers, and the restoration of relative water content and photosynthesis. Moreover, a tentative role of silicon-containing compounds in the osmotic adjustment should also be considered. Water leakage is mostly restricted by the formation of hydrophobic barriers in both aboveground and belowground organs. The fortification of cuticle by an additional silica layer remarkably decreases the rates of cuticular transpiration. Transpiration is then mostly restricted to stomata and should be projected in the increase of stomatal conductance (Ma 2004; Shen et al. 2010b), and an increase in total transpiration rates was also observed quite often (Shen et al. 2010a; Silva et al. 2012; Liu et al. 2014; Yin et al. 2014; Amin et al. 2016). In oil palm seedlings, silicon supplementation was reported to reduce even the stomatal density (Putra et al. 2015), and in banana it increases the thickness of leaf epidermis (Asmar et al. 2013). The silica–cuticle bilayer was reported in many species, such as rice (Yoshida et al. 1962), magnolia (Postek 1981), cucumber (Samuels et al. 1993), and ficus (Davis 1987), but it is very likely present in other plants also.

In roots, silicon often fortifies the barrier structures of endodermal and exodermal cell layers. In this case, it prevents water leakage from the central cylinder toward the soil or other surrounding substrate with lower water potential than is present in the roots (Lux et al. 2002). Generally, the deposition of silicon in cell walls, extracellular spaces, or hydrophobic barrier structures significantly improves water retention, and thus it also improves the maintaining of the plant water balance (Lux et al. 1999; Gao et al. 2004; Ma 2004).

Several studies reported that the content of osmotic adjustment compounds was decreased after silicon treatment even under drought stress (Kaya et al. 2006; Gunes et al. 2008; Shen et al. 2010b; Silva et al. 2012; Mauad et al. 2016). These observations indicate that a role of silicon in the osmotic adjustment should be suggested. Therefore, energy costs on the synthesis of osmotically active solutes should be utilized for other processes to avoid further damage caused by drought stress. However, several contrasting results were observed as well with increased levels of osmolytes such as proline (Crusciol et al. 2009; Pereira et al. 2013; Khattab et al. 2014). Water uptake adjustment by the upregulation of several aquaporin genes was reported in sorghum after silicon supplementation during drought (Liu et al. 2014), and in rice, silicon treatment upregulated several drought-response transcription factors (Khattab et al. 2014). On the other hand, in tomato, which accumulates Si in much lower amounts, no effect of silicon treatment on root osmotic adjustment and the expression of aquaporins was observed; nonetheless, the root hydraulic conductance was improved (Shi et al. 2016). In sorghum, silicon affected also the balance of polyamines and reduced the decrease in the content of both free and conjugated polyamines caused by drought stress. Since the polyamine metabolism is related to ethylene biosynthesis by 1-aminocyclopropane-1-carboxylic acid as a common precursor, the silicon-induced stress alleviation provided via polyamine metabolism was suggested. Decreasing the levels of ethylene while increasing the polyamine content inhibits the senescence of photosynthetic apparatus; moreover, polyamines modulate also the activity of several ion channels (Yin et al. 2014). In white lupin, Si was observed to increase the levels of auxins, gibberellins, cytokinins, and abscisic acid (Abdalla 2011). In soybean, silicon reduced the drought-induced increase in salicylic acid levels and increased the content of gibberellins (Hamayun et al. 2010).

Changes in the antioxidant enzyme activity after silicon supplementation under drought stress vary greatly. As mentioned before, in order to interpret plant stress responses adequately, we must take into account at least the differences between individual species and the duration and intensity of stress exposure. However, since the experiments provided differ not only in the species used, but also in silicon doses, drought simulation character, and cultivation conditions, it is almost impossible to generalize the effect of silicon treatment on the antioxidant machinery. The information on silicon-induced changes in the responses of antioxidant system during drought stress includes species such as cucumber (Jafari et al. 2015), chickpea (Gunes et al. 2007c), maize (Li et al. 2007), rice (Mauad et al. 2016), soybean (Shen et al. 2010b), sunflower (Gunes et al. 2008), tomato (Shi et al. 2014, 2016; Cao et al. 2015), canola (Habibi 2015), wheat (Gong et al. 2005, 2008; Ma et al. 2016), and white lupin (Abdalla 2011). Despite the differences in ROS-scavenging responses, silicon supplementation mostly resulted in the decrease in hydrogen peroxide content and in reduced lipid peroxidation (Gong et al. 2005, 2008; Gunes et al. 2008; Jafari et al. 2015; Ma et al. 2016). Moreover, in rice and white lupin, it was observed that silicon treatment supports the production of phenols that might improve plant antioxidant capacity as well (Abdalla 2011; Emam et al. 2014).

The alleviation of growth reduction by silicon supplementation under water deficit probably results mostly from the preservation of plant water balance (Gong et al. 2003; Ahmed et al. 2016). Improving the leaf relative water content during water deficiency ameliorates both transport and metabolic processes, and thus also supports photosynthesis (Liang et al. 2007; Ahmed et al. 2013; Yin et al. 2014; Amin et al. 2016; Maghsoudi et al. 2016). For example, Si was reported to reduce the negative effects of drought on the chlorophyll content and light use efficiency in wheat and tomato (Cao et al. 2015; Maghsoudi et al. 2015; Ma et al. 2016). Moreover, in tomato, silicon alleviated the disintegration of thylakoid membranes caused by drought (Cao et al. 2015). Photosynthetic assimilation should then provide sufficient supplementation to plant organisms with both matter and energy, and thus contribute to their viability even under persisting stress exposure.

1.5 SILICON AND HIGH-SALINITY STRESS

Soil salinity is another important environmental factor reducing the agricultural productivity. It is estimated to affect about 800×10^6 ha of arable land (almost 20% of total arable land) and increases by 1%–2% each year (Zhu and Gong 2014; Rizwan et al. 2015). Reduced plant growth and yields under saline conditions often seem to not be directly influenced by salinity itself, but rather by harsh changes in water relations, or by reduced water uptake. Therefore, the developmental and metabolic changes occurring during salt stress are very similar to those observed during water stress (Munns 2002). The elevated concentrations of salts decrease the ability of water uptake and cause impairments of metabolic processes such as photosynthesis. Moreover, excessive amounts of Na^+ and Cl^- ions in plant cells and tissues induce hyperosmotic stress and ion imbalance (Deinlein et al. 2014). The water potential homeostasis is therefore disrupted and may lead to severe damage at the cellular level, growth arrest, or even death of the plant. Similarly, as during water stress, increased amounts of salts may alter the uptake of other nutrients and thus lead to various deficiencies. The generation of ROS is obviously stimulated, too (Zhu 2001; Munns 2002). Furthermore, salt stress causes defects in membrane integrity and protein synthesis (Liu et al. 2015).

Plant salt tolerance mechanisms mitigate the hyperosmotic stress by reducing water loss due to the deposition of hydrophobic substances (cuticle, waxes, lignin, suberin, etc.) and maximizing the water uptake, provided by osmotic adjustment (Deinlein et al. 2014). The toxic effects of salts are mostly reduced by their exclusion from the transpiration stream, by their sequestration from cytoplasm to vacuoles, and by the production of new leaves while replacing the ones damaged by salts. It is worth mentioning that costs to maintain the water relations by synthesizing the osmotic adjustment compounds are much higher than for salt exclusion or compartmentalization. Therefore, adaptations of the salt-tolerant species are predominantly performed by the latter mechanism (Munns 2002).

Since the symptoms of salt stress in a large range overlap with those caused by a water deficit (Munns 2002), the suggested mechanisms of salt stress alleviation by silicon supplementation are mostly discussed in Section 1.4. Maintaining plant water balance is crucial for photosynthetic, transport, and other metabolic processes; therefore, the salt-induced reduction of plant growth and the biomass yields are often mitigated by silicon treatment. The relative water content was increased after silicon application in all the studied species under salt stress, even though they belong to various plant families or occupy very distinct habitats (e.g., Liang 1999; Zhu et al. 2004; Romero-Aranda et al. 2006; Zuccarini 2008; Reezi et al. 2009; Liu et al. 2015). Subsequent restoration of the photosynthetic processes then allows further adjustments of the stress responses and helps to sustain the plant viability under continuous stress exposure.

Under saline conditions, plant roots are exposed to an environment possessing relatively high osmotic potential that impairs both the water uptake and the subsequent water leakage from the roots back to the soil (Steudle 2000). Decreasing the rates of water loss by the fortification of water leakage barriers (Lux et al. 2002) and the improvement of water uptake by the enhanced expression of several aquaporin genes (Liu et al. 2015) or by osmotic adjustment (Tuna et al. 2008), silicon seems to also modulate the undesirable uptake and translocation of sodium ions. Most reports indicate that after silicon treatment, the translocation of Na^+ from roots to aboveground organs was restricted (Matoh et al. 1986; Yeo et al. 1999; Gong et al. 2006; Savvas et al. 2007, 2009; Saqib et al. 2008; Zuccarini 2008; Mahdieh et al. 2015). One of the mechanisms of sodium ion detoxification is its sequestration into vacuoles. Silicon supplementation was reported to stimulate the activities of tonoplast H^+-ATPase, H^+-PPase, and plasma membrane H^+-ATPase that can promote the activity of tonoplast Na^+/H^+ antiporters responsible for sodium translocation into vacuoles (Liang et al. 2005, 2006). In some species, Si treatment was reported even to reduce the uptake of sodium (Batool et al. 2015; Farooq et al. 2015; Li et al. 2015). According to Yin et al. (2016), silicon treatment increased the content of polyamines under salt stress. The modulation of several ion channel activities by polyamines can afterwards affect both the uptake and translocation of sodium in plants (Zhao et al. 2007). Moreover, silicon was observed to support the uptake of potassium that should also reduce the uptake of Na^+ by competitive inhibition (Ghassemi-Golezani and Lotfi 2015; Tantawy et al. 2015; Waqas-ud-Din Khan et al. 2015). This was supported by Muneer and Jeong (2015), who detected the Si-induced upregulation of potassium channel genes in tomato under saline conditions. However, not all of the studies detected such detoxifying mechanism and reported no effect of silicon on sodium translocation to the aboveground organs (Romero-Aranda et al. 2006; Liu et al. 2015; Baby et al. 2016). Additionally, silicon was observed to increase the uptake of other ions, including calcium (Tuna et al. 2008; Habibi et al. 2014; Garg and Bhandari 2015), magnesium (Garg and Bhandari 2016; Li et al. 2016), iron, and zinc (Batool et al. 2015).

Improved water uptake under saline conditions can be performed by osmotic adjustment involving the synthesis of proline, glycine betaine, soluble sugars, or other osmolytes decreasing the water potential of plant cells (Slama et al. 2015). Silicon supplementation was observed to have diverse effects on the content of osmotic solutes in different species, and its interpretation is difficult. Since the synthesis of osmolytes depends on various factors including the availability of nitrogen and sufficient photosynthetic activity (Raven 1985), its efficacy is case-specific. Despite the fact that Si supplementation during salt stress mostly resulted in the decreasing of osmotic solute contents, it supports the improvements of plant water status anyway. The decrease in proline content after Si treatment in saline conditions was observed in wheat (Tuna et al. 2008), Chinese liquorice (Zhang et al. 2015), and sorghum (Yin et al. 2013); the decrease in the content of soluble sugars was observed, for example, in grapevine (Qin et al. 2016). However, since the osmotic adjustment can be performed by various compounds, simultaneously observing a decrease in content of one of those solutes can be misleading. For example, in maize, the levels of proline after Si treatment decreased, but the levels of soluble sugars and free amino acids increased (Latef and Tran 2016). In okra, Si treatment increased the levels of all proline, glycine betaine, free aminoacids, and soluble sugars as well (Abbas et al. 2015). Silicon was observed to affect also the metabolism of nitrogen and to mitigate the negative effects of salt stress on its performance. The uptake of N was increased in chickpea (Garg and Bhandari 2016), sweet pepper (Tantawy et al. 2015), and ryegrass (Mahdavi et al. 2016). Additionally, Si increased the translocation of nitrates from roots to shoots and stimulated the ammonia-detoxifying enzymes (Kochanová et al. 2014; de Souza et al. 2016). The electrolyte leakage is often bound to stress physiology assays by estimating the membrane permeability properties

(Whitlow et al. 1992). Silicon application was often reported to adjust the electrolyte leakage under various stresses, including salt stress (Liang 1999; Tuna et al. 2008).

Since the interpretations of antioxidant enzyme activities are significantly case specific, as we have discussed before, it is probably not possible to generalize the influence of silicon on their behavior under salt stress. Nonetheless, the application of silicon often results in decreasing the potentially damage-causing agents or oxidative damage indicators, such as hydrogen peroxide, thiobarbituric acid reactive substances (TBARSs), or MDA. The species examined include tomato (Al-Aghabary et al. 2004; Romero-Aranda et al. 2006), cucumber (Zhu et al. 2004), rose (Reezi et al. 2009), sunflower (Ali et al. 2013), barley (Liang 1999; Liang et al. 2003), wheat (Tale Ahmad and Haddad 2011; Ali et al. 2012), rice (Kim et al. 2014), and sorghum (Liu et al. 2015).

The role of silicon in the amelioration of salt stress is indisputable. However, it is not clear whether its benefits result mostly from maintaining the water balance, such as during drought stress, or other underlying mechanisms are present, too.

1.6 SILICON AND RADIATION STRESS

The sun is a source of visible, ultraviolet (UV), and infrared (IR) radiation that reaches the earth's atmosphere. Generally, at sea level, about 50% of the radiation energy is visible, and about 40% is IR radiation and 6% UV radiation (Moan 2001). UV-C (200–280 nm) radiation is absorbed by ozone; thus, only UV-B (280–320 nm) and UV-A (320–400 nm) reach the ground level (Stapleton 1992; Moan 2001). It is widely known that all types of UV radiation might damage various plant processes. UV-A radiation is 10–100 times more abundant than UV-B (Moan 2001). Predictions of UV-B levels in the future are uncertain; however, full recovery of the ozone column is not expected before 2050, and increased radiation of UV-B to the earth's surface will continue in future years (Madronich et al. 1998; Schrope 2000; Yao et al. 2011).

For plants, it is important to adapt to the increased level of UV-B radiation, because sunlight is essential for photosynthesis, and it is not possible to avoid UV-B exposure. The plant impairment resulting from UV-B depends on several factors, including plant species, developmental phase, anatomical and morphological characteristics, environmental conditions, and radiation intensity and duration (Rybus-Zając and Kubiś 2010). Damage caused by UV radiation can be classified as damage to DNA (may cause heritable mutations) or damage to physiological processes (Stapleton 1992). The mechanism of UV-B radiation toxicity involves ROS production (hydroxyl radicals, singlet oxygen, superoxide radicals, and hydrogen peroxide) and thus oxidative damage (Mackerness 2000; Yao et al. 2011). The accumulation of ROS in plant tissues is responsible for lipid peroxidation via oxidation of unsaturated fatty acids, leading to membrane damage that is followed by electrolyte leakage (Shen et al. 2010b). In these stressful conditions, plants exhibit protective mechanisms to balance and control ROS toxicity by increasing antioxidant enzyme activity and some low-molecular-weight antioxidant compounds, such as proline, anthocyanins, and flavonoids (Yao et al. 2011). Phenolics, besides their many other functions, act as a UV screen and thus protect plants against UV radiation. Each class of phenolics exhibits distinct absorption characteristics; for example, flavonoids with absorption at 250–270 nm and 335–360 nm act as good UV screens (Lattanzio et al. 2006). The negative effect of increased UV-B radiation is evident not only at the molecular and cellular levels, but also at the tissue and whole plant levels. Decreased growth parameters cause lower biomass production and directly affect yield production as a result of higher UV-B radiation (Lizana et al. 2009; Shen et al. 2010a,b). Most of the reduction of aboveground biomass and yield production losses caused by UV-B are mainly due to decreased leaf area, together with damage to photosynthetic pigments that reduces light capture by the plant (Lizana et al. 2009). Also, the quality of seeds may be affected by UV-B radiation. In maize grains, the levels of protein, sugar, and starch decreased under supplemental UV-B in comparison to control plants (Gao et al. 2004). However, the flour protein concentration in wheat grains was not affected by UV-B, even though changes in grain yield were found (Lizana et al. 2009).

Silicon may increase the plant's tolerance against stress from increased UV-B exposure. Silicon supplementation to growth media to some extent contributes to the higher UV-B tolerance of maize seedlings (Mihaličová Malčovská et al. 2014). The authors reported that an increased H_2O_2 concentration in plants exposed to UV-B exposure was significantly decreased after Si application. Also, the content of TBARSs (products of ROS-mediated oxidation of polyunsaturated membrane lipids) was not affected by UV-B in Si-treated plants despite its significant increase in control plants (Mihaličová Malčovská et al. 2014). Silicon application to plants can be associated with decreases of the phenolic biosynthesis. Increased Si accumulation and decreased cinnamyl alcohol hydrogenase (CAD) activity and p-coumaric acid concentration are closely related to alterations in the UV defense system (Goto et al. 2003). Exposure of UV-B increased the total phenol and anthocyanin concentration in soybean seedlings, and after Si application, the concentration of these UV-absorbing compounds decreased significantly (Shen et al. 2010a). Phenolic secondary metabolites, mainly flavonoids, are important in the absorption of UV radiation, and thus may protect plant tissues (Treutter 2005; Schaller et al. 2013). Silicon accumulation and the recently found reverse content of silicon and phenols suggest a protective function of Si and phenols under UV-B stress (Goto et al. 2003; Schaller et al. 2012, 2013). The Si present in leaf in the form of silica indicates UV-B resistance, because leaf silica has a very low UV absorption, as found in rice (Fang et al. 2006). The silicified leaf structures, such as cuticle, prickle hairs, and epidermis, may scatter incoming visible radiation and reduce its penetration to the leaf tissues, as well as affect the reflection of UV radiation (Klančnik et al. 2014). Thus, Si can affect the optical properties of mature and developing leaves. However, the Si effect on reflectance spectra may be different from that on transmittance spectra in leaves, and it indicates that silicified structures influence the internal light gradient (Klančnik et al. 2014). Agarie et al. (1996) studied the optical properties of leaves and found that silica bodies do not act as a "window" to facilitate light transmission to leaf mesophyll tissue of rice, as was formerly expected. Similarly, another study suggested that silica deposition in the epidermis may disperse or reflect UV-B radiation in order to reduce the internal spectra intensity of UV (Goto et al. 2003). Thus, a silicon double layer present in grass may act as a glass layer that is able to decrease UV-B transmission to the leaf tissues (Gatto et al. 1998; Schaller et al. 2013). However, it is not evident whether this silicon double layer evolved to reflect UV radiation, or is just a side effect of Si deposition. Regardless, it would be an advantage with respect to the energy demand for synthesis of phenol compounds as an alternative protection from increased UV-B exposure (Schaller et al. 2013). Wu et al. (2007) found out that the amount of silica on the adaxial epidermis differed between UV-B-tolerant and UV-B-sensitive rice accessions. The Si content was higher in UV-B-tolerant rice accession; thus, the characteristics of Si accumulation may be responsible for increased UV-B tolerance. The accumulation of Si is related to the expression level of genes coding Lsi transporters responsible for Si uptake and translocation (Ma and Yamaji 2008). A higher *Lsi1* expression level is important for increased tolerance to UV-B irradiation in rice (Fang et al. 2011). The overexpression of *Lsi1* genes leads to upregulation of phenylalanine ammonia lyase (PAL), 4-coumarate-CoA ligase (4CL), and photolyase (PL), and also may regulate other genes responsible for resistance to UV-B. The upregulation of PAL and 4CL results in the increased content of UV-B absorptive substances such as flavonoids and total phenolics, and upregulation of PL plays a role in DNA reparation (Mees et al. 2004; Fang et al. 2011). Overexpression of Lsi1 also causes strengthened photosynthesis and detoxification-related pathways that contribute to the enhanced defense against UV-B irradiation in rice (Fang et al. 2011).

1.7 SILICON AND TEMPERATURE STRESS

The accumulation of Si in leaves is advantageous not only for UV-B irradiation defense, but also for cooling leaves in heat stress conditions. In this situation, biosilicified structures present in epidermal cells are effective in cooling plant leaves by the mechanism of efficient mid-IR thermal radiation; thus, silicon creates a physical mechanism against heat stress (Wang et al. 2005). High-temperature stress limits the growth, metabolism, and productivity of plants. Due to heat stress, plant impairment is represented by oxidative stress (increased ROS production), cellular damage, membrane damage, photosynthesis inhibition, and so forth (Tan et al. 2011; Hasanuzzaman et al. 2013). In an experiment with *Agrostis palustris* growing at 35°C–40°C, the temperature of the leaves was decreased about 3°C–4.14°C in the case of Si treatment in comparison to untreated control plants. Also, Si present in soil substrate reduced heat and was effective in the cooling of plant roots (Wang et al. 2005). Another study suggested that Si may be involved in the thermal stability of cell membranes of plants during heat stress (Agarie et al. 1998). In this study, electrolyte leakage caused by high temperature (42.5°C) decreased in the leaves of plants grown with Si, but not in those without Si. Further studies dealing with high-temperature stress and Si interaction found an increased level of some antioxidant enzymes (SOD, APX, and glutathione peroxidase [GPX]); however, some of them decreased CAT (Soundararajan et al. 2014). This variation may be due to species or cultivar specificity and also the difference in Si uptake level. Silicon supplementation also significantly influenced the protein pattern and total protein content during the high-temperature stress in plants. Overall, Si positively affected plant growth and played a vital role against high-temperature stress (Soundararajan et al. 2014). The importance of Si application under high-temperature stress is also evident in the case of alleviating fertility reduction (Wu et al. 2014). This field study revealed that silicon in various concentrations effectively increased the germinated pollen number, the number of pollen grains per stigma, the pollen germination rate, and other fertility parameters in high-temperature-sensitive and -tolerant rice hybrids (Wu et al. 2014). However, exogenous silicon was more effective for the high-temperature-sensitive combination than for the high-temperature-tolerant one (Wu et al. 2014).

On the other hand, Si also mitigates freezing injury of plants growing in low temperatures. Low temperature (cold or chilling) causes many types of damage at the anatomic, cellular, and molecular levels, depending on the intensity and duration of the stress. Generally, these impairments include growth inhibition, necrosis, further cellular dehydration, increased concentration of intracellular salts, membrane damage, increased level of lipid peroxidation and ROS, and inhibited leaf photosynthesis and water use efficiency (Zhu et al. 2006; Liang et al. 2008; Sanghera et al. 2011). The study of Liang et al. (2008) showed that silicon positively affected the growth of plants under freezing conditions. An ameliorative effect may be attributed to the higher antioxidant defense capacity, resulting in decreased lipid peroxidation and enhanced stability and functions of cell membranes. Foliar Si application also positively affects plants stressed by freezing (Habibi 2015). Similarly, in this study, freezing stress caused membrane damage due to lipid peroxidation. The author indicated that ameliorative effect of Si may be attributed to higher antioxidant defense activity and lower lipid peroxidation through leaf water retention, in addition to its role as a mere physical barrier (Habibi 2015). In the case of chilling stress, other studies also confirmed that enhanced levels of ROS and MDA occurred in plant tissues (Liu et al. 2009). Silicon application to hydroponic media increased several enzymatic and nonenzymatic antioxidants and also decreased ROS and MDA content in chilling-stressed plants. Thus, Si plays a protective role in plants stressed by low temperature (Liu et al. 2009).

1.8 CONCLUSION AND FUTURE PROSPECTS

There is no doubt about the positive role of silicon in the mitigation of various biotic and abiotic stresses in plants. This chapter supplied a relatively large amount of evidence on the beneficial role of silicon in plants that suffer from heavy metals and toxic elements, increased salinity, lack of water, increased radiation doses, or higher or lower temperatures in their environment. The mechanisms applied in mitigation are similar for some stresses; however, some aspects are known only for one type of abiotic stress. In general, decreased ROS generation and changes in activation of the antioxidant system in plants are probably the most common reactions to many abiotic stressors. However, we should keep in mind that Si-induced changes in the activity of enzymatic and nonenzymatic antioxidants vary greatly and are species and stress intensity specific. Therefore, general conclusions that Si activates or depresses the antioxidative machinery in plants cannot be drawn. Another relative aspect that helps plants tolerate negative conditions is improved membrane integrity that favors maintaining the homeostasis of organisms. Among the other mechanisms that have been documented as beneficial in overcoming the harmful effects of abiotic stressors are osmotic adjustment, increased water use efficiency of stressed plants, reduced uptake, and decreased root-to-shoot translocation of heavy metals and toxic elements that is often related to anatomical changes in root tissues and restriction of apolasmic transport. Immobilization of dangerous elements might occur in growth media, as well as in specific plant tissues or cell organelles. Additionally, increased synthesis of certain specific secondary metabolites, like phenolics and anthocyanins, as well as increased light reflection by specific cells and the cuticle–Si bilayer, is also considered beneficial in overcoming some abiotic stresses. To advance all these aspects to the readers, Figure 1.2 summarizes all of the positive aspects of Si discussed in this chapter with respect to abiotic stresses.

Although Si is not considered an essential element in general, its beneficial effects on plant growth and alleviation of abiotic stresses rank it high in food safety issues. However, continued investigation and further research activities will allow us to understand the entire mechanisms of Si functioning in plants, and will help us to discover all hidden aspects of this prosperous element for plants and humans.

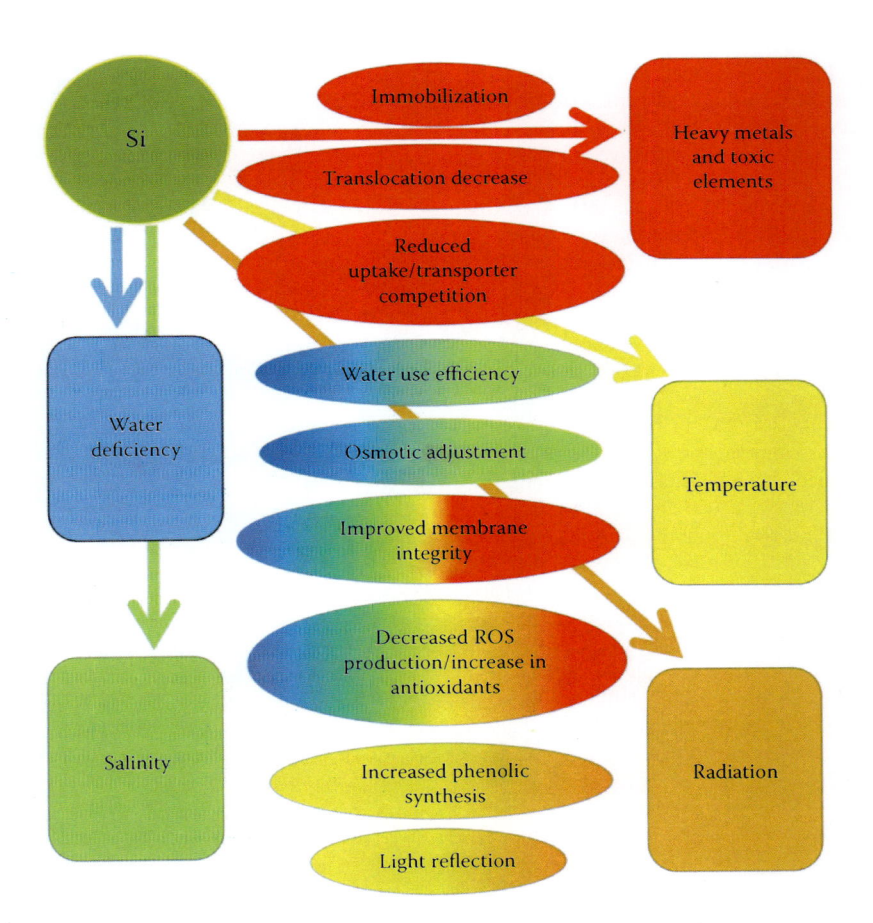

FIGURE 1.2 Overview of plant responses to various forms of abiotic stresses. A response to a particular stress factor is applicable when the same color appears in the ellipse.

ACKNOWLEDGMENTS

This contribution is the result of the project implementation of Comenius University in Bratislava Science Park supported by the Research and Development Operational Programme funded by the European Regional Development Fund (ERDF) (grant ITMS 26240220086).

REFERENCES

Abbas T, Balal RM, Shahid MA, Pervez MA, Ayyub CM, Aqueel MA, Javaid MM. (2015). Silicon-induced alleviation of NaCl toxicity in okra (*Abelmoschus esculentus*) is associated with enhanced photosynthesis, osmoprotectants and antioxidant metabolism. *Acta Physiol Plant* 37: 6.

Abdalla MM. (2011). Beneficial effects of diatomite on the growth, the biochemical contents and polymorphic DNA in Lupinus albus plants grown under water stress. *Agric Biol J N Am* 2: 207–220.

Adrees M, Ali S, Rizwan M, Zia-ur-Rehman M, Ibrahim M, Abbas F, Farid M, Qayyum MF, Irshad MK. (2015). Mechanisms of silicon-mediated alleviation of heavy metal toxicity in plants: A review. *Ecotoxicol Environ Saf* 2015: 186–197.

Agarie S, Agata W, Uchida H, Kubota F, Kaufman PB. (1996). Function of silica bodies in the epidermal system of rice (*Oryza sativa* L.): Testing the window hypothesis. *J Exp Bot* 47: 655–660.

Agarie S, Hanaoka N, Ueno O, Miyazaki A, Kubota F, Agata W, Kaufman PB. (1998). Effects of silicon on tolerance to water deficit and heat stress in rice plants (*Oryza sativa* L.), monitored by electrolyte leakage. *Plant Prod Sci* 1: 96–103.

Ahmed M, Asif M, Hassan F. (2013). Augmenting drought tolerance in sorghum by silicon nutrition. *Acta Physiol Plant* 36: 473–483.

Ahmed M, Quadeer U, Ahmed ZI, Hassan F. (2016). Improvement of wheat (*Triticum aestivum*) drought tolerance by seed priming with silicon. *Arch Agron Soil Sci* 62: 299–315.

Al-Aghabary K, Zhu ZJ, Shi QH. (2004). Influence of silicon supply on chlorophyll content, chlorophyll fluorescence and antioxidative enzyme activities in tomato plants under salt stress. *J Plant Nutr* 27: 2101–2115.

Al Bakheet SA, Attafi IM, Maayah ZH, Abd-Allah AR, Asiri YA, Korashy HM. (2013). Effect of long-term human exposure to environmental heavy metals on the expression of detoxification and DNA repair genes. *Environ Pollut* 181: 226–232.

Ali A, Basra SMA, Iqbal J, Hussain S, Subhani MN, Sarwar M, Haji A. (2012). Silicon mediated biochemical changes in wheat under salinized and non-salinzed solution cultures. *Afr J Biotechnol* 11: 606–615.

Ali S, Farooq MA, Yasmeen T, Hussain S, Arif MS, Abbas F, Bharwana SA, Zhang G. (2013). The influence of silicon on barley growth, photosynthesis and ultra-structure under chromium stress. *Ecotoxicol Environ Saf* 89: 66–72.

Alpaslan M, Gunes A. (2001). Interactive effects of boron and salinity stress on the growth, membrane permeability and mineral composition of tomato and cucumber plants. *Plant Soil* 236: 123–128.

Amin M, Ahmad R, Ali A, Hussain I, Mahmood R, Aslam M, Lee DJ. (2016). Influence of silicon fertilization on maize performance under limited water supply. *Silicon* doi:10.1007/s12633-015-9372-x.

Appenroth KJ. (2010). Soil biology: Soil heavy metals. In Sherameti I, Varma A (eds.), *Soil Heavy Metals, Soil Biology*. Vol. 19. Berlin: Springer-Verlag, 2010, pp. 19–29.

Asher CJ, Reay PF. (1979). Arsenic uptake by barley seedlings. *Aust J Plant Physiol* 6: 459–466.

Ashraf M. (2010). Inducing drought tolerance in plants: Recent advances. *Biotechnol Adv* 28: 169–183.

Asmar SA, Castro EM, Pasqual M, Pereira FJ, Soares JDR. (2013). Changes in leaf anatomy and photosynthesis of micropropagated banana plantlets under different silicon sources. *Sci Hort* 161: 328–332.

Baby T, Collins C, Tyerman SD, Gilliham M. (2016). Salinity negatively affects pollen tube growth and fruit set in grapevines and cannot be ameliorated by silicon. *Am J Enol Vitic* doi:10.5344/ajev.2015.15004.

Batool M, Saqib M, Murtaza G, Basra SMA, Nawaz S. (2015). Silicon application improves Fe and Zn use efficiency and growth of maize genotypes under saline conditions. *Pak J Agri Sci* 52: 445–451.

Baxter I. (2009). Ionomics: Studying the social network of mineral nutrients. *Curr Opin Plant Biol* 12: 381–386.

Baxter I, Dilkes P. (2012). Elemental profiles reflect plant adaptations to the environment. *Science* 336: 1661–1663.

Ben-Gal A, Shani U. (2002). Yield, transpiration and growth of tomatoes under combined excess boron and salinity stress. *Plant Soil* 247: 211–221.

Bertin G, Averbeck D. (2006). Cadmium: Cellular effects, modifications of biomolecules, modulation of DNA repair and genotoxic consequences: A review. *Biochemie* 88: 1549–1559.

Bienert GP, Thorsen M, Schüssler MD, Nilsson HR, Wagner A, Tamás MJ, Jahn TP. (2008). A subgroup of plant aquaporins facilitate the bidirectional diffusion of As(OH)$_3$ and Sb(OH)$_3$ across membranes. *BMC Biol* 6: 26.

Blaylock MJ, Salt DE, Dushenkov S, Zakharova O, Gussman CH, Kapulnik Y, Ensley BD, Raskin I. (1997). Enhanced accumulation of Pb in Indian mustard by soil-applied chelating agents. *Environ Sci Technol* 31: 860–865.

Bogdan K, Schenk MK. (2008). Arsenic in rice (*Oryza sativa* L.) related to dynamics of arsenic and silicic acid in paddy soils. *Environ Sci Technol* 42: 7885–7890.

Bokor B, Vaculík M, Slováková L, Masarovič D, Lux A. (2014). Silicon does not always mitigate zinc toxicity in maize. *Acta Physiol Plant* 36: 733–743.

Brandes EA, Greenaway HT, Stone HEN. (1956). Ductility in chromium. *Nature* 178: 587.

Bray E. (1997). Plant responses to water deficit. *Trends Plant Sci* 2: 48–54.

Broadley MR, White PJ, Hammond JP, Zelko I, Lux A. (2007). Zinc in plants. *New Phytol* 173: 677–702.

Callahan DL, Baker AJM, Kolev SD, Wedd AG. (2006). Metal ion ligands in hyperaccumulating plants. *J Biol Inorg Chem* 11: 2–12.

Camacho-Cristobal JJ, Rexach J, Gonzalez-Fontes A. (2008). Boron in plants: Deficiency and toxicity. *J Integr Plant Biol* 50: 1247–1255.

Cao BL, Ma Q, Zhao Q, Wang L, Xu K. (2015). Effects of silicon on absorbed light allocation, antioxidant enzymes and ultrastructure of chloroplasts in tomato leaves under simulated drought stress. *Sci Hort* 194: 53–62.

Cattivelli L, Rizza F, Badeck FW, Mazzucotelli E, Mastrangelo AM, Francia E, Marè C, Tondelli A, Stanca AM. (2008). Drought tolerance improvement in crop plants: An integrated view from breeding to genomics. *Field Crops Res* 105: 1–14.

Cho-Ruk K, Kurukote J, Supprung P, Vetayasuporn S. (2006). Perennial plants in the phytoremediation of lead-contaminated soils. *Biotechnology* 5: 1–4.

Clarimont KB, Hagar WG, Davis EA. (1986). Manganese toxicity to chlorophyll synthesis in tobacco callus. *Plant Physiol* 80: 291–293.

Clemens S. (2006). Toxic metal accumulation, responses to exposure and mechanisms of tolerance in plants. *Biochemie* 16: 1707–1719.

Collin B, Doelsch E, Keller C, Cazevieille P, Tella M, Chaurand P, Panfili F, Hazemann JL, Meunier JD. (2014). Evidence of sulfur-bound reduced copper in bamboo exposed to high silicon and copper concentrations. *Environ Pollut* 187: 22–30.

Crusciol CA, Pulz AL, Lemos LB, Soratto RP, Lima GP. (2009). Effects of silicon and drought stress on tuber yield and leaf biochemical characteristics in potato. *Crop Sci* 49: 949–954.

Cruz de Carvalho MH. (2008). Drought stress and reactive oxygen species: Production, scavenging and signaling. *Plant Signal Behav* 3: 156–165.

Da Cunha KPV, do Nascimento CWA. (2009). Silicon effects on metal tolerance and structural changes in maize (*Zea mays* L.) grown on a cadmium and zinc enriched soil. *Water Air Soil Poll* 197: 323–330.

Da Cunha KPV, do Nascimento CWA, Silva AJ. (2008). Silicon alleviates the toxicity of cadmium and zinc for maize (*Zea mays* L.) grown on a contaminated soil. *J Plant Soil Sci* 171: 849–853.

Davis RW. (1987). Ultrastructure and analytical microscopy of silicon in the leaf cuticle of *Ficus lyrata* Warb. *Bot Gaz* 148: 318–323.

Deinlein U, Stephan AB, Horie T, Luo W, Xu G, Schroeder JI. (2014). Plant salt-tolerance mechanisms. *Trends Plant Sci* 19: 371–379.

Demirevska-Kepova K, Simova-Stoilova L, Stoyanova Z, Holzer R, Feller U. (2004). Biochemical changes in barely plants after excessive supply of copper and manganese. *Environ Exp Bot* 52: 253–266.

de Souza LC, Lima EGS, de Almeida RF, Neves MG, Nogueira GAS, Neto CFO, da Costa AS, Machado LC, do Nascimento SMC, Brito AEA. (2016). Nitrogen metabolism in sorghum under salinity and silicon treatments in Brazil. *Afr J Agric Res* 11: 199–208.

Doncheva SN, Poschenrieder C, Stoyanova ZL, Georgieva K, Velichkova M, Barceló J. (2009). Silicon amelioration of manganese toxicity in Mn-sensitive and Mn-tolerant maize varieties. *Environ Exp Bot* 65: 189–197.

Ebbs SD, Kochian LV. (1997). Toxicity of zinc and copper to *Brassica* species: Implication for phytoremediation. *J Environ Qual* 26: 776–781.

Emam MM, Khattab HE, Helal NM, Deraz AE. (2014). Effect of selenium and silicon on yield quality of rice plant grown under drought stress. *Aust J Crop Sci* 8: 596–605.

Epstein E. (1999). Silicon. *Annu Rev Plant Physiol Plant Mol Biol* 50: 641–664.

Fang C-X, Wang Q-S, Yu Y, Li Q-M, Zhang H-L, Wu X-C, Chen T, Lin W-X. (2011). Suppression and over-expression of *Lsi1* induce differential gene expression in rice under ultraviolet radiation. *Plant Growth Regul* 65: 1–10.

Fang JY, Wan XC, Ma XL. (2006). Nanoscale silicas in *Oryza sativa* L. and their UV absorption. *Spectrosc Spectral Anal* 26: 2315–2318.

FAO (Food and Agriculture Organization). (2005). Global network on integrated soil management for sustainable use of salt-affected soils. Rome: FAO Land and Plant Nutrition Management Service. http://www.fao.org/ag/agl/agll/spush.

Farooq M, Wahid A, Kobayashi N, Fujita D, Basra SMA. (2009). Plant drought stress: Effects, mechanisms and management. *Agron Sustain Dev* 29: 185–212.

Farooq MA, Saqib ZA, Akhtar J, Bakhat HF, Pasala R-K, Dietz, K-J. (2015). Protective role of silicon (Si) against combined stress of salinity and boron (B) toxicity by improving antioxidant enzymes activity in rice. *Silicon* doi:10.1007/s12633-015-9346-z.

Finnegan PM, Chen W. (2012). Arsenic toxicity: The effects on plant metabolism. *Frontiers Plant Physiol* 3: 182.

Fleck AT, Mattusch J, Schenk MK. (2013). Silicon decreases the arsenic level in rice grain by limiting arsenite transport. *J Plant Nutr Soil Sci* 176: 785–794.

Fleck AT, Schenk MK. (2011). Influence of silicon supply on arsenic concentration in straw and grain fractions of rice. In *Proceedings of the 5th International Conference on Silicon in Agriculture*, Beijing, p. 47.

Foy CD, Weil RR, Coradetti CA. (1995). Differential manganese tolerances of cotton genotypes in nutrient solution. *J Plant Nutr* 18: 685–706.

Gao X, Zou C, Wang L, Zhang F. (2004). Silicon improves water use efficiency in maize plants. *J Plant Nutr* 27: 1457–1470.

Garg N, Bhandari P. (2015). Interactive effects of silicon and arbuscular mycorrhiza in modulating ascorbate-glutathione cycle and antioxidant scavenging capacity in differentially salt-tolerant Cicer arietinum L. genotypes subjected to long-term salinity. *Protoplasma* doi:10.1007/s00709-015-0892-4.

Garg N, Bhandari P. (2016). Silicon nutrition and mycorrhizal inoculations improve growth, nutrient status, K^+/Na^+ ratio and yield of Cicer arietinum L. genotypes under salinity stress. *Plant Growth Regul* 78: 371–387.

Gatto A, Escoubas L, Roche P, Commandre M. (1998). Simulation of the degradation of optical glass substrates caused by UV irradiation while coating. *Opt Commun* 148: 347–354.

Ghassemi-Golezani K, Lotfi R. (2015). The impact of salicylic acid and silicon on chlorophyll a fluorescence in mung bean under salt stress. *Russ J Plant Phys* 62: 611–616.

Gong H, Chen K, Chen G, Wang S, Zhang C. (2003). Effects of silicon on growth of wheat under drought. *J Plant Nutr* 26: 1055–1063.

Gong H, Chen KM, Zhao ZG, Chen GC, Zhou WJ. (2008). Effects of silicon on defense of wheat against oxidative stress under drought at different developmental stages. *Biol Plantarum* 52: 592–596.

Gong H, Randall DP, Flowers TJ. (2006). Silicon deposition in the root reduces sodium uptake in rice (*Oryza sativa* L.) seedlings by reducing bypass flow. *Plant Cell Environ* 29: 1970–1979.

Gong H, Zhu X, Chen K, Wang S, Zhang C. (2005). Silicon alleviates oxidative damage of wheat plants in pots under drought. *Plant Sci* 169: 313–321.

Goto M, Ehara H, Karita S, Takabe K, Ogawa N, Yamada Y, Ogawa S, Yahaya MS, Morita O. (2003). Protective effect of silicon on phenolic biosynthesis and ultraviolet spectral stress in rice crop. *Plant Sci* 164: 349–356.

Grant CA, Sheppard SC. (2008). Fertilizer impacts on cadmium availability in agricultural soils and crops. *Hum Ecol Risk Assess* 14: 210–228.

Guo W, Hou YL, Wang SG, Zhu YG. (2005). Effect of silicate on the growth and arsenate uptake by rice (*Oryza sativa* L.) seedlings in solution culture. *Plant Soil* 272: 173–181.

Guo W, Zhang J, Teng M, Li H, Wang J. (2009). Arsenic uptake is suppressed in a rice mutant defective in silicon uptake. *Plant Nutr Soil Sci* 172: 867–874.

Gunes A, Inal A, Bagci EG, Coban S, Pilbeam DJ. (2007a). Silicon mediates to some physiological and enzymatic parameters symptomatic for oxidative stress in spinach (*Spinacia oleracea* L.) grown under B toxicity. *Sci Hortic* 113: 113–119.

Gunes A, Inal A, Bagci EG, Coban S, Sahin O. (2007b). Silicon increases boron tolerance and reduces oxidative damage of wheat grown in soil with excess boron. *Biol Plant* 51: 571–574.

Gunes A, Pilbeam DJ, Inal A, Bagci EG, Coban S. (2007c). Influence of silicon on antioxidant mechanisms and lipid peroxidation in chickpea (*Cicer arietinum* L.) cultivars under drought stress. *J Plant Interact* 2: 105–113.

Gunes A, Pilbeam DJ, Inal A, Coban S. (2008). Influence of silicon on sunflower cultivars under drought stress. I: Growth, antioxidant mechanisms, and lipid peroxidation. *Commun Soil Sci Plant Anal* 39: 1885–1903.

Guntzer F, Keller C, Meunier J-D. (2012). Benefits of plant silicon for crops: A review. *Agron Sustain Dev* 32: 201–213.

Habibi G. (2015). Exogenous silicon leads to increased antioxidant capacity in freezing-stressed pistachio leaves. *Acta Agric Slovenica* 105: 43–52.

Habibi G, Norouzi F, Hajiboland R. (2014). Silicon alleviates salt stress in pistachio plants. *Prog Biol Sci* 4: 189–202.

Hamayun M, Sohn EY, Khan SA, Shinwari ZK, Khan AL, Lee IJ. (2010). Silicon alleviates the adverse effects of salinity and drought stress on growth and endogenous plant growth hormones of soybean (*Glycine max* L.). *Pak J Bot* 42: 1713–1722.

Hanafy Ahmed AH, Harb EM, Higazy MA, Morgan ShH. (2008). Effect of silicon and boron foliar applications on wheat plants grown under saline soil conditions. *Int J Agric Res* 3: 1–26.

Hasanuzzaman M, Nahar K, Alam MMd, Roychowdhury R, Fujita M. (2013). Physiological, biochemical, and molecular mechanisms of heat stress tolerance in plants. *Int J Mol Sci* 14: 9643–9684.

Hirschi KD, Korenkov VD, Wilganowski NL, Wagner GJ. (2000). Expression of *Arabidopsis* CAX2 in tobacco. Altered metal accumulation and increased manganese tolerance. *Plant Physiol* 124: 125–133.

Hoffmann H, Schenk MK. (2011). Arsenite toxicity and uptake rate of rice (*Oryza sativa* L.) *in vivo*. *Environ Pollut* 159: 2398–2404.

Horst WJ, Marschner H. (1978). Effect of silicon on manganese tolerance of bean plants (*Phaseolus vulgaris* L.). *Plant Soil* 50: 287–303.

Housecroft CE, Sharpe AG. (2008). *Inorganic Chemistry*. 3rd ed. Harlow, UK: Prentice Hall.

Hutchinson TC, Whitby LM. (1977). The effects of acid rainfall and heavy metal particulates on a boreal forest ecosystem near the Sudbury smelting region of Canada. *Water Air Soil Pollut* 7: 421–438.

Inal A, Pilbeam DJ, Gunes A. (2009). Silicon increases tolerance to boron toxicity and reduces oxidative damage in barley. *J Plant Nutr* 32: 112–128.

Isayenkov SV, Maathuis FJM. (2008). The *Arabidopsis thaliana* aquaglyceroporin AtNIP7;1 is a pathway for arsenite uptake. *FEBS Lett* 582: 1625–1628.

Jafari SR, Mohammad S, Arvin J, Kalantari KM. (2015). Response of cucumber (*Cucumis sativus* L.) seedlings to exogenous silicon and salicylic acid under osmotic stress. *Acta Biol Szeged* 59: 25–33.

Kaji M. (2012). Role of experts and public participation in pollution control: The case of itai-itai disease in Japan. *Ethics Sci Environ Pollut* 12: 99–111.

Kar RK. (2011). Plant responses to water stress: Role of reactive oxygen species. *Plant Signal Behav* 6: 1741–1745.

Kaya C, Tuna AL, Guneri M, Asraf M. (2011). Mitigation effects of silicon on tomato plants bearing fruit grown at high boron levels. *J Plant Nutr* 34: 1985–1994.

Kaya C, Tuna L, Higgs D. (2006). Effect of silicon on plant growth and mineral nutrition of maize grown under water-stress conditions. *J Plant Nutr* 29: 1469–1480.

Kaya C, Tuna AL, Sonmez O, Ince F, Higgs D. (2009). Mitigation effects of silicon on maize plants grown at high zinc. *J Plant Nutr* 32: 1788–1798.

Kerkeb L, Mukherjee I, Chatterjee I, Lahner B, Salt DE, Connolly EL. (2008). Iron-induced turnover of the *Arabidopsis* IRON-REGULATED TRANSPORTER1 metal transporter requires lysine residues. *Plant Physiol* 146: 1964–1973.

Kerstiens G. (1996). Cuticular water permeability and its physiological significance. *J Exp Bot* 47: 1813–1832.

Khandekar S, Leisner S. (2011). Soluble silicon modulates expression of *Arabidopsis thaliana* genes involved in copper stress. *J Plant Physiol* 168: 699–705.

Khattab HI, Emam MA, Emam MM, Helal NM, Mohamed MR. (2014). Effect of selenium and silicon on transcription factors NAC5 and DREB2A involved in drought-responsive gene expression in rice. *Biol Plant* 58: 265–273.

Kim YH, Khan AL, Waqas M, Shim JK, Kim DH, Lee KY, Lee IJ. (2014). Silicon application to rice root zone influenced the phytohormonal and antioxidant responses under salinity stress. *J Plant Growth Regul* 33: 137–149.

Kinraide TB, Parker DR, Zobel RW. (2005). Organic acid secretion as a mechanism of aluminium resistance: A model incorporating the root cortex, epidermis, and the external unstirred layer. *J Exp Bot* 56: 1853–1865.

Kitao M, Lei TT, Koike T. (1997). Effects of manganese toxicity on photosynthesis of white birch (*Betula platyphylla* var. *japonica*) seedlings. *Physiol Plant* 101: 249–256.

Kochanová Z, Jašková K, Sedláková B, Luxová M. (2014). Silicon improves salinity tolerance and affects ammonia assimilation in maize roots. *Biologia* 69: 1164–1171.

Klančnik K, Vogel-Mikuš K, Gaberščik A. (2014). Silicified structures affect leaf optical properties in grasses and sedge. *J Photochem Photobiol B* 130: 1–10.

Landridge P, Reynolds MP. (2015). Genomic tools to assist breeding for drought tolerance. *Curr Opin Biotechnol* 32: 130–135.

Lasat MM. (2002). Phytoextraction of toxic metals—A review of biological mechanisms. *J Environ Qual* 31: 109–120.

Latef AAA, Tran LSP. (2016). Impacts of priming with silicon on the growth and tolerance of maize plants to alkaline stress. *Front Plant Sci* 7: 1–10.

Lattanzio V, Lattanzio VMT, Cardinali A. (2006). Role of phenolics in the resistance mechanisms of plants against fungal pathogens and insects. In Imperato F (ed.), *Phytochemistry: Advances in Research.* Trivandrum, Kerala, India: Research Signpost, pp. 23–67.

Lee Ch-H, Huang H-H, Syu Ch-H, Lin T-H, Lee D-Y. (2014). Increase of As release and phytotoxicity to rice seedlings in As-contaminated paddy soils by Si fertilizer application. *J Hazard Mater* 276: 253–261.

Lee CW, Choi JM, Pak CH. (1996). Micronutrient toxicity in seed geranium (*Pelargonium* × *hortorum* Baley). *J Am Soc Horticult Sci* 121: 77–82.

Lewis S, Donkin ME, Depledge MH. (2001). Hsp70 expression in *Enteromorpha intestinalis* (Chlorophyta) exposed to environmental stressors. *Aquat Toxicol* 51: 277–291.

Li LB, Zheng C, Fu YQ, Wu DM, Yang XJ, Shen H. (2012). Silicate-mediated alleviation of Pb toxicity in banana grown in Pb-contaminated soil. *Biol Trace Elem Res* 145: 101–108.

Li P, Song A, Li Z, Fan F, Liang Y. (2015). Silicon ameliorates manganese toxicity by regulating both physiological processes and expression of genes associated with photosynthesis in rice (*Oryza sativa* L.). *Plant Soil* 397: 289–301.

Li QF, Ma CC, Shang QL. (2007). Effects of silicon on photosynthesis and antioxidative enzymes of maize under drought stress. *Ying Yong Sheng Tai Xue Bao* 18: 531–536.

Li RY, Stroud JL, Ma JF, McGrath SP, Zhao FJ. (2009). Mitigation of arsenic accumulation in rice with water management and silicon fertilization. *Environ Sci Technol* 43: 3778–3783.

Li YT, Zhang WJ, Cui JJ, Lang DY, Li M, Zhao QP, Zhang XH. (2016). Silicon nutrition alleviates the lipid peroxidation and ion imbalance of *Glycyrrhiza uralensis* seedlings under salt stress. *Acta Physiol Plant* 38: 96.

Liang Y. (1999). Effects of silicon on enzyme activity and sodium, potassium and calcium concentration in barley under salt stress. *Plant Soil* 29: 217–224.

Liang Y, Chen Q, Liu Q, Zhang W, Ding R. (2003). Exogenous silicon (Si) increases antioxidant enzyme activity and reduces lipid peroxidation in roots of salt-stressed barley (*Hordeum vulgare* L.). *J Plant Physiol* 160: 1157–1164.

Liang Y, Shen Z. (1994). Interaction of silicon and boron in oilseed rape plants. *J Plant Nutr* 17: 415–425.

Liang Y, Sun W, Zhu Y-G, Christie P. (2007). Mechanisms of silicon-mediated alleviation of abiotic stresses in higher plants: A review. *Environ Pollut* 147: 422–428.

Liang Y, Wong JWC, Wei L. (2005). Silicon-mediated enhancement of cadmium tolerance in maize (*Zea mays* L.) grown in cadmium contaminated soil. *Chemosphere* 58: 475–483.

Liang Y, Zhang W, Chen Q, Liu Y, Ding R. (2006). Effect of exogenous silicon (Si) on H^+-ATPase activity, phospholipids and fluidity of plasma membrane in leaves of salt-stressed barley (*Hordeum vulgare* L.). *Environ Exp Bot* 57: 212–219.

Liang Y, Zhu J, Li Z, Chu G, Ding Y, Zhang J, Sun W. (2008). Role of silicon in enhancing resistance to freezing stress in two contrasting winter wheat cultivars. *Environ Exp Bot* 64: 286–294.

Liu J, Zhang H, Zhang Y, Chai T. (2013). Silicon attenuates cadmium toxicity in *Solanum nigrum* L. by reducing cadmium uptake and oxidative stress. *Plant Physiol Biochem* 68: 1–7.

Liu JJ, Lin SH, Xu PL, Wang XJ, Bai JG. (2009). Effects of exogenous silicon on the activities of antioxidant enzymes and lipid peroxidation in chilling-stressed cucumber leaves. *Agric Sci China* 8: 1075–1086.

Liu P, Yin L, Deng X, Wang S, Tanaka K, Zhang S. (2014). Aquaporin-mediated increase in root hydraulic conductance is involved in silicon-induced improved root water uptake under osmotic stress in *Sorghum bicolor* L. *J Exp Bot* 65: 4747–4756.

Liu P, Yin L, Wang S, Zhang M, Deng X, Zhang S, Tanaka K. (2015). Enhanced root hydraulic conductance by aquaporin regulation accounts for silicon alleviated salt-induced osmotic stress in *Sorghum bicolor* L. *Environ Exp Bot* 111: 42–51.

Lizana XC, Hess S, Calderini DF. (2009). Crop phenology modifies wheat responses to increased UV-B radiation. *Agric For Meteorol* 149: 1964–1974.

Lukačová Z, Švubová R, Kohanová J, Lux A. (2013). Silicon mitigates the Cd toxicity in maize in relation to cadmium translocation, cell distribution, antioxidant enzymes stimulation and enhanced endodermal apoplasmic barrier development. *Plant Growth Regul* 70: 89–103.

Lukačová Kuliková Z, Lux A. (2010). Silicon influence on maize, *Zea mays* L., hybrids exposed to cadmium treatment. *Bull Environ Contam Toxicol* 85: 243–250.

Lux A, Luxová M, Hattori T, Inanaga S, Sugimoto Y. (2002). Silicification in sorghum (*Sorghum bicolor*) cultivars with different drought tolerance. *Physiol Plant* 115: 87–92.

Lux A, Luxová M, Morita S, Abe J, Inanaga S. (1999). Endodermal silicification in developing seminal roots of lowland and upland cultivars in rice (*Oryza sativa* L.). *Can J Bot* 77: 955–960.

Lux A, Martinka M, Vaculík M, White PJ. (2011). Root responses to cadmium in the rhizosphere: A review. *J Exp Bot* 62: 21–37.

Lyubenova L, Schröder P. (2010). Soil biology: Soil heavy metals. In Sherameti I, Varma A (eds.), *Soil Heavy Metals, Soil Biology*. Vol. 19. Berlin: Springer-Verlag, p. 492.

Ma D, Sun D, Wang C, Qin H, Ding H, Li Y, Guo T. (2016). Silicon application alleviates drought stress in wheat through transcriptional regulation of multiple antioxidant defense pathways. *J Plant Growth Regul* 35: 1–10.

Ma JF. (2004). Role of silicon in enhancing the resistance of plants to biotic and abiotic stresses. *Soil Sci Plant Nutr* 50: 11–18.

Ma JF, Tamai K, Yamaji N, Mitani N, Konishi S, Katsuhara M, Ishiguro M, Murata Y, Yano M. (2006). A silicon transporter in rice. *Nature* 440: 688–691.

Ma JF, Yamaji N. (2006). Silicon uptake and accumulation in higher plants. *Trends Plant Sci* 11: 392–397.

Ma JF, Yamaji N. (2008). Functions and transport of silicon in plants. *Cell Mol Life Sci* 2008 65: 3049–3057.

Ma JF, Yamaji N, Mitani N, Xu XY, Su YH, McGrath SP, Zhao FJ. (2008). Transporters of arsenite in rice and their role in arsenic accumulation in rice grain. *Proc Natl Acad Sci USA* 105: 9931–9935.

Mackerness SAH. (2000). Plant responses to ultraviolet-B (UV-B: 280–320 nm) stress: What are the key regulators? *Plant Growth Regul* 32: 27–39.

Madronich S, McKenzie RL, Bjorn LO, Caldwell MM. (1998). Changes in biologically active ultraviolet radiation reaching the earth's surface. *J Photochem Photobiol B* 46: 5–19.

Maghsoudi K, Emam Y, Ashraf M. (2015). Influence of foliar application of silicon on chlorophyll fluorescence, photosynthetic pigments, and growth in water-stressed wheat cultivars differing in drought tolerance. *Turk J Bot* 39: 625–634.

Maghsoudi K, Emam Y, Pessarakli M. (2016). Effect of silicon on photosynthetic gas exchange, photosynthetic pigments, cell membrane stability and relative water content of different wheat cultivars under drought stress conditions. *J Plant Nutr* 39: 1001–1015.

Mahdavi S, Kafi M, Fallahi E, Shokrpour M, Tabrizi L. (2016). Water stress, nano silica, and digoxin effects on minerals, chlorophyll index, and growth in ryegrass. *Int J Plant Prod* 10: 251–264.

Mahdieh M, Habibollahi N, Amirjani R, Abnosi MH, Ghorbanpour M. (2015). Exogenous silicon nutrition ameliorates salt-induced stress by improving growth and efficiency of PSII in *Oryza sativa* L. cultivars. *J Soil Sci Plant Nutr* 15: 1050–1060.

Marmiroli M, Pigoni V, Savo-Sardaro ML, Marmiroli N. (2014). The effect of silicon on the uptake and translocation of arsenic in tomato (*Solanum lycopersicum* L.) *Environ Exp Bot* 99: 9–17.

Marschner H. (1995). *Mineral Nutrition of Higher Plants*. 2nd ed. London: Academic Press.

Masarovič D, Slováková Ľ, Bokor B, Bujdoš M, Lux A. (2012). Effect of silicon application on *Sorghum bicolor* exposed to toxic concentration of zinc. *Biologia* 67: 706–712.

Mateos-Naranjo E, Gallé A, Florez-Sarasa I, Perdomo JA, Galmés J, Ribas-Cardó M, Flexas J. (2015). Assessment of the role of silicon in the Cu-tolerance of the C_4 grass *Spartina densiflora*. *J Plant Physiol* 178: 74–83.

Matoh T. (1997). Boron in plant cell walls. *Plant Soil* 193: 59–70.

Matoh T, Kairusmee P, Takahashi E. (1986). Salt-induced damage to rice plants and alleviation effect of silicate. *Soil Sci Plant Nutr* 32: 295–304.

Mauad M, Crusciol CAC, Nascente AS, Grassi Filho H, Lima, GPP. (2016). Effects of silicon and drought stress on biochemical characteristics of leaves of upland rice cultivars. *Rev Ciênc Agron* 47: 532–539.

Mees A, Klar T, Gnau P, Hennecke U, Eker APM, Carell T, Essen L. (2004). Crystal structure of a photolyase bound to a CPD-Like DNA lesion after in situ repair. *Science* 306: 1789–1793.

Meharg AA. (2004). Arsenic in rice—Understanding a new disaster for South-East Asia. *Trends Plant Sci* 9: 415–417.

Meharg AA, Hartley-Whitaker J. (2002). Arsenic uptake and metabolism in arsenic resistant and nonresistant plant species. *New Phytol* 154: 29–43.

Meharg AA, Naylor J, Macnair MR. (1994). Phosphorus nutrition of arsenate tolerant and nontolerant phenotypes of velvet grass. *J Environ Qual* 23: 234–238.

Mihaličová Malčovská S, Dučaiová Z, Maslaňáková I, Bačkor M. (2014). Effect of silicon on growth, photosynthesis, oxidative status and phenolic compounds in maize (*Zea mays* L.) grown in cadmium excess. *Water Air Soil Pollut* 225: 2056.

Mithofer A, Schuze B, Boland W. (2004). Biotic and heavy metal stress response in plants: Evidence for common signals. *FEBS Lett* 566: 1–5.

Moan J. (2001). Visible light and UV radiation. In Brune D, Hellborg R, Persson BRR, Pääkkönen R (eds.), *Radiation at Home, Outdoors and in the Workplace*. Oslo: Scandinavian Science Publisher, pp. 69–85.

Müller U. (2007). *Inorganic Structural Chemistry*. 2nd ed. Chichester, UK: John Wiley.

Muneer S, Jeong BR. (2015). Proteomic analysis of salt-stress responsive proteins in roots of tomato (*Lycopersicon esculentum* L.) plants towards silicon efficiency. *Plant Growth Regul* 77: 133–146.

Munns R. (2002). Comparative physiology of salt and water stress. *Plant Cell Environ* 25: 239–250.

Nable RO, Lance RCM, Cartwright B. (1990). Uptake of boron and silicon by barley genotypes with differing susceptibilities to boron toxicity. *Ann Bot* 66: 83–90.

Nieboer E, Richardson DHS. (1980). The replacement of the nondescript term "heavy metals" by a biologically and chemically significant classification of metal ions. *Environ Pollut Series B Chem Phys* 1: 3–26.

Nordstrom DK. (2002). Public health—Worldwide occurrences of arsenic in ground water. *Science* 296: 2143–2145.

Nowakowski W, Nowakowska J. (1977). Silicon and copper interaction in the growth of spring wheat seedlings. *Biol Plant* 39: 463–466.

Palmer CM, Guerinot ML. (2009). Facing the challenges of Cu, Fe and Zn homeostasis in plants. *Nat Chem Biol* 5: 333–340.

Pearson RG. (1968). Hard and soft acids HSAB. 1. Fundamental principles. *J Chem Educ* 45: 581–587.

Pereira TS, da Silva Lobato AK, Tan DKY, da Costa DV, Uchoa EB, do Nascimento Ferreira R, Silva Guedes EM. (2013). Positive interference of silicon on water relations, nitrogen metabolism, and osmotic adjustment in two pepper (*Capsicum annuum*) cultivars under water deficit. *Aust J Crop Sci* 7: 1064–1071.

Peer WA, Baxter IR, Richards EL, Freeman JL, Murphy AS. (2006). Phytoremediation and hyperaccumulator plants. *Top Curr Genet* 14: 299–340.

Postek MT. (1981). The occurrence of silica in the leaves of *Magnolia grandijlora* L. *Bot Gaz* 142: 124–134.

Prasad MNV. (1995). Cadmium toxicity and tolerance in vascular plants. *Environ Exp Bot* 35: 525–545.

Puig S, Peñarrubia L. (2009). Placing metal micronutrients in context: Transport and distribution in plants. *Curr Opin Plant Biol* 12: 299–306.

Putra ETS, Issukindarsyah, Taryono, Purwanto BH. (2015). Physiological responses of oil palm seedlings to the drought stress using boron and silicon applications. *J Agron* 14: 49–61.

Qin L, Kang W, Qi Y, Zhang Z, Wang N. (2016). The influence of silicon application on growth and photosynthesis response of salt stressed grapevines (*Vitis vinifera* L.). *Acta Physiol Plant* 38: 68.

Raven JA. (1985). Regulation of ph and generation of osmolarity in vascular plants: A cost-benefit analysis in relation to efficiency of use of energy, nitrogen and water. *New Phytol* 101: 25–77.

Reezi S, Babalar M, Kalantari S. (2009). Silicon alleviates salt stress, decreases malondialdehyde content and affects petal color of salt stressed cut rose (*Rosa xhybrida* L.) 'Hot Lady'. *Afr J Biotechnol* 8: 1502–1508.

Reid R. (2014). Understanding the boron transport network in plants. *Plant Soil* 385: 1–13.

Requejo R, Tena M. (2006). Maize response to acute arsenic toxicity as revealed by proteome analysis of plant shoots. *Proteomics* 6: 156–162.

Rizwan M, Ali S, Ibrahim M, Farid M, Adrees M, Bharwana SA, Zia-ur-Rehman M, Qayyum MF, Abbas F. (2015). Mechanisms of silicon-mediated alleviation of drought and salt stress in plants: A review. *Environ Sci Pollut Res* 22: 15416–15431.

Romero-Aranda MR, Jurado O, Cuartero J. (2006). Silicon alleviates the deleterious salt effect on tomato plant growth by improving plant water status. *J Plant Physiol* 163: 847–855.

Rybus-Zając M, Kubiś J. (2010). Effect of UV-B radiation on antioxidative enzyme activity in cucumber cotyledons. *Acta Biol Cracov Bot* 52: 97–102.

Samuels AL, Glass ADM, Ehret DL, Menzies JG. (1993). The effects of silicon supplementation on cucumber fruit: Changes in surface characteristics. *Ann Bot (London)* 72: 433–440.

Sanghera GS, Wani SH, Hussain W, Singh NB. (2011). Engineering cold stress tolerance in crop plants. *Curr Genomics* 12: 30–43.

Sanglard LMVP, Martins SCV, Detmann KC, Silva PEM, Lavinsky AO, Silva MM, Detmann E, Araújo WL, DaMatta FM. (2014). Silicon nutrition alleviates the negative impacts of arsenic on the photosynthetic apparatus of rice leaves: An analysis of the key limitations of photosynthesis. *Physiol Plant* 152: 355–366.

Saqib M, Zörb C, Schubert S. (2008). Silicon-mediated improvement in the salt resistance of wheat (*Triticum aestivum*) results from increased sodium exclusion and resistance to oxidative stress. *Funct Plant Biol* 35: 633–639.

Savvas D, Giotis D, Chatzieustratiou E, Bakea M, Patakioutas G. (2009). Silicon supply in soilless cultivations of zucchini alleviates stress induced by salinity and powdery mildew infections. *Environ Exp Bot* 65: 11–17.

Savvas D, Gizas G, Karras G, Salahas G, Papadimitriou M, Tsouka N. (2007). Interactions between silicon and NaCl-salinity in a soilless culture of roses in greenhouse. *Eur J Hort Sci* 72: 73–79.

Seyfferth AL, Fendorf S. (2012). Silicate mineral impacts on the uptake and storage of arsenic and plant nutrients in rice (*Oryza sativa* L.). *Environ Sci Technol* 46: 13176–13183.

Schaller J, Brackhage C, Bäucker E, Dudel EG. (2013). UV screening of grasses by plant silica layer? *J Biosci* 38: 413–416.

Schaller J, Brackhage C, Dudel E. (2012). Silicon availability changes structural carbon ratio and phenol content of grasses. *Environ Exp Bot* 77: 283–287.

Shanker AK, Cervantes C, Loza-Tavera H, Avudainayagam S. (2005). Chromium toxicity in plants. *Environ Int* 31: 739–753.

Sharma DC, Sharma CP, Tripathi RD. (2003). Phytotoxic lesions of chromium in maize. *Chemosphere* 51: 63–68.

Sharma P, Dubey RS. (2005). Lead toxicity in plants. *Braz J Plant Physiol* 17: 35–52.

Shen X, Li X, Li Z, Li J, Duan L, Eneji AE. (2010a). Growth, physiological attributes and antioxidant enzyme activities in soybean seedlings treated with or without silicon under UV-B radiation stress. *J Agron Crop Sci* 196: 431–439.

Shen X, Zhou Y, Duan L, Li Z, Egrinya Eneji A, Li J. (2010b). Silicon effects on photosynthesis and antioxidant parameters of soybean seedlings under drought and ultraviolet-B radiation. *J Plant Physiol* 167: 1248–1252.

Shi Q, Bao Z, Zhu Z, He Y, Qian Q, Yu J. (2005). Silicon-mediated alleviation of Mn toxicity in *Cucumis sativus* in relation to activities of superoxide dismutase and ascorbate peroxidase. *Phytochemistry* 66: 1551–1559.

Shi Y, Zhang Y, Han W, Feng R, Hu Y, Guo J, Gong H. (2016). Silicon enhances water stress tolerance by improving root hydraulic conductance in *Solanum lycopersicum* L. *Front Plant Sci* 7: 1–15.

Shi Y, Zhang Y, Yao H, Wu J, Sun H, Gong H. (2014). Silicon improves seed germination and alleviates oxidative stress of bud seedlings in tomato under water deficit stress. *Plant Physiol Biochem* 78: 27–36.

Schrope M. (2000). Successes in fight to save ozone layer could close holes by 2050. *Nature* 408: 627.

Schützendübel A, Schwanz P, Teichmann T, Gross K, Langenfeld Heyser R, Godbold DL, Polle A. (2001). Cadmium-induced changes in antioxidative systems, hydrogen peroxide content, and differentiation in Scots pine roots. *Plant Physiol* 127: 887–898.

Silva ON, Lobato AKS, Ávila FW, Costa RCL, Oliveira Neto CF, Santos Filho BG, Martins Filho AP et al. (2012). Silicon-induced increase in chlorophyll is modulated by the leaf water potential in two water-deficient tomato cultivars. *Plant Soil Environ* 58: 481–486.

Singh UM, Sareen P, Sengar RS, Kumar A. (2013). Plant ionomics: A newer approach to study mineral transport and its regulation. *Acta Physiol Plant* 35: 2641–2653.

Slama I, Abdelly C, Bouchereau A, Flowers T, Savouré A. (2015). Diversity, distribution and roles of osmoprotective compounds accumulated in halophytes under abiotic stress. *Ann Bot* 115: 433–447.

Soylemezoglu G, Demir K, Inal A, Gunes A. (2009). Effect of silicon on antioxidant and stomatal response of two grapevine (*Vitis vinifera* L.) rootstocks grown in boron toxic, saline and boron toxic-saline soil. *Sci Hortic* 123: 240–246.

Song A, Li P, Li P, Fan F, Nikolic M, Liang Y. (2011). The alleviation of zinc toxicity by silicon is related to zinc transport and antioxidative reaction in rice. *Plant Soil* 344: 319–333.

Song A, Li Z, Zhang J, Xue G, Fan F, Liang Y. (2009). Silicon-enhanced resistance to cadmium toxicity in *Brassica chinensis* L. is attributed to Si-supressed cadmium uptake and transport and Si-enhanced antioxidant defense capacity. *J Hazard Mater* 172: 74–83.

Soundararajan P, Sivanesan I, Jana S, Jeong BR. (2014). Influence of silicon supplementation on the growth and tolerance to high temperature in salvia splendens. *Hortic Environ Biotechnol* 55: 271–279.

Srivastava S, Tripathi RD, Dwivedi UN. (2004). Synthesis of phytochelatins and modulation of antioxidants in response to cadmium stress in *Cuscuta reflexa*—An angiospermic parasite. *J Plant Physiol* 161: 655–674.

Stapleton AE. (1992). Ultraviolet radiation and plants: Burning questions. *Plant Cell* 4: 1353–1358.

Steudle E. (2000). Water uptake by roots: An integration of views. *Plant Soil* 226: 45–56.

Steudle E, Peterson CA. (1998). How does water get through roots? *J Exp Bot* 49: 775–788.

Styk J. (2001). *Problém t'ažkých kovov (kadmium, olovo, med', zinok) v pôdach Štiavnických vrchov a ich príjem trávnymi porastami.* Bratislava: VÚPOP.

Takano J, Wada M, Ludewig U, Schaaf G, von Wiren N, Fujiwara T. (2006). The *Arabidopsis* major intrinsic protein NIP5;1 is essential for efficient boron uptake and plant development under boron limitation. *Plant Cell* 18: 1498–1509.

Tale Ahmad S, Haddad R. (2011). Study of silicon effects on antioxidant enzyme activities and osmotic adjustment of wheat under drought stress. *Czech J Genet Plant* 1: 17–27.

Tamai K, Ma JF. (2003). Characterization of silicon uptake by rice roots. *New Phytol* 158: 431–436.

Tan W, Meng QW, Brestic M, Olsovska K, Yang X. (2011). Photosynthesis is improved by exogenous calcium in heat-stressed tobacco plants. *J Plant Physiol* 168: 2063–2071.

Tangahu BV, Abdullah SRS, Basri H, Idris M, Anuar N, Mukhlisin M. (2011). A review on heavy metals (As, Pb, and Hg) uptake by plants through phytoremediation. *Int J Chem Eng* 2011: 1–31.

Tantawy AS, Salama YAM, El-Nemr MA, Abdel-Mawgoud AMR. (2015). Nano silicon application improves salinity tolerance of sweet pepper plants. *Int J ChemTech Res* 8: 11–17.

Tao SSH, Bolger PM. (1998). Dietary arsenic intakes in the United States: FDA total diet study, September 1991–December 1996. *Food Addit Contam* 16: 465–472.

Treutter D. (2005). Significance of flavonoids in plant resistance and enhancement of their biosynthesis. *Plant Biol* 7: 581–591.

Tripathi DK, Singh VP, Prasad SM, Chauhan DK, Dubey NK, Rai AK. (2015). Silicon-mediated alleviation of Cr(VI) in wheat seedlings as evidenced by chlorophyll fluorescence, laser induced breakdown spectroscopy and anatomical changes. *Ecotoxicol Environ Saf* 113: 133–144.

Tripathi P, Tripathi RD, Singh RP, Dwivedi S, Goutam D, Shri M, Trivedi PK, Chakrabarty D. (2013). Silicon mediates arsenic tolerance in rice (*Oryza sativa* L.) through lowering of arsenic uptake and improved antioxidant defence system. *Ecol Eng* 52: 96–103.

Tripathi RD, Tripathi P, Dwivedi S, Dubey S, Chatterjee S, Chakrabarty D, Trivedi PK. (2012). Arsenomics: Omics of arsenic metabolism in plants. *Frontiers Plant Physiol* 3: 275.

Tuna AL, Kaya C, Higgs D, Murillo-Amador B, Aydemir S, Girgin AR. (2008). Silicon improves salinity tolerance in wheat plants. *Environ Exp Bot* 62: 10–16.

Ullrich-Eberius CI, Sanz A, Novacky AJ. (1989). Evaluation of arsenate- and vanadate-associated changes of electrical membrane potential and phosphate transport in *Lemna gibba*-G1. *J Exp Bot* 40: 119–128.

Vaculík M, Landberg T, Greger M, Luxová M, Stoláriková M, Lux A. (2012). Silicon modifies root anatomy, and uptake and subcellular distribution of cadmium in young maize plants. *Ann Bot* 110: 433–443.

Vaculík M, Lux A, Luxová M, Tanimoto E, Lichtscheidl I. (2009). Silicon mitigates cadmium inhibitory effects in young maize plants. *Environ Exp Bot* 67: 52–58.

Vaculík M, Pavlovič A, Lux A. (2015). Silicon alleviates cadmium toxicity by enhanced photosynthetic rate and modified bundle sheath's cell chloroplasts ultrastructure in maize. *Ecotoxicol Environ Saf* 120: 66–73.

Vajpayee P, Tripathi RD, Rai UN, Ali MB, Singh SN. (2000). Chromium accumulation reduces chlorophyll biosynthesis, nitrate reductase activity and protein content in *Nympaea alba* L. *Chemosphere* 41: 1075–1082.

Van Bockhaven J, De Vleesschauwer D, Hofte M. (2013). Towards establishing broad-spectrum disease resistance in plants: Silicon leads the way. *J Exp Bot* 64: 1281–1293.

Vatehová Z, Kollárová K, Zelko I, Richterová-Kučerová D, Bujdoš M, Lišková D. (2012). Interaction of silicon and cadmium in *Brassica juncea* and *Brassica napus*. *Biologia* 67: 498–504.

Vert G, Briat JF, Curie C. (2003). Dual regulation of the *Arabidopsis* high affinity root iron uptake system by local and long-distance signals. *Plant Physiol* 132: 796–804.

Vojtáš J. (2000). *Analýza hygienického stavu pôd na Slovensku a návrh doplnenia kódu bonitovaných pôdnoekologických jednotiek pre kontaminované pôdy.* Bratislava: VÚPOP.

Wang L, Nie Q, Li M, Zhang F, Zhuang J, Yang W, Zhang F, Zhuang J, Yang W, Li T, Wang Y. (2005). Biosilicified structures for cooling plant leaves: A mechanism of highly efficient midinfrared thermal emission. *Appl Phys Lett* 87: 194105.

Waqas-ud-Din Khan, Aziz T, Waraich EA, Khalid M. (2015). Silicon application improves germination and vegetative growth in maize grown under salt stress. *Pak J Agri Sci* 52: 937–944.

Whitlow TH, Bassuk NL, Ranney TG, Reichert DL. (1992). An improved method for using electrolyte leakage to assess membrane competence in plant tissues. *Plant Physiol* 98: 198–205.

Williams L, Salt DE. (2009). The plant ionome coming into focus. *Curr Opin Plant Biol* 12: 247–249.

Williams PN, Price AH, Raab A, Hossain SA, Feldmann J, Meharg AA. (2005). Variation in arsenic speciation and concentration in paddy rice related to dietary exposure. *Environ Sci Technol* 39: 5531–5540.

Williams PN, Villada A, Deacon C, Raab A, Figuerola J, Green AJ, Feldmann J, Meharg AA. (2007). Greatly enhanced arsenic shoot assimilation in rice leads to elevated grain levels compared to wheat and barley. *Environ Sci Technol* 41: 6854–6859.

Wójcik M, Tukiendorf A. (2004). Phytochelatin synthesis and cadmium localization in wild type of *Arabidopsis thaliana*. *Plant Growth Regul* 44: 71–80.

Wojtaszek P. (1997). Oxidative burst an early plant response to pathogen infection. *Biochem J* 322: 681–692.

Wu CY, Yao YM, Shao P, Wang Y, Wang ZW, Tian XH. (2014). Exogenous silicon alleviates spikelet fertility reduction of hybrid rice induced by high temperature under field conditions. *Chin J Rice Sci* 28: 71–77.

Wu S. (1994). Effect of manganese excess on the soybean plant cultivated under various growth conditions. *J Plant Nutr* 17: 993–1003.

Wu XC, Lin WX, Huang ZL. (2007). Influence of enhanced ultraviolet-B radiation on photosynthetic physiologies and ultrastructure of leaves in two different resistivity rice cultivars [in Chinese with English abstract]. *Acta Ecol Sinica* 27: 0554–0564.

Xu XY, McGrath SP, Meharg AA, Zhao FJ. (2008). Growing rice aerobically markedly decreases arsenic accumulation. *Environ Sci Technol* 42: 5574–5579.

Yadav SK. (2010). Heavy metals toxicity in plants: An overview on the role of glutathione and phytochelatins in heavy metal stress tolerance of plants. *S Afr J Bot* 76: 167–179.

Yan YH, Zheng ZCh, Li TX, Zhang XZ, Wang Y. (2014). Effect of silicon on translocation and morphology distribution of lead in soil-tobacco system. *Yingyong Shengtai Xuebao* 25: 2991–2998.

Yao X, Chu J, Kunzheng C, Liu L, Shi J, Geng W. (2011). Silicon improves the tolerance of wheat seedlings to ultraviolet-B stress. *Biol Trace Elem Res* 143: 507–517.

Yeo AR, Flowers SA, Rao G, Welfare K, Senanayake N, Flowers TJ. (1999). Silicon reduces sodium uptake in rice (*Oryza sativa* L.) in saline conditions and this is accounted for by a reduction in the transpirational bypass flow. *Plant Cell Environ* 22: 559–565.

Yin L, Wang S, Li J, Tanaka K, Oka M. (2013). Application of silicon improves salt tolerance through ameliorating osmotic and ionic stresses in the seedling of Sorghum bicolor. *Acta Physiol Plant* 35: 3099–3107.

Yin L, Wang S, Liu P, Wang W, Cao D, Deng X. (2014). Silicon-mediated changes in polyamine and 1-aminocyclopropane-1-carboxylic acid are involved in silicon-induced drought resistance in *Sorghum bicolor* L. *Plant Physiol Biochem* 80: 268–277.

Yin L, Wang S, Tanaka K, Fujihara S, Itai A, Den X, Zhang S. (2016). Silicon-mediated changes in polyamines participate in silicon-induced salt tolerance in *Sorghum bicolor* L. *Plant Cell Environ* 39: 245–258.

Yoshida S, Ohnishi Y, Kitagishi K. (1962). Histochemistry of silicon in rice plant. *Soil Sci Plant Nutr* 8: 1–5.

Yu LJ, Luo YF, Liao B, Xie LJ, Chen L, Xiao S, Li JT, Hu SN, Shu WS. (2012). Comparative transcriptome analysis of transporters, phytohormone and lipid metabolism pathways in response to arsenic stress in rice (*Oryza sativa*). *New Phytol* 195: 97–112.

Zeng F, Zhao F, Qiu B, Ouyang Y, Wu F, Zhang G. (2011). Alleviation of chromium toxicity by silicon addition in rice plants. *Agric Sci China* 10: 1188–1196.

Zhang Q, Yan Ch, Liu J, Lu H, Wang W, Du J, Duan H. (2013). Silicon alleviates cadmium toxicity in *Avicennia marina* (Forsk.) Vierh. seedlings in relation to root anatomy and radial oxygen loss. *Mar Pollut Bull* 76: 187–193.

Zhang XH, Zhou D, Cui JJ, Ma HL, Lang DY, Wu XL, Wang ZS, Qiu HY, Li M. (2015). Effect of silicon on seed germination and the physiological characteristics of *Glycyrrhiza uralensis* under different levels of salinity. *J Hortic Sci Biotech* 90: 439–443.

Zhao F, Song C, He JQ, Zhu H. (2007). Polyamines improve K+/Na+ homeostasis in barley seedlings by regulating root ion channel activities. *Plant Physiol* 145: 1061–1072.

Zhao FJ, McGrath SP, Meharg AA. (2010). Arsenic as a food chain contaminant: Mechanisms of plant uptake and metabolism and mitigation strategies. *Ann Rev Plant Biol* 61: 535–559.

Zhu J, Liang YC, Ding YF, Li ZJ. (2006). Effects of silicon on photosynthesis and its related physiological parameters in two winter wheat cultivars under cold stress. *Sci Agric Sinica* 39: 1780–1788.

Zhu J-K. (2001). Plant salt tolerance. *Trends Plant Sci* 6: 66–71.

Zhu Y, Gong H. (2014). Beneficial effects of silicon on salt and drought tolerance in plants. *Agron Sustain Dev* 34: 455–472.

Zhu Z, Wei G, Li J, Qian Q, Yu J. (2004). Silicon alleviates salt stress and increases antioxidant enzymes activity in leaves of salt-stressed cucumber (*Cucumis sativus* L.). *Plant Sci* 167: 527–533.

Zuccarini P. (2008). Effects of silicon on photosynthesis, water relations and nutrient uptake of *Phaseolus vulgaris* under NaCl stress. *Biol Plantarum* 52: 157–160.

2 Role of Silicon under Nutrient Deficiency
Recent Advances and Future Perspective

Muhammad Zia-ur-Rehman, Fatima Akmal, Noorani Mir, Muhammad Rizwan, Asif Naeem, Muhammad Sabir, and Zahoor Ahmad

CONTENTS

ABSTRACT

Silicon is an element that ranks second on the whole Earth in terms of its abundance. It is accumulated by plants at a level equivalent to that of macronutrient elements such as calcium, magnesium, and phosphorous. Instead of their high accumulation in plants, Si does not come in the category of essential elements of higher plants because its essentiality has not been proven yet. However, the reason behind the increasing attention toward silicon is due to its beneficial effects on plant growth and development, particularly for those plants that are subjected to various biotic (plant diseases and pests) and abiotic stresses (salt and heavy metal toxicity). Moreover, those plants that are growing in soil medium uptakes silicon, but their concentrations vary among the plants around 0.1%–10% (w/w). In addition to this, plants are majorly categorized into three types: (i) excluder species, (ii) intermediate type, and (iii) silicon accumulator. Relatively more attention has been paid to the roles of silicon in inducing disease resistance and alleviating heavy metal toxicity. However, its role under nutrient-deficient conditions remains less focused. Therefore, in this chapter we review the role of Si in enhancing the availability of various nutrients, considering both the extrinsic and intrinsic factors, to mitigate their deficiency in plants.

2.1   SILICON AS MICRONUTRIENT

After carbon, silicon (Si) is the second most abundant element on the earth's surface, as it is present in the earth crust and is the constituent of clay minerals having a rich presence in soil, ranging from 50 to 400 g Si kg^{-1} (Kovda, 1973). In soil, Si is mostly present in the form of silica (SiO_2), which comprises 50%–70% of its mass (Ma and Yamaji, 2006). In soil solution, Si is present in the form of silicic acid [$Si(OH)_4$], which is absorbed by plant roots at <9 pH. Si has not yet been declared an essential element for higher plants, but its application to several crops, including sugarcane and rice, has shown noticeably better growth and yield production. Extensive cropping systems and indiscriminate use of commercial fertilizers have resulted in depletion of Si from the soil (Ma and Yamaji, 2006), which requires its frequent application as a fertilizer. It is very important in biotic and abiotic stress tolerance of plants, thereby protecting them from growth inhibition and yield loss (Epstein, 1999; Liang et al., 2007). Regarding absorption from soil and transport to shoots, plants may be categorized as (1) active Si accumulators, (2) passive Si accumulators, and (3) Si rejecters (Takahashi et al., 1990). Depending on the category in which they fall, plants can uptake Si ranging from 0.1% to 10% of its dry weight (Epstein, 1994, 1999; Ma and Yamaji, 2008).

2.1.1   MECHANISM OF SI UPTAKE BY PLANTS

The mechanism of Si uptake varies from plant to plant. Research was conducted to study the mechanism to Si uptake in rice, cucumber, and tomato plant. Mainly, Si uptake is usually done by the same transporter in all these plants; just the density of the transporters varies among plants (Mitani and Ma, 2005). Si is usually accumulated at different sites of plasma membrane, but not detected in vacuoles because the tonoplast lacks the Si transporters (Moore et al., 2014).

2.1.2   ROLE OF SILICON IN SOILS

Soil application of Si could affect the nutrient availability by influencing the following external factors:

- Optimize soil fertility
- Make nutrients more available to plants (Meena et al., 2014)
- Improve soil cation exchange capacity (CEC) (Smyth and Sanchez, 1980)

- Enhances uptake of phosphorus in plants (Matichenkov and Bocharnikova, 2001)
- Reduce aluminum (Al) toxicity by binding it to form hydroxyaluminosilicates (Wada and Wada, 1980; Cocker et al., 1998), and binding monomeric Al in solution in the presence of Si to reduce soluble Al concentrations (Barcelo et al., 1993).

2.1.3 Role of Si in Plants

Silicon accumulates in plants beneath the cuticle and forms a double layer, which limits fungal mycelium access and thus protects plants from infection (Yoshida et al., 1962; Bowen et al., 1992). Furthermore, Chérif et al. (1994) reported that under *Pythium* infection, Si supply to cucumber increased the activities of chitinase, polyphenol oxidase, and peroxidase. It was reported that Si shielded wheat and barley from attack by *Blumeria graminis* and rice from *Pyricularia oryzae* (Fawe et al., 1998; Fauteux et al., 2005). Among abiotic stresses, Si not only produces resistance against fungal diseases, but also protects plants from pests like the stem borer and plant hopper (Ma and Yamaji, 2006). That silicon induces a defense against different insects and pests, including sugarcane stalk borer, leaf spider, stem borer, green leaf hopper, brown plant hopper, and mites, has been witnessed by many researchers (Tanaka and Park, 1966; Ota et al., 1975; Yoshida, 1975; Maxwel et al., 1977; Coulibaly, 1990; Sawant et al., 1994; Savant et al., 1997).

Among abiotic stresses, Si induces resistance against heat, drought, heavy metals, and radiation (Bélanger et al., 2003; Côté-Beaulieu et al., 2009; Epstein, 2009). For example, under heavy metal (Al, Cd, Cr) stress, a deficiency of indole acetic acid (IAA) occurs, due to which the concentration of IAA oxidase increases (Morgan et al., 1976). On the other hand, the addition of Si to growing media increases the Zn content in plants, which is mandatory for IAA production. Thus, Si increases the IAA concentration by increasing the Zn content in plants under heavy metal stress. Similarly, it improves plant uptake of other nutrients to antagonize heavy metal uptake (Singh et al., 2011; Tripathi et al., 2012a,b). It is reported that under Cr, Cd, and Cu stress, the addition of Si to a hydroponic solution increases the uptake of both micro- and macronutrients, that is, Ca, Mg, P, K, Zn, Fe, and Mn (Rizwan et al., 2012; Keller et al., 2015; Tripathi et al., 2015). Tuna et al. (2008) found that in the presence of a high concentration of NaCl in the root zone, the addition of Si results in a decrease in the Ca and K content of plants and improves their growth. One of the intrinsic mechanisms is that Si could immobilize heavy metals in a nontoxic form inside the plant body (Liang et al., 2007).

Under a water shortage, Si could help plants grow for a longer time without any yield loss by preventing excessive loss of water through transpiration. For example, Gao et al. (2004, 2006) reported that Si addition in maize plants lowers the transpiration loss of water of plant leaves and increases the water use capability. It has been reported by several researchers that Si not only provides a physical barrier in plants, but also involves different physiological processes of plants, especially under stress conditions (Datnoff et al., 2007; Liang et al., 2007). Silicon indirectly influences biological systems and makes other elements available to plants, and thus improves plant yield (Birchall et al., 1989; Perry and Keeling-Tuker, 1998).

According to Ma and Yamaji (2006), repeated cropping and the constant application of chemical fertilizers such as nitrogen, phosphorus, and potassium have depleted the amount of plant-available Si in soil. An awareness of Si deficiency in soil is now recognized as being a limiting factor for crop production. There is very little information available on Si alleviation of nutrient deficiency, but recently this has become the focal point of researchers. Several researchers have studied the ameliorative effect of Si in nutrient deficiency (Gonzalo et al., 2013; Pavlovic et al., 2013; Bityutskii et al., 2014).

2.2 IRON NUTRITION AS A FUNCTION OF SILICON

2.2.1 Iron as a Micronutrient and Its Deficiency in Plants

Iron (Fe) is an essential micronutrient and plays a crucial role in plant growth and development. A variety of cellular functions, such as respiration chlorophyll biosynthesis, photosynthetic electron transport, and hexose biosynthesis, are significantly impaired under Fe-deficient conditions (Guerinot, 2000; Briat and Vert, 2004). A deficiency of iron leads to changes in the morphology and physiology of plants (Briat, 2007). Common symptoms in Fe-stressed plants include chlorosis of young leaves, stunting of leaves and roots, and reduced yields (Welch and Graham, 2004; Christin et al., 2009; Kabir et al., 2012). The first symptom of Fe deficiency is chlorosis of the leaves, which is characterized by a sharp decline in chlorophyll concentration (Pestana et al., 2005). Typically, Fe chlorosis begins as a pale leaf color, followed by yellowing of the interveinal areas, while the veins remain green. In severe cases, the entire leaf may turn chlorotic and develop necrotic spots. Iron chlorosis leads to retarded growth and death of the plant in extreme situations.

2.2.2 CAUSES OF FE DEFICIENCY

In the earth's crust, iron (Fe) is the fourth most abundant element, even though it is less soluble in calcareous soils because of being present in the form of oxides and hydroxide compounds. In calcareous soils, bicarbonate $\left(HCO_3^-\right)$ is present in very high quantities, which leads to a rise in pH and consequently unsolubilization of iron (Römheld and Marschner, 1986). Iron solubility in soils is highly pH dependent, decreasing by a factor of 1000 per unit of pH increase (Lindsay, 1979). This explains Fe deficiency in high pH and calcareous soils of arid regions. Excessive moisture and poor aeration are thus conducive to Fe deficiency in plants because a greater HCO_3^- content is caused by CO_2 accumulation. Moreover, iron tends to be poorly available to plants due to its fixation with soil minerals (Marschner, 1995). Since plant-available Fe exists primarily as organic complexes, low-organic-matter soils are low in available Fe. Low soil temperature slows down mineralization of soil organic matter and reduces the Fe supply to plants, which aggravates its deficiency. High levels of P, Mn, Cu, Mo, and Zn decrease Fe iron availability and uptake by plants. Similar is the effect of soil erosion and land leveling that expose the subsoil, which contains less organic matter and more $CaCO_3$ than topsoil.

2.2.3 FE SPECIATION IN SOILS

Fe is a transition metal element and is distinct from the elements at either end of the periodic table in that an electron can be added or removed from the inner electron orbital. Under aerobic conditions, Fe associate strongly with O_2^- and OH^- ions and tend to precipitate as insoluble hydroxide or as minor component of insoluble aluminosilicates. In anaerobic soils, Fe reduces to lower oxidation states; lower oxidation states are more water soluble, but they still tend to precipitate as carbonates and sulfides, or associate with organic matter, thus reducing their movement but increasing their plant availability.

Total soil Fe is always in excess of crop requirement, but the amount of soluble Fe is much less than the total Fe content. Soluble inorganic forms include Fe^{3+}, $Fe(OH)^{2+}$, $Fe(OH)^{2+}$, and Fe^{2+}. In well-aerated soils, however, Fe^{2+} contributes little to the total soluble inorganic Fe. Iron solubility is largely controlled by the solubility of hydrous Fe(III) oxides. These give rise to Fe^{3+} and its hydrolysis species:

$$Fe^{3+} + 3OH^- \rightarrow Fe(OH)_3 \text{ (solid)}$$

The equilibrium is very much in favor of $Fe(OH)_3$ precipitation and is highly pH dependent, with the activity of Fe^{3+} falling with increasing pH. Acidic soils are thus relatively higher in soluble inorganic Fe than calcareous soils, where its level can be extremely low (Lindsay, 1979). When soils are waterlogged, as in the case of flooded rice, a reduction of Fe^{3+} to Fe^{2+} takes place, accompanied by an increase in Fe solubility.

$$Fe(OH)_3 + e^- + 3H^+ \rightarrow Fe^{2+} + 3H_2O$$

2.2.4 ROLE OF SI TO COMBAT FE DEFICIENCY IN PLANTS

Silicon's influence on Fe deficiency is mainly a consequence of altered Fe distribution within the plant body, as no increase in Fe uptake has been described after Si addition. Iron storage in root (apoplastic or plaque) pools seems to be favored by Si addition, and its remobilization seems to be the main factor of the beneficial effect of Si on Fe deficiency. It has been reported that Si delays chlorophyll degradation in different plant species and under different abiotic stresses (Al-aghabary et al., 2004; Feng et al., 2010; Gottardi et al., 2012). There are two possible mechanisms: the first one is related to the structural protection of the chloroplast membranes by Si (Al-aghabary et al., 2004; Feng et al., 2010), and the second one is related to the effect of Si on strengthened cell walls, which might contribute to a more favorable position of leaves to intercept light and increase the photosynthesis rate (Liang, 1998; Ma and Takahasi, 2002). Also, Si could contribute to avoiding chlorophyll degradation by the activation of several enzymatic systems (Bybordi, 2012; Gottardi et al., 2012).

In a study by Gonzalo et al. (2013), the effects of silicon on the alleviation of iron deficiency of soybean (*Glycine max* L. cv. Klaxon), an iron-inefficient and chlorosis-susceptible plant, and cucumber (*Cucumis sativus* L. cv. Ashley), an iron-efficient plant, were evaluated in hydroponic solution using sodium silicate hydrate as the silicon source. Silicon addition prevented chlorophyll degradation, decelerated growth reductions, and maintained leaf iron content of the iron-inefficient soybean variety under iron deficiency conditions. In cucumber, silicon additions delayed reductions in stem dry weight, stem length, number of nodes, and root and shoot iron content.

Silicon supplied via nutrient solution ameliorated Fe deficiency symptoms of cucumber (*Cucumis sativus* L. cv. Semkross) and partially mitigated Zn or Mn deficiency symptoms (Bityutskii et al., 2014). The physiological effects of silicon under Fe deficiency included increases in leaf Fe content and tissue accumulation of organic acids and phenolic compounds. The benefits of silicon were attributed to increased distribution of Fe toward apical shoot portions and accumulation of Fe mobilizing compounds (e.g., citrate in leaves and roots or cathechin in roots). However, Si did not appear to affect mobility and tissue distribution of either Zn or Mn under deficiency conditions.

Si probably contributes to maintain a balance of other micronutrients, such as the Fe/Mn ratio (Pich et al., 1994), which is also beneficial in enhancing chlorophyll synthesis, and provides a possible explanation for the stimulation in growth of Fe-deficient plants supplied with Si (Gonzalo et al., 2013; Pavlovic et al., 2013; Bityutskii et al., 2014). The protective effect of Si in the amelioration of Fe chlorosis was more evident for soybean than for cucumber after a long period of deficiency (Gonzalo et al., 2013). Thus, Si's influence on iron deficiency might depend on the plant species. Pavlovic et al. (2013) concluded that for cucumber, the root responses to Fe deficiency (e.g., root morphological changes and reductive Fe uptake) were indirectly influenced by Si as a consequence of the improved Fe status in the whole plant, rather than directly caused by Si-modulated expression of strategy I–responsive genes (FRO2, IRT1, and HA1). Silicon accumulation in cucumber leaves and stems was, respectively, two and three times higher than in soybean (Gonzalo et al., 2013). The process that controls this accumulation was supposed to depend on plant transpiration rates, giving a positive correlation between the Si content in plants and the transpiration rates. If Fe interacts with Si in the apoplast and forms the apoplastic deposits already mentioned, Fe might follow the Si movement inside the plant and be effectively remobilized from the source to the sink. For both species, the highest Si content was found in the leaves (around 90% of the total Si in the plant) (Gonzalo et al., 2013). In the case of soybean, this Si content resembled the leaves' Fe content, which may imply a direct relationship between the Fe remobilization and transport from root to shoots and the Si content under Fe deficiency, which could improve the chlorosis resistance of the plant. Silicon treatment enhanced citrate concentration in plants. Citrate is one of the molecules that joined Fe on its transport through the xylem (Rellán-Alvarez et al., 2010). The increase in citrate concentration in leaf and root tissues and in the xylem sap of Si-treated cucumber plants, especially the first day after Fe elimination from the nutrient solution (Pavlovic et al., 2013; Bityutskii et al., 2014), should facilitate long-distance transport and Fe utilization in leaves.

However, a high citrate concentration has also been shown to inhibit Fe(III) citrate reduction in leaves (Nikolic and Römheld, 1999).

Another hypothesis has been tested in cucumber, as the Si influences the Fe chelate reductase and the expression profile of the strategy I–related genes FRO2, IRT1, and HA1. Due to the Fe deficiency in nutrient solution, apoplastic Fe pool is gradually decreased, and at the end of the experiment, it was almost threefold lower in the plants receiving Si as compared to those grown without Si. Thus, Si played a vital role in acquisition, mobilization of root Fe compound, and expression of Fe-related genes (Pavlovic et al., 2013).

A proposed hypothesis points out that Si application enhances the oxidation capacity of rice roots and augmented Fe oxidations to ferric insoluble compounds, which will form the Fe plaque (Okuda and Takahashi, 1962; Fu et al., 2012). Although oxide bound ions in soils is often relatively unavailable to plants, the root plaque–bound ions may become available for plant uptake if the plant exudes phytosiderophores that chelate the metal or simply dissolve the iron oxide, making the metal soluble again in the immediate vicinity of the roots (Zhang et al., 1998). In addition to this, accumulation of the iron oxides is also observed in the apoplastic regions (Fu et al., 2012). Moreover, treatment of plants with iron was also observed showing no effect on Si uptake and transport, although silicon facilitates enhanced transportation of Fe from roots to shoots. It was also suggested that the increased expression of Si transporters after Si addition might influence Fe uptake and translocation (Fu et al., 2012) and will benefit Fe nutrition under deficiency conditions.

2.2.5 FACTORS AFFECTING FE AVAILABILITY IN THE PRESENCE OF SI

In (nutrient or soil) solution, all forms of solid silica, crystalline, and amorphous are dissolved to a limited extent in water around pH 7.0, at different rates depending on the phase and surface area. At this pH, the equilibrium concentration of monomeric silicic acid (H_4SiO_4 or $Si(OH)_4$) is less than 2 mM, and above this, polycondensation takes place and colloidal particles could be formed (Birchall, 1990). The pH of nutrient solutions used to study Fe and other micronutrient deficiencies is normally around 7.5–8.0, in contrast to the acidic pH used in metal toxicity studies (pH 5.0–6.5); therefore, only Si concentrations below 2 mM could be tested. Usually, Fe(III)-EDTA and $FeSO_4$ are used as Fe sources in nutrient solution, but the low-stability constants make them inappropriate to be used at pH values above 6.5 and stronger chelates are required. Coupled with this, a decrease in Fe from the nutrient solution was expected due to its coprecipitation with Si or its adsorption onto the silicate surfaces (Gonzalo et al., 2013). Moreover, nutrient soil solution redox potential plays a fundamental role in Fe stability in solution. When $FeSO_4$ is added to pots, part of the Fe could be readily absorbed by the plant as Fe^{2+}, but as the nutrient solution is constantly aerated or due to the radial oxygen loss detected in rice roots under anaerobic conditions (Evans, 2004), Fe(II) is transformed into Fe(III) forms, and the precipitation of iron as oxy-hydroxides is expected. Conversely, Fe(III)-EDTA maintains enough Fe in solution at pH 6.5, but its stability is low at higher pH, and Si addition probably increases pH and may lead to Fe precipitation. As mentioned, Fe and Si could also coprecipitate as Fe(III) silicates or Si may precipitate as negatively charged silica particles (Currie and Perry, 2007) in which the positively charged Fe may bind, resulting in Fe depletion from the solution. All of this may cause an additional increase in Fe deficiency and the effect of Si addition could be disguised. In contrast, when Fe is supplied as a stronger chelate, such as Fe(III)-HBED (N,N-bis(2-hydroxybenzyl) ethylenediamine-N,N-diacetic acid), no Fe depletion in the nutrient solution is observed at any Si rate (Gonzalo et al., 2013).

2.2.6 RESPONSE OF DICOTS AND NONGRAMINACEOUS PLANTS TO FE DEFICIENCY

Two different strategies are used by the efficient plants to overcome the iron deficiency. Strategy I plants, which comprise the dicots and nongrass monocots, develop morphological and biochemical changes, as the enhancement of Fe(III)'s reduction capacity is mediated by a Fe reductase or the rhizosphere's acidification by proton excretion via a plasmalemma H^+-ATPase (Bienfait et al., 1985; Marschner and Römheld, 1994). Dicots and nongraminaceous plants increase the roots' Fe(III) reduction power through an Fe(III) chelate reductase coupled with an increase in the biosynthesis of a Fe(II) transporter; rhizosphere acidification and the release of phenolic compounds were also described as tools to increase the available Fe in the rhizosphere (strategy I) (Hindt and Guerinot, 2012). In *Arabidopsis* (strategy I plant species), iron is absorbed in three steps: rhizosphere acidification through H^+-ATPases, ferric reductase oxidase 2 (FRO_2)–meditated reduction of Fe(III) to Fe(II), and Fe(II) intake by iron-regulated transporter (IRT_1).

2.2.7 RESPONSE OF MONOCOT PLANT TO FE DEFICIENCY

Strategy II plants, which are represented by graminaceous plant species, acquire iron from soil through secretion of high-affinity Fe(III) chelator phytosiderophores (Hindt and Guerinot, 2012). Graminaceous plants (strategy II) release the rhizosphere-specific Fe chelating agents, called phytosiderophores, to solubilize Fe from the soils (Hindt and Guerinot, 2012). In plants from both strategies, insoluble Fe pools might be detected in the apoplast (Bienfait et al., 1985; Briat et al., 1995) and genotypes that form larger Fe apoplastic pools are less susceptible to develop Fe chlorosis symptoms (Longnecker and Welch, 1990). As an example, *Arabidopsis* plants growing with different Fe concentrations in the nutrient solution try to maintain the same Fe concentration in shoots (Baxter et al., 2008) by the remobilization of stored Fe pools within the plant (Jin et al., 2007; García-Mina et al., 2013). Like ferritin in the cell, Fe in these pools could be mobilized when needed and some of the mechanisms previously described in strategies I and II are involved in their solubility (Bienfait et al., 1985). The role of phenolics on the root's Fe pool remobilization in red clover was studied (Jin et al., 2007), and it was found that it delayed the activation of Fe(III) chelate reductase and diminished rhizospheric acidification (Jin et al., 2007), probably due to its contribution to the iron pool remobilization.

2.2.8 ROLE OF PLANT SPECIES AGAINST FE DEFICIENCY IN THE PRESENCE OF SI

Higher plants have been classified based on their ability to take up Si as active Si accumulators, passive accumulators, and Si rejective plants (Takahashi et al., 1990) and differ greatly in their ability to accumulate Si, ranging from 0.1% to 10.0% Si (dry weight) (Epstein, 1999). The differences in Si accumulation between species have been attributed to differences in the Si uptake ability of the roots. The mechanism that controls this uptake has been studied in different plants, such as rice; a typical Si active accumulator cucumber, which, among other dicots, accumulates higher Si concentrations in its shoots (Liang et al., 2005a,b); and tomato, which accumulates low levels of Si (Mitani and Ma, 2005). It was found that transportation of Si from the external solution to the cortical cells is mediated by a similar energy-dependent transporter in all three species, suggesting that the plant species' differing Si accumulations are due to the different densities of the transporter among plant species (Mitani and Ma, 2005). The concerted work of these two transporters produces an effective flow of Si in the exodermis and endodermis to override the barrier of Casparian strips (Ma et al., 2006, 2007).

2.3 ZINC ESSENTIAL AS MICRONUTRIENT

Zinc (Zn) is one of the essential micronutrients required by plants for various physiological processes. It is a constituent of many important enzymes that are used for energy transfer, protein synthesis, nitrogen metabolism, and protein, carbohydrate, lipid, and nucleic acid catabolism (Marschner, 1995; Pedas et al., 2009). Zinc is the stabilizer of the plasma membrane and is necessary for biological membrane stability; thus, lipids and proteins of membranes are protected from oxidative damage (Aravind and Prassad, 2004; Zhao et al., 2005). The average total Zn concentration in soil is approximately 50 mg kg^{-1}, with a range of 10–300 mg kg^{-1} (Mortvedt, 2000). At these concentrations, Zn is an essential element for plant metabolism and growth.

Zinc plays a vital role in the healthy growth and development of plants. Usually in leaf, the Zn requirement is approximately 15–20 mg kg^{-1} dry weight in most crops (Broadley et al., 2007). Zinc not only is important for plant growth, but also plays a very significant role for animals and humans. Thus, its deficiency not only affects plant yield and production, but also has a severe effect on humans. Worldwide, about 3 billion people are severely affected by Fe and Zn deficiency, especially people dependent on cereal-based foods, because in cereals Zn deficiency is the most common mineral disorder (Cakmak et al., 1999; Cakmak, 2002; Graham et al., 2001). Thus, zinc supply is very important for the accomplishment of various metabolic processes in plants. Worldwide, more than half of the agricultural soil is Zn deficient (Cakmak et al., 1999).

2.3.1 REASON FOR LOW-PLANT-AVAILABLE SOIL ZN

No doubt, Zn is present in all part of the world, but its availability to plants is limited. The presence of Zn in soil depends on the weathering of rocks and various soil-forming factors like climate, type, and intensity of weathering (Saeed and Fox, 1997). In spite of that, the limited plant availability of this element is also because of high pH values, high $CaCO_3$ contents, low organic matter contents, lower clay contents, and increased phosphorous applications (Imtiaz, 1999). Thus, calcareous, sandy, and peat soils, as well as soils with high phosphorous concentrations, are expected to have Zn deficiency. In submerged conditions, Zn is precipitated as $Zn(OH)_2$ and Zn-S in sulfur-rich and alkaline soils (Alloway, 2004, 2008). As the soil pH increases from 5 to 8, the Zn concentration in soil solution decreases from 10^{-4} to 10^{-10} M (Kiekens, 1995). Due to repeated application of phosphorous fertilizer, Zn's availability to plants is reduced; this is known as P-induced Zn deficiency (Singh et al., 1986).

2.3.2 ZN DEFICIENCY SYMPTOMS IN PLANTS

Zinc plays a vital role in plant growth and is involved in many enzymatic activities; its insufficient supply severely affects crop production. Deficiency of Zn leads to stunted growth, chlorosis, an expanded maturity period of crops, a reduced number of tillers, smaller leaves, and low quality of harvested crops (Hafeez et al., 2013). Due to low superoxide dismutase (SOD) activity in Zn-deficient plants, oxidative stress is induced, and that leads to reactive oxygen species (ROS) production. These ROS have a devastating impact on the biological membrane and its protein, enzymes, and chlorophyll. As a result, photosynthesis is inhibited (Cakmak, 2000). Thus, for optimal growth of plants, an adequate quantity of Zn must be supplied (Marschner et al., 1996). The reproductive and vegetative growth of plants is very much affected by Zn deficiency, as reported in rice plants in which a deficiency of Zn results in the reduction of dry weight, grain weight (grams per panicle), weight of 1000 seeds, and yield of rice plant (Mehrabanjoubani et al., 2014).

Common deficiency symptoms include stunting of plants, shortened petioles and internodes, dwarf leaves, necrotic spots on leaves, leaves cupping upward and developing interveinal chlorosis, bronzing of leaves, and rosetting of leaves (Snowball and Robson, 1986). Zinc-deficient plants are unthrifty, lack vigor, and give a patchy appearance with short and thin stems. In young plants, interveinal areas have dark brown necrotic lesions (Benton, 2003).

2.3.3 STRATEGIES TO OVERCOME ZN DEFICIENCY

The composition and concentration of growth media determine the uptake of Zn by plants. In the rhizosphere, Zn solubility should be increased by either acidification of growth media or the root releasing low-molecular-weight chelating agents (Sinclair and Krämer, 2012). Zinc is absorbed by roots as a divalent cation through mass flow or as complexes with low-molecular-weight ligands.

Zinc movement inside the root xylem is both symplastic and apoplastic. In phloem, mass flow movement is also reported (Erenoglu et al., 2010). A high concentration of metal binding compounds and pH is overcome by the release of chelaters that bind Zn and help it to move long distances inside the phloem. Inside plants, Zn homeostasis is controlled by chelators. Nicotianamine (NA) is a well-known chelator having a high affinity for binding Zn, and it ensures the movement of Zn inside the phloem (Sinclair and Krämer, 2012).

In Zn-deficient plants, there is dramatic decrease in the concentration of citrate, an organic compound that, along with malate, binds Zn in plants (Sinclair and Krämer, 2012). Excess Zn is stored in cell vacuoles as organic complexes, and it is remobilized in the case of deficiency (Sinclair and Krämer, 2012).

2.3.4 ROLE OF SI IN ZN DEFICIENCY

Cucumber plants grown in Zn-free solution showed no considerable change in chlorophyll, while they develop necrotic spots on their leaves. Silicon supply reduces necrotic spots of the leaves. The concentration of citrate in roots decreased due to Zn deficiency, but Si increased the citrate concentration in Zn-deficient plants. In the same way, fumarate and shikimate root concentrations decrease in micronutrient-deficient plants, whereas the addition of Si increases the fumarate concentration in Zn-deficient plants (Bityutskii et al., 2014).

Zn deficiency results in the production of ROS. It is reported that under abiotic stresses such as salt stress (Liang, 1999; Liang et al., 2003; Zhu et al., 2004), freezing stress (Liang et al., 2008), and drought stress (Gong et al., 2005), the Si mediates the antioxidant defense system in plants. Thus, it is supposed that the Si-mediated antioxidative defense in Zn-deficient plants helps to prevent the development of necrotic spots, as in cucumber, caused by hydroxyl radicals and superoxide anions (Bityutskii et al., 2014). At higher concentrations, the Zn distribution showed different patterns in the presence of Si in both soil and hydroponic experiments (Gu et al., 2011, 2012; Song et al., 2011). Translocation of Zn from roots to shoots is hindered by Si, which strengthens the root cell wall (Gong et al., 2006; Currie and Perry, 2007; Huang et al., 2009; Peleg et al., 2010). Zinc precipitates with Si as zinc silicate around the root epidermis (Gu et al., 2011), which may hinder Zn translocation in the xylem (da Cunha and do Nascimento, 2009). When zinc silicate breaks, the Zn is stored in an unknown form in the cell vacuole (Neumann and Zur Nieden, 2001). As silicate precipitates increases on the cell wall (Currie and Perry, 2007), the binding sites for Zn^{2+} increase (Wang et al., 2000). They bind the Zn^{2+} and limit its movement in the plant (Hodson and Sangster, 1999; Shi et al., 2005b; da Cunha and do Nascimento, 2009). A silicon supply leads to an increase in the bound fraction of Zn in the stems, roots, leaves, and sheaths of the rice plant (Gu et al., 2012). The Si-bound Zn could be remobilized under its deficiency, but the correlation between silicon supplementation and Zn remobilization within plants under Zn-deficient conditions has not yet been studied. Silicon addition increases citrate concentration, which plays a role in the more efficient redistribution of Zn. That is why the Zn content increases in seeds and fruits: efficient Zn use under deficient conditions (Hernandez-Apaolaza, 2014).

Silicon application under Zn treatments ranging from deficiency (1 µg L^{-1} Zn) to moderate toxicity (100 µg L^{-1} Zn) improved both the vegetative and reproductive growth of rice. A maximum growth of rice plant by Si addition was observed at 10 µg L^{-1} Zn treatment. At 100 µg Zn L^{-1}, Si showed a positive effect by alleviating its toxicity. Silicon caused an increase in K^+ and Fe content that might have played a role in ameliorating Zn toxicity. The authors concluded that Si maintained the optimal level of Zn in rice plants (Mehrabanjoubani et al., 2014).

2.4 MANGANESE

Manganese is the 12th most important element, which is present at 0.098% in the biosphere of the earth (Siegel and Siegel, 2000). It is present in the soil as well as the water and sediments and is essential for both humans and plants. It exists in different forms, including oxides, silicates, phosphates, and borates (Gerber et al., 2002; Howe et al., 2004). The most commonly occurring manganese-bearing minerals are manganite, pyrolusite, and rhodonite.

2.4.1 Fate of Mn in Soils

Manganese is released in soil as a result of mineral weathering and atmospheric deposition, originating from both natural and anthropogenic sources. There are three possible oxidation states of manganese in soil: Mn(II), Mn(III), and Mn(IV). The divalent ion is the only form that is stable in soil solution, while Mn(III) and Mn(IV) are only stable in the solid phase of soil (Mcbride, 1994). Manganese mobility in soil is extremely sensitive to soil conditions such as acidity, wetness, organic matter content, and biological activity. The solubility of soil manganese is thus controlled by redox potential and soil pH, where low pH or low redox potential favors the reduction of insoluble manganese oxides, resulting in increased manganese mobility. At soil pH values above 6, manganese binds with organic matter, oxides, and silicates, whereby its solubility decreases. Manganese availability and solubility are thus generally low at high pH values and high organic matter contents, while in acidic soils with low organic matter contents its availability is high. The solubility of manganese is also high in anaerobic conditions at pH values above 6, as well as in aerobic conditions at pH values below 5.5 (Mcbride, 1994; Kabata-Pendias and Pendias, 2001). The availability of Mn is enhanced by nitrates, sulfates, and chlorides (Moore, 1991).

2.4.2 Role of Mn in Plants

Mn is an important element that activates various enzymes in plants during their growth and development and is also a cofactor for various enzymes (Mukhopadhyay and Sharma, 1991). Manganese is an essential component of Mn containing superoxide dismutase (Mn-SOD), which protects plants from ROS such as singlet oxygen and hydrogen peroxide, converting them into H_2O_2 and then water. Mn^{2+} is its plant-available form. Foliar Mn is essential only for citrus and in other tree crops, to cure its deficiency.

2.4.3 Plants' Responses to Mn Deficiency

Plant species and cultivars within species differ considerably in their susceptibility to Mn deficiency. For example, oat, wheat, soybean, and peaches are very susceptible, whereas maize and rye are much less susceptible (Reuter et al., 1988). In cereals, such as barley and wheat, Mn deficiency leads to significant yield reductions and may cause the complete loss of crops during the winter. Below the critical deficiency level, dry matter production, net photosynthesis, and chlorophyll content decline rapidly. Thylakoid structure and chlorophyll degradation occurs under Mn deficiency (Papadakis et al., 2007), leading to the development of characteristic interveinal leaf chlorosis, the most distinct symptom of Mn deficiency. In cereals, greenish gray spots on the more basal leaves (gray specks) are the major visual symptoms. Soils known to be low in plant-available Mn are typically sandy and calcareous, which, in combination with a high content of soil organic matter, favors oxidation of soluble Mn^2 to plant-unavailable MnO_2 (Husted et al., 2005). Soil acidity, or low soil pH, as indicated by an abundance of hydrogen ions (H^+), can cause mineral Mn oxide to dissolve and release enough manganese ions (Mn^{2+}) into the soil solution to make the soil toxic to many plants.

2.4.4 Role of Si in Mitigating Mn Deficiency in Plants

The interaction of manganese and Si in different plants such as rice (Okuda and Takahashi, 1962), barley (Williams and Vlamis, 1957; Horiguchi and Morita, 1987), bean (Horst and Marschner, 1978), and pumpkin (Iwasaki and Matsumura, 1999) grown in hydroponics has been well described.

Generally, $MnSO_4$ is used as a Mn source in nutrient solution or added as a fertilizer in the soil. But well-aerated conditions favor the formation of insoluble oxidized Mn species—Mn(III) and Mn(IV) (Lindsay, 1979)—that might decrease the solution to very low concentrations. Thermodynamically, oxidation of Mn^2 to Mn^4 is favored in aerated soils above pH 4, but the high activation energy required for Mn^2 oxidation inhibits or postpones oxidation processes (Gilkes and McKenzie, 1988). These facts promote the deposition of Mn oxides on root surfaces.

In rice, Si contributes to enhancing the roots' oxidizing capacity, giving a higher Mn oxidation rate in the rhizosphere and increasing the precipitation outside the plant (Okuda and Takahashi, 1962). As mentioned before for Fe and Zn, these deposits could be used under Mn deficiency and could ameliorate Mn deficiency symptoms for a while. In barley, cucumber, and bean (Williams and Vlamis, 1957; Horst and Marschner, 1978; Horiguchi and Morita, 1987; Shi et al., 2005a), Si contributed to a Mn homogeneous distribution in the leaf instead of its concentration in necrotic spots, but the mechanism that controls this is still unknown. On the contrary, Mn absorption by pumpkin was not affected by Si addition, but promoted a Mn accumulation on the trichome's base (Iwasaki and Matsumura, 1999). In cowpea, a lower amount of Mn in the apoplast was obtained when Si was added to the media. This could be explained through the metal adsorption on cell walls mediated by the Si deposits, as for Zn (Horst et al., 1999), but also because soluble Si in the apoplast may affect the Mn oxidation state, promoting its precipitation (Iwasaki et al., 2002).

Cucumber plants treated with Si presented more than 90% of the Mn bound to the cell wall, in comparison with untreated plants, in which 50% of the Mn was in the symplast and 50% in the apoplast (Rogalla and Römheld, 2002; Wiese et al., 2007). Recently, it has been concluded that the Si supply to the nutrient solution did not seem to improve the Mn deficiency symptoms of cucumber (Bityutskii et al., 2014), but as mentioned above for Zn, plants were grown with any Mn in the nutrient solution; therefore, the possibility of pool formation was eliminated. Mn is either present as a precipitate in the apoplast or bound to cell walls; plants might use it in the case of deficiency, although the amount of Mn used in such a case might not be enough to completely suppress symptoms. Also, the mechanism that controls this mobilization, the location of Mn deposits, the Mn availability from deposits, and the Mn speciation on them require further research and the utilization of sophisticated image techniques.

The beneficial effect of Si in cucumber was also explained by its contribution to decrease the membrane lipid peroxidation through the increment of different antioxidants (Shi et al., 2005a). Dragišić et al. (2007) concluded that Si contributes indirectly to a decrease in OH^- in the leaf apoplast by decreasing the free apoplastic Mn^{2+}, thus regulating the Fenton reaction and protecting the plant against Mn toxicity. This mechanism could also explain plant protection against deficiency. Williams and Vlamis (1957) discovered for the first time that total Mn in the leaves was unaffected by Si, but Si caused Mn to be more evenly distributed instead of being concentrated in discrete necrotic spots. This finding has been supported by subsequent experiments (Horst and Marschner, 1978; Horiguchi and Morita, 1987; Shi et al., 2005b). Horst et al. (1999) observed that Si lowered the apoplastic Mn concentration in cowpea, suggesting that Si may modify the cation binding capacity of the cell wall.

2.5 COPPER

Copper (Cu) is an essential element for plant growth. Copper concentrations in soils range between 3 and 100 mg kg^{-1}, but only about 1%–20% exists in a free, readily bioavailable form; the majority is bound to organic matter. Copper is fairly immobile between cell tissues, and deficiency symptoms first appear in the newly formed, younger cells and the reproductive parts. This trace element plays important roles in CO_2 assimilation and ATP synthesis and is a component of various proteins, particularly those involved in both the photosynthetic (plastocyanin) and respiratory (cytochrome oxidase) electron transport chain (Demirevska-kepova et al., 2004). The tendency of uptake and accumulation of Copper from soil to root system depends on the total amount of copper available in the soil and ability of plants to cross the interface of soil and root (Agata and Ernest, 1998). Copper is both a micronutrient for plants and a heavy metal capable of stress induction.

2.5.1 Cu-Associated Deficiency Symptoms

Symptoms of deficiency include impaired photosynthetic electron transport, reduced respiration, and stunted growth due to defects in apical meristems, as well as the rolling up and wilting of leaves (Marschner, 1995).

2.5.2 Interaction of Si and Cu

There are fewer studies that review the Si-Cu interaction in plants (Frantz et al., 2011; Khandekar and Leisner, 2011). In *Arabidopsis thaliana*, Cu toxicity symptoms such as chlorosis on leaves and reduction of shoot and root biomass were diminished by Si addition to the nutrient solution (Khandekar and Leisner, 2011); similar findings were observed for wheat (Nowakowski and Nowakowska, 1997). But leaf Cu concentration did not significantly change as a result of Si addition, suggesting that Si influenced the distribution or bioavailability of Cu within leaves under Cu stress (Li and Leisner, 2008). The formation of Si deposits on the cell wall increased the Cu binding sites and avoided the impact of such high Cu on plant cells that had been proposed to explain this fact, similar to those proposed for the toxicity of other micronutrients (Wang et al., 2000; Rogalla and Römheld, 2002; Ma and Yamaji, 2006; Liang et al., 2007; Frantz et al., 2011). But this could not be the sole mechanism to explain the plant behavior upon Si application against Cu toxicity, because high levels of molecules that bind Cu, as a strategy to minimize its toxic effect, were maintained or even increased when Si was added; therefore, more than one response should be activated to tolerate Cu toxicity (Khandekar and Leisner, 2011).

Erica andevalensis is the plant species that is growing in acidic and metal-rich soil medium and was treated with silicon and toxic element copper to observe the effect of silicon (Oliva et al., 2011). Silicon improved plant growth and reduced the water loss associated with plant death from excess Cu. Leaf Cu concentrations were reduced (up to 32%), while root Cu was increased with Si. Moreover, the phytolith leaves containing Si deposits not only associated with copper but also from other nutrients of plants like calcium (Ca), potassium (K), and phosphorus (P). The silicon-induced Cu tolerance was determined to be a result of inhibited Cu transport from roots to shoots. It was also suggested that the Si phytoliths in leaves may have contributed to Cu toxicity tolerance by immobilizing or deactivating Cu. Si doses were 0.1, 1.7, and 3.4 mM, and Cu concentrations ranged between 1.5 and 150 µM Cu, at pH 5.7. Keller et al. (2015) reported the effect of silicon with copper exposed 0–30 µM to *Triticum turgidum* L. in a hydroponic medium and found that the toxicity of copper can be alleviated by silicon.

Keller et al. (2015) also propounded three major mechanisms of limiting uptake of copper and their translocation to shoots from the root:

1. Alleviation of copper adsorption onto the outer surface of leaf and consequently their immobilization in the epidermis of root
2. Increased Cu complexation by both inorganic and organic anions such as aconitate
3. Restriction of translocation in the thickened silicon-loaded endodermis areas

REFERENCES

Agata, F., and B. Ernest. 1998. Meta-metal interactions in accumulation of V^{5+}, Ni^{2+}, Mo^{6+}, Mn^{2+} and Cu^{2+} in under and above ground parts of *Sinapis alba*. *Chemosphere* 36:1305–1317.

Al-aghabary, K., Z. Zhu, and Q. Shi. 2004. Influence of silicon supply on chlorophyll content, chlorophyll fluorescence, and antioxidative enzyme activities in tomato plants under salt stress. *J. Plant Nutr.* 27:2101–2115.

Alloway, B.J. 2004. *Zinc in Soil and Crop Nutrition*. International Zinc Association and International Fertilizer Industry Association, Brussels.

Alloway, B.J. 2008. Micronutrients and crop production. In B.J. Alloway (ed.), *Micronutrient Deficiencies in Global Crop Production*, pp. 1–39. Springer Science Business Media BV.

Aravind, P., and M.N.V. Prassad. 2004. Zinc protects chloroplasts and associated photochemical functions in cadmium exposed *Ceratophyllum demersum* L., a fresh water macrophyte. *Plant Sci.* 166:1321–1327.

Barcelo, J., P. Guevara, and C. Poschenrieder. 1993. Silicon amelioration of aluminium toxicity in teosinte (*Zea mays* L. ssp. Mexicana). *Plant Soil* 154:249–255.

Baxter, I.R., O. Vitek, B. Lahner, B. Muthukumar, M. Borghi, J. Morrissey, M.L. Guerinot, and D.E. Salt. 2008. The leaf ionome as a multivariate system to detect plant's physiological status. *Proc. Natl. Acad. Sci. U.S.A.* 105:12081–12086.

Bélanger, R.R., N. Benhamou, and J.G. Menzies. 2003. Cytological evidence of an active role of silicon in wheat resistance to powdery mildew (*Blumeria graminis* f. sp. tritici). *Phytopathology* 93:402–412.

Benton, J.J. 2003. *Agronomic Handbook; Management of Crops, Soils and Their Fertility*. CRC Press, Boca Raton, FL.

Bienfait, H.F., W. Vandenbriel, and N.T. Meslandmul. 1985. Free space iron pools in roots generation and mobilization. *Plant Physiol.* 78:596–600.

Birchall, J.D. 1990. Role of silicon in biology. *Chem. Br.* 26:141–144.

Birchall, J.D., C. Exley, J.S. Chappell, and M.J. Philips. 1989. Acute toxicity of aluminium to fish eliminated in silicon-rich acid waters. *Nature* 338:146–148.

Bityutskii, N., J. Pavlovic, K. Yakkonen, V. Maksimovi, and M. Nikolic. 2014. Contrasting effect of silicon on iron, zinc and manganese status and accumulation of metal mobilizing compounds in micronutrient-deficient cucumber. *Plant Physiol. Biochem.* 74:205–211.

Bowen, P., J. Menzies, D. Ehret, L. Samuels, and A.D.M. Glass. 1992. Soluble silicon sprays inhibit powdery mildew development on grape leaves. *J. Am. Soc. Hortic. Sci.* 117:906–912.

Briat, J.F. 2007. Iron dynamics in plants. In J.C. Kader and M. Delseny (eds.), *Advances in Botanical Research: Incorporating Advances in Plant Pathology*, pp. 138–169. Academic Press, London.

Briat, J.F., I. Fobis-Loisy, N. Grignon, S. Lobreaux, N. Pascal, G. Savino, S. Thoiron, N. von Wiren, and O. Van Wuytswinkel. 1995. Cellular and molecular aspects of iron metabolism in plants. *Biol. Cell.* 84:69–81.

Briat, J.F., and G. Vert. 2004. Acquisition et gestion du fer par les plantes. Cahier d'e´tudes et de recherches francophones. *Agriculture* 13:183–201.

Broadley, M.R., P.J. White, J.P. Hammond, I. Zelko, and A. Lux. 2007. Zinc in plants. *New Phytol.* 173:677–702.

Bybordi, A. 2012. Effect of ascorbic acid and silicium on photosynthesis, antioxidant enzyme activity, and fatty acid contents in canola exposure to salt stress. *J. Integr. Agric.* 11:1610–1620.

Cakmak, I. 2002. Plant nutrition research: Priorities to meet human needs for food in sustainable ways. *Plant Soil* 247:3–24.

Cakmak, I., M. Kalayci, H. Ekiz, H.J. Braun, and A. Yilmaz. 1999. Zinc deficiency as an actual problem in plant and human nutrition in Turkey: A NATO-science for stability project. *Field Crop Res.* 60:175–188.

Chérif, M., A. Asselin, and R.R. Bélanger. 1994. Defense responses induced by soluble silicon in cucumber roots infected by *Pythium* spp. *Phytopathology* 84:236–242.

Christin, H., P. Petty, K. Ouertani, S. Burgado, C. Lawrence, and M.A. Kassem. 2009. Influence of iron, potassium, magnesium, and nitrogen deficiencies on the growth and development of sorghum (*Sorghum bicolor* L.) and sunflower (*Helianthus annuus* L.) seedlings. *J. Biotechnol. Res.* 1:64–71.

Cocker, K.M., D.E. Evans, and M.J. Hodson. 1998. The amelioration of aluminium toxicity by silicon in higher plants: Solution chemistry or an in planta mechanism? *Physiol. Plant.* 104:608–614.

Côté-Beaulieu, C., F. Chain, J.G. Menzie, S.D. Kinrade, and R.R. Bélanger. 2009. Absorption of aqueous inorganic and organic silicon compounds by wheat and their effect on growth and powdery mildew control. *Environ. Exp. Bot.* 65:155–161.

Coulibaly, K. 1990. Influence of nitrogen and silicon fertilization on the attack on sugarcane by the stalk borer (*Eldana saccharina* Walker). Sugarcane spring supplement cucumber roots infected by *Pythium* spp. *Phytopathology* 84:236–242.

Currie, H.A., and C.C. Perry. 2007. Silica in plant biological, biochemical and chemical studies. *Ann. Bot.* 100:1383–1389.

da Cunha, K.P.V., and C.W.A. do Nascimento. 2009. Silicon effects on metal tolerance and structural changes in maize (*Zea mays* L.) grown on a cadmium and zinc enriched soil. *Water Air Soil Pollut.* 197:323–330.

Datnoff, L.E., F.A. Rodrigues, and K.W. Seebold. 2007. Silicon and plant disease. In L.E. Datnoff, W.H. Elmer, and D.M. Huber (eds.), *Mineral Nutrition and Plant Disease*, pp. 233–246. APS Press, St. Paul, MN.

Demirevska-kepova, K., L. Simova-Stoilova, Z. Stoyanova, R. Holzer, and U. Feller. 2004. Biochemical changes in barley plants after excessive supply of copper and manganese. *Environ. Exp. Bot.* 52:253–266.

Dragišić, J., J. Bogdanović, V. Maksimović, and M. Nikolic. 2007. Silicon modulates the metabolism and utilization of phenolic compounds in cucumber (*Cucumis sativus* L.) grown at excess manganese. *J. Plant Nutr. Soil Sci.* 170:739–744.

Epstein, E. 1994. The anomaly of silicon in plant biology. *Proc. Natl. Acad. Sci. U.S.A.* 91:11–17.

Epstein, E. 1999. Silicon. *Ann. Rev. Plant Physiol. Plant Mol. Biol.* 50:641–664.

Epstein, E. 2009. Silicon: Its manifold roles in plants. *Ann. Appl. Biol.* 155:155–160.

Erenoglu, E.B., U.B. Kutman, Y. Ceylan, B. Yildiz, and I. Cakmak. 2010. Improved nitrogen nutrition enhances root uptake, root-to-shoot translocation and remobilization of zinc (65Zn) in wheat. *New Phytol.* 189:438–448.

Evans, D.E. 2004. Aerenchyma formation. *New Phytol.* 161:35–49.

Fauteux, F., W. Remus-Borel, J.G. Menzies, and R.R. Belanger. 2005. Silicon and plant disease resistance against pathogenic fungi. *FEMS Microbiol Lett.* 249:1–6.

Fawe, A., M. Abou-Zaid, J.G. Menzies, and R.R. Bélanger. 1998. Silicon-mediated accumulation of flavonoid phytoalexins in cucumber. *Biochem. Cell Biol.* 88:396–401.

Feng, J., Q. Shi, X. Wang, M. Wei, F. Yang, and H. Xu. 2010. Silicon supplementation ameliorated the inhibition of photosynthesis and nitrate metabolism by cadmium (Cd) toxicity in *Cucumis sativus* L. *Sci. Hortic.* 123:521–530.

Frantz, J.M., S. Khandekar, and S. Leisner. 2011. Silicon differentially influences copper toxicity response in silicon-accumulator and non-accumulator species. *J Am SocHortSci* 136:329–338.

Fu, Y.Q., H. Shen, D.M. Wu, and K.Z. Cai. 2012. Silicon-mediated amelioration of Fe^{2+} toxicity in rice (*Oryza sativa* L.) roots. *Pedosphere* 22:795–802.

Gao, X., C. Zou, L. Wang, and F. Zhang. 2004. Silicon improves water use efficiency in maize plants. *J. Plant Nutr.* 27:1457–1470.

Gao, X., C. Zou, L. Wang, and F. Zhang. 2006. Silicon decreases transpiration rate and conductance from stomata of maize plants. *J. Plant Nutr.* 29:1637–1647.

García-Mina, J.M., E. Bacaicoa, M. Fuentes, and E. Casanova. 2013. Fine regulation of leaf iron use efficiency and iron root uptake under limited iron bioavailability. *Plant Sci.* 198:39–45.

Gerber, G.B., A. Léonard, and P. Hantson. 2002. Carcinogenicity, mutagenicity and teratogenicity of manganese compounds. *Crit. Rev. Oncol. Hematol.* 42:25–34.

Gilkes, R.J., and R.M. McKenzie. 1988. Geochemistry and mineralogy of manganese in soils. In R. Graham, R.J. Hannan, and N.C. Uren (eds.), *Manganese in Soils and Plants*, pp. 23–35. Kluwer Academic Publishers, Dordrecht.

Gong, H.J., D.P. Randall, and T.J. Flowers. 2006. Silicon deposition in the root reduces sodium uptake in rice (*Oryza sativa* L.) seedlings by reducing bypass flow. *Plant Cell Environ.* 29:1970–1979.

Gong, H.J., X.Y. Zhu, K.M. Chen, S.M. Wang, and C.L. Zhang. 2005. Silicon alleviates oxidative damage of wheat plants in pots under drought. *Plant Sci.* 169:313–321.

Gonzalo, M.J., J.J. Lucena, and L. Hernández-Apaolaza. 2013. Effect of silicon addition on soybean (*Glycine max*) and cucumber (*Cucumis sativus*) plants grown under iron deficiency. *Plant Physiol. Biochem.* 70:455–461.

Gottardi, S., F. Iacuzzo, N. Tomasi, G. Cortella, L. Manzocco, R. Pinton, V. Römheld, T. Mimmo, M. Scanpicchio, L. Dalla Cost, and S. Cesco. 2012. Beneficial effects of silicon on hydroponically grown corn salad (*Valerianella locusta* (L.) Laterr) plants. *Plant Physiol. Biochem.* 56:14–23.

Graham, R.D., R.M. Welch, and H.E. Bouis. 2001. Addressing micronutrients malnutrition through enhancing the nutritional quality of staple foods principles, perspectives and knowledge gaps. *Adv. Agron.* 70:77–142.

Gu, H.H., H. Qiu, T. Tian, S.S. Zhan, T.H.B. Deng, R.L. Chaney, S.Z. Wang, Y.T. Tang, J.L. Morel, and R.L. Qiu. 2011. Mitigation effects of silicon rich amendments on heavy metal accumulation in rice (*Oryza sativa* L.) planted on multimetal contaminated acidic soil. *Chemosphere* 83:1234–1240.

Gu, H.H., S.S. Zhan, S.Z. Wang, Y.T. Tang, R.L. Chaney, X.H. Fang, X.D. Cai, and R.L. Qiu. 2012. Silicon-mediated amelioration of zinc toxicity in rice (*Oryza sativa* L.) seedlings. *Plant Soil* 350:193–204.

Guerinot, M.L. 2000. Improving rice yields—Ironing out the details. *Nat. Biotechnol.* 19:417–418.

Hafeez, F.Y., M. Abaid-Ullah, and M.N. Hassan. 2013. Plant growth promoting rhizobacteria as zinc mobilizers: A promising approach for cereals biofortification. In *Bacteria in Agrobiology: Crop Productivity*, pp. 217–235. Springer, New York.

Hernandez-Apaolaza, L. 2014. Can silicon partially alleviate micronutrient deficiency in plants? A review. *Planta* 240:447–458.

Hindt, M.N., and M.L. Guerinot. 2012. Getting sense for signals. Regulation of the plant iron deficiency response. *Biochim. Biophys. Acta* 1823:1521–1530.

Hodson, M.J., and A.G. Sangster. 1999. Aluminium/silicon interactions in conifers. *J. Inorg. Biochem.* 76:89–98.

Horiguchi, T., and S. Morita. 1987. Mechanism of manganese toxicity and tolerance of plants. IV. Effect of silicon on alleviation of manganese toxicity of barley. *J. Plant Nutr.* 10:2299–2310.

Horst, W.J., M. Fecht, A. Naumann, A.H. Wissemeier, and P. Maier. 1999. Physiology of manganese toxicity and tolerance in *Vigna unguiculata* (L.) Walp. *J. Plant Nutr. Soil Sci.* 162:263–274.

Horst, W.J., and H. Marschner. 1978. Effect of silicon on manganese tolerance of bean plants (*Phaseolus vulgaris* L.). *Plant Soil* 50:287–303.

Howe, P., H. Malcolm, and S. Dobson. 2004. *Manganese and Its Compounds: Environmental Aspects*. World Health Organization, Geneva.

Huang, C.F., N. Yamaji, M. Nishimura, S. Tajima, and J.F. Ma. 2009. A rice mutant sensitive to Al toxicity is defective in the specification of root outer cell layers. *Plant Cell Physiol.* 50:976–985.

Husted, S., M.U. Thomsen, M. Mattsson, and J. Schjoerring. 2005. Influence of nitrogen and sulphur form on manganese acquisition by barley (Hordeum vulgare). *Plant Soil* 268:309–317.

Imtiaz, M. 1999. Zn deficiency in cereals. PhD thesis, Reading University, Reading, UK.

Iwasaki, K., P. Maier, M. Fecht, and W.J. Horst. 2002. Effect of silicon supply on apoplastic manganese concentrations in leaves and their relation to manganese tolerance in cowpea (*Vigna unguiculata* L.) Walp. *Plant Soil* 238:281–288.

Iwasaki, K., and A. Matsumura. 1999. Effect of silicon on alleviation of manganese toxicity in pumpkin (*Cucurbita moschata* Duch cv. Shintosa). *Soil Sci. Plant Nutr.* 45:909–920.

Jin, C.W., G.Y. You, Y.F. He, C. Tang, P. Wu, and S.J. Zheng. 2007. Iron deficiency-induced secretion of phenolics facilitates the reutilization of root apoplastic iron in red clover. *Plant Physiol.* 144:278–285.

Kabata-Pendias, A., and H. Pendias. 2001. *Trace Elements in Soils and Plants*. 3rd ed. CRC Press, Boca Raton, FL.

Kabir, A.H., N.G. Paltridge, A.J. Able, J.G. Paull, and J.C.R. Stangoulis. 2012. Natural variation for Fe-efficiency is associated with up regulation of strategy I mechanisms and enhanced citrate and ethylene synthesis in *Pisum sativum* L. *Planta* 235:1409–1419.

Keller, C., M. Rizwan, J.C. Davidian, O.S. Pokrovsky, N. Bovet, P. Chaurand, and J.D. Meunier. 2015. Effect of silicon on wheat seedlings (*Triticum turgidum* L.) grown in hydroponics and exposed to 0 to 30 μM Cu. *Planta* 241:847–860.

Khandekar, S., and S. Leisner. 2011. Soluble silicon modulates expression of *Arabidopsis thaliana* genes involved in copper stress. *J. Plant Physiol.* 168:699–705.

Kiekens, L. 1995. Zinc. In B.J. Alloway (ed.), *Heavy Metals in Soils*, pp. 284–303. Blackie Academic and Professional, London.

Kovda, V.A. 1973. *The Bases of Learning about Soils*. 2 vols. Moscow: Nayka.

Li, J., and S.M. Leisner. 2008. Alleviation of copper toxicity in *Arabidopsis thaliana* by silicon addition to hydroponic solutions. *J. Am. Soc. Hortic. Sci.* 133(5):670–677.

Liang, Y. 1998. Effects of Si on leaf ultrastructure, chlorophyll content and photosynthetic activity in barley under salt stress. *Pedosphere* 8:289–296.

Liang, Y., J. Si, and V. Römheld. 2005a. Silicon uptake and transport is an active process in *Cucumis sativus*. *New Phytol.* 167:797–804.

Liang, Y.C. 1999. Effects of silicon on enzyme activity, and sodium, potassium and calcium concentration in barley under salt stress. *Plant Soil* 209:217–224.

Liang, Y.C., Q. Chen, Q. Liu, W.H. Zhang, and R.X. Ding. 2003. Exogenous silicon (Si) increases antioxidant enzyme activity and reduces lipid peroxidation in roots of salt-stressed barley (*Hordeum vulgare* L.). *J. Plant Physiol.* 160:1157–1164.

Liang, Y.C., W.C. Sun, Y.G. Zhu, and P. Christie. 2007. Mechanisms of silicon-mediated alleviation of abiotic stresses in higher plants: A review. *Environ. Pollut.* 47:422–428.

Liang, Y.C., J.W.C. Wong, and L. Wei. 2005b. Silicon-mediated enhancement of cadmium tolerance in maize (*Zea mays* L.) grown in cadmium contaminated soil. *Chemosphere* 58:475–483.

Liang, Y.C., J. Zhu, Z.J. Li, G.X. Chu, Y.F. Ding, J. Zhang, and W.C. Sun. 2008. Role of silicon in enhancing resistance to freezing stress in two contrasting winter wheat cultivars. *Environ. Exp. Bot.* 64:286–294.

Lindsay, W.L. 1979. *Chemical Equilibria in Soils*. Wiley, New York.

Longnecker, N., and R.M. Welch. 1990. Accumulation of apoplastic iron in plant roots. *Plant Physiol.* 92:17–22.

Ma, J.F., and E. Takahasi. 2002. *Soil, Fertilizer, and Plant Silicon Research in Japan*. Elsevier, Amsterdam.

Ma, J.F., K. Tamai, N. Yamaji, N. Mitani, S. Konishi, M. Katsuhara, M. Ishiguro, Y. Murata, and M. Yano. 2006. A silicon transporter in rice. *Nature* 440:688–691.

Ma, J.F., and N. Yamaji. 2006. Silicon uptake and accumulation in higher plants. *Trends Plant Sci.* 11:392–397.

Ma, J.F., and N. Yamaji. 2008. Functions and transport of silicon in plants. *Cell Mol. Life Sci.* 65:3049–3057.

Ma, J.F., N. Yamaji, N. Mitani, K. Tamai, S. Konishi, T. Fujiwara, M. Katsuhara, and M. Yano. 2007. An efflux transporter of silicon in rice. *Nature* 448:209–212.

Marschner, H. 1995. *Mineral Nutrition of Higher Plants.* 2nd ed. Academic Press, London.

Marschner, H., E.A. Kirkby, and I. Cakmak. 1996. Effect of nutrition mineral status on shoot-root partitioning of photo assimilates and cycling of mineral nutrient. *J. Exp. Bot.* 47:1255–1263.

Marschner, H., and V. Römheld. 1994. Strategies of plants for acquisition of iron. *Plant Soil* 165:261–274.

Matichenkov, V.V., and E.A. Bocharnikova. 2001. The relationship between silicon and soil physical and chemical properties. In L.E. Datnoff, G.H. Snyder, and G.H. Korndörfer (eds.), *Silicon in Agriculture: Studies in Plant Science*, pp. 209–219. Elsevier Science, Amsterdam.

Maxwel, F.G., N. Jenkins, and W.L. Parrott. 1977. Resistance of plants to insects. *Adv. Agron.* 24:187–265.

Mcbride, M.B. 1994. *Environmental Chemistry of Soils.* 1st ed. Oxford University Press, Oxford.

Meena, V.D., M.L. Dotaniya, V. Coumar, S. Rajendiran, S. Kundu, and A.S. Rao. 2014. A case for silicon fertilization to improve crop yields in tropical soils. *Proc. Indian Natl. Sci. Acad. B Biol. Sci.* 84:505–518.

Mehrabanjoubani, P., A. Abdolzadeh, H.R. Sadeghipour, and M. Aghdasi. 2014. Impacts of silicon nutrition on growth and nutrient status of rice plants grown under varying zinc regimes. *Theor. Exp. Plant Physiol.* 27:19–29.

Mitani, N., and J.F. Ma. 2005. Uptake system of silicon in different plant species. *J. Exp. Bot.* 56:1255–1261.

Moore, J.W. 1991. *Inorganic Contaminants of Surface Water: Research and Monitoring Priorities.* Springer-Verlag, Berlin.

Moore, K.L., A.M.L. Chen, Y. van de Meene, L. Hughes, W.J. Liu, T. Geraki, F. Mosselmans, S.P. McGrath, C. Grovenor, and F.J. Zhao. 2014. Combined nano SIMS and synchrotron x-ray fluorescence reveals distinct cellular and subcellular distribution patterns of trace elements in rice tissues. *New Phytol.* 201:104–115.

Morgan, P.W., D.M. Taylor, and H.E. Joham. 1976. Manipulations of IAA oxidase activity in and auxin deficiency symptoms in intact cotton plants with manganese nutrition. *Physiol. Plant.* 37:149–156.

Mortvedt, J.J. 2000. Bioavailability of micronutrients. In M.E. Sumner (ed.), *Handbook of Soil Science*, pp. 71–88. CRC Press, Boca Raton, FL.

Mukhopadhyay, M.J., and A. Sharma. 1991. Manganese in cell metabolism of higher plants. *Bot. Rev.* 57:117–149.

Neumann, D., and U. Zur Nieden. 2001. Silicon and heavy metal tolerance of higher plants. *Phytochemistry* 56:685–692.

Nikolic, M., and V. Römheld. 1999. Mechanism of Fe uptake by the leaf symplast: Is Fe inactivation in leaf a cause of Fe deficiency chlorosis? *Plant Soil* 215:229–237.

Nowakowski, W., and J. Nowakowska. 1997. Silicon and copper interaction in growth of spring wheat seedlings. *Biol. Plant.* 39(3):463–466.

Okuda, A., and E. Takahashi. 1962. Effect of silicon supply on the injuries due to excessive amounts of Fe, Mn, Cu, As, AI, Co of barley and rice plant. *Jpn. J. Soil Sci. Plant Nutr.* 33:1–8.

Oliva, S.R., M.D. Mingorance, and E.O. Leidi. 2011. Effects of silicon on copper toxicity in *Erica andevalensis* Cabezudo and Rivera: A potential species to remediate contaminated soils. *J. Environ. Monit.* 13:591–596.

Ota, M., H. Kobayshi, and Y. Kawaguchi. 1975. Effect of slag on paddy rice part 2. Influence of different nitrogen and slag levels on growth and composition of rice plant. *Soil Plant Food* 3:104–107.

Papadakis, I.E., A. Giannakoula, I.N. Therios, A.M. Bosabalidis, M. Moustakas, and A. Nastou. 2007. Mn-induced changes in leaf structure and chloroplast ultrastructure of *Citrus volkameriana* L. plants. *J. Plant Physiol.* 164:100–103.

Pavlovic, J., J. Samardzic, V. Maksimovic, G. Timotijevic, N. Stevic, K.H. Laursen, T.H. Hansen, S. Husted, J.K. Schjoerring, Y. Liang, and M. Nikolic. 2013. Silicon alleviates iron deficiency in cucumber by promoting mobilization of iron in the root apoplast. *New Phytol.* 198:1096–1107.

Pedas, P., J.K. Schjoerring, and S. Husted. 2009. Identification and characterization of zinc-starvation-induced ZIP transporters from barley roots. *Plant Physiol. Biochem.* 47:377–383.

Peleg, Z., Y. Saranga, T. Fahima, A. Aharoni, and R. Elbaum. 2010. Genetic control over silica deposition in wheat awns. *Physiol. Plant.* 140:10–20.

Perry, C.C., and T. Keeling-Tuker. 1998. Aspects of the bioinorganic chemistry of silicon in conjunction with the biometals calcium, iron and aluminium. *J. Inorg. Chem.* 69:181–191.

Pestana, M., A. de Varennes, J. Abadía, and E.A. Faria. 2005. Differential tolerance to iron deficiency of citrus rootstocks grown in nutrient solution. *Sci. Hortic.* 104:25–36.

Pich, A., G. Scholz, and U.W. Stephan. 1994. Iron-dependent changes of heavy metals, nicotianamine, and citrate in different plant organs and in xylem exudate of two tomato genotypes. Nicotianamine as possible copper translocator. *Plant Soil* 165:189–196.

Rellán-Alvarez, R., J. Giner-Martínez-Sierra, T. Orduna, I. Orera, J.A. Rodríguez-Castrillón, J.I. Garcia-Alonso, J. Abadí, and A. Alvarez-Fernandez. 2010. Identification of a tri-iron (III), tri-citrate complex in the xylem sap of iron deficient tomato resupplied with iron. New insights into plant iron long distance transport. *Plant Cell Physiol.* 51:91–102.

Reuter, D.J., A.M. Alston, and J.D. McFarlane. 1988. Occurrence and correction of manganese deficiency in plants. In R.D. Graham (ed.), *Manganese in Soils and Plants*, pp. 205–204. Kluwer Academic, Dordrecht.

Rizwan, M., J.D. Meunier, H. Miche, and C. Keller. 2012. Effect of silicon on reducing cadmium toxicity in durum wheat (*Triticum turgidum* L. cv. Claudio W.) grown in a soil with aged contamination. *J. Hazard Mater.* 209–210:326–334.

Römheld, V., and H. Marschner. 1986. Evidence of a specific uptake system for iron phytosiderophores in roots of grasses. *Plant Physiol.* 78:175–180.

Rogalla, H., and V. Römheld. 2002. Role of leaf apoplast in silicon mediated manganese tolerance of *Cucumis sativus* L. *Plant Cell Environ.* 25:549–555.

Saeed, M., and R.L. Fox. 1977. Relation between suspension pH and Zn solubility in acid and calcareous soils. *Soil Sci.* 124:199–204.

Savant, N.K., G.H. Snyder, and L.E. Datnoff. 1997. Silicon management and sustainable rice production. *Adv. Agron.* 58:151–199.

Sawant, A.S, V.H. Patil, and N.K. Savant. 1994. Rice hull ash applied to seebold reduces dead hearts in transplanted rice. *Int. Rice Res. Notes* 19:21–22.

Shi, Q., Z. Bao, Z. Zhu, Y. He, Q. Qian, and J. Yu. 2005a. Silicon mediated alleviation of Mn toxicity in *Cucumis sativus* in relation to activities of superoxide dismutase and ascorbate peroxidase. *Phytochemistry* 66:1551–1559.

Shi, X., C. Zhang, H. Wang, and F. Zhang. 2005b. Effect of Si on the distribution of Cd in rice seedlings. *Plant Soil* 272:53–60.

Siegel, A., and H. Siegel. 2000. *Metal Ions in Biological Systems: Manganese and Its Role in Biological Processes*, p. 37. CRC Press, Boca Raton, FL.

Sinclair, S.A., and U. Krämer. 2012. The zinc homeostasis network of land plants. *Biochim. Biophys. Acta* 1823:1553–1567.

Singh, J.P., R.E. Karamonas, and J.W.B. Stewart. 1986. Phosphorus-induced zinc deficiency in wheat on residual phosphorus plots. *Agron. J.* 78:668–675.

Singh, R., N. Gautam, A. Mishra, and R. Gupta. 2011. Heavy metal and living systems: An overview. *Ind. J. Pharmacol.* 43:246–253.

Smyth, T.J., and P.A. Sanchez. 1980. Effects of lime, silicate, and phosphorus applications to an oxisol on phosphorus sorption and ion retention. *Soil Sci. Soc. Am. J.* 44:500–505.

Snowball, K., and A.D. Robson. 1986. *Symptoms of Nutrient Deficiencies: Lupins.* University of Western Australia Press, Nedlands, Australia.

Song, A.L., P. Li, Z.J. Li, F.L. Fan, M. Nikolic, and Y.C. Liang. 2011. The alleviation of zinc toxicity by silicon is related to zinc transport and antioxidative reactions in rice. *Plant Soil* 344:319–333.

Takahashi, E., J.F. Ma, and Y. Miyake. 1990. The possibility of silicon as an essential element for higher plants. *Comments Agric. Food Chem.* 2:99–122.

Tanaka, A., and Y.D. Park. 1966. Significance of the absorption and distribution of silica in the rice plant. *Soil Sci. Plant Nutr.* 12:191–195.

Tripathi, D.K., V.P. Singh, D. Kumar, and D.K. Chauhan. 2012a. Impact of exogenous silicon addition on chromium uptake, growth, mineral elements, oxidative stress, antioxidant capacity, and leaf and root structures in rice seedlings exposed to hexavalent chromium. *Acta Physiol. Plant.* 34:279–289.

Tripathi, D.K., V.P. Singh, D. Kumar, and D.K. Chauhan. 2012b. Rice seedlings under cadmium stress: Effect of silicon on growth, cadmium uptake, oxidative stress, antioxidant capacity and root and leaf structures. *Chem. Ecol.* 28:281–291.

Tripathi, D.K., V.P. Singh, S.M Prasad, D.K. Chauhan, N.K. Dubey, and A.K. Rai. 2015. Silicon-mediated alleviation of Cr (VI) toxicity in wheat seedlings as evidenced by chlorophyll florescence, laser induced breakdown spectroscopy and anatomical changes. *Ecotoxicol. Environ. Saf.* 113:133–144.

Tuna, A.L., C. Kaya, D. Higgs, B. Murillo-Amador, S. Aydemir, and A.R. Girgin. 2008. Silicon improves salinity tolerance in wheat plants. *Environ. Exp. Bot.* 62:10–16.

Wada, S., and K. Wada. 1980. Formation, composition and structure of hydroxy-aluminosilicate ions. *Eur. J. Soil Sci.* 31:457–467.

Wang, L.J., Y.H. Wang, Q. Chen, W.D. Cao, M. Li, and F.S. Zhan. 2000. Silicon induced cadmium tolerance of rice seedlings. *J. Plant Nutr.* 23:1397–1406.

Welch, R.M., and R.D. Graham. 2004. Breeding for micronutrients in staple food crops from a human nutrition perspective. *J. Exp. Bot.* 55:353–364.

Wiese, H., M. Nikolic, and V. Römheld. 2007. Silicon in plant nutrition. Effect of zinc, manganese and boron leaf concentrations and compartmentation. In B. Sattelmacher and W.J. Horst (eds.), *The Apoplast of Higher Plants: Compartment of Storage, Transport and Reactions*, pp. 33–47. Springer, Dordrecht.

Williams, D.E., and J. Vlamis. 1957. The effect of silicon on yield and manganese-54 uptake and distribution in the leaves of barley grown in culture solutions. *Plant Physiol.* 32:404–409.

Yoshida, S. 1975. *The Physiology of Silicon in Rice*, vol. 25: *Food FERT*. Technology Center, Taipei, Taiwan.

Yoshida, S., Y. Ohnishi, and K. Kitagishi. 1962. Histochemistry of silicon in rice plant. I. A new method for determining the localization of silicon within plant tissues. *Soil Sci. Plant Nutr.* 8:30–35.

Zhang, X., F. Zhang, and D. Mao. 1998. Effect of iron plaque outside roots on nutrient uptake by rice (*Oryza sativa* L.). Zinc uptake by Fe deficient rice. *Plant Soil* 202:33–39.

Zhao, Z.Q., Y.G. Zhu, R. Kneer, and S.E. Smith. 2005. Effect of zinc on cadmium toxicity induced oxidative stressing winter wheat seedlings. *J. Plant Nutr.* 28:1947–1959.

Zhu, Z.J., G.Q. Wei, J. Li, Q.Q. Qian, and J.Q. Yu. 2004. Silicon alleviates salt stress and increases antioxidant enzymes activity in leaves of salt-stressed cucumber (*Cucumis sativus* L.). *Plant Sci.* 167:527–533.

3 Regulatory Mechanisms by Silicon to Overcome the Salinity-Induced Imbalance of Essential Nutrient Elements

Prabhakaran Soundararajan, Abinaya Manivannan, and Byoung Ryong Jeong

CONTENTS

ABSTRACT

Plants require essential elements such as macronutrients (K, Ca, N, P, S, and Mg) and micronutrients (B, Fe, Cu, Zn, Mo, Ni, and Cl) at certain levels for their growth, development, and reproduction. However, abiotic stresses such as salt, water, and temperature stress create an ionic imbalance in the growth medium. They disintegrate the water availability, affecting the ionic imbalance, which in turn disturbs the uptake and translocation of essential elements. On the other hand, salinity induces a few essential elements to be taken up at phytotoxic levels. These abnormalities cause ionic, oxidative, and osmotic stress, followed by lethal effects in the case of longer exposure to the stress. Integral nutrient management was proposed as an effective strategy to overcome the abiotic stresses. Among those, the application of silicon, a nonessential beneficial element, was proved to enhance the resistance against various kinds of stresses in several plant species. Increases in tolerance by Si in the plants are mainly achieved by regulating the nutrient balance. A Si-mediated mechanism involved in alleviating salinity stress is the deposition of Si in the root wall to block the excessive uptake of sodium (Na), increase the uptake of macronutrients such as potassium (K) and calcium (Ca), and inhibit the toxic ions by formation of a Si–metal complex. A double layer of cuticular Si reduces transpiration. The soluble Si presents in the upper part of the plant, inducing stress-related biochemical changes such as enhanced antioxidant activity, nonenzymatic antioxidants, and phenols to scavenge the excessively generated reactive oxygen species (ROS) group. Expressions of stress-related proteins and transporters are varied based on the addition of Si. Compartmentalization, distribution, vacuolization, and cell wall integrity are additional Si-involved mechanisms for avoiding the toxicity of ions. All the above changes are based on the availability of essential nutrients at the optimal level and in available forms as macro- and micronutrients supports the building blocks of plants. Hence, this chapter deals with the mechanisms of nutrient regulation by Si in mobilization and utilization during salt stress. In particular, the deleterious effects of NaCl on the deficiency of K, Ca, and N and either or both the deficiency and toxicity of P, B, Fe, Cu, Zn, and Mn are covered in this chapter.

3.1 INTRODUCTION

3.1.1 ESSENTIAL ELEMENTS

Elements present in the soil are classified as essential, beneficial, and toxic according to their roles in the plants. Inorganic and organic compounds that are needed by plants for their growth, development, and reproduction are called essential elements. Each element listed among the essential nutrients has a nonreplaceable role in plants and is directly involved in plant metabolism. Plants cannot complete their life cycle in the absence of essential mineral elements (Arnon and Stout, 1939; Asher, 1991; Epstein and Bloom, 2005). There are 17 mineral elements that are considered to be essential. Among them, carbon (C), hydrogen (H), and oxygen (O) are included as the basic nutrients, as they are derived from water and air. The rest of the elements are categorized into macronutrients (nitrogen [N], phosphorus [P], potassium [K], calcium [Ca], magnesium [Mg], and sulfur [S]) and micronutrients (boron [B], chlorine [Cl], copper [Cu], iron [Fe], manganese [Mn], sodium [Na], zinc [Zn], molybdenum [Mo], and nickel [Ni]) according to their content required by plants. Among the various factors that influence plant growth, nutrient elements play an important role in the life cycle of plants.

3.1.2 STRESS

Stress is changes that tend to inhibit the function of the normal life cycle either directly or indirectly. Relationships between abiotic stress and mineral nutrients are more complex (Ali et al., 2012). Numerous essential plant nutrients are involved in the regulatory mechanism under stress conditions. Stress factors simultaneously interact in both antagonistic and synergistic manners on the nutrient present in the growth medium and in the absorption sites of the root, transportation, or distribution (Alam, 1999). Water and nutrient availability during stress conditions varies due to changes in the ionic strength and water availability in the growth medium (Chinnusamy et al., 2005). An imbalance in nutrients caused by stress creates a concentration gradient inside the plants and causes various oxidative and osmotic stresses. As essential nutrients are needed for fundamental cellular roles, toxic ions affect and injure plants, disrupt the normal cellular mechanism, cause deficiency or toxicity with essential elements, and also inactivates the enzymes (Epstein, 1972). Hence, changes in uptake, deficient and toxic types, and the oxidized form of nutrients during stress play an important role in the symptoms that occur in plants during stress conditions. Plants require some nutrients at an excess level to overcome stress; meanwhile, a few nutrients are indirectly induced to the above-optimum level, causing toxicity in plants (Grattan and Grieve, 1999). An ionic imbalance occurs in plants due to an excessive accumulation of Na^+, and Cl^- reduces the uptake of other essential elements (Lutts et al., 1999). Alam (1999) suggests that the reduction of moisture in the root-absorbing surface decreased the nutrients' diffusion rate. Many studies have reported that salt stress can be overcome by restricting the uptake of Na by plants (Greenway and Munns, 1980) and improving the K in wheat (Akram et al., 2007), Ca (Awada et al., 1995), and N in *Phaseolus vulgaris* (Wagenet et al., 1983). In fact, essential elements act as the building blocks of plant cells, cofactors, or activators of enzymes, and integral compound(s) for the stability of nucleic acid and proteins (Marschner, 1971, 1995). In short notes, N is the basic essential element for organic compounds, amino acids, enzymes, and nucleic acids such as DNA and RNA; P, similar to nitrogen, is also an integral component of nucleic acids and proteins. It is involved in energy transfer reactions on adenosine triphosphate and diphosphate (ATP and ADP); K is involved in osmotic and ionic regulation. Ca is required in the maintenance of cell membrane integrity and cell division; Mg is a component of chlorophyll; S is essential for plant lipid synthesis; Fe is important for redox state, heme-, and iron-sulfur proteins; Zn is an important component for dehydrogenase, proteinase, and peptidase; Cu is a constituent of lactase, cytochrome oxidase, polyphenol oxidase, and ascorbic acid oxidase; Mn is a component of arginase and phosphotransferase and is also involved in the O_2-evolving system of photosynthesis; B is involved in carbohydrate metabolism; Mo is required for nitrogen assimilation and is also an integral part of nitrogenase, nitrate reductase, and aldehyde oxidase; Cl acts as an activator of enzymes involved in the splitting of water, which is essential for photosynthesis; Ni is required for enzyme urease to break down into urea to liberate the nitrogen into a usable form for plants, iron absorption, seed germination, and seed sets. The availability or uptake of mineral nutrition plays an important role in the susceptibility or resistance of plants to environmental stress in order to avoid or tolerate excessive toxic compounds in the medium.

3.1.3   IMPORTANCE OF SILICON TO OVERCOMING STRESS

In addition to the macro- and micronutrients, some elements are essential only for certain plants species and promote plant growth particularly under stress; they are called beneficial elements (Marschner, 1995). Though silicon (Si) is not included as an essential element, its application becomes fundamentally important for the amelioration of abiotic and biotic stresses in plants (Datnoff et al., 2001). Silicon is often represented as a quasi-essential element. Its deficiency causes various abnormalities, including improper transportation of nutrients in rice and sugarcane (Epstein, 1994; Takahashi et al., 1990). The beneficial effects of Si are apparently visible in both abiotic and biotic stress conditions in most plants. However, its mechanism for overcoming stress varies between the accumulators and non-accumulators. Miyake and Takahashi (1978, 1982) reported that tomato and cucumber growing in nutrient solution without an additional Si supply showed symptoms such as leaf chlorosis and reduced the growth. The effects of Si are apparently visible during stress conditions (Ma and Yamaji, 2006). During stress, Si-mediated changes at the biochemical and molecular levels are highly associated with its regulation of the essential nutrient elements to overcome stress and alleviate its effects (Ali et al., 2012). Studies on the modulation of plant physiology, photosynthesis, and biochemical processes, such as the modulation of antioxidants and defense-related enzymes, have reported anatomical changes by Si in many plants, such as rice (Agarie et al., 1992; Gong et al., 2006), wheat (Ahmad et al., 1992; Tuna et al., 2008), barley (Liang, 1998), tomato (Al-Aghabary et al., 2004), spinach (Gunes et al., 2007a,b), cucumber (Zhu et al., 2004), salvia (Prabhakaran et al., 2014), and carnation (Prabhakaran et al., 2015). Previously, salt stress has been alleviated by enhanced exogenous application of K (Akram et al., 2007), N (Wagenet et al., 1983), and Ca (Awada et al., 1995). Reasonable amounts of both K^+ and Ca^{2+} are required to maintain cell membrane integrity and function (Wei et al., 2003). Tuna et al. (2008) and Liang (1999) reported that the Si can increase K, Ca, P, and N and optimize the deficient or excessive uptake of micronutrients such as B, Fe, Cu, Zn, and Mn (Mass et al., 1972). Under the saline condition, water stress also occurred due to the reduction in water uptake by root as well as increased transpiration rate and higher water leakage (Munns, 2002). Water stress in plants is also due to low external water potential imposed by a higher salt concentration (Romero-Aranda et al., 2001). Romero-Aranda et al. (2006) reported that Si enhances water storage capacity and is able to dilute the presence of Na in the cell. Since Si has been reported to be involved in the alleviation of stress, it is important to understand the multiple roles played by Si on nutrient regulation to ameliorate salt stress, for example, (1) enhancement of essential nutrient uptake, (2) optimization of the toxic level uptake of nutrients, and (3) regulation of nutrients based on biochemical and molecular mechanisms to overcome the stress effects.

3.2 SALT STRESS

3.2.1 NUTRIENT IMBALANCE CAUSED BY NaCl

A mixture of soluble salts present at an optimal level in the growth medium is essential for plants. However, plant growth has been suppressed at excessive salt concentrations (Francois and Maas, 1994). Initially, the larger amount of salts decreases the free energy, solubility, and availability of water to the plants (Alam, 1999). Growth retardation occurs as a result of the influence of salinity on various vital facets of plant metabolism, such as osmotic adjustment, photosynthesis, respiration, gas exchange, uptake, transportation, compartmentalization, utilization, and maintenance of ions in available forms, nucleic acid and protein synthesis, enzyme activity, and plant hormone regulation (Maas and Nieman, 1978; Lutts et al., 1999). The salt tolerance is mainly considered by restriction of Na^+ and Cl^- ion uptake and translocation (Greenway and Munns, 1980). Generally, Na^+ is required at less than the micronutrient level (Tyler, 2004) and Cl^- ion at the the micronutrient level (Broyer et al., 1954; Xu et al., 1999). Sodium is involved in the osmotic and ionic balance in plants. It maintains the turgor pressure and influences photosynthesis by altering water balances in the root zones. In addition Na is essential for the C4 photosynthesis and also crassulacean acid metabolism (CAM) (Marschner, 1971). Chlorine plays a role in photosystem II. It is associated with the activation of the oxygen-evolving enzyme and is involved in polypeptides for splitting the water complex in photosystem II. Along with K, Cl regulates the opening and closing of stomata. For the activity of enzymes such as asparagines synthethase, amylase, and ATPase, Cl is required. The excessive presence or uptake of Na^+ and Cl^- to the supraoptimal level in plants causes a negative effect on the growth and yield (Pardo and Quintero, 2002). The salinity sensitivity and tolerance are often correlated with blocking the uptake and translocation of Na^+ and Cl^- ions. A higher amount of Na^+ and Cl^- shows a competitive uptake with the other elements, making nutrients into unavailable forms and replacing the other elements, such as K (Akram et al., 2007) and Ca (Awada et al., 1995) toxicity. Hence, it is necessary to alleviate ionic stress caused by the saline condition.

3.2.2 ANOMALY OF SILICON DURING SALINE CONDITION

Numerous mechanisms have been proposed for Si to overcome the salinity in plants. Matoh et al. (1986) suggested that an amorphous form of silica deposition limits the transpiration. This was later supported by Yeo et al. (1999), who found that any reduction in ion transportation could also reduce Na uptake by partial blockage of transpiration bypass flow. Salt enters into plants mostly by the apoplastic pathway (Yadav et al., 1996). The reduction in transportation of Na through trisodium-8-hydroxy-1,3,6-pyrenetrisulphonic acid (PTS) involves blockage of the bypass flow across the rice root. According to Ahmad et al. (1992), the complex formed by Si with Na in the root can also reduce the transportation of salt from the root to the shoot, and partly has an inhibitory effect on the transpiration rate. Higher level of salt presence in the solution or soil causes the loss of internal water and creates osmotic stress (Tester and Davenport, 2003). Sangster (1978) reported that excess leakage of water from a plant to its external surface can be reduced by the physical barrier of Si deposition in the inner tangential wall. This deposition gives mechanical strength to the root apices before completion of rhizodermal and endodermal differentiation. Formation of colloidal silica to silica gel or polysilicic acid throughout the root apoplast by polymerization of Si blocks the apoplastic movements of toxic ions into the plants (Epstein, 1994, 1999). Silicon immobilizes the free Na^+ ion in the plant and inhibits its movement from the root to the aerial part of the plant by forming a complex with Na (Ma et al., 2001). The reduction of toxic Na^+ is paralleled with an increase in root elongation (Tester and Davenport, 2003; Prabhakaran et al., 2015). In contrast, Lu and Neumann (1999) state that deposition of Si on the cell wall could restrict the extensibility of roots. Gong et al. (2005) reported that a possible mechanism for coprecipitation of Si with Na involves the reduction of apoplastic transport across the root. Other mechanisms adapted by plants to overcome salt stress are homeostasis, compartmentalization, and counteruptake or reduction of ions lesser than toxic levels regard Si supplementation will be seen details in below sections.

3.3 REGULATION OF MACRONUTRIENTS BY Si DURING SALT STRESS

3.3.1 POTASSIUM

Potassium is a monovalent cation, and it has the high mobility. Potassium is required for high-yield crop production (Liebersbach et al., 2004). It is well known for its maintenance of osmotic strength. Potassium also plays a vital role in protein synthesis, glycolytic enzyme activity, and photosynthesis. To store energy in the plants via carbohydrate metabolism, K is needed for regulating enzymes to produce carbohydrates and assimilate nitrogen. For the synthesis of tRNA and ribulose-1,5-bisphosphatecarboxylase/oxygenase (RuBisCO), K ion plays an important role (Demmig and Gimmler, 1983). Abundant Na causes a toxic effect to cells (Asch et al., 2000), and an increase in K concentration alleviates the deleterious effects of salt in rice and tomato (Song and Fujiwara, 1996). A deficiency of K highly affects cell expansion, mechanical stability, and turgor-driven movements (Marschner, 1995). The evidence from barley (Liang et al., 1996) and cucumber (Zhu et al., 2004) suggests that Si maintains membrane integrity against Si. The presence of a high amount of Na in the solution or soil leads to the impairment of K acquisition and also excess accumulation of Na^+ in plant cells in many species (Suhaya et al., 1990; Hu and Schmidhalter, 2005). However, intercellular distribution of K and competitive uptake between Na and K modulated by Si are considered as an important factors for enhancing the salt tolerance in wheat (Tuna et al., 2008). Marschner (1995) reported that K is essential for at least 50 major enzymes present in plants. The competitive uptake of K against the Na is frequently important for salt tolerance in plants and has been achieved in barley (Liang, 1998), tomato (Al-Aghabary et al., 2004), and rice (Gong et al., 2006).

Deposition of Si in the lower epidermis of the stomata and increases in K content in the blueberry have been reported by Morikawa and Saigusa (2004). It helps to maintain the gas exchange. For the stomatal response, abscisic acid (ABA) generated during salt stress closes the stomata during the increased efflux of K^+ from the guard cells; ABA-mediated mechanisms can be elucidate by late embryogenesis abundant (LEA) type proteins (Natacha et al., 2008).

The reduction in stomatal closure is due to the K^+ efflux; changes in the guard cell turgor depend on the K^+ concentration in the cell (Marschner, 1995). The Si-associated stomatal movement could be presumably mediated by phenomena such as signal perception, electrochemical gradients across guard cell membranes, transport of ions for osmotic adjustment, and hydraulic maintenance (Agarie et al., 1998). Especially during the competitive uptake of Na, K uptake is invariably less (Marschner, 1995). The K^+ transporters do not discriminate between K^+ and Na^+ because of their similar structure and charge. Liang et al. (1996) state that Si increased the K/Na selectivity ratio. A positive antagonistic uptake of K against Na is reported in the plant-selective transporter mechanism of Na and K (Liang, 1999; Tuna et al., 2008). Miao et al. (2010) found that Si facilitates greater K accumulation in K-deficient conditions than K-sufficient treatment. The addition of Si increased superoxide dismutase (SOD) activity and decreased the malondialdehyde (MDA) concentration in barley; this is associated with a higher uptake of K under Si treatment (Zhu et al., 2004). The enhanced membrane integrity directly decreased the plasma membrane permeability and improved the ultrastructure of chloroplasts (Liang, 1998). Compared with the other elements, K^+ ion is present in the highest concentration in the cytoplasm and chloroplast (Marschner, 1995). Toxic-level Na ions change the cell pH level, and K^+ serves to maintain the pH between 7 and 8, which is necessary for enzymatic activity in plants.

The excessive Na^+ is compartmentalized from the cytoplasm into the vacuole by a proton pump mechanism. This process prevents the cellular components from ionic imbalance caused by excessive Na^+ and helps to maintains all metabolic activities in the symplast (Silva and Geros, 2009). Activation of H^+-ATPases and H^+-PPases must be stimulated (Hasegawa et al., 2000a,b) to compartmentalize the excessive Na ions in the vacuole. Liang (1999) reported the involvement of Si in barley and the compartmentalization of the Na^+ ion in the Na^+/H^+ antiporter. This process is activated and operated by the energy obtained from the H^+-ATPase and H^+-PPase maintained by Si. Furthermore, the H^+ electrochemical potential gradient obtained from the proton pumps acts as an electromotive force (EMF) and the antiport of excessive ions (Blumwald, 2000). Cell wall extension is synergistically dependent on the vacuole size. Formation of a larger vacuole was achieved by plant for the proper maintenance of osmosis (Marschner, 1995). For salt-tolerant mechanisms, a Na^+/H^+ antiporter named AtNHX1 (*Arabidopsis thaliana*) serves to maintain the ion homeostasis (Silva and Geros, 2009) and can be linked with the Si-mediated vacuolation of Na.

3.3.2 CALCIUM

Similar to the K, Ca also plays an integral part in maintaining membrane integrity. It is involved in the senescence process and counteracts the harmful effects against Na (Lahaye and Epstein, 1971). Calcium is an integral component of the cell wall, contributing to the maintenance of the structure and function of the membrane. Calcium is needed preferentially in actively growing tissues (Marschner, 1995). Although Ca has low mobility, small changes in the active pools of Ca within the cytoplasm are reflected in the plant physiology. A high Na concentration inhibiting Ca absorption was reported in many plants (Awada et al., 1995; Liang et al., 1999). However, it can also be readily displaced from its extracellular binding sites by the cations (Na^+) and impaired by less Ca availability in the root growth of Na conditions (Solomon et al., 1989). As Ca is involved in the cell division and cell elongation processes, normal plant growth is dependent on the availability of Ca in the cell. Calcium enters into the plant via apoplasts and binds at the exterior surface of the plasma membrane in an exchangeable form. Cachorro et al. (1994) state that the additional supply of Ca under saline conditions effluxed the Na absorption and accelerated the K absorption. According to Ma et al. (2001), Si-deprived rice plants have less Ca than the Si-added nutrient solution. A combined increase in K and Ca decreases the Na uptake and plays an important role in cell membrane permeability and selectivity (Song and Fujiwara, 1996). McLaughlin and Wimmer (1999) highlighted the role of Ca in the movement of water and solute across the membrane structures and stomatal function, respiratory metabolism, translocation, cell wall synthesis, and signaling roles in plants against stress.

The Ca-mediated linkage response controls the changes in the cell wall. To maintain the cell membrane stability, saline requires conditions of higher Ca. A high Na concentration in the soil or solution inhibits the uptake of Ca and has been reversed by supplementation of Si (Liang et al., 1999). Ion interactions, precipitation, and increased ionic strength together are involved in the decrease of Ca availability and uptake. Higher Ca is needed for the protection of cell membranes and minimizes the leakage of cytosolic K during osmotic or oxidative stress (Sharma et al., 2012). The salinity-derived Ca-deficient condition, which increases the leakage of low-molecular-weight solutes from cells, is avoided by supplemental Si treatment to barley (Liang, 1999). An additional supply of Si alleviates impairment to the membrane structure and cell compartmentation. Under salinity, Tuna et al. (2008) reported that K and Ca are improved by Si and show resistance against NaCl. Furthermore, the membrane permeability impairment that occurs in the lower Ca level, caused by the presence of Na, is alleviated by higher Ca content induction on treated Si (Lahaye and Epstein, 1971; Solomon et al., 1989; Cachorro et al., 1994).

Stabilization of the membrane by Ca is associated with the bridging phosphate and carboxylate groups of phospholipids (Caldwell and Haug, 1982). The cytosolic Ca mediates the osomotic stress alleviation by inducing the stress tolerance genes during salinity. Improvement of the Ca level in the cell by Si helps to improve growth against the stress conditions (Cachorro et al., 1994) because younger leaves require more Ca than older ones (Marschner, 1995). The additional supply of Si, which decreases Ca, might be due to the decrease in transpiration rate (Liang et al., 1999). It is thought, however, that the interaction of Si varies between species during stress, Si increases the Ca content and helps the plant to withstand stress (Solomon et al., 1989; Liang and Shen, 1994).

Ko and Lee (1996) and Kim et al. (2014) reported that Si decreases ABA synthesis (Kim et al., 2014) and avoids the closing of stomata in rice. ABA-induced stomatal closures are partially mediated by Ca (Wilkinson et al., 2001) and involved in the protection of leaves from dehydration. A signaling role of intracellular Ca^{2+} has been implicated in the transduction of salt stress signals for osmoregulation by modulating proteins such as Ca^{2+}-dependent protein kinases (Urao et al., 1994) and in hyperosmotic shock induction of putative Ca^{2+}-binding proteins (Ko and Lee, 1996). Calmodulin can bind with free Ca reversibly and regulate it in cytosol (Snedden and Fromm, 1998). It helps in the activation of phospholipase and nicotinamide adenine dinucleotide (NAD) kinase. The variation in mitochondria upon salt stress is associated with the Ca level because mitochondria are rich in Ca. The calmodulin-mediated release of Ca from mitochondria and vacuoles is responsible for the turgor-related changes in the cell and can be triggered by Si to enhance the cell integrity during ionic imbalance and oxidative stress created by excess Na ions.

3.3.3 NITROGEN

Nitrogen is the vital component for proteins, hormones, chlorophyll, vitamins, and enzymes. Nitrogen metabolism is a major deciding factor for plant growth (Marschner, 1995). Nitrogen is the basic component for protein. A lower nitrogen supply results in the leaf growth being limited by a low photosynthetic rate or insufficient cell expansion (Fallah, 2012). During the high-salinity conditions, N uptake is inhibited (Alva and Syversten, 1991). The formation of white heads under conditions of high nitrogen and salinity is rectified by application of Si. According to Avila et al. (2010), Si increased the level of nitrogen and is correlated with a higher chlorophyll content in rice.

During the alleviation of salt stress, the improvement of nitrogen assimilation is an important process. Both nitrate and nitrite reductase play important roles in the conversion of nitrogen to ammonia. Ammonium and ammonia are highly toxic for plants even in low concentrations. The glutamine synthetase (GS) and glutamine oxoglutarate amino transferase (GOGAT) cycle and glutamate dehydrogenase (GDH) are involved in the conversion of ammonia to active amino acids such as glutamine, arparagine, and arginine. Nitrogen fixation pathway deeply affected by the salinity was reversed due to Si supplementation (Avila et al., 2010; Fallah, 2012). Pereira et al. (2013) reported that the increase in total protein content followed with the changes in the nitrogen assimilation, glycinebetaine, and proline upon Si treatment in pepper. The total soluble content of the salt-sensitive plant is enhanced more than the salt-tolerant genotype by the addition of Si. The maintenance of osmosis and membrane integrity in the cell by K and Ca directly influences protein formation, translocation, and maturation. Proline, an amino acid synthesized from glutamate or arginine, plays important role in the maintenance of water potential in the plant tissues and has been upregulated by Si in potato (Crusciol et al., 2009). An increased concentration of proline inside the cell can also act as a carbon and nitrogen source (Dubey and Pessarakli, 1995) and free radical scavenger (Jain et al., 2001). Similarly, the utilization of Si-induced proline has been reported for the rapid recovery of *Borango officimalis* L. from stress (Shahnaz et al., 2011). In normal conditions, supplemented Si increases the amino acids in both the leaf and phloem sap of rice plants. These results suggest that asparagine synthase and GS are active key enzymes for the nitrogen assimilation modulated by Si (Watanabe et al., 2001). Upon exposure of maize seedlings to NaCl, Si treatment significantly increased GS and GDH activities in the roots (Kochanová et al., 2014) for a defensive reaction against the salt stress conditions. The reduction in the activity of the nitrate and nitrite reductase inhibits protein synthesis. The ROS accumulation and lipid peroxidation deeply affect the cellular membranes, an integral part of amino acid synthesis, and translocation of peptides across the membrane for protein maturation.

3.3.4 PHOSPHOROUS

Phosphorous is crucial in the energy metabolism of cells. It is the constituent of nucleic acids, phospholipids, dinucleotides, and adenosine triphosphate (Bieleski, 1973). Papadopoulos and Rendig (1983) observed that $H_2PO_4^-$ uptake has been suppressed by Cl^-. Unlike competitive uptake of K and the replacement of Ca by Na, the presence of Cl^- in the solution determines the level of P in plants (Hu and Schmidhalter, 2005; Sharpley et al., 1992). The uptake of salinity in plants is based on a higher affinity (Vmax) and lower affinity (Km) toward P in the growth medium and the capacity of roots to regulate the availability of P (Navarro et al., 2001). A reduction in P availability in nutrient solution is due to the high ionic strength created by Na^+ and Cl^- ions (Navarro et al., 2001). In Si, an increased P content helps the plant to overcome stress (Ma and Takahashi, 1991; Ali et al., 2012).

According to Roy et al. (1971), application of Si increases P when it is low and decreases P when it is abundant. However, Si improves the utilization of internal P present in rice (Ma and Takahashi, 1990a, 1991) when available P is low. After uptake of P by plants, it is translocated and redistributed as an inorganic P. Although inorganic P is necessary for metabolism, higher concentration of P inhibits the enzyme reactions and create osmotic pressure (Yoneyama, 1988). Ma and Takahashi (1990a) demonstrate that Si prevents excessive accumulation of P in plants. The excessive inorganic P inside cells has a negative effect on plant growth (Nagaoka, 1998). Both the deficiency (Dong et al., 2004) and abundant level (Shane et al., 2004) of P cause malfunction in the plant cells. When the availability of P is greater, the uptake of Si is reduced and the range of the optimal inorganic level of P inside the cell is broadened (Ma and Takahashi, 1990b).

A larger availability of internal P is achieved by decreasing excess Fe and Mn. Chlorosis or necrosis in leaves occurs during excess P due to the deficiency in availability of Fe and Zn (Ma and Takahashi, 1990a,b). The deposition of Si in the root or decreased transpiration might be responsible for the apoplastic barriers against the radial movement of excess P across the root (Lux et al., 2003; Ma, 2004). The external P strongly regulates the system of multiple transporters in the plasma membrane, and the presence of NaCl has a negative impact on its transportation. In some cases, cytosolic alkylation conditions stimulate a higher uptake of P during salt stress. Mistrik and Ullrich (1996) proposed a co-transport mechanism of $H^+/H_2PO_4^-$. The saline condition lowers P in both the apoplasm and vacuolar under low P, and in P-sufficient or higher conditions, vacuoles act as a reservoir of P. Remobilization of P from the vacuoles inhibited by salt leads to P toxicity in cells. Any H^+ electrochemical gradient changes in the plasma membrane affect the P uptake (Marschner, 1995). Apart from the reduced transpiration or physical barrier mechanism, activation of P-ATPase is due to silica gel formation by acidification of silicic acids (Yeo et al., 1999). As P is important in the activity of ADP-glucose-pyrophosphorlyase, excess could inhibit the formation of starch and stimulate the triosephosphates (Flugge, 1999).

3.4 REGULATION OF MICRONUTRIENTS BY Si DURING SALT STRESS

3.4.1 BORON

Similar to Ca, B is a vital constituent for the stabilization of the cell wall and cell membrane (Whittington, 1957). Boron is involved in cell wall lignification and the differentiation of xylem. Boron is capable of forming stable mono- and di-esters with cis-diols by forming a complex with OH groups. The cis-diol–borate complexes can interact with the membrane components (Marschner, 1995). The sugar–borate complex esters are an integral part of the hemicellulose layer. Stable borate–ester formation between the boric acid and phenolic acids is important for phenol metabolism and lignin synthesis (Brenchley and Warington, 1972). Boron plays an essential role either directly or indirectly in the cell elongation, cell division and cell wall biosynthesis, membrane function, nitrogen metabolism, leaf photosynthesis, and structural stability of uracil (Zhao and Oosterhuis, 2002). The level of boron is accompanied with the salinity of the soil (Alpaslan and Gunes, 2001; Ben-Gal and Shani, 2002). In saline conditions, boron sodicity occurs in a large amount and becomes toxic to plants (Gunes et al., 2007b). Application of Si reduces the uptake of B, Na, and Cl in tomato and spinach.

Silicon and B share similar chemical properties, such as weak, undissociated acids in aqueous solution, and are able to form a complex with polyhydroxy compounds (Brown et al., 1999). Distribution of B in plants depends on the transpiration stream. Polster and Schwenk (1992) suggested that the Si supply increases the B concentration in between the critical deficiency and toxicity level in lily. Liang and Shen (1994) reported that Si enhances B uptake and accumulation under B deficiency, but depresses B uptake at sufficient or elevated B levels. Boron is often found in high concentrations in associatioin with saline-sodic soils, and reduced growth in tomato and cucumber was documented by Alpaslan and Gunes (2001). Si treatment decreasing the B concentration might be due to the formation of a B-Si complex in the soil or solution and leads to lower B availability (Gune et al., 2007a). Deposition in cell walls and the lumen can also reduces the translocation of B, along with Na and Cl. The enhanced tolerance against B under higher NaCl levels is associated with decreased membrane damage. Under a deficiency of B, Si enhances the accumulation of all other nutrients, such as P, K, Ca, Mg, S, Fe, Zn, Mg, Cu, and Mo in oilseed rape (Savic and Marjanović-Jeromela, 2013). The toxic B level is mostly reduced by forming B-Si complexes, which inactivates or immobilizes the B to the shoot (Rogalla and Römheld, 2002). By preventing transpiration-driven water flow, Si can block excess B accumulation (Richmond and Sussman, 2003). The decrease in B concentration is attributed to a possible antagonism between B and Si. In the plants, both B and Si are passively taken up by transpiration-driven water flow (Ben-Gal and Shani, 2002). Stomatal resistance was increased by Si under toxic B in barley (Gunes et al., 2007b).

Supplementation of Si optimized the uptake rate of B between the critical deficiency and toxicity levels in cucumber (Rogalla and Römheld, 2002). The higher activity of peroxidase (POD) and polyphenol oxidase (PPO) under B deficiency released the B-bound POD and PPO from the cell walls, affecting the integrity of the plasma membrane. In B-sufficient conditions, Si reduced the free B and also decreased the higher POD and PPO activities. In both B-sufficient and -deficient conditions, Si treatment increased the total ascorbate. Apart from the modulation of ROS and increased activity of lipoxygenase (LOX), an iron-containing enzyme could break down the polyunsaturated fatty acids released by the membrane and damage could be controlled by Si during B toxicity (Gunes et al., 2007b).

3.4.2 IRON

Iron is an integral component of a number of proteins and enzymes in the plant metabolic process. Mainly, Fe acts as a catalyst for the synthesis of chlorophyll (Marschner, 1995). The deprivation of iron in the medium shows that chlorosis symptoms are due to a blockage in protein synthesis that downregulates the photosystem (Bienfait, 1985). It appears to be mainly bound to the chelators and forms

Fe^{3+}–citrate or an iron–peptide complex (Rodríguez et al., 2005). High-salinity conditions facilitate the Fe to a deficient level (Talei et al., 2012) and toxic level (Achakzai et al., 2010). Silicon supplementation (Chalmardi et al., 2014) alleviates Fe toxicity by reducing the Fe concentration and its transportation from root to shoot. Although Fe taken by plants is mainly used for the electron transport system, excessive Fe is undesirable. A toxic level of Fe causes a malfunction in O_2 depletion and electron donors and acceptors in plants. Iron plays a major role in the redox system of plants. In particular, cytochrome, a hemoprotein, is responsible for the electron transporting chain in the chloroplast and the mitochondria. Cytochrome is the intermediate for electrons in nitrogen assimilation. Ferredoxin, an iron-sulfur protein, acts as a carrier in the electron transport chain (Marschner, 1995). Changes in nitrogen assimilation could be related to the Fe deficiency and the recovery mechanism by Si. Muneer et al. (2014) showed that in tomato, supplementation of silicon induced cytochrome b6/f to alleviate the salinity stress. Catalase (CAT) and POD are important antioxidant enzymes in ROS scavenging belongs to the heme-iron-containing enzymes. The alleviation of Fe toxicity and deficiency by Si is supported by improved antioxidant enzymes such as CAT, ascorbate POD, and cell wall POD (Chalmardi and Zadeh, 2013; Chalmardi et al., 2014). It is absorbed by the roots as either Fe^{2+} or Fe^{3+}. Mostly, it is taken up as Fe^{2+} and the large amount of free Fe^{2+} accelerates the formation of ROS such as superoxide ($O_2^{\cdot-}$), hydrogen peroxide (H_2O_2), and hydroxyl radical (OH^-) (Becana et al., 1998).

Supplementation of Si suppressed the symptoms of Fe deficiency in cucumber (Pavlovic et al., 2013). Supplemental Si delayed leaf chlorosis during Fe deficiency and increased the chlorophyll pigment. Under Fe deficiency obscured upon Si treatment, the plants remain green throughout the experiment. The addition of Si led to significantly higher Fe accumulation in the apoplast of cucumber roots and is associated with the cell wall components, including polysaccharides, lignins, and proteins (Currie and Perry, 2007). Schwarz (1973) stated that Si can strongly bind to pectins rich in galacturonate residues, which can contribute to the cross-linking of cell wall structures. da Cunha and do Nascimento (2009) reported that thickening of the endodermis, xylem, and pericycle cell walls by lignin and Si deposition might account for mitigation of Fe toxicity. The formation of iron plaque on rice roots is associated with occupation of the binding site of Fe by Si (Qiang et al., 2012).

Furthermore, Iler (1979) reported that polysilicate during the polymerization of orthosilicic acid can bind with chelatelike complexes such as the extension of Fe-binding sites in the root apoplast. Phenol exudates from the root play an important role in facilitating the reutilization of apoplastic Fe. The polymerization of phenols to lignin was catalyzed by the cell wall–bound PODs. It causes the reduction-based uptake of Fe by root cells and translocation to the upper part of plants. Si-fed cucumber plants showed higher activities of polyphenol oxidase and cell wall POD, possibly leading to the higher polymerization of polyphenols. This process helps to avoid the browning caused by free phenols and enhances the direct chelation or trapping of Fe at a toxic level (Lavid et al., 2001). Increased lignin in rice could enhance the absorption of free Fe present in cells (Chalmardi and Zadeh, 2013). The improved Fe transportation across a plasma membrane (PM) via divalent cation transporters such as iron regulated transporter (IRT) and myo-inositol transporter (ITR1) were proved by the increase of citrate in the xylem sap of Si-added Fe-deficiency treatment for long distance transportation and utilization of Fe by leaf cells. This was correlated with the enhanced expression of the tricarboxylic acid (TCA) cycle and phosphoenolpyruvate carboxylase (PEPC)–related transcripts in Fe-deficient conditions (Pavlovic et al., 2013). For adequate Fe, a strong enhancement effect was observed on Fe-deficient plants. The expression of genes involved in Fe acquisition, such as ferric chelate reductase (FRO2), ITR1, and H^+-ATPase (HA1) treated with Si for Fe deficiency and without Si for adequate Fe, is observed in Pavlovic et al. (2013).

3.4.3 COPPER

Copper is a divalent cation and is taken up by the plant as Cu^{2+} or as a copper chelate complex (Delas, 1963). Similar to Fe, the functions of Cu are based on the participation of enzymatically bound Cu in redox reactions (Chaignon et al., 2002). In sunflower, Cu toxicity occurs during salt stress (Achakzai et al., 2010). Cu is phytotoxic at higher concentrations (Kovacik et al., 2009). The level of Cu was induced to be higher in *Andrographis paniculata* Nees (Talei et al., 2012) during salt stress. In the redox reactions of the terminal oxidases, copper enzymes react directly with molecular oxygen. The Cu-Zn SOD is involved in the neutralization of $O_2^{\cdot-}$ accumulated during photorespiration (Tanaka et al., 1996). Interestingly, Cu-Zn SOD is mostly found in the stroma of chloroplasts. Excess accumulation of $O_2^{\cdot-}$ has been detoxified by Si-induced SOD activity, and its expression has been reported in *A. thaliana* (Khandekar and Leisner, 2011). Most of the $O_2^{\cdot-}$ and H_2O_2 was formed in chloroplasts (90%) and a smaller portion in mitochondria (4%–5%). Plastocyanins are the components in the electron transport chain of photosystem I. The Cu deficient condition affects the function of the two polypeptides in the chloroplast membrane, which are necessary to maintain the plastoquinone (Marschner, 1995). The involvement of Cu on the mitochondrial electron transport chain has also been proved. The destruction of the thylakoid membrane and the concentrate on the lipid peroxidation cause chlorosis. Shoot growth is mostly affected by the higher availability of Cu than root growth.

The toxicity of Cu in *Arabidopsis* was alleviated by Si through active metal transporter regulation (Li et al., 2008). In zinnia, roots are changed from white to tan brown or deep orange, shoots are stunted, and leaves display interveinal chlorosis (similar to Fe deficiency) at a toxic level of copper; with the addition of Si, boron toxicity was recovered (Frantz et al., 2011). The activity of phenylalanine ammonia lyase (PAL) and POD was reduced by Si during toxic Cu indicates that the leaves are unstressed at an elevated Cu level. This might also be due to the additional Cu-binding sites generated by Si to sequester the metals (Frantz et al., 2011). This effect reduces the ROS generation during stress. Copper is fairly immobile between cell tissues, and deficiency symptoms are clearly visible as chlorosis formation in new leaves and reproductive parts. Li et al. (2008) suggest that Si influences the distribution or bioavailability of Cu within leaves under Cu stress. Chloroplast is very sensitive to Cu, and Cu acts as a cofactor in the ethylene signaling pathway, which could intensify leaf chlorosis and aging (Abdel-Ghany et al., 2005). The suppression of free Cu as a consequence of the formation of Si cross-linked binding pools on the cell wall is analogous to that in the studies on Mn toxicity (Iwasaki et al., 2002).

According to Rogalla and Römheld (2002), the Si–phenolic complex reduces the free phenol, and consequently, the expression of oxidative browning enzymes is reduced in cucumber. Deposition of Si in the shoots differs from that in the roots when ions are at a toxic level. In roots, the Si deposition on the endodermal layer can act as a natural barrier for Cu uptake (Kirkham, 2006). As Cu flow to the shoots is blocked, epidermal and cortical cells in the roots induce a stress response as a result of Cu deposition. A possible explanation from this is that Si helps to stimulate the accumulation of polymerized phenolics in roots by stimulating PAL activity (Cherif et al., 1992) and resistance against Cu toxicity (Li et al., 2008). Two major Cu-binding molecules, such as metallothioneins (MTs) and pytochelatins (PCs), are involved in maintaining the intercellular level of Cu ion. The former helps to keep the free Cu concentration low in cytoplasm, and the latter is involved in preventing free Cu in the cytoplasm and transport to the vacuole. At a higher Cu level, the expression of MTs and PCs indicates that Si efficiently maintains Cu by increasing the free radical metabolizing enzymes (Khandekar and Leisner, 2011). Similarly, during Cu toxicity, the higher level of PAL enzyme and the copper transporter 1 (COPT1) and heavy metal ATPase subunit 5 (HMA5) expressions were reduced by Si (Li et al., 2008).

3.4.4 ZINC

Zinc acts as the metal component of enzymes or as a functional, structural, or regulatory cofactor of a large number of enzymes (Vallee and Auld, 1990). Zinc is involved in the synthesis of protein, carbohydrate, and indole acetic acid (IAA) metabolism, and internodal elongation (Cakmak et al., 1989). Zinc deficiency is one of the critical micronutrient deficiencies in saline or sodic soils (Weisany et al., 2012). The concentration of zinc is also increased under salinity (Taibei et al., 2012).

There has been some discussion that Si can increase the physiological availability of zinc in leaf tissue (Datnoff et al., 2001). At a high P with low Zn supply condition without Si, severe chlorosis has been omitted by the addition of Si (Ma and Takahashi, 1991). The excess Zn in the plant reduced the capacity of phloem to load the sucrose (Rauser and Samarakoon, 1980) and transport other metals (Welch and Shuman, 1999). According to the report of Neumann and Zur Nieden (2001), Zn-silicate formation in the cell wall was involved in the detoxification of Zn. Vacuoles are the main storage compartment for Zn. Neumann and Zur Nieden (2001) found that Zn and Si are colocalized in the vacuolar vesicles. Heavy metals are covalently bound or chelated during their transport through the cytoplasm to a complex with phytochelatins. However, Zn cannot covalently form a thermodynamically stable complex with phytochelatins (Leopold et al., 1999). The Zn is a precipitate such as Zn-silicate, as a transient storage compound for the metal. Later, Si is precipitated in the cytoplasm and Zn is translocated into the vacuole (Neumann and Zur Nieden, 2001).

The Zn complexed SOD is involved and detoxifies $2O_2^- + 2H^+$ to H_2O_2 and O_2 and oxygen in cucumber (Zhu et al., 2004). Zinc is involved in nitrogen formation with histidine and helps to store CO_2 in reverse as HCO_3^-. It can be also used as a RuBisCo substrate. Protein synthesis is affected when the zinc level is low, as Zn is an important component for protein structural stability. An inverse proportion of Zn and RNAse was observed; that is, RNAase activity is higher at a lower Zn ratio. A low availability of Zn also disturbs tryptophan, a precursor for IAA (Singh, 1981). Zn as a component of glutamic, alcohol, and lactic dehydrogenases and carbonic anhydrase is influenced by Si. The toxicity of Zn has been alleviated by precipitation, membrane integrity, and defense capacity.

3.4.5 MANGANESE

Manganese in plants is involved in the photosystem, bound with several metalloproteins (Loneragan, 1988). Mn-SOD is an antioxidant enzyme for scavenging the excessive generation of O_2^-. Formation of scorching in the leaves during salinity is associated with accumulation of Mn (Graham et al., 1988; Horst et al., 1999). In tomato and soybean (Mass et al., 1972), similar to Cu, Mn deficiency changes the chloroplast structure. The occurrence of brown spots in leaves during salinity is associated with the accumulation of oxidized Mn (Mn(IV)) and phenols (Wissemeier et al., 1992). This process has been catalyzed by the presence of PODs in the apoplast (Fecht-Christoffers et al., 2003). This oxidation process of Mn^{2+} and phenols by POD leads to Mn toxicity. It could form Mn^{3+} and pheoxy radicals (Horst et al., 1999). Previously, Fuhrs et al. (2009), Rogalla and Römheld (2002), and Shi et al. (2005) reported the alleviation of Mn toxicity by Si. Accumulation of Mn in the necrotic spot in the absence of Si on the leaves of cucumber and *Vigna unguiculata* was reversed by the presence of Si. The formation of phytotoxic Mn^{3+} and pheoxy has been avoided (Horst et al., 1999). Maksimović et al. (2012) suggest that if Si is able to form a Mn-Si complex, similar to Zn-Si, this could explain the metal-polysilicate involved in increasing cell wall–bound Mn. The Si-enhanced Mn binding could be considered a universal mechanism for the alleviation of Mn toxicity.

Iwasaki et al. (2002) mentioned the precipitation of Mn in the apoplast to avoid its oxidation in pumpkin. During the addition of Si to cucumber, 90% of Mn was bound in the cell wall, and in the absence of Si, Mn was evenly distributed in the apoplast and symplast (Rogalla and Römheld, 2002; Wiese et al., 2007). Moreover, Si and Mn together accumulated in the trichomes (Iwasaki and Matsumura, 1999). The excess Mn is localized by Si in the metabolically inactive form. Horiguchi (1988) states that Si contributes to the distribution of Mn in the root and decreases its content in the leaf. This process was modulated by enhancement of apoplastic NADH-POD and phenolic compounds such as p-coumaric acid, vanillic acid, benzoic acid, and ferulic acid (Fuhrs et al., 2009).

3.5 CONCLUSIONS

To overcome the unlimited stress in the world, several approaches, such as transformation and genetic engineering, have been carried out. However, most of these measures limited by either environmental or economical concern. Although exogenous application of macronutrients has been proved to alleviate stress, it can possibly minimize the solubility of other ions in the medium. The presence of one ion can complicate the availability of other ions in the uptake form. However, supplementation of Si has become a successful method for alleviating salt stress by maintaining proper water levels and enhancing nutrient uptake. The uptake of essential elements from the deficit to the toxic level was the main criterion to improve the resistance by activating the antioxidants and protein-related mechanisms. Controlling the transporter expressions of metals such as Fe and Cu gives an additional dimension on the involvement of Si in the remediation of metal ions. Precipitation, transportation, and the metal binding property of Si are complex mechanisms involved in the compartmentalization, distribution, and vacuolation of excessive elements. To understand the exact mechanisms on involvement of Si in improved tolerance based on nutrient regulation against abiotic stress, studies using cutting-edge genomics and proteomics approaches are needed. Hence, recent advances in technology are helping us to gain more knowledge on nutrient regulation of Si during abiotic stress.

ACKNOWLEDGMENTS

Abinaya Manivannan and Prabhakaran Soundararajan were supported by a scholarship from the BK21 Plus Program, Ministry of Education, South Korea.

REFERENCES

Abdel-Ghany, S.E., P. Müller-Moulé, K.K. Niyogi, M. Pilon, and T. Shikanai. 2005. Two P-type ATPases are required for copper delivery in *Arabidopsis thaliana* chloroplasts. *Plant Cell* 17:1233–1251.

Achakzai, A.K.K., S.A. Kayani, and A.Z.H.A.R. Hanif. 2010. Effect of salinity on uptake of micronutrients in sunflower at early vegetative stage. *Pak. J. Bot.* 42:129–139.

Agarie, S., N. Hanaoka, O. Ueno, A. Miyazaki, F. Kubota, W. Agata, and P.B. Kaufman. 1998. Effects of silicon on tolerance to water deficit and heat stress in rice plants (*Oryza sativa* L.) monitored by electrolyte leakage. *Plant Production Sci.* 1:96–103.

Agarie, S., W. Agata, F. Kubota, and P.B. Kaufman. 1992. Physiological role of silicon in photosynthesis and dry matter production in rice plants. *J. Crop Sci.* 61:200–206.

Ahmad, R., S.H. Zaheer, and S. Ismail. 1992. Role of Si in salt tolerance of wheat (*Triticum aestivum* L.). *Plant Sci.* 85:43–50.

Akram, M.S., H.R. Athar, and M. Ashraf. 2007. Improving growth and yield of sunflower (*Helianthus annuus* L.) by foliar application of potassium hydroxide (KOH) under salt stress. *Pak. J. Bot.* 39:2223–2230.

Al-Aghabary, K., Z. Zhu, and Q. Shi. 2004. Influence of silicon supply on chlorophyll content, chlorophyll fluorescence and anti-oxidative enzyme activities in tomato plants under salt stress. *J. Plant Nutr.* 27:2101–2115.

Alam, S.M. 1999. Nutrient uptake by plants under stress conditions. In M. Pessarakli (ed.), *Handbook of Plant and Crop Stress*. Marcel Dekker, New York, pp. 285–314.

Ali, A., S.M.A. Basra, S. Hussain, and J. Iqbal. 2012. Salt stress alleviation in field crops through nutritional supplementation of silicon. *Pak. J. Nutr.* 11:637–655.

Alpaslan, M., and A. Gunes. 2001. Interactive effects of boron and salinity stress on the growth, membrane permeability and mineral composition of tomato and cucumber plants. *Plant Soil* 236:123–128.

Alva, A.K., and J.P. Syversten. 1991. Irrigation water salinity affects soil nutrient distribution, root density, and leaf nutrient levels of citrus under drip fertigation. *J. Plant Nutr.* 14:715–727.

Arnon, D.I., and P.R. Stout. 1939. The essentiality of certain elements in minute quantity for plants with special reference to copper. *Plant Physiol.* 14(2):371–375.

Asch, F., M. Dingkuhn, K. Miezan, and K. Dörffling. 2000. Leaf K/Na ratio predicts salinity induced yield loss in irrigated rice. *Euphytica* 113:109–118.

Asher, C.J. 1991. Beneficial elements, functional nutrients, and possible new essential elements. In J.J. Mortvedt, F.R. Cox, L.M. Shuman, and R.M. Welch (eds.), *Micronutrients in Agriculture*. 2nd ed. Soil Science Society of America, Madison, WI, pp. 703–723.

Avila, F.W., D.P. Baliza, V. Faquin, J. Araujo, and S.J. Ramos. 2010. Silicon-nitrogen interaction in rice cultivated under nutrient solution. *Rev. Cienc. Agron.* 41:184–190.

Awada, S., W.F. Campbell, L.M. Dudley, J.J. Jurinak, and M.A. Khan. 1995. Interactive effects of sodium chloride, sodium sulfate, calcium sulfate and calcium chloride on snapbean growth, photosynthesis and ion uptake. *J. Plant Nutr.* 18:889–900.

Becana, M., J.F. Moran, and L. Iturbe-Ormaetxe. 1998. Iron-dependent oxygen free radical generation in plants subjected to environmental stress: Toxicity and antioxidant protection. *Plant Soil* 201:137–147.

Ben-Gal, A., and U. Shani. 2002. Yield, transpiration and growth of tomatoes under combined excess boron and salinity stress. *Plant Soil* 247:211–221.

Bieleski, R.L. 1973. Phosphate pools, phosphate transport, and phosphate availability. *Ann. Rev. Plant Physiol.* 24:225–252.

Bienfait, H.F. 1985. Regulated redox processes at the plasmalemma of plant root cells and their function in iron uptake. *J. Bioenerg. Biomembr.* 17:73–83.

Blumwald, E. 2000. Sodium transport and salt tolerance in plants. *Curr. Opin. Cell Biol.* 12:431–434.

Brenchley, W.E., and K. Warington. 1927. The role of boron in the growth of plants. *Ann. Bot.* 37:629–672.

Brown, P.H., N. Bellaloui, H. Hu, and A. Dandekar. 1999. Transgenically enhanced sorbitol synthesis facilitates phloem boron transport and increases tolerance of tobacco to boron deficiency. *Plant Physiol.* 119:17–20.

Broyer, T.C., A.B. Carlton, C.M. Johnson, and P.R. Stout. 1954. Chlorine—A micronutrient element for higher plants. *Plant Physiol.* 29:526–532.

Cachorro, P., A. Ortiz, and A. Cerda. 1994. Implications of calcium on the response of *Phaseolus vulgaris* L. to salinity. *Plant Soil* 159:205–212.

Cakmak, I., H. Marschner, and F. Bangerth. 1989. Effect of zinc nutritional status on growth, protein metabolism and levels of indole-3-acetic acid and other phytohormones in bean (*Phaseolus vulgaris* L.). *J. Exp. Bot.* 40:405–412.

Caldwell, C.R., and A. Haug. 1982. Divalent cation inhibition of barley root plasma membrane-bound Ca^{2+}-ATPase activity and its reversal by monovalent cations. *Physiol. Plant.* 54:112–118.

Chaignon, V., D. Di Malta, and P. Hinsinger. 2002. Fe deficiency increases Cu acquisition by wheat cropped in a Cu-contaminated vineyard soil. *New Phytol.* 154:121–130.

Chalmardi, Z.K., A. Abdolzadeh, and H.R. Sadeghipour. 2014. Silicon nutrition potentiates the antioxidant metabolism of rice plants under iron toxicity. *Acta Physiol. Plant.* 36:493–502.

Chalmardi, Z.K., and A.A. Zadeh. 2013. Role of silicon in alleviation of iron deficiency and toxicity in hydroponically-grown rice (*Oryza sativa* L.) plants. *J. Sci. Technol. Greenhouse Cult.* 3:1–12.

Cherif, M., N. Benhamou, J.G. Menzies, and R.R. Belanger. 1992. Silicon-induced resistance in cucumber plants against *Pythium ultimum*. *Physiol. Mol. Plant Pathol.* 41:411–425.

Chinnusamy, V., A. Jagendorf, and J.K. Zhu. 2005. Understanding and improving salt tolerance in plants. *Crop Sci.* 45:437–448.

Crusciol, C.A., A.L. Pulz, L.B. Lemos, R.P. Soratto, and G.P. Lima. 2009. Effects of silicon and drought stress on tuber yield and leaf biochemical characteristics in potato. *Crop Sci.* 49:949–954.

Currie, H.A., and C.C. Perry. 2007. Silica in plants: Biological, biochemical and chemical studies. *Ann. Bot.* 100:1383–1389.

da Cunha, K.P.V., and C.W.A. do Nascimento. 2009. Silicon effects on metal tolerance and structural changes in maize (*Zea mays* L.) grown on a cadmium and zinc enriched soil. *Water Air Soil Pollut.* 97:323–330.

Datnoff, L.E., G.H. Snyder, and G.H. Korndorfer. 2001. *Silicon in Agriculture*. Elsevier Science, Amsterdam.

Delas, J. 1963. The toxicity of copper accumulated in soils. *Agrochemica* 7:258–288.

Demmig, B., and H. Gimmler. 1983. Properties of the isolated intact chloroplast at cytoplasmic K^+ concentrations. I. Light-induced cation uptake into intact chloroplasts is driven by an electrical potential difference. *Plant Physiol.* 73:169–174.

Dong, D.F., X.X. Peng, and X.L. Yan. 2004. Organic acid exudation induced by phosphorus deficiency and/or aluminium toxicity in two contrasting soybean genotypes. *Physiol. Plant.* 122:190–199.

Dubey, R.S., Pessarakli, M. 1995. Physiological mechanisms of nitrogen absorption, and assimilation in plants under stressful conditions. In M. Pessarakli (ed.), *Handbook of Plant and Crop Physiology*. CRC Press, Boca Raton, FL, pp. 605–625.

Epstein, E. 1972. *Mineral Nutrition of Plants: Principles and Perspectives*. Wiley, New York.

Epstein, E. 1994. The anomaly of silicon in plant biology. *Proc. Natl. Acad. Sci. U.S.A.* 91:11–17.

Epstein, E. 1999. Silicon. *Ann. Rev. Plant Physiol. Plant Mol. Biol.* 50:641–664.

Epstein, E., and A.J. Bloom. 2005. *Mineral Nutrition of Plants: Principles and Perspectives*. 2nd ed. Sinauer Associates, Sunderland, MA.

Fallah, A. 2012. Study of silicon and nitrogen effects on some physiological characters of rice. *Int. J. Agri. Crop Sci.* 4:238–241.

Fecht-Christoffers, M.M., P. Maier, and W.J. Horst. 2003. Apoplastic peroxidases and ascorbate are involved in manganese toxicity and tolerance of *Vigna unguiculata*. *Physiol. Plant.* 117:237–244.

Flugge, U.I. 1999. Phosphate translocators in plastids. *Annu. Rev. Plant Physiol. Plant Mol. Biol.* 50:27–45.

Francois, L.E., and E.V. Maas. 1994. Crop response and management on salt-affected soils. In *Handbook of Plant and Crop Stress*. CRC Press, Boca Raton, FL, pp. 149–181.

Frantz, J.M., S. Khandekar, and S. Leisner. 2011. Silicon differentially influences copper toxicity response in silicon-accumulator and non-accumulator species. *J. Am. Soc. Hortic. Sci.* 136:329–338.

Fuhrs, H., S. Gotze, A. Specht et al. 2009. Characterization of leaf apoplastic peroxidases and metabolites in *Vigna unguiculata* in response to toxic manganese supply and silicon. *J. Exp. Bot.* 60:1663–1678.

Gong, H.J., D.P. Randall, and T.J. Flowers. 2006. Silicon deposition in the root reduces sodium uptake in rice (*Oryza sativa* L.) seedlings by reducing bypass flow. *Plant Cell Environ.* 29:1970–1979.

Gong, H., X. Zhu, K. Chen, S. Wang, and C. Zhang. 2005. Silicon alleviates oxidative damage of wheat plants in pots under drought. *Plant Sci.* 169:313–321.

Graham, R.D., R.J. Hannam, and N.C. Uren. 1988. *Manganese in Soils and Plants*. Kluwer Academic, Amsterdam.

Grattan, S.R., and C.M. Grieve. 1999. Salinity-mineral nutrient relations in horticultural crops. *Sci. Hortic.* 78:127–157.

Greenway, H., and R. Munns. 1980. Mechanism of salt tolerance in non-halophytes. *Ann. Rev. Plant Physiol.* 31:149–190.

Gunes, A., A. Inal, and E.G. Bagci. 2007a. Silicon-mediated changes of some physiological and enzymatic parameters symptomatic for oxidative stress in spinach and tomato grown in sodic-B toxic soil. *Plant Soil* 290:103–114.

Gunes, A., A. Inal, E.G. Bagci, S. Coban, and D.J. Pilbean. 2007b. Silicon mediates changes to some physiological and enzymatic parameters symptomatic for oxidative stress in spinach (*Spinacia oleracea* L.) grown under B toxicity. *Sci. Hortic.* 113:113–119.

Hasegawa, P.M., R.A. Bressan, and J.M. Pardo. 2000a. The dawn of plant salt to tolerance genetics. *Trends Plant Sci.* 5:317–319.

Hasegawa, P.M., R.A. Bressan, J.K. Zhu, and H.J. Bohnert. 2000b. Plant cellular and molecular responses to high salinity. *Annu. Rev. Plant Physiol. Plant Mol. Biol.* 51:463–499.

Horiguchi, T. 1988. Mechanism of manganese toxicity and tolerance of plants. IV. Effects of silicon on alleviation of manganese toxicity of rice plants. *Soil Sci. Plant Nutr.* 34:65–73.

Horst, W.J., M. Fecht, A. Naumann, A.H. Wissemeier, and P. Maier. 1999. Physiology of manganese toxicity and tolerance in *Vigna unguiculata* (L.) Walp. *J. Plant Nutr. Soil Sci.* 162:263–274.

Hu, Y., and U. Schmidhalter. 2005. Drought and salinity: A comparison of their effects on mineral nutrition of plants. *J. Plant Nutr. Soil Sci.* 168:541–549.

Iler, R.K. 1979. *The Chemistry of Silica*. John Wiley, New York.

Iwasaki, K., M. Fecht, P. Maier, and W.J. Horst. 2002. Can leaf apoplastic manganese and Si concentrations explain Si-enhanced manganese tolerance of *Vigna unguiculata* (L.) Walp.? *Dev. Plant Soil Sci.* 92:246–247.

Iwasaki, K., and A. Matsumura. 1999. Effect of silicon on alleviation of manganese toxicity in pumpkin (*Cucurbita moschata* Duch cv. Shintosa). *Soil Sci. Plant Nutr.* 45:909–920.

Jain, M., G. Mathur, S. Koul, and N.B. Sarin. 2001. Ameliorative effects of proline on salt stress-induced lipid peroxidation in cell lines of groundnut (*Arachis hypogea* L.). *Plant Cell Rep.* 20:463–468.

Khandekar, S., and S. Leisner. 2011. Soluble silicon modulates expression of *Arabidopsis thaliana* genes involved in copper stress. *J. Plant Physiol.* 168:699–705.

Kim, Y.H., A.L. Khan, M. Waqas, J.K. Shim, D.H. Kim, K.Y. Lee, and I.J. Lee. 2014. Silicon application to rice root zone influenced the phytohormonal and antioxidant responses under salinity stress. *J. Plant Growth Regul.* 33:137–149.

Kirkham, M.B. 2006. Cadmium in plants on polluted soils: Effect of soil factors, hyperaccumulation, and amendments. *Geoderma* 137:19–32.

Ko, J.H., and S.H. Lee. 1996. Biochemical studies of purified 23 kD calcium-binding protein in *Dunaliella salina* and its cDNA cloning. *Plant Physiol.* 111:714–720.

Kochanová, Z., K. Jašková, B. Sedláková, and M. Luxová. 2014. Silicon improves salinity tolerance and affects ammonia assimilation in maize roots. *Biologia* 69:1164–1171.

Kovacik, J., B. Klejdus, J. Hedbavny, F. Štork, and M. Bačkor. 2009. Comparison of cadmium and copper effect on phenolic metabolism, mineral nutrients and stress-related parameters in *Matricaria chamomilla* plants. *Plant Soil* 320:231–242.

Lahaye, P.A., and E. Epstein. 1971. Calcium and salt tolerance by bean plants. *Plant Physiol.* 25:213–218.

Lavid, N., A. Schwartz, E. Lewinsohn, and E. Tel-Or. 2001. Phenols and phenol oxidases are involved in cadmium accumulation in the water plants *Nymphoides peltata* (Menyanthaceae) and Nymphaeae (Nymphaeaceae). *Planta* 214:189–195.

Leopold, I., D. Gunther, J. Schmidt, and D. Neumann. 1999. Phytochelatins and heavy metal tolerance. *Phytochemistry* 50:1323–1328.

Li, J., S.M. Leisner, and J.M. Frantz. 2008. Alleviation of copper toxicity in *Arabidopsis thaliana* by silicon addition to hydroponic solutions. *J. Am. Soc. Hortic. Sci.* 133:670–677.

Liang, Y. 1999. Effects of silicon on enzyme activity and sodium, potassium and calcium concentration in barley under salt stress. *Plant Soil* 209:217–224.

Liang, Y., and Shen, Z. 1994. Interaction of silicon and boron in oilseed rape plants. *J. Plant Nutr.* 17:415–425.

Liang, Y.C. 1998. Effects of Si on leaf ultrastructure, chlorophyll content and photosynthetic activity in barley under salt stress. *Pedosphere* 8:289–296.

Liang, Y.C., Q.R. Shen, Z.G. Shen, and T.S. Ma. 1996. Effects of silicon on salinity tolerance of two barley cultivars. *J. Plant Nutr.* 19:173–183.

Liebersbach, H., B. Steingrobe, and N. Claassen. 2004. Roots regulate ion transport in the rhizosphere to counteract reduced mobility in dry soil. *Plant Soil* 260:79–88.

Loneragan, J.F. 1988. Distribution and movement of manganese in plants. In *Manganese in Soils and Plants.* Springer, Amsterdam, pp. 113–124.

Lu, Z., and P.M. Neumann. 1999. Low cell-wall extensibility can limit maximum leaf growth rates in rice. *Crop Sci.* 36:126–130.

Lutts, S., J. Bouharmont, and J.M. Kinet. 1999. Physiological characterization of salt-resistant rice (*Oryza sativa* L.) varieties. *Aust. J. Bot.* 47:1843–1852.

Lux, A., M. Luxová, J. Abe, E. Tanimoto, T. Hattori, and S. Inanaga. 2003. The dynamics of silicon deposition in the sorghum root endodermis. *New Phytol.* 158:437–441.

Ma, J.F. 2004. Role of silicon in enhancing the resistance of plants to biotic and abiotic stresses. *Soil Sci. Plant Nutr.* 50:11–18.

Ma, J.F., Y. Miyake, and E. Takahashi. 2001. Silicon as a beneficial element for crop plants. *Studies Plant Sci.* 8:17–39.

Ma, J.F., and E. Takahashi. 1990a. Effect of silicon on the growth and phosphorus uptake of rice. *Plant Soil* 126:115–119.

Ma, J.F., and E. Takahashi. 1990b. The effect of silicic acid on rice in a P-deficient soil. *Plant Soil* 126:121–125.

Ma, J.F., and E. Takahashi. 1991. Effect of silicate on phosphate availability of rice in a P-deficient soil. *Plant Soil* 133:151–155.

Ma, J.F., and N. Yamaji. 2006. Silicon uptake and accumulation in higher plants. *Trends Plant Sci.* 11:392–397.

Maas, E.V., and R.H. Nieman. 1978. Physiology of plant tolerance to salinity. In G.A. Jung (ed.), *Crop Tolerance to Suboptimal Land Conditions.* American Society of Agronomy, Madison, WI, pp. 277–299.

Maksimović, J.D., M. Mojović, V. Maksimović, V. Römheld, and M. Nikolic. 2012. Silicon ameliorates manganese toxicity in cucumber by decreasing hydroxyl radical accumulation in the leaf apoplast. *J. Exp. Bot.* 63:2411–2420.

Marschner, H. 1971. Why can sodium replace potassium in plants. *Potassium Biochem. Physiol.* 1:50–63.

Marschner, H. 1995. *Mineral Nutrition of Higher Plants.* 2nd ed. Academic Press, San Diego.

Mass, E.V., G. Ogata, and M.J. Garber. 1972. Influence of salinity on Fe, Mn, and Zn uptake by plants. *Am. Soc. Agron.* 64:793–795.

Matoh, T., P. Kairusmee, and E. Takahashi. 1986. Salt induced damage to rice plants and alleviation effect of silicate. *J. Plant Nutr.* 32:295–311.

McLaughlin, S.B., and Wimmer, R. 1999. Calcium physiology and terrestrial ecosystem processes. *New Phytol.* 142:373–417.

Miao, B.H., X.G. Han, and W.H. Zhang. 2010. The ameliorative effect of silicon on soybean seedlings grown in potassium-deficient medium. *Ann. Bot.* 105:967–973.

Mistrik, I., and C.I. Ullrich. 1996. Mechanism of anion uptake in plant roots: Quantitative evaluation of H^+/NO_3^- and $H^+/H_2PO_4^-$ stoichiometries. *Plant Physiol. Biochem.* 34:629–636.

Miyake, Y., and E. Takahashi. 1978. Silicon deficiency of tomato plants. *Soil Sci. Plant Nutr.* 24:175–189.

Miyake, Y., and E. Takahashi. 1982. Effect of silicon on the growth of cucumber plants in a solution culture. *Jpn. J. Soil Sci. Plant Nutr.* 53:15–22.

Morikawa, C.K., and M. Saigusa. 2004. Mineral composition and accumulation of silicon in tissues of blueberry (*Vaccinum corymbosus* cv. Bluecrop) cuttings. *Plant Soil* 258:1–8.

Muneer, S., Y.G. Park, M. Abinaya, S. Prabhakaran, and B.R. Jeong. 2014. Physiological and proteomic analysis in chloroplasts of *Solanum lycopersicum* L. under silicon efficiency and salinity stress. *Int. J. Mol. Sci.* 15(12):21803–21824.

Munns, R. 2002. Comparative physiology of salt and water stress. *Plant Cell Environ.* 25:238–250.

Nagaoka, K. 1998. Study on interaction between P and Si in rice plants. Graduation thesis, Kinki University.

Natacha, B.E., P.G. Comella, A. Debures et al. 2008. Inventory, evolution and expression profiling diversity of the LEA (late embryogenesis abundant) protein gene family in *Arabidopsis thaliana*. *Plant Mol. Biol.* 67:107–124.

Navarro, J.M., M.A. Botella, A. Cerdá, and V. Martinez. 2001. Phosphorus uptake and translocation in salt-stressed melon plants. *J. Plant Physiol.* 158:375–381.

Neumann, D., and U. Zur Nieden. 2001. Silicon and heavy metal tolerance of higher plants. *Phytochemistry* 56:685–692.

Papadopoulos, I., and V.V. Rendig. 1983. Interactive effects of salinity and nitrogen on growth and yield of tomato plants. *Plant Soil* 73:47–57.

Pardo, J.M., and F.J. Quintero. 2002. Plants and sodium ions: Keeping company with the enemy. *Genome Biol.* 3:1–4.

Pavlovic, J., J. Samardzic, V. Maksimović et al. 2013. Silicon alleviates iron deficiency in cucumber by promoting mobilization of iron in the root apoplast. *New Phytol.* 198:1096–1107.

Pereira, T.S., A.K. da Silva Lobato, D.K.Y. Tan et al. 2013. Positive interference of silicon on water relations, nitrogen metabolism, and osmotic adjustment in two pepper (*Capsicum annuum*) cultivars under water deficit. *Aus. J. Crop Sci.* 7:1064–1071.

Polster, J., and M. Schwenk. 1992. The role of boron, silicon and nucleic bases on pollen tube growth of *Lilium longiflorum* (L.). *Z. Naturforsch.* 47:102–108.

Prabhakaran, S., M. Abinaya, Y.G. Park, S. Muneer, and B.R. Jeong. 2015. Silicon alleviates salt stress by modulating antioxidant enzyme activities in *Dianthus caryophyllus* 'Tula'. *Hortic. Environ. Biotechnol.* 56:233–239.

Prabhakaran, S., I. Sivanesan, S. Jana, and B.R. Jeong. 2014. Influence of silicon on growth and tolerance to high temperature in *Salvia splendens*. *Hortic. Environ. Biotechnol.* 55:271–279.

Qiang, F.Q., S. Hong, W.D. Ming, and K.C. Zheng. 2012. Silicon mediated amelioration of Fe^{2+} toxicity in rice (*Oryza sativa* L.) roots. *Pedosphere* 22:795–802.

Rauser, W.E., and A.B. Samarakoon. 1980. Vein loading in seedlings of *Phaseolus vulgaris* exposed to excess cobalt, nickel, and zinc. *Plant Physiol.* 65:578–583.

Richmond, K.E., and M. Sussman. 2003. Got silicon? The non-essential beneficial plant nutrient. *Curr. Opin. Plant Biol.* 6:268–272.

Rodríguez, N., N. Menéndez, J. Tornero, R. Amils, and V. De La Fuente. 2005. Internal iron biomineralization in *Imperata cylindrica*, a perennial grass: Chemical composition, speciation and plant localization. *New Phytol.* 165:781–789.

Rogalla, H., and V. Römheld. 2002. Role of leaf apoplast in silicon-mediated manganese tolerance of *Cucumis sativus* L. *Plant Cell Environ.* 25:549–555.

Romero-Aranda, M.R., O. Jurado, and J. Cuartero. 2006. Silicon alleviates the deleterious salt effect on tomato plant growth by improving plant water status. *J. Plant Physiol.* 163:847–855.

Romero-Aranda, M.R., T. Soria, and J. Cuartero. 2001. Tomato plant-water uptake and plant-water relationships under saline growth conditions. *Plant Sci.* 160:265–272.

Roy, A.C., M.Y. Ali, R.L. Fox, and J.A. Silva. 1971. Influence of calcium silicate on phosphate solubility and availability in Hawaiian latosols. *In* Symposium Soil Fertility Evaluation. 1:756–765.

Sangster, A.G. 1978. Silicon in the roots of higher plants. *Am. J. Bot.* 65:929–935.

Savic, J., and A. Marjanović-Jeromela. 2013. Effect of silicon on sunflower growth and nutrient accumulation under low boron supply. *Helia* 36:61–68.

Schwarz, K. 1973. A bound form silicon in glycosaminoglycans and polyuronides. *Proc. Natl. Acad. Sci. U.S.A.* 70:1608–1612.

Shahnaz, G., E. Shekoofeh, D. Kourosh, and B. Moohamadbagher. 2011. Interactive effects of silicon and aluminum on the malondialdehyde (MDA), proline, protein and phenolic compounds in *Borago officinalis* L. *J. Med. Plants Res.* 5:5818–5827.

Shane, M.W., M.E. McCully, and H. Lambers. 2004. Tissue and cellular phosphorus storage during development of phosphorus toxicity in *Hakea prostrata* (Proteaceae). *J. Exp. Bot.* 55(399):1033–1044.

Sharma, P., A.B. Jha, R.S. Dubey, and M. Pessarakli. 2012. Reactive oxygen species, oxidative damage, and antioxidative defense mechanism in plants under stressful conditions. *J. Bot.* 2012:1–26.

Sharpley, A.N., J.J. Meisinger, J.F. Power, and D.L. Suarez. 1992. Root extraction of nutrients associated with long-term soil management. In B. Steward (ed.), *Advances in Soil Science*. Springer-Verlag, Berlin, pp. 151–217.

Shi, Q., Y. Bao, Y. Zhu, Y. He, Q. Qian, and J. Yu. 2005. Silicon-mediated alleviation of Mn toxicity in *Cucumis sativus* in relation to activities of superoxide dismutase and ascorbate peroxidase. *Phytochemistry* 66: 1551–1559.

Silva, P., and H. Geros. 2009. Regulation by salt of vacuolar H^+-ATPase and H^+-pyrophosphatase activities and Na^+/H^+ exchange. *Plant Sig. Behav.* 4:718–726.

Singh, M. 1981. Effect of zinc, phosphorus and nitrogen on tryptophan concentration in rice grains grown on limed and unlimed soils. *Plant Soil* 62:305–308.

Snedden, W.A., and H. Fromm. 1998. Calmodulin, calmodulin-related proteins and plant responses to the environment. *Trends Plant Sci.* 3:299–304.

Solomon, M., R. Ariel, A.M. Mayer, and R.P. Mayber. 1989. Reversal by calcium of salinity induced growth inhibition in excised pea roots. *Isr. J Bot.* 38:65–69.

Song, J.Q., and H. Fujiwara. 1996. Ameliorative effect of potassium on rice and tomato subjected to sodium salinization. *Soil Sci. Plant Nutr.* 42:493–501.

Suhaya, C.G., J.L. Giannini, D.P. Briskin, and M.C. Shannon. 1990. Electostratic changes in *Lycopersicon esculentum* root plasma membrane resulting from salt stress. *Plant Physiol.* 93:471–478.

Takahashi, E., J.F. Ma, and Y. Miyake. 1990. The possibility of silicon as an essential element for higher plants. *Comments Agric. Food Chem.* 2:99–122.

Talei, D., M.A. Kadir, M.K. Yusop, A. Valdiani, and M.P. Abdullah. 2012. Salinity effects on macro and micronutrients uptake in medicinal plant King of Bitters (*Andrographis paniculata* Nees.). *Plant Omics J.* 5:271–278.

Tanaka, K., S. Takio, I. Yamamoto, and T. Satoh. 1996. Purification of the cytosolic CuZn-superoxide dismutase (CuZn-SOD) of *Marachantia paleacea* var. *diptera* and its resemblance to CuZn-SOD from chloroplasts. *Plant Cell Physiol.* 37:523–529.

Tester, N., and R. Davenport. 2003. Na^+ tolerance and Na^+ transport in higher plants. *Ann. Bot.* 91:1–5.

Tuna A.L., C. Kaya, D. Higgs, B.M. Amador, S. Aydemir, and A.R. Girgin. 2008. Silicon improves salinity tolerance in wheat plants. *Environ Exp. Bot.* 62:10–16.

Tyler, G. 2004. Rare earth elements in soil and plant system—A review. *Plant Soil* 267:191–206.

Urao, T., T. Katagiri, T. Mizoguchi, K. Yamaguchi-Shinozaki, N. Hayashida, and K. Shinozaki. 1994. Two genes that encode Ca^{2+}-dependent protein-kinases are induced by drought and high-salt stresses in *Arabidopsis thaliana*. *Mol. Gen. Gen.* 244:331–340.

Vallee, B.L., and D.S. Auld. (1990). Zinc coordination, function, and structure of zinc enzymes and other proteins. *Biochemistry* 29:5647–5659.

Wagenet, R.J., R.R. Rodriguez, W.F. Campbell, and D.L. Turner. 1983. Fertilizer and salty water effects on *Phaseolus. Agron. J.* 75:161–166.

Watanabe, S., T. Fujiwara, T. Yoneyama, and H. Hayashi. 2001. Effects of silicon nutrition on metabolism and translocation of nutrients in rice plants. *Dev. Plant Soil Sci.* 92:174–175.

Wei, W.X., P.E. Bilsborrow, P. Hooley, D.A. Fincham, E. Lombi, and B.P. Forster. 2003. Salinity induced differences in growth, ion distribution and partitioning in barley between the cultivar Maythorpe and its derived mutant golden promise. *Plant Soil* 250:183–191.

Weisany, W., Y. Sohrabi, G. Heidari, A. Siosemardeh, and K. Ghassemi-Golezani. 2012. Changes in antioxidant enzymes activity and plant performance by salinity stress and zinc application in soybean (*Glycine max* L.). *Plant Omics J.* 5:60–67.

Welch, R.M., and L. Shuman. 1995. Micronutrient nutrition of plants. *Crit. Rev. Plant Sci.* 14:49–82.

Whittington, W.J. 1957. The role of boron in plant growth. I. The effect on general growth, seed production and cytological behaviour. *J. Exp. Bot.* 8:353–367.

Wiese, H., M. Nikolic, and V. Romheld. 2007. Silicon in plant nutrition. Effect of zinc, manganese and boron leaf concentrations and compartmentation. In Sattelmacher, B. and W.J. Horst (eds.), *The Apoplast of Higher Plants: Compartment of Storage, Transport and Reactions*. Springer, The Netherlands, pp. 33–47.

Wilkinson, S., A.L. Clephan, and W.J. Davies. 2001. Rapid low temperature induced stomatal closure occurs in cold tolerant *Commelina cummunis* leaves but not in cold sensitive tobacco leaves via a mechanism that involves apoplastic calcium but not abscisic acid. *Plant Physiol.* 126:1566–1578.

Wissemeier, A.H., A. Diening, A. Hergenröder, W.J. Horst, and G. Mix-Wagner. 1992. Callose formation as parameter for assessing genotypical plant tolerance of aluminium and manganese. *Plant Soil* 146:67–75.

Xu, G., H. Magen, J. Tarchitzky, and V. Kafkafi. 1999. Advances in chloride nutrition. *Adv. Agron.* 68:96–150.

Yadav, R., T.J. Flowers, and A.R. Yeo. 1996. The involvement of the transpirational bypass flow in sodium uptake by high- and low-sodium-transporting lines of rice developed through intravarietal selection. *Plant Cell Environ.* 19:329–336.

Yeo, A.R., S.A. Flowers, G. Rao, K. Welfare, N. Senanayake, and T.J. Flowers. 1999. Silicon reduces sodium uptake in rice (*Oryza sativa* L.) in saline conditions and this is accounted for by a reduction in the transpirational bypass flow. *Plant Cell Environ.* 22:559–565.

Yoneyama, T. 1988. Natural abundance of 15N in root nodules of pea and broad bean. *J. Plant Physiol.* 132:59–62.

Zhao, D., and D.M. Oosterhuis. 2002. Cotton carbon exchange, nonstructural carbohydrates, and boron distribution in tissues during development of boron deficiency. *Field Crops Res.* 78:75–87.

Zhu, Z., G. Wei, J. Li, Q. Qian, and J. Yu. 2004. Silicon alleviates salt stress and increases antioxidant enzymes activity in leaves of salt-stressed cucumber (*Cucumis sativus* L.). *Plant Sci.* 167:527–533.

4 Silicon: A Potential Element to Impart Resistance to Photosynthetic Machinery under Different Abiotic Stresses

Gausiya Bashri, Durgesh Kumar Tripathi, Vijay Pratap Singh, Sheo Mohan Prasad, and Devendra Kumar Chauhan

CONTENTS

ABSTRACT

Abiotic stress limits the crop growth and productivity of plants. Indeed, abiotic stress decreases the growth of plants by inhibiting the process of photosynthesis. Among the elements, silicon (Si) ranks second in abundance on earth crust. Unfortunately, so far, it is not found to be essential in higher plant growth processes (Epstein 1999). However, the application of Si has enhanced the tolerance of plants to various abiotic and biotic stresses. Si is also known to improve the process of photosynthesis in plants under different abiotic stresses, such as drought, salinity, heavy metal toxicity, and ultraviolet B (UV-B), and provide tolerance to plants under these conditions. There were several works on the role of Si on plant growth by defense mechanisms, but they did not specifically address the effect of Si on photosynthetic machinery. Therefore, in this chapter we discuss the tolerance mechanism of plants with special reference to the photosynthetic machinery subjected to different abiotic stresses.

4.1 INTRODUCTION

Development is necessary for the growing population and should be achieved without disturbing the environment. However, rapid enlargement in industrialization, urbanization, and other human activities have disturbed the whole environmental system and caused various kinds of pollution in the environment. Plants, being sessile organisms, are frequently exposed to a wide range of environmental stresses, such as drought, salinity, heavy metals, and ultraviolet B (UV-B). Abiotic stresses result in an approximately 70% yield reduction in the world (Acquaah 2007). One of the important reasons for yield reduction is the inhibition of the growth and photosynthetic abilities of plants grown under different abiotic stresses. The main target of abiotic stresses is chloroplasts, and these stresses can heavily affect the photosynthetic pigment content, maximum yield of photosystem (PS) II photochemical reactions (Fv/Fm), and photosynthetic apparatus: PSs I and II and light harvesting complexes (LHCs) I and II in plants (Chen et al. 2011; Song et al. 2014; Tripathi et al. 2015). In recent years, chlorophyll (Chl) a fluorescence and gas exchange characteristics were used as a sensitive technique to determine the changes in photosynthetic apparatus under different abiotic stress conditions (Feng et al. 2010; Bashri and Prasad 2015). Chl a fluorescence (JIP-test) has been developed to measure the phenomenological and biophysical expressions of PS II (Govindjee 1995; Strasser et al. 2000).

Silicon (Si) is found mostly as silicic acid ($Si(OH)_4$) in soil solution, and it is the second most abundant element of the earth's crust (Lindsay 1979; Epstein 1994; Gong et al. 2006). However, a great number of Si compounds are insoluble in soil and not available for plants because they are mostly found as oxides or silicates (Richmond and Sussman 2003). Silicon is an essential element for the diatoms (Ketchum 1954; Volcani 1978; Round et al. 1990; Kinrade et al. 2001, 2002). But it is not found essential for higher plants, although it showed many positive effects on growth and development under different environmental stress conditions (Liang et al. 2003, 2006). Plants take Si in the form of silicic acid, $Si(OH)_4$ (Weiss and Herzog 1978; Epstein 1999, 2002; Richmond and Sussman 2003; Epstein and Bloom 2005) and then polymerized into silica gel in shoots ($SiO_2 \cdot nH_2O$). The amount of Si varies between plant species and ranges from 0.1% to 10.0% (on dry weight basis)

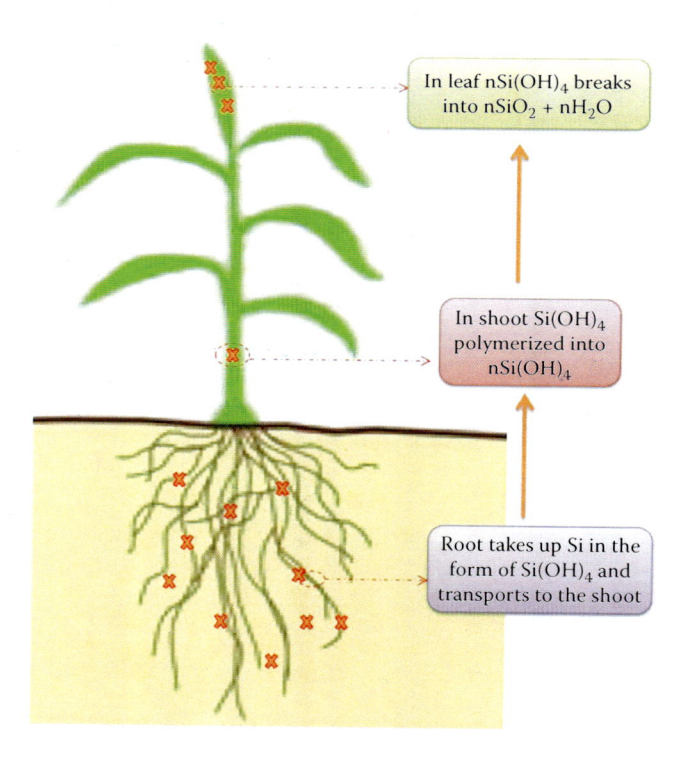

In leaf $nSi(OH)_4$ breaks into $nSiO_2 + nH_2O$

In shoot $Si(OH)_4$ polymerized into $nSi(OH)_4$

Root takes up Si in the form of $Si(OH)_4$ and transports to the shoot

FIGURE 4.1 Graphical representation of the movement of Si from soil to leaf tips. (Modified from Janislampi KW, Effect of silicon on plant growth and drought stress tolerance, All Graduate Thesis and Dissertations, 2012, Paper 1360, http://digitalcommons.usu.edu/etd/1360.)

(Jones and Handreck 1967; Ma et al. 2001; Richmond and Sussman 2003). Si is transported passively by transpiration stream, and it accumulates at the sites of high transpiration rate (Wiese et al. 2007). It was suggested that, after water transpires from the leaf, silicic acid converts into colloidal silicic acid, then it forms amorphous silica ($SiO_2 \cdot nH_2O$) (also referred to as silica gel, phytoliths, or opal), which polymerizes at high concentrations (>2 mM), and creates a polymer in the plant (Jones and Handreck 1967; Gao et al. 2006; Ma and Yamaji 2006). Figure 4.1 shows a graphical representation of the movement of Si from soil to leaf tips (Janislampi 2012). After accumulation inside the plant, it provides protection to the plants (Van Soest 2006).

So far, Si has not been considered an essential element for plants, but in recent years it has been shown to have many effects on growth and development (Epstein 1994, 1999; Ma et al. 2001; Liang et al. 2007). On the whole, Si can be used for improving the tolerance of plants to different abiotic stresses, that is, drought, salinity, heavy metals, and UV-B. For these reasons, Si has received increasing attention from scientists. Many scientists have suggested that Si may be part of certain structural or functional attributes of plants and can affect the metabolic or physiological processes of plants exposed to different environmental stress (Liang et al. 2003, 2006). It has been reported that Si increases plant tolerance to drought (Gong et al. 2005; Chen et al. 2011), salinity (Moussa 2006; Ali et al. 2012; Xie et al. 2015), heavy metals (Feng et al. 2010; Song et al. 2014; Tripathi et al. 2015), and UV-B radiation (Shen et al. 2010, 2014). Moreover, Si has also been reported to improve photosynthetic performance by increasing the content of photosynthetic pigments and photosynthesis, gas exchange characteristics, and PSII activity in plants (Detmann et al. 2012; Pilon et al. 2013; Harizanova et al. 2014). Therefore, in this chapter, the effects of Si application on the photosynthetic machinery of plants are addressed under different abiotic stress conditions.

4.2 BENEFITS OF SILICON TO PHOTOSYNTHETIC MACHINERY UNDER DIFFERENT ABIOTIC STRESSES

Silicon has shown a number of beneficial effects through pot, hydroponic, and field experiments in plants (Jones and Handreck 1967; Savant et al. 1997; Epstein 1999; Datnoff et al. 2001; Datnoff and Rodrigues 2005). Silicon application works at several levels for the plant growth and provides tolerance to the plant (Liang et al. 2003, 2006; Tripathi et al. 2011). Further, Si has also been reported to improve the photosynthetic performance of plants under different abiotic stress conditions, that is, drought (Chen et al. 2011; Maghsoudi et al. 2015), salinity (Ali et al. 2012; Xie et al. 2015), heavy metals (Feng et al. 2010; Song et al. 2014; Tripathi et al. 2015), and UV-B radiation (Shen et al. 2010, 2014). Therefore, in this chapter, we address the benefits of Si application to the photosynthetic system of plants under different abiotic stresses. Table 4.1 presents a brief summary of Si application under different abiotic stresses in plants based on recent studies, and below we discuss this in detail.

4.2.1 UNDER DROUGHT STRESS

At present, water scarcity is a severe problem and its limitation in plants is called drought stress. It has been predicted that drought stress will frequently increase in the whole world due to current climatic conditions (IPCC 2007). Drought stress severely affects the plant growth by inhibiting stomatal movements, leaf expansion, and stem elongation (Engelbrecht et al. 2007). Drought stress negatively affects a number of processes in plants, and the most significant inhibitory effect can be observed on photosynthesis (Chen et al. 2011). Drought stress also reduces chlorophyll contents and damages the photosynthetic apparatus: photosystems (PSs) I and II, light harvesting complexes (LHCs) I and II, and the oxygen evolving complex (Iturbe-Ormaetxe et al. 1998; Baker and Rosenqvist 2004; Naumann et al. 2010; Signarbieux and Feller 2011). On the whole, drought stress inhibits the PSII photochemistry and reaction centers in plants (Naumann et al. 2010). Such alternations by drought stress may be ascribed to the degradation of structural proteins of PS II reaction

TABLE 4.1

Effect of Si Application on Photosynthetic Machinery of Plants under Drought, Salinity, Heavy Metals, and UV-B Stress

Abiotic Stress	Crop Plant	Effect of Silicon on Photosynthetic Activity	References
Drought	*Triticum aestivum* L.	Increased chlorophyll and carotenoid contents	Gong et al. 2005;
		Improved gas exchange characteristics: *Pn* and *Ci*	Yao et al. 2011;
		Increased the values F_0, Fm, Fv/Fm, ΦPSII, ETR, and qP while decreasing qN and F_0/Fm	Maghsoudi et al. 2015
	Zea mays L.	Increased chlorophyll content	Kaya et al. 2006
	Cucumis sativus L.	Increased net photosynthesis	Hattori et al. 2008
	Glycine max L.	Increased chlorophyll content	Shen et al. 2010
		Increased *Pn, E, Gs*, and *Ci*	
	Oriza sativa L.	Increased *Pn* and *E*	Chen et al. 2011
		Improved PSII activity: Fv/Fm and Fv/F_0	
	Lycopersicum esculentum L.	Increased chlorophyll content, *E*, and *Gs*	Silva et al. 2012
	Pistacia vera L.	Increased *Gs* and Fv/Fm	Habibi and Hajiboland 2013
	Kentucky bluegrass	Increased *Pn, E*, and *Gs*	Saud et al. 2014
Salinity	*Zea mays* L.	Increased chlorophyll content and $^{14}CO_2$ fixation	Moussa 2006;
		Increased *Pn, Ci*, and *Gs*	Xie et al. 2015
	Phaseolus vulgaris L.	Increased *Pn* and *Gs*	Zuccarini 2008
	Vigna unguiculata and *Phaseolus vulgaris* L.	Increased *Pn* and *Ci*	Murillo-Amador et al. 2007
	Sorghum bicolour	Increased *Pn, Tr*, and *Ci*	Yin et al. 2013
	Brassica napus L.	Increased chlorophyll content and *Pn*	Hashemi et al. 2010; Nezami and Bybordi 2011; Farshidi et al. 2012
	Lycopersicum esculentum Mill.	Increased chlorophyll content, Fv/Fm, and ΦPSII	Romero-Aranda et al. 2006
		Increased *Pn, E*, and *Gs*	
	Solanum lycopersicum L.	Increased chlorophyll content, *Pn, E*, and *Gs*	Haghighi and Pessarakli 2013; Muneer et al. 2014
	C. pepo	Increased *Pn*	Savvas et al. 2009
	Oryza sativa L.	Increased *Pn, E*, and *Gs*	Shi et al. 2013
	Spartina densiflora	Increased chlorophyll content, *Pn, Gs*, Fv/Fm, and ΦPSII	Mateos-Naranjo et al. 2013
		Decreased qN and Fm	
	Triticum aestivum L.	Increased chlorophyll content	Tuna et al. 2008; Ali et al. 2012; Tahir et al. 2012
Heavy metal Cadmium	*Gossypium hirsutum* L.	Increased chlorophyll content, *Pn*, and *Gs*	Farooq et al. 2013
	Zea mays	Increased chlorophyll and carotenoid content and Fv/Fm	Malčovská et al. 2014a

(Continued)

TABLE 4.1 (CONTINUED)

Effect of Si Application on Photosynthetic Machinery of Plants under Drought, Salinity, Heavy Metals, and UV-B Stress

Abiotic Stress	Crop Plant	Effect of Silicon on Photosynthetic Activity	References
	Oryza sativa L.	Increased *Pn*, Fv/Fm, and qP	Nwugo and Huerta 2008
	Cucumis sativus L.	Increased chlorophyll content Improved *Pn*, *Gs*, *E*, *Ci*, and Fv/Fm	Feng et al. 2010
Manganese	*Zea mays* L.	Increased chlorophyll content, *Pn*, *Gs*, *Ci*, Fv/Fm, and ΦPSII	Feng et al. 2009
Chromium	*Oryza sativa* L.	Increased chlorophyll content Improved *Pn*, *Gs*, *E*, *Ci*, and Fv/Fm	Ali et al. 2013; Tripathi et al. 2015
Zinc	*Gossypium hirsutum* L.	Increased chlorophyll content, *Pn*, *Gs*, and *E*	Anwaar et al. 2015
	Oryza sativa L.	Protected chloroplast ultrastructure Upregulated the photosynthetic genes PsbY, PsaH, PetC, and PetH and expressions of Os09g26810 and Os04g38410	Song et al. 2014
Arsenic	*Oryza sativa* L.	Increased *Pn* and *Gs*	Sanglard et al. 2014
UV-B	*Triticum aestivum* L.	Increased chlorophyll content	Yao et al. 2011
	Glycine max L	Increased chlorophyll content Increased *Pn*, *Gs*, *E*, *Ci*, and Fv/Fm	Shen et al. 2010, 2014
	Zea mays L.	Increased chlorophyll and carotenoid content	Malčovská et al. 2014b

Note: *Ci*, intercellular CO_2 concentration; *E*, transpiration rate; F_0, minimum fluorescence; Fm, maximum fluorescence; Fv/Fm, maximum yield of PSII photochemical reactions; *Gs*, stomatal conductance; *Pn*, net photosynthetic rate; qN, nonphotochemical quenching; qP, photochemical quenching; ΦPSII, effective quantum yield of PSII.

centers (Ohashi et al. 2006). Further, it also inhibits the CO_2 fixation rate by decreasing the activities of enzymes in the Calvin cycle (Monakhova and Chernyad'ev 2002).

Silicon plays a crucial role in plants under drought stress (Habibi and Hajiboland 2013). Gong et al. (2005) suggested that Si increases plant tolerance to drought by increasing photosynthetic activity. In another study, Hattori et al. (2005) showed that under drought stress, Si-treated sorghum plants have a better photosynthetic rate and stomatal conductance than Si-untreated plants. Further, Hattori et al. (2008) showed that the photosynthetic rate and intercellular CO_2 concentration were higher in cucumber seedlings supplied with Si under osmotic stress. Similarly, effects of Si application were reported in pepper plants under water deficit conditions (Lobato et al. 2009). In soybean seedlings, Si application increased photosynthesis by 18.3% under drought (Shen et al. 2010). Silicon application also increases the drought resistance in rice plants by increasing the chlorophyll content, rate of photosynthesis and transpiration, basal quantum yield (Fv/F_0), and maximum quantum yield of PSII photochemistry (Fv/Fm) (Chen et al. 2011). Further, Silva et al. (2012) observed a linear positive correlation between leaf water potential and chlorophyll levels, and application of Si improved the chlorophyll level in tomato seedlings. Si addition alleviated the drought stress in Kentucky bluegrass by increasing the stomatal conductance, leaf water content, relative growth rate, water use efficiency, and rates of photosynthesis and transpiration (Saud et al. 2014). Recently, Maghsoudi et al. (2015) showed that foliar application of Si improved the growth, chlorophyll content, and values of minimal fluorescence (F_0), maximal fluorescence (Fm), Fv/Fm, effective quantum yield of PSII (ΦPSII), photochemical quenching (qP), and photosynthetic electron transport rate (ETR) under water stress condition. In contrast to this, Si application decreased the drought-induced increase in nonphotochemical quenching (qN) and F_0/Fm. These results suggest that application of Si ameliorated the drought-induced effects by dropping nonphotochemical quenching while enhancing the value of Fv/Fm and qP (Maghsoudi et al. 2015).

4.2.2 UNDER SALT STRESS

Salinity has always been known as an abstruse problem that affects agricultural production (Nezami and Bybordi 2011). The response of all plants to high salinity depends on the saline concentration in their tissues, the time to imposition of salinity stress, and the form of salts (Liang 1999). Salinity effects can be seen in plants at different levels, ranging from a decrease in yield to complete death of plants. It affects all processes of plants, including growth, photosynthesis, protein synthesis, and fat and energy metabolisms (Ma 2004; Parida and Das 2005). Salinity stress can close the stomata, reduce the cell membrane permeability for CO_2, and ultimately reduce the photosynthetic rate (Kafi and Rahimi 2011). Proteomics analysis by two-dimensional blue native polyacrylamide gel electrophoresis (2D-BN-PAGE) of chloroplasts revealed that photosynthetic apparatus—photosystems (PSs) I and II—are highly sensitive to salt stress (Muneer et al. 2014), and it also reduces the content of large and small subunits of ribulose-1,5-bisphosphate carboxylase/oxygenase (Rubisco), as well as other enzymes required for photosynthesis and chlorophyll biosynthesis (Kim et al. 2005; Parker et al. 2006). Si application has been reported to improve salt stress in several plant species, such as wheat (Tuna et al. 2008; Ali et al. 2012; Tahir et al. 2012), barley (Liang et al. 2003), maize (Moussa 2006; Xie et al. 2015), and tomatoes (Haghighi and Pessarakli 2013; Muneer et al. 2014).

The addition of Si also increases photosynthesis in saline conditions (Nezami and Bybordi 2011). Si application in the form of silicate alleviates the reduction in growth, pigment contents, stomatal conductance, and net photosynthesis caused by salt in rice and legume seedlings, respectively (Yeo et al. 1999; Murillo-Amador et al. 2007; Shi et al. 2013). Further, Moussa (2006) indicated that Si ameliorated the salt-induced toxicity and increased the tolerance of maize plants through improved chlorophyll content and photosynthetic activity. Similarly, Romero-Aranda et al. (2006) documented that salinized tomato plants, upon Si application, exhibited increased leaf turgor potential and net photosynthesis rates by 42% and 20%, respectively, compared to Si-untreated salinized tomato plants. Similar to this, Savvas et al. (2009) showed that supplementation of 1 mM Si through nutrient solution attenuated the negative impact of salinity on net photosynthesis, and this effect was related to decreased Na and Cl translocation to the shoot of tomato plants through their roots. In *Phaseolus vulgaris* seedlings Si application increased the growth, stomatal conductance, and net photosynthetic rate under NaCl (Zuccarini 2008). Wheat plants also showed improved growth, higher relative water content (RWC), increased chlorophyll fractions and ratios, and stimulated activities of superoxide dismutase (SOD) and catalase (CAT) upon supplementation of Si in the solution culture (Tuna et al. 2008; Ali et al. 2012; Tahir et al. 2012). In recent studies, Yin et al. (2013) reported that Si alone had no effect on sorghum growth, while it partly reversed the salt-induced reduction in plant growth and photosynthesis. Further, Haghighia and Pessarakli (2013) observed the effect of bulk Si and nano-Si and reported that application of both improved the plant's fresh weight, chlorophyll content, photosynthetic rate, leaf water content, mesophyll conductance, and water use efficiency, and offset the negative impact of saline stress. Similarly, Mateos-Naranjo et al. (2013) reported that Si improved the growth by increasing the photosynthetic rate and water-use efficiency (WUE), and they also suggested that increasing the photosynthetic rate at high salinity might occur due to increased chlorophyll content and stomatal conductance under salt stress. Si-untreated plants showed a significant decrease in PSII activity at the highest salinity treatment (680 mM NaCl), while the addition of Si offset this effect. Later, Muneer et al. (2014) reported that addition of Si increased the photosynthetic pigment content, net photosynthesis, transpiration rate, and stomatal conductance, and also improved the cytochrome *b6/f* and ATP–synthase complex in salt stress; moreover, it rapidly upregulated the light harvesting complexes (LHCs). On the basis of these results, Muneer et al. (2014) suggested that Si plays an important role in attenuating damage to chloroplasts and their metabolism in salt stress conditions. Further, Al-aghabary et al. (2005)

reported that Si alleviates NaCl-induced toxicity and increases the tolerance of tomato plants to NaCl by increasing chlorophyll content and the maximum yield of PSII photochemistry (Fv/Fm). Xie et al. (2015) showed that the values of the net photosynthetic rate, intercellular CO_2 concentration, and stomatal conductance of maize were significantly enhanced by Si (used as fertilizers) under salt stress. Overall, it can be said that Si plays a significant role in salinity tolerance through various protective effects on the photosynthetic apparatus.

4.2.3 UNDER HEAVY METAL STRESS

Heavy metal pollution is a global problem due to environmental reasons. These metals may be introduced into the environment by various anthropogenic activities, such as fertilizer use, mining, electroplating, and use of metal-based pesticides and batteries (Sanitá di Toppi and Gabrielli 1999; Benavides et al. 2005). Heavy metals can create many physiological disturbances, including reduced biomass production, photosynthesis, and nutrient uptake. Phytotoxicity of heavy metals in plants can be seen as leaf chlorosis, diminutive growth, and the inhibition of key physiological processes together with photosynthesis (Atal et al. 1991; Moya et al. 1995; Das et al. 1998; Chugh and Sawhney 1999; Sanitá di Toppi and Gabrielli 1999; Kukier and Chaney 2002; Popova et al. 2009).

Previously, it has been well documented that Si application increases plant tolerance to metals toxicity, that is, aluminum (Al), boron (B), cadmium (Cd), chromium (Cr), manganese (Mn), and zinc (Zn) (Liang et al. 2001; Neumann and Zur Nieden 2001; Iwasaki et al. 2002a,b; Shi et al. 2005; Gunes et al. 2007; Nwugo and Huerta 2008; Inal et al. 2009; Kaya et al. 2009; Song et al. 2009; Tripathi et al. 2015). Silicon application significantly attenuated the negative effects of Cd stress on photosynthesis and Fv/Fm and qP in rice seedlings (Nwugo and Huerta 2008). Later, Feng et al. (2010) documented that Si application alleviated Cd-induced damage like chlorosis and significantly increased the pigments contents and chlorophyll *a* fluorescence parameters (Fv/Fm) by protecting the chloroplastin in cucumber plants. Further, Farooq et al. (2013) showed that Si application enhanced plant growth, gas exchange characteristics (net photosynthetic rate, *Pn*; stomatal conductance, *Gs*; transpiration rate, *E*; water use efficiency, *Pn/E*), and chlorophyll and carotenoid contents, as well as the performance of antioxidant enzymes in cotton plants under Cd stress. Recently, Malčovská et al. (2014a) documented that the supplementation of silicon successfully ameliorated the toxic effect of Cd on maize plants and enhanced the growth and some of the photosynthetic parameters, like Fv/Fm, and reduced the level of oxidative stress. Manganese significantly repressed the stomatal conductance, net photosynthetic rate, Fv/Fm, and quantum yield of PSII electron transport (PSII-ETR), and application of Si offset the negative effects of Mn toxicity in maize seedlings (Feng et al. 2009). Lead (Pb) is a highly toxic metal and causes many direct and indirect effects on plants, animals, and humans, like other heavy metals. Lead decreases the gas exchange parameters—*Pn*, *Gs*, *E*, and *Pn/E*—and exogenous application of Si alleviates the inhibitory effect induced by lead stress on these parameters in maize seedlings (Bharwana et al. 2013). Similarly, Ali et al. (2013) have reported that application Si enhances plant growth and attenuates Cr toxicity, which evidences a significant increase in photosynthetic parameters in rice seedlings: single photon avalanche diode (SPAD) value, *Pn*, *Ci*, *Gs*, *E*, and Fv/Fm. Further, the results of Tripathi et al. (2015) suggest that Si addition significantly ameliorates the Cr-induced decline in chlorophyll fluorescence parameters—Fv, Fm, Fv/F_0, Fv/Fm, Fm/F_0, and qP—and also reduces the Cr-mediated enhancement in F_0 and qN non-photochemical quenching (NPQ) in rice seedlings. Further, high Zn also decreases the chlorophyll content, *Pn*, *Ci*, *Gs*, *E*, and chlorophyll fluorescence in rice (Song et al. 2014). High-Zn stress also damages the chloroplast structure, which is seen as disintegrated and misplaced thylakoid membranes and uneven swelling (Song et al. 2014). However, all negative effects of high-Zn stress were attenuated by the supplementation of Si (1.5 mM). Furthermore, Si application also increased the expression levels of Os08g02630 (PsbY, polyprotein of PSII), Os05g48630 (PsaH, subunit of PSI located in the center of the PSI dimer), Os07g37030 (PetC, associated with the cytochrome's biological processes in photosynthesis), Os03g57120 (PetH), Os09g26810, and Os04g38410 under

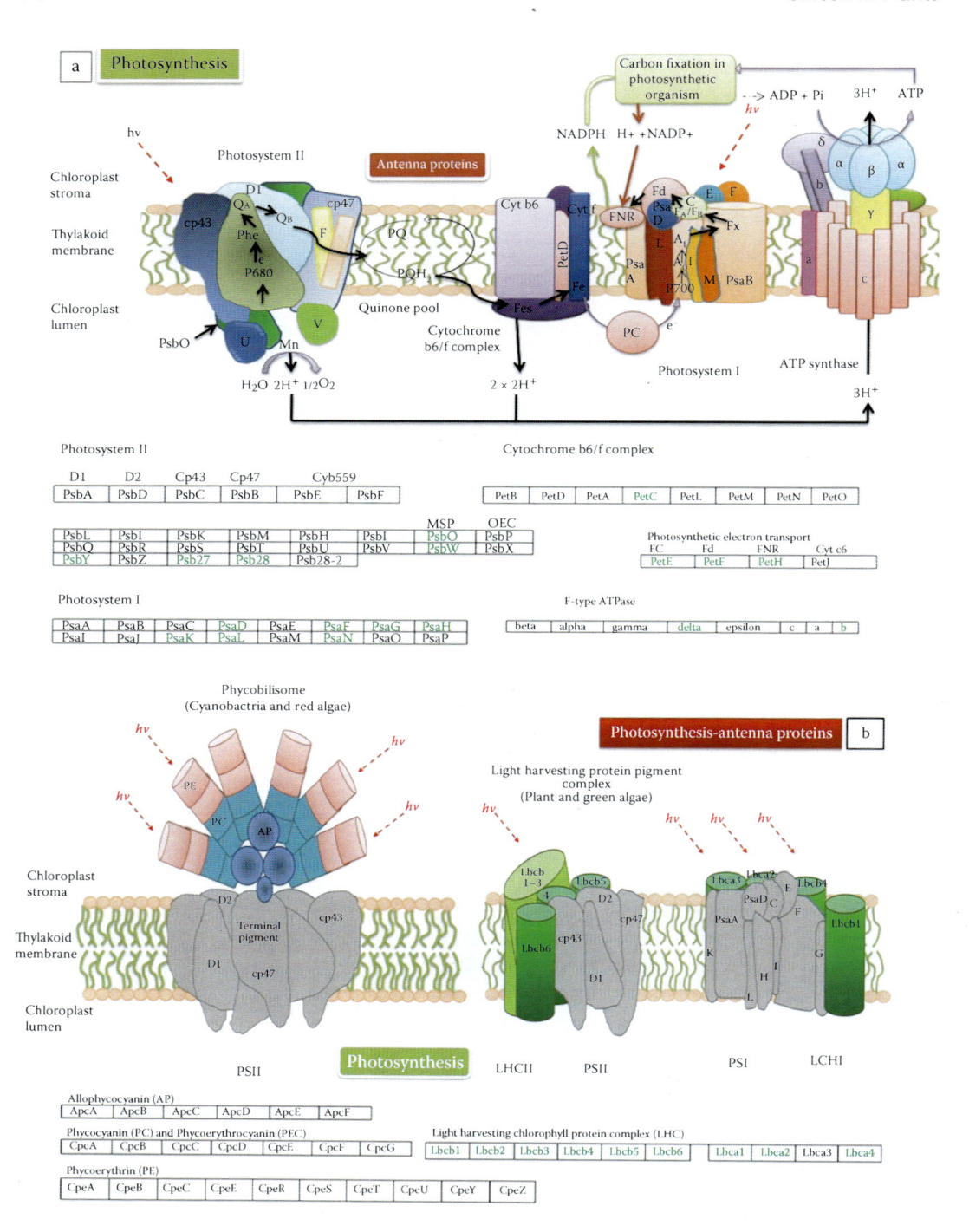

FIGURE 4.2 Effect of silicon on photosynthetic machinery genes under high-Zn stress in rice seedlings. Green font represents the upregulated genes of photosynthetic machinery. (Adapted from Song A et al., *PLoS One* 9(11): e113782, 2014. http://dx.doi.org/10.1371/journal.pone.0113782. With permission.)

high-Zn stress (Song et al. 2014). Cotton plants are also damaged by increased concentrations of Zn and generate oxidative damage, which is evident from elevated levels of hydrogen peroxide (H_2O_2), electrolyte leakage, and malondialdehyde (MDA), and consequently inhibit cotton growth, biomass, chlorophyll pigments, and the photosynthetic process. These effects were reversed by the application of Si (Anwaar et al. 2015). Similarly, Hu et al. (2013) reported that Si supply increases the net photosynthetic rate and stomatal conductance under arsenic (As) toxicity in rice. Furthermore, Sanglard et al. (2014) also documented the effect of Si nutrition on photosynthetic performance (Fv/Fm and qP) in rice plants under As (arsenite) toxicity. They reported that Si considerably reverted As-induced inhibition in carbon fixation and leaf conductance at the stomata and mesophyll levels, and photosynthetic rates showed a strong correlation with both conductances: stomatal ($r^2 = 0.90$) and mesophyll ($r^2 = 0.95$). In common, Si protected the photosynthetic machinery of plants by improving the gas exchange parameters and upregulating the gene expression of photosynthetic apparatus, as shown in Figure 4.2, in rice seedlings (Song et al. 2014); this might improve the photosynthetic performance of plants under heavy metal toxicity.

4.2.4 Under UV-B Stress

Solar UV-B (280–320 nm) radiation is increasing in the earth's surface due to the depletion of the stratospheric ozone layer (Madronich et al. 1998). Plants, being sessile organisms, are frequently exposed to UV-B radiation, and the response to UV-B radiation varies from plant to plant. Some plants are tolerant, some showed stimulated growth, but most of them are sensitive and showed altered metabolism (Kakani et al. 2003). Exposure to UV-B radiation has been shown to decrease leaf chlorophyll contents (Prasad and Zeeshan 2005; Mishra et al. 2008), photosynthesis (Mishra et al. 2008), Rubisco activity (Keiller et al. 2003), and ultimately the productivity of the plant (Kakani et al. 2003). The sensitivity of UV-B radiation to crops was also influenced by other factors, like water regime and nutrient status (Balakumar et al. 1993).

Silicon accumulates maximally in leaf tissues of the plants. For this reason, it may provide protection from UV-B radiation and also protect the photosynthetic machinery of plants. Shen et al. (2010, 2014) reported that application of Si increased the chlorophyll content and net photosynthetic rate in soybean seedlings under UV-B radiation. Further, Malčovská et al. (2014b) reported that maize seedlings, upon UV-B exposure, showed enhanced generation of reactive oxygen species (ROS) contents (hydrogen peroxide and superoxide radical) and thiobarbituric acid reactive substances (TBARSs), while Si-treated plants showed less generation of ROS. This may be due to increased production of phenolic metabolites (total phenols and flavonoids), important for their screening function on Si application under radiation stress, which may contribute to higher UV-B tolerance of maize seedlings and ultimately increased growth and photosynthetic pigment content (chlorophyll and carotenoids).

4.3 SUMMARY

Abiotic stress causes huge losses in plant productivity worldwide. Therefore, researchers aimed to overcome abiotic stresses by the use of different growth regulating substances. A number of studies support the role of Si withstanding different abiotic stress conditions. These reports suggest that Si has certain physiological functions in plants. Its role becomes more important under adverse environmental conditions. Si supplementation increased the SiO_2 content in leaf tissues of plants, and the presence of Si in the cell walls of plants increases their strength against drought, salinity, toxic metals, and UV-B radiation. On the basis of literature, we can say that Si improves plant growth, increases the maximum quantum yield of PSII (Fv/Fm) and photochemical quenching (qP), and protects the chloroplast ultrastructure from disorganization, as well as increases the expression of genes associated with photosynthesis in plants under different abiotic stresses. Researchers also suggest that Si could be used as a potential growth regulating substance to improve plant growth

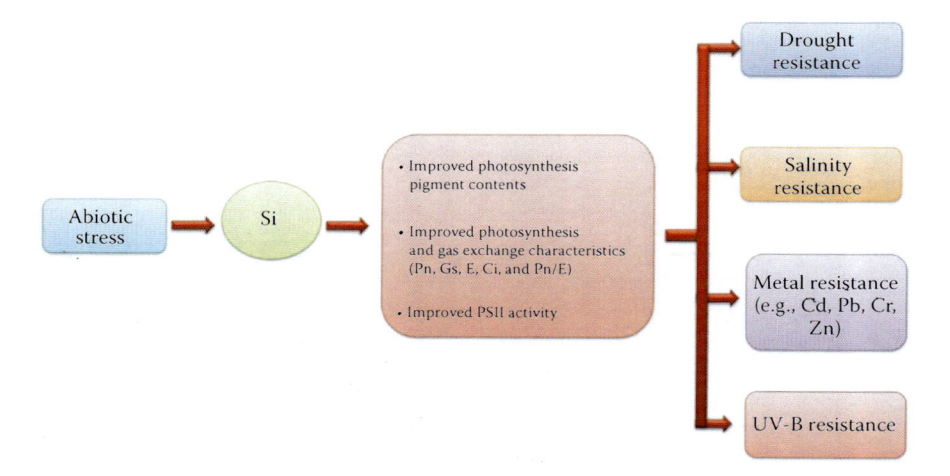

FIGURE 4.3 Schematic representation of Si-induced resistance to different abiotic stresses by improving the photosynthetic machinery of plants.

by improving photosynthetic machinery and providing resistance under drought, salinity, heavy metals, and UV-B stress (Figure 4.3). This may be a promising new strategy for the improvement of plant growth in the agricultural field.

4.4 FUTURE PERSPECTIVES

Current knowledge of silicon's function in individual plant species under abiotic stress conditions (salinity, metal toxicity, drought, and UV radiation) is developing, but there is also the need to better understand its function in more species and under other stress conditions. In natural conditions, plants are subjected to a number of environmental stress factors, that is, low or high temperature, water deficiency, excessive photosynthetically active radiation, UV radiation, salinity, and heavy metals, simultaneously that alter the plant physiology. Taking into account the ability of Si to influence the photosynthetic efficiency in plants under stress (drought, salinity, metal toxicity, and UV radiation) it may be said that Si will have a protective function in other environmental stress conditions. Unfortunately, there is a lack of molecular mechanisms of Si uptake in plants (except some gramineous plant species); hence, Si uptake genes in other plant species need to be isolated and characterized. Furthermore, most dicot plants do not have the ability to accumulate enough Si; for this reason, Si supplementation is not fruitful for these plants. So, there is a need for biotechnology to insert the desired gene of Si uptake so that plants can accumulate more Si and become tolerant to both biotic and abiotic stresses. However, further studies are needed for understanding the signaling behavior of Si in plants under these stress conditions. Understanding these processes should be the objective of future experiments. Such studies will broaden the knowledge of plant adaptation to adverse environmental conditions and can be used in farming practice to increase crop yields.

REFERENCES

Acquaah G. (2007). *Principles of Plant Genetics and Breeding*. Blackwell, Oxford, p. 780.
Al-aghabary K, Zhu Z, Shi Q. (2005). Influence of silicon supply on chlorophyll content, chlorophyll fluorescence, and antioxidative enzyme activities in tomato plants under salt stress. *J Plant Nutr* 27: 2101–2115.
Ali S, Cai S, Zeng F, Qiu B, Zhang GP. (2012). The effect of salinity and chromium stresses on uptake and accumulation of mineral elements in barley genotypes differing in salt tolerance. *J Plant Nutr* 35: 827–839.

Ali S, Farooq MA, Yasmeen T, Hussain S, Arif MS, Abbas F, Bharwana SA, Zhang G. (2013). The influence of silicon on barley growth, photosynthesis and ultra-structure under chromium stress. *Ecotoxicol Environ Saf* 89: 66–72.

Anwaar SA, Ali S, Ali S, Ishaque W, Farid M, Farooq MA, Najeeb U, Abbas F, Sharif M. (2015). Silicon (Si) alleviates cotton (*Gossypium hirsutum* L.) from zinc (Zn) toxicity stress by limiting Zn uptake and oxidative damage. *Environ Sci Poll Res* 22: 3441–3450.

Atal N, Saradhi PP, Mohanty K. (1991). Inhibition of the chloroplast photochemical reactions by treatment of wheat seedlings with low concentrations of Cd: Analysis of electron transport activities and changes in fluorescence yield. *Plant Cell Physiol* 32: 943–951.

Baker NR, Rosenqvist E. (2004). Application of chlorophyll fluorescence can improve crop production strategies: An examination of future possibilities. *J Exp Bot* 55: 1607–1621.

Balakumar T, Hani BVV, Paliwal K. (1993). On the interaction of UV-B radiation (280–315 nm) with water stress in crop plants. *Physiol Plant* 87: 217–222.

Bashri G, Prasad SM. (2015). Indole acetic acid modulates changes in growth, chlorophyll a fluorescence and antioxidant potential of *Trigonella foenum-graecum* L. grown under cadmium stress. *Acta Physiol Plant* 37: 49.

Benavides MP, Gallego SM, Tomaro ML. (2005). Cadmium toxicity in plants. *Braz J Plant Physiol* 17: 21–34.

Bharwana SA, Ali S, Farooq MA, Iqbal N, Abbas F, Ahmad MSA. (2013). Alleviation of lead toxicity by silicon is related to elevated photosynthesis, antioxidant enzymes suppressed lead uptake and oxidative stress in cotton. *J Bioremediat Biodegrad* 4: 1–11.

Chen W, Yao X, Cai K, Chen J. (2011). Silicon alleviates drought stress of rice plants by improving plant water status, photosynthesis and mineral nutrient absorption. *Biol Trace Elem Res* 142: 67–76.

Chugh LK, Sawhney SK. (1999). Photosynthetic activities of *Pisum sativum* seedlings grown in presence of cadmium. *Plant Physiol Biochem* 37: 297–303.

Das P, Samantaray S, Rout R. (1998). Studies on cadmium toxicity in plants: A review. *Environ Pollut* 98: 29–36.

Datnoff LE, Rodrigues FA. (2005). The role of silicon in suppressing rice diseases. American Phytopathological Society, February. http://www.apsnet.org/publications/apsnetfeatures/Pages/SiliconInRiceDiseases.aspx (accessed October 2, 2007).

Datnoff LE, Snyder GH, Korndörfer GH. (2001). *Silicon in Agriculture*. Elsevier Science, Amsterdam.

Detmann KC, Araujo WL, Martins SCV, Sanglard LMVP, Ries JV, Detmann E, Rodrigues FA, Nunes-Nesi A, Fernie AR, DaMatta FM. (2012). Silicon nutrition increases grain yield, which, in turn, exerts a feed-forward stimulation of photosynthetic rates via enhanced mesophyll conductance and alters primary metabolism in rice. *New Phytol* 196: 752–762.

Engelbrecht BMJ, Comita LS, Condit R, Kursar TA, Tyree MT, Turner BL et al. (2007). Drought sensitivity shapes species distribution patterns in tropical forests. *Nature* 447: 80–82.

Epstein E. (1994). The anomaly of silicon in plant biology. *Proc Natl Acad Sci USA* 91: 11–17.

Epstein E. (1999). Silicon. *Annu Rev Plant Physiol Plant Mol Biol* 50: 641–664.

Epstein E. (2002). Silicon in plant nutrition. In *Second Silicon in Agriculture Conference*, Tsuruoka, Japan, August 22–26.

Epstein E, Bloom AJ. (2005). *Mineral Nutrition of Plants: Principles and Perspectives*. 2nd ed. Sinauer Associates, Sunderland, MA.

Farooq MA, Ali S, Hameed A, Ishaque W, Mahmood K, Iqbal Z. (2013). Alleviation of cadmium toxicity by silicon is related to elevated photosynthesis, antioxidant enzymes; suppressed cadmium uptake and oxidative stress in cotton. *Ecotoxicol Environ Saf* 96: 242–249.

Farshidi M, Abdolzadeh A, Sadeghipour HR. (2012). Silicon nutrition alleviates physiological disorders imposed by salinity in hydroponically grown canola (*Brassica napus* L.) plants. *Acta Physiol Plant* 34: 1779–1788.

Feng J, Shi Q, Wang X. (2009). Effects of exogenous silicon on photosynthetic capacity and antioxidant enzyme activities in chloroplast of cucumber seedlings under excess manganese. *Agric Sci China* 8: 40–50.

Feng J, Shi Q, Wang X, Wei M, Yang F, Xu H. (2010). Silicon supplementation ameliorated the inhibition of photosynthesis and nitrate metabolism by cadmium (Cd) toxicity in *Cucumis sativus* L. *Sci Hort* 123: 521–530.

Gao X, Zou C, Wang L, Zhang F. (2006). Silicon decreases transpiration rate and conductance from stomata of maize plants. *J Plant Nutr* 29: 1637–1647.

Gong H, Zhu X, Chen K, Wang S, Zhang C. (2005). Silicon alleviates oxidative damage of wheat plants in pots under drought. *Plant Sci* 169: 313–321.

Gong HJ, Randall DP, Flowers TJ. (2006). Silicon deposition in the root reduces sodium uptake in rice (*Oryza sativa* L.) seedlings by reducing bypass flow. *Plant Cell Environ* 29: 1970–1979.

Govindjee R. (1995). Sixty-three years since Kautsky: Chlorophyll a fluorescence. *Aust J Plant* Physiol 22: 131–160.

Gunes A, Inal A, Bagci EG, Coban S, Sahin O. (2007). Silicon increases boron tolerance and reduces oxidative damage of wheat grown in soil with excess boron. *Biol Plant* 51: 571–574.

Habibi G, Hajiboland R. (2013). Alleviation of drought stress by silicon supplementation in pistachio (*Pistacia vera* L.) plants. *Folia Horticulturae* 25: 21–29.

Haghighi M, Pessarakli M. (2013). Influence of silicon and nano-silicon on salinity tolerance of cherry tomatoes (*Solanum lycopersicum* L.) at early growth stage. *Sci Hort* 161: 111–117.

Harizanova A, Zlatev Z, Koleva L. (2014). Effect of silicon on activity of antioxidant enzymes and photosynthesis in leaves of cucumber plants (*Cucumis sativus* L.). *Turk J Agr Nat Sci* Special Issue 2.

Hashemi A, Abdolzadeh A, Sadeghipour HR. (2010). Beneficial effects of silicon nutrition in alleviating salinity stress in hydroponically grown canola, *Brassica napus* L., plants. *Soil Sci Plant Nutr* 56: 244–253.

Hattori T, Inanagaa S, Arakib H. (2005). Application of silicon enhanced drought tolerance in *Sorghum bicolor. Physiol Plant* 123: 459–466.

Hattori T, Sonobe K, Inanaga S, An P, Morita S. (2008). Effects of silicon on photosynthesis of young cucumber seedlings under osmotic stress. *J Plant Nutr* 31: 1046–1058.

Hu H, Zhang J, Wang H, Li R, Pan F, Wu J, Liu Q. (2013). Effect of silicate supplementation on the alleviation of arsenite toxicity in 93-11 (*Oryza sativa* L. indica). *Environ Sci Pollut Res* 20: 8579–8589.

Inal A, Pilbeam DJ, Gunes A. (2009). Silicon increases tolerance to boron toxicity and reduces oxidative damage in barley. *J Plant Nutr* 32: 112–128.

IPCC (Intergovernmental Panel on Climate Change). 2007. Climate change 2007: Synthesis report. Contributions of Working Groups I, II, and III to the Fourth Assessment Report of the Intergovernmental Panel on Climate Change (AR4). Geneva: IPCC.

Iturbe-Ormaetxe I, Escudero PR, Arrese-Igor C, Becana M. (1998). Oxidative damage in pea plants exposed to water deficit or paraquat. *Plant Physiol* 116: 173–181.

Iwasaki K, Maier P, Fecht M, Horst WJ. (2002a). Leaf apoplastic silicon enhances manganese tolerance of cowpea (*Vigna unguiculata*). *J Plant Physiol* 15: 167–173.

Iwasaki K, Maier P, Fecht M, Horst WJ. (2002b). Effects of silicon supply on apoplastic manganese concentrations in leaves and their relation to manganese tolerance in cowpea (*Vigna unguiculata* L. Walp.). *Plant Soil* 238: 281–288.

Janislampi KW. (2012). Effect of silicon on plant growth and drought stress tolerance. All Graduate Theses and Dissertations. Paper 1360. http://digitalcommons.usu.edu/etd/1360.

Jones LHP, Handreck KA. (1967). Silica in soils, plants and animals. *Adv Agron* 19: 107–149.

Kafi M, Rahimi Z. (2011). Effect of salinity and silicon on root characteristics, growth, water status, proline contents and ion accumulation of purslane (*Portulaca oleracea* L.). *Soil Sci Plant Nutr* 57: 341–347.

Kakani V, Reddy G, Zhao KR, Mohammed AR. (2003). Effects of ultraviolet-B radiation on cotton (*Gossypium hirsutum* L.) morphology and anatomy. *Ann Bot* 91: 817–826.

Kaya C, Tuna L, Higgs D. (2006). Effect of silicon on plant growth and mineral nutrition of maize grown under water-stress conditions. *J Plant Nutr* 29: 1469–1480.

Kaya C, Tuna AL, Sonmez O, Ince F, Higgs D. (2009). Mitigation effects of silicon on maize plants grown at high zinc. *J Plant Nutr* 32: 1788–1798.

Keiller DR, Mackerness SAH, Holmes MG. (2003). The action of a range of supplementary ultraviolet (UV) wavelengths on photosynthesis in *Brassica napus* L. in the natural environment: Effects on PSII, CO_2 assimilation and level of chloroplast proteins. *Photosynth Res* 75: 139–150.

Ketchum BH. (1954). Mineral nutrition of phytoplankton. *Annu Rev Plant Physiol* 5: 55–64.

Kim DW, Rakwal R, Agrawal GK, Jung YH, Shibato J, Jwa NS et al. (2005). A hydroponic rice seedling culture model system for investigating proteome of salt stress in rice leaf. *Electrophoresis* 26: 4521–4539.

Kinrade SD, Gillson AME, Knight CTG. (2002). Silicon-29 NMR evidence of a transient hexavalent silicon complex in the diatom *Navicula pelliculosa*. *J Chem Soc Dalton Tran* 3: 307–309.

Kinrade SD, Hamilton RJ, Schach, Knight CTG. (2001). Aqueous hypervalent silicon complexes with aliphatic sugar acids. *J Chem Soc Dalton Tran* 2001: 961–963.

Kukier U, Chaney RL. (2002). Growing rice grain with controlled cadmium concentrations. *J Plant Nutr* 25: 1793–1820.

Liang Y. (1999). Effects of silicon on enzyme activity and sodium, potassium and calcium concentration in barley under salt stress. *Plant Soil* 209: 217–224.

Liang Y, Yang C, Shi H. (2001). Effects of silicon on growth and mineral composition of barley grown under toxic levels of aluminum. *J Plant Nutr* 24: 229–243.

Liang YC, Chen Q, Liu Q, Zhang WH, Ding RX. (2003). Exogenous silicon (Si) increases antioxidant enzyme activity and reduces lipid peroxidation in roots of salt-stressed barley (*Hordeum vulgare* L.). *J Plant Physiol* 160: 1157–1164.

Liang YC, Hua H, Zhu YG, Zhang J, Cheng C, Romheld V. (2006). Importance of plant species and external silicon concentration to active silicon uptake and transport. *New Phytol* 172: 63–72.

Liang YC, Sun WC, Zhu YG, Christie P. (2007). Mechanisms of silicon-mediated alleviation of abiotic stresses in higher plants. A review. *Environ Pollut* 147: 422–428.

Lindsay WL. (1979). *Chemical Equilibria in Soil*. John Wiley & Sons, New York.

Lobato AKS, Coimbra GK, Neto MAM, Costa RCL, Santos Filho BG, Oliveira Neto CF et al. (2009). Protective action of silicon on water relations and photosynthetic pigments in pepper plants induced to water deficit. *Res J Biol Sci* 4: 617–623.

Ma JA. (2004). Role of silicon in enhancing the resistance of plants to biotic and abiotic stresses. *Soil Sci Plant Nutr* 50: 11–18.

Ma JF, Goto S, Tamai K, Ichii M. (2001). Role of root hairs and lateral roots in silicon uptake by rice. *Plant Physiol* 127: 1773–1780.

Ma JF, Yamaji N. (2006). Silicon uptake and accumulation in higher plants. *Trends Plant Sci* 11: 392–397.

Madronich SRL, McKenzie LO Björn, Caldwell MM. (1998). Changes in biologically active ultraviolet radiation reaching the earth's surface. *J Photochem Photobiol B* 6: 5–19.

Maghsoudi K, Emam Y, Ashraf M. (2015). Influence of foliar application of silicon on chlorophyll fluorescence, photosynthetic pigments, and growth in water-stressed wheat cultivars differing in drought tolerance. *Turk J Bot* 39: 1407–1411.

Malčovská SM, Dučaiová Z, Maslaňáková I, Bačkor M. (2014a). Effect of silicon on growth, photosynthesis, oxidative status and phenolic compounds of maize (*Zea mays* L.) grown in cadmium excess. *Water Air Soil Pollut* 225: 2056.

Malčovská SM, Dučaiová Z, Bačkor M. (2014b). Impact of silicon on maize seedlings exposed to short-term UV-B irradiation. *Biologia* 69(10): 1349–1355.

Mateos-Naranjo E, Andrades-Morenoa L, Davy AJ. (2013). Silicon alleviates deleterious effects of high salinity on the halophytic grass *Spartina densiflor*. *Plant Physiol Biochem* 63: 115–121.

Mishra V, Srivastava G, Prasad SM, Abraham G. (2008). Growth, photosynthetic pigments and photosynthetic activity on seedling stage of *Vigna unguiculata* in response to UV-B and dimethoate. *Pestic Biochem Physiol* 92: 30–37.

Monakhova OF, Chernyad'ev II. (2002). Protective role of kartolin-4 in wheat plants exposed to soil drought. *Appl Environ Microbiol* 38: 373–380.

Moussa HR. (2006). Influence of exogenous application of silicon on physiological response of salt-stressed maize (*Zea mays* L.). *Int J Agric Biol* 8: 293–297.

Moya JL, Ros R, Picazo I. (1995). Heavy metal-hormone interactions in rice plants: Effects on growth, net photosynthesis, and carbohydrate distribution. *J Plant Growth Regul* 14: 61–67.

Muneer S, Park YG, Manivannan A, Prabhakaran S, Jeong BR. (2014). Physiological and proteomic analysis in chloroplasts of *Solanum lycopersicum* L. under silicon efficiency and salinity stress. *Int J Mol Sci* 15: 21803–21824.

Murillo-Amador B, Yamada S, Yamaguchi T, Rueda-Puente E, Avila-Serrano N, Garcia-Hernandez JL, Lopez-Aguilar R, Troyo-Dieguez E, Nieto-Garibay A. (2007). Influence of calcium silicate on growth, physiological parameters and mineral nutrition in two legume species under salt stress. *J Agron Crop Sci* 193: 413–421.

Naumann JC, Bissett SN, Young DR, Edwards J, Anderson JE. (2010). Diurnal patterns of photosynthesis, chlorophyll fluorescence, and PRI to evaluate water stress in the invasive species, *Elaeagnusum bellata* Hub. *Trees* 24: 237–245.

Neumann D, Zur Nieden U. (2001). Silicon and heavy metal tolerance of higher plants. *Phytochemistry* 56: 685–692.

Nezami MT, Bybordi A. (2011). Effects of silicon on photosynthesis and concentration of nutrients of *Brassica napus* L. in saline-stressed conditions. *J Food Agric Environ* 9: 655–659.

Nwugo CC, Huerta AJ. (2008). Effects of silicon nutrition on cadmium uptake, growth and photosynthesis of rice seedlings (*Oryza sativa* L.) exposed to long term low level cadmium. *Plant Soil* 311: 73–86.

Ohashi Y, Nakayama N, Saneokai H, Fujita K. (2006). Effects of drought stress on photosynthetic gas exchange, chlorophyll fluorescence and stem diameter of soybean plants. *Biol Plant* 50: 138–141.

Parida AK, Das AB. (2005). Salt tolerance and salinity effects on plants: A review. *Ecotoxicol Environ Saf* 60: 324–349.

Parker R, Flowers T, Moore A, Harpham N. (2006). An accurate and reproducible method for proteome profiling of the effects of salt stress in the rice leaf lamina. *J Exp Bot* 57: 1109–1118.

Pilon C, Soratto RP, Moreno LA. (2013). Effects of soil and foliar application of soluble silicon on mineral nutrition, gas exchange, and growth of potato plants. *Crop Sci* 53: 1605–1614.

Popova LP, Maslenkova LT, Yordanova RY, Ivanova AP, Krantev AP, Szalai G, Janda T. (2009). Exogenous treatment with salicylic acid attenuates cadmium toxicity in pea seedlings. *Plant Physiol Biochem* 47: 224–231.

Prasad SM, Zeeshan M. (2005). UV-B radiation and cadmium induced changes in growth, photosynthesis, and antioxidant enzymes of cyanobacterium *Plectonema boryanum*. *Biol Plant* 49: 229–236.

Richmond KE, Sussman M. (2003). Got silicon? The nonessential beneficial plant nutrient. *Curr Opin Plant Biol* 6: 268–272.

Romero-Aranda MS, Jurado O, Cuartero J. (2006). Silicon alleviates the deleterious salt effect on tomato plant growth by improving plant water status. *J Plant Physiol* 163: 847–855.

Round FE, Crawford RM, Mann DG. (1990). *The Diatoms: Biology and Morphology of the Genera*. Cambridge University Press, Cambridge.

Sanglard LM, Martins SC, Detmann KC, Silva PE, Lavinsky AO, Silva MM, Detmann E, Araújo WL, Damatta FM. (2014). Silicon nutrition alleviates the negative impacts of arsenic on the photosynthetic apparatus of rice leaves: An analysis of the key limitations of photosynthesis. *Physiol Plant* 152: 355–366.

Sanitá di Toppi L, Gabrielli R. (1999). Response to cadmium in higher plants. *Environ Exp Bot* 41: 105–130.

Saud S, Li X, Chen Y, Zhang L, Fahad S, Hussain S, Sadiq A, Chen Y. (2014). Silicon application increases drought tolerance of Kentucky bluegrass by improving plant water relations and morphophysiological functions. *Sci World J* 2014: 368694. http://dx.doi.org/10.1155/2014/368694.

Savant NK, Snyder GH, Datnoff LE. (1997). Silicon management and sustainable rice production. *Adv Agron* 58: 151–199.

Savvas D, Papastavrou D, Ntatsi G, Ropokis A, Olympios C, Hartmann H, Schwarz D. (2009). Interactive effects of grafting and manganese supply on growth, yield, and nutrient uptake by tomato. *Hortscience* 44: 1178–1982.

Shen X, Xiao X, Dong Z, Chen Y. (2014). Silicon effects on antioxidative enzymes and lipid peroxidation in leaves and roots of peanut under aluminum stress. *Acta Physiol Plant* 36: 3063–3069.

Shen XF, Zhou YY, Duan LS, Li ZH, Eneji AE, Li JM. (2010). Silicon effects on photosynthesis and antioxidant parameters of soybean seedlings under drought and ultraviolet-B radiation. *J Plant Physiol* 167: 1248–1252.

Shi XH, Zhang CC, Wang H, Zhang FS. (2005). Effect of Si on the distribution of Cd in rice seedlings. *Plant Soil* 272: 53–60.

Shi Y, Wang YC, Flowers TJ, Gong HJ. (2013). Silicon decreases chloride transport in rice (Oryza sativa L.) in saline conditions. *J Plant Physiol* 170: 847–853.

Signarbieux C, Feller U. (2011). Non-stomatal limitations of photosynthesis in grassland species under artificial drought in the field. *Environ Exp Bot* 71: 192–197.

Silva ON, Lobato AKS, Ávila FW, Costa RCL, Oliveira Neto CF, Santos Filho BG et al. (2012). Silicon-induced increase in chlorophyll is modulated by the leaf water potential in two water-deficient tomato cultivars. *Plant Soil Environ* 5: 481–486.

Song A, Li P, Fan F, Li Z, Liang Y. (2014). The effect of silicon on photosynthesis and expression of its relevant genes in rice (*Oryza sativa* L.) under high-zinc stress. *PLoS One* 9(11): e113782. http://dx.doi.org/10.1371/journal.pone.0113782.

Song A, Li Z, Zhang J, Xue G, Fan F, Liang Y. (2009). Silicon-enhanced resistance to cadmium toxicity in *Brassica chinensis* L. is attributed to Si-suppressed cadmium uptake and transport and Si-enhanced antioxidant defense capacity. *J Hazard Mater* 172: 74–83.

Strasser RJ, Srivastava A, Tsimilli-Michael M. (2000). The fluorescence transient as a tool to characterize and screen photo-synthetic samples. In M Yunus, U Pathre, P Mohanty (eds.), *Probing Photosynthesis: Mechanisms, Regulation and Adaptation*. Taylor & Francis, London, pp. 445–483.

Tahir MA, Aziz T, Farooqc M, Sarwara G. (2012). Silicon-induced changes in growth, ionic composition, water relations, chlorophyll contents and membrane permeability in two salt-stressed wheat genotypes. *Arch Agron Soil Sci* 5: 247–256.

Tripathi DK, Kumar R, Chauhan DK, Rai AK, Bicanic D. (2011). Laser-induced breakdown spectroscopy for the study of the pattern of silicon deposition in leaves of *Saccharum* species. *Instrum Sci Technol* 39: 510–521.

Tripathi DK, Singh VP, Prasad SM, Chauhan DK, Dubey NK, Rai AK. (2015). Silicon-mediated alleviation of Cr (VI) toxicity in wheat seedlings as evidenced by chlorophyll florescence, laser induced breakdown spectroscopy and anatomical changes. *Ecotoxicol Environ Saf* 113: 133–144.

Tuna AL, Kaya C, Higgs D, Murillo-Amador B, Girgin AR, Aydemir S. (2008). Silicon improves salinity tolerance in wheat plants. *Environ Exp Bot* 62: 10–16.

Van Soest PJ. (2006). Rice straw, the role of silica and treatments to improve quality. *Anim Feed Sci Technol* 130: 137–171.

Volcani BE. (1978). Role of silicon in diatom metabolism and silicification. In G Bendz, J Lindquist (eds.), *Biochemistry of Silicon and Related Problems*. Plenum, New York, pp. 177–206.

Weiss A, Herzog A. (1978). Isolation and characterization of a silicon organic complex from plants. In G Bendz, J Lindquist (eds.), *Biochemistry of Silicon and Related Problems*. Plenum, New York, pp. 109–127.

Wiese H, Nikolic M, Römheld V. (2007). Silicon in plant nutrition. In B Sattelmacher, WJ Horst (eds.), *The Apoplast of Higher Plants: Compartment of Storage, Transport and Reactions*. Springer, Amsterdam, pp. 33–47.

Xie Z, Song R, Shao H, Song F, Xu H, Lu Y. (2015). Silicon improves maize photosynthesis in saline-alkaline soils. *Sci World J* 2015: 245072. http://dx.doi.org/10.1155/2015/245072.

Yao X, Chu J, Kunzheng C, Liu L, Shi J, Geng W. (2011). Silicon improves the tolerance of wheat seedlings to ultraviolet-B stress. *Biol Trace Elem Res* 143: 507–517.

Yeo AR, Flowers SA, Rao G, Welfare K. (1999). Silicon reduces sodium uptake in rice (*Oryza Sativa* L.) in saline condition and this is accounted for by a reduction in the transpirational bypass flow. *Plant Cell Environ* 22: 559–565.

Yin L, Wang S, Li J, Tanaka K, Oka M. (2013). Application of silicon improves salt tolerance through ameliorating osmotic and ionic stresses in the seedling of *Sorghum bicolor*. *Acta Physiol Plant* 35: 3099–3107.

Zuccarini P. (2008). Effects of silicon on photosynthesis, water relations and nutrient uptake of *Phaseolus vulgaris* under NaCl stress. *Biol Plant* 52: 157–160.

5 Silicon and Heavy Metal Tolerance of Plants

*Clístenes Williams Araújo do Nascimento
and Karina Patrícia Vieira da Cunha*

CONTENTS

ABSTRACT

Silicon (Si) is the second most abundant element in the earth's crust. Although not considered an essential element, Si has many benefits to plants, including in the control of diseases and in the alleviation of saline, water, and heavy metal stresses. The alleviation of heavy metal phytotoxicity by the application of Si has been well documented. However, the mechanisms by which Si increases the tolerance of plants are not completely understood. Increasing plant tolerance to heavy metals may have a significant impact on the phytostabilization of metals in the soil and may also enhance metal phytoextraction through a greater accumulation of metals in the aerial part of the plant due to the alleviation of stress. This chapter provides a synthesis of the current knowledge regarding Si-mediated alleviation of metal stress in plants and aims to integrate the effects of Si in the soil, as well as physiological reactions mediated by Si in plants exposed to heavy metal stresses.

5.1 INTRODUCTION

Silicon (Si) is the second most abundant element in the earth's crust, after oxygen. In soil, the silicon concentration can range from <1 to 45 dag kg^{-1} in dry weight. Silicon is present in the soil solution as monosilicic acid $(Si(OH)_4)$, mostly in the nondissociated form, which is readily available to plants. As a result of the desilication caused by the intense weathering and leaching of tropical soils, Si is mostly found in the form of quartz (SiO_2), opal $(SiO_2 \cdot nH_2O)$, and other forms that are not available to plants. The Si forms that are chemically active in the soil are monosilicic acid (both soluble and weakly adsorbed), polysilicic acid, and organosilicon compounds (Matichenkov and Calvert, 2002). There are many processes and transformations that influence Si concentration in the soil (Figure 5.1). Monosilicic acid is at the center of these interactions and transformations as a product of the dissolution of Si-rich minerals (Lindsay, 1979).

The principal sources of monosilicic acid in soil are as follows: organic residue decomposition, polysilicic acid dissociation, desorption of Si from Fe and Al oxides and hydroxides, dissolution of crystalline and noncrystalline minerals, addition of silicate fertilizers, and irrigation water. The main losses include the precipitation of Si in solution, polymerization of monosilicic acid into

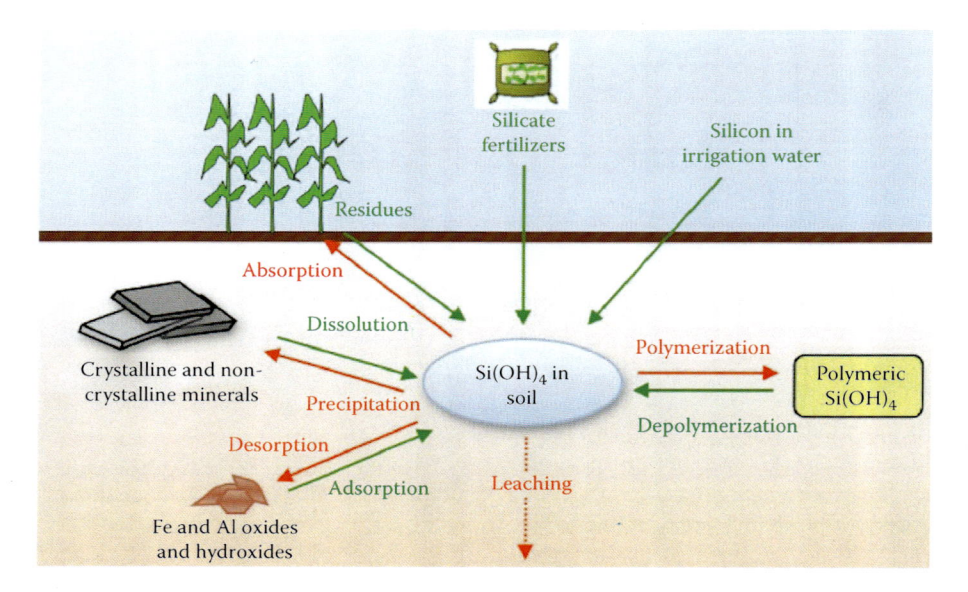

FIGURE 5.1 Transformations and processes that influence silicon concentration in the soil.

polysilicic acid, leaching, and adsorption into Fe and Al oxides and hydroxides, as well as absorption by plants.

Tropical soils typically present low pH, high Al^{3+} content, low base saturation, and high capacity for phosphorus fixation. Especially in these soils, silicate fertilizers can influence crops in two ways: (1) by improving the chemical properties and soil fertility and (2) by directly affecting plant growth and development (Matichenkov and Calvert, 2002).

In the soil, silicates show behaviors similar to those of calcium carbonates and magnesium, as they are able to elevate pH and neutralize the exchangeable Al. They are also associated with the increase of soluble Si availability and exchangeable Ca and Mg contents, which promote greater base saturation and lower Al saturation (Epstein, 1999; Savant et al., 1999).

Silicate fertilizers are generally neutral to slightly alkaline (Lindsay, 1979). According to Savant et al. (1999), the corrective effect on soil acidity, promoted by silicates, occurs through the reactions of SiO_3^{-2} anions with H^+ protons in the soil:

$$CaSiO_3 + H_2O \leftrightarrow Ca^{2+} + SiO_3^{2-} + H_2O$$

$$SiO_3^{2-} + 2H^+ \leftrightarrow H_2SiO_3$$

$$H_2SiO_3 + H_2O \leftrightarrow H_4SiO_4$$

Another important aspect in investigating the effects of silicate on soil properties concerns its interaction with P (Matichenkov and Calvert, 2002) and with nitrogen, phosphorus, and potassium (NPK) fertilizer (Lima-Filho et al., 1999). Lima-Filho et al. (1999) affirmed that the use of silicate fertilizers increases the efficiency of the NPK fertilizer. The silicates present good adsorption properties and promote lower leaching of K and other mobile nutrients from the soil surface. According to Matichenkov and Calvert (2002), silicate fertilizers adsorb P, which decreases its leaching by 40%–90%, without decreasing its availability to plants because P is held in exchangeable forms on the silicate surface. Lima-Filho et al. (1999) also reported that with the addition of silicates in soil, chemical reactions occur in which Ca, Al, and Fe silicates and phosphates are exchanged, consequently forming their respective silicates with the concurrent release of phosphate ions to the soil (Matichenkov and Calvert, 2002). In addition to these phosphate exchange reactions, the silicate anion can displace the phosphate anion from oxide adsorption sites, or preferentially occupy them. The exchange reactions between Ca, Al, and Fe silicates and phosphates are as follows:

$$CaHPO_4 + Si(OH)_4 \leftrightarrow CaSiO_3 + H_2O + H_3PO_4$$

$$2Al(H_2PO_4)_3 + 2Si(OH)_4 + 5H^+ \leftrightarrow Al_2Si_2O_5 + 5H_3PO_4 + 5H_2O$$

$$2FePO_4 + Si(OH)_4 + 2H^+ \leftrightarrow Fe_2SiO_4 + 2H_3PO_4$$

Depending on the chemical composition of the soil, monosilicic acids can be polymerized and polysilicic acid formed, influencing the texture, adsorption capacity, and stability of the soil's organic matter (Savant et al., 1999; Matichenkov and Calvert, 2002).

Plants absorb Si as monosilicic acid ($Si(OH)_4$), the nondissociated form present in soil at concentrations of 0.1–0.6 mmol L^{-1}, and, in lower quantities, as $Si(OH)_3O^-$, the ionic form predominant in pH > 9 (Epstein, 1994). Accumulation of Si in plant tissue varies between 0.1% and 10% of dry matter (Ma and Takahashi, 2002; Currie and Perry, 2007). This wide variation in plant tissue Si concentration is principally attributed to the absorption and transport characteristics of Si by plants (Epstein, 1994, 2009; Ma and Yamaji, 2006). Most plants, especially dicots, are not able to absorb large quantities of Si from the soil (Sahebi et al., 2014). The majority of species absorbs Si by passive diffusion, so that Si arrives at the xylem and reaches the aerial part while accompanying the transpiration stream. However, species from the families Poaceae, Equisetaceae, and Cyperaceae, which present a high accumulation of Si (>4% Si in dry weight), absorb Si actively (Currie and Perry, 2007). In this case, Si is absorbed via membrane-specific proteins, which ensures Si accumulation by the plant, regardless of the concentration gradient (Ma and Yamaji, 2006; Epstein, 2009).

The Si absorbed by the roots is transported to the aerial part and deposited intra- or extracellularly in the plant tissues as hydrated silica ($SiO_2 \cdot nH_2O$). In Graminaceae species, including maize, rice, and sorghum, silica is deposited in the form of silica bodies, mainly in the epidermal, siliceous, and bulliform cells, as well as in the stoma and trichome of leaves (Currie and Perry, 2007). In many species, a dense layer formed by the deposition of silica can be found below the cuticle. The development of this layer has been of great importance in conditions of biotic and abiotic stress. This layer contributes to reduced water loss by transpiration and increased efficiency (Nwugo and Huerta, 2008) and acts as a mechanical barrier to pathogen penetration and herbivore mastication (Epstein, 1999; Savant et al., 1999).

In this chapter, the role of silicon in the tolerance of abiotic stresses caused by heavy metals in plants is discussed, contextualizing these interactions and suggesting future directions for research that may have a significant impact on phytotechnologies, such as phytoextraction and revegetation of areas contaminated by metals and phytoremediation programs.

5.2 SILICON IN THE ALLEVIATION OF STRESSES CAUSED BY HEAVY METALS IN PLANTS

Although not considered an essential element, Si has shown many benefits to plants, especially in the control of diseases and in the alleviation of saline, water, and heavy metal stresses. The alleviation of heavy metal phytotoxicity by the application of Si has been well documented (Table 5.1). However, the mechanisms by which Si increases the tolerance of plants are not completely

TABLE 5.1

Principal Studies on the Alleviation of Heavy Metals in Plants by the Application of Silicon

Metal	Plant	Reference
Manganese	Rice	Horiguchi (1988)
	Bean	Horst and Marschner (1978)
	Barley	Williams and Vlamis (1957), Horiguchi and Morita (1987)
	Cucumber	Rogalla and Romheld (2002), Shi et al. (2005a), Maksimovic et al. (2007)
	Cowpea	Horst et al. (1999), Iwasaki et al. (2002a,b)
	Pumpkin	Iwasaki and Matsumura (1999)
	Soybean	Miyake and Takahashi (1985)
	Sorghum	Galvez et al. (1987)
Cadmium	Rice	Wang et al. (2000), Shi et al. (2005b), Nwugo and Huerta (2008), Zhang et al. (2008), Tripathi et al. (2012), Kim et al. (2014), Ma et al. (2015), Wang et al. (2015)
	Maize	Liang et al. (2005), Cunha et al. (2008b), Cunha and Nascimento (2009)
	Strawberry	Treder and Cieslinski (2005)
	Eucalyptus	Accioly et al. (2009)
Zinc	*Minuartia verna*	Neumann et al. (1997)
	Cardaminopis halleri	Neumann and Nieden (2001)
	Minuartia	Lichtenberger and Neumann (1996)
	Maize	Cunha et al. (2008b), Cunha and Nascimento (2009)
	Rice	Song et al. (2014)
	Eucalyptus	Accioly et al. (2009)
Chromium	Rice	Wang et al. (2015)
Copper	Wheat	Keller et al. (2015)
	Bamboo	Collin et al. (2014)
	Rice	Kim et al. (2014)
Arsenic	Rice	Guo et al. (2005)
	Maize	Kidd et al. (2001), Wang et al. (2004)
	Soybean	Baylis et al. (1994)
	Sorghum	Galvez et al. (1987), Hodson and Sangster (1993)

understood. In analyzing the literature, the role of silicon in relieving the toxicity of Cd and Mn is evident, while metals such as Pb and Cr have been poorly studied, as reported Wu et al. (2013).

It is known that Si acts both on the soil, by reduction in metal bioavailability, and inside plants (Figure 5.2). In the plant, the ameliorating effect of Si can be attributed to the following: (1) metal retention in the roots and inhibition of translocation to the aerial part (Shi et al., 2005b); (2) SiO_2 deposition in the root apoplasm and leaf surface, constructing a barrier to the apoplastic flow of metallic ions and to the transpiration stream, respectively (Shi et al., 2005b); (3) coprecipitation of Si–metal complexes in the cell wall (Neumann and Nieden, 2001); (4) compartmentalization of metals bound to organic acids in the vacuole (Neumann and Nieden, 2001); (5) a more homogenous distribution of metals and the formation of Si–polyphenol complexes in the tissues (Williams and Vlamis, 1957; Horst and Marschner, 1978; Shi et al., 2005a; Maksimovic et al., 2007); and (6) a reduction in the peroxidation of lipid membranes by the stimulation of enzymatic and nonenzymatic antioxidants (Shi et al., 2005a).

The role of Si in the immobilization of heavy metals in soil is well known (Liang et al., 2005, 2007). Some authors assert that Si alleviates metal phytotoxicity by decreasing the bioavailability of these elements due to a rise in soil pH (Paim et al., 2006; Liang et al., 2007; Accioly et al., 2009), whereas other authors defend the immobilization as a consequence of the direct effect of Si in solution, regardless of changes in soil pH (Baylis et al., 1994; Liang et al., 2005; Cunha et al., 2008b).

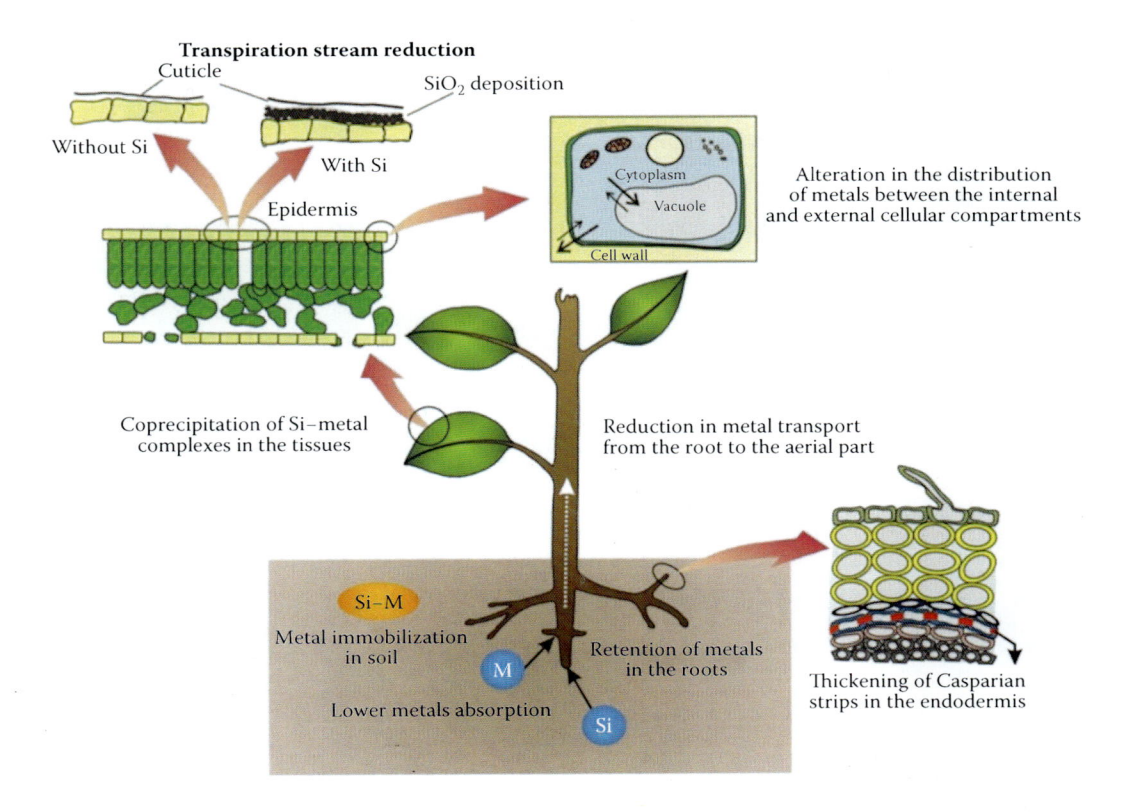

FIGURE 5.2 Schematic representation of the possible mechanisms mediated by silicon in the alleviation of heavy metal toxicity in plants.

Generally, the pH shows an inverse relationship with the bioavailability of heavy metals. The rise in pH that results from the application of silicate promotes an increase in adsorption sites in the soil, reducing the bioavailable content of the metals. In addition to the increase in adsorption, higher precipitation occurs with the rise in pH, contributing to the immobilization of metals in the soil.

Similar to the carbonate ion $\left(CO_3^{2-}\right)$, the SiO_3^{2-} anions combine with H^+ ions in the soil, decreasing chemical activity by the precipitation of metal in the form of hydroxides. However, besides the precipitation in the form of hydroxides, the addition of Si promotes the precipitation of metals in the form of silicates (Dietzel, 2000), which is an added and advantageous benefit in relation to limestone (Paim et al., 2006).

To investigate the effect of Si on the immobilization of Cd in soil, Liang et al. (2007) conducted two parallel experiments that differed only in the dose of Si applied to the soil: 400 mg kg^{-1} Si in the first experiment and 50 mg kg^{-1} Si in the second experiment. With the application of 400 mg kg^{-1} Si to contaminated soil, the authors concluded that a reduction in Cd toxicity in maize was due to a significant rise in soil pH and the consequent reduction of the available Cd content. However, in the second experiment, the Cd detoxification in maize plants could not be attributed to soil effects, since there were no significant alterations observed in the pH, in the bioavailability, and in the distribution of Cd between soil fractions (Table 5.2). Even though there were no effects on the soil, the plants treated with Si showed biomass production up to 138% greater than plants without Si; this was in addition to the 50% increase in total Cd accumulation in the aerial part, without toxicity symptoms being observed. This result indicates the potential of Si in phytoextraction in studies that aim to increase the removal of metals by greater accumulation in the aerial part.

TABLE 5.2

pH and Cadmium Fractions of Soil Experimentally Contaminated with Cadmium as a Result of the Application of Silicon

Treatment	pH	Cd Fraction (%)			
		Exchangeable	Specifically Adsorbed	Fe and Mn Oxides	Organic Matter
First Experiment					
Control	4.09b	nd	nd	nd	nd
Cd20	4.04b	84.9a	9.6b	4.7b	0.8b
Cd40	3.91b	84.4a	10.1b	4.8b	0.7b
Si400 + Cd20	5.24a	69.8c	13.2a	13.0a	3.9a
Si400 + Cd40	5.10a	72.9b	12.1a	12.7a	2.4a
Second Experiment					
Control	3.97a	nd	nd	nd	nd
Cd20	3.99a	91.1a	8.2a	0.7a	nd
Cd40	3.95a	91.6a	6.9a	1.5a	nd
Si50 + Cd20	4.13a	86.4a	10.3a	3.3a	nd
Si50 + Cd40	4.19a	89.9a	7.4a	2.7a	nd

Source: Liang, Y. et al., *Chemosphere*, 58: 475–483, 2005.

Note: Means followed by the same letters in the same column and the same experiment are not significantly different according to Duncan's test ($p < 0.05$). nd, not detected.

Cunha et al. (2008b) observed that the alteration of the soil pH was not the most important factor in the reduction of bioavailability of the metals Cd and Zn after treatment of contaminated soil with silicates. These authors verified significant reductions of the content of bioavailable metals after the addition of increasing doses of Si, even without significant changes in the soil pH. In this case, the reduction in bioavailability may result from the precipitation of metals in the form of silicates, which occur independently from the soil pH alteration (Dietzel, 2000; Sommer et al., 2006).

This difference in results, regarding the effect of Si on the immobilization of metals in the soil, can be attributed to the different doses and sources of Si. Liang et al. (2005) found a significant rise in pH when 400 mg kg^{-1} Si was applied to the soil; this was not observed with 50 mg kg^{-1} Si dose (Table 5.2). Accioly et al. (2009) and Paim et al. (2006) did not have the opportunity to observe the behavior of metals in soil independent of the influence of Si on pH because the lowest doses used by these authors were 450 and 500 mg kg^{-1} Si, respectively, which promoted significant increases in soil pH. In the experiment conducted by Cunha et al. (2008b), the highest dose added to the soil was 200 mg kg^{-1} Si, which resulted in a bioavailability reduction of roughly 24% and 41% in Cd and Zn, respectively.

According to Dietzel (2000), in the normal range of soil pH, the incorporation of silicates promotes the polymerization of silicate compounds that are potential ligands to the formation of insoluble complexes with heavy metals (metal silicates). In fact, silicic acid can exist as isolated molecules, which are referred to as monosilicic acid [Si(OH)$_4$], and as polymers, which are called polysilicic acid. These polymers consist of tetrahedrons that are connected by Si-O-Si bonds capable of forming dimeric silica $\left(\text{Si}_2\text{O}_3(\text{OH})_4^{2-}\right)$, trimeric silica $\left(\text{Si}_3\text{O}_5(\text{OH})_5^{3-}\right)$, tetrameric silica $\left(\text{Si}_4\text{O}_8(\text{OH})_4^{4-}\right)$, and so on. The formation of polysilicic acids can be generically described according to the following equation:

$$HO-\underset{\underset{OH}{|}}{\overset{\overset{OH}{|}}{Si}}-OH \;+\; HO-\underset{\underset{OH}{|}}{\overset{\overset{OH}{|}}{Si}}-OH \quad\leftrightarrow\quad HO-\underset{\underset{OH}{|}}{\overset{\overset{OH}{|}}{Si}}-O-\underset{\underset{OH}{|}}{\overset{\overset{OH}{|}}{Si}}-OH$$

Monosilicic acid   Polysilicic acid

The application of doses greater than 100 mg dm^{-3} of Si to the soil promotes a change in Si coordination from four to six, which is the starting point for its polymerization and consequent formation of insoluble Si–metal complexes in the soil (Dietzel, 2000; Paim et al., 2006). Paim et al. (2006) observed the presence of whitish corpuscles on soil particles with a sharp increase in the Si dose from 6040 to 13,660 mg dm^{-3} of Si. Furthermore, the decrease in pH, from approximately 7.5 to 6.0 observed by these authors for higher doses of Si (>8030 mg dm^{-3}), suggests a decrease in the activity of SiO_3^{-2} anions as bases and an increase in its participation as Si polymers, acting as ligands for the formation of insoluble Si complexes with metals in the soil.

Regarding the Si potential in studies of phytoextraction, Araújo (2009) reported an increase in the efficiency of Pb phytoextraction assisted by NTA and citric acid after the addition of increasing Si doses. In the experiments with assisted phytoextraction, many authors report a reduction in the production of dry matter in plants cultivated in contaminated soil after treatment with chelants. This indicates that the phytotoxicity of chelants may represent a limitation to the application of higher doses of chelants because they reduce biomass and consequently lower the net efficiency of heavy metal phytoextraction (Melo et al., 2006; Nascimento and Xing, 2006; Freitas and Nascimento, 2009; Freitas et al., 2009). In the work developed by Araújo (2009), when soil was enriched with Si before planting and phytoextraction induction, the same result was not observed, even when using doses of 30 mmol kg^{-1} citric acid and 10 mmol kg^{-1} NTA, which are greater than those used by Freitas and Nascimento (2009) and Freitas et al. (2009). Araújo (2009) found average increases of 22% (30 mmol kg citric acid) and 8% (10 mmol kg NTA) in the production of dry matter when compared to the chelant treatment without Si addition to soil. This result demonstrates the alleviating action of Si, expressed not only as an increase in the production of dry matter, but also by the reduction of chelant toxicity to plants, which is of great value for the development and improvement of assisted phytoextraction techniques. As reported by Araújo (2009), the beneficial effect of Si application in Pb-contaminated soil was greater when the stress was established, that is, with the application of chelants and the resulting increase of Pb solubility in the soil. This finding supports the idea presented by Epstein (2009), in which the alleviating action of Si can especially be observed when the plants are under stress conditions; in beneficial conditions, the role of Si is often underestimated.

Alterations in the distribution of metals between soil fractions are also reported to be an effect of Si in contaminated soil. After soil treatment with Si, Liang et al. (2005) and Cunha et al. (2008b) found an increase in the content of metals bound to Fe oxides and the concomitant reduction of concentration in the exchangeable fraction. Sommer et al. (2006) reported the sorption of Si polymers on the surface of Fe and Al oxides, in which these polymers showed greater stability at pH < 8 and in the presence of cations in solution, especially divalent cations, according to the following order: Cu > Zn > Sr > Ca > Mg > Na > K (Dietzel, 2000). Furthermore, the bonding of organosilicic complexes to the surface of oxides was reported by Sommer et al. (2006) and may be involved in the complexation of heavy metals in soil. After Si addition, higher concentrations of metals bonded to organic matter were reported by Cunha et al. (2008b). Even without alterations in soil pH, these authors found that Cd and Zn passed into the less available fractions in the presence of Si. Without Si application, 47% and 70% of total Cd and Zn, respectively, were in the exchangeable fraction. After the addition of 200 mg kg^{-1} Si, the participation of Cd and Zn in the exchangeable fraction was reduced to 32% and 40%, respectively.

5.2.1 Silicon and Tolerance of Cellular Mechanisms to Heavy Metals

The relationship of Si to the heavy metal tolerance of cellular mechanisms has been well studied. One of the first studies associating the effect of Si to the alleviation of heavy metal toxicity was developed by Williams and Vlamis (1957). These authors observed a reduction in Mn^{2+} toxicity symptoms in barley plants after the addition of a small quantity of Si (357 μmol L^{-1} Si) in a nutrient solution. Although the concentrations of Mn in leaf tissues were the same in treatments with and without Si, the plants were severely injured in the absence of Si. Using stained Mn (^{54}Mn), these authors observed that Mn showed a more homogenous distribution in the leaves in the presence of Si, avoiding the formation of lesions that are typical of Mn toxicity in plants. The visual symptoms of Mn toxicity in plants are characterized by the development of brown and necrotic spots on leaves as a result of Mn point concentration and the oxidation of phenolic compounds in leaf tissue (Rogalla and Romheld, 2002). Subsequently, Horst and Marschner (1978) observed similar alterations in the distribution of Mn in the presence and absence of Si in the leaves of bean plants. These results were recently confirmed by Shi et al. (2005a) in cucumber plants exposed to Mn toxicity and treated with Si.

In contrast with some authors (Williams and Vlamis, 1957; Horiguchi and Morita, 1987), Horst and Marschner (1978) observed a greater accumulation of Mn in tissues after Si treatment, whereas the toxic level of Mn in bean plant leaves was 100 mg kg^{-1} in the absence of Si. After Si supplementation in nutrient solution, this level rose to more than 1000 mg kg^{-1}.

Changes in the speciation and distribution of metals between cellular compartments are also described as an effect of Si on plants. The addition of Si in nutrient solutions alleviates Mn toxicity by increasing the adsorption of Mn in the cell wall of bean plants and by decreasing the concentration of soluble Mn in the apoplasm (Horst et al., 1999; Iwasaki et al., 2002a,b). According to Iwasaki et al. (2002a,b), when supplied at the same time, 50 μmol L^{-1} Mn and 1.44 mmol L^{-1} Si alleviated the symptoms of Mn toxicity without reducing the total concentration of Mn in the aerial part of the plant. On the other hand, the symptoms were not completely alleviated when cowpea plants were only pretreated with 1.44 mmol L^{-1} Si and then exposed to 50 μmol L^{-1} Mn without the concomitant supply of Si. This fact indicates that Si must be continually present in the culture medium during stress for a more favorable response. Although the total Mn concentrations in the tissues were similar, both pretreated plants and plants subjected to a continuous supply of Si exhibited lower concentrations of soluble Mn in the apoplasm extract and higher Mn concentrations adsorbed in the cell wall structure when compared to plants without Si (Iwasaki et al., 2002a).

Corroborating the results of Iwasaki et al. (2002a,b) and Rogalla and Romheld (2002) verified that although the absorption and concentration of Mn in cucumber leaves were not affected after the addition of Si, significant alterations in the distribution of Mn between the internal and external cellular compartments were observed.

It is important to stress that in addition to the change in cellular distribution of Mn between the apoplasm and the symplasm, Si promoted a change in the Mn species present in the apoplasm. Thus, soluble Mn, the predominant form in apoplasm in the absence of Si, became Mn adsorbed in the cell wall after the addition of Si to the solution. This result is important because Mn present in plants treated with Si is less soluble and, consequently, less toxic to cellular metabolism. Therefore, both Iwasaki et al. (2002a,b) and Rogalla and Romheld (2002) concluded that the Mn tolerance mediated by Si in cucumber and cowpea plants is a consequence of the strong bond of Mn to the cell wall and the reduction of Mn concentrations in the intracellular environment. In contrast, Horst and Marschner (1978) observed greater Mn concentrations in the vacuoles of bean leaves after treatment with Si. In an experiment with rice plants exposed to Cd toxicity, Shi et al. (2005a) found no alteration in the distribution of Cd between the symplast and the apoplast of rice roots exposed to 50 μmol L^{-1} $CdSO_4$ after treatment with 1.8 mmol L^{-1} Si. Even in the presence of Si, the majority of Cd (87%) in the root remained in the intracellular environment. This discrepancy in results was attributed to differences between species in response to different heavy metals. Nevertheless, Wang et al. (2000)

studied the effect of Si on the toxicity of Cd in rice plants and found by energy-dispersive x-ray (EDX) analysis that after treatment with Si, Cd became predominantly allocated to the cell wall. This finding corroborates the results obtained by Iwasaki et al. (2002a,b) and Rogalla and Romheld (2002). A recent study suggests that a hemicellulose-bound form of Si with net negative charges is responsible for inhibition of Cd uptake in rice cells by a mechanism of "[Si-hemicellulosematrix] Cd" complexation and subsequent codeposition (Ma et al., 2015).

The effect of Si modifying the bonding capacity of the cell wall promotes the precipitation of Si–metal complexes, reducing the concentration of free metals in the tissues (Wang et al., 2000; Iwasaki et al., 2002a,b; Rogalla and Romheld, 2002; Ma et al., 2015). In fact, several authors have located heavy metals bound to Si in the cell wall. Using the transmission electron microscope technique in association with EDX, Zhang et al. (2008) reported that the leaf distribution of Cd was similar to the distribution of Si, suggesting the coprecipitation of these elements in plant tissue. Neumann and Nieden (2001) considered the occurrence of precipitates of Zn silicate in the vacuole and cell wall of *Cardaminopsis halleri* leaves a mechanism of heavy metal tolerance. Cunha and Nascimento (2009) reported that after the addition of increasing doses of Si to the soil, Cd and Zn, as well as Si, became preferentially deposited in the cell wall of the collenchyma and the leaf epidermis, especially in the bulliform cells (Figure 5.3). Similarly, Neumann et al. (1997) verified that Zn was precipitated as Zn silicate in the cell wall of the epidermis of *Minuartia verna*, a species that is tolerant to Zn. These results support the hypothesis that the Si–metal interaction in the tissue has a significant role in the reduction of metal toxicity mediated by Si, corroborating the involvement of Si in the tolerance of cellular mechanisms to heavy metals.

The coprecipitation of Si–metal in the cell wall has been identified as an important mechanism of heavy metal toxicity alleviation mediated by Si because it prevents concentrations of free metal ions at sites of high sensibility to its toxicity, such as in the interior of the mesophyll (Ma et al., 2015; Vollenweider et al., 2006). Although Si–metal deposits can be detected in the cytoplasm, Neumann and Nieden (2001) showed that these deposits are unstable and transitory, being more common in

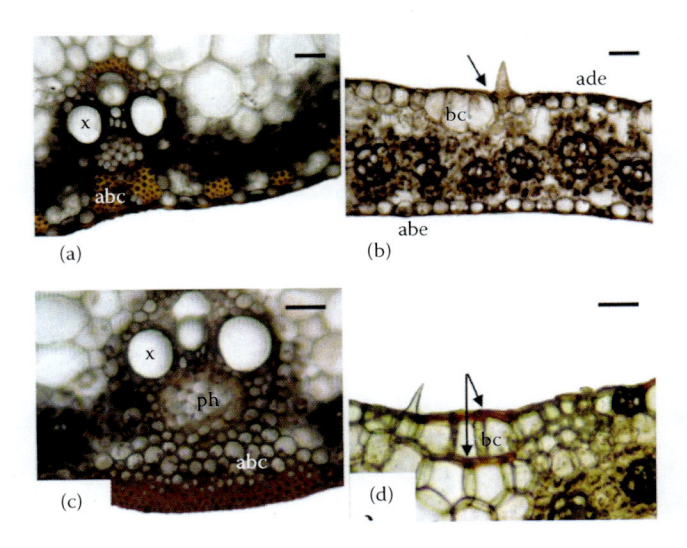

FIGURE 5.3 Locations of silicon, cadmium, and zinc in maize leaves cultivated in soil contaminated with 10 or 100 mg kg^{-1} cadmium or zinc, respectively, after silicon application. Silicon apoplastic deposits in the collenchyma (a), epidermis (b), and in detail on the surface of bulliform cells (arrow, d). Cadmium and zinc in abaxial collenchyma (c) after silicon application. Cadmium and zinc deposits in the internal and external cell walls of bulliform cells (arrow, b). abc, abaxial collenchyma; abe, abaxial epidermis; ade, adaxial epidermis; bc, bulliform cells; ph, phloem; x, xylem. Scale bars: 50 μm. (Adapted from Cunha, K.P.V., and Nascimento, C.W.A., *Water Air Soil Pollut.*, 197: 323–330, 2009. With permission.)

the apoplasm. These authors report that these silicates are slowly degraded so that the resulting SiO_2 is detectable as a thin layer over the tonoplast, while heavy metals are sequestered by organic acids in the vacuoles. In recent research on Si and relief Zn toxicity in rice, the leaf chloroplast structure was disordered under high-Zn stress, including uneven swelling, disintegrated and missing thylakoid membranes, and decreased starch granule size and number; however, all were counteracted by the addition of 1.5 mM Si.

The internal mechanisms by which Si increases the tolerance of plants to heavy metals are contradictory. While some researchers have observed that Si increased tolerance but reduced the absorption of metals (Horiguchi and Morita, 1987; Guo et al., 2005; Nwugo and Huerta, 2008; Keller et al., 2015), other researchers demonstrated that Si promoted greater production of biomass without reducing the absorption and accumulation of metals in the aerial part of the plant (Horst and Marschner, 1978; Horiguchi, 1988; Iwasaki et al., 2002a,b; Rogalla and Romheld, 2002; Maksimovic et al., 2007; Cunha and Nascimento, 2009; Cunha et al., 2008b). It is worth mentioning, however, that the increase in growth of plants treated with Si can cause a dilution effect in the concentration of metals (Shi et al., 2005b). This effect was also reported by Cunha and Nascimento (2009), who observed the linear reduction of Zn content in the aerial part and the root of maize plants (Figure 5.4) without the reduction in metal content after Si addition.

The reduction of metal concentration in the shoots of plants as an effect of Si has been attributed to the retention of metals in the roots, as a result of the deposition of metal silicates in these tissues (Shi et al., 2005b; Treder and Cieslinski, 2005; Zhang et al., 2008; Kim et al., 2014; Keller et al., 2015). As an example of this effect, Shi et al. (2005b) reported that 85% of total Cd in rice plants was retained in the roots after Si addition. Blocking the apoplastic flow by the deposition of SiO_2 or metal silicates in the cell wall of roots, in addition to the thickening of the endodermis, can explain the metal retention in the roots.

Cocker et al. (1998) proposed a model to explain the alleviation of Al phytotoxicity mediated by Si through the formation of aluminosilicates (ASs) and hydroxyaluminosilicates (HASs) in the apoplasm. According to these authors, the formation of these silicates occurs both by the direct interaction between Si and Al with the cell wall of apical roots and by the exudation of organic acid or H^+ by the cells induced by Si. This model is not restricted to Al because Si can similarly interact with other metal ions in the root apoplasm (Shi et al., 2005b; Zhang et al., 2008; Cunha and Nascimento, 2009; Collin et al., 2014; Keller et al., 2015). However, this model only explains detoxification in the external compartment, not accounting for the interaction between Si and metal in the cytoplasm and the vacuoles. The formation of Cd and Zn silicates in the intracellular environment was suggested by Cunha and Nascimento (2009), who found Si and metals in the interior of the endodermis and pericycle cells after the addition of high doses of Si (150–200 mg kg^{-1} Si). This mechanism has been confirmed in bamboos, since the high Si accumulation in the endodermis may influence transpiration bypass flow through the roots, but Cu concentrations in aboveground parts were not significantly affected by Si supplementation (Collin et al., 2014).

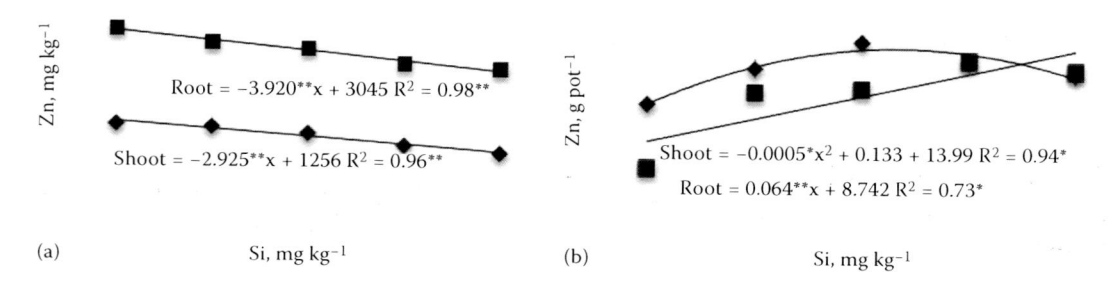

(a)   Si, mg kg^{-1} (b)   Si, mg kg^{-1}

Root = −3.920**x + 3045 R^2 = 0.98**

Shoot = −2.925**x + 1256 R^2 = 0.96**

Shoot = −0.0005*x² + 0.133 + 13.99 R^2 = 0.94*

Root = 0.064**x + 8.742 R^2 = 0.73*

FIGURE 5.4 Relationship between silicon doses in the soil and levels (a) and content (b) of zinc in the shoots and roots of maize plants cultivated in soil contaminated with 100 mg kg^{-1} zinc. (Adapted from Cunha, K.P.V., and Nascimento, C.W.A., *Water Air Soil Pollut.*, 197: 323–330, 2009. With permission.)

In addition to the deposition of SiO_2 or metal silicates in the root apoplasm, some authors reported that a deposition of SiO_2 on the surface of the leaf epidermis also significantly contributes to the reduction of absorption and translocation of metals from the roots to the aerial part of the plant. With the reduction of the transpiration stream, the metal absorption by mass flow is also reduced. In this context, Zhang et al. (2008) reported a reduction of 30%–50% in Cd concentration in the aerial part of rice plants after the addition of Si doses in a nutrient solution.

The deposition of SiO_2 on the surface of the leaf epidermis has been associated with a reduction of water loss by transpiration. In this context, Nwugo and Huerta (2008) found that after Si addition, rice plants exposed to Cd toxicity showed a significant reduction in the rate of stomatal conductance. However, this was without a reduction of CO_2 assimilation, which suggests an increase in the efficiency of water usage by the plants. These plants also showed a greater efficiency in the use of luminous energy, maintaining their photosynthetic activities. Recent studies have shown significant increases in photosynthesis induced by Si in tomato and sorghum plants exposed to salt and water stress, respectively (Al-Aghabary et al. 2004; Hattori et al., 2005). According to Gong et al. (2005), the increase in photosynthetic activity mediated by Si in plants under stress is also associated with the increased activity of the photosynthetic enzymes ribulose–bisphosphate carboxylase and glyceraldehyde-3-phosphate dehydrogenase, increased photosynthetic pigments, and increased assimilation of CO_2. Nwugo and Huerta (2008) report that the effects of Si in the photosynthetic parameters were minimal in healthy plants. They also stress that Si seems to play a significant role in the improvement of these parameters only under stress conditions.

Another important aspect in the alleviation of heavy metal toxicity mediated by Si is its involvement in metabolism and in the utilization of polyphenol compounds by plants submitted to biotic and abiotic stresses. Si increases the production of polyphenols that act as pathogens inhibiting substances in plants that are under biotic stress (Epstein, 1999; Savant et al., 1999). However, this effect of Si in plants exposed to heavy metal toxicity is very complex. In heavy metal toxicity, Si can act by either stimulating or inhibiting the production of polyphenols. According to Maksimovic et al. (2007), the concentrations of phenolic compounds in cucumber plants cultivated with higher concentrations of Mn were reduced after Si addition, suggesting the formation of Si-polyphenol compounds in the cells.

Induction of the polyphenol metabolism by plants has been attributed to the cellular oxidative imbalance resulting from biotic and abiotic stresses (Schützendübel et al., 2001; Chaoui and Ferjani, 2005). Soluble polyphenols increased in wheat leaves as a result of nickel toxicity (Pandolfini et al., 1992) and in cucumber exposed to Cd (Dudjak et al., 2004). The accumulation of insoluble polyphenols, such as lignin, has also been observed after treatment with Cd and Cu in pea leaves (Chaoui and Ferjani, 2005); in pepper leaves as a response to Cu stress (Díaz et al., 2001); and in maize exposed to Cd toxicity (Cunha et al., 2009). According to Vollenweider et al. (2006), the induction of the lignin metabolism seems to be a strategy in plants to increase the bonding sites in the cell wall that are capable of complexing metals and reducing oxidative damage in the cells because polyphenols may present antioxidative activity.

In plants treated with Si, the lignification of cell walls seems to be significantly reduced (Cunha and Nascimento, 2009). According to Boudet (2000), lignin synthesis represents a biosynthesis process of high cellular energy demand, while the incorporation of SiO_2 in the cell wall has a much lower energy cost. Therefore, there is an energetic advantage to the use of SiO_2 by plants as a structural component in cell walls. Consequently, Hossain et al. (2002) defend the assertion that the promotion of plant growth induced by Si in Poaceae leaves is a consequence of the qualitative metabolic changes of the cell wall.

After Si treatment, the production of soluble polyphenols appears to be stimulated, while insoluble polyphenols are inhibited. Kidd et al. (2001) reported a greater production of soluble polyphenols (catechol, catechin, quercetin, and curcumin) after Si application as a strategy to increase the complexation of Al and therefore detoxify cells, which increases the resistance to Al in a maize cultivar.

Inanaga and Okasaka (1995) and Inanaga et al. (1995) reported that, in addition to being deposited as silica gel (inert) in the cell wall, Si is involved in Si–phenol–carbohydrate and Si–lignin–carbohydrate type complexes, as well as stabilizes the bonding of pectins-proteins to the middle lamella, similarly to Ca. This involvement of Si in these bonds may justify the alterations observed in the metabolism of polyphenols.

References to the involvement of Si in cellular oxidative stress have also been made in the literature. It has been reported that Si application decreases the peroxidation of lipids by reducing the concentration of H_2O_2 in cells, in addition to stimulating the activity of the antioxidant defense system (Gong et al., 2005; Shi et al., 2005a; Gunes et al., 2007). In fact, under stress conditions, maintaining or increasing the activity of antioxidant enzymes and nonenzymatic components of the cellular antioxidant system is important for the toleration of heavy metal stress.

While it is not completely clear how plants respond to the presence of metals in cells, there is growing evidence that the phytotoxicity of metals can be attributed, at least in part, to the cellular oxidative stress that damages various cell components, especially the membrane lipids (Chaffai et al., 2007).

In the antioxidant system, the first line of defense against free radicals is the superoxide dismutase (SOD) enzyme, which is crucial for removing the superoxide anion $\left(O_2^-\right)$ from the compartments in which it is formed (Shi et al., 2005a). SOD activity generally decreases significantly in tissues exposed to environmental stresses. In the presence of Si, SOD activity increases and the concentration of malondialdehyde (MDA) significantly decreases (Gunes et al., 2007; Wang et al., 2015). MDA, a product of the polyunsaturated fatty acid peroxidation of the membrane, has been considered a biomarker for lipid peroxidation of the cell membranes under stress conditions (Shi et al., 2005a; Gunes et al., 2007). The reduction of MDA concentrations indicates less lipid peroxidation, which contributes to the integrity and functionality of the cell membranes and is considered an assuaging effect from the oxidative stress mediated by Si (Gunes et al., 2007; Liang et al., 2007). According to Shi et al. (2005a), Si promotes significant increases in SOD activity under normal conditions (10 μmol L^{-1} Mn), as well as in plants exposed to toxic levels of Mn in solution (600 μmol L^{-1} Mn). In contrast to SOD, the guaiacol peroxidase (GPX) activity increased under normal conditions after the addition of Si, but GPX activity reduced to toxic levels after addition of Mn in the nutrient solution (Shi et al., 2005a). Similar results were found by Zhu et al. (2004) and Liang et al. (2007).

The removal of O_2^- by SOD is accompanied by the production of H_2O_2, which is a highly toxic molecule that quickly spreads through the membranes. In plants, the peroxidase enzymes have the important function of decomposing H_2O_2, which is produced during cell metabolism, into water and molecular oxygen. However, under conditions of stress by heavy metals, the function of the peroxidase is more complex. In the toxicity of Mn, the GPX is responsible for catalyzing the production of polyphenols in locations of Mn accumulation in the tissue. These polyphenol accumulations are responsible for necrotic stains on the leaf tissue. In other cases, the peroxidases are involved in the lignification of cell walls, increasing the bonding of metals to their structure by promoting new bonding sites. These enzymes supply necessary H_2O_2 for the polymerization of the lignin precursors, favoring their fusion with proteins and carbohydrates from the cell wall and also reducing the level of H_2O_2 in the cell (Aquino-Bolaños and Mercado-Silva, 2004).

In plants submitted to Mn toxicity, it is common to find high concentrations of GPX, an antioxidant enzyme related to the production of polyphenols; H_2O_2; phenols; and accumulations of oxidated Mn (Iwasaki et al., 2002a; Shi et al., 2005a). Horiguchi and Morita (1987) stated that although the concentration of Mn remained unaltered in the tissue, peroxidase activity was reduced in the leaves after addition of Si. According to Iwasaki et al. (2002a), the increase in soluble Si concentrations in the apoplasm of cowpea plants provided significant reductions in the GPX enzyme activity ($r = -0.72**$), which is involved in the oxidation of polyphenols and Mn. These authors concluded that, in the case of Mn toxicity, Si is maintained in the reduced state of the apoplasm, preventing the oxidation of phenolic precursors and Mn. This significantly reduces the expression of symptoms from the toxicity of this metal (Iwasaki et al., 2002a). Additionally, Maksimovic et al. (2007) stated that the Si–polyphenol interaction in the tissue prevents the oxidation of these compounds, avoiding the development of Mn toxicity symptoms.

5.3 DETECTION OF METALS AND Si IN THE CELLS

Based on the results presented in this chapter, the controversies surrounding the effects of Si in the soil, as well as in the plant and its interactions with metals in the plants, may result from differences in the concentrations used, the exposure time, and plant species, or from the detection methods used.

The interpretation of data on the location of Si or metals in cells remains quite difficult. Conflicting results are obtained even within the same species and using the same method of detection. An example of this can be cited for rice plants exposed to Cd toxicity. Using EDX analysis, Wang et al. (2000) encountered a greater bonding of Cd to the cell wall after Si treatment, whereas Shi et al. (2005b) concluded that there were no alterations in the distribution of Cd between the symplast and the apoplast after the addition of Si.

There are different methods for determining the location and distribution of heavy metals and Si in plant tissues, such as histochemistry, cell fractionation, EDX analysis, and elemental distribution analysis by micro-PIXE (particle-induced X-ray emission) or the nuclear microprobe technique (NMP). According to Wójcik and Tukiendorf (2005), histochemical tests are highly sensitive and allow for the detection of metal even in small concentrations using light and electronic microscopes. However, the reactions may not be specific enough for a determined metal. Therefore, it is impossible to distinguish between deposits of different metals, making it an infeasible technique for the identification of different metal deposits. On the other hand, x-ray analyses, though not always useful for low concentrations of metals (Shi et al., 2005b; Liang et al., 2007), provide more specific signs and are therefore becoming more common.

Liang et al. (2007), in their review of the effects of Si in the alleviation of heavy metal toxicity, stated that they were not able to detect Cd in maize roots containing concentrations of metal as high as 420 mg kg^{-1} using EDX. Yet according to these authors, in the study conducted by Shi et al. (2005b), the EDX analysis could not have been used to identify Cd accumulation sites in the cells after Si treatment because the small concentrations of Cd found in the roots remained below the detection limit of the EDX technique. This possibly resulted in the absence of alterations in the distribution of Cd between the apoplast and symplast. Of course, studies of greater depth are needed to reexamine the microdistribution of metals interacting with Si in tissue using more sensitive methods, such as x-ray microfluorescence (SRXFA) (synchrotron radiation) (Liang et al., 2007). Using electron energy loss spectroscopy (EELS), electron spectroscopic imaging (ESI), and EDX, Neumann and Nieden (2001) safely showed the coprecipitation of Si and Zn in the tissues of *Cardaminopsis halleri*, inferring a concentration gradient for these precipitates between the cell wall, cytoplasm, and vacuoles. Combining the techniques of freeze substitution and high-pressure freezing, Ribeiro et al. (2001) revealed a reduction in the risk of element redistribution in fixed tissue for EDX analysis. This option was successfully used by Belleghem et al. (2007).

According to Belleghem et al. (2007), the most probable source of irregularities found among the results in which EDX is used is the combination of an exposure to elevated concentrations and inadequate preparation of samples. Elevated metal concentrations are often applied to plants as a method of compensating for elements lost during the chemical fixation stage. Although this exposure to elevated dosages is occasionally necessary to allow for detection of intercellular metals through methods such as EDX, they have highly destructive effects on the structure of the cells, preventing the examination of Cd distribution inside the cells (Wójcik and Tukiendorf, 2005). In addition, exposure to elevated dosages of metals is irrelevant, both from the physiological viewpoint and from the ecological viewpoint (Nwugo and Huerta, 2008). For Cd, for example, solution concentrations varying from 0.32 to 1 µmol L^{-1} represent moderate contamination, while concentrations of >5 µmol L^{-1} are considered to be highly contaminated.

5.4 FINAL CONSIDERATIONS

The Si–metal interactions in the soil are complex and deserve more attention from researchers. The effects of Si in soil have remained in the background, while the role of Si in the physiology of plants is being studied more intensely. The importance of studies on the effects of Si in crops includes the possibility of integrating the effects of Si in soil reactions and physiological reactions mediated by Si in plants exposed to the stress of heavy metals. These effects may have a significant impact on the phytostabilization of metals in soils and may increase the phytoextraction process through a greater accumulation of metals in the aerial part of the plant due to the alleviation of stress. Studies on these aspects are still basic, but they show promise due to their environmental importance and the potential commercial implementation of new technologies.

REFERENCES

Accioly, A.M.A., Soares, C.R.F.S., and Siqueira, J.O. 2009. Silicato de cálcio como amenizante da toxidez de metais pesados em mudas de eucalipto. *Pesquisa Agropecuária Bras.*, 44: 1–9.

Al-Aghabary, K., Zhu, Z., and Shi, Q. 2004. Influence of silicon supply on chlorophyll content, chlorophyll fluorescence, and antioxidative enzyme activities in tomato plants under salt stress. *J. Plant. Physiol.*, 27(12): 2101–2115.

Aquino-Bolaños, E.N., and Mercado-Silva, E. 2004. Effects of polyphenol oxidase and peroxidase activity, phenolics and lignin content on the browning of cut jicama. *Postharvest Biol. Technol.*, 33: 275–283.

Araújo, J.C.T. 2009. Efeito do silicato de cálcio na fitoextração induzida em solo contaminado por chumbo. Tese de Doutorado, Universidade Federal Rural de Pernambuco, Recife.

Baylis, A.D., Gragopoulou, C., Davidson, K.J., and Birchall, J.D. 1994. Effects of silicon on the toxicity of aluminum to soybean. *Commun. Soil Sci. Plant Anal.*, 25: 537–546.

Belleghem, F.V., Cuypers, A., Semane, B., Smeets, K., Angronsveld, J., D'haen, J., and Valcke, R. 2007. Subcellular localization of cadmium in roots and leaves of *Arabidopsis thaliana*. *New Phytol.*, 173: 495–508.

Boudet, A.M. 2000. Lignins and lignification: Selected issues. *Plant Physiol. Biochem.*, 38(1/2): 81–96.

Chaffai, R., Elhammadi, M.A., Seybou, N.T., Tekitek, A., Marzouk, B., and Ferjani, E.E. 2007. Altered fatty acid profile of polar lipids in maize seedlings in response to excess copper. *J. Agron. C Sci.*, 193: 207–217.

Chaoui, A., and Ferjani, E.E. 2005. Effects of cadmium and copper on antioxidant capacities, lignification and auxin degradation in leaves of pea (*Pisum sativum* L.) seedlings. *C. R. Biol.*, 328: 23–31.

Cocker, K.M., Evans, D.E., and Hodson, M.J. 1998. The amelioration of aluminium toxicity by silicon in higher plants: Solution chemistry or an in planta mechanism? *Physiol. Plant.*, 104: 608–614.

Collin, B., Doelsch, E., Keller, C., Cazevieille, P., Tella, M., Chaurand, P., Panfili, F., Hazemann, J.L., and Meunier, J.D. 2014. Copper distribution and speciation in bamboo exposed to a high Cu concentration and Si supplementation. First evidence on the presence of reduced copper bound to sulfur compounds in Poaceae. *Environ Pollut*, 187: 22–30.

Cunha, K.P.V., and Nascimento, C.W.A. 2009. Silicon effects on metal tolerance and structural changes in maize (*Zea mays* L.) grown on a cadmium and zinc enriched soil. *Water Air Soil Pollut.*, 197: 323–330.

Cunha, K.P.V., Nascimento, C.W.A., Pimentel, R.M.M., and Ferreira, C.P. 2008a. Cellular localization of cadmium and structural changes in maize plants grown on a cadmium contaminated soil with and without liming. *J. Hazard. Mater.*, 160: 228–234.

Cunha, K.P.V., Nascimento, C.W.A., and Silva, A.J. 2008b. Silicon alleviates the toxicity of cadmium and zinc for maize (*Zea mays* L.) grown on a contaminated soil. *J. Plant Nutr. Soil Sci.*, 171: 849–853.

Currie, H.A., and Perry, C. 2007. Silica in plants: Biological, biochemical and chemical studies. *Ann. Bot.*, 1–7.

Díaz, J., Bernal, A., Pomar, F., and Merino, F. 2001. Induction of shikimate dehydrogenase and peroxidase in pepper (*Capsicum annuum* L.) seedlings in response to copper stress and its relation to lignification. *Plant Sci.*, 161: 179–188.

Dietzel, M. 2000. Dissolution of silicates and the stability of polysilicic acid. *Geochim. Cosmochim. Acta*, 64(19): 3275–3281.

Dudjak, J., Lachman, J., Miholová, D., Kolihová, D., and Pivec, V. 2004. Effect of cadmium on polyphenol content in young barley plants (*Hordeum vulgare* L.). *Plant Soil Environ.*, 50: 471–477.

Epstein, E. 1994. The anomaly of silicon in plant biology. *Proc. Natl. Acad. Sci. U.S.A.*, 91: 11–17.

Epstein, E. 1999. Silicon. *Ann. Rev. Plant Physiol. Plant Mol. Biol.*, 50: 641–664.

Epstein, E. 2009. Silicon: Its manifold roles in plants. *Ann. Appl. Biol.*, 155: 1–6.

Freitas, E.V.S., and Nascimento, C.W.A. 2009. The use of NTA for lead phytoextraction from soil from a battery recycling site. *J. Hazard. Mater.*, 171: 849–853.

Freitas, E.V.S., Nascimento, C.W.A., Biondi, C.M., Silva, J.P.S., and Souza, A.P. 2009. Dessorção e lixiviação de chumbo em espodossolo tratado com agentes quelantes. *Rev. Bras. Cienc. Solo*, 33: 517–525.

Galvez, L., Clark, R.B., Gourley, L.M., and Maranville, J.W. 1987. Silicon interactions with manganese and aluminum toxicity in sorghum. *J. Plant Nutr.*, 10: 1139–1147.

Gong, H.J., Zhu, X.Y., Chen, K.M., Wang, S.M., and Zhang, C.L. 2005. Silicon alleviates oxidative damage of wheat plants in pots under drought. *Plant Sci.*, 169: 313–321.

Gunes, A., Inal, A., Bagci, E.G., Coban, S., and Sahin, O. 2007. Silicon increases boron tolerance and reduces oxidative damage of wheat grown in soil with excess boron. *Biol. Plant.*, 51: 571–574.

Guo, W., Hou, Y.L., Wang, S.G., and Zhu, Y.G. 2005. Effect of silicate on the growth and arsenate uptake by rice (*Oryza sativa* L.) seedlings in solution culture. *Plant Soil*, 72: 173–181.

Hattori, T., Inanaga, S., Araki, H., An, P., Morita, S., Luxová, M., and Lux, A. 2005. Application of silicon enhanced drought tolerance in *Sorghum bicolor*. *Physiol Plant.*, 123(4): 459–466.

Hodson, M.J., and Sangster, A.G. 1993. The interaction between silicon and aluminium in *Sorghum bicolor* (L.) Moench: Growth analysis and x-ray microanalysis. *Ann. Bot.*, 72: 389–400.

Horiguchi, T. 1988. Mechanism of manganese toxicity and tolerance of plants. IV. Effects of silicon on alleviation of manganese toxicity of rice plants. *Soil Sci. Plant Nutr.*, 34(1): 65–73.

Horiguchi, T., and Morita, S. 1987. Mechanism of manganese toxicity and tolerance of plants. VI. Effect of silicon on alleviation of manganese toxicity of barley. *J. Plant Nutr.*, 10: 2299–2310.

Horst, W.J., Fecht, M., Naumann, A., Wissemeier, A.H., and Maier, P. 1999. Physiology of manganese toxicity and tolerance in *Vigna unguiculata* (L.) Walp. *J. Plant Nutr Soil Sci.*, 162: 263–274.

Horst, W.J., and Marschner, H. 1978. Effect of silicon on manganese tolerance of bean plants (*Phaseolus vulgaris* L.). *Plant Soil*, 50: 287–303.

Hossain, M.T., Mori, R., Wakabayashi, K.S.K., Fujii, S.K.S., Yamamoto, R., and Hoson, T. 2002. Growth promotion and an increase in cell wall extensibility by silicon in rice and some other Poaceae seedlings. *J. Plant Res.*, 115: 23–27.

Inanaga, S., and Okasaka, A. 1995. Calcium and silicon binding compounds in cell walls of rice shoots. *Soil Sci. Plant Nutr.*, 41: 103–110.

Inanaga, S., Okasaka, A., and Tanaka, S. 1995. Does silicon exist in association with organic compounds in rice plants? *Soil Sci. Plant Nutr.*, 41: 111–117.

Iwasaki, K., Maier, P., Fecht, M., and Horst, W.J. 2002a. Leaf apoplastic silicon enhances manganese tolerance of cowpea (*Vigna unguiculata*). *J. Plant Physiol.*, 159: 167–173.

Iwasaki, K., Maier, P., Fecht, M., and Horst, W.J. 2002b. Effects of silicon supply on apoplastic manganese concentrations in leaves and their relation to manganese tolerance in cowpea (*Vigna unguiculata* L. Walp.). *Plant Soil*, 238: 281–288.

Iwasaki, K., and Matsumura, A. 1999. Effect of silicon on alleviation of manganese toxicity in pumpkin (*Cucurbita moschata* Duch cv. Shintosa). *Soil Sci. Plant Nutr.*, 45: 909–920.

Keller, C., Rizwan, M., Davidian, J.C., Pokrovsky, O.S., Bovet, N., Chaurand, P., and Meunier, J.D. 2015. Effect of silicon on wheat seedlings (*Triticum turgidum* L.) grown in hydroponics and exposed to 0 to 30 μM Cu. *Planta*, 241: 847–860.

Kidd, P.S., Llugany, M., Poschenrieder, C., Gunse, B., and Barcelo, J. 2001. The role of root exudates in aluminium resistance and silicon-induced amelioration of aluminium toxicity in three varieties of maize (*Zea mays* L.). *J. Exp. Bot.*, 52: 1339–1352.

Kim, Y.-H., Khan, A.L., Kim, D.-H., Lee, S.-Y., Kim, K.-M., Waqas, M., Jung, H.-Y., Shin, J.-H., Kim, J.-G., and Lee, I.-J. 2014. Silicon mitigates heavy metal stress by regulating P-type heavy metal ATPases, *Oryza sativa* low silicon genes, and endogenous phytohormones. *BMC Plant Biology*, 14: 13. doi: 10.1186/1471-2229-14-13.

Liang, Y., Sun, W., Zhu, Y.-G., and Christie, P. 2007. Mechanisms of silicon-mediated alleviation of abiotic stresses in higher plants: A review. *Environ Pollut.*, 147(2): 422–428.

Liang, Y., Wong, J.W.C., and Wei, L. 2005. Silicon-mediated enhancement of cadmium tolerance in maize (*Zea mays* L.) grown in cadmium contaminated soil. *Chemosphere*, 58: 475–483.

Lichtenberger, O., and Neumann, D. 1996. A study Si-L and O-K ELNES in plant material, SiO_2, Ca- and Zn-silicate in *Minuartia*. *J. Microsc.*, 183: 45–52.

Lima-Filho, O.F., Lima, M.T.G., and Tsai, S.M. 1999. O silício na agricultura. *Inform. Agron. Piracicaba*, 87: 1–7.

Lindsay, W.L. 1979. *Chemical Equilibria in Soils*. New York: Wiley-Interscience.

Ma, J., Cai, H., He, C., Zhang, W., and Wang, L. 2015. A hemicellulose-bound form of silicon inhibits cadmium ion uptake in rice (*Oryza sativa*) cells. *New Phytol*, 206: 1063–1074. doi:10.1111/nph.13276.

Ma, J.F., and Takahashi, E. 2002. *Soil, Fertilizer, and Plant Silicon Research in Japan*. Elsevier Science, Amsterdam.

Ma, J.F., and Yamaji, N. 2006. Silicon uptake and accumulation in higher plants. *Trends Plant Sci.*, 11: 392–397.

Maksimovic, J.D., Bogdanovic, J., Maksimovic, V., and Nikolic, M. 2007. Silicon modulates the metabolism and utilization of phenolic compounds in cucumber (*Cucumis sativus* L.) grown at excess manganese. *J. Plant Nutr. Soil Sci.*, 170(6): 739–744.

Matichenkov, V.V., and Calvert, D.V. 2002. Silicon as a beneficial element for sugarcane. *J. Am. Soc. Sugarcane Technol.*, 22: 21–30.

Melo, E.E.C., Nascimento, C.W.A., and Santos, A.C.Q. 2006. Solubilidade, fracionamento e fitoextração de metais pesados após aplicação de agentes quelantes. *Rev. Bras. Cienc. Solo*, 30: 1051–1060.

Miyake, Y., and Takahashi, E. 1985. Effect of silicon on the growth of soybean plants in a solution culture. *Soil Sci. Plant Nutr.*, 31: 625–636.

Nascimento, C.W.A., and Xing, B. 2006. Phytoextraction: A review on enhanced metal availability and plant accumulation. *Sci. Agric.*, 63(3): 299–311.

Neumann, D., and Nieden, U.Z. 2001. Silicon and heavy metal tolerance of higher plant. *Phytochemistry*, 56: 685–692.

Neumann, D., Zur Neiden, U., Lichenberger, O., Leopold, I., and Schwieger, W. 1997. Heavy metal tolerance of *Minuartia verna. J. Plant Physiol.*, 151: 101–108.

Nwugo, C.C., and Huerta, A.J. 2008. Effects of silicon nutrition on cadmium uptake, growth and photosynthesis of rice plants exposed to low-level cadmium. *Plant Soil*, 311: 73–86.

Paim, L.A., Carvalho, R., Abreu, C.M.P., and Guerreiro, M.C. 2006. Estudo dos efeitos do silício e do fósforo na redução da disponibilidade de metais pesados em área de mineração. *Quim. Nova*, 29(1): 28–33.

Pandolfini, T., Gabbrielli, R., and Comparini, C. 1992. Nickel toxicity and peroxidase activity in seedlings of *Triticum aestivum* L. *Plant Cell Environ.*, 15: 719–725.

Ribeiro, K.C., Benchimol, M., and Farina, M. 2001. Contribution of cryofixation and freeze-substitution to analytical microscopy: A study of *Tritrichomonas foetus* hydrogenosomes. *Microsc. Res. Tech.*, 53: 87–92.

Rogalla, H., and Romheld, V. 2002. Role of leaf apoplast in silicon-mediated manganese tolerance of *Cucumic sativus* L. *Plant Cell Environ.*, 25: 549–555.

Sahebi M., Hanafi M.M., Abdullah S.N.A., Rafii M.Y., Azizi P., Nejat N. et al. 2014. Isolation and expression analysis of novel silicon absorption gene from roots of mangrove (*Rhizophora apiculata*) via suppression subtractive hybridization. *Biomed Res Int.*, 2014: 971985 10.1155/2014/971985.

Savant, N.K., Korndörfer, G.H., Datnoff, L.E., and Snyder, G.H. 1999. Silicon nutrition and sugarcane production: A review. *J. Plant Nutr.*, 22: 1853–1903.

Schützendübel, A., Schwanz, P., Teichmann, T., Gross, K., Langenfeld-Heyser, R., Godbold, O.L., and Polle, A. 2001. Cadmium-induced changes in antioxidative systems, H_2O_2 content and differentiation in pine (*Pinus sylvestris*) roots. *Plant Physiol.*, 127: 887–892.

Shi, Q.H., Bao, Z.Y., Zhu, Z.J., He, Y., Qian, Q.Q., and Yu, J.Q. 2005a. Silicon-mediated alleviation of Mn toxicity in *Cucumis sativus* in relation to activities of superoxide dismutase and ascorbate peroxidase. *Phytochemistry*, 66: 1551–1559.

Shi, X.H., Zhang, C.C., Wang, H., and Zhang, F.S. 2005b. Effect of Si on the distribution of Cd in rice seedlings. *Plant Soil.*, 272: 53–60.

Sommer, M., Kaczorek, D., Kuzyakov, Y., and Breuer, J. 2006. Silicon pools and fluxes in soils and landscapes: A review. *J. Plant Nutr. Soil Sci.*, 169: 310–329.

Song, A., Li, P., Fan, F., Li, Z., and Liang, Y. 2014. The effect of silicon on photosynthesis and expression of its relevant genes in rice (*Oryza sativa* L.) under high-zinc stress. *PLoS ONE*, 9(11): e113782. doi:10.1371/journal.pone.0113782.

Treder, W., and Cieslinski, G. 2005. Effect of silicon application on cadmium uptake and distribution in strawberry plants grown on contaminated soils. *J. Plant Nutr.*, 28: 917–929.

Tripathi, D.K., Singh, V.P., Kumar, D., and Chauhan, D.K. 2012. Rice seedlings under cadmium stress: Effect of silicon on growth, cadmium uptake, oxidative stress, antioxidant capacity and root and leaf structures. *Chem Ecol.*, 28(3): 281–291.

Vollenweider, P., Cosio, C., Gunthardt-Goerg, M.S., and Keller, C. 2006. Localization and effects of cadmium in leaves of a cadmium-tolerant willow (*Salix viminalis* L.). Part II. Microlocalization and cellular effects of cadmium. *Environ. Exp. Bot.*, 58: 25–40.

Wang, L., Wang, W., Chen, Q., Cao, W., Li, M., and Zhang, F. 2000. Silicon-induced cadmium tolerance of rice seedlings. *J. Plant Nutr.*, 23: 1397–1406.

Wang, Y., Stass, A., and Horst, W.J. 2004. Apoplastic binding of aluminum is involved in silicon-induced amelioration of aluminum toxicity in maize. *Plant Physiol.*, 136: 3762–3770.

Wang, S., Liu, P., Chen, D., Yin, L., Li, H., and Deng, X. 2015. Silicon enhanced salt tolerance by improving the root water uptake, and decreasing the ion toxicity in cucumber. *Front. Plant Sci.*, 6: 759 10.3389 /fpls.2015.00759.

Williams, D.E., and Vlamis, J. 1957. The effect of silicon on yield and manganese-54 uptake and distribution in the leaves of barley plants grown in culture solutions. *Plant Physiol.*, 32: 404–409.

Wójcik, M., and Tukiendorf, A. 2005. Cadmium uptake, localization and detoxification in *Zea mays*. *Biol. Plant.*, 49: 237–245.

Wu, J.W., Shi, Y., Zhu, Y.X., Wang, Y.C., and Gong, H.J. 2013. Mechanisms of enhanced heavy metal tolerance in plants by silicon: A review. *Pedosphere*, 23: 815–825. 10.1016/S1002-0160(13)60073-9.

Zhang, C., Wang, L. Nie, Q., Zhang, W., and Zhang, F. 2008. Long-term effects of exogenous silicon on cadmium translocation and toxicity in rice (*Oryza sativa* L.). *Environ. Exp. Bot.*, 62: 300–307.

Zhu, Z.J., Wei, G.Q., Li, J., Qian, Q.Q., and Yu, J.Q. 2004. Silicon alleviates salt stress and increases antioxidant enzymes activity in leaves of salt-stressed cucumber (*Cucumis sativus* L.). *Plant Sci.*, 167: 527–533.

6 Silicon and Nanotechnology
Role in Agriculture and Future Perspectives

Anita Singh, Shikha Singh, and Sheo Mohan Prasad

CONTENTS

ABSTRACT

As silicon (Si) is one of the copiously available elements, it has several beneficial impacts on soil and plant systems. It makes plants resistant against diseases, insect attacks, and other biotic and abiotic stresses. It maintains soil fertility by improving physicochemical properties of the soil, which consequently increases the availability of nutrients to plants. Nowadays, technology based on nonmaterials is increasing day by day, which has led to the production of nano-Si. The application of nano-Si from an agricultural perspective has been given a lot of attention by researchers. It has greater surface area than the bulk form. Due to higher solubility and surface reactivity, nano-Si helps in greater absorption of nutrients by plants, and it is also applied for mitigating stress conditions. So, the application of nanosilicon in agriculture systems has provided new solutions to problems in plants and food science to enhance the quality of plant products.

6.1 INTRODUCTION

Silicon (Si) is present in the earth's crust in the form of silicate. After chemical and physical weathering, it is released in the soil and absorbed by plants. The concentration of Si varies with plant species (Hodson et al. 2005). It is found in higher concentrations in monocot plants than in dicot ones.

Si occurs mainly as monosilicic acid (H_4SiVO_4) in soil, and its concentrations varies from 0.1 to 0.6 mM. Plants can absorb Si only in this form (Epstein 1994; Ma and Takahashi 2002). Earlier, Si had not been characterized as a vital element because it is not involved in plant metabolism, which is one criterion for being an essential element (Arnon and Stout 1939), but Epstein and Bloom (2003) have redefined the characteristics required for being an essential element. An element is essential if it is part of an intrinsic component that is vital for plant metabolism, and in its deficiency, the plants show growth disorders and disturbances in reproduction and other developmental processes. Based on this, Si is considered an essential element because its deficiency results in the disturbance of plant growth and development.

Accumulation of Si shows variation in different plant species (Ma et al. 2001). In rice plants, Si can accumulate up to 10% on a dry weight basis in the shoot. Its higher accumulation in rice is responsible for healthy growth and stable production. Therefore, Si can be recognized as an "agronomically essential element," and it is applied in paddy fields in the form of silicate fertilizers. It also helps in the amelioration of various abiotic and biotic stresses, and it is able to increase plant resistance. Silicon is necessary for the proper growth of plants grown under stress conditions (Suriyaprabha et al. 2012). Application of silicon as a silicate fertilizer is one of the sustainable techniques for enhancing nutrient availability and the yield of crops. With the help of silicon fertilizers, plant resistance to disease can also be increased (Romero-Aranda et al. 2006). When silicon fertilizer is applied in soil, it increases the absorption of potassium and decreases sodium absorption. This results in enhancement of the ratio of potassium to sodium. Along with this, it also increases the absorption of nitrogen and sulfur, which improves plant nutrition. In the presence of stress conditions, Si can play an important role by increasing the stability of cell membranes (Romero-Aranda et al. 2006). The reason behind its beneficial aspect is its hydrophilic nature. Si maintains the deterioration of cell membranes in the presence of environmental stress (Azimi et al. 2014). It was reported that in comparison to bulk Si element as oxide salt, nanostructured SiO_2 showed the best results in plants affected by salt stress. Changbai larch (*Larix olgensis*) seedlings grown under 500 µl L^{-1} nanostructured silicon dioxide showed enhancement in plant height, diameter of root collar, and length and number of lateral roots compared to those grown in control and with bulk Si salt (Lin et al. 2004).

Application of Si as nanoparticles in agriculture is a novel approach, as they are more reactive than their bulk counterparts. In order to improve seedling growth and crop protection against disease, silica nanoparticles (SNPs) were applied (Bao-shan et al. 2004; Torney et al. 2007). Nano-Si materials have been given a lot of attention by agricultural researchers due to their tiny size and greater surface area, the physicochemical properties of which can change compared to those of the bulk materials. The nanoform of silicon showed a larger surface area and higher reactivity. Table 6.1 shows different studies related to the beneficial role of Si in bulk and nanoforms in plant systems.

In this context, this chapter deals with the role of silicon and its nanoform in agriculture. Improvement in the quality of crops and tolerance behavior to various stresses with the help of the nanoform of Si is discussed in detail. This chapter also focuses on the application of Si in agriculture systems, providing future perspectives.

6.2 ROLE OF Si IN PLANTS

6.2.1 DISTRIBUTION AND FUNCTIONING

Silicon (Si) acts as a macroelement and plays a very important role in plant functioning. It is the eighth most common element in nature (Sahebi et al. 2015). As discussed above, Si is taken up by the plants in the form of monosilicic acid, and its concentration varies by plant species (Guntzer et al. 2012). Dicotyledonous plants accumulate Si concentrations at the level of 0.1% of dry weight basis in their tissue. Dryland grasses like oats and rye have about a 1% accumulation tendency of

TABLE 6.1

Role of Silica in Bulk and Nanoforms in Agricultural Systems

Concentration/Size	Studied Plant	Effect	Reference
0 and 250 mg L^{-1} silica	*Oryza sativa* L. and *Oryza glaberrima*	Silicon application alleviated iron toxicity symptoms through reduction in iron concentration in leaf tissue	Dufey et al. 2014
4 tons ha^{-1} calcium silicates	*Saccharum officinarum*	Increased tolerance to shoot borer (*Diatraea saccharalis*)	Sidhu et al. 2013
Nutrient solution containing 0 or 2 mM silicon	*Oryza sativa*	Silicon nutrition increased grain yield and harvest index, stimulated photosynthetic rates via enhanced mesophyll conductance, and altered primary metabolism	Detmann et al. 2012
10 μM silica	*Oryza sativa*	Si application ameliorated chromium toxicity	Tripathi et al. 2011
0, 55, 110, and 165 kg ha^{-1} silicon in form of calcium-magnesium silicate	*Diatraea saccharalis*	Added Si increased the amounts of extractable Si in a Quartzapsament soil, as well as increasing the yield and Si uptake in stalks of cultivar SP 89 1115	Camargo et al. 2010
–	*Zea mays*	Silicon ameliorated manganese toxicity in Mn-sensitive and Mn-tolerant maize varieties	Doncheva et al. 2009
50, 100, and 150 mg L^{-1} NaSiO$_3$	*Zinnia elegans*	Si foliar application moderately increased the leaf resistance of zinnias grown under optimal greenhouse conditions	Kamenidou et al. 2009
5000 and 10,000 kg ha^{-1} as calcium silicate	*Saccharum officinarum*	Increased resistance to stalk borer (*E. saccharina*)	Keeping and Meyer 2003
15 metric tons ha^{-1} of silicates	*Saccharum officinarum*	Silica application increased leaf chlorophyll and decreased leaf freckling in plants	Elawad et al. 1982
Potassium silicate at 5 g per pot	*Oryza sativa*	Silica content was an important component of resistance; high silica content resulted in higher insect mortality and lower larval weight	Subbarao and Perraju 1976
–	*Oryza sativa*	Prevented attack by the larvae of the yellow rice borer, *Scirpophaga incertulas*	Panda et al. 1975
Silica nanoparticles (100, 200, and 300 ppm per 4200 m^2)	*Lycopersicon esculentum*	Silica nanoparticles significantly reduced the numbers of *Tuta absoluta* larvae compared with control (without any treatments)	El-Samahy et al. 2014
SiO$_2$ concentrations (0, 5, 20, 40, 60, and 80 mg L^{-1})	*Agropyron elongatum*	Applications of nanomaterial encouraged earlier plant germination, broke seed dormancy, and improved plant production	Azimi et al. 2014
12 nm with corresponding surface area of 200 m^2 g^{-1}	*Lycopersicum esculentum*	Lower concentrations of nanosilicon dioxide (nSiO$_2$; size 12 nm) improved seed germination of tomato	Siddiqui and Al-Whaibi 2014
Si and nano-Si at 4 concentrations each (0, 1, 2, and 3 mM)	*Vicia faba*	Silicon and nanosilicon protected *Vicia faba* plants against the hazardous effect of salinity	Abdul Qados and Moftah 2014
100, 150, 200, 250, 300, 350 ppm	*Lycopersicum esculentum*	Application of nanosilica minimized the problems caused by *Spodoptera littoralis* (Lepidoptera: Noctuidae)	El-Bendary and El-Helaly 2013
Amorphous nanosilica from rice husk (20–40 nm)	*Zea mays*	Properly promoted fungal resistance, especially against *Fusarium oxysporum* and *Aspergillus niger* in maize crop	Suriyaprabha et al. 2013

(Continued)

TABLE 6.1 (CONTINUED)

Role of Silica in Bulk and Nanoforms in Agricultural Systems

Concentration/Size	Studied Plant	Effect	Reference
1 and 2 mM nano-Si	*Lycopersicum esculentum*	1 mM N-Si showed great enhancement of germination characteristics, while 2 mM N-Si showed reduction of germination properties	Haghighi et al. 2012
Nano-SiO$_2$	*Zea mays*	Nano-SiO$_2$ promoted seed coat resistance and improved the nutritional availability to maize plants	Suriyaprabha et al. 2012
15 kg/ha L nano-SiO$_2$	*Zea mays*	Enhancement in seed germination, water use efficiency and chlorophyll content	Yuvakkumar et al. 2011
0, 200, 400, 800, and 1600 mg L^{-1} SiO$_2$ nanoparticles	*Scenedesmus obliquus*	Contents of chlorophyll decreased under moderate and high concentrations (50, 100, and 200 mg/L) of SiO$_2$ nanoparticles, while SiO$_2$ bulk particles (BPs) were found to be nontoxic up to 200 mg/L	Wei et al. 2010
2000, 1000, 500, 250, 125, and 62 μl L^{-1}	*Larix olgensis*	Exogenous application of nano-SiO$_2$ improved seedling growth and quality of seedlings	Bao-shan et al. 2004

their dry weight, and wetland grasses, such as paddy-grown rice, have a 5% or higher accumulation tendency. So, based on these data, it was reported that sugarcane (*Saccharum officinarum*), rice (*Oryza sativa*), and wheat (*Triticum aestivum*) absorb the largest amounts of Si, in the ranges (kg ha^{-1}) of 300–700, 150–300, and 50–150, respectively (Snyder et al. 2006). Plants belonging to the Poaceae family showed higher uptake of Si than other plant species (Jian et al. 2006). The dicotyledon plants, such as cucumbers (*Cucumis sativus*), melons, strawberries, and soybeans (*Glycine max* L. Merr), have a very low tendency to absorb Si from the growth medium (Mitani and Ma 2005). As it was discussed, Si is absorbed as silicic acid in the plants. It enters into plants just like water transport and reaches to transpiration termini.

Each plant differs in its ability to accumulate silica depending on its morphological properties. In plant systems, Si uptake occurs in three different ways: active, passive, or rejective mode. Rice plants have a low-affinity transporter *Lsi1*, which is involved in the Si uptake. This transporter is present on the lateral roots and helps in the uptake of Si in the form of silicic acid (Ma and Yamaji 2006). The genomic analysis of this particular transporter shows that it is mapped at chromosome 2 and has five exons and four introns. It is expressed constitutively and located mainly in the exo- and endodermal cells of roots. Its regulation is based on the availability of silicic acid. It was reported that when cDNA encoding *Lsi1* was injected into *Xenopus laevis* oocytes, it showed increased transportation of silicic acid (Ma and Yamaji 2006). In other plants (cucumber and tomato), studies were also done to investigate the uptake of silicic acid at different levels, that is, at moderate and low levels. In these species, transportation of silicic acid is also mediated by a transporter that shows the same affinity as *Lsi1*, but it has a different density on the lateral roots than *Lsi1*. Along with the higher plants, some photosynthetic organisms such as diatoms (Bacillariophyceae: Heterokontophyta) have silicon involvement. Their cell walls are silicified and provide protection from grazers (Hamm et al. 2003). After transportation of silicic acid, when water is evaporated from plants, it gets precipitated in the form of phytoliths. These silica deposits provide resistance to plants against parasites and grazers, and they also change the posture of the leaves (Datnoff et al. 2001). So overall, the beneficial effects of Si include an increase in growth and fruit yields and the provision of resistance against environmental stress. A mechanism to alleviate stress conditions through Si is discussed in Section 6.2.2.

6.2.2 MITIGATING STRESS CONDITIONS

Agricultural productivity is strongly affected by various stresses, such as heavy metals, drought, and salinity, and also by several plant diseases and pests. There are various reports that show Si's ability to ameliorate these stresses (Ma 2004; Lee and Luan 2012).

6.2.2.1 Mechanism to Mitigate Biotic Stress

Biotic stress includes various pathogens that are responsible for causing different diseases in plants. Si plays an important role in making plants resistant against these disease-causing pathogens. Bélanger et al. (2003) and Rodrigues et al. (2003) have shown that in the presence of Si, the production of glycosylated phenolics and antimicrobial products such as diterpenoid and phytoalexins was increased in wheat and rice plants against pests. It was also found that Si provides resistance against sheath blight diseases in rice plants (Zhang et al. 2006; Cai et al. 2008). The addition of silicon resulted in the hardening of the plant cuticle layer coating by SiO_2 polymerization; this is the only possible mechanism for reducing disease susceptibility by Si (Rodrigues et al. 2003). It also helps in the defense mechanisms by increasing the amounts of phenolic components and the activities of peroxidase, chitinases, polyphenoloxidase, β-1, 3-glucanases, and phenylalanine ammonialyase enzymes (Rodrigues et al. 2003). Treatment with Si resulted in decreases of leaf blast disease in rice cultivars (Seebold et al. 2000). It has also been applied in cucumber shoots to reduce the occurrence rate of powdery mildew disease caused by *Sphaerotheca fuliginea* (Liang et al. 2005). Si plays an important role against the fungi *Pythium aphanidermatum* and *Pythium ultimum* grown in cucumber roots. It also provides resistance against rice green leafhoppers (*Eurymela distinct*), mites (*Lorryia formosa*), and brown plant hoppers (*Nilaparvata lugens*) (Sahebi et al. 2015).

Plants infected with necrotizing pathogens showed enhancement in resistance against the attack of various pathogens; this is known as systemic acquired resistance (SAR). Silicon acts in a similar way in order to show defense responses against pathogens. When plants were treated with Si, they showed higher activity in the protective enzymes (chitinase, peroxidases, and polyphenoloxidases) in the leaves of rice, wheat, and cucumber (Yang et al. 2003; Liang et al. 2005; Cai et al. 2008). These enzymes help in the production and accumulation of antifungal compounds and pathogenesis-related proteins in plants. Application of Si has increased the content of lignin–carbohydrate complexes in rice epidermal cells, and it provides resistance against pathogen attack (Cai et al. 2009). At the molecular level, Si application also shows some changes in plants to develop resistance. Kauss et al. (2003) have shown enhancement in the gene expression responsible for encoding a novel proline-rich protein to provide SAR against pathogen attack in cucumber plant. This protein helps in the cell wall reinforcement at the site of fungal infection in epidermal cells. Si-treated plants react to pathogen inoculation by increasing their regulation of defense and pathogenesis-related genes. This confirms that silicon plays a very important role in increasing resistance against pathogen infections.

6.2.2.2 Mechanism to Mitigate Abiotic Stress

Plant functional activities are dependent on light, water, carbon, and mineral nutrients. Abiotic stress is basically a change in these environmental conditions due to accumulation of certain xenobiotics, like metals and pesticides. There are several strategies that regulate the alleviation of various abiotic stresses in plants. Among them, the application of Si is a safe strategy to mitigate abiotic stresses.

Alleviation of metal toxicity with the help of Si is widely accepted in plants. In various plants, such as barley, bean, pumpkin, cowpea, and cucumber, Si reduces Mn toxicity (Rogalla and Römheld

2002; Shi et al. 2005a; Liang et al. 2006a). It was found that Si-untreated cucumber plants showed higher Mn toxicity than Si-treated plants (Rogalla and Römheld 2002). As in Si-treated plants the Mn toxicity is reduced by its binding with cell wall as well as its accumulation in the symplast. So, its availability will be reduced and it becomes less toxic to the plant (Rogalla and Römheld 2002). Si played an important role in reducing the toxicity and uptake of cadmium (Cd) and Arsenic (As) in rice seedlings (Shi et al. 2005b). A pot experiment showed that addition of Si significantly reduced the toxic effects of Cd treatment in maize plants by increasing the yield. A higher concentration of Si (i.e., 400 mg kg^{-1} Si) in soil resulted in a high pH that consequently decreased Cd uptake in the shoots and roots of maize plants (Liang et al. 2005). In rice plants, it was also found that Cd and copper (Cu) toxicity can be reduced with the help of Si (Kim et al. 2014). Si treatment improved the root function by reducing Cd/Cu concentration, lipid peroxidation, and fatty acid desaturation in plant tissues. Due to reduction in the metal concentration, the mRNA expression of enzymes responsible for encoding metal transporters (*OsHMA2* and *OsHMA3*) in Si metal-treated rice plants is significantly downregulated and the expression of genes for Si transportation (*OsLSi1* and *OsLSi2*) is significantly upregulated (Kim et al. 2014).

Other abiotic stresses are also nullified with the help of Si treatment. Under salt stress, Si enhances dry biomass and the rate of transpiration. It helps in the reduction of salt stress by enhancing water absorption in plant tissues, which contributes to the dilution of solutes in plants (Romero-Aranda et al. 2006; Tuna et al. 2008). Silicon also changes the cell wall anatomy by depositing silica as polymerized silicon dioxide (SiO_2), called opaline phytoliths. It provides tolerance to plants against drought stress (Ma 2004).

In the presence of salt toxicity, Si treatment improves the growth of various crop species, like rice (*Oryza sativa* L.), wheat (*Triticum durum*), and tomato (*Lycopersicon esculentum* Mill.) (Romero-Aranda et al. 2006; Tuna et al. 2008). Yin et al. (2013) have reported that Si treatment reversed the salt-induced reduction in plant growth and photosynthesis. The level of osmolytes like sucrose and fructose was also increased significantly, and the Na$^+$ concentration was decreased in silicon-treated plants under salt stress. This proves that silicon has the potential to improve salt tolerance. The Na$^+$ concentration was reduced by adding Si. It helps in the blockage of the transpiration bypass flow and consequently leads to reduction in the rate of transpiration. In the presence of Si, barley plants grown under salt stress showed stimulation in *adenosine triphosphatase* (H$^+$-ATPase) and *H$^+$ pyrophosphatase* (H$^+$-PPase) activity of the root plasma membrane. It enhances the K$^+$ absorption and reduces the Na$^+$ transportation from roots to shoots (Liang and Ding 2002; Liang et al. 2006b). Si also changes the basic structure and metabolic activity of plasma membranes by affecting the stress-dependent peroxidation of membrane lipids (Al-Aghabary et al. 2004; Zhu et al. 2004). So, a significant number of studies have proved that silicon has a role in providing resistance against biotic and abiotic stresses, as shown in Figure 6.1.

Along with silicon as bulk material, its application as a nanoparticle has also been given a lot of attention by scientists. Due to the greater surface area, its solubility and reactivity tend to be higher than those of bulk. It was reported that nanoparticles of SiO_2 have a useful impact on plants. It showed an ameliorative effect on salt stress in seedlings of tomato (Haghighi et al. 2012). Kalteh et al. (2014) have also found that application of silicon nanoparticles reduced salinity stress by increasing the growth, chlorophyll, and proline content in *Ocimum basilicum*. Compared to bulk silicon, nanosilica is less toxic to bacterial colonies present in soil. It has been found to be very useful in maintaining the growth and other development of plants grown under stress conditions.

The above-discussed beneficial role played by silicon in the bulk form, as well as in the nanoform, is dependent on the size of the nanoparticle. When the size is reduced to a nanometer scale, any substance shows the basic properties of nanoparticles. Its chemical, electronic, magnetic, and mechanical properties are dependent on the surface area and size. The particle size is dependent on the synthesis and characterization techniques. So, synthesis and characterization of nanoparticles are among the essential tasks.

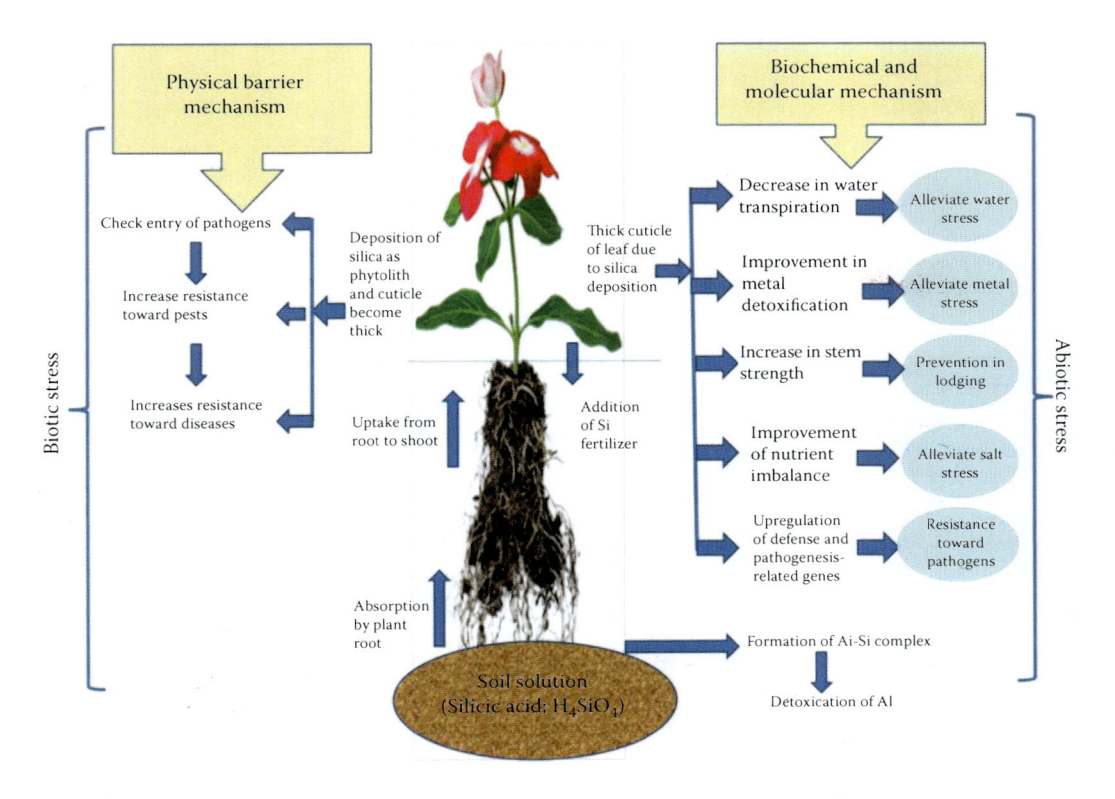

FIGURE 6.1 Si-mediated defense mechanism in plant system under biotic and abiotic stress.

6.3 SYNTHESIS AND CHARACTERIZATION OF NANOFORM OF SILICON

6.3.1 SYNTHESIS OF SILICA NANOPARTICLE

Nanoparticles can pass the cell membrane of crops very easily due to their small size. So, they are more useful than their bulk materials. The nanoform of silicon is the major and frequently used engineered nanoparticle in different fields. Top-down and bottom-up are the two main methods for obtaining nanoforms of silicon (Reverchon and Adami 2006). Top-down approaches include physical processes that utilize special size reduction techniques to reduce the dimension of the original size of the substance, and bottom-up approaches include a chemical process in which atomic and molecular scaling is done to produce nanoparticles. There are various methods to synthesize silica nanoparticles (SNPs), such as reverse microemulsion, flame synthesis, and the sol-gel process.

In reverse microemulsion, a spherical micelle is formed by dissolving the surfactant molecules in organic solvents. The polar head group of surfactants is organized to form microcavities in the presence of water. Microcavities that are formed are called reverse micelles (Tan et al. 2011). With this process, by adding silicon alkoxides and catalyst into the medium, nanosilica is synthesized within microcavities. This approach is not cost-effective, and it also has some difficulties with the removal of surfactants.

SNPs can also be produced with the help of flame synthesis, in which flame decomposition of metal-organic precursors is done in the presence of high temperature. This is also known as chemical vapor condensation (CVC), and in this process silicon tetrachloride ($SiCl_4$) with hydrogen and oxygen reacts to form SNPs. It has one disadvantage: particle size, morphology, and phase composition cannot be controlled in this process (Rahman and Padavettan 2012).

Along with this, to synthesize SNPs, the sol-gel process is among the widely acceptable techniques. It forms homogenous products with the help of hydrolysis and the condensation of metal alkoxides $(Si(OR)_4)$ such as tetra ethyl orthosilicate (TEOS; $Si(OC_2H_5)_4$) or inorganic salts such as sodium silicate (Na_2SiO_3). In this process, mineral acids act as catalysts. Silanol groups are formed after the hydrolysis of TEOS molecules. The siloxane bridge (Si–O–Si), which is responsible for the formation of silica structures, is formed after condensation and polymerization between the silanol groups or between the silanol groups and ethoxy groups. Spherical and monodisperse SNPs can be synthesized with the help of silicon alkoxide in the presence of ammonia as a catalyst. Their size ranges from 50 nm to 1 μm.

Among other nonconventional routes of SNP production, Amutha et al. (2010) extracted SNPs from rice husk (RH), an agricultural waste. Silica is mainly localized around cellulose microcompartments in rice plants (Tabata et al. 2010). RH contains a large amount of silica, and it is produced in large volumes during the agricultural process. One ton of husks is obtained from 5 tons of rice. The RHs produced have different roles, such as fertilizers, fuels, and land filling or paving materials. Due to its high silicon content, RH has become a source for the preparation of nanosilicon. To form nanoparticles, RH is burned in air until ash is produced. This process leads to the decomposition of organic matter, which releases silica as a major remnant. With this process, a pure form of nanoparticles can be prepared. To synthesize SiO_2 nanocomposite sawdust, waste paper, corncob, and sugarcane bagasse are other cellulosic waste materials. In this case, dried cellulosic samples are submerged in silica sol for about 16 h and nanostructured SiO_2 is formed by a calination process (Pang et al. 2011). Figure 6.2 shows different processes involved in the synthesis of nanosilica. After synthesis, characterization of nanoparticles is done.

6.3.2 CHARACTERIZATION TECHNIQUES OF SILICA NANOPARTICLES

There are various techniques through which nanoforms of silicas can be characterized. They can be characterized by using x-ray diffraction (XRD), infrared (IR) spectroscopy, scanning electron microscopy (SEM), energy-dispersive x-ray (EDX) spectroscopy, and dynamic light scattering (DLS). Abou Rida and Harb (2014) have characterized amorphous silica nanoparticles by using these techniques. First, they used an x-ray powder diffraction pattern to characterize precipitated SiO_2. In this case, the spectrum appears as a broad band with the equivalent Bragg angle at $2\theta = 22°$, which indicates the amorphous nature of the synthesized nanoparticle. It did not show an ordered crystalline structure due to the absence of sharp peaks in the prepared silica nanoparticles (Bhavornthanayod and Rungrojchaiporn 2009).

In the case of Fourier transform–infrared (FTIR) spectroscopy, absorption of IR radiation by the nanomaterial is measured against the wavelength. It gives the impression of molecular structures. IR radiation leads to excitation of nonmaterial molecules into a higher vibrational state. The wavelengths that are absorbed by the sample are characteristic of its molecular structure. Due to the presence of various functional groups in the sample, its IR absorption exhibits a wide range. The particular functional groups give characteristic IR absorption at a specific narrow frequency range, but sometimes a functional group gives rise to several characteristic absorptions. So, the whole spectrum should be examined; its interpretation is not dependent on one or two bands only.

With the help of SEM, the nature of synthesized nanosilica particles can be examined. In this process, an electron beam is focused on the surface of the sample, and with the help of a detector, the resulting electrons are measured. This leads to the formation of an image in the form of pixels. Abou Rida and Harb (2014) showed that the majority of primary SiO_2 particles had a uniform size varying from ~50 to ~70 nm. The size of particle is dependent on the chain length of the cationic surfactant. Compared to dodecyltrimethyl ammonium bromide, cetyltrimethyl ammonium bromide surfactant showed better dispersivity and smaller size.

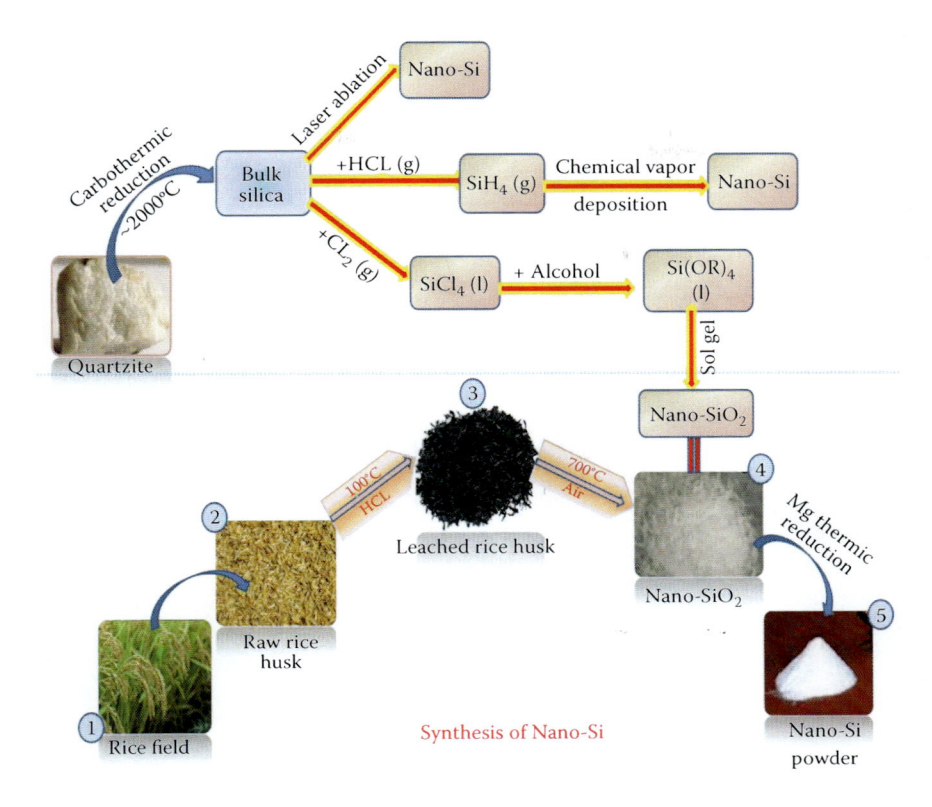

FIGURE 6.2 Schematic representation of Si nanoparticle synthesis. (Modified from Liu N et al., *Sci Rep* 3: 1919–1926, 2013.)

EDX analysis is another technique to characterize nanosilicon. It is also based on the principles of scanning electron microscope. Interaction of an electron beam with nanosilica produces a variety of emissions, including x-rays. To determine the abundance of specific elements, an energy-dispersive detector (EDS) is used. It transforms the characteristic x-rays of different elements into an energy spectrum, and with the help of EDS system software, the energy spectrum is analyzed. It also provides information on the chemical composition of materials and creates element composition maps. So, it helps in providing fundamental compositional information about nanomaterials.

Silicon nanoparticles are also characterized on the basis of their hydrodynamic size and colloidal stability; this is done by the DLS technique. This analytical method provides structural information and aggregation process kinetics. In addition, this technique detects the thickness of the macromolecules adsorbed onto nanoparticles. So overall, it gives an idea of the size, shape, concentration, and surface properties of synthesized nanoparticles (Lim et al. 2013). After synthesis and characterization, these silicon nanoparticles are used in different fields, like industrial, medical, and most importantly, agriculture fields. Here its application in agricultural systems is discussed in detail.

6.4 APPLICATION OF NANOFORM IN AGRICULTURAL SYSTEMS

The application of nanotechnology in agriculture is an emerging approach and increasing rapidly (Duran and Marcato 2013; Giraldo et al. 2014). The applications in the agricultural field are enormous, such as in insect or pest management through the formulation of nanopesticides and insecticides and in the enhancement of plant productivity with the help of nanonutrients. The application of conventional pesticide in agricultural soil causes a decline in soil fertility and an accumulation of toxic metals, which consequently affects human health. Hence, nanotechnology is one of the competent substitutes for managing pest problems in agriculture without any negative impacts. Extensive studies have been done to find the chemical stability, versatility, and biocompatibility of Silica-based nanomaterials (Slowing et al. 2008). Nanosilica is used for other purposes, like to deliver DNA and trigger gene expression (Torney et al. 2007). In the following sections, the application of nanosilicon in agriculture systems is discussed in detail.

6.4.1 Using Nanosilicon as a Nutrient Supplement

Exogenously supplied mineral nutrients play a crucial role in plants against toxicity (Pankovic et al. 2000; Sarwar et al. 2010). The application of the nanoform of silicon (nanosilica) also plays an important role in maintaining the growth and development of plants. Nanosilica is highly reactive and has a high surface-to-volume ratio. The application of nanosilicas in agricultural fields can be done in several ways. It can be used as fertilizers and nutrient supplements. The use of nanosilicon is expected to have an advantage over the usual particle size of Si fertilizer. It is more readily absorbed by plants and more efficient than conventional chemical fertilizers (Ranjbar and Shams 2009). Some studies have been conducted to test the beneficial effects of nanosilica to plants. For instance, Siddiqui and Al-Whaibi (2014) have found that the application of 8 g L^{-1} of nanosilicon dioxide (nSiO$_2$; size 12 nm) increased the germination rate of tomato (*Lycopersicum esculentum* Mill. cv. Super Strain B) seedlings. According to Suriyaprabha et al. (2012), nano-SiO$_2$ increased seed germination by providing better nutrient availability to maize seeds and better pH and conductivity to the growing medium. Bao-shan et al. (2004) applied an exogenous application of nano-SiO$_2$ on Changbai larch (*Larix olgensis*) seedlings. It was found that nano-SiO$_2$ improved the growth and quality of seedlings and also induced the synthesis of chlorophyll. Haghighi et al. (2012) in tomato and Siddiqui et al. (2014) in squash reported that nano-SiO$_2$ enhanced seed germination and stimulated the antioxidant system under NaCl stress. Exogenous application of nano-SiO$_2$ improves seed germination of soybean by increasing nitrate reductase and also by enhancing the seeds' ability to absorb and utilize water and nutrients (Zheng et al. 2005). Under salinity stress, nano-SiO$_2$ improves the leaf fresh and dry weights, chlorophyll contents, and proline accumulation. Wang et al. (2014) performed an experiment on rice plants treated with quantum dots (QDs) coated with silica. It was found that it promoted markedly the root growth of rice plants. Nano-SiO$_2$ enhances plant growth and development by increasing gas exchange and chlorophyll fluorescence parameters, such as the rates of photosynthesis and transpiration, stomatal conductance, photochemical efficiency, and photochemical quenching (Xie et al. 2011; Siddiqui et al. 2014). As we know, photosynthesis is a key process for plants on earth that changes light energy to chemical energy, and its efficiency is being reduced because of various environmental stresses. At the present time, researchers are trying to improve these adverse impacts on the photosynthetic process by using nanobiotechnology. They are trying to develop bionic plants that could have better photosynthetic efficiency and biochemical sensing. Recently, Giraldo et al. (2014) suggested that a nanobionics approach to engineered plants would facilitate new and advanced functional properties in photosynthetic organelles. Noji et al. (2011) reported that a nanomesoporous silica compound (SBA) bound with photosystem II (PS II) and made the activity of a photosynthetic O$_2$ evolving reaction stable, indicating transport of electrons from water to the quinone molecules. They suggested that the PS II–SBA conjugate might have properties to develop for photosensors and artificial photosynthetic systems. Supanjani et al. (2006) have reported that the growth of maize (*Zea mays* L.) seedlings is enhanced by nanosilica application in soil. Amrullah et al. (2015) have also found that the application of nanosized Si in the form of fertilizer having concentrations of 0.09, 0.23, 0.45, 0.68, and 0.9 ppm increased the height, root length, leaf length, canopy, and root fresh and dry weights of rice plants compared to the control. The antioxidant potential of plants has been also found to be increased with the application of nanosilicon. Roohizadeh et al. (2015) have treated seeds of *Vicia faba* by 1.5 and 3 mM nanosilica. They found that the activity of peroxidase in the 1.5 and 3 mM treatments of nanosilica was significantly increased. With the 3 mM nanosilica treatments, the activities of both superoxide dismutase and peroxidase were significantly increased. Thus, nanosilica particles can nullify the oxidative damage via increasing the potential of antioxidant enzymes. So, in order to protect against biotic and abiotic stresses, it improves the physiological and biochemical processes of the plants.

6.4.2 As a Resistance Provider against Pathogens

As discussed in Section 6.4, silicon shows a dynamic potential against pathogen resistance (Datnoff et al. 1997; Carver et al. 1998; Shettya et al. 2012). In agriculture, problems related to insects and pest is of major concern. Silicon acts as a mechanical barrier in leaf tissues to inhibit the entry of fungal pathogens. A number of papers were presented in South Africa on the use of silicon in combating plant diseases. One of the most innovative approaches was presented by the Spanish company Bioiberica. Findings from Bioiberica verified the synergic effect of amino acids and silicon on reduction in the harshness of powdery mildew disease in different plant species, like fruit trees and horticultural crops. There are mainly two modes of action to elucidate the effect of silicon: either by structural reinforcement in the form of deposition below the plant cell wall or by inducing plant defense responses.

In this line, nanotechnology also has potential to become an emerging tool in order to detect plant diseases and their treatment by using nano-based kits, such as (1) nanosensors that enhance the nutrient absorption capability of plants, (2) nanoporous zeolites for maximized crop yield, and (3) nanocapsules for insect pest management (Chaudhry et al. 2008; Rai and Ingle 2012). In recent years, nanotechnology has improved the performance and acceptability of Si as nanosilicon by increasing its effectiveness in the form of nanopesticides and herbicides to provide resistance against disease. It has been found that nanosilicon is useful as a catalyst and, most importantly, has been found to be useful as a nanopesticide. Earlier, Barik et al. (2008) reviewed the use of nanosilica as a nanopesticide. Uses of insecticides and pesticides affect application costs to farmers, as most of the applied pesticides, that is, about 90%, are lost to the air, as well as through water runoff. Along with this indiscriminate usage of pesticide, several problems resulted, such as the enhancement of pathogen resistance, reduction in soil biodiversity and nitrogen fixation, and pesticide accumulation (Ghormade et al. 2011). In order to decrease the level of these problems, pesticide was applied inside nanoparticles that can be released whenever required. Formulations made by encapsulating the insecticides, fungicides, or nematicides with nanoparticles offer effective control of pests and have also resulted in less accumulation of pesticide residues in soil (Goswami et al. 2010). With the help of this technique, the rate of pesticide application can be reduced, as its smaller quantity can provide much better management for pesticide attention. These nanoencapsulated pesticides are released upon the occurrence of an environmental stress such as temperature, humidity, and light. Wang et al. (2007) have reported that a pesticide formulated by nanoemulsions, including oil in water, was effective against various insect pests in agriculture. Among them, validamycin is an effective transfer system for pesticides; porous hollow silica nanoparticles that are soluble in water are filled with it and released under controlled conditions (Barik et al. 2008). Nanosilica (~3–5 nm) that is prepared from silica can also be used as a nanoinsectide. It results in the death of insects by being absorbed into their cuticular lipids. For the protection of water obstruction on their bodies, insects use a diversity of lipids on their cuticles (Liu et al. 2006). Typically, nanosilica is absorbed into cuticular lipids by physiosorption, and by this means it causes the death of the insect. Also with the help of this technique, less herbicide is required to achieve weed eradication (Nair et al. 2010). The potential of nanoparticles for insect pest management has also been reported (Bhattacharyya et al. 2010). The process of nanoencapsulation helps in insect pest control by slowing the release of chemicals to a particular host through various release mechanisms, including dissolution, biodegradation, diffusion, and osmotic pressure with specific pH (Vidyalakshmi et al. 2009). In recent years, nanosilica has been successfully employed to control a range of insects and pests in the agricultural system. Antifungal activities of polymer-based silica-silver nanoparticles against *Bipolaris sorokiniana* and *Magnaporthe grisea* (Jo et al. 2009) have been reported. In an earlier study, Suriyaprabha et al. (2013) found that properly functionalized nanoparticles may be efficiently used to decrease the application rate of fungicides on plant tissues. Further, the results obtained from the use of hydrophobic silica nanoparticles, promoting resistance against fungus in maize crop, will express the uses of SNPs for plant defense applications.

6.5 CONCLUSION AND FUTURE PERSPECTIVES

In soil solution, silicon is basically present in the form of monosilicic acid (H_4SiO_4). It is the only form of silicon that plants can absorb. Its application increases the strength and rigidity of plant cell walls and tissues. In addition, it also decreases the transpiration rate from the stomata, thereby reducing transpiration and increasing resistance to abiotic stresses like drought, high radiation, and temperature. Nowadays, it is also used in the form of nanosilica in agriculture fields, as it is more reactive than its bulk counterpart. Nanosilica improves seedling growth and crop protection against biotic and abiotic stresses more precisely. Applications of nano-Si materials have been given a lot of attention by agricultural researchers due to their tiny size and greater surface area. However, in the future there should be more studies in order to find the interaction mechanism between nanosilica and plants, which can increase its potential use in crop improvement.

ACKNOWLEDGMENTS

The authors acknowledge the University of Allahabad for providing necessary facilities. Dr. Anita Singh is thankful to the Science & Engineering Research Board (SERB) for providing fellowship under Start Up Research Grant SB/YS/LS-228/2013 as a DST–Young Scientist. Ms. Shikha Singh is thankful to the University Grants Commission, New Delhi, for providing project-JRF (project no. 41-460/2012 [SR]).

REFERENCES

Abdul Qados AMS, Moftah AE. (2014). Influence of silicon and nano-silicon on germination, growth and yield of faba bean (*Vicia faba* L.) under salt stress conditions. *Am J Exp Agric* 5: 509–524.

Abou Rida, Harb F. (2014). Synthesis and characterization of amorphous silica nanoparticles from aqueous silicates using cationic surfactants maurice. *J Met Mater Miner* 24: 37–42.

Al-aghabary K, Zhu Z, Shi Q. (2004). Influence of silicon supply on chlorophyll content, chlorophyll fluorescence, and antioxidative enzyme activities in tomato plants under salt stress. *J Plant Nutr* 27: 2101–2115.

Amrullah, Sopandie D, Sugianta, Junaedi A. (2015). Influence of nano-silica on the growth of rice plant (*Oryza sativa* L.) *Asian J Agric Res* 9: 33–37.

Amutha K, Ravibaskar R, Sivakumar G. (2010). Extraction, synthesis and characterization of nanosilica from rice husk ash. *Int J Nanotech Appl* 4: 61–66.

Arnon DI, Stout PR. (1939). The essentiality of certain elements in minute quantity for plants with special reference to copper. *Plant Physiol* 14: 371–375.

Azimi R, Borzelabad MJ, Feizi H, Amin Azimi A. (2014). Interaction of SiO_2 nanoparticles with seed prechilling on germination and early seedling growth of tall wheatgrass (*Agropyron elongatum* L.). *Polish J Chem Technol* 16: 25–29.

Bao-shan L, Shao-qi D, Chun-hui L, Li-jun F, Shu-chun Q, Min Y. (2004). Effects of TMS (nanostructured silicon dioxide) on growth of Changbai larch seedlings. *J Forestry Res* 15: 138–140.

Barik TK, Sahu B, Swain V. (2008). Nanosilica—From medicine to pest control. *Parasitolol Res* 103: 253–258.

Bélanger R, Benhamou N, Menzies J. (2003). Cytological evidence of an active role of silicon in wheat resistance to powdery mildew (*Blumeria graminis* f. sp. *tritici*). *Phytopathology* 93: 402–412.

Bhattacharyya A, Bhaumik A, Rani PU, Mandal S, Epidi TT. (2010). Nanoparticles—A recent approach to insect pest control. *Afr J Biotechnol* 9: 3489–3493.

Bhavornthanayod C, Rungrojchaiporn P. (2009). Synthesis of zeolite A membrane from rice husk ash. *J Met Mater Miner* 19: 79–83.

Cai K, Gao D, Chenand J, Luo S. (2009). Probing the mechanisms of silicon-mediated pathogen resistance. *Plant Signal Behav* 4: 1–3.

Cai K, Gao D, Luo S, Zeng R, Yang J, Zhu X. (2008). Physiological and cytological mechanisms of silicon-induced resistance in rice against blast disease. *Physiol Planta* 134: 324–333.

Camargo MS, Júnior ARG, Wyler P, Korndörfer BGH. (2010). Silicate fertilization in sugarcane: Effects on soluble silicon in soil, uptake and occurrence of stalk borer (*Diatraea saccharalis*). Presented at 19th World Congress of Soil Science, Soil Solutions for a Changing World, Brisbane, Australia, August 1–6, 2010.

Carver TLW, Zeyen RJ, Ahlstrand GG. (1987). The relationship between insoluble silicon and success or failure of attempted primary penetration by powdery mildew (*Erysiphe graminis*) germlings on barley. *Physiol Mol Plant Pathol* 31: 133–148.

Chaudhry Q, Scotter M, Blackburn J, Ross B, Boxall A, Castle L, Watkins R. (2008). Applications and implications of nanotechnologies for the food sector. *Food Addit Contam* 25: 241–258.

Datnoff LE, Brecht M, Stiles C, Rutherford B. (1997). The role of silicon in suppressing foliar diseases in warm-season turf. *Int Turfgrass Soc Res J* 10: 175–179.

Datnoff LE, Snyder GH, Korndörfer GH (eds.). (2001). *Silicon in Agriculture*. Studies in Plant Science 8. Amsterdam: Elsevier.

Detmann KC, Araújo WL, Martins SC, Sanglard L, Reis JV, Detmann E, Rodrigues FA, Nunes-Nesi A, Fernie AR, DaMatta FM. (2012). Silicon nutrition increases grain yield, which, in turn, exerts a feed-forward stimulation of photosynthetic rates via enhanced mesophyll conductance and alters primary metabolism in rice. *New Phytol* 196: 752–762.

Doncheva SN, Poschenrieder C, Stoyanova ZL, Georgieva K, Velichkova M, Barceló b J. (2009). Silicon amelioration of manganese toxicity in Mn-sensitive and Mn-tolerant maize varieties. *Environ Exp Bot* 65: 189–197.

Dufey I, Gheysens S, Ingabire A, Lutts S, Bertin P. (2014). Silicon application in cultivated rice (*Oryza sativa* L. and *Oryza glaberrima* Steud.) alleviates iron toxicity symptoms through the reduction in iron concentration in the leaf tissue. *J Agron Crop Sci* 200: 132–142.

Duran N, Marcato PD. (2013). Nanobiotechnology perspectives. Role of nanotechnology in the food industry: A review. *Int J Food Sci Technol* 48: 1127.

Elawad SH, Gascho GJ, Street JJ. (1982). Response of sugarcane to silicate source and rate. I. Growth and yield. *Agron J* 74: 481–484.

El-bendary HM, El-Helaly AA. (2013). First record nanotechnology in agricultural: Silica nanoparticles: A potential new insecticide for pest control. *Appl Sci Rep* 4: 241–246.

El-Samahy EFM, Asmaa M, El-Ghobary, Khafagy IF. (2014). Using silica nanoparticles and neemoil extract as new approaches to control *Tuta absoluta* (meyrick) in tomato under field conditions. *Int J Plant Soil Sci* 3: 1355–1365.

Epstein E. (1994). The anomaly of silicon in plant biology. *Proc Natl Acad Sci USA* 91: 11–17.

Epstein E, Bloom AJ. (2003). *Mineral Nutrition of Plants: Principles and Perspectives*. 2nd ed. New York: John Wiley & Sons.

Ghormade V, Deshpande MV, Paknikar KM. (2011). Perspectives for nano-biotechnology enabled protection and nutrition of plants. *Biotechnol Adv* 29: 792–803.

Giraldo JP, Landry MP, Faltermeier SM, McNicholas TP, Iverson NM, Boghossian AA, Reuel NF, Hilmer AJ, Sen F, Brew JA, Strano MS. (2014). Plant nanobionics approach to augment photosynthesis and biochemical sensing. *Nat Mater* 13: 400–408.

Goswami A, Roy I, Sengupta S, Debnath N. (2010). Novel applications of solid and liquid formulations of nanoparticles against insect pests and pathogens. *Thin Solid Films* 519: 1252–1257.

Guntzer F, Keller C, Meunier JD. (2012). Benefits of plant silicon for crops: A review. *Agron Sustain Dev* 32: 201–213.

Haghighi M, Afifipour Z, Mozafarian M. (2012). The effect of N-Si on tomato seed germination under salinity levels. *J Biol Environ Sci* 6: 87–90.

Hamm CE, Merkel R, Springer O, Kurkoje P, Maler C, Prechtel K, Smetacek V. (2003). Architecture and material properties of diatom shells provide effective mechanical protection. *Nature* 42: 841–843.

Hodson MJ, White PJ, Mead A, Broadley MR. (2005). Phylogenetic variation in the silicon composition of plants. *Ann Bot* 96: 1027–1046.

Jian FM, Tamai K, Yamaji N. (2006). A silicon transporter in rice. *Nature* 440: 688–691.

Jo YK, Kim BH, Jung G. (2009). Antifungal activity of silver ions and nano-particles on phytopathogenic fungi. *Plant Dis* 93: 1037–1043.

Kalteh M, Alipour ZT, Ashraf S, Aliabadi MM, Nosratabadi AF. (2014). Effect of silica nanoparticles on basil (*Ocimum basilicum*) under salinity stress. *J Chem Health Risks* 4: 49–55.

Kamenidou S, Cavins TJ, Marek S. (2009). Evaluation of silicon as a nutritional supplement for greenhouse zinnia production. *Sci Hortic* 119: 297–301.

Kauss H, Seehaus K, Franke R, Gilbert S, Dietrich RA, Kroger N. (2003). Silica deposition by a strongly cationic proline-rich protein from systemically resistant cucumber plants. *Plant J* 33: 87–95.

Keeping MG, Meyer JH. (2003). Effect of four sources of silicon on resistance of sugarcane varieties to *Eldana saccharina* Walker (Lepidoptera: Pyralidae). *Proc S Afr Sug Technol Assoc* 77: 99–103.

Kim JH, Lee Y, Kim EJ, Gu S, Sohn EJ, Seo YS, An HJ, Chang YS. (2014). Exposure of iron nanoparticles to *Arabidopsis thaliana* enhances root elongation by triggering cell wall loosening. *Environ Sci Technol* 48: 3477–3485.

Lee SC, Luan S. (2012). ABA signal transduction at the crossroad of biotic and abiotic stress responses. *Plant Cell Environ* 35: 53–60.

Liang Y, Sun W, Zhu YG, Christie P. (2006a). Mechanisms of silicon-mediated alleviation of abiotic stresses in higher plants: A review. *Environ Pollut* xx: 1–7.

Liang Y, Zhang W, Chen Q, Liu Y, Ding R. (2006b). Effect of exogenous silicon (Si) on H+-ATPase activity, phospholipids and fluidity of plasma membrane in leaves of salt-stressed barley (*Hordeum vulgare* L.). *Environ Exp Bot* 57: 212–219.

Liang YC, Ding RX. (2002). Influence of silicon on microdistribution of mineral ions in roots of salt-stressed barley as associated with salt tolerance in plants. *Sci China C* 45: 298–308.

Liang YC, Sun WC, Si J, Romheld V. (2005). Effects of foliar and root-applied silicon on the enhancement of induced resistance to powdery mildew in *Cucumis sativus*. *Plant Pathol* 54: 678–685.

Lim J, Yeap SP, Che HX, Low SC. (2013). Characterization of magnetic nanoparticle by dynamic light scattering. *Nanoscale Res Lett* 8: 381.

Lin B, Diao S, Li C, Fang L, Qiao S, Yu M. (2004). Effect of TMS (nanostructured silicon dioxide) on growth of Changbai larch seedlings. *J Forestry Res* 15: 138–140.

Liu F, Wen LX, Li ZZ, Yu W, Sun HY, Chen JF. (2006). Porous hollow silica nanoparticles as controlled delivery system for water soluble pesticide. *Mater Res Bull* 41: 2268–2275.

Liu N, Huo K, McDowell MT, Zhao J, Cui Y. (2013). Rice husks as a sustainable source of nanostructured silicon for high performance Li-ion battery anodes. *Sci Rep* 3: 1919–1926.

Ma JF. (2004). Role of silicon in enhancing the resistance of plants to biotic and abiotic stresses. *Soil Sci Plant Nutr* 50: 11–18.

Ma JF, Miyake Y, Takahashi E. (2001). Silicon as a beneficial element for crop plants. In Datnoff LE, Snyder GH, Korndörfer GH (eds.), *Silicon in Agriculture*. Studies in Plant Science 8. Amsterdam: Elsevier, pp. 17–39.

Ma JF, Takahashi. (2002). *Soil, Fertilizer, and Plant Silicon Research in Japan*. Amsterdam: Elsevier.

Ma JF, Yamaji N. (2006). Silicon uptake and accumulation in higher plants. *Trends Plant Sci* 11: 392–397.

Mitani N, Ma JF. (2005). Uptake system of silicon in different plant species. *J Exp Bot* 56: 1255–1261.

Nair R, Varghese SH, Nair BG, Maekawa T, Yoshida Y. (2010). Nanoparticulate material delivery to plants. *Plant Sci* 179: 154–163.

Noji T, Kamidaki C, Kawakami K, Shen JR, Kajino T, Fukushima Y, Sekitoh T, Itoh S. (2011). Photosynthetic oxygen evolution in mesoporous silica material: Adsorption of photosystem II reaction center complex into 23 nm nanopores in SBA. *Langmuir* 27: 705–713.

Panda N, Pradhan B, Samalo AP, Rao PSP. (1975). Note on the relationship of some biochemical factors with the resistance in rice varieties to yellow rice borer. *Ind J Agric Sci* 45: 499–501.

Pang SC, Chin SF, Yih V. (2011). Conversion of cellulosic waste materials into nanostructured ceramics and nanocomposites. *Adv Mater Lett* 2: 118–124.

Pankovic D, Plesnicar M, Maksimoviic IA, Petrovic N, Sakac Z, Kastori R. (2000). Effects of nitrogen nutrition on photosynthesis in Cd-treated sunflower plants. *Ann Bot Lond* 86: 841–847.

Rahman IA, Padavettan V. (2012). Synthesis of silica nanoparticles by sol-gel: Size-dependent properties, surface modification, and applications in silica–polymer nanocomposites—A review. *J Nanomater* 2012: 132424.

Rai M, Ingle A. (2012). Role of nanotechnology in agriculture with special reference to management of insect pests. *Appl Microbiol Biotechnol* 4: 287–293.

Ranjbar M, Shams GA. (2009). Uses of nanotechnology. *J Environ Green* 3: 29–34.

Reverchon E, Adami R. (2006). Nanomaterials and supercritical fluids. *J Supercritical Fluids* 37: 1–22.

Rodrigues FN, Benhamou L, Datnoff J, Jones, Bélanger R. (2003). Ultrastructural and cytochemical aspects of silicon mediated rice blast resistance. *Phytopathology* 93: 535–546.

Rogalla H, Römheld V. (2002). Role of leaf apoplast in Si mediated manganese tolerance of *Cucumis sativus* L. *Plant Cell Environ* 25: 549–555.

Romero-Aranda MR, Jurado O, Cuartero J. (2006). Silicon alleviates the deleterious salt effect on tomato plant growth by improving plant water status. *J Plant Physiol* 163: 847–855.

Roohizadeh G, Arbabian S, Tajadod G, Majd A, Salimpour F. (2015). The study of nano silica effects on the total protein content and the activities of catalase, peroxidase and superoxide dismutase of *Vicia faba* L. *Trop. Plant Res* 2(1): 47–50.

Sahebi M, Hanafi MM, Abdullah SNA, Rafii MY, Azizi P, Tengoua FF, Azwa JNM, Shabanimofrad M. (2015). Importance of silicon and mechanisms of biosilica formation in plants. *Biol Med Res* 2015: 396010.

Sarwar N, Saifullah, Malhi SS, Zia MH, Naeem A, Bibi S, Farid G. (2010). Role of mineral nutrition in minimizing cadmium accumulation by plants. *J Sci Food Agric* 90: 925–937.

Seebold KW, Datnoff LE, Correa-Victoria FJ, Kucharek TA, Snyder GH. (2000). Effect of silicon rate and host resistance on blast, scald, and yield of upland rice. *Plant Dis* 84: 871–876.

Shettya R, Jensena B, Shettya NP, Hansena M, Hansenb CW, Starkeyb KR, Jorgensen HJL. (2012). Silicon induced resistance against powdery mildew of roses caused by *Podosphaera pannosa*. *Plant Pathol* 61: 120–131.

Shi QH, Bao ZY, Zhu ZJ, He Y, Qian QQ, Yu JQ. (2005a). Silicon-mediated alleviation of Mn toxicity in *Cucumis sativus* in relation to activities of superoxide dismutase and ascorbate peroxidase. *Phytochemistry* 66: 1551–1559.

Shi QH, Zhu ZJ, Xu M, Qian QQ, Yu JQ. (2005b). Effect of excess manganese on the antioxidant systemin *Cucumis sativus* L. under two light intensities. *Environ Exp Bot* 58: 197–2051.

Siddiqui MH, Al-Whaibi MH. (2014). Role of nano-SiO_2 in germination of tomato *Lycopersicum esculentum* seeds Mill.). *Saudi Biol Sci* 21: 13–17.

Siddiqui MH, Al-Whaibi MH, Faisal M, Al Sahli AA. (2014). Nano-silicon dioxide mitigates the adverse effects of salt stress on *Cucurbita pepo* L. *Environ Toxicol Chem* 33(11): 2429–2437.

Sidhu JK, Stout MJ, Blouin DC, Datnoff LE. (2013). Effect of silicon soil amendment on performance of sugarcane borer, *Diatraea saccharalis* (Lepidoptera: Crambidae) on rice. *Bull Entomol Res* 103: 656–664.

Slowing H, Vivero-Escoto JL, Wu CW, Lin VSY. (2008). Mesoporous silica nanoparticle as controlled releases drug delivery and gene transfer carriers. *Adv Drug Deliv Rev* 60: 1278–1288.

Snyder GH, Matichenkov VV, Datnoff LE. (2006). Silicon. In *Plant Nutrition*. Taylor & Francis, Belle Glade, FL, pp. 551–562.

Subbarao DV, Perraju A. (1976). Resistance in some rice strains to first-instar larvae of *Tryporyza incertulas* (Walker) in relation to plant nutrients and anatomical structure of the plants. *Int Rice Res Newsl* 1: 14–15.

Supanjani Han HS, Jung JS, Lee KD. (2006). Rock phosphate-potassium and rock-solubilising bacteria as alternative, sustainable fertilizers. *Agron Sust Dev* 26: 233–240.

Suriyaprabha R, Karunakaran G, Kavitha K, Yuvakkumar R, Rajendran V, Kannan N. (2013). Application of silica nanoparticles in maize to enhance fungal resistance. *IET Nanobiotechnol* 8(3): 133–137.

Suriyaprabha R, Karunakaran G, Yuvakkumar R, Rajendran V, Kannan N. (2012). Silica nanoparticles for increased silica availability in maize (*Zea mays* L) seeds under hydroponic conditions. *Curr Nanosci* 8: 902–908.

Tabata S, Iida H, Horie T, Yamada S. (2010). Hierarchical porous carbon from cell assemblies of rice husk for in vivo applications. *Med Chem Comm* 1: 136–138.

Tan TTY, Liu S, Zhang Y, Han MY, Selvan ST. (2011). Microemulsion preparative method [Overview]. *Comprehensive Nanosci Technol* 5: 399–441.

Torney F, Trewyn BG, Lin VSY, Wang K. (2007). Mesoporous silica nanoparticles deliver DNA and chemicals into plants. *Nat Nanotechnol* 2: 295–300.

Tripathi DK, Singh VP, Kumar D, Chauhan DK. (2011). Impact of exogenous silicon addition on chromium uptake, growth, mineral elements, oxidative stress, antioxidant capacity, and leaf and root structures in rice seedlings exposed to hexavalent chromium. *Acta Physiol Plant* 34: 279–289.

Tuna AL, Kaya G, Higgs D, Murillo-Amador B, Aydemir S, Girgin AR. (2008). Silicon improves salinity tolerance in wheat plants. *Environ Exp Bot* 62: 10–16.

Vidyalakshmi R, Bhakyaraj R, Subhasree RS. (2009). Encapsulation "the future of probiotics"—A review. *Adv Biol Res* 3: 96–103.

Wang A, Zheng Y, Peng F. (2014). Thickness-controllable silica coating of CdTe QDs by reverse microemulsion method for the application in the growth of rice. *J Spectrosc* 2014: 169245. http://dx.doi.org/10.1155/2014/169245.

Wang L, Li X, Zhang G, Dong J, Eastoe J. (2007). Oil-in-water nanoemulsions for pesticide formulations. *J Colloid Interface Sci* 314: 230–235.

Wei C, Zhang Y, Guo J, Han B, Yang X, Yuan J. (2010). Effects of silica nanoparticles on growth and photosynthetic pigment contents of *Scenedesmus obliquus*. *J Environ Sci* 22: 155–160.

Xie Y, Li B, Zhang Q, Zhang C, Lu K, Tao G. (2011). Effects of nano-TiO2 on photosynthetic characteristics of *Indocalamus barbatus*. *J Northeast For Univ* 39: 22–25.

Yang YF, Liang YC, Lou YS, Sun WC. (2003). Influences of silicon on peroxidase, superoxide dismutase activity and lignin content in leaves of wheat *Tritium aestivum* L. and its relation to resistance to powdery mildew. *Sci Agric Sin* 36: 813–817.

Yin L, Wang S, Li J, Tanaka K, Oka M. (2013). Application of silicon improves salt tolerance through ameliorating osmotic and ionic stresses in the seedling of *Sorghum bicolor*. *Acta Physiol Plant* 35: 3099–3107.

Yuvakkumar R, Elango V, Rajendran V, Kannan NS, Prabu P. (2011). Influence of nanosilica powder on the growth of maize crop (*Zea Mays* L.). *Int J Green Nanotechnol* 3: 80–190.

Zhang GL, Dai QG, Zhang HC. (2006). Silicon application enhances resistance to sheath blight (*Rhizoctonia solani*) in rice. *J Plant Physiol Mol Biol* 32: 600–606.

Zheng L, Hong F, Lu S, Liu C. (2005). Effect of nano-TiO_2 on strength of naturally aged seeds and growth of spinach. *Biol Trace Elem Res* 104: 83–91.

Zhu ZJ, Wei GQ, Li J, Qian QQ, Yu JQ. (2004). Silicon alleviates salt stress and increases antioxidant enzymes activity in leaves of salt-stressed cucumber (*Cucumis sativus* L.). *Plant Sci* 167: 527–533.

7 Silicon and Floricultural Crops
Evaluation of Silicon Uptake and Deposition in Floricultural Plants

Yoo Gyeong Park, Chung Ho Ko, Sowbiya Muneer, and Byoung Ryong Jeong

CONTENTS

ABSTRACT

Silicon (Si) is still deemed to be a nonessential nutrient for the majority of plant species, although its uptake has been found to be beneficial for improving resistance to insects, pathogens, drought, and heavy metals, and for crop quality and yield in some plant species. Although not extensively studied, the addition of Si to nutrient solutions has enhanced flower quality in many ornamental plants, such as begonias, carnations, chrysanthemums, gerberas, kalanchoes, marigolds, miniature roses, petunias, phalaenopsis, poinsettias, salvias, sunflowers, torenias, and zinnias. The reported effects vary and depend strongly on plant species. The beneficial effects of Si are usually more apparent under biotic or abiotic stress. Thus, the addition of Si to the nutrient solution of plants has also shown positive effects in floristic crops. However, most floricultural plants studied were low Si accumulators. The low Si uptake by floricultural plants suggests that they may lack homologs to Lsi genes, which encode the Si transporter in rice, and thus may not have the underlying molecular mechanisms to accumulate significant amounts of Si in the tissues. The objective of this chapter is to summarize research on the effects of Si supplementation on plant growth and changes in stress resistance under abiotic and biotic stress conditions, and also to evaluate Si uptake and deposition in floricultural crops.

7.1 INTRODUCTION

Silicon (Si) is the second most abundant element in the earth's crust and is among the major inorganic constituents of higher plants (Epstein, 1999; Ma and Takahashi, 2002). The function of Si in horticultural crops is not well understood, primarily because Si is not considered an element essential for plant growth, as indicated by the "criteria of essentiality" of Arnon and Stout (1939; Epstein, 1994).

Generally, only monocots such as rice and sugarcane can accumulate Si at concentrations higher than many other essential nutrients, and the application of Si is deemed important for their sustainable production (Ma and Yamaji, 2006). However, many floricultural crops are dicots and are usually grown in greenhouses in Si-depleted growth media for either cut flower production or potted ornamentals. Until recent studies by Frantz et al. (2008), Mattson and Leatherwood (2010), and Son et al. (2013), appreciable amounts of Si accumulation in ornamental plants were not well known. Most horticultural plants studied have been shown to be low Si accumulators.

Si promotes plant growth and plays a unique role in conferring tolerance in plants to various abiotic and biotic stresses (Liang et al., 2007). The addition of Si to the nutrient solution enhanced flower quality in gerberas (Savvas et al., 2002; Kamenidou et al., 2010), sunflowers (Kamenidou et al., 2008), roses (Ehret et al., 2005; Hwang et al., 2005), zinnias (Kamenidou et al., 2009), and chrysanthemums (Sivanesan et al., 2013a,b). Furthermore, it has been suggested that the addition of Si-based fertilizers may increase plant resistance to fungal pathogens and insect pests, although the mechanisms associated with Si-mediated resistance are not well understood (Sangster and Hodson, 1986; Bowen et al., 1992; Belanger et al., 1995; Sangster et al., 2001; Ma and Takahashi, 2002; Keeping and Kvedaras, 2008). This chapter describes the effects of Si supplementation on growth and resistance under abiotic and biotic stress, and evaluates Si uptake and deposition in floristic crops.

7.2 Si IN FLORICULTURAL PLANTS

Floristic crops have been studied as models to decipher the effects of Si on growth parameters (plant height, dry matter, stem diameter, etc.) and the quality of flowers or fruits. The following descriptions are reviews of these aspects in some major floristic crops (Table 7.1).

7.2.1 *AJUGA MULTIFLORA*

The genus *Ajuga*, belonging to the mint family (Lamiaceae), comprises about 301 species and occurs in the cooler parts of Europe, Asia, Africa, and Australia (Sivanesan and Jeong, 2014). The addition of Si to NaCl-containing medium enhanced the shoot regeneration and growth of *Ajuga multiflora* by altering the activity of antioxidant enzymes, maintaining the ultrastructure of stomata, and limiting NaCl deposition in leaves (Sivanesan and Jeong, 2014).

7.2.2 BEGONIA

Begonias (*Begonia semperflorens* Link et Otto) are classic bedding plants because they are beautiful for the entire season and are easy to grow (Lim et al., 2012). Although commonly grown as

TABLE 7.1

Physiological Responses of Floricultural Plants Grown with Different Forms of Si and Methods of Application

Physiological Response	Plant	Silicon Form	Application Method	References
Plant height				
Increase	Chrysanthemum	K_2SiO_3	Subirrigation	Sivanesan et al., 2013a,b
	Gerbera	Na_2SiO_3	Foliar spray	Kamenidou et al., 2010
	Kalanchoe	$NaSiO_3$	Foliar spray	Bae et al., 2010b
	Lobelia	Silicate fertilizer	Supplemented medium	Mattson and Leatherwood, 2010
	New Guinea	$KSiO_3$	Subirrigation	Mattson and Leatherwood, 2010
	Portulaca	$KSiO_3$	Subirrigation	Mattson and Leatherwood, 2010
	Sunflower	$KSiO_3$	Subirrigation	Kamenidou et al., 2008
		Substrate	Supplemented medium	Mattson and Leatherwood, 2010
Decrease	Bracteantha	$KSiO_3$	Subirrigation	Mattson and Leatherwood, 2010
	Carnation	Silicate fertilizer	Supplemented medium	Bae et al., 2010a
	Kalanchoe	K_2SiO_3	Subirrigation	Son et al., 2012
	Kalanchoe	Na_2SiO_3	Foliar spray	Son et al., 2012
	Marigold	K_2SiO_3	Subirrigation	Sivanesan et al., 2010
	Nephrolepis	K_2SiO_3	Drip irrigation	Sivanesan et al., 2014
	Zinnia	$KSiO_3$	Subirrigation	Kamenidou et al., 2009
Leaf width				
Increase	Chrysanthemum	K_2SiO_3	Subirrigation	Sivanesan et al., 2013a,b
	Kalanchoe	$CaSiO_3$	Subirrigation	Bae et al., 2010b
		Silicate fertilizer	Supplemented medium	
Chlorophyll content				
Increase	Kalanchoe	$CaSiO_3$	Subirrigation	Son et al., 2012
	Marigold	K_2SiO_3	Subirrigation	Sivanesan et al., 2010
Stem diameter				
Increase	Carnation	Silicate fertilizer	Supplemented medium	Bae et al., 2010a
	Chrysanthemum	Silicate fertilizer	Supplemented medium	Moon et al., 2008
	Gerbera	$NaSiO_3$	Foliar spray	Kamenidou et al., 2010
	Marigold	K_2SiO_3	Subirrigation	Sivanesan et al., 2010
	Rose	K_2SiO_3	Root feeding and foliar spray	Hwang et al., 2005
	Sunflower	$KSiO_3$, $NaSiO_3$	Foliar spray, subirrigation	Kamenidou et al., 2008
	Zinnia	K_2SiO_3	Subirrigation	Kamenidou et al., 2009
Decrease	Kalanchoe	Na_2SiO_3	Foliar spray	Son et al., 2012

(Continued)

TABLE 7.1 (CONTINUED)

Physiological Responses of Floricultural Plants Grown with Different Forms of Si and Methods of Application

Physiological Response		Plant	Silicon Form	Application Method	References
Root length	Increase	Chrysanthemum	$CaSiO_3$	Subirrigation	Sivanesan et al., 2013a,b
		Marigold	K_2SiO_3	Subirrigation	Moon et al., 2008
					Sivanesan et al., 2010
Days to anthesis	Increase	Sunflower	$KSiO_3$	Subirrigation	Kamenidou et al., 2008
	Decrease	Gerbera	$NaSiO_3$	Foliar spray	Kamenidou et al., 2010
Number of flowers	Increase	Chrysanthemum	K_2SiO_3	Subirrigation	Sivanesan et al., 2013a,b
Flower stalk length	Increase	Chrysanthemum	K_2SiO_3, Na_2SiO_3	Foliar spray	Sivanesan et al., 2013b
	Decrease	Kalanchoe	$CaSiO_3$	Subirrigation	Son et al., 2012
Flower diameter	Increase	Calibrachoa	$KSiO_3$	Subirrigation	Mattson and Leatherwood, 2010
		Fuchsia	$KSiO_3$	Subirrigation	Mattson and Leatherwood, 2010
		Gerbera	$NaSiO_3$	Foliar spray	Kamenidou et al., 2010
		Petunia	$KSiO_3$	Subirrigation	Mattson and Leatherwood, 2010
		Portulaca	$KSiO_3$	Subirrigation	Mattson and Leatherwood, 2010
		Sunflower	K_2SiO_3, Na_2SiO_3	Subirrigation	Kamenidou et al., 2008
		Torenia	$KSiO_3$	Subirrigation	Mattson and Leatherwood, 2010
		Zinnia	$NaSiO_3$	Foliar spray	Kamenidou et al., 2009
	Decrease	Argyranthemum	$KSiO_3$	Subirrigation	Mattson and Leatherwood, 2010
		Bacopa	$KSiO_3$	Subirrigation	Mattson and Leatherwood, 2010
		Lobelia	$KSiO_3$	Subirrigation	Mattson and Leatherwood, 2010
		Verbena	$KSiO_3$	Subirrigation	Mattson and Leatherwood, 2010

annuals, they are actually evergreen perennials. In the begonia 'Super Olympia Rose', the structure of stomata was more compact in size in the 300 mg·L^{-1} K$_2$SiO$_3$ treatment than in the control (Lim et al., 2012). It has been reported that many species of begonia have stomata that occur in clusters, which is an unusual character, and have an extremely limited distribution among the higher plants (Lim et al., 2012). Gan et al. (2010) hypothesized that the stomatal clusters could be induced by certain degrees of drought or salt stress and have a different ecological significance according to the type of clustering. The surface of the leaf epidermis appeared to be compact due to Si deposition, which indicated that Si positively affected the growth and biomass production of begonias (Lim et al., 2012).

7.2.3 CALENDULA

Calendulas, commonly known as pot marigolds, are planted widely in gardens and landscapes (Bayat et al., 2012). Si alleviates the negative effects of salt stress on growth, chlorophyll reading values, electrolyte leakage, and flower quality, depending on the concentration of Si. The maximum alleviation of salt stress was found with 100 mg·L^{-1} Si applications (Bayat et al., 2013).

7.2.4 CARNATION

Dianthus caryophyllus, commonly known as the carnation, is used as either a cut flower or potted flower worldwide due to its attractive and colorful inflorescences (Soundararajan et al., 2015). Soundararajan et al. (2015) reported that 50 mg·L^{-1} Si supplementation could be optimal for improved growth *in vitro* and enhanced resistance against salinity in *D. caryophyllus* 'Tula'. Carnations treated with K$_2$SiO$_3$ (100, 150, 200 mg·L^{-1}) produced less ethylene, leading to the lower carbohydrates and higher dry weight in Si-treated flowers (Jamali and Rahemi, 2011).

7.2.5 CHRYSANTHEMUM

Chrysanthemums are among the top-selling cut flowers, potted plants, and herbaceous landscape plants worldwide with high ornamental and commercial interest (Song et al., 2011). All three cultivars of chrysanthemums have been classified as Si accumulators (Ma et al., 2001), as the presence of Si in the leaves of the cultivars was found to be more than 10 g·kg^{-1} dry weight. The addition of potassium silicate to the nutrient solution increases the ability of chrysanthemums to withstand attack by leafminers (Parrella et al., 2007; Jeong et al., 2012), and silicate supplementation to the growing medium increased the number of branches, plant height, number of nodes, and stem diameter in the chrysanthemum 'Backwang' (Moon et al., 2008). However, Sivanesan et al. (2013a,b) reported that the number and size of chrysanthemum flowers increased when the nutrient solution was supplemented with Si, but necrotic lesions developed in adult leaves. The Si leaf content was greater in chrysanthemums that received silicate fertilization, but this increase in Si uptake did not affect any of the phytochemical characteristics evaluated (Carvalho-Zanão et al., 2012).

7.2.6 *COTONEASTER WILSONII*

The genus *Cotoneaster* (Rosaceae) consists of around 300 species of woody plants varying in stature from 0.2 m shrubs to 15–20 m trees, and occurring across Europe, North Africa, and temperate regions of Asia, excluding Japan (Bartish et al., 2001). Hyperhydricity, a common problem reported during its micropropagation, has been significantly reduced by supplementation with Si. The inclusion of 50 or 100 mg·L^{-1} Si to the medium significantly reduced the percentage of hyperhydricity of *C. wilsonii* when compared to the control (Sivanesan et al., 2011). Scanning electron microscope analysis showed that the stomata of nonhyperhydric plants exhibited normal morphology, while hyperhydric leaves displayed abnormal stomata that were elevated above the leaf surface and wide open due to deformed guard cells (Sivanesan et al., 2011).

7.2.7 GERBERA

Gerbera jamesonii belongs to the Asteraceae family and is popularly grown for cut or potted flowers (Jana and Jeong, 2014). Gerberas supplemented with several Si source and rate combinations, particularly NaSiO$_3$ foliar sprays, produced thicker flower peduncles, increased flower diameters, increased height, and flowered earlier than nonsupplemented controls (Kamenidou et al., 2010). Gerberas grown in Si-supplemented media all accumulated higher levels of Si in leaf, peduncle, and flower tissues than nonsupplemented controls (Kamenidou et al., 2010). The inclusion of Si in the nutrient solution of gerberas resulted in improved overall crop quality with thicker flower stems and a higher percentage of flowers when compared to the plants grown with the standard nutrient solution (Savvas et al., 2002). However, there were no significant differences between treatments in the dry weight of flowers and all photosynthetic parameters, such as transpiration, photosynthesis, water use efficiency, and stomatal conductance. In general, nanosilica increased the number of gerbera flowers, but potassium silicate did not improve the studied traits (Barakatain et al., 2013).

7.2.8 KALANCHOE

Kalanchoe blossfeldiana is one of the prettiest and most well-known representatives of the succulent family. Silicate fertilizer (MgCa(SiO$_3$)$_2$) supplementation at 40 g·L^{-1} medium resulted in the greatest plant height, leaf width and length, root length, and root fresh and dry weights in kalanchoe 'Kaluna' and 'Tula' (Bae et al., 2010b). Moreover, transverse section of kalanchoe 'Kaluna' and 'Tula' in roots and leaves tissues, silicate fertilizer supplementation of 40 g·L^{-1} medium were observed to be more compact tissues than the control one (Bae et al., 2010a). CaSiO$_3$ supplied through a subirrigation system improved the plant quality of kalanchoe 'Peperu', making compact potted plants (Son et al., 2012).

7.2.9 MARIGOLD

Tagetes patula L., commonly known as the French marigold, is mainly cultivated for ornamental purposes all over the world (Sivanesan et al., 2010). The application of Si decreases seedling height (Sivanesan et al., 2010).

7.2.10 *NEPHROLEPIS EXALTATA*

The most frequently used indoor ferns are Boston ferns (*Nephrolepis exaltata* and cultivars), which come in a variety of textures (Sivanesan et al., 2014). Sivanesan et al. (2014) reported that a high amount of Si deposition may reduce the transpiration rate and thereby enhance high-temperature tolerance in *Nephrolepis exaltata* 'Corditas'.

7.2.11 PHALAENOPSIS

The growth of the silica body was increased by increasing the concentration of CaSiO$_3$ from 0.01 to 0.5 mg·L^{-1} and was maximized from 0.5 to 1.0 mg·L^{-1} (Zhou, 1995). The Si concentrations in phalaenopsis orchid leaf tissues increased 3- to 10-fold as Si (KSiO$_3$) application rates increased, compared with a control (no Si fertilization), confirming that phalaenopsis orchid liners are Si accumulators and responsive to Si fertilization (Vendrame et al., 2010). Increased concentration of Silicon resultant in their active transportation in tissues of *Phalaenopsis* orchid (Vendrame et al., 2010).

7.2.12 POINSETTIA

Si sprays significantly reduced the occurrence and severity of bract necrosis in poinsettias (*Euphorbia pulcherrima* Willd. ex. Klotzsch), a physiological disorder caused by calcium deficiency (McAvoy and Bible, 1996). Si application increased the concentration of Si, while stem length decreased and the width of the canopy was decreased proportionally with increasing Si concentration (unpublished data).

7.2.13 ROSE

Mist applications of sodium silicate to rose (*Rosa hybrida* L.) cuttings decreased leaflet drop and increased rooting (Gillman and Zlesak, 2000). Si alleviates salt stress, decreases malondialdehyde content, and affects the petal color of salt-stressed cut roses (Reezi et al., 2009). The addition of Si to recirculated nutrient solution in a closed hydroponic system ameliorated most of the negative effects of recirculation on cut rose (*Rosa hybrida* L. 'Kardinal') production, improving stem quality (Ehret et al., 2005). Hwang et al. (2005) reported that applications of potassium silicate had beneficial effects on the growth and quality of cut flowers of the miniature rose 'Pinocchio' in a rockwool culture system. The incidence of powdery mildew in *Rosa hybrida* 'Remata' from the infection of *Sphaerotheca fuliginea* significantly decreased with 100 mg·L^{-1} K_2SiO_3 applied as foliar sprays, compared to that in the control (0 mg·L^{-1} K_2SiO_3) (Park et al., 2013).

7.2.14 SALVIA SPLENDENS

The genus *Salvia* (Lamiaceae) consists of approximately 900 species, including shrubs, herbaceous perennials, and annuals (Karousou et al., 2000). *In vitro* studies under salt stress have been reported on *Salvia splendens*. At 50 mM NaCl, 50 or 100 mg·L^{-1} K_2SiO_3 helped to overcome the salt effect and maximize the plant growth (Soundararajan et al., 2013). The addition of Si enhanced tolerance and significantly affected plant growth (Soundararajan et al., 2013). Particularly in 'Vista Red' treated with Si, a protein band, approximately 46 kDa, was expressed strongly under temperature stress (Soundararajan et al., 2014). Si not only promoted the growth of salvia, but also played a vital role against temperature stress (Soundararajan et al., 2014).

7.2.15 SUNFLOWER

Helianthus annuus L. belongs to the Asteraceae family and is prized for its ornamental flowers, seed oil, and medicinal value (Khursheed et al., 2009). Thick, straight stems, increased flower and stem diameters, and increased height were observed in some of the Si treatments of ornamental sunflower. In addition, sunflower quality was upgraded in Si-treated plants compared with untreated controls (Kamenidou et al., 2008).

7.2.16 ZINNIA

Zinnia elegans is known for its long vase life and disease-resistant characteristics (Dole, 1999). Potassium silicate applied as weekly drenches at 200 mg·L^{-1} increased apical and basal stem diameter and suppressed plant height (Kamenidou et al., 2009). *Z. elegans* grown in a greenhouse with weekly $NaSiO_3$ foliar spray at 100 mg·L^{-1} Si moderately increased leaf transpiration resistance and flower diameter (Kamenidou et al., 2009). Ranger et al. (2009) reported that nutrient solutions amended with potassium silicate did not affect the prereproductive period and survival of the green peach aphid (*Myzus persicae* Sulzer) reared on *Z. elegans*, whereas total cumulative fecundity and the intrinsic rate of increase were slightly reduced; however, this was only a modest increase in resistance levels. When Si was added to treatments exposed to Cu toxicity, the occurrence of discolored roots diminished, suggesting at least partial alleviation of Cu toxicity symptoms (Frantz et al., 2011).

7.3 Si UPTAKE AND DEPOSITION IN FLORICULTURAL PLANTS

7.3.1 POSITIVE EFFECTS

The positive effects of Si observed in monocots have generated interest for research with floricultural crops as well. The reported effects vary and depend strongly on plant species. Horticultural crops grown in Si-amended substrates exhibit a variety of responses related to abiotic and biotic stresses and morphology (Mattson and Leatherwood, 2010). The effects have been associated mainly with Si deposition in cell walls and as a double layer of polymerized Si in the cuticle, which presumably passively impedes evapotranspiration (Ma and Takahashi, 2002) and provides a mechanical defense against pests and pathogens (Belanger et al., 1995). Si, being dispersed through the plant via the transpiration stream (Samuels et al., 1991), inhibits fungal diseases through modifications of the epidermal layer of the leaves and fruits, as well as by increasing the presence of low-molecular-weight metabolites (Fawe et al., 1998; Gillman et al., 2003). Si alleviated salt stress by modulating antioxidant enzyme activities in *Dianthus caryophyllus* (Soundararajan et al., 2015). Si has many effects, which include improving the cell wall thickness below the cuticle and also the leaf angle, making leaves more erect, and thus reducing self-shading, especially under a high nitrogen rate (Mauad et al., 2003).

7.3.2 NEGATIVE EFFECTS

Si accumulation in natural systems is rarely linked to adverse effects in plants, even in those that have a high Si uptake capacity (Cooke and Leishman, 2011). Montpetit et al. (2012) also reported that Si absorption caused leaf necrosis in *Arabidopsis*, and this symptom was proportional to the external Si concentration. Chrysanthemums with a subirrigational supply of Si developed necrotic lesions in the older leaves at the beginning of the flowering stage, compared to the control and foliar spray of Si (Sivanesan et al., 2013a,b). Further, the addition of Si to the nutrient solution decreased leaf tissue concentrations of Ca, Mg, S, Na, B, Cu, Fe, and Mn in chrysanthemums (Sivanesan et al., 2013b). Si treatment reduced the leaf tissue Ca concentration in sunflowers (Kamenidou et al., 2008). Growth abnormalities of sunflowers were observed when concentrations of 100 and 200 $mg \cdot L^{-1}$ Si were supplied as $KSiO_3$ substrate drenches (Kamenidou et al., 2008).

7.3.3 SI UPTAKE AND DEPOSITION IN FLORICULTURAL PLANTS

Si is the second most abundant element in the earth's crust and soil and is found in significant amounts in all plants (Ma et al., 2006). However, plant species differ greatly in Si accumulation, ranging from 0.1% to 10% of dry weight (Epstein, 1994; Ma and Takahashi, 2002). This difference is attributed to the difference in the ability of roots to take up Si (Takahashi et al., 1990). Si is taken up by the roots in the form of silicic acid ($Si(OH)_4$), an uncharged monomeric molecule, when the solution pH is below 9 (Ma and Takahashi, 2002). Physiological studies have shown that Si uptake by rice roots is mediated by a type of transporter (Tamai and Ma, 2003). After it is taken up, Si is translocated to the shoot in the form of monomeric silicic acid (Casey et al., 2003; Mitani et al., 2005) and is finally deposited on cell wall materials as a polymer of hydrated, amorphous silica, forming silica–cuticle double layers and silica cellulose double layers on the surface of leaves, stems, and hulls (Yoshida, 1965). Si is still not recognized as an essential element for plant growth, but the beneficial effects of this element on growth, development, yield, and disease resistance have been observed in a wide variety of plant species (Ma, 2004). Si mitigates fungal infections and pest attacks, alleviates lodging and other abiotic stresses, improves the light interception ability of plants in a community, and minimizes transpiration losses (Ma and Takahashi, 2002). Soluble Si in plants also has an active function in enhancing host resistance to plant diseases by stimulating one or more defense reaction mechanisms (Fauteux et al., 2005). Two mechanisms for Si-enhanced resistance to diseases have been proposed. One is that

Si acts as a physical barrier. Si is deposited beneath the cuticle to form a cuticle Si double layer (Fauteux et al., 2005). Another mechanism proposed recently is that soluble Si acts as a modulator of host resistance to pathogens. Several studies in monocots (rice and wheat) and dicots (cucumbers) have shown that plants supplied with Si can produce phenolics and phytoalexins in response to fungal infections, such as those causing rice blast and powdery mildew (Fawe et al., 1998; Belanger et al., 2003; Rodrigues et al., 2004; Remus-Borel et al., 2005). The deposition of Si in culms, leaves, and hulls enhances the strength and rigidity of cell walls and decreases transpiration from the cuticle, and thus increases resistance to lodging, low and high temperatures, radiation, UV, and drought stresses. Under drought and salt stresses, Si-alleviated effects have been associated with an increase in antioxidant defense abilities (Liang et al., 2003; Zhu et al., 2004; Gong et al., 2005).

Horticultural crops grown in Si-amended substrates exhibit a variety of responses related to abiotic and biotic stresses (Kamenidou et al., 2002; Gillman et al., 2003; Datnoff et al., 2006; Heine et al., 2007; Larsen, 2008) and morphology. Some plants accumulates more silicon than some essential micronutrients, they known as accumulator species, and they are capable to accumulate more than 1000 mg·kg^{-1} (Epstein, 1999), while those plant species that can accumulate less than 1000 mg·kg^{-1} are known as Si-non accumulating species or Si-excluders (Frantz et al., 2008, 2010; Mattson and Leatherwood, 2010). Si is considered to be a beneficial element to higher plants (Epstein and Bloom, 2004), since the absorption process must be active or passive (Ma et al., 2007), and deposition in cell walls of several organs such as leaves and stems can promote beneficial effects (Cunha et al., 2008). Therefore, Si has been frequently linked to physiological, morphological, nutritional, and molecular aspects in plants (Ma et al., 2004; Ma and Yamaji, 2006; Lobato et al., 2009; Isa et al., 2010). Franz et al. (2008) very easily identified Si deposition based on the lack of wrinkled epidermal cells in various floricultural species. It has been reported in a variety of species that Si accumulates at trichome bases or wherever the xylem ends (Dengler and Lin, 1980; Dayanandan and Kaufman, 1976) and along leaf margins (Frantz et al., 2005). It appears from various reports that the capability of plants to benefit from Si supply is related to Si uptake into the root symplast. This hypothesis has been well documented by many studies with different crops exhibiting varied degrees of Si uptake (Jones and Handreck, 1967; Miyake and Takahashi, 1978; Epstein, 1994).

Si is accumulated by a broad range of bedding and potted plant species (Frantz et al., 2008) (Table 7.2). Voogt and Sonneveld (2001) and Voogt et al. (2005) evaluated the floricultural crops gerberas (*Gerbera* spp.), carnations (*Dianthus caryophyllus*), heath asters (*Aster ericoides*), poinsettias (*Euphorbia pulcherrima*), African violets (*Saintpaulia ionantha*), and roses (*Rosa*) and found that asters, poinsettias, African violets, and roses contained appreciable amounts of Si (more than 25 mmol·kg^{-1} dry mass). Overall, Si-supplemented gerberas accumulated less Si than zinnias and sunflowers supplemented with the same Si treatments (Kamenidou et al., 2008, 2009). All 14 species examined by Frantz et al. (2008) accumulated additional Si in their leaves when supplemented with potassium silicate. Zinnias accumulated Si to more than 1% dry weight, whereas verbenas, impatiens, New Guinea impatiens, and calibrachoas accumulated Si from 0.2 to 0.8% (Frantz et al., 2008). Mattson and Leatherwood (2010) found similar patterns for these species: petunias accumulated the least Si of any species examined; there was much greater accumulation by *Calibrachoa* than the related genus *Petunia*; verbena had among the highest Si concentrations; and New Guinea impatiens (*Impatiens hawkeri*) accumulated more Si than the related species *Impatiens walleriana*. All three cultivars of chrysanthemums were classified as Si accumulators, per Ma et al. (2001), as the concentration of Si in the leaves of the cultivars was found to be more than 10 g·kg^{-1} dry weight. Phalaenopsis orchid liners have been reported to be Si accumulators with an active uptake system (Vendrame et al., 2010).

The uptake mechanisms differ among plant species. In a study using rice, cucumbers, and tomatoes, species that accumulate high, medium, and low levels of Si, respectively, it was found that the transportation of Si from the external solution to the cortical cells is mediated by a similar

TABLE 7.2

Shoot Silicon Concentration in Si-Supplemented Floricultural Plants in a Descending Order

Plant	Si Concentration of Shoot (mg kg⁻¹ dwt)	Reference
Sunflower	18,300	Kamenidou et al., 2008
Zinnia	16,400	Kamenidou et al., 2009
	5365	Hogendorp et al., 2012
Aster	4796	Hogendorp et al., 2012
Torenia	4267	Mattson and Leatherwood, 2010
Argyranthemum	3507	Mattson and Leatherwood, 2010
Verbena	3365	Mattson and Leatherwood, 2010
Phlox	2427	Hogendorp et al., 2012
Calibrachoa	2308	Mattson and Leatherwood, 2010
Coreopsis	2219	Hogendorp et al., 2012
New Guinea impatiens	1976	Mattson and Leatherwood, 2010
Astilbe	1411	Hogendorp et al., 2012
Rosemary	1053	Mattson and Leatherwood, 2010
Sweet basil	1052	Hogendorp et al., 2012
Salvia	1070	Hogendorp et al., 2012
French marigold	624	Hogendorp et al., 2012
Coral flower	555	Hogendorp et al., 2012
Petunia	211	Mattson and Leatherwood, 2010
Gerbera	13	Kamenidou et al., 2010
Chrysanthemum	0.179	Sivanesan et al., 2013b
	0.136	Sivanesan et al., 2013a
Nephrolepis	0.060	Sivanesan et al., 2014

Note: dwt, dry weight.

transporter with a Km value of 0.15 mM in all three species (Mitani and Ma, 2005). However, the Vmax differed among the plant species (i.e., rice > cucumber > tomato), suggesting that the density of the transporter differs among plant species. It seems that this transport process is energy dependent because metabolic inhibitors and low temperatures inhibited transport (Mitani and Ma, 2005). Following uptake by the roots, Si is translocated to the shoot via the xylem. Chemically, silicic acid polymerizes to form silica gel ($SiO_2 \cdot nH_2O$) when the concentration of silicic acid exceeds 2 mM. However, the concentration of Si in the xylem sap is usually much higher than 2 mM in rice and wheat, even though the major form of Si in the xylem has been identified as monomeric silicic acid in these plant species (Casey et al., 2004; Mitani et al., 2005). In the shoots, silicic acid is further concentrated through water loss (transpiration) and is polymerized. The process of Si polymerization converts silicic acid to colloidal silicic acid and finally to silica gel with increasing silicic acid concentration (Ma and Takahashi, 2002). Si is deposited as a 2.5 mm thick layer in the space immediately beneath the thin (0.1 mm) cuticle layer, forming a cuticle–Si double layer in rice leaf blades (Ma and Takahashi, 2002). There are two types of silicified cells in rice leaf blades: silica cells, and silica bodies or silica motor cells (Ma and Takahashi, 2002).

High levels of Si in rice tissues are attributed to the superior ability of the roots to take up this element. Physiological studies have shown that rice takes up Si actively; the uptake is much faster than that of water and is not affected by transpiration (Ma and Takahashi, 2002; Raven, 2003). This high accumulation in the shoots has been attributed to the presence of an influx Si transporter (Low Silicon Rice 1 [Lsi1]) and efflux transporter (Lsi2) in rice roots. Both Lsi1 and Lsi2 of rice are localized at the plasma membrane of both exodermal and endodermal cells of the roots (Ma et al., 2006, 2007). Lsi1 belongs to a nodulin-26-like major intrinsic protein (NIP) subfamily in the aquaporin family. Lsi1 is expressed mainly in the roots, and the encoded protein shows transport activity specific to Si (Ma et al., 2007). The concerted functioning of these two transporters produces an effective flow of Si in the exodermis and endodermis to override the barrier of Casparian strips (Ma et al., 2006, 2007).

7.3.4 Lsi Genes

Transformation was achieved using a transporter gene (Lsi; low Si rice) responsible for high Si accumulation capacity in rice (Ma et al., 2006). Lsi1 belongs to the NIP III subgroup of aquaporins and is an influx transporter of silicic acid in plant cells. Lsi1 is localized in the root exodermis and endodermis, but is on the distal side (Ma et al., 2006, 2007). Lsi2 is mainly expressed in the roots as Lsi1 (Ma and Yamaji, 2008). There is little accumulation of Lsi2 transcripts in the root tip (0–10 mm), but much accumulation in the mature parts of roots (Ma et al., 2007). The expression of Lsi2 is also regulated by ABA, as seen in Lsi1 (Ma and Yamaji, 2008). Like Lsi1, Lsi2 is also localized on the exodermis and endodermis cells of the roots. However, in contrast to Lsi1 localized on the distal side, Lsi2 was localized on the proximal side of these cells (Ma et al., 2007). Si is then transported into the endodermis cells by Lsi1 and released into the stele by Lsi2 (Ma and Yamaji, 2008). The Si transported via Lsi1 and Lsi2 into the stele is then translocated to the shoot by transpirational flow through the xylem. Si is present in the xylem at a high concentration (Casey et al., 2003; Mitani et al., 2005), and over 90% of Si taken up by roots is translocated to the shoot (Ma and Takahashi, 2002). Si is deposited into the cell wall as a polymer of hydrated amorphous silica, forming silica–cuticle double layers, and also deposited in specific shoot cells as silica bodies (Ma and Takahashi, 2002; Prychid et al., 2004). Thus, agrobacterium-mediated transformation using the Lsi1 gene might be a good way of either transporting or translocating Si from the soil to the root. We carried out a study to obtain the Lsi1 and Lsi2 transformants using an economically important horticultural plant, the chrysanthemum, to see if Si uptake and tolerance to stresses are promoted (unpublished data). A plant regeneration system was established from different explants. Then, a transformation protocol was established and applied to chrysanthemum 'Sumi' using Lsi1 and Lsi2 genes. The Lsi1 and Lsi2 transformants increased Si uptake and tolerance to salinity and heat stresses (Table 7.3).

TABLE 7.3

Accumulated Shoot Si Concentrations in Chrysanthemum Plants Not Transformed (Wild-Type) or Transformed with Lsi1 and Lsi2 Genes, and Fed with (+) or without (–) 100 mg·L⁻¹ Si and 200 mM NaCl (Salt Stress-Inducing Agent) from the Nutrient Solution

	Shoot Si Concentration (μg·mg^{-1} dwt)	
Treatment	Lsi1	Lsi2
Not transformed	470c[a]	448c
Transformed	2078a	760b
–Si	435c	413c
+Si	2113a	905a
–NaCl	1275b	718b
+NaCl	1235b	600b

Note: Values are means of the three variables used as the treatments.

[a] Mean separation within columns by Duncan's multiple.

7.4 CONCLUSIONS AND FUTURE PROSPECTS

Si application has been found to be beneficial for improving resistance to insects, pathogens, drought, and heavy metals, and for floristic crop quality and yield. The beneficial effects of Si are usually expressed more clearly under stressed conditions. Thus, Si application in floristic crop production will ensure safe food production and environment protection. However, most horticultural plants studied have been shown to be low Si accumulators. The low Si uptake by horticultural plants suggests that they may lack homologs to the Lsi gene, the gene encoding Si transport in rice, and thus may not have the underlying molecular mechanisms to accumulate significant amounts of Si. Recently, the Lsi1, 2, 3, and 6 genes, which are responsible for active Si uptake, have been identified from a rice mutant that is defective in Si uptake. Si transporter studies conducted by Ma's research group in Japan have so far discovered the genes Lsi1, 2, 3, and 6. Among these, the Lsi1 and 2 genes are expressed mainly in the root and localized on plasma membranes at the proximal side of cells in both the exodermis and endodermis. Therefore, further studies are needed on the Lsi gene using an economically important floristic crop to see if Si uptake and tolerance to stresses are promoted.

REFERENCES

Arnon, D.I., and P.R. Stout. 1939. The essentiality of certain elements in minute quantity for plant with special reference to copper. *Plant Physiol.* 14:371–375.

Bae, M.J., Y.G. Park, and B.R. Jeong. 2010a. Effect of a silicate fertilizer supplemented to a medium on the growth and development of potted plants. *Flowers Res. J.* 18:50–56.

Bae, M.J., Y.G. Park, and B.R. Jeong. 2010b. Effect of a silicate fertilizer supplemented to the medium on rooting and subsequent growth of potted plants. *Hortic. Environ. Biotechnol.* 51:355–359.

Barakatain, L., A. Nikbakht, N. Etemadi, and J.K. Ali. 2013. Effect of source and method of silica application on some of the quantitative and physiological characteristics of *Gerbera jamesonii* L. *J. Sci. Technol. Greenhouse Cult.* 4:39–47.

Bartish, I.V., B. Hylmo, and H. Nybom. 2001. RAPD analysis of interspecific relationships in presumably apomictic *Cotoneaster* species. *Euphytica* 120:273–280.

Bayat, H., M. Alirezaie, and H. Neamati. 2012. Impact of exogenous salicylic acid on growth and ornamental characteristics of calendula (*Calendula officinalis* L.) under salinity stress. *J. Stress Physiol. Biochem.* 8:258–267.

Bayat, H., M. Alirezaie, H. Neamati, and A.A. Saadabad. 2013. Effect of silicon on growth and ornamental traits of salt-stressed calendula (*Calendula officinalis* L.). *J. Ornam. Plants* 3:207–214.

Belanger, R.R., N. Benhamou, and J.G. Menzies. 2003. Cytological evidence of an active role of silicon in wheat resistance to powdery mildew (*Blumeria graminis* f. sp. *tritici*). *Phytopathology* 93:402–412.

Belanger, R.R., P.A. Bowen, D.L. Ehret, and J.G. Menzes. 1995. Soluble silicon: Its role in crop disease management of greenhouse crops. *Plant Dis.* 79:329–336.

Bowen, P., J. Menzies, D. Ehret, L. Samuels, and A.D.M. Glass. 1992. Soluble silicon sprays inhibit powdery mildew development on grape leaves. *J. Am. Soc. Hortic. Sci.* 117:906–912.

Carvalho-Zanão, M.P., L.A. Zanão Júnior, J.G. Barbosa, J.A.S. Grossi, and V.T. de Ávila, 2012. Yield and shelf life of chrysanthemum in response to the silicon application. *Hortic. Brasil.* 30:403–408.

Casey, W.H., S.D. Kinrade, C.T.G. Knight, D.W. Rains, and E. Epstein. 2003. Aqueous silicate complexes in wheat, *Triticum aestivum* L. *Plant Cell Environ.* 27:49–52.

Cooke, J., and M.R. Leishman. 2011. Is plant ecology more siliceous than we realize? *Trends Plant Sci.* 16:61–68.

Cunha, K.P.V., C.W.A. Nascimento, A.M.A. Accioly, and A.J. Silva. 2008. Cadmium and zinc availability, accumulation and toxicity in maize grown in a contaminated soil. *Rev. Bras. Ciênc. Solo* 3:1319–1328.

Datnoff, L.E., T.A. Nell, R.T. Leonard, and B.A. Rutherford. 2006. Effect of silicon on powdery mildew development on miniature potted rose [abstract]. *Phytopathology* 96:S28.

Dayanandan, P., and P.B. Kaufman. 1976. Trichomes of *Cannabis sativa* L. (Cannabaceae). *Am. J. Bot.* 63:578–591.

Dengler, N.G., and E.Y.-C. Lin. 1980. Electron microprobe analysis of the distribution of silicon in the leaves of *Selaginella emmeliana*. *Can. J. Bot.* 58:2459–2466.

Dole, H.C. 1999. Zinnias: Colorful, butterfly-approved. *Butterfly Gardeners Quarterly*.

Ehret, D.L., J.G. Menzies, and T. Helmer. 2005. Production and quality of greenhouse roses in recirculating nutrient systems. *Sci. Hortic.* 106:103–113.

Epstein, E. 1994. The anomaly of silicon in plant biology. *Proc. Nat. Acad. Sci. USA* 94:11–17.

Epstein, E. 1999. Silicon. *Annu. Rev. Plant Physiol. Plant Mol. Biol.* 50:641–664.

Epstein, E., and A.J. Bloom. 2004. *Mineral Nutrition of Plants: Principles and Perspectives.* Sinauer Associates, Sunderland, MA.

Fauteux, F., W. Remus-Borel, J.G. Menzies, and R.R. Belanger. 2005. Silicon and plant disease resistance against pathogenic fungi. *FEMS Microbiol. Lett.* 249:1–6.

Fawe, A., M. Abou-zaid, J.G. Menzies, and R.R. Belanger. 1998. Silicon-mediated accumulation of flavonoid phytoalexins in cucumber. *Phytopathology* 88:396–401.

Frantz, J.M., J.C. Locke, L. Datnoff, M. Omer, A. Widrig, D. Sturtz, L. Horst, and C.R. Krause. 2008. Detection, distribution, and quantification of silicon in floricultural crops utilizing three distinct analytical methods. *Commun. Soil Sci. Plant Anal.* 39:2734–2751.

Frantz, J.M., J.C. Locke, D. Sturtz, and S. Leisner. 2010. Silicon in ornamental crops: Detection, delivery, and function. In F.A. Rodrigues (ed.), *Silicio na agricultura: Anais do V simposio brasileiro sobre silicio na agricultura*, pp. 111–134. Universidade Federal de Vicxosa, Vicxosa, Brazil.

Frantz, J.M., S. Khandekar, and S. Leisner. 2011. Silicon differentially influences copper toxicity response in silicon-accumulator and non-accumulator species. *J. Am. Soc. Hortic. Sci.* 136:329–338.

Frantz, J.M., D.D.S. Pitchay, J.C. Locke, L.E. Horst, and C.R. Krause. 2005. Silicon is deposited in leaves of New Guinea impatiens. *Plant Health Progress.* Available at http://www.Plantmanagementnetwork.org /sub/php/research/2005/silicon.

Gan, Y., L. Zhou, S.J. Zhong, Z.X. Shen, Y.Q. Zhang, and G.X. Wang. 2010. Stomatal clustering, a new marker for environmental perception and adaptation in terrestrial plants. *Bot. Stud.* 51:325–326.

Gillman, J.H., and D.C. Zlesak. 2000. Mist applications of sodium silicate to rose (*Rosa* L. × 'Nearly Wild') cuttings decrease leaflet drop and increase rooting. *Hortscience* 35:773.

Gong, H., X. Zhu, K. Chen, S. Wang, and C. Zhang. 2005. Silicon alleviates oxidative damage of wheat plants in pots under drought. *Plant Sci.* 169:313–321.

Hogendorp, B.K., R.A. Cloyd, and J.M. Swiader. 2012. Determination of silicon concentration in some horticultural plants. *Hortscience* 47:1593–1595.

Hwang, S.J., H.M. Park, and B.R. Jeong. 2005. Effects of potassium silicate on the growth of miniature rose 'Pinocchio' grown on rockwool and its cut flower quality. *J. Jpn. Soc. Hortic. Sci.* 74:242–247.

Isa, M., S. Bai, T. Yokoyama, J.F. Ma, Y. Ishibashi, T. Yuasa, and M. Iwaya-Inoue. 2010. Silicon enhances growth independent of silica deposition in a low-silica rice mutant, *lsi1. Plant Soil* 331:361–375.

Jamali, B., and M. Rahemi. 2011. Carnation flowers senescence as influenced by nickel, cobalt and silicon. *J. Biol. Environ. Sci.* 5:147–152.

Jana, S., and B.R. Jeong. 2014. Silicon: The most under-appreciated element in horticultural crops. *Trends Hortic. Res.* 4:1–19.

Jeong, K.J., Y.S. Chon, S.H. Ha, H.K. Kang, and J.G. Yun. 2012. Silicon application on standard chrysanthemum alleviates damages induced by disease and aphid insect. *Kor. J. Hortic. Sci. Technol.* 30:21–26.

Jones, L.H.P., and K.A. Handreck. 1967. Silica in soils, plants and animals. *Adv. Agron.* 19:107–149.

Kamenidou, S., T.J. Cavins, and S. Marek. 2002. Silicon supplementation affects greenhouse produced cut flowers. MS thesis, Oklahoma State University, Stillwater.

Kamenidou, S., T.J. Cavins, and S. Marek. 2008. Silicon supplements affect horticultural traits of greenhouse-produced ornamental sunflowers. *Hortscience* 43:236–239.

Kamenidou, S., T.J. Cavins, and S. Marek. 2009. Evaluation of silicon as a nutritional supplement for greenhouse zinnia production. *Sci. Hortic.* 119:297–301.

Kamenidou, S., T.J. Cavins, and S. Marek. 2010. Silicon supplements affect floricultural quality traits and elemental nutrient concentrations of greenhouse produced gerbera. *Sci. Hortic.* 123:390–394.

Karousou, R., E. Hanlidou, and S. Kokkini. 2000. The sage plants in Greece: Distribution and infraspecific variation. In S.E. Kintzios (ed.), *Sage: The Genus Salvia*, pp. 27–46. Harwood Academic Publishers, Amsterdam.

Keeping, M.G., and O.L. Kvedaras. 2008. Silicon as a plant defense against insect herbivory: Response to Massay, Ennos, and Hartley. *J. Anim. Ecol.* 77:631–633.

Khursheed, T., M.Y.K. Ansari, and D. Shahab. 2009. Studies on the effect of caffeine on growth and yield parameters in *Helianthus annuus* L. variety Modern. *Biol. Med.* 1:56–60.

Larsen, A.K. 2008. Less mildew in pot roses with silicon. *FlowerTech* 11:18–19.

Liang, Y., Q. Chen, Q. Liu, W. Zhang, and R. Ding. 2003. Exogenous silicon (Si) increases antioxidant enzyme activity and reduces lipid peroxidation in roots of salt-stressed barley (*Hordeum vulgare* L.). *J. Plant Physiol.* 160:1157–1164.

Liang, Y., W. Sun, Y.G. Zhu, and P. Christie. 2007. Mechanisms of silicon-mediated alleviation of abiotic stresses in higher plants: A review. *Environ. Pollut.* 147:422–428.

Lim, M.Y., E.J. Lee, S. Jana, I. Sivanesan, and B.R. Jeong. 2012. Effect of potassium silicate on growth and leaf epidermal characteristics of begonia and pansy grown in vitro. *Kor. J. Hortic. Sci. Technol.* 30:579–585.

Lobato, A.K.S., G.K. Coimbra, M.A.M. Neto, R.C.L. Costa, B.G. Santos Filho, C.F. Oliveira Neto, L.M. Luz et al. 2009. Protective action of silicon on relations and photosynthetic pigments in pepper plants induced to water deficit. *Res. J. Biol. Sci.* 4:617–623.

Ma, J.F. 2004. Role of silicon in enhancing the resistance of plants to biotic and abiotic stresses. *Soil Sci. Plant Nutr.* 50:11–18.

Ma, J.F., N. Mitani, S. Nagao, S. Konishi, K. Tamai, T. Iwashita, and M. Yano. 2004. Characterization of the silicon uptake system and molecular mapping of the silicon transporter gene in rice. *Plant Physiol.* 136:3284–3289.

Ma, J.F., and E. Takahashi. 2002. *Soil, Fertilizer, and Plant Silicon Research in Japan.* Elsevier, Amsterdam.

Ma, J.F., K. Tamai, N. Yamaji, N. Mitani, S. Konishi, M. Katsuhara, M. Ishiguro, Y. Murata, and M. Yano. 2006. A silicon transporter in rice. *Nature* 440:688–691.

Ma, J.F., and N. Yamaji. 2006. Silicon uptake and accumulation in higher plants. *Trends Plant Sci.* 11:392–397.

Ma, J.F., and N. Yamaji. 2008. Functions and transport of silicon in plants. *Cell. Mol. Life Sci.* 65:3049–3057.

Ma, J.F., N. Yamaji, N. Mitani, K. Tamai, S. Konishi, T. Fujiwara, M. Katsuhara, and M. Yano. 2007. An efflux transporter of silicon in rice. *Nature* 448:209–212.

Ma, J.F., Y. Miyake, and E. Takahashi. 2001. Silicon as a beneficial element for crop plant. In L.E. Datnoff, G.H. Korndorfer, and G. Snyder (eds.), *Silicon in Agriculture*, pp. 17–39. Elsevier, Amsterdam.

Mattson, N.S., and W.R. Leatherwood. 2010. Potassium silicate drenches increase leaf silicon content and affect morphological traits of several floriculture crops grown in a peat-based substrate. *Hortscience* 45:43–47.

Mauad, M., C.A.C. Crusciol, H. Grassi Filho, and J.C. Corrêa. 2003. Nitrogen and silicon fertilization of upland rice. *Sci. Agri.* 60:761–765.

McAvoy, R.J., and B.B. Bible. 1996. Silica sprays reduce the incidence and severity of bract necrosis in poinsettia. *Hortscience* 31:1146–1149.

Mitani, N., and J.F. Ma. 2005. Uptake system of silicon in different plant species. *J. Exp. Bot.* 56:1255–1261.

Mitani, N., J.F. Ma, and T. Iwashita. 2005. Identification of silicon form in the xylem of rice (*Oryza sativa* L.). *Plant Cell Physiol.* 46:279–283.

Miyake, Y., and E. Takahashi. 1978. Silicon deficiency of tomato plant. *Soil Sci. Plant Nutr.* 24:175–189.

Montpetit, B., A. Royer, A. Langlois, P. Cliche, A. Roy, N. Champollion, G. Picard, F. Domine, and R. Obbard. 2012. New shortwave infrared albedo measurements for snow specific surface area retrieval. *J. Glaciol.* 58:941.

Moon, H.H., M.J. Bae, and B.R. Jeong. 2008. Effect of silicate supplemented medium on rooting of cutting and growth of chrysanthemum. *Flower Res. J.* 16:107–111.

Park, Y.G., I. Sivanesan, and B.R. Jeong. 2013. Effect of silicon source and application method on growth and development, and incidence of powdery mildew (*Sphaerotheca pannosa* var. *rosae*) in potted *Rosa hybrida* 'Apollo' and 'Remata'. *Flower Res. J.* 21:56–62.

Parrella, M.P., T.P. Costamagna, and R. Kaspi. 2007. The addition of potassium silicate to the fertilizer mix to suppress *Liriomyza leafminers* attacking chrysanthemum. *Acta Hortic.* 747:365–370.

Prychid, C.J., P.J. Rudall, and M. Gregory. 2004. Systematics and biology of silica bodies in monocotyledons. *Bot. Rev.* 69:377–440.

Ranger, C.M., A.P. Singh, J.M. Frantz, L. Canas, J.C. Locke, M.E. Reding, and N. Vorsa. 2009. Influence of silicon on resistance of *Zinnia elegans* to *Myzus persicae* (Hemiptera: Aphididae). *Environ. Entomol.* 38:129–136.

Raven, J.A. 2003. Cycling silicon—The role of accumulation in plants. *New Phytol.* 158:419–430.

Reezi, S., M. Babalar, and S. Kalantari. 2009. Silicon alleviates salt stress, decreases malondialdehyde content and affects petal color of salt stressed cut rose (*Rosa* × *hybrida* L.) Hot Lady. *Afr. J. Biotechnol.* 8:1502–1508.

Remus-Borel, W., J.G. Menzies, and R.R. Belanger. 2005. Silicon induces antifungal compounds in powdery mildew-infected wheat. *Physiol. Mol. Plant Pathol.* 66:108–115.

Rodrigues, F.A., D.J. McNally, L.E. Datnoff, J.B. Jones, C. Labbe, N. Benhamou, J.G. Menzies, and R.R. Belange. 2004. Silicon enhances the accumulation of diterpenoid phytoalexins in rice: A potential mechanism for blast resistance. *Phytopathology* 94:177–183.

Samuels, A.L., A.D.M. Glass, D.M. Ehret, and J.G. Menzies. 1991. Mobility and deposition of silicon in cucumber plants. *Plant Cell Environ.* 14:485–92.

Sangster, A.G., and M.J. Hodson. 1986. Silica in higher plants. *CIBA Found. Symp.* 121:90–111.

Sangster, A.G., M.J. Hodson, and H.J. Tubb. 2001. Silicon deposition in higher plants. In L.E. Datnoff, G.H. Snyder, and G.H. Korndörfer (eds.), *Silicon in Agriculture*, pp. 85–113. Elsevier, Amsterdam.

Savvas, D., G. Manos, A. Kotsiras, and S. Souvaliotis. 2002. Effects of silicon and nutrient-induced salinity on yield, flower quality and nutrient uptake of gerbera grown in a closed hydroponic system. *J. Appl. Bot.* 76:153–158.

Sivanesan, I., and B.R. Jeong. 2014. Silicon promotes adventitious shoot regeneration and enhances salinity tolerance of *Ajuga multiflora* bunge by altering activity of antioxidant enzyme. *Sci. World J.* 2014:1–10.

Sivanesan, I., M.S. Son, J.P. Lee, and B.R. Jeong. 2010. Effects of silicon on growth of *Tagetes patula* L. 'Boy Orange' and 'Yellow Boy' seedlings cultured in an environment controlled chamber. *Propag. Ornam. Plants* 10:136–140.

Sivanesan, I., M.S. Son, J.Y. Song, and B.R. Jeong. 2013a. Silicon supply through the subirrigation system affects growth of three chrysanthemum cultivars. *Hortic. Environ. Biotechnol.* 54:14–19.

Sivanesan, I., M.S. Son, P. Soundararajan, and B.R. Jeong. 2013b. Growth of chrysanthemum cultivars as affected by silicon source and application method. *Kor. J. Hortic. Sci. Technol.* 31:544–551.

Sivanesan, I., M.S. Son, P. Soundararajan, and B.R. Jeong. 2014. Effect of silicon on growth and temperature stress tolerance of *Nephrolepis exaltata* 'Corditas'. *Kor. J. Hortic. Sci. Technol.* 32:142–148.

Sivanesan, I., J.Y. Song, S.J. Hwang, and B.R. Jeong. 2011. Micropropagation of *Cotoneaster wilsonii* Nakai-A rare endemic ornamental plant. *Plant Cell Tiss. Org. Cult.* 105:55–63.

Son, M.S., H.J. Oh, J.Y. Song, M.Y. Lim, I. Sivanesan, and B.R. Jeong. 2012. Effect of silicon source and application method on growth of kalanchoe 'Peperu'. *Kor. J. Hortic. Sci. Technol.* 30:250–255.

Son, M.S., J.Y. Song, M.Y. Lim, I. Sivanesan, G.S. Kim, and B.R. Jeong. 2013. Silicon uptake level of six potted plants from a potassium silicate-supplemented hydroponic solution. *Kor. J. Hortic. Sci. Techonol.* 31:153–158.

Song, J.Y., N.S. Mattson, and B.R. Jeong. 2011. Efficiency of shoot regeneration from leaf, stem, petiole and petal explants of six cultivars of *Chrysanthemum morifolium*. *Plant Cell Tiss. Org. Cult.* 107:295–304.

Soundararajan, P., A. Manivannan, Y.G. Park, S. Muneer, and B.R. Jeong. 2015. Silicon alleviates salt stress by modulating antioxidant enzyme activities in *Dianthus caryophyllus* 'Tula'. *Hortic. Environ. Biotechnol.* 56:233–239.

Soundararajan, P., I. Sivanesan, S. Jana, and B.R. Jeong. 2014. Influence of silicon supplementation on the growth and tolerance to high temperature in *Salvia splendens*. *Hortic. Environ. Biotechnol.* 55:271–279.

Soundararajan, P., I. Sivanesan, E.H. Jo, and B.R. Jeong. 2013. Silicon promotes shoot proliferation and shoot growth of *Salvia splendens* under salt stress in vitro. *Hortic. Environ. Biotechnol.* 54:311–318.

Takahashi, E., J.F. Ma, and Y. Miyake. 1990. The possibility of silicon as an essential element for higher plants. *Comments Agric. Food Chem.* 2:99–122.

Tamai, K., and J.F. Ma. 2003. Characterization of silicon uptake by rice roots. *New Phytol.* 158:431–436.

Vendrame, W.A., A.J. Palmateer, A. Pinares, K.A. Moore, and L.E. Datnoff. 2010. Silicon fertilization affects growth of hybrid phalaenopsis orchid liners. *Horttechnology* 20:603–607.

Voogt, W., and C. Sonnenveld. 2001. Silicon in horticultural crops grown in soilless culture. In L.E. Datnoff, G.H. Snyder, and G.H. Korndörfer (eds.), *Silicon in Agriculture: Studies in Plant Science*, pp. 115–131. Elsevier Science, Amsterdam.

Voogt, W., J.P. Wubben, and N.A. Straver. 2005. The effect of silicon application on some ornamental plants. In Proceedings of the III Silicon in Agriculture Conference, Uberlandia, Brazil, October 22–26, p. 128 (abstract).

Yoshida, S. 1965. Chemical aspects of the role of silicon in physiology of the rice plant. *Bull. Natl. Inst. Agric. Sci. B* 15:1–58.

Zhou, T. 1995. The detection of the accumulation of silicon in *Phalaenopsis* (Orchidaceae). *Ann. Bot.* 75:605–607.

Zhu, Z., G. Wei, J. Li, Q. Qian, and J. Yu. 2004. Silicon alleviates salt stress and increases antioxidant enzymes activity in leaves of salt-stressed cucumber (*Cucumis sativus* L.). *Plant Sci.* 167:527–533.

8 Silicon and Alleviation of Salt Stress in Crop Genotypes Differing in Salt Tolerance

Muhammad Shahzad, Abdul Qadir, and Sajid Masood

CONTENTS

ABSTRACT

Soil salinity, a white plague, affects the plant growth and ultimately yield of important agricultural crops throughout the world. The most common channel of salt stress injury is to disrupt crop plants via cellular ionic and osmotic imbalance. Albeit silicon (Si) is generally considered nonessential for plant growth and development, nevertheless, it has been found beneficial for improving crop tolerance to both biotic (e.g., diseases and pathogens) and abiotic (e.g., salinity, drought, and nutrient toxicity or deficiency) stresses. Si is ranked second regarding its abundance in the earth's crust, after oxygen. Regardless of the fact that its role in plant biology is incompletely understood, Si application has been widely suggested to be used for the improvement of crop production under adverse climate and soil conditions. A number of possible mechanisms have been explained where Si confers tolerance mechanisms to plants against environmental stresses, especially salinity stress. However, the responses of different crop genotypes have been found to vary with Si addition under salinity stress.

The underlying mechanisms of Si for improving the resistance in plants to salt stress include maintenance of plant water status, reduction in Na^+ uptake, improvement of the K^+/Na^+ ratio, maintenance of the nutrient balance, promotion of the photosynthetic rate, regulation of the levels of endogenous plant hormones, alteration of the transpiration rate, and mitigation of the toxicity of other heavy metals. Si application has also been found to improve the defensive system of plants by producing antioxidants, which sequentially detoxify the reactive oxygen species (ROS). These physiological and morphological improvements in plants have proved the Si role in the mitigation of salt stress. Recent studies with monocots have shown that Si accumulation in plants is mediated by specific transporters. Here, in this chapter, we summarize the agronomic, physiological/biochemical, and genetic/proteomic responses of different crop genotypes in the alleviation of salt stress as a result of Si supply. Further, advancements with a combination of molecular and genetic studies on Si transport and functions in different crop genotypes will help advance understanding of its complete role under abiotic stress conditions like salinity.

8.1 INTRODUCTION

8.1.1 SALINITY AND PLANT GROWTH

Salt stress badly influences plant growth and limits the agricultural productivity of the world. Climate change and agronomic practices throughout the world have contributed to increased soil salinity, which currently affects an estimated 45 million hectares of irrigated land (Rengasamy 2010). The problem with salinity is aggravated as a result of the burgeoning population and intensive use of natural resources, especially in developing countries. To increase food quality and quantity, plants should utilize different strategies to overcome the adverse effects of salinity. For food security, high priority should be given to minimize the harmful effects of salinity. However, the genetic approaches can increase the plant resistance against salt toxic effects.

Understanding the mechanism of plant growth under salt stress conditions has been found to be complex, especially because the way it is affected by salinity stress is not fully understood, as the response of plants to undue salinity is versatile and comprised of changes in the morphology, physiology, and metabolism of the plants (Rhoades 1993; Hilal et al. 1998). A lower transpiration rate, coupled with reduced ion uptake by the roots, or reduced xylem loading, may cause poor supply via the xylem. It is possible that an inadequate supply of ions to the expanding region may restrict the cell division or expansion when the plants are grown at high levels of NaCl (Bernstein et al. 1995). Salinity affects the plant physiology by changing the water and ionic status of the cells (Muscolo et al. 2003; Masood et al. 2012; Shahzad et al. 2012). Ionic imbalance takes place in the cells due to excessive buildup of Na^+ and Cl^-, which affects the uptake of other mineral nutrients (Wang and Han 2007; Shahzad et al. 2013). High Na^+ disturbs the K^+ nutrition and inhibits many enzymes, activities (Jaleel et al. 2007; Wang and Han 2007). Under salt stress, plants must expend supplementary cellular resources to maintain high concentrations of cytosolic K^+ and low concentrations of Na^+. Excessive salinity normally disturbs the cellular electron transport within the different subcellular compartments and results in the generation of ROS, which activates phytotoxic reactions such as lipid peroxidation, protein degradation, and DNA mutation (Ali and Alqurainy 2006).

8.1.2 Silicon and Plant Growth

Si is generally considered nonessential for plant growth and development; nevertheless, it can improve crop growth and production under adverse climate and soil conditions. According to Epstein and Bloom (2005), Si should be added to the list of essential elements for higher plants, because in the absence of Si, abnormal plant growth is observed, compared with Si-treated plants. Despite the abundance of Si in soil, most of it cannot be absorbed directly by plants. Si is generally taken up by the plant root in the form of soluble silicic acid, $Si(OH)_4$, which ranges from 0.1 to 0.6 mM in normal soil solution (Mitani and Ma 2005). However, many factors, like soil pH, temperature, water status, presence of cations, and organic compounds, eventually affect the availability of Si to plants (Liu et al. 2002).

Increasing evidence suggests that species of plants and their variety highly varies against the resistance, in terms of salinity (Wahid and Rasul 1997). As mentioned in the above section, a high salinity level alters the morphology, physiology, and metabolism of the plants; however, these changes have been partially overcome by the addition of Si in different crop genotypes (Table 8.1). Furthermore, Si is not damaging to plants when it is present in surplus (Epstein 1994; Ma et al. 2001). Earlier findings have demonstrated that there are great differences even among genotypes within the same species (Deren 2001; Ma and Takahashi 2002; Hodson et al. 2005). Moreover, the absorption of Si by selected species or cultivars, such as banana, rice, or tomato, were different in various Si concentrations (Ma and Takahashi 2002; Henriet et al. 2006). This suggested that the availability of Si also regulates the amount of Si absorbed by the plants.

A recent review by Zhu and Gong (2014) presented literature related to an efficient uptake system of Silicon, that is mediated by both influx (Lsi1) and efflux (Lsi2) transporters and accounts for Si accumulation in monocots, including wheat (Montpetit et al. 2012), rice (Ma et al. 2006, 2007), barley (Chiba et al. 2009; Yamaji et al. 2012), maize (Mitani et al. 2009a,b), and the dicot pumpkin (*Cucurbita moschata* Duch.) (Mitani et al. 2011). According to Ma et al. (2006) and Ma (2010), aquaporin-like protein (influx transporter) is involved in the transport of Si from the outside solution to the exodermal cells and then to the endodermal cells, which is encoded by Lsi1. Lsi2 (efflux transporter) is a plasma membrane–localized transporter that functions to release Si from exodermal cells to the apoplast and then to the stele (Ma et al. 2007).

8.2 SILICON AND SALT STRESS IN MONOCOTS

8.2.1 Wheat

Wheat is generally considered a moderately salt-sensitive crop, as the growth is adversely affected by salt stress (Zhu et al. 2004). Better growth of crops, especially from the Poaceae and Cyperaceae families, along with certain other crops, has been observed after Si application, because they tend to accumulate great amounts of Si (Mitani and Ma 2005). However, according to Matichenkov and Kosobrukhov (2004), Si is hypothetically deposited more in the tissues of graminous plants than in other species. Wheat belongs to the family Poaceae and has a tendency to accumulate Si. Therefore, the use of Si can be beneficial to the growth of wheat against salt stress, as it is cost-efficient and eco-friendly to the environment. Si has been widely suggested by scientists to ameliorate the adverse effects of salinity and recover the development of crops exposed to saline conditions. As salinity causes a substantial drop in the development, phytochemical contents, and harvest of wheat, however, Si application has been found to mitigate salinity stress through different ways (Sharaf 2010). It is evident from the literature that adding a silicate supplement to the nutrient solution of wheat improves the plant's resistance to salinity stress. According to Ahmad et al. (1992), Si application improved the shoot dry weight; however, the root dry weight of wheat plants remained unaffected. Similarly, wheat leaf area, leaf thickness, and leaf water potential were improved with the addition of Si in saline conditions (Gong et al. 2003, 2005). In Ahmad et al.'s (1992) study, Si

TABLE 8.1

Various Aspects (Agronomic, Physiological/Biochemical, and Genetic/Proteomic) Studied with Si Addition in Different Monocot and Dicot Plants under Salinity Stress

Plant Flowering Group	Crop	Observed Parameters	References
Monocots	Wheat	Agronomic	Ahmad et al. (1992), Sharaf (2010), Tahir et al. (2012), Ahmad (2014)
		Physiological/biochemical	Ahmad et al. (1992), Matichenkov and Kosobrukhov (2004), Mali and Aery (2008), Tuna et al. (2008), Saqib et al. (2008), Sharaf (2010), Tahir et al. (2012), Ahmad (2014)
	Rice	Agronomic	Gong et al. (2006), Lekklar and Chaidee (2011)
		Physiological/biochemical	Matoh et al. (1986), Yeo et al. (1999), Gong et al. (2006), Shi et al. (2013)
	Barley	Physiological/biochemical	Liang et al. (1996, 2003, 2005), Liang (1999), Liang and Ding (2002), Gunes et al. (2007a), Matichenkov and Kosobrukhov (2004)
	Maize	Physiological/biochemical	Moussa (2006), Rohanipoor et al. (2013), Moussa and El-Galad (2015)
	Sugarcane	Physiological/biochemical	Ashraf et al. (2010a,b)
	Sorghum	Physiological/biochemical	Kafi et al. (2011), Yin et al. (2013)
	KBG	Agronomic	Chai et al. (2010), Bae et al. (2012)
		Physiological/biochemical	Chai et al. (2010), Bae et al. (2012)
Dicots	Tomato	Agronomic	Muneer et al. (2014), Haghighi et al. (2012)
		Physiological/biochemical	Al-aghabary et al. (2005), Romero-Aranda et al. (2006), Gunes et al. (2007b)
		Genetic/proteomic	Muneer et al. (2014)
	Cucumber	Physiological/biochemical	Zhu et al. (2004)
	Spinach	Physiological/biochemical	Gunes et al. (2007a)
	Strawberry	Physiological/biochemical	Wang and Galletta (1998)
	Alfalfa	Physiological/biochemical	Wang and Han (2007)
	Beans	Physiological/biochemical	Shahzad et al. (2013), Zuccarini (2008), Miao et al. (2010), Lee et al. (2010), Hamayun et al. (2010)
	Grapevine	Physiological/biochemical	Soylemezoglu et al. (2009)
	Canola	Agronomic	Solatni et al. (2012)
		Physiological/biochemical	Hashemi et al. (2010)
	Purslane	Agronomic	Kafi et al. (2011)
		Physiological/biochemical	Kafi et al. (2011)
	Zucchini	Agronomic	Savvas et al. (2009)
		Physiological/biochemical	Savvas et al. (2009)
	Tobacco	Agronomic	Hajiboland and Cheraghvareh (2014)
		Physiological/biochemical	Hajiboland and Cheraghvareh (2014)
	Clover	Physiological/biochemical	Abdalla (2011)
		Genetic/proteomic	Abdalla (2011)
	Lettuce	Physiological/biochemical	Milne et al. (2012)
	Sunflower	Physiological/proteomics	Asilan et al. (2013)
	Lentil	Agronomic	Sabaghnia and Janmohammadi (2015)
	Pea	Physiological/biochemical	Shahid et al. (2015)

supplement augmented the tiller number in both saline and nonsaline conditions. Si significantly increased the growth of two wheat genotypes differing in salt tolerance. In another study, Si supplement decreased the Na^+ contents, while enhancing K^+, and subsequently improved the plant biomass, 100-seed weight, length of ear, and rate of photosynthesis. The use of a salt-tolerant wheat cultivar (Pishgam) treated with appropriate Si treatment at the booting stage was suggested as an advantageous approach to attain greater grain yield of wheat plants grown on saline lands (Ahmad 2014). Furthermore, Si supplement significantly increased the biomass of wheat plants both in the control and in a saline environment. However, an increase in the biomass was more under saline conditions than in normal nutrient solution.

If we consider the physiological aspects of wheat studied with Si supply under salinity stress, it has been found that Si positively affects the wheat plants. Beneficial impacts of Si were observed via a significant decrease in Na^+ contents of the flag leaves and plant roots of wheat under salt stress (Ahmad et al. 1992). Similarly, Na^+ uptake in both plant shoots and roots of the wheat crop was decreased with the addition of Si under salinity stress (Saqib et al. 2008; Tuna et al. 2008). Si significantly alleviates the toxic effects of Na^+ by reducing its uptake and translocation with the improvement in plant growth, particularly in salt-sensitive genotypes. The decreased Na^+ contents in Si-treated wheat plants have also been reported previously (Ali et al. 2012). The selective uptake of K^+ reduces the uptake of toxic Na^+ ion, which could be the mechanism of salt tolerance in wheat. Si may also increase the cell wall Na^+ binding and, as a result, decrease the potential Na^+ toxicity (Saqib et al. 2008). Adding Si significantly decreased the Na^+ contents, whereas it improved the K^+ contents and shoot K^+/Na^+ ratio of both salt-sensitive and salt-tolerant genotypes of wheat. Decreased Na^+ uptake and enhanced K^+ uptake by the treatment of Si was suggested as a reason for the improved salinity tolerance and better plant growth (Ali et al. 2012). Si supplement significantly increased the K^+ concentration in five wheat genotypes; however, the increase was more pronounced under salt stress (Tahir et al. 2012). The K^+/Na^+ ratio was significantly enhanced in various experiments when Si was applied to the medium. Improved salinity tolerance by Si addition was suggested to be due to increased K^+ uptake in wheat plants (Mali and Aery 2008), which in turn enhanced the K^+/Na^+ ratio and decreased the Na^+ translocation in shoots (Na^+ exclusion at the root level). In another study under saline conditions, proline accumulation and Na^+ content in the wheat plant tissues were augmented, while K^+ uptake was decreased (Ahmad 2014). When treated with Si-amended highly saline water, wheat plants usually have high selectivity of potassium over Na^+. Si-treated plants improved the membrane stability index in leaves compared to non-Si-treated plants (Ali et al. 2012). Si addition enhanced the relative water content in salt-stressed wheat, but remained insignificant for the control plants (Tuna et al. 2008). The water restoration in salt-stressed plants was suggested to be due to the hydrophilic nature of Si (Romero-Aranda et al. 2006). However, further experiments are needed to explore whether and how exogenous Si is involved in regulation of water transport even at salt stress conditions application of silicon in wheat was observed to be completely restored the level of chlorophyll, even under the salt stress conditions, not only this, the level of chlorophyll was also observed to be enhanced as compared to control (Tuna et al. 2008). According to Liang et al. (2007), Si possesses the characteristics of functioning as a physical or mechanical barrier in plants. When plants are subjected to salinity stress, Si functions are not limited to its deposition in the cell wall; however, it actively participates in the physiological and metabolic changes in plants. Chlorophyll and water contents were little increased in the Si-treated plants' leaves (Ali et al. 2012). This study (Ali et al. 2012) concluded that in wheat plants under salt stress, Si contributes to the metabolic, physiological, or structural activity. Si enhances the photosynthetic activity in plants by enhancing the plant water and chlorophyll content under salt stress.

Considering the antioxidant aspects of wheat plant shoots under salinity stress, it has been found that superoxide dismutase (SOD), peroxidase (POX), catalase (CAT), and glutathione reductase (GR) show a significant increase in their activities (Sharaf 2010; Masood et al. 2012). Interestingly, Si addition under salinity stress enhanced the antioxidative activity of various enzymes (Sharaf 2010). With Si application under treatments without salinity, the activities of both SOD and CAT were markedly higher; however, POX and GR were significantly reduced under the same conditions. Moreover, Si treatment in wheat plants under salinity stress distinctly decreased the rise in SOD, POX, CAT, and GR activities. Endogenous phytohormone contents were also markedly changed with Si, salinity, and their interactions. The conclusion from Sharaf (2010) was that under salinity stress, Si plays a favorable regulatory role in wheat plants.

8.2.2 RICE

The effects of Si on rice plants grown under salinity have been widely studied and proved to enhance the yield. The Na^+ silicate application in rice resulted in a higher grain yield while increasing the Si accumulation in plants (Winslow 1992). Si supplement also led to improved rice tolerance to salt stress (Matoh et al. 1986). Salt stress induced a pronounced inhibition in shoot and root growth after 7 days of treatment, following a significant reduction of about 50% in photosynthetic pigment contents after 14 days (Lekklar and Chaidee 2011). After 7 days, salt stress was completely inhibited by the exogenous supply of Silicon in shoot and root. When the stress period was extended to 14 days, the root growth of salt-stressed seedlings was not improved by Si application, while shoot growth was enhanced. Moreover, the root/shoot ratio was drastically declined after 14 days of salt stress regardless of Si (Lekklar and Chaidee 2011). Interestingly, chlorophyll a, b, and carotenoid contents of salt-stressed seedlings were maintained at the control level in the presence of Si. It was concluded that exogenous Si improves the shoot growth of rice seedlings by protecting the photosynthetic apparatus against salt stress.

A study on rice (*Oryza sativa* L.) seedlings (using three cultivars) in regard to chloride transport under salinity stress found enhanced plant shoot growth with Si supplement; however, root growth was not affected. Moreover, salt conditions in all three cultivars reduced the shoot and root fresh biomass. In culture solution, the same results were observed where Si enhanced the biomass of shoots with no effect on the roots (Shi et al. 2013). Friesen et al. (1994) determined that up to a 40% yield of upland rice was restrained by Si scarcity. However, when Si was added, it significantly increased the production of rice. According to Deren et al. (1994), Si supplement increases the yield due to an increase in grain number per panicle, while the 100-seed weight and panicles in each square meter exhibit less variation. Si treatment was also reported to significantly increase rice production on calcareous lands (Liang et al. 1994). Si treatment increased the grain number per unit area, mature grain percentage, and eventually the production of rice (Ando et al. 1999). The effect of addition of Si on the growth and grain discoloration of rice plants was also noticed (Korndörfer et al. 1999). Si supplementation increases total grain weight and significantly reduces grain discoloration, independent of soil type. An increasing amount of Si fertilizer increases Si concentration in the leaves. The effect of Si treatment on the growth of rice plants in the fields was examined by Wang and Galletta (1998). Si supplement increased phosphorus content, productive panicle number, and grain number in each panicle. The study resulted in a 5% improvement in the production of rice with silicate treatment. Rooting ability and early growth by Si addition was observed by the use of silica gel (Mori et al. 1999). The findings concluded that increased dry biomass and an improved ratio of dry biomass to plant height were influenced by the silica gel.

The rice plant growth was reduced by the 50 mM NaCl treatment (Gong et al. 2006); however, Si addition significantly enhanced the shoot growth by improving fresh and dry shoot weights and plant height. Moreover, Si treatment did not show any effect on the rice roots. Further, reduced Na^+ uptake in shoots was attributed to the enhanced rice growth with a Si supplement under salinity stress. Silicate-deficient plants under salinity stress showed almost 50% higher Na^+ uptake in rice (Matoh et al. 1986). Coprecipitation of Na^+ ion and Si elements in roots was proposed as a probable mechanism responsible for the reduced Na^+ uptake even under saline conditions (Gong et al. 2006). Si precipitation in the exodermis and endodermis decreases the Na^+ translocation through reduced apoplastic transport in the root. Yeo et al. (1999) also suggested partial blockage of transpirational bypass flow as a reason for reduced Na^+ contents in rice plants. A decrease in the transpiration rate with Si supplement increased stomata conductance through improved CO_2 intake, which resulted after Na^+ uptake was reduced. Yeo et al.'s (1999) findings further suggested that silicate can decrease Na^+ uptake as a result of a reduced transpiration rate; this condition leads to improved stomatal conductance of rice plants that are otherwise facing net photosynthesis and growth reduction under salinity stress.

Necrotic spots in leaves of non-Si-treated rice plants were observed, which might be called deficiency symptoms. Abnormal growth and sterility proportion were significantly higher in non-Si-treated rice plants. Si supplement prominently enhanced rice production (Mitsui and Takatoh 1963). Si treatment also sustains photosynthetic activity (Agarie et al. 1992), which was attributed to the increased dry biomass in rice. The enhanced rate of Si nutrition in the field promotes the dry biomass, leaf area, and leaf chlorophyll contents of rice plants. In rice, Gong et al. (2006) suggested that the mechanism by which Si mediates salt tolerance in plants can be a Si-induced physical barrier in roots. A study further investigated the depositing of Si on the exodermis and endodermis of rice roots, which dramatically decreases apoplastic transport (transpirational bypass flow) and therefore Na^+ accumulation. Similarly, addition of Si also decreases the Cl^- level in shoots (Shi et al. 2013). Faiyue et al. (2010) suggested that the lateral root may play a role in the bypass flow, because it lacks an exodermis, whereas Si can enhance exodermal development in rice (Fleck et al. 2011). Therefore, in rice, both Si-enhanced exodermal development and Si deposition on the exodermis contribute to decreased loading of salt ions into the xylem of roots, resulting in decreased salt ion accumulation in shoots. It was suggested by Gong et al. (2006) that although Si-induced reduction in transpirational bypass flow (and therefore Na^+ transport) contributes to salt tolerance in rice, this mechanism may not work in other plants, such as grapevine. Stronger barriers to radical oxygen loss are believed to reduce Na^+ uptake and improve salt tolerance in rice. The contradictory results obtained from canola and rice were attributed to the differences in their abilities to accumulate Si, and further investigations were proposed (Krishnamurthy et al. 2009). Si addition considerably reduced the chloride concentrations in plant shoots under salinity stress, which also resulted in an improved K^+/Cl^- ratio; however, chloride concentrations remained the same under the influence of Si added to the roots. Si supplement in rice plant shoots reduced the transpiration bypass flow, as displayed by the apoplastic tracer transport (Shi et al. 2013), which was attributed to the reduced chloride concentration in rice shoots under salinity stress. In addition, Si treatment enhanced the photosynthetic rate and stomata conductance, which suggested that the decrease in Cl^- (and Na^+) uptake was not only through a lower transpiration rate. Further investigation was recommended to investigate the possible deposition of Si in exodermis and endodermis and the role of Si in suberization of the exodermis and lignification of the sclerenchyma in lateral roots (Shi et al. 2013).

8.2.3 BARLEY

In salt-stressed barley roots, application of Si decreases both Na^+ and Cl^- levels but increases K^+, with Na^+ and K^+ being more uniformly distributed over the whole root section and proposed to be the key mechanism of Si-enhanced salt tolerance in this species (Liang and Ding 2002). In barley grown in sodic-B toxic soil, the addition of Si decreases the concentrations of Na^+ and boron in shoots (Gunes et al. 2007a). Salt tolerance mechanisms adopted by plants include active Na^+ efflux, reduced Na^+ influx, and Na^+ scavenging into the vacuole (Marschner 2011). However, during the scavenging process, transport of Na^+ from the cytoplasm to the vacuole is facilitated by proton pumps to guard the cellular compartment and keep all metabolic events in the symplasm. According to Liang et al. (2005), any factor that activates the proton pumps will definitely favor the compartmentation of salt toxic ions. The salinity tolerance by Si application was also attributed to the selective uptake and transport of K^+ and Na^+ by plants as they compete with each other at the root and shoot level (Liang et al. 2005). Si treatment was reported to decrease the Na^+ and enhance the K^+ concentration in salt-sensitive and -tolerant barley genotypes (Liang 1999). Supplemental Si prominently reduced the Na^+ contents and translocation from roots to the shoots in barley plants. Si treatment under salinity stress did not greatly influence the Ca^{2+} concentrations in barley shoots. In another greenhouse experiment, Si supplement moderately reduced the salinity impacts by facilitating the higher K^+ uptake, especially in roots. However, Si supplement had no influence on the Cl^- concentrations in both root and shoot under salinity stress (Liang 1999). It was concluded that the growth improvement of salt-stressed barley plants is attributed to the decreased Na^+ uptake in leaves, hence mitigating the salinity impacts.

The Si-mediated decrease in lipid peroxidation helps to maintain the membrane integrity and decrease the plasma membrane permeability under salt stress, as observed in barley (Liang et al. 1996, 2003). In barley leaf cell organelles, Si reportedly increased the photosynthetic activity and chlorophyll content either with or without salt stress (Liang et al. 1996; Liang 1997). Earlier the resistance of barley plants against salinity stresses was inclined by exogenous supply of active silicon either in the form of liquid or solid (Matichenkov and Kosobrukhov 2004).

Improvement in the K^+/Na^+ ratio in the barley shoot under salinity stress was due to the increased H^+-ATPase activity by Si treatment that increased the transport of K^+ and reduced the Na^+ uptake (Liang et al. 2005). However, the energy-dependent Na^+/H^+ antiports get Electromotive Force (EMF) by H^+ electrochemical potential gradient which is mainly induced by both H^+-ATPase and H^+-PPase (Blumwald 2000). According to Liang et al. (1996), the activities of antioxidant enzymes such as CAT, SOD, and POD were increased by Si treatment in salt-sensitive and salt-tolerant barley plant roots when exposed to salinity stress. These findings with Si treatment were proposed to protect the plants from salt injury and later improved the barley plant growth under salinity stress. Furthermore, it was suggested that the influence of Si is more clear with the growth stages.

About 80% Si deposition was observed in hulled barley when compared to hull-less barley with less Si contents (Ma et al. 2003). It was proposed that accumulation of Si in barley grain is a genetically controlled mechanism. Liang (1999) observed a decrease in the activities of plasma membrane H^+-ATPase in barley roots under salt stress, but the activities were increased significantly when Si was added to plants. The increase in the activities of plasma membrane H^+-ATPases may facilitate the Na^+ export from the cell. Liang et al. (2005) reported that the activities of barley root tonoplast H^+-ATPase and H^+-PPase were also considerably stimulated by the addition of Si under salt stress, which may have facilitated the Na^+ compartmentation in vacuoles through the tonoplast Na^+/H^+ antiporter. Si has been reported to decrease the concentration of madondialdehyde, the end product of membrane lipid peroxidation, in barley under salinity, suggesting that addition of Si can decrease the lipid peroxidation (Liang et al. 2003). Liang et al. (2003) observed an increase in the CAT activity under salt-stressed barley at day 2 when compared with the control, in spite of adding or not adding Si. On days 4 and 6 of salt treatment, CAT activity decreased, but Si addition significantly alleviated the decrease. Future studies can target the further genetic and proteomic aspects to understand the role of Si in improving the salt tolerance of barley plants.

8.2.4 MAIZE

Si has been proposed to mitigate salt stress by reducing the Na^+ influx in plants; however, it also reduces the movement into the aerial parts of plants by decreasing the free Na^+ transport. When compared to physiological impacts of Si on plants, Ma et al. (2001) concluded that the Na^+ and Si complex formation in the solution is more dominant in mitigating the harmful impacts of salinity stress. Si addition reduced the Na^+ toxicity, and this was also accompanied by enhanced root length. Moussa (2006) found that supplemental application of Si increased the dry biomass in all parts of maize (*Zea mays* L.) under saline conditions, and dry biomass of leaf and total plant was significantly higher. Added Si in the nutrient solution alleviated the NaCl-induced growth inhibition. In another study, by Rohanipoor et al. (2013), salinity stress significantly decreased the fresh and dry weights of shoot and root, stem length, leaf area, chlorophyll content, and relative water content of maize plants, whereas application of Si significantly ameliorated the negative impacts of salinity on plant growth.

Moussa (2006) reported that chlorophyll *a* and *b* were significantly reduced under salinity stress, and these were recovered after 60 days of treatment with Si in maize plants. Proper Si nutrition can increase salt resistance in maize plants (Rohanipoor et al. 2013). Xie et al. (2015) planned research to determine Si effects on the photosynthetic characteristics of maize grown on saline-alkaline soil. Experimental results showed that the values of the photosynthetic rate, stomatal conductance, and intercellular CO_2 concentration of maize were significantly enhanced, while decreasing the transpiration rate. This indicated that Si application with proper doses significantly increased the photosynthetic efficiency of maize in different growth stages under the environment of saline-alkaline soil. The positive effects of Si on the physiological parameters were observed in plants grown under salt stress conditions (Rohanipoor et al. 2013). Similarly, in a hydroponic study, growth and gas exchange attributes in maize plants were improved by the addition of silicon ($SiOH_4$) under salinity stress (Parveen and Ashraf 2015). Application of Si to the salt-stressed plants increased the concentrations of N, P, and K in maize plants (Giza 2 and Trihybrid 321), compared to salt-stressed plants alone (Moussa and El-Galad 2015).

Moussa (2006) examined the role of Si supplement on the physiological/biochemical response of maize plants under salinity stress. The results indicated that Si partially offset the negative impacts and increased the tolerance of maize to NaCl stress by increasing the activities of SOD and CAT, chlorophyll content, and photosynthetic activity. In contrast, salt stress considerably increased the H_2O_2, free proline level, and MDA concentration, whereas Si addition significantly reduced them in stressed maize leaves. Maize plants (Giza 2 and Trihybrid 321) grown under saline conditions showed a decrease in the total protein contents, enzyme activity of ribulose-1,5-bisphosphate-carboxylase/oxygenase (RuBPCase), photosynthetic activity, and macronutrient level (N, P, and K) compared to the control plants. However, the findings with Si treatment partially offset the negative impacts and increased the tolerance to salinity stress by enhancing the above parameters (Moussa and El-Galad 2015).

In a recent study by Rohanipoor et al. (2013), the effects of Si on some physiological properties of maize under salt stress revealed that potassium silicate (K_2SiO_3) supplement increases the salt resistance in maize plants. Results showed that salinity significantly decreased the fresh and dry biomass of shoots and roots length of stems, leaf areas, chlorophyll contents, and relative water content of maize plant, whereas the application of Si significantly improved the measured parameters. However, studies dealing with genetics and proteomics are lacking to better understand the Si contribution to the enhanced salinity tolerance of maize plants.

8.2.5 SUGARCANE

Many researchers have experienced the beneficial impacts of Si on the growth of sugarcane (Anderson and Sosa 2001; Matichenkov and Calvert 2002). However, the significant role of Si in the alleviation of salinity stress was reported by Ashraf et al. (2010a,b) in different salt-sensitive and salt-tolerant sugarcane genotypes. Si application significantly decreased the Na^+, while increasing the K^+ concentration in shoots. Juice quality characteristics, such as brix, sucrose, commercial cane sugar, and sugar recovery, were significantly improved in both salt-sensitive and salt-tolerant sugarcane genotypes with the supplementation of silicates. Studies with this particular crop focusing on the antioxidant, proteomic, and genetic changes with Si treatment under salinity stress are scarce compared to those for other crops.

8.2.6 SORGHUM

Yin et al. (2013) reported that short-term application of Si decreases leaf Na^+ concentration in the seedling of sorghum (*Sorghum bicolor* L.), but does not enhance K^+ content, which is in agreement with the observations in salt-stressed rice (Gong et al. 2006). Yin et al. (2013) also found that short-term application of Si could significantly increase the levels of sucrose and fructose in sorghum under salt stress, suggesting that Si can alleviate salt-induced osmotic stress. However, the evidences about the relationship between Si addition and compatible solute metabolism and water transport is very few (Kafi et al. 2011).

It was reported that Si had no effect on the seedling growth under normal conditions (Yin et al. 2014). Under salt stress, the photosynthesis and transpiration rate were decreased, but these reductions were alleviated by Si application. In addition, water content and elongation rates in leaves were maintained at higher levels with Si than without Si. The root hydraulic conductance of the seedlings was inhibited by salt, but Si application alleviated this inhibition. Moreover, transcript levels of several aquaporin genes were upregulated by Si (Yin et al. 2014). Under salt stress, Si inhibited the increase in the root H_2O_2 levels and enhanced the activities of antioxidant enzymes.

In a recent study, Yin et al. (2015) showed that the salt tolerance of *Sorghum bicolor* L. was enhanced by Si treatment as a result of decreased Na^+ concentration. In the same study, polyamine synthesis genes were reported to be upregulated with Si amendment under salinity stress. As a result, PA levels were found to be increased and, at the same time, ethylene precursor (1-aminocyclopropane-1-carboxylic acid [ACC]) concentrations were decreased. The study concluded that Si ameliorated the salinity impacts by controlling the PA and ACC in sorghum plants.

8.2.7 KENTUCKY BLUE GRASS

Kentucky blue grass (KBG) had shown lower salinity tolerance when compared with perennial ryegrass and creeping bent grass (Dai et al. 2009). Furthermore, when exposed to 1.5% NaCl, earlier investigations found that KBG plants can survive only 10–15 days (Fei et al. 2005). Under saline conditions, abnormalities occur in germination of turf grass, and later survival of the seedling is also hard (Daniel and Freeborg 1987). Salinity stress resulted in the reduced germination rate along with 12%–35% reduction in canopy coverage, number of tillers, and height of plant when compared to control (Chai et al. 2010).

Beneficial effects of Si have been reported in many species of Poacea and Cyperaceae. In Chai et al.'s (2010) study, Si supplement enhanced the canopy coverage, number of tillers, and germination rate of KBG under saline conditions. Furthermore, Si addition increased the K^+ transfer and reduced the N^+ transport from roots to shoots, which not only lessened the harmful impacts of salinity, but also improved the quality of turf even under salt stress.

A study by Bae et al. (2012) was planned to investigate the effects of Si supplement on the salt-sensitive 'Perfection' and 'Midnight' Kentucky bluegrass (*Poa pratensis* L.). The 400 mM NaCl treatment reduced the shoot length, fresh and dry weights of shoots and roots, relative water content, chlorophyll, and carotenoid contents in both cultivars. The NaCl caused an increase in electrolyte leakage, malondialdehyde (MDA) concentration, and H_2O_2 in both cultivars. It was concluded that these physiological interferences and visually noticeable disturbances in Kentucky bluegrass can be mitigated by the addition of Si under salinity stress.

It can be concluded from the investigations with Si supplements on the monocots under salinity stress that Si induced improvements in the growth of all studied crops. Beneficial impacts found on the antioxidant, water, and ionic statuses with Si addition under salinity stress suggest that its role must be further examined using advanced proteomic and genetic approaches to better understand the Si-induced salt tolerance mechanism.

8.3   SILICON AND SALT STRESS IN DICOTS

8.3.1 TOMATO

Seed germination percentage, germination rate, and mean germination time of *Lycopersicon esculentum* L. were inhibited severely by increasing salt concentrations. On the other hand, Si supplement enhanced the above-mentioned germination characteristics (Haghighi et al. 2012). Si almost completely alleviated germination characteristics of tomato seeds, even under adverse salinity stress. Moreover, in an earlier investigation, Al-aghabary et al. (2005) found that Si treatment alleviates the harmful salinity impacts on the growth by increasing the dry biomass of tomato plants. Increased water influx in the symplast of tomato seedlings was attributed to the increase in growth by Si addition under salinity stress (Romero-Aranda et al. 2006). When compared to non-Si-treated plants, the Si-treated plants showed a 42% and 20% improvement in the leaf turgor pressure and rates of net photosynthesis, respectively. Under salt stress, water content was reduced by 54% in tomato plants; however, when Si was added to the salt-treated plants, this effect was reversed by 40% (Romero-Aranda et al. 2006). In the same study, NaCl-treated plants with Si addition revealed a 42% increase in the values of turgor potential compared to the plants treated with only NaCl. Si decreases the translocations of Na^+, Cl^-, and boron from roots to shoots of tomato plants grown in sodic-boron toxic soil (Gunes et al. 2007b). Si-mediated salt tolerance in plants is not always accompanied by a decrease in tissue Na^+ or Cl^- levels. In tomato, inclusion of Si reportedly has no significant effect on Na^+ and Cl^- concentrations in leaves, but improves water storage in plants. This higher water content contributes to salt dilution, thereby reducing the toxicity of salt and improving plant growth (Romero-Aranda et al. 2006). Si has beneficial effects on the photochemical apparatus and photosynthetic pigment in tomato and *Spartina densiflora*. However, under salinity, the beneficial effects of Si on the photosynthetic apparatus, and later the increased photosynthetic activities, may be a result of Si's role in decreasing the Na^+ uptake and increasing the K^+ uptake while enhancing the antioxidant defense (Al-aghabary et al. 2005; Mateos-Naranjo et al. 2013). Improvement of chloroplast proteome expression and photosynthetic metabolism with Si addition under salinity stress concluded an overall protective effect on tomato seedlings (Muneer et al. 2014). However, this study pointed to one major disadvantage for further investigations with tomato plants by comparing them to the rice plants, which have a superior ability to accumulate Si.

Another positive aspect of Si found on tomato plants under salinity stress is the improvement of the tolerance via augmentation of SOD, CAT activities, and leaf soluble protein contents (Al-aghabary et al. 2005). APX activity was slightly decreased by the addition of Si, as the concentrations of H_2O_2 and MDA were significantly reduced in plant leaves when compared with the salt-treated plants. According to Santa-Cruz et al. (1997), total polyamines are reduced to a greater extent in the salt-sensitive tomato species (*Lycopersicon esculentum*) than the salt-tolerant species (*Lycopersicon pennellii*). This shows that accumulation of polyamines, especially spermidine and spermine, may contribute to the salt tolerance mechanism. Proteomics analysis of chloroplasts by Muneer et al. (2014) under salinity stress by second-dimensional blue native polyacrylamide-gel electrophoresis (2D-BN-PAGE) showed an increased sensitivity of multiprotein complex proteins. Si addition significantly reduced the ATP-synthase complex during salt condition, whereas the rapid upregulation of the light harvesting complex (LHC) trimers and monomers was observed. With these findings, it was suggested that damage to chloroplasts and their metabolism under salinity stress is crucially restrained by Si addition (Muneer et al. 2014). Their findings indicated that the *Lsi1* gene is involved in tomato plants for the transport of Si in subcellular localizations to overcome salinity stress. The other two Si efflux genes, *Lsi2* and *Lsi3*, were also proposed to be characterized in the same plants.

8.3.2 Cucumber

Si supplement under salinity stress in cucumber plants significantly reduced the electrolyte leakage, because of limited lipid peroxidation production, reduced H_2O_2, and reactive substances like thiobarbituric acid (Zhu et al. 2004). Increased activities of SOD, dehydroascorbate reductase (DHAR), ascorbate peroxidase (APX), guaiacol peroxidase (GPX), and GR with Si addition suggested the alleviation of cucumber plants to withstand an oxidative damage and ultimately enhanced the tolerance against salinity stress. A significant increase in antioxidant enzymes of salt-stressed leaves by Si addition projected that Si may be involved in the metabolic or physiological activity in cucumber exposed to salt stress. Furthermore, in cucumber plants, a decrease in lipid peroxidation by the addition of Si was suggested to maintain membrane integrity and decrease plasma membrane permeability under salt stress. However, Si mechanisms in mitigating the salt stress in cucumber plants need to be investigated at the genetic and proteomic levels.

8.3.3 SPINACH

Si addition in the growth medium of spinach plants grown on sodic-B toxic soil resulted in reduced H_2O_2 production, which was increased under salinity stress (Gunes et al. 2007b). In this study, Si addition under salinity stress also improved the chlorophyll content, photosynthetic activity, and ribulose bisphosphate carboxylase activity of the spinach leaves. Under salinity stress, Si supplement prohibited the cell membrane's physical and functional weakening by improving the lipid steadiness in the cell membrane. The findings determined that metabolic or physiological changes are influenced by Si addition under a saline environment. However, Si's role in mediating salt tolerance at the genetic and proteomic levels in spinach plants could be the focus of future investigations.

8.3.4 STRAWBERRY

Foliar spray of potassium silicate induced metabolic changes, like inclined levels of citric and malicacids and declined glucose, sucrose, fructose and myo-inositol contents, and also increased chlorophyll content in strawberry leaves under salinity stress (Wang and Galletta 1998). Higher proportions of fatty acid unsaturation in glycolipids and phospholipids and noticeable amounts of lipids in membrane were observed in Si-treated strawberry plants. The salinity tolerance of Si-treated strawberry plants was attributed to the degree and contents of unsaturated lipids that facilitate the maintenance of the leaf chloroplast and protection from salt injury.

8.3.5 ALFALFA

Wang and Han (2007) studied exogenous Si application in alfalfa (*Medicago sativa* L.) under salinity stress. The findings revealed that exogenous Si decreased the Na^+ content in alfalfa roots, but not in shoots, when compared to salt treatment alone. However, when compared to salt-treated plants, application of Si to alfalfa notably increased the K^+ content in shoots.

8.3.6 BEAN

Shahzad et al. (2013) demonstrated that high salinity causes the Na^+ concentration to increase in the extracellular space of the field bean (*Vicia faba* L.) leaves, and this was significantly lowered by Si treatment. Moreover, in the apoplastic fluid extracted from the field bean leaves, Si addition significantly decreased the Na^+ concentration under salinity stress. Si supplement under salinity stress increased the shoot biomass, which was attributed to the significant reduction in Na^+ in the leaves. Earlier, common bean (*Phaseolus vulgaris*) growth was found to be improved by enhanced photosynthesis, water, and K^+ status in the plants with Si addition under salt stress (Zuccarini 2008). The same crop was treated with various levels of Si treatments, along with various levels of salinity, to investigate the effect of Si on the yield and component of yield of beans in salt stress conditions (Parande et al. 2013). The findings revealed that an increase in salinity caused a significant reduction in the number of seeds in the pod, number of fertile pods, and dry matter. Si application significantly increased the 100-seed weight and yield of beans under saline environments. Finally, it was concluded that use of Si in salinity conditions reduced harmful effects of salinity on the yield component in bean plants. Experimental evidence has shown that the presence of abscisic acid largely inhibits Na^+ and Cl^- transport to the shoot in intact bean seedlings (Karmoker and Van Steveninck

1979). However, Lee et al. (2010) found that abscisic acid content increases in soybean plants under salt stress, but decreases upon addition of Si. After investigating the growth and physiohormonal attributes, ameliorative effects by added Si against salinity stress were also observed by Hamayun et al. (2010) in soybean plants. The application of Si to potassium (K^+)-deficient medium markedly alleviated K^+ deficiency–induced inhibition of both root and shoot growth in soybean seedlings (Miao et al. 2010). In addition, Si significantly enhanced K^+ accumulation in soybean plants exposed to K-deficient medium, thus alleviating K^+ deficiency–induced membrane lipid peroxidation and oxidative stress by modulating antioxidant enzymes (Miao et al. 2010). These findings were suggested to have important implications in agronomic practice to enhance the efficiency of nutrient by application of Si fertilizer under nutrient-deficient conditions (Miao et al. 2010). Unfortunately, studies are lacking that employ genetic and proteomic approaches to understand the complete role of Si in mediating salinity tolerance in bean plants.

8.3.7 GRAPEVINE

Soylemezoglu et al. (2009) reported that Si improves the combined salt and boron tolerance of grapevine grown under saline, boron toxic, and boron toxic and saline conditions, by studying membrane-related parameters and antioxidant responses. Si reportedly decreases the concentration of MDA, the end product of membrane lipid peroxidation, in grapevine rootstocks under salinity, suggesting that addition of Si can decrease lipid peroxidation. However, in salt-stressed grapevine, addition of Si did not affect the activity of SOD, and it decreased the CAT activity, whereas the activity of APX was increased or unchanged, depending on cultivars.

8.3.8 CANOLA

Hashemi et al. (2010) found that addition of Si decreases the lignin content in canola. The ability of Si to reduce lignification in tissues may facilitate cell wall loosening and extensibility and promote further plant growth under stress conditions (Hattori et al. 2003). However, further investigations were proposed after the contradictory results obtained for canola and rice, which were related to the differences in their abilities to accumulate Si. Solatni et al. (2012) observed the effect on canola, cv. Talayeh seedlings after the exposure of Silicon and salinity stress. Treatment of Si showed the highest values for germination percentage and germination ratio and the lowest one for mean germination time compared to other treatments. Moreover, Si with a concentration of 60 mM NaCl led to the highest lengths of the radicle and plumule. Salinity had a significantly negative effect on all studied characteristics, but application silicon supplement improved those traits.

8.3.9 PURSLANE

In an experiment with purslane, salinity influenced a decrease in the shoot weight and root development (area, volume, diameter, length, and dry root biomass) (Kafi and Rahimi 2011). Proline and Na^+ contents were augmented; however, the K^+ content decreased in both the leaf and root of purslane under salt stress. Si supplement enhanced root characteristics and also the K^+ uptake in root and leaf even under salinity stress. The findings also revealed reduced Na^+ contents, which enhanced the K^+/Na^+ ratio in purslane leaves. Kafi and Rahimi (2011) suggested that purslane, having medicinal and vegetable uses, can be grown on saline lands using Si treatment. It was presumed after findings that Si is more advantageous in shoots than roots in purslane. It was further explained in this study that safe growth of purslane under salinity stress is due to increased solute uptake that is needed to encourage cell expansion, since in the plant tissue this may facilitate sustained pressure potential (Munns et al. 1983; Flowers and Colmer 2008).

8.3.10 Zucchini

The role of added silicon in the nutrient solution against salinity tolerance in zucchini squash (*Cucurbita pepo* L. cv. 'Rival') plants was tested by Savvas et al. (2009). Treatments include one at the highest level of salinity (6.2 dS m^{-1}, 35 mM NaCl) in combination with a 1.0 mM Si level in the nutrient solution that was supplied to the crop. Significantly affected growth and fruit yield under salt treatment alone was recovered by the addition of 1 mM Si in the salinized nutrient solution. Growth suppression under salinity was attributed to restriction in the net photosynthesis and increased Na$^+$ and Cl$^-$ in the plant tissues; however, results with addition of 1 mM Si showed ameliorative effects under salinity stress.

8.3.11 Tobacco

Salt-stressed tobacco (*Nicotiana rustica* L.) plants were supplemented with Si to examine the influence of Si on growth and some physiological and biochemical parameters (Hajiboland and Cheraghvareh 2014). Si application caused a reduction of Na$^+$ concentration, while it increased the K$^+$ and Ca^{2+} significantly and resulted in a higher K$^+$/Na$^+$ ratio in the leaves, stem, and roots under salinity stress. After findings, it was proposed that tolerance to salt stress in tobacco was improved due to an enhancement of photosynthesis, accumulation of organic osmolytes, and improvement of K$^+$/Na$^+$ selectivity, but not limiting water loss.

8.3.12 Clover

Two Egyptian clover (*Trifolium alexandrinum*) cultivars differing in salinity tolerance were used (Helaly, salt sensitive; Sarw1, salt tolerant) to investigate the protective role of Si under salinity stress (Abdalla 2011). Addition of diatomite at an upgraded rate alone or combined with both concentrations of NaCl significantly enhanced the complete considered plant development parameters (height and fresh and dry biomass of fodder or pot), photosynthetic rate, percentage of relative water content, percentage of membrane stability index, total photosynthetic pigment, and contents of magnesium, potassium, phosphorus, and calcium, respectively. Additionally, diatomite fertilization with and without salinity stress induced two distinctive protein electrophoratic bands (233 and 25 kD), which were absent in either the control or salinity-stressed cultivars (Abdalla 2011). Diatomite application with and without salinity induced various different types of distinguished amplified DNA fragments in both clover cultivars using polymerase chain reaction-random amplified polymorphic DNA (PCR-RAPD) analysis, although the number of induced polymorphic DNA fragments was greater in Helaly than in Sarw1. According to Abdalla (2011), these results indicate that diatomite recovered and improved the morphologic, metabolic, and biochemical status of both cultivars under salinity stress and especially the sensitive line (Helaly).

8.3.13 Lettuce

The antioxidant content, capacity, and certain oxidative stress parameters were investigated in salinity-stressed *Lactuca sativa* L. with the addition of exogenous Si (Milne 2012). Applications of Si significantly increased oxygen radical absorbance capacity (ORAC) analysis values, and reduced glutathione under salinity stress (Milne 2012). The activity of CAT was significantly decreased with the addition of Si at 30 mM NaCl. In the same study, lipid peroxidation was not influenced by the different Si concentrations. With these findings, it was suggested by Milne (2012) that applications of Si could be beneficial with regards to the modulation of oxidative stress in salinity-stressed lettuce and should be considered an addition when cultivating lettuce. Moreover, low salinity levels (maximum 60 mM NaCl) were used in this study, and it was recommended that investigations dealing with the effects of Si application on antioxidant status will reveal clear results in lettuce when treated with higher-salinity concentrations. However, there is a need to completely understand the mechanism behind Si's role in mitigating salt stress through genetic and proteomic studies.

8.3.14 SUNFLOWER

Ali et al. (2013) observed that silicon can alleviate the salinity stress in sunflower plants (*Helianthus annuus L.*). They found that the treatment with 100 mM NaCl improved the protein content as well as increased the malondialdehyde (MDA) content in sunflower plants. In addition, NaCl stress enhanced the endogenous, nonenzymatic antioxidants and the activity of antioxidant enzymes, such as POX, SOD, and CAT. When these plants were treated with a combination of NaCl and 2 mM Si, the inhibitory effects of NaCl stress were declined by alleviation of protein content and antioxidant enzyme activity by detoxifying the oxidative injury induced by salinity. Results suggested that Si has a key role in the enhancement of plant antioxidant potentials and resistance to salinity in sunflower plants at the seedling stage. In conclusion, addition of Si could ameliorate salt stress in sunflower (Ali et al. 2013). In the same year, in a greenhouse experiment growth and ornamental characteristics were observed with Si addition as a foliar spray on salt-stressed calendula (Bayat et al. 2013). The findings presented that growth, number of flowers per plant, soil plant analysis development (SPAD) values, and diameter of flowers were reduced under high salt conditions. Conversely, Si treatment as foliar spray on calendula plants under saline conditions revealed increased shoot and root total dry biomass, leaf area, and plant height. In comparison to salt treatment alone, Si increased flower number per plant and diameter of flower under salinity stress. Salinity stress increased electrolyte leakage in calendula plants; moreover, Si foliar spray under salt stress significantly decreased electrolyte leakage. Outcomes with calendula plants suggested that the growth and ornamental characteristics can be enhanced by Si foliar spray under salt stress (Bayat et al. 2013).

8.3.15 LENTIL

Sabaghnia and Janmohammadi (2015) conducted research to investigate the use of nano-Si dioxide (nSiO$_2$) supplements on seed germination of lentils under different salt concentrations. Their findings revealed that addition of 1 mM nSiO$_2$ could noticeably lessen the harmful impacts of salinity stress by improving germination percentage of lentil seed, shoot and root length, weight of seedling, mean germination time, seedling vigor index, and seed reserve mobilization. After findings of this study, for enhanced production of lentil under saline conditions, the use of an optimal concentration of economic nSiO$_2$ was considered beneficial. Sabaghnia and Janmohammadi (2015), before making practical recommendations, suggested investigation of the effects of other nanoparticles on different crops under both biotic and abiotic conditions. They further commented that this interesting topic will be considered in detail in future investigations.

8.3.16 PEA

Improvements with Si, leaf extract of *Melia azedarchta*, and root extract of sugar beet (applied alone or in combination) addition in pea (*Pisum sativum L.*) were tested under salinity stress (Shahid et al. 2015). Salinity stress significantly inhibited the growth, different gas exchanges attributes, stability index of membrane, content of total phenol, and productivity. Moreover, salinity significantly enhanced production of lipid peroxidation, electrolyte leakage, H$_2$O$_2$ content, and activities of antioxidant, leaf free proline, and glycinebetaine contents (Shahid et al. 2015). Si and phytoextract addition markedly reduced the harmful impacts of salinity on plant growth, enzymatic activities, gas exchange attributes, osmolytes, and later productivity. The phytoextracts and Si suppressed the lipid peroxidation, electrolyte leakage, and H$_2$O$_2$ content by strengthening the enzymatic and nonenzymatic (proline and glycinebetaine) antioxidant defense system (Shahid et al. 2015). The phytoextracts and Si application also checked the root or leaf Na$^+$ and chloride contents, but improved the Si contents (Shahid et al. 2015). With the above findings, Shahid et al. (2015) proposed that it is very effective to add Si and both phytoextracts to improve the physiological processes in pea plants even under salinity stress. Furthermore, these are inexpensive and a certainly manageable solution to minimize the detrimental salinity effects, and they can result in improved productivity with saline groundwater in marginal saline lands. We were unable to find studies that use genetic and proteomic approaches to examine the mechanisms behind the positive role of Si in inducing salinity tolerance in pea plants.

8.4 CONCLUSIONS AND FUTURE PROSPECTS

Salinity markedly decreases agriculture productivity throughout the world after disrupting plant physiology and biochemistry. Different crop genotypes show contrasting growth responses to salinity stress either in the presence of Si or without Si addition. Si presence in the cell wall of plants has led to the maintenance of tensile strength and conferring of the tolerance mechanism in plants to salinity stress. From many research findings of Si performance under salinity stress, many scientists have suggested agronomic improvements with respect to Si-induced modifications of the physiological/biochemical functions in plants. In recent years, studies of the characterization of Si transporters have emerged with an objective to understand Si role's under stress conditions; however, the genetic approaches are still in their nascent phase. Although Si's role has been tested in both monocot and dicot plants under salinity stress alone, salinity, along with other stress conditions (combined stresses), is also an important but largely ignored aspect to be studied. Optimal dosage and duration of Si treatment and the appropriate application procedures should be studied more precisely. In conclusion, studies focusing on a combination of molecular and genetic approaches are desired to reveal the mechanisms behind the plant adaptations, together with Si under salt stress conditions.

REFERENCES

Abdalla MM. (2011). Impact of diatomite nutrition on two *Trifolium alexandrinum* cultivars differing in salinity tolerance. *Int J Plant Physiol Biochem* 3: 233–246.

Agarie S, Agata W, Kubota F, Kaufman PB. (1992). Physiological roles of silicon in photosynthesis and dry matter production in rice plants. I. Effects of silicon and shading treatments. *Jpn J Crop Sci* 61: 200–206.

Ahmad B. (2014). Interactive effects of silicon and potassium nitrate in improving salt tolerance of wheat. *J Integr Agric* 13: 1889–1899.

Ahmad R, Zaheer SH, Ismail S. (1992). Role of silicon in salt tolerance of wheat (*Triticum aestivum* L.). *Plant Sci* 85: 43–50.

Al-aghabary K, Zhu Z, Shi Q. (2005). Influence of silicon supply on chlorophyll content, chlorophyll fluorescence, and antioxidative enzyme activities in tomato plants under salt stress. *J Plant Nutr* 27: 2101–2115.

Ali A, Alqurainy F. (2006). Activities of antioxidants in plants under environmental stress. In Motohashi N (ed.), *The Lutein-Prevention and Treatment for Diseases*. Transworld Research Network, Kerala, India, pp. 187–256.

Ali A, Basra SM, Hussain S, Iqbal J. (2012). Increased growth and changes in wheat mineral composition through calcium silicate fertilization under normal and saline field conditions. *Chil J Agric Res* 72: 98–103.

Ali MAM, Ramezani A, Far SM, Asilan KS, Moradi-Ghahderijani M, Jamian SS. (2013). Application of silicon ameliorates salinity stress in sunflower (*Helianthus annuus* L.) plants. *Int J Agri Crop Sci* 6: 1367.

Anderson DL, Sosa O. (2001). Effect of silicon on expression of resistance to sugarcane borer (*Diatraea saccharalis*). *J Am Soc Sugar Cane Technol* 21: 43–50.

Ando H, Fujii H, Hayasaka T, Yokoyama K, Mayum H. (1999). New silicon source for rice cultivation. 3. Growth and yield of wetland rice with reference to silica gel application. In *Conference "Silicon in Agriculture,"* Fort Lauderdale, FL, September 26–30, p. 32.

Ashraf M, Afzal M, Ahmed R, Mujeeb F, Sarwar A, Ali L. (2010a). Alleviation of detrimental effects of NaCl by silicon nutrition in salt-sensitive and salt-tolerant genotypes of sugarcane (*Saccharum officinarum* L.). *Plant Soil* 326: 381–391.

Ashraf M, Ahmad R, Bhatti A, Afzal M, Sarwar A, Maqsood M, Kanwal S. (2010b). Amelioration of salt stress in sugarcane (*Saccharum officinarum* L.) by supplying potassium and silicon in hydroponics. *Pedosphere* 20: 153–162.

Bae EJ, Lee KS, Huh MR, Lim CS. (2012). Silicon significantly alleviates the growth inhibitory effects of NaCl in salt-sensitive 'Perfection' and 'Midnight' Kentucky bluegrass (*Poa pratensis* L.). *Hortic Environ Biotechnol* 53: 477–483.

Bayat H, Alirezaie M, Neamati H, Saadabad AA. (2013). Effect of silicon on growth and ornamental traits of salt-stressed calendula (*Calendula officinalis* L.). *J Ornam Plants* 3: 207–214.

Bernstein N, Silk WK, Läuchli A. (1995). Growth and development of sorghum leaves under conditions of NaCl stress: Possible role of some mineral elements in growth inhibition. *Planta* 196: 699–705.

Blumwald E. (2000). Sodium transport and salt tolerance in plants. *Curr Opin Cell Biol* 12: 431–434.

Chai Q, Shao X, Zhang J. (2010). Silicon effects on *Poa pratensis* responses to salinity. *Hortscience* 45: 1876–1881.

Chiba Y, Mitani N, Yamaji N, Ma JF. (2009). *HvLsi1* is a silicon influx transporter in barley. *Plant J* 57: 810–818.

Dai J, Huff DR, Schlossberg MJ. (2009). Salinity effects on seed germination and vegetative growth of greens-type relative to other cool-season turfgrass species. *Crop Sci* 49: 696–703.

Daniel WH, Freeborg R. (1987). *The Turf Managers' Handbook*. APPA Association of Higher Education, Alexandria, VA.

Deren C. (2001). Chapter 8 Plant genotype, silicon concentration, and silicon-related responses. *Stud Plant Sci* 8: 149–158.

Deren C, Datnoff L, Snyder G, Martin F. (1994). Silicon concentration, disease response, and yield components of rice genotypes grown on flooded organic histosols. *Crop Sci* 34: 733–737.

Epstein E. (1994). The anomaly of silicon in plant biology. *Proc Natl Acad Sci USA* 91: 11–17.

Epstein E, Bloom A. (2005). Inorganic components of plants. In *Mineral Nutrition of Plants: Principles and Perspectives*. 2nd ed. Sinauer Associates, Sunderland, MA, pp. 44–45.

Faiyue B, Al-Azzawi MJ, Flowers TJ. (2010). The role of lateral roots in bypass flow in rice (*Oryza sativa* L.). *Plant Cell Environ* 33: 702–716.

Fei Y-J, Wang Y, Li Q-J, Xiao F, Deng J. (2005). Assessment on salt tolerance of several grass cultivars. *Sichuan Cao Yuan* 5: 20–24.

Fleck AT, Nye T, Repenning C, Stahl F, Zahn M, Schenk MK. (2011). Silicon enhances suberization and lignification in roots of rice (*Oryza sativa*). *J Exp Bot* 62: 2001–2011.

Flowers TJ, Colmer TD. (2008). Salinity tolerance in halophytes. *New Phytol* 179: 945–963.

Friesen D, Sanz J, Correa F, Winslow M, Okada K, Datnoff L, Snyder G. (1994). Silicon deficiency of upland rice on highly weathered savanna soils in Colombia. I. Evidence of a major yield constraint. In *IX Conferência Internacional de arroz para a América Latina e para o Caribe*, Goiania, Brazil, pp. 21–25.

Gong H, Randall D, Flowers T. (2006). Silicon deposition in the root reduces sodium uptake in rice (*Oryza sativa* L.) seedlings by reducing bypass flow. *Plant Cell Environ* 29: 1970–1979.

Gong H, Zhu X, Chen K, Wang S, Zhang C. (2005). Silicon alleviates oxidative damage of wheat plants in pots under drought. *Plant Sci* 169: 313–321.

Gong HJ, Chen Km, Chen Gc, Wang Sm, Zhang Cl. (2003). Effects of silicon on growth of wheat under drought. *J Plant Nutr* 26: 1055–1063.

Gunes A, Inal A, Bagci EG, Coban S. (2007a). Silicon-mediated changes on some physiological and enzymatic parameters symptomatic of oxidative stress in barley grown in sodic-B toxic soil. *J Plant Physiol* 164: 807–811.

Gunes A, Inal A, Bagci EG, Pilbeam DJ. (2007b). Silicon-mediated changes of some physiological and enzymatic parameters symptomatic for oxidative stress in spinach and tomato grown in sodic-B toxic soil. *Plant Soil* 290: 103–114.

Haghighi M, Afifipour Z, Mozafarian M. (2012). The effect of N-Si on tomato seed germination under salinity levels. *J Biol Environ Sci* 6: 87–90.

Hajiboland R, Cheraghvareh L. (2014). Influence of Si supplementation on growth and some physiological and biochemical parameters in salt-stressed tobacco (*Nicotiana rustica* L.) plants. *J Sci Islamic Repub Iran* 25: 205–217.

Hamayun M, Sohn E-Y, Khan SA, Shinwari ZK, Khan AL, Lee I-J. (2010). Silicon alleviates the adverse effects of salinity and drought stress on growth and endogenous plant growth hormones of soybean (*Glycine max* L.). *Pak J Bot* 42: 1713–1722.

Hashemi A, Abdolzadeh A, Sadeghipour HR. (2010). Beneficial effects of silicon nutrition in alleviating salinity stress in hydroponically grown canola, *Brassica napus* L., plants. *Soil Sci Plant Nutr* 56: 244–253.

Hattori T, Inanaga S, Tanimoto E, Lux A, Luxová M, Sugimoto Y. (2003). Silicon-induced changes in visco-elastic properties of sorghum root cell walls. *Plant Cell Physiol* 44: 743–749.

Henriet C, Draye X, Oppitz I, Swennen R, Delvaux B. (2006). Effects, distribution and uptake of silicon in banana (*Musa* spp.) under controlled conditions. *Plant Soil* 287: 359–374.

Hilal M, Zenoff AM, Ponessa G, Moreno H, Massa EM. (1998). Saline stress alters the temporal patterns of xylem differentiation and alternative oxidase expression in developing soybean roots. *Plant Physiol* 117: 695–701.

Hodson M, White PJ, Mead A, Broadley M. (2005). Phylogenetic variation in the silicon composition of plants. *Ann Bot* 96: 1027–1046.

Jaleel CA, Gopi R, Sankar B, Manivannan P, Kishorekumar A, Sridharan R, Panneerselvam R. (2007). Studies on germination, seedling vigour, lipid peroxidation and proline metabolism in *Catharanthus roseus* seedlings under salt stress. *S Afr J Bot* 73: 190–195.

Kafi M, Nabati J, Zare Mehrjerdi M. (2011). Effect of salinity and silicon application on oxidative damage of sorghum (*Sorghum bicolor* L. moench). *Pak J Bot* 43(5): 2457–2462.

Kafi M, Rahimi Z. (2011). Effect of salinity and silicon on root characteristics, growth, water status, proline content and ion accumulation of purslane (*Portulaca oleracea* L.). *Soil Sci Plant Nutr* 57: 341–347.

Karmoker J, Van Steveninck R. (1979). The effect of abscisic acid on sugar levels in seedlings of *Phaseolus vulgaris* L. cv. Redland Pioneer. *Planta* 146: 25–30.

Korndörfer G, Datnoff L, Corrêa G. (1999). Influence of silicon on grain discoloration and upland rice grown on four savanna soils of Brazil. *J Plant Nutr* 22: 93–102.

Krishnamurthy P, Ranathunge K, Franke R, Prakash H, Schreiber L, Mathew M. (2009). The role of root apoplastic transport barriers in salt tolerance of rice (*Oryza sativa* L.). *Planta* 230: 119–134.

Lee S, Sohn E, Hamayun M, Yoon J, Lee I. (2010). Effect of silicon on growth and salinity stress of soybean plant grown under hydroponic system. *Agroforest Syst* 80: 333–340.

Lekklar C, Chaidee A. (2011). Roles of silicon on growth of Thai jasmine rice (*Oryza sativa* L. cv. KDML105) under salt stress. *Agric Sci J* 42: 45–48.

Liang Y. (1997). Effect of silicon on leaf ultrastructure, chlorophyll content and photosynthetic activity of barley under salt stress. *Pedosphere* 8: 289–296.

Liang Y. (1999). Effects of silicon on enzyme activity and sodium, potassium and calcium concentration in barley under salt stress. *Plant Soil* 209: 217–224.

Liang Y, Chen Q, Liu Q, Zhang W, Ding R. (2003). Exogenous silicon (Si) increases antioxidant enzyme activity and reduces lipid peroxidation in roots of salt-stressed barley (*Hordeum vulgare* L.). *J Plant Physiol* 160: 1157–1164.

Liang Y, Ding R. (2002). Influence of silicon on microdistribution of mineral ions in roots of salt-stressed barley as associated with salt tolerance in plants. *Sci China C Life Sci* 45: 298–308.

Liang Y, Shen Q, Shen Z, Ma T. (1996). Effects of silicon on salinity tolerance of two barley cultivars. *J Plant Nutr* 19: 173–183.

Liang Y, Sun W, Zhu Y-G, Christie P. (2007). Mechanisms of silicon-mediated alleviation of abiotic stresses in higher plants: A review. *Environ Pollut* 147: 422–428.

Liang Y, Zhang W, Chen Q, Ding R. (2005). Effects of silicon on H+-ATPase and H+-PPase activity, fatty acid composition and fluidity of tonoplast vesicles from roots of salt-stressed barley (*Hordeum vulgare* L.). *Environ Exp Bot* 53: 29–37.

Liang YC, Ma TS, Li FJ, Feng YJ. (1994). Silicon availability and response of rice and wheat to silicon in calcareous soils. *Commun Soil Sci Plan* 25: 2285–2297.

Liu W, Wang L, Bai Y. (2002). Research progress in the beneficial elements—Silicon for plants. *Acta Bot Boreali Occidental Sin* 23: 2248–2253.

Ma J, Miyake Y, Takahashi E. (2001). Silicon as a beneficial element for crop plants. *Stud Plant Sci* 8: 17–39.

Ma JF. (2010). Silicon transporters in higher plants. In *MIPs and Their Role in the Exchange of Metalloids*. Springer, Berlin, pp. 99–109.

Ma JF, Higashitani A, Sato K, Takeda K. (2003). Genotypic variation in silicon concentration of barley grain. *Plant Soil* 249: 383–387.

Ma JF, Takahashi E. (2002). *Soil, Fertilizer, and Plant Silicon Research in Japan*. Elsevier Science, Amsterdam.

Ma JF, Tamai K, Yamaji N, Mitani N, Konishi S, Katsuhara M, Ishiguro M, Murata Y, Yano M. (2006). A silicon transporter in rice. *Nature* 440: 688–691.

Ma JF, Yamaji N, Mitani N, Tamai K, Konishi S, Fujiwara T, Katsuhara M, Yano M. (2007). An efflux transporter of silicon in rice. *Nature* 448: 209–212.

Mali M, Aery N. (2008). Influence of silicon on growth, relative water contents and uptake of silicon, calcium and potassium in wheat grown in nutrient solution. *J Plant Nutr* 31: 1867–1876.

Marschner H. (2011). *Marschner's Mineral Nutrition of Higher Plants*. 3rd ed. Elsevier/Academic Press, Amsterdam, p. 684.

Masood S, Saleh L, Witzel K, Plieth C, Mühling KH. (2012). Determination of oxidative stress in wheat leaves as influenced by boron toxicity and NaCl stress. *Plant Physiol Biochem* 56: 56–61.

Mateos-Naranjo E, Andrades-Moreno L, Davy AJ. (2013). Silicon alleviates deleterious effects of high salinity on the halophytic grass *Spartina densiflora*. *Plant Physiol Biochem* 63: 115–121.

Matichenkov V, Calvert D. (2002). Silicon as a beneficial element for sugarcane. *J Am Soc Sugar Cane Technol* 22: 21–30.

Matichenkov V, Kosobrukhov A. (2004). Si effect on the plant resistance to salt toxicity. In *Proceedings of the 13th International Soil Conservation Organization Conference (ISCO)*, Brisbane, Australia, pp. 287–295.

Matoh T, Kairusmee P, Takahashi E. (1986). Salt-induced damage to rice plants and alleviation effect of silicate. *Soil Sci Plant Nutr* 32: 295–304.

Miao B-H, Han X-G, Zhang W-H. (2010). The ameliorative effect of silicon on soybean seedlings grown in potassium-deficient medium. *Ann Bot* 105(6): 967–973.

Milne CJ. (2012). The alleviation of salinity induced stress with the application of silicon in soilless grown *Lactuca sativa* L. 'Eish!' *Int J Phys Sci* 7(5): 735–742.

Mitani N, Chiba Y, Yamaji N, Ma JF. (2009a). Identification and characterization of maize and barley *Lsi2*-like silicon efflux transporters reveals a distinct silicon uptake system from that in rice. *Plant Cell* 21: 2133–2142.

Mitani N, Ma JF. (2005). Uptake system of silicon in different plant species. *J Exp Bot* 56: 1255–1261.

Mitani N, Yamaji N, Ago Y, Iwasaki K, Ma JF. (2011). Isolation and functional characterization of an influx silicon transporter in two pumpkin cultivars contrasting in silicon accumulation. *Plant J* 66: 231–240.

Mitani N, Yamaji N, Ma JF. (2009b). Identification of maize silicon influx transporters. *Plant Cell Physiol* 50: 5–12.

Mitsui S, Takatoh H. (1963). Nutritional study of silicon in graminaceous crops (Part 1). *Soil Sci Plant Nutr* 9: 7–11.

Montpetit J, Vivancos J, Mitani-Ueno N, Yamaji N, Rémus-Borel W, Belzile F, Ma JF, Bélanger RR. (2012). Cloning, functional characterization and heterologous expression of *TaLsi1*, a wheat silicon transporter gene. *Plant Mol Biol* 79: 35–46.

Mori R, Wakabayashi K, Hoson T, Fujii S, Yamamoto R, Kamisaka S. (1999). Mechanism of growth promotion in rice third leaves by silicon [in Japanese]. In *Proceedings of the Annual Meeting of the Botanical Society of Japan*, Tohoku, Japan, p. 120.

Moussa HR. (2006). Influence of exogenous application of silicon on physiological response of salt-stressed maize (*Zea mays* L.). *Int J Agric Biol* 8: 293–297.

Moussa, HR, El-Galad MAE. (2015). Comparative response of salt tolerant and salt sensitive maize (*Zea mays* L.) cultivars to silicon. *Eur J Acad Essays* 2(1): 1–5.

Muneer S, Park YG, Manivannan A, Soundararajan P, Jeong BR. (2014). Physiological and proteomic analysis in chloroplasts of *Solanum lycopersicum* L. under silicon efficiency and salinity stress. *Int J Mol Sci* 15: 21803–21824.

Munns R, Greenway H, Kirst G. (1983). Halotolerant eukaryotes. In *Physiological Plant Ecology* III. Springer, Berlin, pp. 59–135.

Muscolo A, Panuccio MR, Sidari M. (2003). Effects of salinity on growth, carbohydrate metabolism and nutritive properties of kikuyu grass (*Pennisetum clandestinum* Hochst). *Plant Sci* 164: 1103–1110.

Parande S, Zamani GR, Syyari Zahan M, Ghaderi M. (2013). Effects of silicon application on the yield and component of yield in the common bean (*Phaseolus vulgaris*) under salinity stress. *Int J Agron Plant Prod* 4: 1574–1579.

Parveen N, Ashraf M. (2010). Role of silicon in mitigating the adverse effects of salt stress on growth and photosynthetic attributes of two maize (*Zea mays* L.) cultivars grown hydroponically. *Pak J Bot* 42(3): 1675–1684.

Rengasamy P. (2010). Soil processes affecting crop production in salt-affected soils. *Funct Plant Biol* 37: 613–620.

Rhoades J. (1993). Electrical conductivity methods for measuring and mapping soil salinity. In Sparks DL (ed.), *Advances in Agronomy*. Vol. 49. Academic Press, San Diego, pp. 201–251.

Rohanipoor A, Norouzi M, Moezzi A, Hassibi P. (2013). Effect of silicon on some physiological properties of maize (*Zea mays*) under salt stress. *J Biol Environ Sci* 7: 71–79.

Romero-Aranda MR, Jurado O, Cuartero J. (2006). Silicon alleviates the deleterious salt effect on tomato plant growth by improving plant water status. *J Plant Physiol* 163: 847–855.

Sabaghnia N, Janmohammadi M. (2015). Effect of nano-silicon particles application on salinity tolerance in early growth of some lentil genotypes. *Ann UMCS Biol* 69: 39–55.

Santa-Cruz A, Acosta M, Pérez-Alfocea F, Bolarin MC. (1997). Changes in free polyamine levels induced by salt stress in leaves of cultivated and wild tomato species. *Physiol Plant* 101: 341–346.

Saqib M, Zörb C, Schubert S. (2008). Silicon-mediated improvement in the salt resistance of wheat (*Triticum aestivum*) results from increased sodium exclusion and resistance to oxidative stress. *Funct Plant Biol* 35: 633–639.

Savvas D, Giotis D, Chatzieustratiou E, Bakea M, Patakioutas G. (2009). Silicon supply in soilless cultivations of zucchini alleviates stress induced by salinity and powdery mildew infections. *Environ Exp Bot* 65: 11–17.

Shahid MA, Balal RM, Pervez MA, Abbas T, Aqeel MA, Riaz A, Mattson NS. (2015). Exogenous 24-Epibrassinolide elevates the salt tolerance potential of pea (*Pisum sativum* L.) by improving osmotic adjustment capacity and leaf water relations. *J Plant Nutr* 38: 1050–1072.

Shahzad M, Witzel K, Zörb C, Mühling K. (2012). Growth-related changes in subcellular ion patterns in maize leaves (*Zea mays* L.) under salt stress. *J Agron Crop Sci* 198: 46–56.

Shahzad M, Zörb C, Geilfus CM, Mühling K. (2013). Apoplastic Na⁺ in *Vicia faba* leaves rises after short-term salt stress and is remedied by silicon. *J Agron Crop Sci* 199: 161–170.

Sharaf A. (2010). Improvement growth, and yield of wheat plants grown under salinity stress by using silicon. *J Am Sci* 6: 559–566.

Shi Y, Wang Y, Flowers TJ, Gong H. (2013). Silicon decreases chloride transport in rice (*Oryza sativa* L.) in saline conditions. *J Plant Physiol* 170: 847–853.

Solatni Z, Shekari F, Jamshidi K, Fotovat R, Azimkhani R. (2012). The effect of silicon on germination and some growth characteristics of salt-stressed canola seedling. *Int J Agric Agric R* 2: 12–21.

Soylemezoglu G, Demir K, Inal A, Gunes A. (2009). Effect of silicon on antioxidant and stomatal response of two grapevine (*Vitis vinifera* L.) rootstocks grown in boron toxic, saline and boron toxic-saline soil. *Sci Hortic—Amsterdam* 123: 240–246.

Tahir MA, Aziz T, Farooq M, Sarwar G. (2012). Silicon-induced changes in growth, ionic composition, water relations, chlorophyll contents and membrane permeability in two salt-stressed wheat genotypes. *Arch Agron Soil Sci* 58: 247–256.

Tuna AL, Kaya C, Higgs D, Murillo-Amador B, Aydemir S, Girgin AR. (2008). Silicon improves salinity tolerance in wheat plants. *Environ Exp Bot* 62: 10–16.

Wahid A, Rasul E. (1997). Identification of salt tolerance traits in sugarcane lines. *Field Crops Res* 54: 9–17.

Wang S, Galletta G. (1998). Foliar application of potassium silicate induces metabolic changes in strawberry plants. *J Plant Nutr* 21: 157–167.

Wang XS, Han JG. (2007). Effects of NaCl and silicon on ion distribution in the roots, shoots and leaves of two alfalfa cultivars with different salt tolerance. *Soil Sci Plant Nutr* 53: 278–285.

Winslow MD. (1992). Silicon, disease resistance, and yield of rice genotypes under upland cultural conditions. *Crop Sci* 32: 1208–1213.

Xie Z, Song R, Shao H, Song F, Xu H, Lu Y. (2015). Silicon improves maize photosynthesis in saline-alkaline soils. *Sci World J* 2015: 245072.

Yamaji N, Chiba Y, Mitani-Ueno N, Ma JF. (2012). Functional characterization of a silicon transporter gene implicated in silicon distribution in barley. *Plant Physiol* 160: 1491–1497.

Yeo A, Flowers S, Rao G, Welfare K, Senanayake N, Flowers T. (1999). Silicon reduces sodium uptake in rice (*Oryza sativa* L.) in saline conditions and this is accounted for by a reduction in the transpirational bypass flow. *Plant Cell Environ* 22: 559–565.

Yin L, Wang S, Li J, Tanaka K, Oka M. (2013). Application of silicon improves salt tolerance through ameliorating osmotic and ionic stresses in the seedling of *Sorghum bicolor*. *Acta Physiol Plant* 35: 3099–3107.

Yin L, Wang S, Liu P, Wang W, Cao D, Deng X, Zhang S. (2014). Silicon-mediated changes in polyamine and 1-aminocyclopropane-1-carboxylic acid are involved in silicon-induced drought resistance in *Sorghum bicolor* L. *Plant Physiol Biochem* 80: 268–277.

Yin L, Wang S, Tanaka K, Fujihara S, Itai A, Den X, Zhang S. (2015). Silicon-mediated changes in polyamines participate in silicon-induced salt tolerance in *Sorghum bicolor* L. *Plant Cell Environ* 39: 245–258.

Zhu Y, Gong H. (2014). Beneficial effects of silicon on salt and drought tolerance in plants. *Agron Sustain Dev* 34: 455–472.

Zhu Z, Wei G, Li J, Qian Q, Yu J. (2004). Silicon alleviates salt stress and increases antioxidant enzymes activity in leaves of salt-stressed cucumber (*Cucumis sativus* L.). *Plant Sci* 167: 527–533.

Zuccarini P. (2008). Effects of silicon on photosynthesis, water relations and nutrient uptake of *Phaseolus vulgaris* under NaCl stress. *Biol Plant* 52: 157–160.

9 Silicon-Mediated Modulations of Genes and Secondary Metabolites in Plants
A Comprehensive Overview

Abinaya Manivannan, Prabhakaran Soundararajan, and Byoung Ryong Jeong

CONTENTS

ABSTRACT

Silicon (Si) is considered as a quasi-essential element for the plant system, but its beneficial effects on growth and stress tolerance have been recorded in a wide variety of agricultural and horticultural crops. However, Si-mediated molecular regulation of physiological processes by the mediation of gene expression, particularly under stressed conditions, has been dealt with in a limited number of plants. Similarly, the prophylactic mechanism of Si against several pathogen attacks has been attributed to the secondary metabolism in plants. Silicon triggers vital secondary metabolite biosynthetic pathways such as the phenylpropanoid and terpenoid pathways for the accumulation of antimicrobial compounds or phytoalexins. In this chapter, the Si-mediated regulations of genes and secondary metabolites under abiotic and biotic stress are discussed.

9.1 INTRODUCTION

9.1.1 SILICON: A MODEST ELEMENT WITH VERSATILE ROLES

Geographically, silicon (Si) is the second most abundant element in the earth's crust, covering 27.70% in the lithosphere. However, the occurrence of Si in its pure form is rare, and it exists in the form of silicate minerals that cover 90% of the earth's crust (Mitra, 2015). Broadly, Si is distributed as quartz (SiO_2), sand, and sandstone (Rédei, 2008). In biological organisms, Si occurs as amorphous silica ($SiO_2 \cdot nH_2O$) and soluble silicic acid ($Si(OH)_4$), accumulated and metabolized by some prokaryotes (Das and Chattopadhyay, 2000), and stimulates the growth of fungi (Wainwright et al., 1997). It is well known that Si is essential for diatoms (Martin-Jézéquel et al., 2000) and algae (Currie and Perry, 2007). In mammals, Si is vital for bones, cartilage, connective tissue formation, enzymatic activities, and lymphocyte proliferation (Carlisle, 1988, 1997).

Under biological pH ranges, Si occurs as a silicic acid and is taken up by plants in the same form (Knight and Kinrade, 2001). According to Ingri (1978), monosilicic acid has the ability to combine with a hydroxyl functional group containing organic molecules. Likewise, biosilica can associate with biological molecules such as proteins and carbohydrates; especially Si in a hypervalent form has been found to be in complexes with derivatives of sugars (Kinrade et al., 1999, 2001; Bond and McAliffe, 2003). Irrespective of the geographic occurrence and uptake of Si, the first evidence of *in vivo* formation of organosilicon compound has been discovered in the diatom *Navicula pelliculosa* (Kinrade et al., 2002). In diatoms, Si implies that the phosphorylation of proteins is involved in the synthesis of DNA and specific mRNA (Reeves and Volcani, 1984). In plants, the organosilicon was identified to bind proteins, cellular polyphenols for lignification, lipids, cellulose, and pectin (Kolesnikov et al., 2001).

Thus, the distribution and physiological importance of Si have been recorded in several biological systems. However, in this chapter, the versatile role of Si in plant biology, in particular with the molecular regulation of genes and secondary metabolites, is discussed.

9.1.2 MULTIFACET ROLE OF SI IN THE PLANT KINGDOM

The essential roles of Si in the plant system have been extensively studied by numerous plant biologists for several years. Yet to date, by definition, Si is considered as a quasi-essential or non-essential element for the plant because most plant species can complete their life cycle without it (Arnon and Stout, 1939). However, there are a number of hypotheses concerning the physiological functions of Si in monocots and dicots. In addition, ample positive effects of Si on growth and biomass (Egrinya Eneji et al., 2008; Soundararajan et al., 2014; Zhang et al., 2015), productivity or yield (Epstein, 1999), stress tolerance (Liang, 1966; Ma, 2004; Liang et al., 2007; Zhu et al., 2007; Muneer et al., 2014), macro- and micronutrient management (Tripathi et al., 2014), pest and pathogen resistance (Lanning, 1966; Coors, 1987; Fawe et al., 1998; Cookson et al., 2007), provision of structural rigidity (Epstein, 1994), nutrient transportation (Tripathi et al., 2012), enhancement of light absorption (Li et al., 2004), resistance to lodging (Ma et al., 2006), and ion homeostasis (Mehrabanjoubani et al., 2015) have been confirmed in the plant kingdom. For instance, Si was reported to alleviate abiotic and biotic stresses in various agricultural and horticultural plants (Liang et al., 1966; Epstein, 1999; Ma, 2004; Liang et al., 2007; Soundararajan et al., 2014). Although Si biology in the plant was studied for decades, the mechanisms behind the physiological responses or regulation by Si under normal and stressed conditions are not well elucidated. The enhancement of disease resistance against pathogenic fungi in plants could be resultant from the Si-mediated induced resistance (Fawe et al., 1998). The Si-induced resistance (Si-IR) mechanism can be attributed to either the form of mechanical barrier for penetration of host tissue due to the specific accumulation and polymerization of $Si(OH)_4$ in cell walls (Yoshida et al., 1962) or the means of chemical resistance, like the accumulation of antifungal compounds (Fawe et al., 1998). Evidently, Chérif et al. (1992) illustrated that induced resistance by Si resulted in a significant increase in the activities of chitinases, peroxidases (POXs), and polyphenoloxidases (PPOs) against *Pythium ultimum* in cucumber plants. In addition, the microscopic studies on the cells affected by *P. ultimum* or *Sphaerotheca fuliginea* (Schlechtend.: Fr.) Pollacci revealed the accumulation of phenolic compounds (in a manner similar to that of phytoalexins) toxic to fungal structures (Chérif et al., 1992).

Under abiotic stress such as salinity, Si application resulted in the alleviation of stress and enhancement of plant growth (Soundararajan et al., 2014; Li et al., 2015). For instance, in our recent study, Si supplementation significantly mitigated the detrimental effects of salinity and improved the growth and antioxidant enzyme activity of salt-stressed carnation in *in vitro* conditions (Soundararajan et al., 2015). During salt stress, Si decreases Na^+ and Cl^- transport due to silicon deposition (Gong et al., 2006; Shi et al., 2013).

According to Zhu and Gong (2014), the silicon-mediated mechanism behind the alleviation of salt stress includes the following aspects: (1) maintenance of optimal water, (2) enhancement of photosynthesis and curbing of the transpiration rate, (3) limiting the oxidative stress by ion toxicity alleviation, and (4) biosynthetic regulation of solutes and plant hormones. In line with other researchers, Al-Aghabary et al. (2005) observed increased activities of antioxidant enzymes and enhanced photochemical efficiency of the PS II under salt stress. Also noteworthy, Si decreases the toxicity of aluminum (Al) (Wang et al., 2004), boron (B) (Gunes et al., 2007), cadmium (Cd) (Liang et al., 2005), chromium (Cr) (Tripathi et al., 2012), copper (Cu) (Li et al., 2008), and zinc (Zn) (Neumann and zur Nieden, 2001). Similarly, the applications of fertilizers fortified with Si have been reported to reduce the electrolyte leakage and promote photosynthesis in rice under drought stress conditions (Chen et al., 2011).

Taken together, the inclusion of Si for plant growth becomes an indispensable solution for abiotic and biotic stresses. Although some information is available on the positive roles of Si, as mentioned earlier, few studies have attempted to elucidate the molecular rationale behind the Si-mediated regulation of genes and secondary metabolites. As a comprehensive overview, in this chapter Si-mediated regulation of gene expression and secondary metabolites is discussed.

9.2   Si-MEDIATED REGULATION OF GENE EXPRESSION

The first initiative for the identification of gene regulation by Si in rice was led by Watanabe et al. (2004). Microarray analysis of rice leaf demonstrated the differential expression of genes upon Si nutrition. The results of the first microarray experiment on rice revealed the downregulation of four genes and upregulation of a single gene under limited supply of Si (Wantanabe et al., 2004). The transcript levels of chlorophyll *a/b* binding protein, metallothione-like protein, *Xa21* gene family member, and carbonic anhydrase homologue were downregulated by Si supplementation (Wantanabe et al., 2004). However, the application of Si upregulated the expression of a zinc finger protein homologue. Apparently, the zinc finger protein plays a vital role as the major transcription factor for stress-responsive genes, and thus the results of Wantanabe et al. (2004) suggested the regulation of stress-related genes by Si. Different types of genes regulated by Si amendment, under biotic and abiotic stresses, have been provided in Table 9.1.

9.2.1   REGULATION OF PHOTOSYNTHESIS-RELATED GENES BY SI

Oxidative stress caused by both abiotic and biotic stress factors retards plant growth by targeting major physiological processes such as photosynthesis. For instance, the metal toxicity, salinity, and drought stress affect the vital enzymes involved in the Calvin cycle and the photosynthetic electron transport chain (Nwugo and Huerta, 2008; Gong et al., 2005; Muneer et al., 2014). In particular, the enzymes Ribulose-1,5-bisphosphate carboxylase/oxygenase (RuBisCO), fructose-1,6-bisphosphate, and NADP-glyceraldehyde-3-phosphate dehydrogenase are negatively affected (Nwugo and Huerta, 2008). Moreover, the photosynthetic efficiency of plants was degenerated by the disruption of thylakoid membranes by Cd toxicity (Souza et al., 2005). Consequently, the perturbation in the photosynthetic process negatively affects the cascade of other physiological process. Amendment of Si increased the photosynthetic efficiency of agricultural as well as horticultural plants under both normal and stressed conditions (Muneer et al., 2014). One of the stress-combating strategies utilized by Si was enhancing the photosynthesis of plants.

TABLE 9.1

Genes Regulated by Si Amendment under Different Environmental Stresses

Gene	Description	Regulation	Biological Process	Stress	Reference
Os01g0713200	β-1,3-Glucanase precursor	Up	Defense-related gene	*Magnaporthe oryzae*	Brunings et al. (2009)
Os02g0584800	Heavy metal transport/ detoxification protein domain-containing protein	Up	Defense-related gene	*Magnaporthe oryzae*	Brunings et al. (2009)
Os02g0585100	Heavy metal transport/ detoxification protein domain-containing protein	Up	Defense-related gene	*Magnaporthe oryzae*	Brunings et al. (2009)
Os04g0469000	Heavy metal transport/ detoxification protein domain-containing protein	Up	Defense-related gene	*Magnaporthe oryzae*	Brunings et al. (2009)
Os04g0610400	Pathogenesis-related transcriptional factor and ERF domain-containing protein	Up	Defense-related gene	*Magnaporthe oryzae*	Brunings et al. (2009)
Os07g0104100	Peroxidase 27 precursor (EC 1.11.1.7) (Atperox P27) (PRXR7) (ATP12a)	Up	Defense-related gene	*Magnaporthe oryzae*	Brunings et al. (2009)
Os08g0539700	PibH8 protein	Up	Photosynthesis	*Magnaporthe oryzae*	Brunings et al. (2009)
Os08g02630	PsbY	Up	Photosynthesis	Zinc toxicity	Song et al. (2014)
Os05g48630	PsaH	Up	Photosynthesis	Zinc toxicity	Song et al. (2014)
Os07g37030	PetC	Up	Photosynthesis	Zinc toxicity	Song et al. (2014)
Os03g57120	PetH	Up	Photosynthesis	Zinc toxicity	Song et al. (2014)
Os09g26810	Chlorophyll *a/b* binding protein	Up	Photosynthesis	Zinc toxicity	Song et al. (2014)
Os04g38410	Chlorophyll *a/b* binding protein CP24, chloroplast precursor	Up	Photosynthesis	Zinc toxicity	Song et al. (2014)
AU068650	Zinc finger protein	Up	Transcription factor–encoding gene	–	Watanabe et al. (2004)
AF048750	Metallothionein-like protein	Down	Metal binding	–	Watanabe et al. (2004)
D00641	Chlorophyll *a/b* binding protein	Down	Photosynthesis	–	Watanabe et al. (2004)
U37133	Xa21 gene family member	Down	Disease resistance	–	Watanabe et al. (2004)
AF182806	Carbonic anhydrase	Down	Photosynthesis	–	Watanabe et al. (2004)
Os11g0692500	Bacterial blight resistance protein	Up	Defense-related gene	*Magnaporthe oryzae*	Brunings et al. (2009)

(Continued)

TABLE 9.1 (CONTINUED)
Genes Regulated by Si Amendment under Different Environmental Stresses

Gene	Description	Regulation	Biological Process	Stress	Reference
Os01g0963200	Peroxidase BP 1 precursor	Up	Defense-related gene	*Magnaporthe oryzae*	Brunings et al. (2009)
Os01g0378100	Peroxidase precursor (EC 1.11.1.7)	Up	Defense-related gene	*Magnaporthe oryzae*	Brunings et al. (2009)
Os03g0379300	Basic helix–loop–helix dimerization region bHLH domain-containing protein	Up	Transcription factor–encoding genes	*Magnaporthe oryzae*	Brunings et al. (2009)
Os01g0633400	CBS domain-containing protein	Up	Transcription factor–encoding genes	*Magnaporthe oryzae*	Brunings et al. (2009)
Os02g0731200	Transcription factor MADS57	Up	Transcription factor–encoding genes	*Magnaporthe oryzae*	Brunings et al. (2009)
Os01g0871600	TGF-β receptor, type I/II extracellular region family protein	Up	Transporter-encoding genes	*Magnaporthe oryzae*	Brunings et al. (2009)
Os02g0550800	Ammonium transporter	Up	Transporter-encoding genes	*Magnaporthe oryzae*	Brunings et al. (2009)
Os12g0227400	Allyl alcohol dehydrogenase	Up	Housekeeping genes	*Magnaporthe oryzae*	Brunings et al. (2009)
Os03g0226400	Cation diffusion facilitator 8	Up	Housekeeping genes	*Magnaporthe oryzae*	Brunings et al. (2009)
Os01g0898500	Patatin-like protein 2 (fragment)	Up	Housekeeping genes	*Magnaporthe oryzae*	Brunings et al. (2009)
Os11g0608300	Barley stem rust resistance protein	Down	Defense-related gene	*Magnaporthe oryzae*	Brunings et al. (2009)
Os11g0673600	Disease resistance protein family protein	Down	Defense-related gene	*Magnaporthe oryzae*	Brunings et al. (2009)
Os03g0266300	Heat shock protein Hsp20 domain-containing protein	Down	Defense-related gene	*Magnaporthe oryzae*	Brunings et al. (2009)
Os03g0235000	Peroxidase (EC 1.11.1.7)	Down	Defense-related gene	*Magnaporthe oryzae*	Brunings et al. (2009)
Os12g0491800	Terpene synthase-like domain-containing protein	Down	Defense-related gene	*Magnaporthe oryzae*	Brunings et al. (2009)
Os10g0191300	Type 1 pathogenesis-related protein	Down	Defense-related gene	*Magnaporthe oryzae*	Brunings et al. (2009)
Os09g0417800	DNA-binding WRKY domain-containing protein	Down	Transcription factor–encoding genes	*Magnaporthe oryzae*	Brunings et al. (2009)
Os08g0332700	Trans-acting transcriptional protein ICP0 (Immediate–early protein IE110)	Down	Transcription factor–encoding genes	*Magnaporthe oryzae*	Brunings et al. (2009)
Os02g0695200	P-type R2R3 Myb protein (fragment)	Down	Transcription factor–encoding genes	*Magnaporthe oryzae*	Brunings et al. (2009)
Os09g0110300	Putative cyclase family protein	Down	Housekeeping genes	*Magnaporthe oryzae*	Brunings et al. (2009)

(Continued)

TABLE 9.1 (CONTINUED)
Genes Regulated by Si Amendment under Different Environmental Stresses

Gene	Description	Regulation	Biological Process	Stress	Reference
Os08g0112300	Transferase family protein	Down	Housekeeping genes	*Magnaporthe oryzae*	Brunings et al. (2009)
Os10g0154700	Cyclophilin Dicyp-2	Down	Housekeeping genes	*Magnaporthe oryzae*	Brunings et al. (2009)
Os08g0155700	DNA-directed RNA polymerase (EC 2.7.7.6) largest chain (isoform B1)-like protein	Down	Housekeeping genes	*Magnaporthe oryzae*	Brunings et al. (2009)
Os11g0194800	DNA-directed RNA polymerase II 7.6 kDa polypeptide	Down	Housekeeping genes	*Magnaporthe oryzae*	Brunings et al. (2009)
Os11g0106700	Ferritin 1, chloroplast precursor (ZmFer1)	Down	Housekeeping genes	*Magnaporthe oryzae*	Brunings et al. (2009)
Os12g0258700	Multicopper oxidase, type 1 domain-containing protein	Down	Housekeeping genes	*Magnaporthe oryzae*	Brunings et al. (2009)
Os01g0770200	Tyrosine decarboxylase 1 (EC 4.1.1.25) (ELI5) (fragment)	Down	Housekeeping genes	*Magnaporthe oryzae*	Brunings et al. (2009)
Os01g0627800	Cytochrome P450 monooxygenase CYP72A5 (fragment)	Down	Cytochrome P450–encoding genes	*Magnaporthe oryzae*	Brunings et al. (2009)
Os02g0241100	Protein kinase domain-containing protein	Down	Kinase/phosphatase-encoding genes	*Magnaporthe oryzae*	Brunings et al. (2009)
CHS	Chalcone synthase	Up	Defense-related gene	*Magnaporthe oryzae*	Rahman et al. (2015)
PAL	Phenylalanine ammonia lyase	Up	Defense-related gene	*Magnaporthe oryzae*	Rahman et al. (2015)
POX	Peroxidase	Up	Defense-related gene	*Magnaporthe oryzae*	Rahman et al. (2015)
PR-1	Pathogenesis-related protein transcript-1	Up	Defense-related gene	*Magnaporthe oryzae*	Rahman et al. (2015)
Os08g0539700	PibH8 protein	Up	Defense-related gene	*Magnaporthe oryzae*	Brunings et al. (2009)
Os04g0385600	Cyclin-like F-box domain-containing protein	Up	Transcription factor–encoding gene	*Magnaporthe oryzae*	Brunings et al. (2009)
Os11g0665600	Helix–turn–helix, Fis-type domain-containing protein	Up	Transcription factor–encoding gene	*Magnaporthe oryzae*	Brunings et al. (2009)
Os04g0669200	Ethylene response factor 3	Up	Phytohormone-related gene	*Magnaporthe oryzae*	Brunings et al. (2009)
Os05g0497300	Ethylene-responsive element-binding factor 5 (AtERF5)	Up	Phytohormone-related gene	*Magnaporthe oryzae*	Brunings et al. (2009)
Os01g0675800	No apical meristem (NAM) protein domain-containing protein	Up	Phytohormone-related gene	*Magnaporthe oryzae*	Brunings et al. (2009)

(Continued)

TABLE 9.1 (CONTINUED)
Genes Regulated by Si Amendment under Different Environmental Stresses

Gene	Description	Regulation	Biological Process	Stress	Reference
Os07g0181100	C4-dicarboxylate transporter/ malic acid transport protein family protein	Up	Housekeeping gene	*Magnaporthe oryzae*	Brunings et al. (2009)
Os02g0807000	Phosphoenolpyruvate carboxylase kinase	Up	Housekeeping gene	*Magnaporthe oryzae*	Brunings et al. (2009)
Os01g0554100	RNA-directed DNA polymerase (reverse transcriptase) domain-containing protein	Up	Housekeeping gene	*Magnaporthe oryzae*	Brunings et al. (2009)
Os03g0803500	2OG-Fe(II) oxygenase domain-containing protein	Up	Housekeeping gene	*Magnaporthe oryzae*	Brunings et al. (2009)
Os10g0559500	2OG-Fe(II) oxygenase domain-containing protein	Up	Housekeeping gene	*Magnaporthe oryzae*	Brunings et al. (2009)
Os09g0432300	AAA ATPase, central region domain-containing protein	Up	Housekeeping gene	*Magnaporthe oryzae*	Brunings et al. (2009)
Os06g0676700	High pI α-glucosidase	Up	Housekeeping gene	*Magnaporthe oryzae*	Brunings et al. (2009)
Os08g0190100	Oxalate oxidase-like protein or germin-like protein (germin-like 8, germin-like 12)	Up	Housekeeping gene	*Magnaporthe oryzae*	Brunings et al. (2009)
Os05g0495600	P-type ATPase (fragment)	Up	Housekeeping gene	*Magnaporthe oryzae*	Brunings et al. (2009)
Os03g0405500	PDI-like protein	Up	Housekeeping gene	*Magnaporthe oryzae*	Brunings et al. (2009)
Os01g0227700	Cytochrome P450 family protein	Up	Cytochrome P450–encoding gene	*Magnaporthe oryzae*	Brunings et al. (2009)
Os02g0822900	Protein kinase domain-containing protein	Up	Kinase/phosphatase-encoding gene	*Magnaporthe oryzae*	Brunings et al. (2009)
Os07g0538300	Serine/threonine kinase receptor-like protein	Up	Kinase/phosphatase-encoding gene	*Magnaporthe oryzae*	Brunings et al. (2009)
Os04g0634700	Diacylglycerol kinase	Up	Kinase/phosphatase-encoding gene	*Magnaporthe oryzae*	Brunings et al. (2009)
WRKY-IId5 (AY157064)	WRKY group II transcription factor	Ns	Transcription factor–encoding gene	*Ralstonia solanacearum*	Ghareeb et al. (2011)
JERF3 (AY383630)	Jasmonate and ethylene-responsive factor 3	Up	Transcription factor–encoding gene	*Ralstonia solanacearum*	Ghareeb et al. (2011)
TSRF1 (AF494201)	Tomato stress-responsive factor	Up	Defense-related gene	*Ralstonia solanacearum*	Ghareeb et al. (2011)
ACCO (X04792)	1-Aminocyclopropane-1-carboxylate oxidase	Up	Phytohormone-related gene	*Ralstonia solanacearum*	Ghareeb et al. (2011)

(Continued)

TABLE 9.1 (CONTINUED)
Genes Regulated by Si Amendment under Different Environmental Stresses

Gene	Description	Regulation	Biological Process	Stress	Reference
PR1 (M69247)	Pathogenesis-related protein 1	Ns	Defense-related gene	*Ralstonia solanacearum*	Ghareeb et al. (2011)
FD-I (Z75520)	Ferredoxin I	Up	Photosynthesis-related gene	*Ralstonia solanacearum*	Ghareeb et al. (2011)
GLU 1,3 (M80604)	β-Glucanase	Up	Defense-related gene	*Ralstonia solanacearum*	Ghareeb et al. (2011)
CHI-II (U30465)	Chitinase class II	Up	Defense-related gene	*Ralstonia solanacearum*	Ghareeb et al. (2011)
POD (X94943)	Peroxidase	Up	Defense-related gene	*Ralstonia solanacearum*	Ghareeb et al. (2011)
PAL (M83314)	Phenylalanine ammonia lyase	Up	Defense-related gene	*Ralstonia solanacearum*	Ghareeb et al. (2011)
AGP-1g (X99147)	Arabinogalactan protein	Up	Defense-related gene	*Ralstonia solanacearum*	Ghareeb et al. (2011)
PGIP (L26529)	Polygalacturonase inhibitor protein	Up	Defense-related gene	*Ralstonia solanacearum*	Ghareeb et al. (2011)
OsRDCP1	RING domain-containing protein family members	Up	Transcription factor–encoding genes	Drought	Khattab et al. (2014)
OsDREB2	Dehydration-responsive element-binding protein	Up	Transcription factor–encoding genes	Drought	Khattab et al. (2014)
OsCMO	Rice choline monooxygenase	Up	Transcription factor–encoding genes	Drought	Khattab et al. (2014)
OsNAC5	NAC regulons (NAM, *Arabidopsis thaliana* activating factor [ATAF], and cup-shaped cotyledon [CUC])	Up	Transcription factor–encoding genes	Drought	Khattab et al. (2014)
OsRab16b	Dehydrin	Up	Transcription factor–encoding genes	Drought	Khattab et al. (2014)
SAMDC02	S-Adenosyl-L-methionine decarboxylase	Up	Polyamine synthesis–related gene	Salinity	Yin et al. (2015)
SAMDC04	S-Adenosyl-Met-decarboxylase	Up	Polyamine synthesis–related gene	Salinity	Yin et al. (2015)
SAMDC06	S-Adenosyl-Met-decarboxylase	Up	Polyamine synthesis–related gene	Salinity	Yin et al. (2015)
ODC1	Ornithine decarboxylase	Ns	Polyamine synthesis–related gene	Salinity	Yin et al. (2015)
ODC2	Ornithine decarboxylase	Ns	Polyamine synthesis–related gene	Salinity	Yin et al. (2015)

(Continued)

TABLE 9.1 (CONTINUED)
Genes Regulated by Si Amendment under Different Environmental Stresses

Gene	Description	Regulation	Biological Process	Stress	Reference
ODC3	Ornithine decarboxylase	Ns	Polyamine synthesis–related gene	Salinity	Yin et al. (2015)
ADC	Arginine decarboxylase	Up	Polyamine synthesis–related gene	Salinity	Yin et al. (2015)
CPA	N-Carbamoyl putrescine amidohydrolase	Up	Polyamine synthesis–related gene	Salinity	Yin et al. (2015)
ACS1	1-Aminocyclopropane-1-carboxylic acid synthase	Ns	Polyamine synthesis–related gene	Salinity	Yin et al. (2015)
ACS2	1-Aminocyclopropane-1-carboxylic acid synthase	Down	Polyamine synthesis–related gene	Salinity	Yin et al. (2015)
SPDS	Spermidine synthase	Ns	Polyamine synthesis–related gene	Salinity	Yin et al. (2015)
MCO	Multicopper oxidase type I family protein	Up	Heavy metal toxicity	Copper toxicity	Li and Leisner (2008)
ELT	Esterase lipase thioesterase family protein	Up	Heavy metal toxicity	Copper toxicity	Li and Leisner (2008)
COPT1	Copper transporter	Up	Transporter gene	Copper toxicity	Li and Leisner (2008)
HMA5	Heavy metal–transporting P-type ATPases	Ns	Transporter gene	Copper toxicity	Li and Leisner (2008)

Note: Up and down represent Si-mediated up- and downregulation of the corresponding genes; Ns represents nonsignificant expression regulation.

Although several reports recorded the positive effects of Si on photosynthesis only, few have studied the molecular mechanism behind the gene expression upon Si addition, especially in rice. Recently, Song et al. (2014) illustrated the effect of Si on the expression of photosynthesis-relevant genes under zinc toxicity. The results revealed variations in the gene expression upon Zn toxicity and Si supplementation. In their study, the expression of photosynthesis-responsible genes, such as *PsbY*, *PsaH*, *PetC*, *PetH*, *Os03g57120*, and *Os09g26810* (chlorophyll *a/b* binding protein), has been dealt in detail.

Si has upregulated the level of *PsbY* (*Os08g02630*), an important polyprotein of photosystem II (PS II), whereas the higher concentration of Zn downregulated the same. *PsbY* is one of the low-molecular-mass subunits of the oxygen-evolving complex of the PS II with manganese-binding polypeptide with L-arginine metabolizing enzyme activity (Kawakami et al., 2007). Moreover, the authors suggested that the induction of *PsbY* transcripts by Si might trigger the manganese-binding capacity and water oxidation, and ultimately increase the activity of the PS II and electron transfer rate.

Similarly, the addition of Si has increased the expression of the *PsaH* gene that encodes 13 polypeptides and is located in the middle of the PS I dimer. The *PsaH* is encoded by nuclear genes, and its position of location contacts the *PsaA* and *PsaD* genes (Pfannschmidt and Yang, 2012). The major role of the *PsaH* gene is to maintain the stability of photosynthetic *co8y* (Naver et al., 1999). The *PsaH* knockout resulted in the impairment of the light harvesting complex (LHC) - II complex, leading to the energy transition delay between PS II and PS I (Lunde et al., 2000). Thus, the ability of Si to enhance the capacity of the PS I system was demonstrated once again.

The *PetC* expression level decreased by Zn toxicity was increased by Si supplementation. The *PetC* encodes for the Rieske Fe-S center-binding polypeptide of cytochrome *bf* complex and is associated with the cytochrome's biological processes in photosynthesis (Breyton et al., 1994). By upregulating the expression of *PetC*, Si protected the structural integrity of the chloroplast (Song et al., 2014). In addition, the transcripts of *PetH* were also observed to be in a trend similar to that of *PetC*. *PetH* encodes for ferredoxin NADP$^+$ reductase, a key enzyme involved in the generation of NADPH in the final step of the photosynthetic electron transport chain, and it also maintains the reducible glutathione content in the cells (Song et al., 2014). Apart from the above-mentioned genes, Si has upregulated the expression of light-harvesting complex genes (*Os03g57120* and *Os09g26810*). Overall, the supplementation of Si has induced the vital genes involved in PS I and PS II to protect the photosynthetic efficiency of plants, especially under a stressful environment.

9.2.2 REGULATION OF HOUSEKEEPING GENES BY SI

The augmentation of Si regulates the expression of "constitutive genes" (housekeeping genes or reference genes) that are globally expressed in all cells, irrespective of their pathophysiological conditions, and are necessary for the maintenance of fundamental cellular functions. Even though the housekeeping genes are considered for constant expression, several reports question the stability of their expressions under biotic and abiotic stresses (Selvey et al., 2001; Volkov et al., 2003; Nicot et al., 2005; Olsvik et al., 2005; Jain et al., 2006).

In rice, the amendment of Si differentially regulated the expression of genes involved in cellular housekeeping processes (Brunings et al., 2009). Under normal conditions, Si supplementation downregulated most of the housekeeping genes, whereas upon pathogen infection, upregulation of the genes was noted by Brunings et al. (2009). Likewise, the Si-mediated enhancement of housekeeping genes was evident in tomato plants infected with *Ralstonia solanacearum*, an aerobic gram negative bacterium (Ghareeb et al., 2011). According to Ghareeb et al. (2011), the addition of Si has induced the levels of housekeeping genes encoding actin (*ACT*), α-tubulin (*TUB*), and phosphoglycerate kinase (*PGK*), in bacteria-infected tomato. Jarosch et al. (2005) proposed that actin cytoskeleton conferred the basal resistance against pathogens and the colonization of *R. solanacearum* can be facilitated by enfeebling the host's actin cytoskeleton (Ghareeb et al., 2011). Therefore, the increase in the expression level of the actin gene in bacterial infected plants by Si has induced the host resistance. Moreover, the results supported the Si-IR-mediated reduction of disease incidence in *R. solanacearum*–infected plants (Ghareeb et al., 2011). Overall, the application of silicon resulted in significantly increased expression stability of the housekeeping genes, illustrating the alleviation of the stress imposed by the pathogen.

9.2.3 REGULATION OF TRANSCRIPTION FACTORS BY SI

Numerous kinds of genes are provoked by environmental onslaughts in plants (Shinozaki and Yamaguchi-Shinozaki, 2000; Rabbani et al., 2003). These genes play multiple roles, including stress tolerance, maintenance of gene expression, and signal transduction in response to stress (Xiong et al., 2002; Shinozaki et al., 2003).

Among the stress-stimulated genes, transcription factors (TFs) function as the key regulators of the downstream genes important for plant tolerance against biotic and abiotic stresses (Gao et al., 2007; Lucas et al., 2011). The TFs are mediated by specific cis-elements present in the promoter region of the target gene called regulons (Nakashima et al., 2009; Qin et al., 2011). A plant system encompasses a variety of regulons that are activated by different kinds of stresses. For instance, during heat and drought stress, the dehydration-responsive element-binding protein (*DREB2*) regulons come into action (Mizoi et al., 2012). Likewise, the binding factor *NAC* regulons (no apical meristem [NAM], *Arabidopsis thaliana* activating factor [ATAF], and cup-shaped cotyledon [CUC]) operate under osmotic stress (Nakashima et al., 2009; Fujita et al., 2011).

Further, the higher expressions of TFs can regulate a wide range of signaling pathways leading to stress tolerance (Chaves and Oliveira, 2004; Umezawa et al., 2006). Recently, Si-mediated upregulation of TFs involved in drought-responsive gene expression in rice was extensively dealt by Khattab et al. (2014). According to Khattab et al. (2014), plants pretreated with Si plants displayed enhanced levels of *DREB2A*, *NAC5*, RING domain-containing *OsRDCP1* gene, *OsCMO* coding rice choline monooxygenase (CMO), and dehydrin *OsRAB16b* (Khattab et al., 2014).

The *OsDREB* TFs in rice mediating the expression of downstream stress-responsive genes in an abscisic acid (ABA)–independent manner are vital for osmotic stress tolerance (Dubouzet et al., 2003; Hussain et al., 2011). The expression of *OsDREB2A* induced by drought in rice resulted in stress resistance (Chen et al., 2008; Wang et al., 2008). The NACs are plant-specific TFs with multiple roles in plant development and stress response (Tran et al., 2010). Around 140 putative *NAC* or *NAC*-like genes are identified in the rice genome; among them, 20 genes are identified as stress-responsive genes, including *OsNAC5* (Fang et al., 2008). The products of these genes are involved in various functions, such as osmolytes, detoxification, redox homeostasis, and fortification of macromolecules (Hu et al., 2008). An increase in the transcripts level of *OsNAC5* in rice plants resulted in the accumulation of proline and soluble sugars, and also prevented the lipid peroxidation and production of an excess amount of hydrogen peroxide (H_2O_2). These metabolic changes protect plants from dehydration and oxidative damage under stressed conditions (Takasaki et al., 2010; Song et al., 2011). Moreover, the upregulation of the *OsNAC5* enhanced the stress tolerance of the rice cultivars by increasing the expression levels of stress-inducible rice genes such as *LEA3* (Takasaki et al., 2010).

Ubiquitin (Ub)-26S proteasome pathway regulates the turnover of several eukaryotic proteins. The multiple Ub chains are linked to the target proteins by ubiquitin ligases E1, E2, and E3 (Kraft et al., 2005; Stone et al., 2005). Previous findings revealed that the RING E3 Ub ligases play an important role in drought stress responses in rice (Bae et al., 2011; Ning et al., 2011; Park et al., 2011). In rice, five homologues of *Oryza sativa RING* domain-containing protein family members (*OsRDCP*) were identified to contain a single *RING* motif in their N-terminal regions (Khattab et al., 2014). Among the homologues, the *OsRDCP1*, which counteracts the dehydration stress in rice plants, was enhanced by Si treatment (Bae et al., 2011; Khattab et al., 2014). In addition, the CMO gene product ferredoxin-dependent enzyme CMO catalyzes the initial step in the biosynthesis of glycine betaine, an important quaternary ammonium compound with an osmolytic role in higher plants, which leads to the abiotic stress tolerance (Burnet et al., 1995).

Further, *OsRAB16b* is one *LEA* gene that is expressed in vegetative and reproductive tissues in response to abiotic stresses (Tunnacliffe and Wise, 2007; Bies-Ethève et al., 2008). These *LEA* genes encode *LEA* proteins that confer the adaptation property to the plants during stressful conditions and render stress tolerance (Lenka et al., 2011).

Thus, the Si-mediated regulation of genes encoding TFs, which in turn are involved in expression of stress tolerance genes, can be considered one of the stress-combating tactics attributed to Si.

9.2.4 REGULATION OF AQUAPORIN, PHYTOHORMONES, AND POLYAMINE BIOSYNTHESIS GENES BY SI

Aquaporin belongs to integral membrane proteins involved in the regulation of water uptake by roots and facilitates the radial transportation of water across cell membranes, especially under abiotic stresses (Boursiac et al., 2005). During abiotic stress, the uptake of water mainly occurs via a "cell-to-cell" pathway mediated by aquaporin (Boursiac et al., 2005; Horie et al., 2011; Sutka et al., 2011). A variety of stimuli, such as ABA, calcium ions, free radicals, and ethylene, affect the activity of aquaporin (Azad et al., 2004; Parent et al., 2009; Hu et al., 2012). Moreover, the short-term salt stress–induced H_2O_2 tends to limit the aquaporin activity by oxidant gating, phosphorylation status, and aquaporin relocalization, which prevents water uptake (Boursiac et al., 2008).

According to Romero-Aranda et al. (2006), the supplementation of Si improves the water status of plants under salt stress conditions. However, the molecular mechanism behind the water status enhancement by Si remained unclear. Recent findings by Liu et al. (2015) have shed light onto the Si-mediated regulation of aquaporin genes for the improvement of water uptake in *Sorghum bicolor* under a salt stress condition. Their results revealed that the enhancement of aquaporins by Si amendment alleviated the root hydraulic conductance. Further, the improvement of aquaporin activity was ascribed by an increase in the expression level of aquaporin genes. Under salt stress, the upregulation of the genes such as *SbPIP1;6, SbPIP2;2*, and *SbPIP2;6* encoding plasma membrane intrinsic protein (PIP), the copious aquaporin in the root plasma membrane upon Si nutrition, has been illustrated by Liu et al. (2015). Furthermore, the overexpression of aquaporin genes leads to rapid uptake of water, and dilutes the excess Na^+ concentration (Gao et al., 2010). In line with the existing benefits of aquaporin upregulation, Sutka et al. (2011) proposed that a significant increase in the expression of aquaporin genes in the roots compensates the soil water uptake reduction under water-deficit conditions.

Under abiotic stress, Si not only controls the regulation of aquaporin genes, but also is involved in the molecular mediation of ABA biosynthesis. As evidence, the illustrations made by Kim et al. (2014a) for rice aid in the understanding of the transcriptional regulation of genes involved in ABA biosynthesis, such as *ZEP, NCED1, NCED3*, and *NCED5*, by Si addition. Previous reports suggested that zeaxanthin epoxidase (ZEP or ABA1) and 9-cis-epoxycarotenoid dioxygenase (NCED1 and NCED3) are the vital enzymes involved in the ABA biosynthesis pathway in *Arabidopsis* and other plant species (Qin and Zeevaart, 1999; Seo et al., 2000; Xiong et al., 2002; Xiong and Zhu 2003). In general, abiotic stresses like salinity increase the regulation of ABA biosynthesis genes (Qin and Zeevaart 1999; Xiong et al., 2002). However, Si treatment has downregulated the transcripts of *ZEP, NCED1, NCED3*, and *NCED5* under salinity (Kim et al., 2014b). Consequently, Si supplementation has alleviated the negative effect of salinity stress in the rice root zone by regulating the genes responsible for ABA synthesis. However, the molecular rationale behind the phytohormonal regulation of Si has to be addressed in the future.

Novel insight into the molecular regulation of polyamines (PAs) by Si under abiotic stress has been comprehensively investigated recently in *Sorghum bicolor* (Yin et al., 2015). According to the report, Si improved the expression of the S-adenosyl-L-methionine decarboxylase (*SAMDC*) gene, encoding an important enzyme in PA and ethylene biosynthesis. PAs such as Put (putrescine), Spd (spermidine), and Spm (spermine) play a vital role in plant salt tolerance (Yamaguchi et al., 2006; Zhao et al., 2007; Kuznetsov and Shevyakova, 2010; Pottosin and Shabala, 2014). Plants with elevated PA levels exhibited resistance to salt stress, and exogenous PAs mitigated the deleterious effects of salt toxicity (Liu et al., 2006; Chai et al., 2010; Quinet et al., 2010). Moreover, several reports state that the overexpression of PA synthesis genes stimulated salt stress tolerance (Roy and Wu, 2001; Tang et al., 2007; Wen et al., 2008). Based on the account of the previous knowledge on PAs, it can be firmly considered that PAs play indispensable roles against salinity. Correspondingly, ethylene, a gaseous hormone with an important role in leaf senescence, is associated with the regulation of PAs (Spd, Spm), and ethylenes share a common precursor, S-adenosyl-L-methionine (SAM) (Pandey et al., 2000). Yin et al. (2015) hypothesized that PAs and ethylene are involved in Si-mediated salt tolerance in sorghum. As mentioned, the amendment of Si has upregulated the genes responsible for PA biosynthesis; on the other hand, Si significantly decreased the level of ethylene (Yin et al., 2015). Further, the interpretations by Yin et al. (2015) demonstrated that the promotion of PA biosynthesis and inhibition of ethylene precursor, 1-aminocyclopropane-1-1-carboxylic acid (ACC), generation may be involved in Si-induced salt tolerance. Since Spd, Spm, and ethylene share a common precursor, SAM, it is most often considered to be the existence of a competitive demand among the PAs and ethylene (Pandey et al., 2000). Consequently, Si influenced the accumulation of ACC by the regulation of PA synthesis and retarded the ethylene synthesis, ultimately protecting the salt-induced leaf senescence. Based on the aforementioned information, the regulation of gene expressions related to aquaporins, phytohormones, and PAs by Si under abiotic stresses can be highly appreciated.

9.2.5 Regulation of Defense-Responsive Genes by Si

Numerous reports evidence the defensive role of Si in plants. In particular, the protection mechanism of Si against powdery mildew and blast diseases was studied in depth. The increase in resistance against *Magnaporthe grisea* by Si amendment was elaborately dealt with in both the biochemical and molecular level by Rodrigues et al. (2005). Si supplementation has enhanced the expression of defense-related genes, such as chalcone synthase (CHS), phenylalanine ammonia lyase (PAL), pathogenesis-related protein (PR1), peroxidase (POX), chitinases, and β-1,3-glucanases (Rodrigues et al., 2005). CHS is a vital enzyme involved in the flavonoid biosynthesis pathway and regulates the accumulation of phytoalexin-like sakuranetin (Rodrigues et al., 2005). Similarly, PAL determines the production of defensive phenolic compounds via the phenylpropanoid pathway, whereas the POXs are involved in the biosynthesis of lignin (a major cell wall component and a well-known mechanical barrier) (Rhodes, 1994).

Likewise, the induction of antimicrobial PR1 proteins along with genes involved in secondary metabolism is the primary consequence of the plant defense response (Zeier et al., 2004). In addition, the Si-induced molecular signaling mechanism conferred resistance against *M. oryzae* in rice by altering the defense gene expressions (Brunings et al., 2009). Apart from the genes involved in secondary metabolism, Si has differentially regulated the heavy metal transport/detoxification protein-related genes (*Os01g0713200*, *Os02g0584800*, *Os02g0585100*, and *Os04g0469000*) in rice, which might shed light on the rationale behind the heavy metal toxicity alleviation of Si (Brunings et al., 2009).

On the whole, Si appears to be involved in the regulation of genes responsible for vital physiological functions, especially in a stressed environment.

9.3 SECONDARY METABOLITES: A BRIEF OUTLINE

The striking discovery of secondary metabolites was made in the eighteenth century by Kossel (1891) with the first remarkable notion that secondary metabolites are different from the large-scale primary metabolites. After a decade, the concept of secondary metabolites was further fine-tuned by Czapek (1921). According to him, the secondary metabolites are defined as the end products derived from nitrogen metabolism that undergo secondary modifications, like deamination. In addition, secondary metabolites, in contrast to primary metabolites, are characterized by lower abundance and narrow distribution, and are stored in specialized cells or organelles (Rhodes, 1994). All of the above-mentioned discoveries have led to the birth of a new discipline, phytochemistry. After the development of chromatographic methods and molecular biology techniques, the vital role of secondary metabolites with respect to plant environmental adaptations was demonstrated extensively. Plant-based secondary metabolites are generally categorized into three large families: phenolics, terpenes and steroid, and alkaloids (Luckner, 1984). However, phenolics are attributed to a major class of secondary metabolites, especially in higher plants, because of their role in the lignin biosynthesis pathway (Rhodes, 1994).

In general, secondary metabolites immensely contribute to the fitness of the plants by interacting with the environment. Therefore, the classic functions of secondary metabolites include antifungal, antibiotic, antiviral, and phytoalexins (Rhodes, 1994). Moreover, the secondary metabolites also aid in the absorption of harmful ultraviolet (UV) radiations and protect from leaf injuries in plants (Li et al., 1993). Thus, the major function of plant secondary metabolites is to protect plants from onslaughts by insects, herbivores, and pathogens, or to survive other biotic and abiotic stresses. They constitute the main elements of chemical defense in plants. Since the roles of secondary metabolites are numerous, in this chapter the regulation of secondary metabolites by Si is discussed.

9.3.1 DOES SI REGULATE SECONDARY METABOLITES?

The Si-mediated prophylactic effects are evident from several previous reports, especially the enhanced accumulation of antifungal compounds similar to phytoalexins in cucumber against powdery mildew (Fawe et al., 1998). According to Fawe et al. (1998), cucumber encloses a pool of conjugated compounds stimulated or modified by Si treatment, giving rise to new antifungal metabolites. Furthermore, the Si-induced compounds were not detected in control plants, suggesting that they are derived essentially from neosynthesized conjugates. The nuclear magnetic resonance (NMR)–based structural analysis revealed the novel antifungal compound as rhamnetin (3,5,3′,4′-tetrahydroxy-7-O-methoxyflavone), which was observed only on the Si-treated cucumber plants. Similarly, in the same study, several other antifungal compounds induced by Si have been identified as a postinfection response. The flavanol rhamnetin was synthesized via the phenylpropanoid pathway, an important biosynthetic pathway through which numerous phenols and flavonoids are synthesized (Currie and Perry, 2007).

Biosynthesis of polyphenols entails a cascade of central enzyme-regulated reactions from which several tributaries emanate toward different secondary metabolites. In general, the phenylpropanoid pathway in plants is associated with the shikimate pathway, which is involved in the synthesis of aromatic amino acids from central carbon metabolisms (Herrmann, 1995). In the phenylpropanoid pathway, PAL is the first committed enzyme; it diverts the central flux of carbon from the primary metabolism and leads to the synthesis of myriad phenolics (Currie and Perry, 2007; Rahman et al., 2015). The activity of PAL is stimulated by a variety of environmental cues, including pathogen attacks, tissue wounding, UV irradiation, exposure to heavy metals, low temperatures, and low levels of nitrogen, phosphate, or ions (Dixon and Paiva, 1995; Weisshaar and Jenkins, 1998). The molecular regulations of PAL enzyme activity, especially in stress responses, elucidate its sophisticated regulatory control. For instance, the upregulation of PAL transcripts by Si has been recently documented in ryegrass (Rahman et al., 2015) and also discussed in Section 9.2.5. Moreover, the Si-induced significant increase in PAL enzyme activity has been well documented in several plant species, particularly in stressful environments (Shetty et al., 2011; Rahman et al., 2015).

The stimulation of antimicrobial secondary metabolites by Si as a prophylactic response has been recorded in miniature potted roses. In roses, the application of Si has tremendously alleviated the negative effects imposed by *Podosphaera pannosa* (Shetty et al., 2011). High-performance liquid chromatography (HPLC) and liquid chromatography–mass spectrometry (LC-MS) analysis of Si-treated rose leaf extracts revealed significant enhancement in the phenolic acids and flavonol glycosides. In particular, phenolic acids such as chlorogenic acid and neochlorogenic acid, along with an unknown phenolic acid in higher concentrations, were identified in Si treatments (Shetty et al., 2011). A similar pattern of increase in the same phenolic acids by Si has been observed recently in ryegrass against *Magnaporthe oryzae* (Rahman et al., 2015). As mentioned earlier, Si primed the rose plants with flavonol glycosides such as quercetin, rutin, and kaempferol, the reputed antimicrobial components against *P. pannosa* (Shetty et al., 2011). Both chlorogenic acid and rutin prevented the conidial germination of *P. pannosa* and also suppressed the appressoria formation, thus reducing the severity of the pathogen (Shetty et al., 2011). This denotes that Si might play a positive regulatory role in the secondary metabolites and strengthen the chemical defense of the plant system against pathogen attack.

Likewise, the amendment of Si elicited the synthesis of momilactones (A and B) in rice plants infected with *M. grisea* (Rodrigues et al., 2004). Momilactones are the subset of secondary metabolites with allelopathic effects. These compounds are synthesized from the terpenoid pathway with the geranylgeranyl phosphate as the precursor (Rhodes, 1994). The terpenoid pathway is characterized as a principal pathway resulting in the synthesis of metabolites such as gibberellins, ABA, and membrane sterols (Rhodes, 1994). Moreover, this pathway is considered a highly branched pathway with several intermediates that are channeled at different levels to produce terpenoids such as monoterpene compounds like essential oils and sesquiterpenoid phytoalexins like rishitin (Rhodes, 1994). Thus, the report by Rodrigues et al. (2004) shed light on the Si-mediated regulation of terpenoid biosynthesis in response to blast disease in rice. The molecular rationale behind the enhancement of secondary metabolites by Si has been hypothesized to be functionally similar to the systemic-acquired resistance (SAR) process.

In contrast, the Si supplementation retarded the phenolic compounds, such as conoferyl alcohol and coumaric acid in cucumber grown under excess Mn (Dragišić et al., 2007). The toxic effect of surplus Mn has induced oxidative stress perturbing the phenolic metabolism and triggered the activity of phenol metabolizing enzymes like PPO. The PPO catalyzes the oxidation of phenols to harmful quinone, which causes the browning of leaves (Dragišić et al., 2007). However, the added Si rectified the problem of leaf browning by regulating the excess phenol synthesis during the Mn stress by regulating the vital rate-limiting enzymes in the phenylpropanoid pathway (Dragišić et al., 2007). However, the mechanism behind the regulation of secondary metabolites is still under debate.

Apart from the chemical defense mediation, it has been well defined that the binding of silicic acid to phenolic and carbohydrate constituents of the cell walls results in phenol–carbohydrate complexes of different types (Inanaga and Okasaka, 1995). Moreover, the polymerization of phenols induces the process of Si-mediated lignification (Fang et al., 2003). Noteworthy is that the phenolic compounds, ferulic acid, p-coumaric acid, and coniferyl alcohol, are the vital precursors in lignin biosynthesis. Thus, it has been demonstrated that the supplemented Si channeled the above-mentioned phenolics for lignin synthesis and prevented their accumulation in leaf tissues (Goto et al., 2003). A schematic representation of Si-mediated secondary metabolites for the improvement of stress tolerance in plants is portrayed in Figure 9.1.

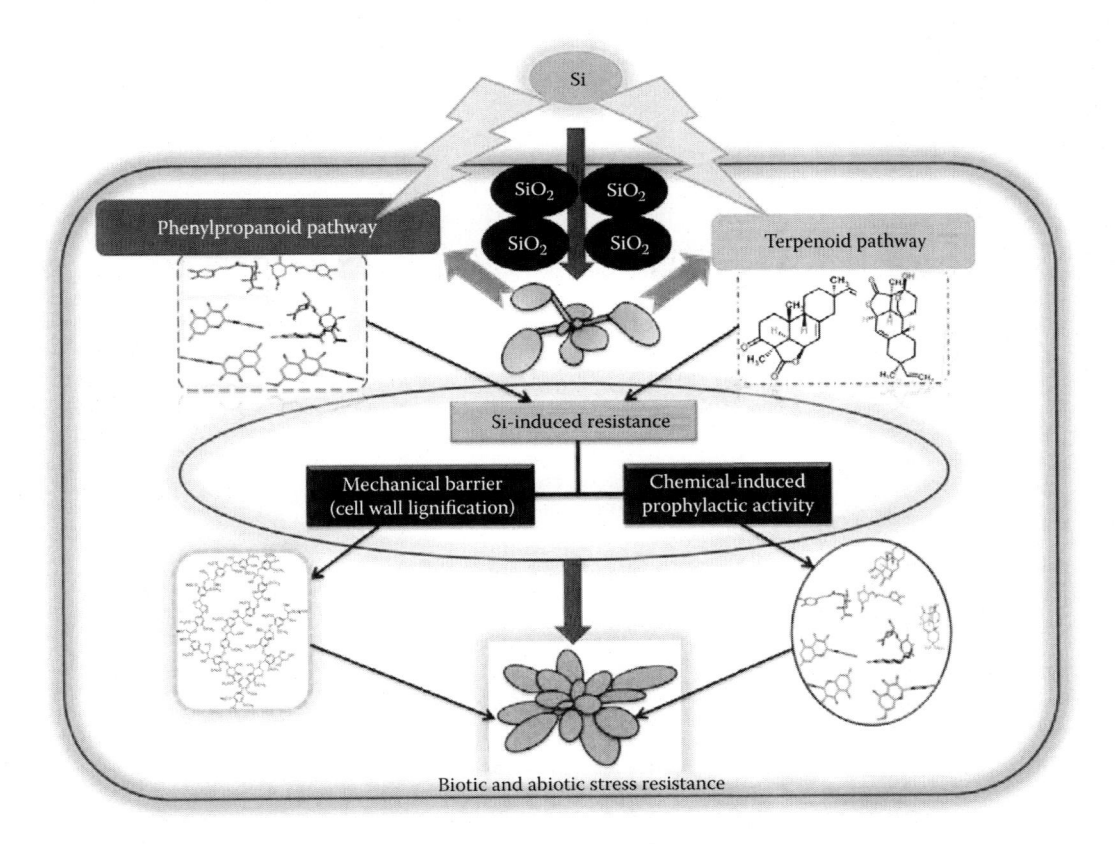

FIGURE 9.1 Schematic representation of the role of Si in mediation of secondary metabolites to render stress tolerance in plants.

In conclusion, Si nutrition to plants enhances the regulation of genes involved in physiological processes to improve the growth and stress tolerance against abiotic and biotic stresses. Si also mediates the synthesis and accumulation of important secondary metabolites, such as phenols and flavonoids, by triggering various genes involved in the phenylpropanoid pathway in plants under a stressed environment.

ACKNOWLEDGMENTS

Abinaya Manivannan and Prabhakaran Soundararajan were supported by a scholarship from the BK21 Plus Program, Ministry of Education, South Korea. The authors are thankful to Dr. Sowbiya Muneer, Institute of Agriculture and Life Science, Gyeongsang National University, for reviewing the chapter.

REFERENCES

Al-aghabary, K., Z. Zhu, and Q. Shi. Influence of silicon supply on chlorophyll content, chlorophyll fluorescence, and antioxidative enzyme activities in tomato plants under salt stress. *Journal of Plant Nutrition* 27, no. 12 (2005): 2101–2115.

Arnon, D. I., and P. R. Stout. The essentiality of certain elements in minute quantity for plants with special reference to copper. *Plant Physiology* 14, no. 2 (1939): 371.

Azad, A. K., Y. Sawa, T. Ishikawa, and H. Shibata. Phosphorylation of plasma membrane aquaporin regulates temperature-dependent opening of tulip petals. *Plant and Cell Physiology* 45, no. 5 (2004): 608–617.

Bae, H., S. K. Kim, S. K. Cho, B. G. Kang, and W. T. Kim. Overexpression of OsRDCP1, a rice RING domain containing E3 ubiquitin ligase, increased tolerance to drought stress in rice (*Oryza sativa* L.). *Plant Science* 180, no. 6 (2011): 775–782.

Bies-Etheve, N., P. Gaubier-Comella, A. Debures, E. Lasserre, E. Jobet, M. Raynal, R. Cooke, and M. Delseny. Inventory, evolution and expression profiling diversity of the LEA (late embryogenesis abundant) protein gene family in *Arabidopsis thaliana*. *Plant Molecular Biology* 67, no. 1–2 (2008): 107–124.

Bond, R. and J. C. McAuliffe. Silicon biotechnology: New opportunities for carbohydrate science. *Australian Journal of Chemistry* 56, no. 1 (2003): 7–11.

Boursiac, Y., J. Boudet, O. Postaire, D. T. Luu, C. Tournaire-Roux, and C. Maurel. Stimulus-induced down regulation of root water transport involves reactive oxygen species-activated cell signaling and plasma membrane intrinsic protein internalization. *The Plant Journal* 56, no. 2 (2008): 207–218.

Boursiac, Y., S. Chen, D.-T. Luu, M. Sorieul, N. van den Dries, and C. Maurel. Early effects of salinity on water transport in *Arabidopsis* roots. Molecular and cellular features of aquaporin expression. *Plant Physiology* 139, no. 2 (2005): 790–805.

Breyton, C., C. de Vitry, and J.-L. Popot. Membrane association of cytochrome b6f subunits. The Rieske iron-sulfur protein from *Chlamydomonas reinhardtii* is an extrinsic protein. *Journal of Biological Chemistry* 269, no. 10 (1994): 7597–7602.

Brunings, A. M., L. E. Datnoff, J. F. Ma, N. Mitani, Y. Nagamura, B. Rathinasabapathi, and M. Kirst. Differential gene expression of rice in response to silicon and rice blast fungus *Magnaporthe oryzae*. *Annals of Applied Biology* 155, no. 2 (2009): 161–170.

Burnet, M., P. J. Lafontaine, and A. D. Hanson. Assay, purification, and partial characterization of choline monooxygenase from spinach. *Plant Physiology* 108, no. 2 (1995): 581–588.

Carlisle, E. M. Silicon as a trace nutrient. *Science of the Total Environment* 73, no. 1 (1988): 95–106.

Carlisle, E. M. Silicon. In *Handbook of Nutritionally Essential Mineral Elements*, ed. B. L. O'Dell and R. A. Sunde, pp. 603–608. Marcel Dekker, New York, 1997.

Chai, Y. Y., C. D. Jiang, L. Shi, T. S. Shi, and W. B. Gu. Effects of exogenous spermine on sweet sorghum during germination under salinity. *Biologia Plantarum* 54, no. 1 (2010): 145–148.

Chaves, M. M., and M. M. Oliveira. Mechanisms underlying plant resilience to water deficits: Prospects for water-saving agriculture. *Journal of Experimental Botany* 55, no. 407 (2004): 2365–2384.

Chen, J.-Q., X.-P. Meng, Y. Zhang, M. Xia, and X.-P. Wang. Over-expression of OsDREB genes lead to enhanced drought tolerance in rice. *Biotechnology Letters* 30, no. 12 (2008): 2191–2198.

Chen, W., X. Yao, K. Cai, and J. Chen. Silicon alleviates drought stress of rice plants by improving plant water status, photosynthesis and mineral nutrient absorption. *Biological Trace Element Research* 142, no. 1 (2011): 67–76.

Cherif, M., N. Benhamou, J. Ge Menzies, and R. R. Belanger. Silicon induced resistance in cucumber plants against *Pythium ultimum*. *Physiological and Molecular Plant Pathology* 41, no. 6 (1992): 411–425.

Cookson, L. J., D. K. Scown, K. J. McCarthy, and N. Chew. The effectiveness of silica treatments against wood boring invertebrates. *Holzforschung* 61, no. 3 (2007): 326–332.

Coors, J. G. Resistance to the European corn borer, *Ostrinia nubilalis* (Hubner), in maize, *Zea mays* L., as affected by soil silica, plant silica, structural carbohydrates, and lignin. In *Genetic Aspects of Plant Mineral Nutrition*, pp. 445–456. Springer, Berlin, 1987.

Currie, H. A., and C. C. Perry. Silica in plants: Biological, biochemical and chemical studies. *Annals of Botany* 100, no. 7 (2007): 1383–1389.

Czapek, F. *Spezielle Biochemie, Biochemie der Pflanzen*, p. 369. Vol. 3. G. Fischer Jena, Thuringia, Germany, 1921.

Das, S., and U. K. Chattopadhyay. Role of silicon in modulating the internal morphology and growth of *Mycobacterium tuberculosis*. *Indian Journal of Tuberculosis* 47 (2000): 87–91.

Dixon, R. A., and N. L. Paiva. Stress-induced phenylpropanoid metabolism. *Plant Cell* 7, no. 7 (1995): 1085.

Dragišić M., J. Bogdanović, V. Maksimović, and M. Nikolic. Silicon modulates the metabolism and utilization of phenolic compounds in cucumber (*Cucumis sativus* L.) grown at excess manganese. *Journal of Plant Nutrition and Soil Science* 170, no. 6 (2007): 739–744.

Dubouzet, J. G., Y. Sakuma, Y. Ito, M. Kasuga, E. G. Dubouzet, S. Miura, M. Seki, K. Shinozaki, and K. Yamaguchi-Shinozaki. OsDREB genes in rice, *Oryza sativa* L., encode transcription activators that function in drought-, high-salt- and cold-responsive gene expression. *Plant Journal* 33, no. 4 (2003): 751–763.

Egrinya Eneji, A., S. Inanaga, S. Muranaka, J. Li, T. Hattori, P. An, and W. Tsuji. Growth and nutrient use in four grasses under drought stress as mediated by silicon fertilizers. *Journal of Plant Nutrition* 31, no. 2 (2008): 355–365.

Epstein, E. The anomaly of silicon in plant biology. *Proceedings of the National Academy of Sciences* 91, no. 1 (1994): 11–17.

Epstein, E. Silicon. *Annual Review of Plant Biology* 50, no. 1 (1999): 641–664.

Fang, J., H. Wang, Y. Chen, and F. Zhang. Silica nanospheres formation induced by peroxidase-catalyzed phenol polymerization. *Progress in Natural Science* 13, no. 7 (2003): 501–504.

Fang, Y., J. You, K. Xie, W. Xie, and L. Xiong. Systematic sequence analysis and identification of tissue specific or stress-responsive genes of NAC transcription factor family in rice. *Molecular Genetics and Genomics* 280, no. 6 (2008): 547–563.

Fawe, A.-Z., J. G. Menzies, and R. R. Belanger. Silicon-mediated accumulation of flavonoid phytoalexins in cucumber. *Phytopathology* 88, no. 5 (1998): 396–401.

Fujita, Y., M. Fujita, K. Shinozaki, and K. Yamaguchi-Shinozaki. ABA-mediated transcriptional regulation in response to osmotic stress in plants. *Journal of Plant Research* 124, no. 4 (2011): 509–525.

Gao, J.-P., D.-Y. Chao, and H.-X. Lin. Understanding abiotic stress tolerance mechanisms: Recent studies on stress response in rice. *Journal of Integrative Plant Biology* 49, no. 6 (2007): 742–750.

Gao, Z., X. He, B. Zhao, C. Zhou, Y. Liang, R. Ge, Y. Shen, and Z. Huang. Overexpressing a putative aquaporin gene from wheat, TaNIP, enhances salt tolerance in transgenic *Arabidopsis*. *Plant and Cell Physiology* 51, no. 5 (2010): 767–775.

Ghareeb, H., Z. Bozso, P. G. Ott, and K. Wydra. Silicon and *Ralstonia solanacearum* modulate expression stability of housekeeping genes in tomato. *Physiological and Molecular Plant Pathology* 75, no. 4 (2011): 176–179.

Gong, H., X. Zhu, K. Chen, S. Wang, and C. Zhang. Silicon alleviates oxidative damage of wheat plants in pots under drought. *Plant Science* 169, no. 2 (2005): 313–321.

Gong, H. J., D. P. Randall, and T. J. Flowers. Silicon deposition in the root reduces sodium uptake in rice (*Oryza sativa* L.) seedlings by reducing bypass flow. *Plant, Cell and Environment* 29, no. 10 (2006): 1970–1979.

Goto, M., H. Ehara, S. Karita, K. Takabe, N. Ogawa, Y. Yamada, S. Ogawa, M. Sani Yahaya, and O. Morita. Protective effect of silicon on phenolic biosynthesis and ultraviolet spectral stress in rice crop. *Plant Science* 164, no. 3 (2003): 349–356.

Gunes, A., A. Inal, E. G. Bagci, S. Coban, and O. Sahin. Silicon increases boron tolerance and reduces oxidative damage of wheat grown in soil with excess boron. *Biologia Plantarum* 51, no. 3 (2007): 571–574.

Herrmann, K. M. The shikimate pathway: Early steps in the biosynthesis of aromatic compounds. *Plant Cell* 7, no. 7 (1995): 907.

Horie, T., T. Kaneko, G. Sugimoto, S. Sasano, S. K. Panda, M. Shibasaka, and M. Katsuhara. Mechanisms of water transport mediated by PIP aquaporins and their regulation via phosphorylation events under salinity stress in barley roots. *Plant and Cell Physiology* 52, no. 4 (2011): 663–675.

Hu, H., J. You, Y. Fang, X. Zhu, Z. Qi, and L. Xiong. Characterization of transcription factor gene SNAC2 conferring cold and salt tolerance in rice. *Plant Molecular Biology* 67, no. 1–2 (2008): 169–181.

Hu, W., Q. Yuan, Y. Wang, R. Cai, X. Deng, J. Wang, and S. Zhou. Overexpression of a wheat aquaporin gene, TaAQP8, enhances salt stress tolerance in transgenic tobacco. *Plant and Cell Physiology* 53, no. 12 (2012): 2127–2141.

Hussain, S. S., M. A. Kayani, and M. Amjad. Transcription factors as tools to engineer enhanced drought stress tolerance in plants. *Biotechnology Progress* 27, no. 2 (2011): 297–306.

Inanaga, S., and A. Okasaka. Calcium and silicon binding compounds in cell walls of rice shoots. *Soil Science and Plant Nutrition* 41, no. 1 (1995): 103–110.

Ingri, N. Aqueous silicic acid, silicates and silicate complexes. In *Biochemistry of silicon and related problems*, ed. G. Bendz and I. Lindqvist. pp. 3–51. Springer US, 1978.

Jain, M., A. Nijhawan, A. K. Tyagi, and J. P. Khurana. Validation of housekeeping genes as internal control for studying gene expression in rice by quantitative real-time PCR. *Biochemical and Biophysical Research Communications* 345, no. 2 (2006): 646–651.

Jarosch, B., N. C. Collins, N. Zellerhoff, and U. Schaffrath. RAR1, ROR1, and the actin cytoskeleton contribute to basal resistance to *Magnaporthe grisea* in barley. *Molecular Plant-Microbe Interactions* 18, no. 5 (2005): 397–404.

Kawakami, K., M. Iwai, M. Ikeuchi, N. Kamiya, and J.-R. Shen. Location of PsbY in oxygen-evolving photosystem II revealed by mutagenesis and x-ray crystallography. *FEBS Letters* 581, no. 25 (2007): 4983–4987.

Khattab, H. I., M. A. Emam, M. M. Emam, N. M. Helal, and M. R. Mohamed. Effect of selenium and silicon on transcription factors NAC5 and DREB2A involved in drought-responsive gene expression in rice. *Biologia Plantarum* 58, no. 2 (2014): 265–273.

Kim, Y.-H., A. L. Khan, M. Waqas, H.-J. Jeong, D.-H. Kim, J. S. Shin, J.-G. Kim, M.-H. Yeon, and I.-J. Lee. Regulation of jasmonic acid biosynthesis by silicon application during physical injury to *Oryza sativa* L. *Journal of Plant Research* 127, no. 4 (2014a): 525–532.

Kim, Y. H., A. L. Khan, M. Waqas, J. K. Shim, D. H. Kim, K. Y. Lee, and I. J. Lee. Silicon application to rice root zone influenced the phytohormonal and antioxidant responses under salinity stress. *Journal of Plant Growth Regulation* 33, no. 2 (2014b): 137–149.

Kinrade, S. D., J. W. Del Nin, A. S. Schach, T. A. Sloan, K. L. Wilson, and C. T. G. Knight. Stable five- and six-coordinated silicate anions in aqueous solution. *Science* 285, no. 5433 (1999): 1542–1545.

Kinrade, S. D., A.-M. E. Gillson, and C. T. G. Knight. Silicon-29 NMR evidence of a transient hexavalent silicon complex in the diatom *Navicula pelliculosa*. *Journal of the Chemical Society, Dalton Transactions* 3 (2002): 307–309.

Kinrade, S. D., R. J. Hamilton, A. S. Schach, and C. T. G. Knight. Aqueous hypervalent silicon complexes with aliphatic sugar acids. *Journal of the Chemical Society, Dalton Transactions* 7 (2001): 961–963.

Knight, C. T. G., and S. D. Kinrade. A primer on the aqueous chemistry of silicon. *Studies in Plant Science* 8 (2001): 57–84.

Kolesnikov, M. P., and V. K. Gins. Forms of silicon in medicinal plants. *Applied Biochemistry and Microbiology* 37, no. 5 (2001): 524–527.

Kossel, H. Ueber die chemische Zusammensetzung der Zelle. *Archives of Analytical Physiology, Physiol Abteilung* 181–186 (1891).

Kraft, E., S. L. Stone, L. Ma, N. Su, Y. Gao, O.-S. Lau, X.-W. Deng, and J. Callis. Genome analysis and functional characterization of the E2 and RING-type E3 ligase ubiquitination enzymes of *Arabidopsis*. *Plant Physiology* 139, no. 4 (2005): 1597–1611.

Kuznetsov, V. V., and N. I. Shevyakova. Polyamines and plant adaptation to saline environments. In *Desert Plants*, pp. 261–298. Springer, Berlin, 2010.

Lanning, F. C. Relation of silicon in barley to disease, cold, and pest resistance. *Journal of Agricultural and Food Chemistry* 14, no. 6 (1966): 636–638.

Lenka, S. K., A. Katiyar, V. Chinnusamy, and K. C. Bansal. Comparative analysis of drought-responsive transcriptome in Indica rice genotypes with contrasting drought tolerance. *Plant Biotechnology Journal* 9, no. 3 (2011): 315–327.

Li, H., Y. Zhu, Y. Hu, W. Han, and H. Gong. Beneficial effects of silicon in alleviating salinity stress of tomato seedlings grown under sand culture. *Acta Physiologiae Plantarum* 37, no. 4 (2015): 1–9.

Li, J., S. M. Leisner, and J. Frantz. Alleviation of copper toxicity in *Arabidopsis thaliana* by silicon addition to hydroponic solutions. *Journal of the American Society for Horticultural Science* 133, no. 5 (2008): 670–677.

Li, J., T.-M. Ou-Lee, R. Raba, R. G. Amundson, and R. L. Last. *Arabidopsis* flavonoid mutants are hypersensitive to UV-B irradiation. *Plant Cell Online* 5, no. 2 (1993): 171–179.

Li, W.-B., X.-H. Shi, H. Wang, and F.-S. Zhang. Effects of silicon on rice leaves resistance to ultraviolet-B. *Acta Botanica Sinica—English Edition* 46, no. 6 (2004): 691–697.

Liang, Y., S. Qirong, S. Zhenguo, and T. Ma. Effects of silicon on salinity tolerance of two barley cultivars. *Journal of Plant Nutrition* 19, no. 1 (1996): 173–183.

Liang, Y., J. W. C. Wong, and L. Wei. Silicon-mediated enhancement of cadmium tolerance in maize (*Zea mays* L.) grown in cadmium contaminated soil. *Chemosphere* 58, no. 4 (2005): 475–483.

Liang, Y., W. Sun, Y.-G. Zhu, and P. Christie. Mechanisms of silicon-mediated alleviation of abiotic stresses in higher plants: A review. *Environmental Pollution* 147, no. 2 (2007): 422–428.

Liu, J.-H., K. Nada, C. Honda, H. Kitashiba, X.-P. Wen, X.-M. Pang, and T. Moriguchi. Polyamine biosynthesis of apple callus under salt stress: Importance of the arginine decarboxylase pathway in stress response. *Journal of Experimental Botany* 57, no. 11 (2006): 2589–2599.

Liu, P., L. Yin, S. Wang, M. Zhang, X. Deng, S. Zhang, and K. Tanaka. Enhanced root hydraulic conductance by aquaporin regulation accounts for silicon alleviated salt-induced osmotic stress in *Sorghum bicolor* L. *Environmental and Experimental Botany* 111 (2015): 42–51.

Lucas, S., E. Durmaz, B. A. Akpınar, and H. Budak. The drought response displayed by a DRE-binding protein from *Triticum dicoccoides*. *Plant Physiology and Biochemistry* 49, no. 3 (2011): 346–351.

Luckner, M. *Secondary Metabolism in Microorganisms, Plants and Animals*. Springer Verlag, Berlin, 1984.

Lunde, C., P. E. Jensen, A. Haldrup, J. Knoetzel, and H. V. Scheller. The PSI-H subunit of photosystem I is essential for state transitions in plant photosynthesis. *Nature* 408, no. 6812 (2000): 613–615.

Ma, J. F. Role of silicon in enhancing the resistance of plants to biotic and abiotic stresses. *Soil Science and Plant Nutrition* 50, no. 1 (2004): 11–18.

Ma, J. F., K. Tamai, N. Yamaji, N. Mitani, S. Konishi, M. Katsuhara, M. Ishiguro, Y. Murata, and M. Yano. A silicon transporter in rice. *Nature* 440, no. 7084 (2006): 688–691.

Martin-Jezequel, V., M. Hildebrand, and M. A. Brzezinski. Silicon metabolism in diatoms: Implications for growth. *Journal of Phycology* 36, no. 5 (2000): 821–840.

Mehrabanjoubani, P., A. Abdolzadeh, H. R. Sadeghipour, and M. Aghdasi. Silicon affects transcellular and apoplastic uptake of some nutrients in plants. *Pedosphere* 25, no. 2 (2015): 192–201.

Mitra, G. N. *Regulation of Nutrient Uptake by Plants: A Biochemical and Molecular Approach*. Springer, Berlin, 2015.

Mizoi, J., K. Shinozaki, and K. Yamaguchi-Shinozaki. AP2/ERF family transcription factors in plant abiotic stress responses. *Biochimica et Biophysica Acta (BBA)—Gene Regulatory Mechanisms* 1819, no. 2 (2012): 86–96.

Muneer, S., Y. G. Park, A. Manivannan, P. Soundararajan, and B. R. Jeong. Physiological and proteomic analysis in chloroplasts of *Solanum lycopersicum* L. under silicon efficiency and salinity stress. *International Journal of Molecular Sciences* 15, no. 12 (2014): 21803–21824.

Nakashima, K., Y. Ito, and K. Yamaguchi-Shinozaki. Transcriptional regulatory networks in response to abiotic stresses in *Arabidopsis* and grasses. *Plant Physiology* 149, no. 1 (2009): 88–95.

Naver, H., A. Haldrup, and H. V. Scheller. Cosuppression of photosystem I subunit PSI-H in *Arabidopsis thaliana* efficient electron transfer and stability of photosystem I is dependent upon the PSI-H subunit. *Journal of Biological Chemistry* 274, no. 16 (1999): 10784–10789.

Neumann, D. and U. Zur Nieden. Silicon and heavy metal tolerance of higher plants. *Phytochemistry* 56, no. 7 (2001): 685–692.

Nicot, N., J.-F. Hausman, L. Hoffmann, and D. Evers. Housekeeping gene selection for real-time RT-PCR normalization in potato during biotic and abiotic stress. *Journal of Experimental Botany* 56, no. 421 (2005): 2907–2914.

Ning, Y., C. Jantasuriyarat, Q. Zhao, H. Zhang, S. Chen, J. Liu, L. Liu et al. The SINA E3 ligase OsDIS1 negatively regulates drought response in rice. *Plant Physiology* 157, no. 1 (2011): 242–255.

Nwugo, C. C., and A. J. Huerta. Effects of silicon nutrition on cadmium uptake, growth and photosynthesis of rice plants exposed to low-level cadmium. *Plant and Soil* 311, no. 1–2 (2008): 73–86.

Olsvik, P. A., K. K. Lie, A.-E. O. Jordal, T. O. Nilsen, and I. Hordvik. Evaluation of potential reference genes in real-time RT-PCR studies of Atlantic salmon. *BMC Molecular Biology* 6, no. 1 (2005): 21.

Pandey, S., S. A. Ranade, P. K. Nagar, and N. Kumar. Role of polyamines and ethylene as modulators of plant senescence. *Journal of Biosciences* 25, no. 3 (2000): 291–299.

Parent, B., C. Hachez, E. Redondo, T. Simonneau, F. Chaumont, and F. Tardieu. Drought and abscisic acid effects on aquaporin content translate into changes in hydraulic conductivity and leaf growth rate: A trans-scale approach. *Plant Physiology* 149, no. 4 (2009): 2000–2012.

Park, J.-J., J. Yi, J. Yoon, L.-H. Cho, J. Ping, H. J. Jeong, S. K. Cho, W. T. Kim, and G. An. OsPUB15, an E3 ubiquitin ligase, functions to reduce cellular oxidative stress during seedling establishment. *Plant Journal* 65, no. 2 (2011): 194–205.

Pfannschmidt, T., and C. Yang. The hidden function of photosynthesis: A sensing system for environmental conditions that regulates plant acclimation responses. *Protoplasma* 249, no. 2 (2012): 125–136.

Pottosin, I., and S. Shabala. Polyamines control of cation transport across plant membranes: Implications for ion homeostasis and abiotic stress signaling. *Frontiers in Plant Science* 5 (2014).

Qin, F., K. Shinozaki, and K. Yamaguchi-Shinozaki. Achievements and challenges in understanding plant abiotic stress responses and tolerance. *Plant and Cell Physiology* 52, no. 9 (2011): 1569–1582.

Qin, X., and J. A. D. Zeevaart. The 9-cis-epoxycarotenoid cleavage reaction is the key regulatory step of abscisic acid biosynthesis in water-stressed bean. *Proceedings of the National Academy of Sciences* 96, no. 26 (1999): 15354–15361.

Quinet, M., A. Ndayiragije, I. Lefevre, B. Lambillotte, C. C. Dupont-Gillain, and S. Lutts. Putrescine differently influences the effect of salt stress on polyamine metabolism and ethylene synthesis in rice cultivars differing in salt resistance. *Journal of Experimental Botany* 61, no. 10 (2010): 2719–2733.

Rabbani, M. A., K. Maruyama, H. Abe, M. A. Khan, K. Katsura, Y. Ito, K. Yoshiwara, M. Seki, K. Shinozaki, and K. Yamaguchi-Shinozaki. Monitoring expression profiles of rice genes under cold, drought, and high-salinity stresses and abscisic acid application using cDNA microarray and RNA gel-blot analyses. *Plant Physiology* 133, no. 4 (2003): 1755–1767.

Rahman, A., C. Wallis, and W. Uddin. Silicon induced systemic defense responses in perennial ryegrass against infection by *Magnaporthe oryzae*. *Phytopathology* 105, no. 6 (2015): 748–757.

Redei, G. P. *Encyclopedia of Genetics, Genomics, Proteomics, and Informatics*. Vol. 2. Springer Science & Business Media, Berlin, 2008.

Reeves, C. D., and B. E. Volcani. Role of silicon in diatom metabolism. *Archives of Microbiology* 137, no. 4 (1984): 291–294.

Rhodes, M. J. C. Physiological roles for secondary metabolites in plants: Some progress, many outstanding problems. *Plant Molecular Biology* 24, no. 1 (1994): 1–20.

Rodrigues, F. A., D. J. McNally, L. E. Datnoff, J. B. Jones, C. Labbe, N. Benhamou, J. G. Menzies, and R. R. Belanger. Silicon enhances the accumulation of diterpenoid phytoalexins in rice: A potential mechanism for blast resistance. *Phytopathology* 94, no. 2 (2004): 177–183.

Rodrigues, F. Á., W. M. Jurick, L. E. Datnoff, J. B. Jones, and J. A. Rollins. Silicon influences cytological and molecular events in compatible and incompatible rice-*Magnaporthe grisea* interactions. *Physiological and Molecular Plant Pathology* 66, no. 4 (2005): 144–159.

Romero-Aranda, M. R., O. Jurado, and J. Cuartero. Silicon alleviates the deleterious salt effect on tomato plant growth by improving plant water status. *Journal of Plant Physiology* 163, no. 8 (2006): 847–855.

Roy, M., and R. Wu. Arginine decarboxylase transgene expression and analysis of environmental stress tolerance in transgenic rice. *Plant Science* 160, no. 5 (2001): 869–875.

Selvey, S., E. W. Thompson, K. Matthaei, R. A. Lea, M. G. Irving, and L. R. Griffiths. β-Actin—An unsuitable internal control for RT-PCR. *Molecular and Cellular Probes* 15, no. 5 (2001): 307–311.

Seo, M., H. Koiwai, S. Akaba, T. Komano, T. Oritani, Y. Kamiya, and T. Koshiba. Abscisic aldehyde oxidase in leaves of *Arabidopsis thaliana*. *Plant Journal* 23, no. 4 (2000): 481–488.

Shetty, R., X. Frette, B. Jensen, N. Prasad Shetty, J. D. Jensen, H. J. L. Jørgensen, M.-A. Newman, and L. P. Christensen. Silicon-induced changes in antifungal phenolic acids, flavonoids, and key phenylpropanoid pathway genes during the interaction between miniature roses and the biotrophic pathogen *Podosphaera pannosa*. *Plant Physiology* 157, no. 4 (2011): 2194–2205.

Shi, Y., Y. Wang, T. J. Flowers, and H. Gong. Silicon decreases chloride transport in rice (*Oryza sativa* L.) in saline conditions. *Journal of Plant Physiology* 170, no. 9 (2013): 847–853.

Shinozaki, K., and K. Yamaguchi-Shinozaki. Molecular responses to dehydration and low temperature: Differences and cross-talk between two stress signaling pathways. *Current Opinion in Plant Biology* 3, no. 3 (2000): 217–223.

Shinozaki, K., K. Yamaguchi-Shinozaki, and M. Seki. Regulatory network of gene expression in the drought and cold stress responses. *Current Opinion in Plant Biology* 6, no. 5 (2003): 410–417.

Song, A., P. Li, F. Fan, Z. Li, and Y. Liang. The effect of silicon on photosynthesis and expression of its relevant genes in rice (*Oryza sativa* L.) under high-zinc stress. *PloS One* 9, no. 11 (2014): e113782.

Song, S.-Y., Y. Chen, J. Chen, X.-Y. Dai, and W.-H. Zhang. Physiological mechanisms underlying OsNAC5-dependent tolerance of rice plants to abiotic stress. *Planta* 234, no. 2 (2011): 331–345.

Soundararajan, P., A. Manivannan, Y. Gyeong Park, S. Muneer, and B. R. Jeong. Silicon alleviates salt stress by modulating antioxidant enzyme activities in *Dianthus caryophyllus* 'Tula'. *Horticulture, Environment, and Biotechnology* 56, no. 2 (2015): 233–239.

Soundararajan, P., I. Sivanesan, S. Jana, and B. R. Jeong. Influence of silicon supplementation on the growth and tolerance to high temperature in *Salvia splendens*. *Horticulture, Environment, and Biotechnology* 55, no. 4 (2014): 271–279.

Souza, J. F., H. Dolder, and A. L. Cortelazzo. Effect of excess cadmium and zinc ions on roots and shoots of maize seedlings. *Journal of Plant Nutrition* 28, no. 11 (2005): 1923–1931.

Stone, S. L., H. Hauksdottir, A. Troy, J. Herschleb, E. Kraft, and J. Callis. Functional analysis of the RING-type ubiquitin ligase family of *Arabidopsis*. *Plant Physiology* 137, no. 1 (2005): 13–30.

Sutka, M., G. Li, J. Boudet, Y. Boursiac, P. Doumas, and C. Maurel. Natural variation of root hydraulics in *Arabidopsis* grown in normal and salt-stressed conditions. *Plant Physiology* 155, no. 3 (2011): 1264–1276.

Takasaki, H., K. Maruyama, S. Kidokoro, Y. Ito, Y. Fujita, K. Shinozaki, K. Yamaguchi-Shinozaki, and K. Nakashima. The abiotic stress-responsive NAC-type transcription factor OsNAC5 regulates stress inducible genes and stress tolerance in rice. *Molecular Genetics and Genomics* 284, no. 3 (2010): 173–183.

Tang, W., R. J. Newton, C. Li, and T. M. Charles. Enhanced stress tolerance in transgenic pine expressing the pepper CaPF1 gene is associated with the polyamine biosynthesis. *Plant Cell Reports* 26, no. 1 (2007): 115–124.

Tran, L. S. P., R. Nishiyama, K. Yamaguchi-Shinozaki, and K. Shinozaki. Potential utilization of NAC transcription factors to enhance abiotic stress tolerance in plants by biotechnological approach. *GM Crops* 1, no. 1 (2010): 32–39.

Tripathi, D. K., V. P. Singh, S. Gangwar, S. M. Prasad, J. N. Maurya, and D. K. Chauhan. Role of silicon in enrichment of plant nutrients and protection from biotic and abiotic stresses. In *Improvement of Crops in the Era of Climatic Changes*, pp. 39–56. Springer, New York, 2014.

Tripathi, D. K., V. P. Singh, D. Kumar, and D. K. Chauhan. Impact of exogenous silicon addition on chromium uptake, growth, mineral elements, oxidative stress, antioxidant capacity, and leaf and root structures in rice seedlings exposed to hexavalent chromium. *Acta Physiologiae Plantarum* 34, no. 1 (2012): 279–289.

Tunnacliffe, A., and M. J. Wise. The continuing conundrum of the LEA proteins. *Naturwissenschaften* 94, no. 10 (2007): 791–812.

Umezawa, T., M. Fujita, Y. Fujita, K. Yamaguchi-Shinozaki, and K. Shinozaki. Engineering drought tolerance in plants: Discovering and tailoring genes to unlock the future. *Current Opinion in Biotechnology* 17, no. 2 (2006): 113–122.

Volkov, R. A., I. I. Panchuk, and F. Schoffl. Heat-stress-dependency and developmental modulation of gene expression: The potential of house-keeping genes as internal standards in mRNA expression profiling using real-time RT-PCR. *Journal of Experimental Botany* 54, no. 391 (2003): 2343–2349.

Wainwright, M., K. Al-Wajeeh, and S. J. Grayston. Effect of silicic acid and other silicon compounds on fungal growth in oligotrophic and nutrient-rich media. *Mycological Research* 101, no. 08 (1997): 933–938.

Wang, Y., A. Stass, and W. J. Horst. Apoplastic binding of aluminum is involved in silicon-induced amelioration of aluminum toxicity in maize. *Plant physiology* 136, no. 3 (2004): 3762–3770.

Wang, Q., Y. Guan, Y. Wu, H. Chen, F. Chen, and C. Chu. Overexpression of a rice OsDREB1F gene increases salt, drought, and low temperature tolerance in both *Arabidopsis* and rice. *Plant Molecular Biology* 67, no. 6 (2008): 589–602.

Watanabe, S., E. Shimoi, N. Ohkama, H. Hayashi, T. Yoneyama, J. Yazaki, F. Fujii et al. Identification of several rice genes regulated by Si nutrition. *Soil Science and Plant Nutrition* 50, no. 8 (2004): 1273–1276.

Weisshaar, B., and G. I. Jenkins. Phenylpropanoid biosynthesis and its regulation. *Current Opinion in Plant Biology* 1, no. 3 (1998): 251–257.

Wen, X.-P., X.-M. Pang, N. Matsuda, M. Kita, H. Inoue, Y.-J. Hao, C. Honda, and T. Moriguchi. Over-expression of the apple spermidine synthase gene in pear confers multiple abiotic stress tolerance by altering polyamine titers. *Transgenic Research* 17, no. 2 (2008): 251–263.

Xiong, L., K. S. Schumaker, and J.-K. Zhu. Cell signaling during cold, drought, and salt stress. *Plant Cell Online* 14, suppl. 1 (2002): S165–S183.

Xiong, L., and J.-K. Zhu. Regulation of abscisic acid biosynthesis. *Plant Physiology* 133, no. 1 (2003): 29–36.

Yamaguchi, K., Y. Takahashi, T. Berberich, A. Imai, A. Miyazaki, T. Takahashi, A. Michael, and T. Kusano. The polyamine spermine protects against high salt stress in *Arabidopsis thaliana*. *FEBS Letters* 580, no. 30 (2006): 6783–6788.

Yin, L., S. Wang, K. Tanaka, S. Fujihara, A. Itai, X. Den, and S. Zhang. Silicon-mediated changes in poly-amines participate in silicon-induced salt tolerance in *Sorghum bicolor* L. *Plant, Cell and Environment* 39, no. 2 (2015): 245–258.

Yoshida, S., Y. Ohnishi, and K. Kitagishi. Histochemistry of silicon in rice plant III. The presence of cuticle-silica double layer in the epidermal tissue. *Soil Science and Plant Nutrition* 8 (1962): 1–5.

Zeier, J., M. Delledonne, T. Mishina, E. Severi, M. Sonoda, and C. Lamb. Genetic elucidation of nitric oxide signaling in incompatible plant-pathogen interactions. *Plant Physiology* 136, no. 1 (2004): 2875–2886.

Zhang, Q., J. Liu, H. Lu, S. Zhao, W. Wang, J. Du, and C. Yan. Effects of silicon on growth, root anatomy, radial oxygen loss (ROL) and Fe/Mn plaque of *Aegiceras corniculatum* (L.) Blanco seedlings exposed to cadmium. *Environmental Nanotechnology, Monitoring and Management* 4 (2015): 6–11.

Zhao, F., C.-P. Song, J. He, and H. Zhu. Polyamines improve K+/Na+ homeostasis in barley seedlings by regu-lating root ion channel activities. *Plant Physiology* 145, no. 3 (2007): 1061–1072.

Zhu, Z., G. Wei, J. Li, Q. Qian, and J. Yu. Silicon alleviates salt stress and increases antioxidant enzymes activ-ity in leaves of salt-stressed cucumber (*Cucumis sativus* L.). *Plant Science* 167, no. 3 (2004): 527–533.

Zhu, Y., and H. Gong. Beneficial effects of silicon on salt and drought tolerance in plants. *Agronomy for Sustainable Development* 34, no. 2 (2014): 455–472.

10 Role of Silicon under Heavy Metal and Toxic Element Stress
An Emphasis on Root Biology

Marek Vaculík and Miroslava Vaculíková

CONTENTS

ABSTRACT

Contamination of land by heavy metals and toxic elements has raised a serious environmental problem in recent decades. This generates a great problem for food and feed production worldwide. Therefore, the mechanisms that allow plants to cope with these threats are intensively studied. There is much evidence that application of silicon (Si) ameliorates the negative effects of various environmental stresses, including heavy metals and toxic elements. These contaminants are frequently present in the soil; therefore, roots often represent the first point of contact with them. This chapter demonstrates the role of Si in plant roots and their response to the elevated concentration of heavy metals, such as Cd, Zn, Cr, and Cu, and the toxic concentrations of other metals and metalloids, such as Al, Mn, As, and Sb.

10.1 INTRODUCTION

Although investigation of the effects of silicon (Si) on plants has dramatically increased in recent years, it still has not been proved that this mysterious element is somehow involved in metabolic processes in plants, and therefore it cannot be listed among the essential elements (Epstein and Bloom 2003). However, its beneficial effects are well known, and Si is applied in the form of Si fertilizers in agricultural praxis in various parts of the world (e.g., Guntzer et al. 2012). The positive effects of Si in enhancing the resistance of plants against pathogenic fungi and insects are relatively well described (Epstein 2002; Reynolds et al. 2009). Another form of stressful conditions for plants is abiotic stress. There are several factors that influence crop production worldwide. Among them, drought, increased salinity, temperature changes, and radiation are especially the focus of many researchers. However, probably the largest amount of abiotic stress studies investigating the effect of Si have been performed in the area of plant nutrition and element toxicity (Balakhina and Borkowska 2013).

For optimal growth and development, plants need to be exposed to several elements present in soil solution. Some of these elements have a so-called dual function—they are required in certain concentrations (mostly as microelements). However, when present in excess, they might be toxic for plants. This is the case with essential zinc (Zn), copper (Cu), manganese (Mn), and chromium (Cr). On the other hand, there are also elements that can be taken up by plants that have no known function; moreover, they are also toxic for plants. This is the case with nonessential cadmium (Cd), lead (Pb), arsenic (As), and antimony (Sb).

Soil contamination by metals and metalloids increased enormously in the last decades, mainly due to extensive mining and ore processing, industrial activities, and increasing traffic. Contaminants negatively influence agriculture and cause yield losses. Therefore, there is a serious need to decrease the uptake of dangerous elements in plants and eliminate their harmful effects. For this purpose, Si has been shown as one possible solution. There are many studies confirming the positive effect of Si on the growth of plants exposed to elevated concentrations of heavy metals and other toxic elements (e.g., Ma 2004; Liang et al. 2007; Wu et al. 2013; Adrees et al. 2015). Moreover, Si has been shown to improve nutrient deficiency and mineral imbalance in some plants, holding more potential for agriculture in the twenty-first century (Pavlovic et al. 2013; Bityutskii et al. 2014; Hernandez-Apaolaza 2014).

The roots are the first plant parts to deal with the elements present in the soil solution. Sometimes these plant organs are partially neglected because of their "hidden life" in the earth; however, their role in the essential processes, such as water uptake, mineral nutrition, element storage, and translocation, cannot be ignored. This present contribution is an overview of the current knowledge about the effects of Si on plants exposed to various heavy metals and toxic elements, with special emphasis on roots, which also need our attention.

10.2 SILICON AND CADMIUM TOXICITY

Cadmium is one of the most dangerous threats to all living organisms. This heavy metal is a hazardous contaminant of food and through the food chains enters the human body as a cumulative poison. Contamination of agricultural land by Cd represents a serious problem in many countries and ranks high in food safety issues (Benavides et al. 2005; Hasan et al. 2009; Nagajyoti et al. 2010). Cadmium is mostly taken up by plant roots and retained in this organ; only a small part is usually translocated to the aerial plant organs, with the exception of hyperaccumulating plants (Martinka et al. 2014). In general, Cd in plants causes a reduction of root and shoot growth, and in high concentrations, it can lead to cell death and destruction of the whole plant (Lux et al. 2011; Clemens et al. 2013).

There are many records documenting that addition of Si can ameliorate the negative effects of Cd on plant growth. Liang et al. (2005) conducted a series of experiments with maize (*Zea mays*) grown in soil containing different concentrations of Si and Cd, and found that Cd negatively influenced the growth of plants and reduced root and shoot length. However, plants that were grown in soils containing both Cd and Si had significantly improved root and shoot biomass as compared with plants that were grown in soils not containing Si. Da Cunha et al. (2008) described that addition of Si (200 mg kg^{-1}) into soil polluted by Cd (10 mg kg^{-1}) and Zn induced a significant increase in maize biomass. The root biomass increased about 82% and the shoot biomass about 67% compared with plants treated without Si. This increase in biomass production was related to the decrease in Cd and Zn bioavailability and detoxification of these metals within plant tissues (da Cunha et al. 2008).

Also, from other experiments it is evident that Si enhanced the growth of plants exposed to Cd. The roots of hydroponically cultivated maize were considerably longer (about 33%) than the roots of Cd-treated plants. Similarly, the fresh weight of the belowground part was about 77% higher in Cd + Si treatment than in Cd alone, and the difference in the dry weight between these two treatments was not as high (about 11%), but it was still significant. In aboveground plant parts, the difference in dry biomass between maize plants treated with Cd and Cd + Si was about 20% (Vaculík et al. 2009). Lukačová Kuliková and Lux (2010) reported that Si either increased or decreased the root length and dry weight of five various Cd-treated maize hybrids. Similarly, Lukačová et al. (2013) found that Si increased the length of the root of maize plants exposed to various Cd concentrations, and negative effects of Cd on fresh and dry root biomass were mitigated by Si only at lower Cd doses.

There is also evidence of a beneficial effect of Si on biomass production on other plant species exposed to elevated Cd. Shi et al. (2005b) found that after the application of both Cd and Si to Yoshida nutrient hydroponics medium, the root and shoot biomass significantly increased in rice (*Oryza sativa*) plants. Similarly, Zhang (2008) demonstrated that Si enhanced the root and shoot biomass in rice plants treated with 2 µM Cd, and the beneficial effect of Si decreased in plants treated with a double Cd concentration (4 µM Cd). Nwugo and Huerta (2008) observed no significant changes in root and shoot length, dry weight, or total leaf area in rice plants treated with Cd and Si together from the first day of experiment. However, all these observed parameters were significantly improved in plants treated from the first day with Cd and from the twentieth day also with Si (Nwugo and Huerta 2008). Gu et al. (2011) found that silicon-rich amendment mitigated the negative effects of metals, including Cd, in rice grown on multi-metal-contaminated acidic soil.

From other than monocotyledonous species, the effect of Si on biomass production in plants suffering from Cd toxicity was investigated, for example, by Song et al. (2009) on pakchoi (*Brassica chinensis*). They found that application of Si increased the fresh and dry weight of roots and shoots in plants treated with Cd + Si compared with plants treated without Si. Similarly, Feng et al. (2010) reported enhanced growth of roots and shoots of cucumber (*Cucumis sativus*) as the effect of 1 mM Si addition into the nutrient medium containing 100 µM Cd. The same alleviating effect of Si was observed by Shi et al. (2010) on a Cd-tolerant and Cd-sensitive cultivar of peanut (*Arachis hypogaea*). Several species of mangroves were also investigated for the effect of Si on their growth when exposed to heavy metals. For example, Zhang et al. (2013) observed that addition of Si to Cd-treated mangrove *Avicennia marina* improved the growth of the roots and shoots. Similarly, Zhang et al. (2015) found that Si improved the root and shoot biomass in *Aegiceras corniculatum* when grown in Cd excess. Farooq et al. (2013) observed a positive effect of Si on the length of the root and root biomass of Cd-treated cotton (*Gossypium hirsutum*) plants. Similarly, Liu et al. (2013b) found a stimulatory effect of Si on the biomass and root length of *Solanum nigrum* grown in Cd exposure. Therefore, we conclude that there is no doubt about the role of Si in enhancing the root and shoot biomass of various plants suffering from Cd toxicity, and this effect might be attributed to the alleviating effect of Si on the growth of Cd-treated plants.

In general, there is knowledge about the detoxifying mechanisms of Cd in the soil–plant system. These mechanisms are mostly based on a reduction of available Cd via its immobilization in the media through an increase in its pH. For example, Chen et al. (2000) observed that furnace slag that contained a lot of Si was more efficient in restricting Cd uptake in rice and wheat plants than waste from mining of Ca carbonates and iron ores. Also, Cheng and Hseu (2002) found that Si-containing slag decreased the concentration of soil-available Cd and Pb and decreased the concentration of these elements in pakchoi (*Brassica chinensis*) plants. Therefore, it was suggested that a higher pH and an increased availability of Si in the slag contributed to reduced Cd uptake in the plants.

Later, Liang et al. (2005) found that an elevated Si concentration in the soil (400 mg kg^{-1}) significantly increased the soil pH, and oppositely decreased the availability of Cd, which resulted in a decreased Cd concentration in the roots and shoots and a decrease of the total Cd content in the maize shoot. However, opposite results were obtained when Si was applied at a lower concentration in the soil (50 mg kg^{-1}). This Si concentration did not influence the soil pH and Cd availability, and the Cd concentration in the shoot significantly decreased when compared with plants treated with only Cd. Although the amount of xylem sap increased, the Cd concentration in the sap decreased in Cd + Si compared with Cd-treated plants. Therefore, Liang et al. (2005) suggested that a Si-induced increase in Cd tolerance can be attributed to not only Cd immobilization due to increased pH in the medium, but also detoxification of Cd effects directly in plants. This suggestion was confirmed by Gu et al. (2011), who found that Si-rich amendments were able to mitigate the negative effects of heavy metals, including Cd, on rice grown in multi-metal-contaminated acidic soil, and the reduction in Cd uptake can be attributed to *in situ* immobilization of heavy metals in soil and also Si-mediated alleviative effects on plants.

Vaculík et al. (2009) showed that Si induced an increase in Cd concentration in roots and shoots, as well as total Cd content per plant in maize plants treated with Cd + Si when compared with plants treated with Cd (5 µM Cd) and without Si. Lukačová Kuliková and Lux (2010) compared the uptake of Cd and Si in five different maize hybrids. In three hybrids, the root Cd concentration significantly decreased, and in two hybrids, the shoot Cd concentration significantly decreased upon Si supplementation. However, only in one of five hybrids did the Cd concentration decrease simultaneously in both the roots and shoots after Cd + Si treatment. Lukačová et al. (2013) later found that Si increased the concentration of Cd in roots exposed to lower Cd doses (5 and 10 µM Cd), and the opposite was observed in those roots exposed to high Cd stress (100 µM Cd). In contrast, Mihaličová Malčovská et al. (2014) showed that Si decreased the concentration of Cd only in those roots that were exposed to higher doses of Cd (Cd 50 µM), and Cd concentration did not change after Si addition to roots exposed to lower Cd (Cd 5 µM). Dresler et al. (2015) reported that the roots of maize plants that were exposed to elevated Cd together with Si accumulated less Cd, and decreased the phytochelatin content, than roots exposed only to Cd, although no differences were observed on the shoot level. This indicated that the effect of Si on the uptake and concentration of Cd within one species might be variable and probably depends on the cultivar or hybrid and concentration of Cd used in the experiments.

There is knowledge on the effect of Si on Cd uptake in other species, especially in rice. Shi et al. (2005b) found that the total amount of Cd decreased about 24% in shoots, but increased about 21% in roots of Cd-treated rice grown with Cd + Si, and the highest amount of Cd remained in the root. Therefore, they suggested that Si inhibited the transport of Cd from root to shoot. At the same time, the growth of Cd-treated plants was enhanced due to Si application. This resulted in a decrease in shoot Cd concentration of about 67% and a decrease in root Cd concentration of about 35% in plants treated with Cd + Si when compared with Cd-treated plants without Si application. Higher biomass production probably dilutes the metal concentration in tissues, by which a negative effect of metals on plants is eliminated (Shi et al. 2005b). This effect of Si on biomass production, together with the accumulation potential of some plants, might be used in phytoextraction technologies (do Nascimento et al. 2006). Shi et al. (2005b) also found that Si reduced the transport of 8-hydroxy-1,3,6-pyrenetrisulphonic acid (PTS), a fluorescent marker of apoplasmic transport from root to shoot. They suggested that increased Si deposition in the endodermis is related to the physical blockage of apoplasmic transport through root tissues.

Zhang (2008) studied the long-term effect of Si on Cd toxicity in rice. They found that Si decreased the Cd concentration in both roots and shoots during the entire cultivation period (105 days of treatment), and the same was observed by Nwugo and Huerta (2008) on the same species. Gu et al. (2011) found that Si-rich amendments reduced the Cd accumulation in rice root, stem, and leaf.

It is known that addition of Si to soil inhibited Cd uptake in strawberry plants. A lower Cd concentration, as the result of Si addition, was observed in stems, leaves, and fruits, but not in roots, although foliar application of Si had no effect on Cd concentration in strawberry (Treder and Cieslinski 2005). Song et al. (2009) compared the effect of additional Si supplementation (1.4 mM) on Cd-treated pakchoi cultivars tolerant or sensitive to Cd. They found that the addition of Si significantly decreased the shoot Cd concentration in both cultivars. On the contrary, the addition of Si considerably increased the root Cd concentration in both cultivars, especially in the tolerant one

(Song et al. 2009). Shi et al. (2010) found that 1.8 mM Si pretreatment for two weeks decreased the Cd concentration in the roots and shoots of peanut plants exposed an additional three weeks to 200 μM Cd. Zhang et al. (2013) observed that most Cd is accumulated in the roots of mangrove *Avicennia marina*. However, after Si addition, the concentration of Cd decreased not only in roots, but also in the stem and leaves. Opposite results were found by Rizwan et al. (2012) in durum wheat grown in historically contaminated soil. They found that the addition of Si increased the concentration of Cd in roots; however, it decreased in shoots. Therefore, they suggested that Si in soil probably decreased the root-to-shoot translocation and retained the Cd in roots (Rizwan et al. 2012). Liu et al. (2013b) reported that Si enhanced tolerance of *Solanum nigrum* by decreased Cd uptake in roots and decreased root-to-shoot translocation.

Da Cunha and do Nascimento (2009) investigated the influence of various soil Si concentrations on structural modifications of maize exposed to a negative effect of Cd and Zn. Although the presence of heavy metals usually reduces the number and diameter of xylem vessels, resulting in decreased uptake and transport of water (e.g., Barceló and Poschenrieder 1990; Gong et al. 2005), da Cunha and do Nascimento (2009) found an increase in xylem vessel diameter due to Si application when compared with plants grown in soils containing Cd and Zn without Si addition. Besides the enlargement of xylem vessels, the biomass of roots and shoots also increased. The central xylem vessel in the leaf, as well as the width of the leaf mesophyll, increased after Si application. Therefore, da Cunha and do Nascimento (2009) suggested that the Si-induced increase in diameter and conductance of xylem elements, as well as enlargement of the leaf mesophyll, is a structural change responsible for alleviation of Cd and Zn toxicity in maize.

In roots, the application of Si increased the thickness of the xylem and pericycle cell walls, as well as the width of the Casparian bands in the endodermis of plants grown in soil contaminated with Cd + Zn when compared with plants grown without Si addition (da Cunha and do Nascimento 2009). Similarly, Zhang et al. (2013) observed that Si induced an increase in the xylem area, thickness of the epidermal and exodermal layer, and width of the Casparian band in the roots of the Cd-treated mangrove *Avicennia marina*. Additionally, Tripathi et al. (2012b) observed that Si prevented a decrease in the frequency of root hairs of rice plants grown under Cd excess.

There are also some reports that Si influenced the development of apoplasmic barriers, root suberization, and lignification, as well as modified root anatomy in plants exposed to Cd. A silicon-induced delay in the endodermis suberinization was observed on hydroponically cultivated Cd-treated maize when compared with plants treated without Si addition (Vaculík et al. 2009). A later study showed that Si delayed apoplasmic barrier formation and xylem element lignification only when maize plants were exposed to a lower Cd concentration (5 μM), and in roots exposed to a higher dose of Cd (50 μM), no differences in apoplasmic barrier formation were observed. This is probably related to the Cd uptake characteristics of the investigated maize hybrid (Vaculík et al. 2012). On the contrary, there are also reports documenting premature suberization and lignification due to Si in roots exposed to Cd. For example, Vatehová et al. (2012), as well as Zhang et al. (2013, 2015), reported that Si enhanced the development of apoplasmic barriers in maize and two mangrove species, respectively. Lukačová et al. (2013) obtained different results with maize exposed to various Cd concentrations. At a lower Cd concentration, the addition of Si enhanced the endodermis development, and the opposite was observed at higher Cd concentrations. Therefore, it seems that endodermal suberization might depend on the species and concentration of the metal applied to plants, and the exact role of Si in the processes of formation of root apoplasmic barriers in still not fully understood.

Da Cunha and do Nascimento (2009) found that in maize root grown in soil contaminated with Cd and Zn, these metals accumulated mainly in the cell wall of the endodermis, pericycle, xylem, phloem, epidermis, exodermis, and cell wall of the cortical parenchyma. In the same plants, the Si deposits were detected in the cell wall of the epidermis, exodermis, endodermis, pericycle, and xylem. Therefore, they suggested that in those tissues where Si and Cd + Zn are deposited, a coprecipitation of Si with metals might occur. Similarly, Seregin and Ivanov (2001) found Si deposits in the cell wall of the endodermis along thickened Casparian bands, and therefore they supposed that the Si deposition is directly related to inhibition of apoplasmic transport of metal into the inner root tissues. Shi et al. (2005b) found that in rice roots grown in medium containing Cd + Si, Cd is predominantly accumulated in the surrounding endodermis and epidermis, and Si was localized mostly in the endodermis. They suggested that a similar localization of Cd and Si was probably related to coprecipitation of these elements in the surrounding endodermis in the roots of rice (Shi et al. 2005b).

In maize leaves, da Cunha and do Nascimento (2009) found codeposition of Si and Cd with Zn in cell walls of the sclerenchyma and adaxial and abaxial epidermis. In the epidermis, these elements were localized mostly in bulliform cells. Another study was performed by Zhang (2008), who found phytolits in rice leaves as a place of codeposition of Cd and Si. Da Cunha and do Nascimento (2009) later detected Si–metal complexes in the vacuoles and cytoplasm of maize leaf mesophyll cells, suggesting that directly in these parts of cells, the coprecipitation of Si with metals occurred. This probably represents one of the key detoxifying mechanisms of Si in plants exposed to Cd and Zn (da Cunha and do Nascimento 2009).

Most of Cd (87%) accumulated in the root of rice plants treated with Cd + Si (Shi et al. 2005b) localized in the symplasm. However, opposite results were obtained by Wang et al. (2000) and Liang et al. (2007). According to Wang et al. (2000), a considerably higher amount of Cd was detected in the cytoplasm, vacuoles, and cell organelles in Cd-treated plants, and only a minimal amount of Cd was detected in the same structures in plants treated with Cd + Si. Therefore, they suggested that the cell wall in plants treated with Si probably contains specialized colloidal Si layers with high absorption potential, and in this way restricts the release of Cd inside plant cells (Wang et al. 2000). Liang et al. (2007) found that most of the Cd was bound to cell walls and less was detected in the symplasm of maize roots treated with Cd and Si. Therefore, Liang et al. (2007) suggested that the apoplasm plays a key role in Cd detoxification, similarly as Si-induced mechanisms of Mn tolerance in plants (Iwasaki et al. 2002; Rogalla and Römheld 2002). Also, Ye et al. (2012) observed the effect of Si on Cd compartmentation in the root tip of *Kandelia obovata*. They found that Si enhanced the ratio of apoplasmic-bound Cd and decreased the ratio of symplasmically bound Cd. On the other hand, Vaculík et al. (2012), using radioactively labeled Cd^{109} isotopes, detected the highest amount of Cd in a soluble fraction and organelle-rich fraction of maize root cells. The addition of Si did not influence the distribution of Cd within different cell compartments in roots; however, in shoots, significantly more Cd was bound to the cell wall fraction. Later, Lukačová et al. (2013), on another maize hybrid, found that Si increased the deposition of Cd in the cell wall fractions of both roots and shoots. Liu et al. (2013a) reported a decreased net influx of Cd ions into rice protoplasts after Si addition and attributed this to potential binding of Cd ions with the organosilicon matrix in the cell walls. Recently, Ma et al. (2015) found that most of the Cd is colocalized with a hemicellulose fraction of cell walls and suggested that a hemicellulose-bound form of Si with net negative charges is responsible for inhibition of the net Cd uptake in rice cells by a mechanism of (Si–hemicellulose matrix)–Cd complexation and codeposition.

In leaves, Shi et al. (2005b) found that application of Si to Cd-treated rice plants reduced the symplasmic and increased the apoplasmic content of Cd. Also, it was found that the addition of Si decreased the Cd content in the organelle fraction of Cd-sensitive cultivars and in the cell wall fraction of Cd-tolerant cultivars of peanut (Shi et al. 2010b).

Another mechanism that could be responsible for Si-alleviated growth might be the influence of Si on the retention of Cd in the Fe-Mn plaque on the root surface. Zhang et al. (2013) observed that addition of Si prompted the accumulation of Mn plaque on the root surface of Cd-treated mangrove *Avicennia marina* and suggested that this might be responsible for improved Cd tolerance of this species. Similarly, Zhang et al. (2015) observed that addition of Si to Cd-treated mangrove *Aegiceras corniculatum* reduced Fe plague and, oppositely, increased Mn plaque. At the same time, a higher amount of Cd was detected in the Fe-Mn plaque of those plants treated with Si when compared with those treated only with Cd. Therefore, they suggested that Fe-Mn plaque after Si addition could possibly block the absorption of Cd from the growth media to the root tissues, and thus improve plant growth (Zhang et al. 2015).

There are also some reports that Si might influence the oxidative status of plants that are exposed to an elevated concentration of Cd. For example, Lukačová et al. (2013) observed that addition of Si stimulated the activity of root peroxidase. The activities of root superoxide dismutase (SOD) and catalase (CAT) were not uniform and depended on the concentration of Cd in the medium. Similarly, Farooq et al. (2013) observed an increased protein content and enhanced activity of ascorbate peroxidase (APX), guaiacol peroxidase (GPX), SOD, and CAT due to the addition of Si in Cd-treated cotton roots. This positive Si effect might be attributed to decreased free radicals in the plant cells. Tripathi et al. (2012b) reported that the elevated concentration of Cd increased both production of hydrogen peroxide and lipid peroxidation in rice roots, and application of Si decreased all these parameters. Mihaličová Malčovská et al. (2014) reported that production of hydrogen peroxide and lipid peroxidation (thiobarbituric acid relative species) increased in maize plants that were exposed to an elevated concentration of Cd. After application of Si, the formation of hydrogen peroxide decreased, and this probably also coincided with improved membrane integrity, as lipid peroxidation decreased after Si addition in maize roots, too.

10.3 SILICON AND ZINC TOXICITY

Zinc is an essential element, and it plays an important role in plants. An appropriate concentration of Zn is required for optimal plant growth and development. Zinc is involved in various metabolic pathways and acts as a cofactor of many enzymes and other proteins (Broadley et al. 2007; Marschner and Marschner 2012). However, excess Zn is dangerous for plants, and various symptoms of Zn cytotoxicity, for example, damage to the chloroplast ultrastructure, vacuolation, an increased number of peroxisomes, and mitochondria with dilated cristae, have been described as well (Martinka et al. 2014).

There are several reports documenting the effect of Si on plants that grow in higher Zn conditions. Da Cunha et al. (2008) observed a positive effect of Si on the roots of maize plants grown in Zn-contaminated soil. The root biomass increased about 80% in soil treated with Zn and Si when compared with soil treated only with Zn. Although soil Zn bioavailability was lower in Zn-Si-treated soils, the root and shoot Zn concentrations increased. Kaya et al. (2009) reported alleviation of Zn toxicity on hydroponically cultivated maize. The biomass of roots and shoots was improved after Si addition. Additionally, the concentration of Zn, as well as P and Fe, decreased in roots after Si addition. Song et al. (2011) investigated the effect of Si on Zn toxicity in rice plants. They observed significant changes in the morphology and biomass of the root system; the length of the roots, number of root tips, root surface, and biomass significantly increased when Si was present in the medium containing high Zn. They also reported that addition of Si stimulated Zn accumulation in roots. On the other hand, they observed that in roots not showing symptoms of oxidative damage, the concentration of hydrogen peroxide was lower after Si addition. This might be due to increased activity of antioxidant enzymes (APX, CAT, and SOD) in roots treated with Zn and Si. Moreover, they found that peroxidation of membrane lipids decreased after Si addition, indicating that Si improved the integrity of membranes of root cells (Song et al. 2011). A similar study was conducted by Gu et al. (2012) on rice. Also They found a stimulating effect of Si on the root biomass of rice treated by higher Zn. On the other hand, Gu et al. (2012) observed a decrease in Zn concentration in roots treated by two various Si doses, which did not correspond to the previous findings of Song et al. (2011). It should be noted, however, that Gu et al. (2012) found that the concentration of Zn decreased in xylem exudates with increasing dose of Si application. Anwaar et al. (2015) recently investigated the effect of Si on the high toxicity of Zn in cotton plants. They found that root length and biomass were improved after Si addition. This might be a result of lower Zn uptake, as the concentration of Zn in roots, stems, and leaves decreased after Si addition. They also found that the protein level, as well as the activity of antioxidative enzymes, increased more in those plants treated with Zn + Si than in those treated only with Zn. In contrast, the content of hydrogen peroxide and peroxidation of lipids in cotton roots decreased, which is in agreement with the previous findings of Song et al. (2011).

There is also contrasting knowledge of the influence of Si on plants treated with higher Zn. Masarovič et al. (2012) investigated sorghum plants grown under Zn toxicity and found that Si stimulated the growth of primary seminal roots; however, the root biomass did not improve. Silicon did not influence the uptake and translocation of Zn from roots to shoots. They also observed a contrasting effect of Si on the activity of various antioxidative enzymes (Masarovič et al. 2012). Similarly, Bokor et al. (2014) found that root length, branching, and biomass did not improve with several doses of Si applied to highly Zn-treated maize plants, although the concentration of Zn decreased in all Si treatments. Si decreased the activity of investigated antioxidative enzymes and lowered root lignification in maize roots exposed to high Zn (Bokor et al. 2014). Recently, Bokor et al. (2015) reported that Zn alone, as well as in combination with Si, greatly imbalanced the maize ionome, while the root ionome was more affected than the shoot ionome. Mineral elements Mn, Fe, Ca, P, Mg, Ni, Co, and K significantly decreased, and Se increased in roots treated with Zn and Si + Zn. Additionally, Zn alone and in combination with Si downregulated the expression level of Si influx and efflux transporters (*ZmLsi1* and *ZmLsi2*) in maize roots (Bokor et al. 2015).

There are a few works describing the colocalization of Zn and Si in plant tissues, although most of them investigated mainly aboveground plant parts. In roots, da Cunha and do Nascimento (2009) showed that precipitation of Si in the root endodermis and pericycle seems to play a role in detoxification of Zn and Cd in maize plants. Gu et al. (2012), using various microscopy techniques, localized most of the Zn in less metabolically active tissues in rice roots, especially in sclerenchyma layers, and in the area of exodermal and endodermal cells and the stele. Addition of Si decreased the intensity of the signal, especially in the stele. Tissue fractionation studies indicated that due to Si application, more Zn is bound to the cell walls and less Zn can be found in free apoplasmic and symplasmic space (Gu et al. 2012). In leaves, Neumann and zur Nieden (2001) investigated the distribution of Si and some elements toxic in excess, especially electron-dense precipitates containing Zn and Si, that were localized in the intercellular space, cytoplasm, and nucleus. Using electron energy loss spectroscopy (EEL-Spectra), they defined these precipitates as silicates with bound Zn. Moreover, they found colocalization of Zn and Si in vacuolar vesicles. Zinc silicates, a temporary storage place for metals in vacuoles, are degraded to SiO_2 precipitates, which are stored in the cytoplasm. During the formation of the above-mentioned vacuolar SiO_2 deposits, Zn is transported into the vacuoles. Also, the presence of uncommon Zn silicate invaginations in the vacuolar direction was determined in the parenchyma cells of a leaf mesophyll in *Cardaminopsis halleri* (Neumann and zur Nieden 2001). Similarly, da Cunha and do Nascimento (2009) observed a simultaneous presence of Si, Cd, and Zn deposits in maize mesophyll cells, suggesting the precipitation of Si–metal complexes in the cytoplasm and vacuoles.

It is interesting that no Zn silicates were found in *Silene cucubalis* and *Minuartia verna* grown in lower polluted soils, although these plants species are considered to be heavy metal tolerant, and it is known that they accumulate Si at higher amounts than other dicotyledonous plants (Neumann et al. 1997; Bringezu et al. 1999). However, these Zn-Si precipitates were detected in the cytoplasm of those plants grown in highly polluted soils. Therefore, Neumann and zur Nieden (2001) suggested that the formation of Zn-Si precipitates was a mechanism partially responsible for the alleviation of Zn toxicity in *Cardaminopsis halleri*.

10.4 SILICON AND MANGANESE TOXICITY

Manganese is an essential nutrient that is required for the optimal growth and development of plants. It is involved in many enzymatic reactions running in plants. The role of Mn is to scavenge reactive oxygen species when present as a cofactor of enzyme SOD (Shenker et al. 2004; Marschner and Marschner 2012). However, increased levels of Mn can be toxic. Symptoms of Mn toxicity and relative toxic tissue concentration vary between the plant species, as phytotoxic mechanisms are involved in different biochemical pathways in various plant species and genotypes (El-Jaoual and Cox 1998).

It is known that Si positively influenced the growth of plants exposed to Mn excess, although almost all papers exclusively deal with the positive effects of Si on Mn toxicity in leaves, and very limited knowledge is available about roots. For the first time, this was observed in barley (Williams and Vlamis 1957a,b). Although these authors did not record a Si-induced decrease in the Mn content in leaves, they found that Si caused a homogenous distribution of Mn in leaf tissues. This was different in leaves treated without Si, where, as the effect of high Mn, several necrotic areas in leaf mesophyll occurred. A similar effect of Si on Mn toxicity was also described by Horst and Marschner (1978), Horiguchi and Morita (1987), and Shi et al. (2005a). Nonhomogenous distribution of Mn with characteristic spotted accumulation was observed in the absence of Si. These spots were dark brown, contained oxidized Mn and polyphenols, and chlorotic and necrotic zones were formed in their surroundings. A positive effect of Si was attributed to the homogenous deposition of Mn, which prevents the occurrence of these Mn spot deposits. Therefore, a Si-induced increase in Mn resistance was not related to a decrease of Mn uptake, but to a different pattern of Mn deposition within the leaf tissues (Horst and Marschner 1978).

Already in 1972, J.E. Bowen noticed that simultaneous treatment of various plants with 5 mg L^{-1} of Si caused reduction of Mn, as well as Zn, Cu, and Fe uptake, and in some plants a Mn deficiency, as the result of Si addition (Bowen 1972). Horst et al. (1999) observed a decrease in apoplasmic Mn concentration in *Vigna unguiculata* as the result of Si addition into the cultivation medium. Based on these results, they supposed that Si might influence the cell wall binding capacity of cations. Iwasaki et al. (2002) found that application of 1.44 mM Si into the medium containing 50 μM Mn alleviated the symptoms of Mn toxicity, which were already observed at 10 μM Mn without Si addition. They also found a higher amount of Mn to be localized in the apoplasm of plants treated with Mn + Si when compared with plants treated only with Mn. Finally, they supposed that the alleviative effect of Si on Mn toxicity consisted not only of increased Mn adsorption to cell walls, but also of active detoxifying of Mn excess through the soluble Si in the apoplasm (Iwasaki et al. 2002).

It was found that Mn toxicity is related to the intercellular concentration of this metal, and by increasing the Si concentration, these toxicity symptoms disappeared in cucumber plants. When 1.8 mM Si was applied, these symptoms disappeared completely. Moreover, it was observed that Mn is deposited homogenously in the cell wall (50%) and in symplasm (50%). When plants were treated with additional Si, more than 90% of Mn was localized in the cell wall. Therefore, it is probable that Si induced binding of Mn into the cell wall, which positively affected the impact of Mn toxicity on plants (Rogalla and Römheld 2002).

According to the results published by Shi et al. (2005a), Si is involved in the mitigation of Mn toxicity effects mainly through decreasing lipid peroxidation and enhancing enzymatic (SOD, APX, and glutathione reductase [GR]) and nonenzymatic (ascorbate [AsA] and glutathione) antioxidants. Doncheva et al. (2009) attributed the beneficial effect of Si on Mn toxicity to a Si-induced increase in the content of chlorophylls and carotenoids, as well as to an increase in the leaf and epidermis thickness. Concerning the roots, Feng et al. (2009) described that application of Si to cucumber plants increased both the dry biomass of roots and aerial parts. Zanao Junior et al. (2010) found that the concentration of Mn increased in rice roots after Si application. Therefore, they suggested that Si probably ameliorated the negative effect of Mn toxicity by decreasing the root-to-shoot translocation of Mn in rice plants.

10.5 SILICON AND ALUMINUM TOXICITY

Aluminum toxicity represents a serious problem for cultivation of agricultural plants on acidic soils, as well as for forest production because of the increase in soil Al content caused by acid rains. The binding of Al to a pectic matrix of root cell walls and apoplasmic face of plasma membranes of root apical cells seems to be a major factor leading to Al-induced inhibition of root elongation. The protection of the root apoplasm, as well as release of organic acid anions complexing Al, is one of the most important mechanisms of plant Al resistance (Horst et al. 2010). However, it is also known that Si-Al interaction, which decreases the negative effect of Al, is one of the external mechanisms of Al detoxification in plants (Hiradate et al. 1988). Silicon with Al forms stabile inert hydroxyaluminosilicates at neutral pH. At a lower pH, Al is released from the minerals to the soil solution, and that causes contamination of soils by this toxic metal. This is also the principle of soil contamination by Al via acid rain (Hodson and Sangster 2002a).

Investigation of Al-Si interaction is probably the best studied feature between the topics dealing with the interaction of Si and other heavy metals or toxic elements. Hodson and Evans (1995) reported the beneficial effect of Si on the mitigation of Al toxicity in plants. According to Liang et al. (2007), several various plant species have been investigated, and a positive effect of Si on Al toxicity was documented for some of them: sorghum (*Sorghum bicolor*), barley (*Hordeum vulgare*), teosinte (*Zea mays* ssp. *mexicana*), and soya (*Glycine max*). Only a minimal or no effect of Si on Al toxicity was observed on rice, wheat (*Triticum aestivum*), cotton, and pea (*Pisum sativum*). Similarly, Baylis et al. (1994) showed that a mitigating effect of Si on Al toxicity depended on the pH of the cultivation media.

Hammond et al. (1995) found that Si inhibited root Al uptake, and a positive effect of Si on an increase in root and shoot growth was visible mainly at higher Al levels. However, an increase in Al shoot concentration as the effect of Si application was reported by Birchall (1990), suggesting that Si probably enhanced the transport of Al from roots to shoots where hydroxyaluminosilicates were produced and stored. Barceló et al. (1993) found that a small concentration of Si can reduce Al uptake and toxicity symptoms on teosinte. They also recorded an increase in malate and formic acid in plants grown on Si-Al-containing medium. Corrales et al. (1997) recorded a positive influence of Si on the mitigation of Al toxicity effects on the growth of maize roots. In wheat, a positive effect of Si on Al toxicity was attributed to internal mechanisms that influenced various physiological processes in plants (Cocker et al. 1998b).

Aluminum significantly reduces plant growth and increases the root/shoot ratio. In experiments with Al-treated sorghum, the highest Al concentration was detected in the outer tangential walls of the root epidermis. Similarly, Al was detected in the cell walls of the hypodermis and no Al was detected in the stele. It is known that Si is usually deposited in the inner tangential walls of the endodermis. In roots exposed to Si and Al, most of the Si was deposited in the inner tangential walls, and also, atypical Si deposition, together with Al, in the outer tangential wall of the epidermis was observed. These Si-Al deposits on the outer epidermis cell wall are probably related to the mechanisms responsible for the enhancement of sorghum root growth in excess Al (Hodson and Sangster 1993, 2002a). Hard soluble aluminosilicates or hydroxyaluminosilicates were formed in cell walls, which are part of an apoplasmic space, and in this way reduced the content of Al^{3+} in the symplasm (Cocker et al. 1998a). Hodson and Sangster (1999) did not find an elevated amount of Al in the shoots of various species from Graminae, while in conifers the content of Al increased intensively. Therefore, they supposed the existence of some mechanism blocking Al transportation from roots to shoots. Similar results were obtained by Carnelli et al. (2002) by paleoecological study of opaline phytoliths of biogenic origin in the Alps. An identical tissue localization was observed for Al and Si in pine (*Pinus strobus*) needles, which is in relation to the supposed codeposition and Al sequestration in needles (Hodson and Sangster 2002b). This is further supported by findings of Britez et al. (2002), who showed that Si-Al complex formation in leaves and stems of *Faramea marginata* (Rubiaceae) contributes to internal detoxification. In experiments with maize root apical segments, more than 85% of Al was bound to the cell walls. The addition of Si did not influence the content of Al in symplasmic and apoplasmic space. Aluminum induced a higher Si accumulation in the cell walls, which reduced the mobility of apoplasmic Al. Therefore, it was supposed that Si, together with Al, formed the hydroxyaluminosilicates in the root apoplasmic space with high detoxifying importance (Wang et al. 2004).

Kidd et al. (2001) found that short-term pretreatment of maize with 1 mM Si, which is later exposed to Al toxicity, induced the production of catechine and quercetine in roots. These exudates of phenolic compounds with flavonoid character probably play an important role in amelioration of Al toxicity in roots (Kidd et al. 2001). Pragabar et al. (2011) investigated the effect of Si on Al-treated suspension cultures of Norway spruce (*Picea abies*). They found that all the negative effects of Al on the cell wall thickening, degree of vacuolization, and degeneration of Golgi bodies, nucleus, mitochondria, and endoplasmic reticule were ameliorated by Si addition. Singh et al. (2011) observed that on the level of root anatomy, Al reduced the length and frequency of root hairs. Similarly, the negative effects of Al resulted in a decrease of stomata frequency and length of epidermal cells and caused a disorder in the mesophyll cells. All negative aspects of Al toxicity were alleviated by Si addition (Singh et al. 2011). Shen et al. (2014) investigated the effect of Si application on various characteristics of peanut roots exposed to an elevated concentration of Al. They found that application of Al resulted in enhanced accumulation of this metal in the root and less in the leaves and stem. However, addition of Si decreased the root Al concentration about 41% compared with plants treated with only Al and not Si. Similarly, the root and shoot biomass decreased with Al, and this was alleviated by Si addition. A decrease in Al tissue concentration and an increase in biomass probably resulted in the improvement of photosynthesis. Additionally, the production of malondialdehyde (MDA), which is commonly used as an indicator of membrane damage, decreased after Si addition, about 28% and 25% in the roots and shoot of peanut plants, respectively, suggesting that addition of Si helps to maintain the integrity of the membranes when plants are grown in Al stress (Shen et al. 2014).

10.6 SILICON AND COPPER TOXICITY

Copper belongs to the essential micronutrients, like Zn and Mn. However, an already slightly elevated concentration of Cu can induce visible toxicity symptoms, and higher doses of Cu are seriously toxic for all living organisms, including plants (Marschner and Marschner 2012). There is relatively limited knowledge about the influence of Si on plants exposed to Cu toxicity, although some new evidence has appeared in recent years.

Nowakowski and Nowakowska (1997) found that Si actively ameliorated the negative Cu influence on biomass production on wheat (*Triticum aestivum*). Plants grown in a medium containing Cu + Si accumulated less Cu in their below- and aboveground plant tissues compared with plants treated with Cu without Si. Similarly, the below- and aboveground parts of plants growing in Cu + Si medium contained considerably more water than relevant tissues of plants grown in Cu treatment without Si addition (Nowakowski and Nowakowska 1997). Also, Li et al. (2008) found a positive effect of Si on the mitigation of Cu toxicity on *Arabidopsis thaliana*. Copper-mediated leaf chlorosis was reduced, and shoot and root biomass increased when Si was applied together with Cu. Also, a reduction in the activity of PAL in leaves was observed in plants treated with Cu + Si when compared with Cu-treated ones, although the activity of PAL in roots considerably increased. The authors explain this as a possible mechanism by which Si activated defense mechanisms in those plant parts directly exposed to Cu, and suggested a different role of Si on phenolics synthesis and metabolism in roots and shoots (Li et al. 2008). Additionally, Li et al. (2008) observed that expressions of multicopper oxidase type I family protein (*MOC*), esterase lipase thioesterase family protein (*TLE*), copper transporter 1 (*COPT1*), and heavy metal ATPase 5 (*HMA5*) genes in *Arabidopsis* roots were downregulated due to the presence of Si in media containing elevated Cu. Khandekar and Leisner (2011) observed increased SOD activity in plants treated with Cu + Si when compared with Cu-treated plants. They suggested that Si apparently allows plants to more efficiently respond to Cu toxicity by maintaining or upregulating Cu binding molecules and increasing expression of free radical metabolizing enzymes, and through these, other mechanisms as well, helping to reduce the stress of plants (Khandekar and Leisner 2011). Collin et al. (2014) investigated the proportion of Cu ligands in bamboo roots after Si supplementation that were determined from the linear combination fitting of the plant's Cu K-edge extended x-ray absorption fine-structure spectroscopy (EXAFS) data. They found that the best component fits were obtained with 48% Cu(II)-malate and 52% Cu(II)-histidine for the Cu100 treatment and 66% Cu(II)-malate and 34% Cu(II)-histidine for the Cu100Si treatment (Collin et al. 2014). Recently, Keller et al. (2015) investigated the effect of Si on growth of durum wheat exposed to elevated Cu concentrations. They found no differences in root length between Cu and Cu + Si plants at lower applied Cu doses; however, at higher Cu stress, Si positively enhanced the root length of wheat plants. Additionally, they observed that with increasing Cu concentration in the media, more Si was stored in roots, and at the same time, the concentration of Cu in roots decreased. At high Cu exposure, the concentration of chelating agents like malate, citrate, and aconitate increased after Si addition. Regarding the distribution of Cu within roots, Keller et al. (2015) reported that in the absence of Si, Cu was mainly localized in the central cylinder of root, while in the presence of Si, Cu was mainly localized in the vicinity of the root epidermis. Mateos-Naranjo et al. (2015) reported that the addition of Si to Cu-stressed C4 grass *Spartina densiflora* resulted in an increase in root biomass. Although the root Cu concentration was not affected, addition of Si decreased the root-to-shoot Cu translocation, suggesting that this could be one of the mechanisms by which Si mitigates Cu toxicity in plants (Mateos-Naranjo et al. 2015).

10.7   SILICON AND CHROMIUM TOXICITY

Chromium (Cr) is the second most common contaminant of soils and groundwater. This nonessential element is highly toxic for all organisms, including plants. Chromium affects the uptake of water and imbalance nutrient status of plants; it negatively influences photosynthesis, destroys membranes, and causes oxidative stress (Pal-Singh et al. 2013). In recent years, some reports have been published on the possibility of the mitigation of negative effects of this nonessential and dangerous metal on the growth of plants using Si.

Zeng et al. (2011) compared the effect of external Si application to various rice cultivars suffering from Cr toxicity. They investigated whether the effect of Si depended on the cultivar. In general, they attributed the positive effect of Si to a reduced Cr concentration within root tissues, a reduced root-to-shoot translocation of Cr, and an improved antioxidant capacity of the plants, especially an increase in the activity of CAT and POX in roots that were decreased due to Cr toxicity (Zeng et al. 2011). Similarly, Tripathi et al. (2012a) investigated the effect of Cr excess on rice plants and found that addition of Si increased root growth, enhanced root biomass, and increased protein content in roots. At the same time, the concentration of Cr decreased and the concentration of Si increased in roots of those plants treated with both Cr and Si when compared with those roots exposed only to Cr. Plants treated with Si coped with Cr toxicity by improved antioxidant capacity, enhanced content of total phenolics, and decreased lipid peroxidation (Tripathi et al. 2012a). Ali et al. (2013) investigated the influence of Si on the growth and ultrastructure of barley plants exposed to Cr. They found that Si improved the root biomass and decreased the concentration of Cr in roots. Negative influences of Cr on the ultrastructural levels as disruption of the nucleus, disappearance of the nucleolus, and disruption of the nuclear membrane were partially alleviated by addition of Si to barley root tip cells. Ding et al. (2013) compared the influence of various doses of Si on the growth of pakchoi exposed to Cr and found that lower application of Si increased the total root Cr content. In contrast, at higher Si treatment the total root content of Cr decreased. It has also been observed that Si can ameliorate the negative effects of Cr on the growth of wheat plants (Tripathi et al. 2015). The root length, as well as root fresh and dry biomass, increased after Si addition to Cr-exposed plants. Cr caused also several modifications in root anatomy; root hairs were shorter and had a thick-walled tip, and root contained a lot of aerenchymous spaces in the cortical tissues. The addition of Si ameliorated this, and Si promoted root lignification and suberization when compared with roots of those plants treated only by Cr (Tripathi et al. 2015).

10.8 SILICON AND ARSENIC TOXICITY

Arsenic is a dangerous metalloid that is present in agricultural soils in many places throughout the world. Several studies indicated that growing crop plants on As-polluted soils accounts for a potential health risk for humans because of accumulation of As in plant biomass (Antosiewicz et al. 2008; Bergqvist and Greger 2012; Drličková et al. 2013). In recent years, interesting results were obtained regarding the role of Si in the uptake and detoxification of As in plants.

Guo et al. (2005) tested three various Si concentrations and proposed 1 mM Si as the best for mitigation of As toxicity in rice. They also found a decrease in the root and shoot As concentration with increasing Si concentration in the medium. Moreover, they observed a significant decrease in the phosphorus concentration in plants growing in hydroponics with increasing As and Si concentration. Silicon did not influence the As translocation from root to shoot. With increasing concentration of As in the medium, As translocation from root to shoot decreased, and with increasing concentration of Si in the medium, the total content of As in plants significantly decreased (Guo et al. 2005). Later, it was discovered that Si and As share the same transport pathways. Ma et al. (2006) identified OSNIP2;1 (also called Lsi1) as a Si transporter, which is also the main entrance of As in the form of arsenite (As(III)) into the rice roots. Mutation in this gene led to 60% inhibition of As(III) uptake in rice roots when compared with the control (Ma et al. 2006). With respect to these findings, Fleck et al. (2013) compared the uptake of different forms of As in various aboveground organs and plant parts of rice. They found that Si addition to soil solution did not affect the arsenate (As(V)) and dimethylarsinic acid (DMA) concentration; however, it significantly decreased the As(III) concentration, which correlates with previous findings that As(III) and Si share the same transport mechanisms. On the other hand, Liu et al. (2014) observed in rice that Si attenuated the uptake of inorganic forms of As, however enhanced the uptake of methylated forms (DMA), although the latter ones are less dangerous, as they are less toxic. Another problem in As accumulation might represent cultivar variability. Marmiroli et al. (2014) showed that root and shoot uptake of As can vary between various tomato cultivars after addition of Si. In general, the concentration of As(III) decreased in roots of most tomato cultivars, whereas the concentration of As(V) mostly increased in tomato roots, although not significantly.

Tripathi et al. (2013) also observed a positive effect of Si on As in rice plants, although their research was focused on shoots and not on roots. They attributed the ameliorative effect of Si to reduced As uptake, increased antioxidants, and enhanced synthesis of glutathione and nonprotein thiols. Similarly, positive effects on photosynthesis were recorded by Sanglard et al. (2014) on rice and by da Silva et al. (2015) on maize. Recently, Lou et al. (2015) observed the effect of Si on the growth of wheat roots influenced by As. They found that an alleviating effect of Si occurred only at a certain Si concentration in the growth media (960 μmol L^{-1}), and lower applied doses of Si did not influence the growth of plants. Roots grown in the presence of As and Si (960 μmol L^{-1}) were longer, and the root As concentration significantly decreased when compared with those roots treated only by As. At the same time, they also observed that Si addition decreased As(V) uptake and decreased As(III) efflux from roots (Lou et al. 2015). Lee et al. (2014) pointed out the problem of increased bioavailability of As by Si application in the soils highly contaminated by As. They found that Si added in lower doses to the soils might enhance As availability and increase root and shoot As concentration and decrease root length. Therefore, they supposed that foliar Si application instead of soil fertilization might improve the growth of crops in soils with a high content of As (Lee et al. 2014).

10.9 SILICON AND ANTIMONY TOXICITY

Antimony, similar to arsenic, belongs to the group of metalloids. It is a nonessential element that can be dangerous for living organisms when present in excess. Natural soil concentrations of Sb are relatively small; however, it is present in high concentrations, especially in mining areas and in the vicinity of smelters. A considerably high amount of Sb enters the soil by military activities, and traffic is also an important source (Tschan et al. 2009b). It is known that considerably increased levels of Sb can be found in plants colonizing old mining sites (Bech et al. 2012; Vaculík et al. 2013). Accumulation of Sb in agricultural plants might represent a threat for crop production; Sb was found to decrease yield, reduce photosynthesis and respiration, and affect root and shoot anatomy (Tschan et al. 2009a; Vaculík et al. 2015).

Recently, there has also been some evidence for the ameliorative role of Si on Sb toxicity in plants. Huang et al. (2012) investigated the effect of Sb on wild-type rice and mutants differing in the uptake of Si. Although they observed no change in plant biomass, application of Si reduced the uptake of Sb in aerial parts and increased the concentration of Sb in roots, especially in wild-type rice. Vaculíková et al. (2014) investigated the effect of Si on an elevated concentration of Sb on maize roots. Although they did not observe changes in root length and root biomass after Si addition, they found significant differences in the antioxidant status of plants. Sb increased the activity of investigated antioxidative enzymes (APX and CAT), which was ameliorated by addition of Si. This was probably due to lower oxidative stress in maize roots, as the peroxidation of lipids and synthesis of proline, considered a suitable marker of abiotic stress, decreased when Si was added to the medium. These results suggested that although Si did not directly improve root length and biomass, changes in plant biochemistry indicated a possible mitigation role of Si on Sb toxicity in plants (Vaculíková et al. 2014). Recently, in another study, Vaculíková et al. (2016) found that Sb(V) did not influence root biomass, although it reduced the growth of the root. The addition of Si ameliorated the negative effect of Sb on root growth and architecture, and enhanced aerenchyma formation that was suppressed by Sb. Additionally, Vaculíková et al. (2016) found that Si did not affect root Sb concentrations but modified Sb translocation to the shoot, and therefore suggested that the interactions between Sb and Si occurred through different types of mechanisms.

10.10   CONCLUSION AND PERSPECTIVES

A previous literature survey clearly demonstrated the beneficial effects of Si on the growth of various monocot and dicot species exposed to elevated concentrations of dangerous heavy metals or toxic elements. An alleviative role of Si has been proved in the case of Cd, Zn, Mn, Cu, Al, Pb, Cr, Ni, As, and Sb. However, we may suppose that the list of elements is not complete. We have shown that there are several Si-induced mechanisms involved at the root level that facilitate the plants' coping with element toxicity. Restriction of root uptake and immobilization of metals in the rhizosphere are probably among the first defense mechanisms. It is known that some elements form complexes with Si directly in the soil; a positive role of Si in enhancing root exudation has also been observed. Additionally, retention of toxic elements in the root apoplasm might enhance the resistance and vigor of plants. This is supported by evidence that Si actively influenced the development of root apoplasmic barriers, as was shown, for example, in the case of Cd. Formation of insoluble complexes between Si and metals and their storage within cell walls help plants to decrease the available metal concentration and restrict symplasmic uptake. This was demonstrated in the cases of Al, Mn, and Cd. There is no doubt that Si influences the oxidative status of plants; either an increase or a decrease in the activity of antioxidants has been observed in the cases of Cd, Zn, Cu, As, and Sb. Therefore, it is difficult to generalize these responses, and results of the experiments should be evaluated in the particular research context. One of the other possible mechanisms explaining the strong ability of plants to cope with element toxicity is Si-mediated enhancement of the permeability of cell membranes.

Usually, the ameliorative effect of Si tends to be connected to reduced metal uptake in plant tissues. However, it was documented that Si can stimulate the uptake of dangerous metals and, at the same time, improve the growth of plants. These plants should be carefully supervised mainly for the production of food and feed; however, at the same time, they might be interesting for phytoremediation purposes. Additionally, it should also be noted that there are some reports documenting a vague effect of Si on alleviation of metal toxicity, especially in the case of Zn. Despite this, there are many more records about the ameliorative role of Si in the toxicity of various heavy metals and toxic elements. Therefore, further research focused on individual plant species and their cultivars or hybrids, as well as a deep focus on the root and rhizosphere interaction of Si with heavy metals and toxic elements, would better explain the mitigating role of Si in plants.

ACKNOWLEDGMENT

This work was supported by Slovak Grant Agency VEGA No. VEGA 2/0022/13, and was part of the project supported by the Slovak Research and Development Agency under Contract APVV SK-SRB-2013-0021 and APVV SK-AT-2015-0009.

REFERENCES

Adrees M, Ali S, Rizwan M, Zia-ur-Rehman M, Ibrahim M, Abbas F, Farid M, Qayyum MF, Irshad MK. (2015). Mechanisms of silicon-mediated alleviation of heavy metal toxicity in plants: A review. *Ecotoxicol Environ Saf* 119: 186–197.

Ali S, Farooq MA, Yasmeen T, Hussain S, Arif MS, Abbas F, Bharwana SA, Zhang G. (2013). The influence of silicon on barley growth, photosynthesis and ultra-structure under chromium stress. *Ecotoxicol Environ Saf* 89: 66–72.

Anwaar SA, Ali S, Ali S, Ishaque W, Farid M, Farooq MA, Najeeb U, Abbas F, Sharif M. (2015). Silicon (Si) alleviates cotton (*Gossypium hirsutum* L.) from zinc (Zn) toxicity stress by limiting Zn uptake and oxidative damage. *Environ Sci Pollut Res* 22: 3441–3450.

Antosiewicz DM, Escudě-Duran C, Wierzbowska E, Skłodowska A. (2008). Indigenous plant species with the potential for the phytoremediation of arsenic and metals contaminated soil. *Water Air Soil Pollut* 193: 197–210.

Balakhina T, Borkowska A. (2013). Effects of silicon on plant resistance to environmental stresses: A review. *Int Agrophys* 27: 225–232.

Barceló J, Guevara P, Poschenrieder C. (1993). Silicon amelioration of aluminium toxicity in teosinte, *Zea mays* L. ssp. *mexicana*. *Plant Soil* 154: 249–255.

Barceló J, Poschenrieder C. (1990). Plant water relations as effected by heavy metal stress: A review. *J Plant Nutr* 13: 1–37.

Baylis AD, Gragopoulou C, Davidson KJ, Birchall JD. (1994). Effects of silicon on the toxicity of aluminum to soybean. *Commun Soil Sci Plant Anal* 25: 537–546.

Bech J, Corrales I, Tume P, Barceló J, Duran P, Roca N, Poschenrieder C. (2012). Accumulation of antimony and other potentially toxic elements in plants around a former antimony mine located in the Ribes Valley (Eastern Pyrenees). *J Geochem Explor* 113: 100–105.

Benavides MP, Gallego SM, Tomaro ML. (2005). Cadmium toxicity in plants. *Braz J Plant Physiol* 17: 21–34.

Bergqvist C, Greger M. (2012). Arsenic accumulation and speciation in plants from different habitats. *Appl Geochem* 27: 615–622.

Birchall JD. (1990). The role of silicon in biology. *Chem Britain* 26: 141–144.

Bityutskii N, Pavlovic J, Yakkonen K, Maksimovic V, Nikolic M. (2014). Contrasting effect of silicon on iron, zinc and manganese status and accumulation of metal-mobilizing compounds in micronutrient-deficient cucumber. *Plant Physiol Biochem* 74: 205–211.

Bokor B, Bokorová S, Ondoš S, Švubová R, Lukačová Z, Hýblová M, Szemes T, Lux A. (2015). Ionome and expression level of Si transporter genes (Lsi1, Lsi2, and Lsi6) affected by Zn and Si interaction in maize. *Environ Sci Pollut Res* 22: 6800–6811.

Bokor B, Vaculík M, Slováková Ľ, Masarovič D, Lux A. (2014). Silicon does not always mitigate zinc toxicity in maize. *Acta Physiol Plant* 36: 733–743.

Bowen JE. (1972). Manganese-silicon interaction and its effect on growth of Sudan grass. *Plant Soil* 37: 577–588.

Bringezu K, Lichtenberger O, Leopold I, Neumann D. (1999). Heavy metal tolerance of *Silene vulgaris*. *J Plant Physiol* 154: 534–546.

Britez RM, Watanabe T, Jansen S, Reissmann CB, Osaki M. (2002). The relationship between aluminium and silicon accumulation in leaves of *Faramea marginata* (Rubiaceae). *New Phytol* 156: 437–444.

Broadley MR, White PJ, Hammond JP, Zelko I, Lux A. (2007). Zinc in plants. *New Phytol* 173: 677–702.

Carnelli A, Madella M, Theurillat J-P, Ammann B. (2002). Aluminium in the opal silica reticule of phytoliths: A new tool in palaeoecological studies. *Am J Bot* 89: 346–351.

Chen HM, Zheng CR, Tu C, She ZG. (2000). Chemical methods and phytoremediation of soil contaminated with heavy metals. *Chemosphere* 41: 229–234.

Cheng S, Hseu Z. (2002). *In-situ* immobilization of cadmium and lead by different amendment in two contaminated soils. *Water Air Soil Pollut* 140: 73–84.

Clemens S, Aarts MGM, Thomine S, Verbruggen N. (2013). Plant science: The key to preventing slow cadmium poisoning. *Trends Plant Sci* 18: 92–99.

Cocker KM, Evans DE, Hodson MJ. (1998a). The amelioration of aluminium toxicity by silicon in higher plants: Solution chemistry or an in plants mechanism? *Physiol Plant* 104: 608–614.

Cocker KM, Evans DE, Hodson MJ. (1998b). The amelioration of aluminium toxicity by silicon in wheat (*Triticum aestivum* L.): Malate exudation as evidence for an in planta mechanism. *Planta* 204: 318–323.

Collin B, Doelsch E, Keller C, Cazevieille P, Tella M, Chaurand P, Panfili F, Hazemann JL, Meunier JD. (2014). Evidence of sulfur-bound reduced copper in bamboo exposed to high silicon and copper concentrations. *Environ Pollut* 187: 22–30.

Corrales I, Poschenrieder C, Barceló J. (1997). Influence of silicon pretreatment on aluminium toxicity in maize roots. *Plant Soil* 190: 203–209.

da Cunha KPV, do Nascimento CWA. (2009). Silicon effects on metal tolerance and structural changes in maize (*Zea mays* L.) grown on a cadmium and zinc enriched soil. *Water Air Soil Pollut* 197: 323–330.

da Cunha KPV, do Nascimento CWA, da Silva AJ. (2008). Silicon alleviates the toxicity of cadmium and zinc for maize (*Zea mays* L.) grown on a contaminated soil. *J Plant Soil Sci* 171: 849–853.

da Silva AJ, do Nascimento CW, da Silva Gouveia-Neto A, da Silva Jr EA. (2015). Effects of silicon on alleviating arsenic toxicity in maize plants. *Rev Bras Cienc Solo* 39: 289–296.

Ding X, Zhang S, Li S, Liao X, Wang R. (2013). Silicon mediated the detoxification of Cr on pakchoi (*Brassica Chinensis* L.) in Cr-contaminated soil. *Proc Environ Sci* 18: 58–67.

do Nascimento CWA, Amarasiriwardena D, Xing B. (2006). Comparison of natural organic acids and synthetic chelates at enhancing phytoextraction of metals from a multi-metal contaminated soil. *Environ Pollut* 140: 114–123.

Doncheva S, Poschenrieder C, Stoyanova Z, Georgieva K, Velichkova M, Barceló J. (2009). Silicon amelioration of manganese toxicity in Mn-sensitive and Mn-tolerant maize varieties. *Environ Exp Bot* 65: 189–197.

Dresler S, Wojcik M, Bednarek W, Hanaka A, Tukiendorf A. (2015). The effect of silicon on maize growth under cadmium stress. *Russ J Plant Physiol* 62: 86–92.

Drličková G, Vaculík M, Matejkovič P, Lux A. (2013). Bioavailability and toxicity of arsenic in maize (*Zea mays* L.) grown in contaminated soils. *Bull Environ Contam Toxicol* 91: 235–239.

El-Jaoual T, Cox DA. (1998). Manganese toxicity in plants. *Planta* 2: 353–386.

Epstein E. (2002). Silicon in plant nutrition. In *Second Silicon in Agriculture Conference*. Tsuruoka, Yamagata, Japan, pp. 1–4.

Epstein E, Bloom AJ. (2003). *Mineral Nutrition of Plants: Principles and Perspectives*. John Wiley & Sons, New York.

Farooq MA, Ali S, Hameed A, Ishaque W, Mahmood K, Iqbal Z. (2013). Alleviation of cadmium toxicity by silicon is related to elevated photosynthesis, antioxidant enzymes; suppressed cadmium uptake and oxidative stress in cotton. *Ecotox Environ Saf* 96: 242–249.

Feng JP, Shi QH, Wang XF. (2009). Effects of exogenous silicon on photosynthetic capacity and antioxidant enzyme activities in chloroplast of cucumber seedlings under excess manganese. *Agric Sci China* 8: 40–50.

Feng J, Shi Q, Wang X, Wei M, Yang F, Xu H. (2010). Silicon supplementation ameliorated the inhibition of photosynthesis and nitrate metabolism by cadmium (Cd) toxicity in *Cucumis sativus* L. *Sci Horticult* 123: 521–530.

Fleck AT, Mattusch J, Schenk MK. (2013). Silicon decreases the arsenic level in rice grain by limiting arsenite transport. *J Plant Nutr Soil Sci* 176: 785–794.

Gong HJ, Zhu XY, Chen KM, Wang S, Zhang Ch. (2005). Silicon alleviates oxidative damage of wheat plants in pots under drought. *Plant Sci* 169: 313–321.

Gu HH, Qiu H, Tian T, Zhan SS, Deng THB, Chaney RL, Wang SZ, Tang YT, Morel JL, Qiu RL. (2011). Mitigation effect of silicon rich amendments on heavy metal accumulation in rice (*Oryza sativa* L.) planted on multi-metal contaminated acidic soil. *Chemosphere* 83: 1234–1240.

Gu HH, Zhan SS, Wang SH, Tang YT, Chaney RL, Fang XH, Cai XD, Qiu RL. (2012). Silicon-mediated amelioration of zinc toxicity in rice. *Plant Soil* 350: 193–204.

Guntzer F, Keller C, Meunier J-D. (2012). Benefits of plant silicon for crops: A review. *Agron Sustain Dev* 32: 201–213.

Guo W, Hou YL, Wang SG, Zhu YG. (2005). Effect of silicate on the growth and arsenate uptake by rice (*Oryza sativa* L.) seedlings in solution culture. *Plant Soil* 272: 173–181.

Hammond KE, Evans DE, Hodson MJ. (1995). Aluminium/silicon interactions in barley (*Hordeum vulgare* L.) seedlings. *Plant Soil* 173: 89–95.

Hasan SA, Fariduddin Q, Ali B, Hayat S, Ahmad A. (2009). Cadmium: Toxicity and tolerance in plants. *J Environ Biol* 30: 165–174.

Hernandez-Apaolaza L. (2014). Can silicon partially alleviate micronutrient deficiency in plants? *Planta* 240: 447–458.

Hiradate S, Taniguchi S, Sakurai K. (1988). Aluminum speciations in aluminum-silica solutions and potassium chloride extracts of acidic soils. *Soil Sci Soc Am J* 62: 630–636.

Hodson MJ, Evans DE. (1995). Aluminium/silicon interactions in higher plants. *J Exp Bot* 46: 161–171.

Hodson MJ, Sangster AG. (1993). The interaction between silicon and aluminium in *Sorghum bicolor* (L.) Moench: Growth analysis and x-ray microanalysis. *Ann Bot* 72: 389–400.

Hodson MJ, Sangster AG. (1999). Aluminium/silicon interactions in conifers. *J Inorg Biochem* 76: 89–98.

Hodson MJ, Sangster AG. (2002a). Silicon and abiotic stress. In *Second Silicon in Agriculture Conference*, Tsuruoka, Yamagata, Japan, pp. 99–104.

Hodson MJ, Sangster AG. (2002b). X-ray microanalytical studies of mineral localization in the needles of white pine (*Pinus strobus* L.). *Ann Bot* 89: 367–374.

Horiguchi T, Morita S. (1987). Mechanism of manganese toxicity and tolerance of plants. IV. Effect of silicon on alleviation of manganese toxicity of barley. *J Plant Nutr* 10: 2299–2310.

Horst WJ, Fecht M, Naumann A, Wissemeier AH, Maier P. (1999). Physiology of manganese toxicity and tolerance in *Vigna unguiculata* (L.) Walp. *J Plant Nutr Soil Sci* 162: 263–274.

Horst WJ, Marschner H. (1978). Effect of silicon on manganese tolerance of bean plants (*Phaseolus vulgaris* L.). *Plant Soil* 50: 287–303.

Horst WJ, Wang Y, Eticha D. (2010). The role of the root apoplast in aluminium-induced inhibition of root elongation and in aluminium resistance of plants: A review. *Ann Bot* 106: 185–197.

Huang YZ, Zhang WQ, Zhao LJ. (2012). Silicon enhances resistance to antimony toxicity in the low-silica rice mutant, lsi1. *Chem Ecol* 28: 341–354.

Iwasaki K, Maier P, Fecht M, Horst WJ. (2002). Effects of silicon supply on apoplastic manganese concentrations in leaves and their relation to manganese tolerance in cowpea (*Vigna ustulata* (L.) Walp.). *Plant Soil* 238: 281–288.

Kaya C, Tuna AL, Sonmez O, Ince F, Higgs D. (2009). Mitigation effects of silicon on maize plants grown at high zinc. *J Plant Nutr* 32: 1788–1798.

Keller C, Rizwan M, Davidian JC, Pokrovsky OS, Bovet N, Cahurand P, Meunier JD. (2015). Effect of silicon on wheat seedlings (*Triticum turgidum* L.) grown in hydroponics and exposed to 0 to 30 μM Cu. *Planta* 241: 847–860.

Khandekar S, Leisner S. (2011). Soluble silicon modulates expression of *Arabidopsis thaliana* genes involved in copper stress. *J Plant Physiol* 168: 699–705.

Kidd PS, Llugany M, Poschenreider C, Gunsé B, Barceló J. (2001). The role of root exudates in aluminum resistance and silicon-induced amelioration of aluminium toxicity in three varieties of maize (*Zea mays* L.). *J Exp Bot* 52: 1339–1352.

Lee Ch-H, Huang H-H, Syu Ch-H, Lin T-H, Lee D-Y. (2014). Increase of As release and phytotoxicity to rice seedlings in As-contaminated paddy soils by Si fertilizer application. *J Hazard Mater* 276: 253–261.

Li J, Frantz J, Leisner S. (2008). Alleviation of copper toxicity in *Arabidopsis thaliana* by silicon addition to hydroponic solutions. *J Am Soc Horticult* Sci 133: 670–677.

Liang Y, Sun W, Zhu YG, Christie P. (2007). Mechanisms of silicon-mediated alleviation of abiotic stresses in higher plants: A review. *Environ Pollut* 147: 422–428.

Liang Y, Wong JWC, Wei L. (2005). Silicon-mediated enhancement of cadmium tolerance in maize (*Zea mays* L.) grown in cadmium contaminated soil. *Chemosphere* 58: 475–483.

Liu J, Ma J, He C, Li X, Zhang W, Xu F, Lin Y, Wang L. (2013a). Inhibition of cadmium ion uptake in rice (*Oryza sativa*) cells by a wall-bound form of silicon. *New Phytol* 200: 691–699.

Liu J, Zhang H, Zhang Y, Chai T. (2013b). Silicon attenuates cadmium toxicity in *Solanum nigrum* L. by reducing cadmium uptake and oxidative stress. *Plant Physiol Biochem* 68: 1–7.

Liu WJ, McGrath SP, Zhao FJ. (2014). Silicon has opposite effects on the accumulation of inorganic and methylated arsenic species in rice. *Plant Soil* 376: 423–431.

Lou LQ, Shi GL, Wu JH, Zhu S, Qian M, Wang HZ, Cai QS. (2015). The influence of phospohorus on arsenic uptake/efflux and As toxicity to wheat roots in comparison with sulphur and silicon. *J Plant Growth Regul* 34: 242–250.

Lukačová Z, Švubová R, Kohanová J, Lux A. (2013). Silicon mitigates the Cd toxicity in maize in relation to cadmium translocation, cell distribution, antioxidant enzymes stimulation and enhanced endodermal apoplasmic barrier development. *Plant Growth Regul* 70: 89–103.

Lukačová Kuliková Z, Lux A. (2010). Silicon influence on maize, *Zea mays* L., hybrids exposed to cadmium treatment. *Bull Environ Contam Toxicol* 85: 243–250.

Lux A, Martinka M, Vaculík M, White PJ. (2011). Root responses to cadmium in the rhizosphere: A review. *J Exp Bot* 62: 21–37.

Ma J, Cai H, He C, Zhang W, Wang L. (2015). A hemicelluloses-bound form of silicon inhibits cadmium ion uptake in rice (*Oryza sativa*) cells. *New Phytol* 206: 1063–1074.

Ma JF. (2004). Role of silicon in enhancing the resistance of plants to biotic and abiotic stresses. *Soil Sci Plant Nutr* 50: 11–18.

Ma JF, Tamai K, Yamaji N, Mitani N, Konishi S, Katsuhara M, Ishiguro M, Murata Y, Yano M. (2006). A silicon transporter in rice. *Nature* 440: 688–691.

Marmiroli M, Pigoni V, Savo-Sardaro ML, Marmiroli N. (2014). The effect of silicon on the uptake and translocation of arsenic in tomato (*Solanum lycopersicum* L.) *Environ Exp Bot* 99: 9–17.

Marschner H, Marschner P. (2012). *Marschner's Mineral Nutrition of Higher Plants*. 3rd ed. Academic Press, London.

Martinka M, Vaculík M, Lux A. (2014). Plant cell responses to cadmium and zinc. In Nick P, Opatrný Z (eds.), *Applied Plant Cell Biology*. Plant Cell Monograph 22. Springer Berlin, pp. 209–246.

Masarovič D, Slováková Ľ, Bokor B, Bujdoš M, Lux A. (2012). Effect of silicon application on *Sorghum bicolor* exposed to toxic concentration of zinc. *Biologia* 67: 706–712.

Mateos-Naranjo E, Gallé A, Florez-Sarasa I, Perdomo JA, Galmés J, Ribas-Cardó M, Flexas J. (2015). Assessment of the role of silicon in the Cu-tolerance of the C_4 grass *Spartina densiflora*. *J Plant Physiol* 178: 74–83.

Mihaličová Malčovská S, Dučaiová Z, Maslaňáková I, Bačkor M. (2014). Effect of silicon on growth, photosynthesis, oxidative status and phenolic compounds of maize (*Zea mays* L.) grown in cadmium excess. *Water Air Soil Pollut* 225: 2056.

Nagajyoti PC, Lee KD, Sreekanth TVM. (2010). Heavy metals, occurrence and toxicity for plants: A review. *Environ Chem Lett* 8: 199–216.

Neumann D, zur Nieden U. (2001). Silicon and heavy metal tolerance of higher plants. *Phytochemistry* 56: 685–692.

Neumann D, zur Nieden U, Schwieger W, Leopold I, Lichtenberger O. (1997). Heavy metal tolerance of *Minuartia verna*. *J Plant Physiol* 151: 101–108.

Nowakowski W, Nowakowska J. (1997). Silicon and copper interaction in the growth of spring wheat seedlings. *Biol Plant* 39: 463–466.

Nwugo ChC, Huerta AJ. (2008). Effects of silicon nutrition on cadmium uptake, growth and photosynthesis of rice exposed to low-level cadmium. *Plant Soil* 311: 73–86.

Pal-Singh H, Mahajan P, Kaur S, Batish DR, Kohli RK. (2013). Chromium toxicity and tolerance in plants. *Environ Chem Lett* 11: 229–254.

Pavlovic J, Samardzic J, Maksimovic V, Timotijevic G, Stevic N, Laursen KH, Hansen TH, Husted S, Schjoerring JK, Liang Y, Nikolic M. (2013). Silicon alleviates iron deficiency in cucumber by promoting mobilization of iron in the root apoplast. *New Phytol* 198: 1096–1107.

Pragabar S, Hodson MJ, Evans DE. (2011). Silicon amelioration of aluminium toxicity and cell death in suspension cultures of Norway spruce (*Picea abies* (L.) Karst.). *Environ Exp Bot* 70: 266–276.

Reynolds OL, Keeping MG, Meyer JH. (2009). Silicon-augmented resistance of plants to herbivorous insects: A review. *Ann Appl Biol* 155: 171–186.

Rizwan M, Meunier JD, Miche H, Keller C. (2012). Effect of silicon on reducing cadmium toxicity in durum wheat (*Triticum turgidum* L. cv. Claudio W.) grown in a soil with aged contamination. *J Hazard Mater* 209–210: 326–334.

Rogalla H, Römheld V. (2002). Role of apoplast in silicon-mediated manganese tolerance of *Cucumis sativus* L. *Plant Cell Environ* 25: 549–555.

Sanglard LMVP, Martins SCV, Detmann KC, Silva PEM, Lavinsky AO, Silva MM, Detmann E, Araújo WL, DaMatta FM. (2014). Silicon nutrition alleviates the negative impacts of arsenic on the photosynthetic apparatus of rice leaves: An analysis of the key limitations of photosynthesis. *Physiol Plant* 152: 355–366.

Seregin IV, Ivanov VB. (2001). Physiological aspects of cadmium and lead toxic effects on higher plants. *Russ J Plant Physiol* 48: 523–544.

Shen XF, Xiao XM, Dong ZX, Chen Y. (2014). Silicon effects on antioxidative enzymes and lipid peroxidation in leaves and roots of peanut under aluminum stress. *Acta Physiol Plant* 36: 3063–3069.

Shenker M, Plessner OE, Tel-Or E. (2004). Manganese nutrition effects on tomato growth, chlorophyll concentration, and superoxide dismutase activity. *J Plant Physiol* 161: 197–202.

Shi G, Cai Q, Liu C, Wu L. (2010). Silicon alleviates cadmium toxicity in peanut plants in relation to cadmium distribution and stimulation of antioxidative enzymes. *Plant Growth Regul* 61: 45–52.

Shi QH, Bao ZY, Zhu ZJ, He Y, Qian QQ, Yu JQ. (2005a). Silicon-mediated alleviation of Mn toxicity in *Cucumis sativus* in relation to activities of superoxide dismutase and ascorbate peroxidase. *Phytochemistry* 66: 1551–1559.

Shi XH, Zhang CC, Wang H, Zhang FS. (2005b). Effect of Si on the distribution of Cd in rice seedlings. *Plant Soil* 272: 53–60.

Singh VP, Tripathi DK, Kumar D, Chauhan DK. (2011). Influence of exogenous silicon addition on aluminium tolerance in rice seedlings. *Biol Trace Elem Res* 144: 1260–1274.

Song A, Li P, Li P, Fan F, Nikolic M, Liang Y. (2011). The alleviation of zinc toxicity by silicon is related to zinc transport and antioxidative reaction in rice. *Plant Soil* 344: 319–333.

Song A, Li Z, Zhang J, Xue G, Fan F, Liang Y. (2009). Silicon-enhanced resistance to cadmium toxicity in *Brassica chinensis* L. is attributed to Si-supressed cadmium uptake and transport and Si-enhanced antioxidant defense capacity. *J Hazard Mater* 172: 74–83.

Treder W, Cieslinski G. (2005). Effect of silicon application on cadmium uptake and distribution in strawberry plants grown on contaminated soils. *J Plant Nutr* 28: 917–929.

Tripathi DK, Singh VP, Kumar D, Chauhan DK. (2012a). Impact of exogenous silicon addition on chromium uptake, growth, mineral elements, oxidative stress, antioxidant capacity, and leaf and root structures in rice seedlings exposed to hexavalent chromium. *Acta Physiol Plant* 34: 279–289.

Tripathi DK, Singh VP, Kumar D, Chauhan DK. (2012b). Rice seedlings under cadmium stress: Effect of silicon on growth, cadmium uptake, oxidative stress, antioxidant capacity and root and leaf structures. *Chem Ecol* 28: 281–291.

Tripathi DK, Singh VP, Prasad SM, Chauhan DK, Dubey NK, Rai AK. (2015). Silicon-mediated alleviation of Cr(VI) in wheat seedlings as evidenced by chlorophyll fluorescence, laser induced breakdown spectroscopy and anatomical changes. *Ecotoxicol Environ Saf* 113: 133–144.

Tripathi P, Tripathi RD, Singh RP, Dwivedi S, Goutam D, Shri M, Trivedi PK, Chakrabarty D. (2013). Silicon mediates arsenic tolerance in rice (*Oryza sativa* L.) through lowering of arsenic uptake and improved antioxidant defence system. *Ecol Eng* 52: 96–103.

Tschan M, Robinson B, Nodari M, Schulin R. (2009a). Antimony uptake by different plant species from nutrient solution, agar and soil. *Environ Chem* 6: 144–152.

Tschan M, Robinson B, Schulin R. (2009b). Antimony in the soil-plant system—A review. *Environ Chem* 6: 106–115.

Vaculík M, Jurkovič L, Matejkovič P, Molnárová M, Lux A. (2013). Potential risk of arsenic and antimony accumulation by medicinal plants naturally growing on old mining sites. *Water Air Soil Pollut* 224: 1546.

Vaculík M, Landberg T, Greger M, Luxová M, Stoláriková M, Lux A. (2012). Silicon modifies root anatomy, and uptake and subcellular distribution of cadmium in young maize plants. *Ann Bot* 110: 433–443.

Vaculík M, Lux A, Luxová M, Tanimoto E, Lichtscheidl I. (2009). Silicon mitigates cadmium inhibitory effects in young maize plants. *Environ Exp Bot* 67: 52–58.

Vaculík M, Mrázová A, Lux A. (2015). Antimony (SbIII) reduces growth, declines photosynthesis and modifies leaf tissue anatomy in sunflower (*Helianthus annuus* L.). *Environ Sci Pollut Res* 22: 18699–18706.

Vaculíková M, Vaculík M, Šimková L, Fialová I, Kochanová Z, Sedláková B, Luxová M. (2014). Influence of silicon on maize roots exposed to antimony—Growth and antioxidative response. *Plant Physiol Biochem* 83: 279–284.

Vaculíková M, Vaculík M, Tandy S, Luxová M, Schulin R. (2016). Alleviation of antimonate (SbV) toxicity in maize by silicon (Si). *Environ Exp Bot* 128: 11–17.

Vatehová Z, Kollárová K, Zelko I, Richterová-Kučerová D, Bujdoš M, Lišková D. (2012). Interaction of silicon and cadmium in *Brassica juncea* and *Brassica napus*. *Biologia* 67: 498–504.

Wang LJ, Wang YH, Chen Q, Cao WD, Li M, Zhang FS. (2000). Silicon induced cadmium tolerance of rice seedlings. *J Plant Nutr* 23: 1397–1406.

Wang YX, Stass A, Horst WJ. (2004). Apoplastic binding of aluminium is involved in silicon-induced amelioration of aluminium toxicity in maize. *Plant Physiol* 136: 3762–3770.

Williams DE, Vlamis J. (1957a). Manganese toxicity in standard culture solutions. *Plant Soil* 8: 183–193.

Williams DE, Vlamis J. (1957b). The effect of silicon on yield and manganese-54 uptake and distribution in leaves of barley plants grown in culture solutions. *Plant Physiol* 32: 404–409.

Wu JW, Shi Y, Zhu YX, Wang YC, Gong HJ. (2013). Mechanisms of enhanced heavy metal tolerance in plants by silicon: A review. *Pedosphere* 23: 815–825.

Ye J, Yan Ch, Liu J, Lu H, Liu T, Song Z. (2012). Effects of silicon on the distribution of cadmium compartmentation in root tips of *Kandelia obovata* (S., L.) Young. *Environ Pollut* 162: 369–373.

Zanao Junior LA, Ferreira Fontes RL, Lima Neves JC, Körndorfer GH, de Ávila VT. (2010). Rice grown in nutrient solution with doses of manganese and silicon. *Rev Bras Cienc Solo* 34: 1629–1639.

Zeng FR, Zhao FS, Qiu BY, Ouyang YN, Wu FB, Zhang GP. (2011). Alleviation of chromium toxicity by silicon addition in rice plants. *Agric Sci China* 10: 1188–1196.

Zhang F. (2008). Long-term effects of exogenous silicon on cadmium translocation and toxicity in rice (*Oryza sativa* L.). *Environ Exp Bot* 62: 300–307.

Zhang Q, Liu J, Lu H, Zhao S, Wang W, Du J, Yan Ch. (2015). Effects of silicon on growth, root anatomy, radial oxygen loss (ROL) and Fe/Mn plaque of *Agiceras cornicultaum* (L.) Blanco seedlings exposed to cadmium. *Environ Nanotechnol Monitor Manag* 4: 6–11.

Zhang Q, Yan Ch, Liu J, Lu H, Wang W, Du J, Duan H. (2013). Silicon alleviates cadmium toxicity in *Avicennia marina* (Forsk.) Vierh. seedlings in relation to root anatomy and radial oxygen loss. *Marine Pollut Bull* 76: 187–193.

11 Silicon Uptake and Translocation in Plants
Recent Advances and Future Perspectives

Anuradha Patel, Jitendra Kumar, Madhulika Singh, Vijay Pratap Singh, and Sheo Mohan Prasad

CONTENTS

ABSTRACT

Silicon (Si) plays a crucial role during growth and developmental processes of plants. It retains in the plant tissues and shows constructive effects by regulating the negative impacts of numerous abiotic and biotic stresses. Si abundance in the earth's crust is a huge matter of concern for the scientific world and is sometimes remarked as a stress alleviator. At the same time, frequent application of Si-based fertilizers is reported to increase yield, growth, and quality of crops. In this chapter, an attempt has been made to provide insight and knowledge on distribution pattern of Si in soil, uptake of Si, its translocation at different parts of the plant, and consequences at different stages of growth with an emphasis on future Si research.

11.1 INTRODUCTION

After evaluating composition of the earth's crust, it was found that silicon (Si) is the second most copious element to oxygen. Soils have a Si range between 100 and 500 µM in monosilicic acid (H_4SiO_4) form (Sommer et al., 2006) and in silicon dioxide that covers about 50%–70% of the soil mass (Ma et al., 2011). Moreover, in soil after silicate minerals experience chemical and physical weathering, which further decide the fate of Si. It can form clay minerals by mixing with numerous elements present in the earth's crust or utilized by crops, or it can be easily lost by runoff and join streams, rivers, and oceans. Although Si is abundant, however, it is never found independently and has a great affinity toward other elements of the crust, and forming oxides or silicates (Ma, 2009). At the same time Si is considered as a nonessential element, it plays a beneficial role in growth and developmental processes of plants (Tripathi et al., 2013a). Silicon usually enters plants in a neutralized silicic acid state and is eventually irrevocably discharged throughout the plant as amorphous silica. Plants retain Si in their tissues and show constructive effects by ameliorating the negative impacts caused by different biotic as well as abiotic stresses, including diseases caused by numerous pathogens (Ma, 2009; Tripathi et al., 2015a). Furthermore, evidence shows that most silicon resources are insoluble, and therefore easily available to plant uptake (Richmond and Sussman, 2003). The proficient roles of silicon against numerous stresses have been well documented (Epstein, 1999; Richmond and Sussman, 2003; Singh et al., 2011; Tripathi et al., 2012a,b, 2015a,b, 2016). From ongoing research, amazing facts about silicon have been illuminated. However, frequent use of synthetic fertilizers based on nitrogen, phosphorus, and potassium, and repeated cropping practices easily demolish the Si content in soil to an extent that makes it unable for plants to their use. In this way, in Si-deficient land, a new type of threat may arise for Si-accumulating plants such as rice and sugarcane (Ma and Yamaji, 2006). Although silicon is widely dispersed in the environment, its deficiency may contribute to a new kind of stress and provide a new direction for researchers. In addition, Si is also recognized as a stress alleviator. Numerous of biotic and abiotic stresses, consisting of diseases caused by pests and other pathogens, gravity, UV-B radiation, and metal stresses, can be successfully ameliorated by the application of Si. Silicon performs its function by polymerizing silicic acid, and after polymerization, it forms dense amorphous hydrated silica, which plays a pivotal role in the arrangement of organic defense compounds. In this chapter, we provide insight and knowledge on the distribution pattern of silicon in soil, uptake of Si, its translocation in different parts of plants, and consequences.

11.2 SILICON CONTENT IN SOIL

Silicon is introduced into a soil environment via the action of different natural events, which include the weathering of mineral rocks, runoff, and capillary ascension from the water table (Savant et al., 1999). Studies show that, on the basis of weight, soil contains about 32% Si (Lindsay, 1979), and percentages may reach up to 50%–70% due to various anthropogenic activities (Ma et al., 2011). The concentration of Si in soil can be easily determined with the difference between Si leaching and its uptake by plants (Kittrick, 1969). Although the occurrence of Si in soil is very high, however, its essentiality as a micronutrient for higher plants is very difficult to prove. Si availability in soil is efficiently regulated by chemical kinetics, rather than by thermodynamics (Hallmark et al., 1982).

11.3 SILICON CONTENT IN PLANTS

The concentration of Si in plants is so heterogeneous, on the basis of dry weight, that it covers approximately 1–100 g Si kg^{-1}, and plants often absorb it in the range between 50 and 200 kg Si ha^{-1} (Matichenkov et al., 1997). Usually plants acquire Si from the soil in suspension form as monosilicic acid, sometimes known as orthosilicic acid (H_4SiO_4). Depending on the concentration of Si in the shoot, the plants may be classified in three types are as follows

- Si accumulator: The plant contains more than 1.0% Si, for instance, species of Gramineae and Cyperaceae.
- Intermediate type: The plant contains between 0.5% and 1.0% Si.
- Si excluder: The plant contains less than 0.5% Si (Takahashi et al., 1990).

The solubility of Si in the soil is totally dependent on ionization of monosilicic acid, which is regulated by standard pH limits in the range of 2 and 8.5 (Savant et al., 1999). The ionization of monosilicic acid is demonstrated below:

$$Si(OH)_4 + OH \rightarrow Si(OH)_3O^- + H_2O$$

$$H_4SiO_4 + OH \rightarrow H_3SiO_4 + H_2O$$

After assaying silicon concentration in plants, its uptake and distribution pattern can be easily understood. At the same time, there are also some factors, like fluctuations in humidity and disturbance of the wetting–drying regime in the soil, that seriously hinder the uptake of silicon.

11.3.1 ROLE OF SILICON IN PLANTS

Although silicon is not recognized as an essential element, however, it has advantageous performance in the plant development process. To date, research has shown that exogenous application of Si successfully ameliorates negative consequences of different biotic and abiotic stresses (Epstein and Bloom, 2005; Singh et al., 2011; Tripathi et al., 2012a,b, 2015a,b, 2016). At the same time, Si also stimulates yield of crops by providing resistance to plants against diseases caused by numerous pathogens and bacteria. The deposition of silica confers rigidity and strength to culms and erectness to leaves, thereby facilitating nonlodging and enhancing interception of sunlight and photosynthesis. Evidence shows that Si supplementation enhances photoincorporation of carbon, and it also stimulates incorporated carbon in the rice panicle (Takahashi and Miyake, 1982; Meena et al., 2014). Moreover, in rice leaves cuticle layer, which becomes highly thick due to the deposition of silica, makes it difficult for pests and other pathogenic fungi to penetrate, suck, chew, and feed. Similarly, silicon nutrition may also reduce the transpiration rate in plants and improve the water economy (Gao et al., 2004), chaffiness, and shattering of grain, and therefore enhance the grain yield and productivity (Jones and Handreck, 1967). The cuticular layer of epidermal cells of the leaf contains stomata, which are responsible for the gaseous exchange between the atmosphere and the leaf. There are bulliform cells, which are arranged in a fan shape on both sides of the leaf veins. Under dry weather conditions, excessive water loss through transpiration causes bulliform cells to shrink and the leaf to roll up in order to lessen transpiration loss. When the atmosphere is well laden with moisture, the bulliform cells absorb moisture and expand. The leaf then reopens. The cuticle on the leaf surface plays an important role in reducing transpiration loss and also prevents pest and disease attacks. The accumulation of silicon forms a bulky layer of silica on the leaf surface, and it efficiently controls the cuticular transpiration (Meena et al., 2014). Studies have shown that after Si application, the transpiration rate of plants can be obstructed by around 20%–30% (Kudinova, 1974).

11.4 SILICON UPTAKE AND ITS TRANSLOCATION

The accumulation of Si in plants is efficiently regulated by the capability of roots to take it up. Si is mostly present in the range between 0.1 and 0.6 mM (Epstein, 1994) as neutralized monomeric orthosilicic acid: H_4SiO_4 (Ma et al., 2002; Casey et al., 2004). However, the accumulation and translocation of Si between plants show a heterogeneous response; sometimes it adopt to a passive process that is regulated by transpiration (Richmond and Sussman, 2003; Ma and Yamaji, 2015), which occurs via the xylem and water (mass flow), and other times it adopt to an active process that includes transporters located in the plasma membrane of root cells (Takahashi, 1996; Rodrigues et al., 2011). The translocation of silicic acid initiates from the root, towards the shoot, and through the xylem, with loss of water from transpiration, and then it is polymerized into silica (SiO_2) and ultimately retained in different tissues. Uptake and translocation occur using monosilicic acid (Barber and Shone, 1966), and deposition occurs as amorphous silica gel (SiO_nH_nO). Silicon uptake is most likely a typical process and gets handled via the action of multiple genes. In this way, the translocation of Si from the external soil solution to the roots, and then to different tissues, requires different transporter genes.

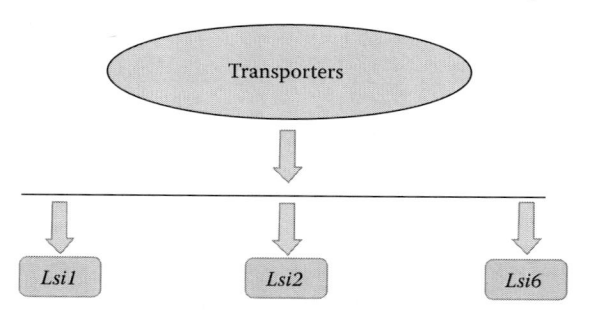

11.4.1 INFLUX TRANSPORTERS

Si influx transporters are coded by two genes (*Lsi1* and *Lsi2*), and these two genes have been acknowledged in rice roots (Ma et al., 2006, 2007). These two sets of genes (*Lsi1* and *Lsi2*) play a pivotal role during the translocation of Si from the epidermis toward steles of the root, and subsequently shift toward the shoot by transpirational water flow via the xylem, where it is then further polymerized and retained in plant tissues as silica (Ma et al., 2006). Along with these transporter genes, another gene, *Lsi6*, is also involved in Si distribution, which was earlier reported in rice shoots (Yamaji and Ma, 2007). *Lsi6* is a homolog of *Lsi1* and also has outstanding transportation potential for silicic acid, though it is different from *Lsi1* and *Lsi2* (Ma, 2009). *Lsi6* is extrinsically found at the root edges, and the adaxial side of the xylem parenchyma in the leaf sheath and leaf blades (Yamaji et al., 2008; Ma, 2009). More than 90% of Si that is taken up by the root further moves toward the shoots (Ma and Takahashi, 2002). Mainly, the monosilicic acid form of Si is gathered in the xylem (Casey et al., 2004; Mitani et al., 2005). These influx transporter genes are isolated by adopting a mutant approach; the *Lsi1* gene is isolated by following the method based on map cloning, causing many facts to come to light. For example, *Lsi1* belongs to a Nod26-like major intrinsic protein (NIP), which is a subfamily of aquaporin-like proteins (Ma, 2009). Moreover, the translocation potential of these transporter genes is not affected much by temperature fluctuation (Mitani et al., 2008). Higher expression of the *Lsi1* gene was found in the root region, making it more vulnerable for metal uptake; however, application of Si suppresses the intake of metal by plants (Ma et al., 2008). In addition, a study done by Yamaji and Ma (2007) showed that the expression of the *Lsi1* gene is maximally found at the heading stage in the rice plant. Another study done regarding the expression of *Lsi1* at different growth stages, it was revealed that 67% of total Si was taken up during the reproductive stage (Ma et al., 1989). The Si starvation during this growth stage drastically reduced the crop productivity (Ma, 2009); therefore, in order to achieve high crop productivity, it is necessary to add a surplus amount of Si on some lands. A study showed that higher expression of the *Lsi1* gene could be achieved via supplementation of Si (Ma et al., 2006). Usually, the existence of the *Lsi1* gene can be seen at both the exodermis and endodermis, along with the Casparian strips of the cell. Altogether, *Lsi1* is accountable for the translocation of silicic acid between the external solution and root cells. Moreover, several other transporter genes, such as *ZmLsi1* and *HvLsi1*, have also been introduced in maize and barley seedlings, respectively, that have great potential for the movement of Si in plants (Chiba et al., 2009; Mitani et al., 2009). These transporter genes basically exist at the epidermal, hypodermal, and cortical cells. *HvLsi1* and *ZmLsi1* also exhibited great potential for Si influx activity, like *Lsi1*, but their configuration and existence patterns are distinct from that of *Lsi1* (Ma, 2009).

11.4.2 EFFLUX TRANSPORTERS

The gene encoding the efflux transporter of Si was also cloned by using a novel rice mutant *Lsi2* (Ma et al., 2007). The *Lsi2* gene is recognized as a membrane protein with 11 transmembrane domains, which belong to an anion transporter family and are efficient for the movement of silicon from outside of the cells (Ma, 2009). This influx and efflux of silicon is necessary for the translocation and distribution of silicon. The expression of the *Lsi2* gene maximally responds at a lower pH in an external medium (Ma et al., 2007). The silicon translocation by *Lsi2* is a force-dependent active process, and this energy is obtained by the fluctuation of the proton gradient. In addition, the *Lsi2* gene is mainly found in the roots, as in the case of *Lsi1* (Ma, 2009). However, the expression of these transporter genes is inhibited by temperature fluctuations, as well as three protonophores inhibitors: 2,4-dinitrophenol (DNP), carbonyl cyanide 3-chlorophenylhydrazone (CCCP), and carbonyl cyanide p-(trifluoromethoxy) phenyl hydra zone (FCCP) (Okuda and Takahashi, 1962; Ma, 2009).

11.4.3 MECHANISMS INVOLVED IN COUPLING OF BOTH GENES (*LSI1* AND *LSI2*)

Both transporter genes (*Lsi1* and *Lsi2*) are familiar with exodermis and endodermis, along with the Casparian strips of the cell (Figure 11.1). The Casparian strips block the passage of solutes and prevent them from moving independently between the soil solution and stele. Moreover, in plants that possess aerenchyma in their roots, demolition of all cortical cells occurs, except the exodermis and endodermis of the cells (Ma, 2009). Therefore, Si initiates traveling from the exodermis cells by *Lsi1* and is released into the apoplast by *Lsi2*. Further, Si moves toward the endodermis cells by *Lsi1* and is released into the stele by *Lsi2* (Ma and Yamaji, 2008) (Figure 11.1). Therefore, coupling of *Lsi1* and *Lsi2* is essential for the efficient transport of Si across cells into the stele. Silicon translocation is affected by several factors, such as temperature, pH, and metabolic inhibitors (Ma, 2009). A study done by Okuda and Takahashi (1962) characterized NaCN (sodium cyanide) and DNP (2,4-dinitrophenol) as restricting agents for Si uptake in rice plants (Figure 11.2).

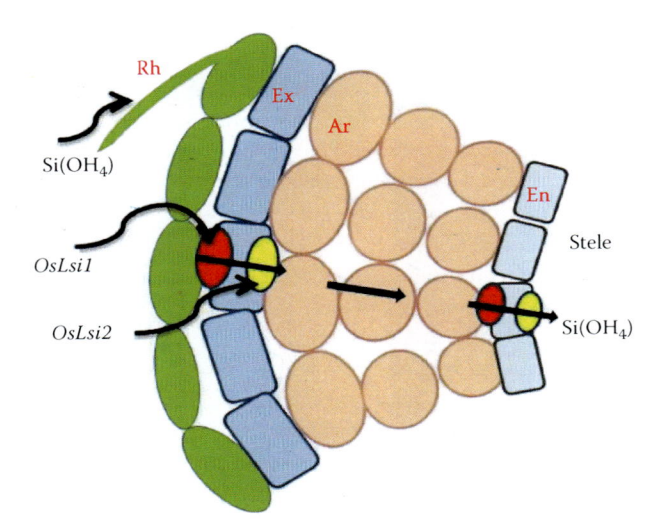

FIGURE 11.1 Schematic representation of Si uptake system in rice roots through transporters. Si is absorbed in the form of silicic acid ($Si(OH)_4$) through exodermal cells of the root by help of the transporter, i.e., *OsLsi1*, and released from exodermal cells to the apoplast, by transporter, *OsLsi2*, and then transported into the stele by both *OsLsi1* and *OsLsi2*, again at the endodermal cells. Ar, aerenchyma; En, endodermis; Ex, exodermis; Rh, root hair. (Modified from Ma JF et al., *Proc Jpn Acad Ser B* 87: 377–385, 2011.)

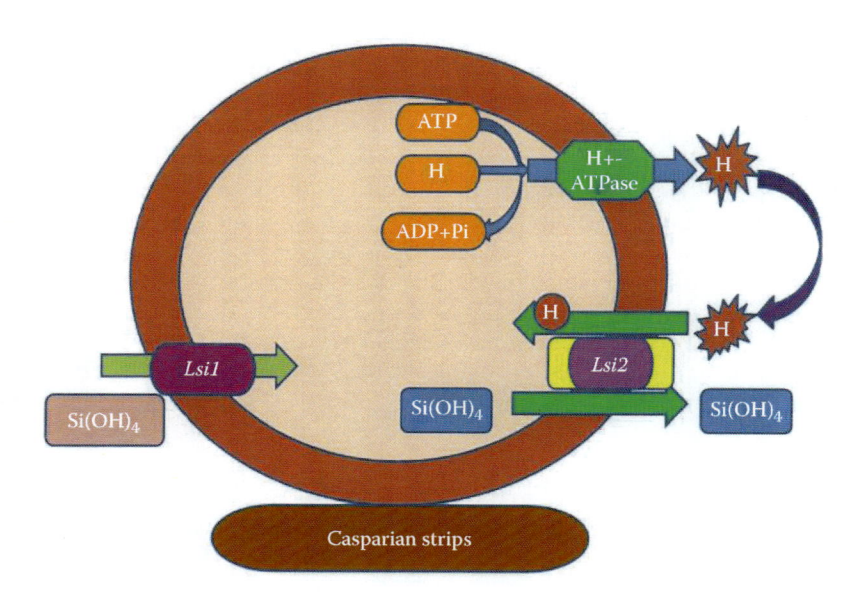

FIGURE 11.2   Schematic representation of genes involved in translocation of Si in rice plants. The existence of *Lsi* genes at the exodermis and endodermis, along with Casparian strips, accelerates the uptake of Si within the cell.

11.5   IMPACT OF SILICON ON GROWTH OF PLANTS

Due to the ever-increasing pollution and contamination of the environment, yield and productivity of crops are being affected more and more. These declines in crop productivity are associated with improper nutrition. Nutrition in plants is achieved by the application of numerous chemical fertilizers. Among different fertilizers used in crop fields, silicon proves to be beneficial in enhancing crop productivity (Marschner, 1995; Marschner and Hengge 1997; Iwasaki et al., 2002). Evidence has shown that the routine application of fertilizers based on Si stimulates the productivity of rice and sugarcane (Chérif et al., 1992; Lux et al., 2002). Common sources of Si fertilizer include blast furnace slag (calcium silicate and magnesium silicate), silicomanganese slag (calcium silicate and manganese silicate), fused magnesium phosphate, calcium silicate, and potassium silicate. This beneficial effect of Si is attributed to its nature of deposition in various tissues, thereby producing a physical blockade that stimulates the potency and rigidity of tissues (Yoshida et al., 1976; Ma and Yamaji, 2006; Chauhan et al., 2011; Tripathi et al., 2012a,b, 2013b, 2014). Si also has a role in the hull formation of rice, and therefore improves the grain quality (Savant et al., 1997). In addition to this, plants growing in silicon-deficient soil are prone to diseases like blast and brown spots, which are becoming more widespread (Bélanger, 2003; Rodrigues et al., 2004; Rémus-Borel et al., 2005). Further, in Si-deficient soil, the rice leaves become soft and droopy, causing mutual shading and finally moving towards an inefficient photosynthetic rate (Ma et al., 2002). According to an International Rice Research Institute report, under silicon-deprived conditions, the yield of panicles, as well as grains, significantly decreases, and at the same time, such land types are also susceptible to lodging (Epstein et al., 1999; Ma and Takahashi, 2002; Tripathi et al., 2013a). In contrast to this, through rational application of silicon fertilizer, the negative consequences of nitrogen fertilizer, such as flaccidity of the leaves, can be rectified. A study showed that exogenous supplementation of Si fertilizers significantly enhanced the dry mass of barley by 21% and 54% over 20 and 30 days of growth, respectively (Kudinova, 1974). The poor-quality hull and milky white grains (kernels) are associated with a deficient level of silicon, which is highly correlated with the silicon concentration in the rice straw (Aleshin et al., 1978; Meena et al., 2014). An experimental study

TABLE 11.1
Silicon Fertilizers Used to Supplement the Yield in Different Crops

Plants	Si Fertilizers	References
Rice	Calcium silicate, Ca-Mg silicate, sodium metasilicate, silicate slag, rice hull, rice hull ash, rice straw, Poha industry waste, lignite fly ash, coal fly ash	Rodrigues et al. (2011)
Sugarcane	Calcium silicate, calcium metasilicate, Ca-Mg silicate, silicate slag, electric furnace slag, steel slag, Tennessee Valley Authority (TVA) slag, bagasse furnace ash, basalt, cement	Savant et al. (1999)
Maize	Coal fly ash	Moussa et al. (2006)
Soybean	Poha industry waste	Rodrigues et al. (2011)
Potato	Lignosilicon	Liang et al. (2005a)
Mustard	Coal fly ash	Rodrigues et al. (2011)
Cucumber	Sodium metasilicate	Samuels et al. (1991)

reported that barley grains that were harvested from a silicon-enriched land had superior capability for germination compared to other seeds collected from silicon-deficient land (Matichenkov, 1990). Matichenkov et al. (2000) showed that nutrition with a silicon-based fertilizer accelerated citrus growth by 30%–80%. The study thus revealed that calcium silicate, a discharge material of steel and the phosphorus-producing industry, may provide a good source for silicon fertilizers (Gascho et al., 2001). Another Si-containing fertilizer is the potassium silicate that is hydrophilic in nature, and therefore can be used in hydroponic culture. Several other options for Si-based fertilizers, such as calcium silicate hydrate, silica gel, and thermophosphate, have now been widely accepted at the commercial level (Gascho et al., 2001). Table 11.1 shows the various sources of silicon fertilizers that actively participate in enriching growth and development of plants.

11.5.1 Role of Silicon against Environmental Stress

The productivity of crop plants is severely troubled by numerous biotic and abiotic stresses, such as pests and plant diseases, salinity, drought, cold, UV, and heavy metals, raising the question of the survivability of plants (Parihar et al., 2014; Kumar et al., 2015) (Table 11.2). Si has many roles such as upgrading of the nutrient imbalance, diminution in mineral toxicities, and stimulation of resistance to different stresses (Ma, 2004; Ma and Yamaji, 2006). The advantageous role of Si is more evident against different biotic and abiotic stress conditions, as it efficiently protects plants by modifying cell wall properties (Figure 11.3).

11.5.2 Role of Silicon against Heavy Metal Toxicity

Nowadays, increasing pollution and exposure to heavy metals have become major limiting factors for plant growth and productivity. Heavy metals such as lead, arsenic, selenium, cadmium, copper, zinc, uranium, mercury, and nickel have been found to be responsible for posing severe health risks, as it is easier for them to enter into the food chain at high levels (Kim et al., 2014). Heavy metals are well known to cause detrimental effects on plant growth by arresting the proper cellular tasks of proteins and lipids, as well as of the rudimentary components of thylakoid membranes, which have resulted in major physiological disorders, including reduced biomass production, photosynthesis inhibition, and troubling nutrient uptake (Ali et al., 2012, 2013; Guntzer et al., 2012). Numerous studies have shown that through the application of Si, the concentration of heavy metals may be successfully minimized in plants (Liang et al., 2005a,b, 2007). Silicon interacts with metals such as Al, As, and Cd, and ameliorates their effects within the plant through stimulation of the antioxidant system, metal ion chelation, immobilization of heavy

TABLE 11.2

Summary of Si-Mediated Abiotic Stress Tolerance in Plants

Plants	Silicon-Enhanced Tolerance against Abiotic Stress		Work	Reference
Triticum turgidum, Oryza sativa	Heavy metal stress	Copper	Alleviates Cu toxicity; regulates P-type heavy metal ATPases, reduced uptake of metals in the roots; modulates the signaling of phytohormones involved in responses to stress and host defense, such as abscisic acid, jasmonic acid, and salicylic acid	Kim et al. (2014), Keller et al. (2015)
Avicennia marina, Amaranthus hypochondriacus, Amaranthus tricolor, Brassica oleracea, Oryza sativa, Zea mays		Cadmium	Suppresses Cd uptake from soil solution by increasing the relative dissolved concentrations of competing cations and changes in root morphology	Zhang et al. (2014), Lu et al. (2014), Kim et al. (2014), Malčovská et al. (2014a)
Oryza sativa		Arsenic	Reduces As accumulation	Guo et al. (2007), Li et al. (2009)
Oryza sativa		Chromium	Alleviates Cr toxicity mainly through inhibiting the uptake and translocation of Cr and enhancing the capacity of defense against oxidative stress induced by Cr toxicity	Zeng et al. (2011)
Aloe vera, Vicia faba, Solanum lycopersicum	Salt stress		Improves activity of plasma membrane ATPase, tonoplast membrane ATPase, and H^+-pyrophosphates of aloe root; significantly increases the contents of P and K, and the K/Na ratio; decreases Na ion contents and also increases the antioxidant system	Xu et al. (2015), Qados and Moftah (2015), Li et al. (2015)
Triticum aestivum	Drought		Alleviates oxidative damage and increases growth	Gong et al. (2005)
Triticum aestivum, Zea mays	UV-B		Improves the tolerance by increasing antioxidant compounds' content and also UV-B screening compounds	Yao et al. (2011), Malčovská et al. (2014b)

metals during plant growth, and compartmentalization of heavy metals in vacuoles, cytoplasm, and cell wall (Liang et al., 2007). Amelioration against heavy metal toxicity can be achieved by any of the following methods:

- Si inhibits heavy metal translocation to the shoots, as it is trapped in the roots.
- Si reduces plasma membrane lipid peroxidation (Shi et al., 2005).
- Si forms a barrier to metal ion in apoplastic and transpirational flows (Lux et al., 2002) by accumulating SiO_2 in the root apoplasm and leaf surface.
- The metals are compartmentalized by organic acids in vacuoles, and by coprecipitation of Si–metal complexes in the cell wall.
- Si–polyphenol complexes are formed in tissues (Maksimovic et al., 2007).

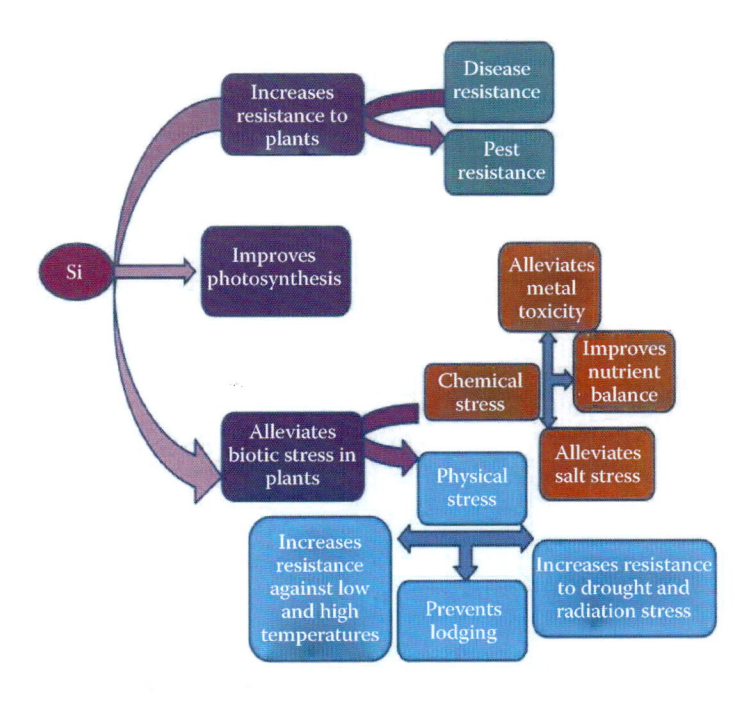

FIGURE 11.3 Role of Si in rice plants under different abiotic and biotic stress conditions.

Similarly, Si also plays a pivotal role against mineral stresses, including manganese and aluminum toxicity, as well as phosphorus deficiency (Ma and Takahashi, 1990; Ma et al., 2002). The soluble form of silicon derivatives, including monosilicic acid and polysilicic acid, efficiently modify the physicochemical properties of the soil. Further, due to a higher affinity of monosilicic acid to different metals (Lindsay, 1979), it chemically reacts with aluminum, iron, and manganese and can form slightly soluble silicates (Horiguchi, 1988; Lumsdon and Farmer, 1995). The following reactions show how metals react with silicon:

$$Al_2Si_2O_5 + 2H + 3H_2O = 2Al^3 + 2H_4SiO_4$$

$$Al_2Si_2O_5(OH)_4 + 6H = 2Al^3 + 2H_4SiO_4 + H_2O$$

$$Fe_2SiO_4 + 4H^+ = 2Fe^2 + 2H_4SiO_4$$

$$MnSiO_3 + 2H^+ + H_2O = Mn^{2+} + H_4SiO_4,$$

$$Mn_2SiO_4 + 4H^+ = 2Mn^{2+} + H_4SiO_4,$$

The least concentration of monosilicic acid has great potential to react with heavy metals in soil and form complex soluble compounds that are easily taken up by plants (Schindler et al., 1976). However, when reacting with heavy metals, an excess amount of monosilicic acid forms less soluble complex compounds that are not easily available to the plant (Lindsay, 1979; Cherepanov et al., 1994):

$$ZnSiO_4 \ 4H \ 2Zn^{2+} \ H_4SiO_4, \ \log Ko \ 13.15$$

$$PbSiO_4 \ 4H \ 2Pb^{2+} \ H_4SiO_4, \ \log Ko \ 18.45$$

11.5.3 ROLE OF SILICON AGAINST SALINITY STRESS

Salinity has become a major crop-limiting factor and is projected to worsen by the year 2050; as expected, >50% of the arable land will suffer from severe salinity problems (Vinocur and Altman 2005; Kumar et al., 2015). Salinity-induced damage to plasma membranes is associated with the accretion of a surplus amount of Na and Cl ions in plant tissues (Liu et al., 1987; Singha and Choudhuri, 1990; Hernandez et al., 1993). Further, these ions may cause severe oxidative stress in cells, caused by a disturbance in the balance between various toxic oxygen derivatives (oxidants) and the antioxidants (Liu et al., 1987; Singha and Choudhuri, 1990; Hernandez et al., 1993). However, plants have numerous efficient antioxidant enzymes to scavenge these oxygen derivatives for their survivability. Studies have shown that under hypersaline conditions, antioxidant enzymes such as superoxide dismutase (SOD) and catalase (CAT) decrease noticeably in plants (Singha and Choudhuri, 1990; Hernandez et al., 1993), which in turn stimulates lipid peroxidation via overproduction of ROS (Hernandez et al., 1993; Lutts et al., 1996; Fadzilla et al., 1997) which cause damage to the lipids and membranes. In this way, evidence shows that Si plays a crucial role in barley (Liang et al., 1996, 2003), cucumber (Adatia and Besford, 1986; Zhu et al., 2004), maize (Moussa et al., 2006), tomato (Romero-Arnada et al., 2006), and wheat (Ahmad et al., 1992; Tuna et al., 2008) in mitigating salinity stress.

The mechanisms through which silicon increases tolerance in plants are as follows:

- Si immobilizes toxic sodium ion, which reduces the uptake of sodium into the plant and stimulates potassium uptake (Hasegawa et al., 2000; Liang et al., 2003, 2005c). Si decreases the membrane permeability of cells (Liang et al., 1996), and considerably improves the configuration of chloroplasts. Si also stimulates the SOD activity in leaves and H^+-ATPase in the membranes, therefore suppressing the products of lipid peroxides induced by salt stress.
- Si presents itself as a physical barrier and activates some defense mechanisms of plants. Studies have revealed that exogenous application of Si astonishingly increases yields of rice and spring wheat (Rafi and Epstein, 1999), sugarcane (Savant et al., 1999), and barley (Liang et al., 2005c). In addition, Liang et al. (2007) also showed that Si nutrition efficiently enhances the productivity of tomato plants against salt stress.

11.5.4 ROLE OF SILICON AGAINST DROUGHT STRESS

Drought has become a common concern among environmentalists, as it restricts plant growth and productivity (Said-Al Ahl et al., 2009; Devkota and Jha, 2011; Singh et al., 2015). Generally, drought is defined as an area with deficient rainfall and a higher evapotranspiration rate, resulting in a water crisis for plants (Ahmad et al., 2013). To overcome drought stress, plants acclimatize to drought or arid circumstances by adjusting their physiomorphic features. In this way, plants have efficient antioxidants that play a key role against drought stress. The study done by Schmidt et al. (1999) showed enhanced SOD activity in bentgrass against drought conditions after exogenous supply of silicates during a high fertilization regime. In another study, Ma et al. (2004) showed that silicon elevated the physiological response of the peroxidase (POD), SOD, and CAT enzymes against drought stress. Under severe stress, these physiological and biochemical responses showed a positive correlation with the supplementation of silicon. Evidence has shown that under drought stress condition, Si plays a key role in the upregulation of stomatal conductance, relative water content, and water potential in wheat plants (Guntzer et al., 2012). Further, Si is responsible for larger and thicker leaves, that reduces the risk of water crisis in cells by adjusting the water consumption and transpiration loss (Gong et al., 2003; Eneji et al., 2005; Hattori et al., 2005).

11.5.5 ROLE OF SILICON AGAINST UV-B RADIATION

During the last few decades, UV-B radiation has attracted much attention due to enhanced depletion of the stratospheric ozone via numerous ozone-depleting substances, like synthetic chlorofluorocarbons (CFCs), chlorocarbons (CCs), and organobromides (OBs), that is why earth is surface experiencing extreme levels of UV-B radiation (290–320 nm) (Gao et al., 2007). UV-B radiation negatively affects plant growth by disrupting the various electron transport chains (photosynthetic and respiratory), as well as producing different reactive oxygen species (ROS) (Zancan et al., 2008; Lizana et al., 2009; Beckmann et al., 2012; Singh et al., 2014). The intense UV-B radiation pierces the plant cells and accelerates the destruction of essential macromolecules of the cell, which results in malfunctioning of photosystem II (PSII) and Rubisco (Ribulose-1,5-bisphosphate carboxylase/oxygenase) activity (Li et al., 2007; Fang et al., 2011). UV-B also affects the stomatal conductance, CO_2 fixation, and integrity of thylakoid membranes (Hollosy, 2002; Lidon et al., 2012; Zlatev et al., 2012). In contrast to this, application of Si successfully ameliorates the negative consequences of UV-B by reducing the production of ROS, and thus prevents the cell to experience the oxidative stress. Apart from this, Si plays numerous positive roles in plants against UV-B stress. Silicon application increases the total biomass and chlorophyll content and also reduces membrane damage in wheat seedlings (*Triticum aestivum* L.) exposed to enhanced UV-B stress (Yao et al., 2011). Similarly, the enhanced activity of antioxidative enzymes in soybean seedlings (*Glycine max* L.) increased the leaf area after silicon supplementation (Shen et al., 2010). A recent study by Shen et al. (2014) showed that the exogenous application of Si stimulated the stem length, photosynthesis, and chlorophyll content in soybean against enhanced levels of UV-B; however, these changes were cultivar dependent. In another study, Wu et al. (2009) showed that silicon nutrition stimulates the accumulation of phenolic compounds, which help to reduce the negative effect of UV-B on rice (*Oryza sativa* L.). Furthermore, the deposition of silica near the epidermis of the leaves may also decrease the penetration of UV radiation in cells.

11.5.6 ROLE OF SILICON AGAINST BIOTIC STRESS

Insects and pests are one of the major biological constraints that limit crop productivity. The exogenous application of Si stimulates the resistance of plants against fungi, pests, and other pathogens. Evidence shows that Si in a soluble form is more efficient than that in a solid form (silica) in plants (Samuels et al., 1991; Chérif et al., 1992). In addition, exogenous supply of Si stimulates the confrontation in rice leaf against neck blast, sheath blight, brown spot, leaf scald, and stem rot (Rodrigues et al., 2005). Similarly, Wang et al. (2004) revealed that Si decreases the incidence of powdery mildew in cucumber, barley, and wheat, as well as provides strength against numerous diseases, such as ring spot in sugarcane, rust in cowpea, leaf spot in Bermuda grass (*Cynodon dactylon*), and gray leaf spot in St. Augustine grass (*Stenotaphrum secundatum*) and perennial ryegrass.

11.6 MECHANISM OF ACTION

There are two mechanisms that have evolved in plants through which Si enhances resistance against diseases. Primarily, Si provides a barrier by gathering beneath the cuticle and forming a cuticle–Si double layer (Sangster et al., 2001). This layer causes disruption of the infection process as it mechanically obstructs penetration by fungi, and thereby soluble Si serves as a modulator for host resistance against pathogens (Ma and Yamaji, 2006). Several studies conducted with rice, wheat, and cucumber have shown that plants nourished with Si produce a significant amount of phenolics and phytoalexins against fungal infection, such as those causing rice blast and powdery mildew (Ma et al., 2001; Sahebi et al., 2015). At the same time, Si also activates some defense mechanisms in plants. In an earlier finding, Chérif et al. (1993) showed that Si stimulates the activity of chitinases, peroxidases, and polyphenoloxydases in cucumber plant against the colonization and infection caused by the *Pythium*.

TABLE 11.3
Summary of Si-Mediated Biotic Stress Tolerance in Plants

Plants	Pathogen	Diseases	References
Rice	*Pyricularia oryzae*	Rice blast	Rodrigues et al. (2004)
Wheat	*Bipolaris, Fusarium*	Grain discoloration	Hayasaka et al. (2008)
Cucumber	*Xanthomonas oryzae* pv. oryzae	Bacterial leaf blight	Fauteux et al. (2005)
Banana	*Erysiphe graminis*	Powdery mildew	Liang et al. (2005a)
Bell pepper	*Mycosphaerella pinodes*	Leaf spot	Guével et al. (2007)
	Sphaerotheca fuliginea phythium spp.	Powdery mildew	Adatia and Besford (1986), Menzies
	Cylindrocladium spathiphylli,	Root disease	et al. (1992), Chérif et al. (1994),
	Fusarium oxysporum	Root disease	Liang et al. (2005a)
	Phytophthora capsici	Fusarium wilt	Vermeire et al. (2011)
		Phytophthora blight	Fortunato et al. (2012)
			French-Monar et al. (2010)

Similarly, in rice plants, exogenous supplementation of Si enhanced the glucanase, peroxidase, and PR-1 transcripts level, and therefore decreased the colonization of *Magnaporthe grisea* in epidermal cells of a vulnerable rice cultivar (Rodrigues et al., 2005). Silicon also stimulates the tolerance capacity of plants against insects and pests, such as the stem borer and plant hopper (Bienert et al., 2008) (Table 11.3). These biochemical responses are only driven by soluble Si, signifying that soluble Si plays a dynamic role in enhancing host resistance to diseases by stimulating the defense reaction. However, the exact nature of the interaction between soluble Si and its biochemical pathways in the plant that leads to disease resistance remains unclear (Fauteux et al., 2005; Ma and Yamaji, 2006).

The beneficial impact of Si on plant yield and productivity in relation to different biotic and abiotic stresses is attributed to its deposition in the plant tissues, which provides a mechanical defense against probing and chewing by insects (Ma and Yamaji, 2006). A secondary mechanism is based on Si-mediated resistance, in which a plant modifies its physiological and biochemical processes to obtain resistance against pathogens (Liang et al., 2005a; Hayasaka et al., 2008; Sun et al., 2010). Similarly, Rodrigues et al. (2004) proposed that Si-mediated formation of phytoalexins may provide resistance in the leaf tissue of rice against rice blast.

11.7 CONCLUSION AND FUTURE PERSPECTIVES

In conclusion, it can be inferred that with the application of Si, we can protect different edible crops against numerous biotic and abiotic stresses. Si may efficiently protect plants against heavy metal by decreasing their accumulation and transportation and increasing mineral elements, as well as antioxidant capacity. At the same time, it can be projected that exogenous supplementation with Si in metal-susceptible areas definitely decreases the metal pressure on edible grains via inhibiting their accumulation and transport in crop plants. Moreover, under other biotic and abiotic stresses, like plant diseases caused by different pathogens, salinity, and UV radiation, Si acts in a pivotal role by stimulating resistance and the formation of phenolic compounds and phytoalexins. These beneficial attributes of Si provide a way to reduce the deleterious impact of metals by restricting their entry into crops. This is important in order to achieve the goal of food security, since metal infectivity of croplands has attracted global concern owing to the improper handling of metal-based manures, fertilizers, and pesticides. In addition, current knowledge of silicon function in individual plant species under abiotic stress conditions (salinity, metal toxicity, drought, and UV radiation) is developing, but there is also the need to better understand its function in more species, under other stress conditions (high and low temperature) and more complex interactions (especially higher temperatures combined with optimal watering, drought, and UV-B radiation).

REFERENCES

Adatia MH, Besford RT. (1986). The effects of silicon on cucumber plants grown in recirculating nutrient solution. *Ann Bot* 58: 343–351.

Ahmad R, Lim CJ, Kwon SY. (2013). Glycine betaine: A versatile compound with great potential for gene pyramiding to improve crop plant performance against environmental stresses. *Plant Biotechnol Rep* 7: 49–57.

Ahmad R, Zaheer SH, Ismail S. (1992) Role of silicon in salt tolerance of wheat. *(Triticum aestivum) J Plant Sci* 85: 43–50.

Aleshin EP, Aleshin NE, Avakian AR. (1978). The effect of various nutrition and gibberellins on SiO_2 content in hulls of rice. *Agrochemistry* 7: 64–68.

Ali B, Wang B, Ali S, Ghani MA, Hayat MT, Yang C, Xu L, Zhou WJ. (2013). 5-Aminolevulinic acid ameliorates the growth, photosynthetic gas exchange capacity, and ultrastructural changes under cadmium stress in *Brassica napus* L. *J Plant Growth Regul* 32: 604–614.

Ali S, Cai S, Zeng F, Qiu B, Zhang GP. (2012). The effect of salinity and chromium stresses on uptake and accumulation of mineral elements in barley genotypes differing in salt tolerance. *J Plant Nutr* 35: 827–839.

Barber DA, Shone MGT. (1966). The absorption of silica from aqueous solutions by plants. *J Exp Bot* 17: 569–578.

Bélanger RR. (2003). Cytological evidence of an active role of silicon in wheat resistance to powdery mildew (*Blumeria graminis* f. sp. *tritici*). *Phytopathology* 93: 402–412.

Beckmann M, Hock M, Bruelheide H, Erfmeier A. (2012). The role of UV-B radiation in the invasion of *Hieracium pilosella*—A comparison of German and New Zealand plants. *Environ Exp Bot* 75: 173–180.

Bienert GP, Schussler MD, Jahn TP. (2008). Metalloids essential, beneficial or toxic. Major intrinsic proteins. *Biochem Sci* 33: 20–26.

Casey WH, Kinrade SD, Knight CTG, Rains DW, Epstein E. (2004). Aqueous silicate complexes in wheat, *Triticum aestivum* L. *Plant Cell Environ* 27: 51–54.

Chauhan DK, Tripathi DK, Kumar R, Kumar Y. (2011). Diversity, distribution and frequency based attributes of phytolith in *Arundo donax* L. *Int J Innov Biol Chem Sci* 1: 22–27.

Cherepanov KA, Chernish GI, Dinelt VM, Suharev JI. (1994). *The Utilization of Secondary Material Resources in Metallurgy*. Metallurgy Elsevier science B.V. 120–145.

Chérif M, Asselin A, Bélanger R. (1994). Defense responses induced by soluble silicon in cucumber roots infected by *Pythium* spp. *Phytopathology* 84: 236–242.

Chérif M, Benhamou N, Bélanger RR. (1993). Occurrence of cellulose and chitin in the hyphal walls of Pythium ultimum: A comparative study with other plant pathogenic fungi. *Can J Microbiol* 2: 213–222.

Chérif M, Benhamou N, Menzies JG, Bélanger RR. (1992). Silicon induced resistance in cucumber plants against *Pythium ultimum*. *Physiol Mol Plant Pathol* 41: 411–425.

Chiba Y, Mitani N, Yamaji N, Ma JF. (2009). *HvLsi1* is a silicon influx transporter in barley. *Plant J* 57: 810–818.

Devkota A, Jha PK. (2011). Influence of water stress on growth and yield of *Centella asiatica*. *Int Agrophys* 25: 211–214.

Eneji E, Inanaga S, Muranaka S, Li J, An P, Hattori T, Tsuji W. (2005). Effect of calcium silicate on growth and dry matter yield of *Chloris gayana* and *Sorghum sudanense* under two soil water regimes. *Grass Forage Sci* 60: 393–398.

Epstein AL, Gussman CD, Blaylock MJ, YermiyahuU, Huang JW, Kapulnik Y, Orser CS. (1999). EDTA and Pb-EDTA accumulation in *Brassica juncea* grown in Pb-amended soil. *Plant Soil* 208: 87–94.

Epstein E. (1994). The anomaly of silicon in plant biology. *Proc Natl Acad Sci USA* 91: 11–17.

Epstein E. (1999). Silicon. *Annu Rev Plant Physiol Plant Mol Biol* 50: 641–664.

Epstein E, Bloom AJ. (2005). *Mineral Nutrition of Plants: Principles and Perspectives*. 2nd ed. Sinauer Associates, Sunderland, MA.

Fadzilla NM, Finch RP, Burdon RH. (1997). Salinity oxidative stress and antioxidant responses in shoot cultures of rice. *J Exp Bot* 48: 325–331.

Fang Ch-X, Wang Q-S, Yu Y, Huang L-K, Wu X-Ch, Lin W-X. (2011). Silicon and its uptaking gene *Lsi1* in regulation of rice UV-B tolerance. *Acta Agron Sin* 37: 1005–1011.

Fauteux F, Remus-Borel W, Menzies JG, Belanger RR. (2005). Silicon and plant disease resistance against pathogenic fungi. *FEMS Microbiol* 249: 1–6.

Fortunato AA, Rodrigues FA, Baroni JCP, Soares GCB, Rodriguez MAD, Pereira OL. (2012). Silicon suppresses *Fusarium* wilt development in banana plants. *J Phytopathol* 160: 674–679.

French-Monar R, Avila F, Korndorfer G, Datnoff L. (2010). Silicon suppresses *Phytophthora* blight development on bellpepper. *J Phytopathol* 158: 554–560.

Gao K, Yu H, Brown MT. (2007). Solar PAR and UV radiation affects the physiology and morphology of the cyanobacterium *Anabaena* sp. *PCC 7120. J Photochem Photobiol B: Biol* 89: 117–124.

Gao X, Zou C, Wang L, Zhang F. (2004). Silicon improves water use efficiency in maize plants. *J Plant Nutr* 27: 1457–1470.

Gascho GJ, Datnoff LE, Snyder GH, Korndorfer GH. (2001). Silicon sources for agriculture. In *Silicon in Agriculture*. Elsevier, Amsterdam, pp. 197–207.

Gong H, Zhu X, Chen K, Wang S, Zhang C. (2005). Silicon alleviates oxidative damage of wheat plants in pots under drought. *Plant Sci* 169: 313–321.

Gong HJ, Chen KM, Chen GC, Wang SM, Zhang CL. (2003). Effects of silicon on growth of wheat under drought. *J Plant Nutr* 26: 1055–1063.

Guével MH, Menzies JG, Belanger RR. (2007). Effect of root and foliar applications of soluble silicon on powdery mildew control and growth of wheat plants. *Eur J Plant Pathol* 119: 429–436.

Guntzer F, Keller C, Meunier J-D. (2012). Benefits of plant silicon for crops: A review. *Agron Sustain Develop* 32: 201–213.

Guo W, Zhu Y-G, Liu W-J, Liang Y-C, Geng C-N, Wang S-G. (2007). Is the effect of silicon on rice uptake of arsenate (AsV) related to internal silicon concentrations, iron plaque and phosphate nutrition? *Environ Pollut* 148: 251–257.

Hallmark CT, Wilding LP, Smeck NE. (1982). Silicon. *Agron Monogr* 9: 263–265.

Hasegawa PM, Bressan RA, Zhu JK, Bohnert HJ. (2000). Plant cellular and molecular responses to high salinity. *Annu Rev Plant Physiol Plant Mol Biol* 51: 463–499.

Hattori T, Inanaga S, Araki H, An P, Morita S, Luxova M, Lux A. (2005). Application of silicon enhanced drought tolerance in *Sorghum bicolor. Physiol Plant* 123: 459–466.

Hayasaka T, Fujii H, Ishiguro K. (2008). The role of silicon in preventing appressorial penetration by the rice blast fungus. *Phytopathology* 98: 1038–1044.

Hernandez JA, Corpass FJ, Gomez M, Rio LA, Sevilla F. (1993). Salt-induced oxidative stress mediated by active oxygen species in pen leaf mitochondria. *Physiol Plant* 89: 103–110.

Hollosy F. (2002). Effect of ultraviolet radiation on plant cells. *Micron* 33: 179–197.

Horiguchi T. (1988). Mechanism of manganese toxicity and tolerance of plant. *Soil Sci Plant Nutr* 34: 65–73.

Iwasaki K, Maier P, Fecth M, Horst WJ. (2002). Leaf apoplastic silicon enhances manganese tolerance of cowpea (*Vigna unguiculata). J Plant Physiol* 159: 167–173.

Jones LHP, Handreck KA. (1967). Silica in soils, plants, and animals. *Adv Agron* 19: 107–149.

Keller C, Rizwan M, Davidian JC, Pokrovsky OS, Bovet N, Chaurand P, Meunier JD. (2015). Effect of silicon on wheat seedlings (*Triticum turgidum* L.) grown in hydroponics and exposed to 0 to 30 lM Cu. *Planta* 241: 847–860.

Kim Y-H, Khan AL, Kim D-H, Lee S-Y, Kim K-M, Waqas M, Jung H-Y, Shin J-H, Kim J-G, Lee I-J. (2014). Silicon mitigates heavy metal stress by regulating P-type heavy metal ATPases, *Oryza sativa* low silicon genes, and endogenous phytohormones. *BMC Plant Biol* 14: 13.

Kittrick JA. (1969). Soil minerals in the Al_2O_3-SiO_2-H_2O system and a theory of their formation. *Clays Clay Miner* 17: 157–167.

Kudinova LI. (1974). The effect of silicon on weight of plant barley. *Sov Soil Sci* 6: 39–41.

Kumar J, Singh VP, Prasad SM. (2015). NaCl-induced physiological and biochemical changes in two cyanobacteria *Nostoc muscorum* and *Phormidium foveolarum* acclimatized to different photosynthetically active radiation. *J Photochem Photobiol B: Biol* 151: 221–232.

Li H, Zhu Y, Hu Y, Han W, Gong H. (2015). Beneficial effects of silicon in alleviating salinity stress of tomato seedlings grown under sand culture. *Acta Physiol Plant* 37: 71.

Li HB, Cheng KW, Wong CC, Fan KW, Chen F, Jiang Y. (2007). Evaluation of antioxidant capacity and total phenolic content of different fractions of selected microalgae. *Food Chem* 102: 771–776.

Li RY, Stroud JL, Ma JF, Mcgrath SP, Zhao FJ. (2009). Mitigation of arsenic accumulation in rice with water management and silicon fertilization. *Environ Sci Technol* 43: 3778–3783.

Liang Y, Sun W, Si J, Romheld V. (2005a). Effects of foliar and root-applied silicon on the enhancement of induced resistance to powdery mildew in *Cucumis sativus. Plant Pathol* 54: 678–685.

Liang Y, Wong JWC, Wei L. (2005b). Silicon-mediated enhancement of cadmium tolerance in maize (*Zea mays* L.) grown in cadmium contaminated soil. *Chemosphere* 58: 475–483.

Liang YC, Chen Q, Liu Q, Zhang W, Ding R. (2003). Exogenous silicon (Si) increases antioxidant enzyme activity and reduces lipid per-oxidation in roots of salt stressed. *J Plant Physiol* 10: 1157–1164.

Liang YC, Shen QR, Shen ZG, Ma TS. (1996). Effects of silicon on salinity tolerance of two barley cultivars. *J Plant Nutr* 19: 173–183.

Liang YC, Sun WC, Zhu YG, Christie P. (2007). Mechanisms of silicon-mediated alleviation of abiotic stresses in higher plants: A review. *Environ Pollu* 147: 422–428.

Liang YC, Zhang WQ, Chen J, Ding R. (2005c). Effect of silicon on H+-ATPase and H+-PPase activity, fatty acid composition and fluidity of tonoplast vesicles from roots of salt-stressed barley (*Hordeum vulgare* L.). *Environ Exp Bot* 53: 29–37.

Lidon FJC, Teixeira M, Ramalho JC. (2012). Decay of the chloroplast pool of ascorbate switches on the oxidative burst in UV-B-irradiated rice. *J Agron Crop Sci* 198: 130–144.

Lindsay WL. (1979). *Chemical Equilibria in Soil*. New York, Wiley.

Liu YL, Mao CL, Wang LJ. (1987). Advances in salt tolerance in plants. *Commun Plant Physiol* 23: 1–7.

Lizana C, Hess S, Calderini DF. (2009). Crop phenology modifies wheat responses to increased UV-B radiation. *Agric Forest Meteorol* 149: 1964–1974.

Lu H, Zhuang P, Li Z, Tai Y, Zou B, Li Y, McBride MB. (2014). Contrasting effects of silicates on cadmium uptake by three dicotyledonous crops grown in contaminated soil. *Environ Sci Pollut Res* 21: 9921–9930.

Lumsdon DG, Farmer VC. (1995). Solubility characteristics of proto-imogolite sols: How silicic acid can detoxify aluminium solutions. *Eur Soil Sci* 46: 179–186.

Lutts S, Kinet JM, Bouharmont J. (1996). NaCl-induced senescence in leaves of rice (*Oryza sativa* L.) cultivars differing in salinity resistance. *Ann Bot* 78: 389–398.

Lux A, Luxová M, Hattori T, Inanaga S, Sugimoto Y. (2002). Silicification in sorghum (*Sorghum bicolor*) cultivars with different drought tolerance. *Physiol Plant* 115: 87–92.

Ma JF. (2004). Role of silicon in enhancing the resistance of plants to biotic and abiotic stresses *Soil Sci Plant Nutr* 50: 11–18.

Ma JF. (2009). Silicon uptake and translocation in plants. In *Proceedings of the International Plant Nutrition Colloquium XVI*, Department of Plant Sciences, UC Davis.

Ma JF, Miyake Y, Takahashi E. (2001). Silicon as a beneficial element for crop plants. *Stud Plant Sci* 8: 17–39.

Ma JF, Nishimura K, Takahashi E. (1989). Effect of silicon on the growth of rice plant at different growth-stages. *Soil Sci Plant Nutr* 35: 347–356.

Ma JF, Takahashi E. (1990). The effect of silicic acid on rice in a P-deficient soil. *Plant Soil* 126: 121–125.

Ma JF, Takahashi E. (2002). *Soil, Fertilizer, and Plant Silicon Research in Japan*. Elsevier, Amsterdam.

Ma JF, Tamai K, Ichii M, Wu GF. (2002). A rice mutant defective in Si uptake. *Plant Physiol* 130: 2111–2117.

Ma JF, Tamai K, Yamaji N, Mitani N, Konishi S, Katsuhara M, Ishiguro M, Murata Y, Yano M. (2006). A silicon transporter in rice. *Nature* 440: 688–691.

Ma JF, Yamaji N. (2006). Silicon uptake and accumulation in higher plants. *Trends Plant Sci* 11: 392–397.

Ma JF, Yamaji N. (2015). A cooperative system of silicon transport in plants. *Trends Plant Sci* 20: 435–442.

Ma JF, Yamaji N, Mitani N, Tamai K, Konishi S, Fujiwara T, Katsuhara M, Yano M. (2007). An efflux transporter of silicon in rice. *Nature* 448: 209–212.

Ma JF, Yamaji N, Mitani N, Xu XY, Su YH, McGrath S, Zhao FJ. (2008). Transporters of arsenite in rice and their role in arsenic accumulation in rice grain. *Proc Natl Acad Sci* 105: 9931–9935.

Ma JF, Yamaji N, Mitani-Ueno N. (2011). Transport of silicon from roots to panicles in plants. *Proc Jpn Acad Ser B* 87: 377–385.

Maksimovic JD, Bogdanovic J, Maksimovic V, Nikolic M. (2007). Silicon modulates the metabolism and utilization of phenolic compounds in cucumber (*Cucumis sativus* L.) grown at excess manganese. *J Plant Nutri Soil Sci* 170: 739–744.

Malčovská SM, Dučaiová Z, Bačkor M. (2014a). Impact of silicon on maize seedlings exposed to short-term UV-B irradiation. *Biologia* 69: 1349–1355.

Malčovská SM, Dučaiová Z, Maslaňáková I, Bačkor M. (2014b). Effect of silicon on growth, photosynthesis, oxidative status and phenolic compounds of maize (*Zea mays* L.) grown in cadmium excess. *Water Air Soil Pollut* 225: 2056.

Marschner C, Hengge E. (1997). Stepwise synthesis of functional polysilane dendrimers. In Auner N, Weis J (eds.), *Organosilicon Chemistry*. Verlag Chemie, Weinheim, Germany, p. 333.

Marschner H. (1995). *Mineral Nutrition of Higher Plants*. 2nd ed. Academic Press, London, pp. 405–435.

Matichenkov VV. (1990). Amorphous oxide of silicon in soddy podzolic soil and its influence on plants. Autoref Diss Cand, Moscow State University, Moscow.

Matichenkov VV, Ammosova YM, Bocharnikova EA. (1997). The method for determination of plant-available silica in soil. *Agrochem* 1: 76–84.

Matichenkov VV, Calvert DV, Snyder GH. (2000). Prospective silicon fertilization for citrus in Florida. *Proc Soil Crop Sci Soc Fla* 59: 137–141.

Meena VD, Dotaniya ML, Vassanda C, Rajendiran S, Ajay, Kundu S, Subba Rao A. (2014). A case for silicon fertilization to improve crop yields in tropical soils. *Proc Natl Acad Sci India Sect B Biol Sci* 84(3): 505–518.

Menzies J, Bowen P, Ehret D, Glass ADM. (1992). Foliar applications of potassium silicate reduce severity of powdery mildew on cucumber, muskmelon, and zucchini squash. *J Am Soc Hortic Sci* 117: 902–905.

Mitani N, Ma JF, Iwashita T. (2005). Identification of the silicon form in xylem sap of rice (*Oryza sativa* L.) *Plant Cell Physiol* 46: 279–283.

Mitani N, Yamaji N, Ma JF. (2008). Characterization of substrate specificity of a rice silicon transporter, *Lsi1*. *Pflugers Arch-Eur J Physiol* 456: 679–668.

Mitani N, Yamaji N, Ma JF. (2009). Identification of maize silicon influx transporters. *Plant Cell Physiol* 50: 5–12.

Moussa TAA, Ahmed AM, Abdelhamid SMS. (2006). Optimization of cultural conditions for biosurfactant production from *Nocardia amarae*. *J Appl Sci Res* 2: 844–850.

Okuda A, Takahashi E. (1962). Studies on the physiological role of silicon in crop plants. Part 9. Effect of various metabolic inhibitors on the silicon uptake by rice plant. *Jpn J Soil Sci Manure* 33: 453–455.

Parihar P, Singh S, Singh R, Singh VP, Prasad SM. (2014). Effect of salinity stress on plants and its tolerance strategies. *Environ Sci Pollut Res* 22: 6.

Qados AMSA, Moftah AE. (2015). Influence of silicon and nano-silicon on germination, growth and yield of *Faba Bean* (*Vicia faba* L.) under salt stress conditions. *Am J Exp Agric* 5: 509–524.

Rafi MM, Epstein E. (1999). Silicon absorption by wheat. *Plant Soil* 211: 223–230.

Rémus-Borel W, Menzies JG, Bélanger RR. (2005). Silicon induces antifungal compounds in powdery mildew-infected wheat. *Physiol Mol Plant Pathol* 66: 108–115.

Richmond KE, Sussman M. (2003). Got silicon? The nonessential beneficial plant nutrient. *Curr Opin Plant Biol* 6: 268–272.

Rodrigues FA, McNally DJ, Datnoff LE, Jones JB, Labbe C, Benhamou N, Menzies JG, Belanger RR. (2004). Silicon enhances the accumulation of diterpenoid phytoalexins in rice: A potential mechanism for blast resistance. *Phytopathology* 94: 177–183.

Rodrigues FA, Oliveira LA, Korndörfer AP, Korndörfer GH. (2011). Silício elemento benéfico e importante às plantas. *Inform Agron* 134: 14–20.

Rodrigues FA, Wayne MJ, Datnoff LE, Jeffrey BJ, Jeffrey AR. (2005). Silicon influences cytological and molecular events in compatible and incompatible rice-*Magnaporthe grisea* interactions. *Physiol Mol Plant Pathol* 66: 144–159.

Romero-Arnada MR, Jourado O, Cuartero J. (2006). Silicon alleviates the deleterious salt effects on tomato plant growth by improving plant water status. *J Plant Physiol* 163: 847–855.

Sahebi M, Hanafi MM, Akmar ASN, Rafii MY, Azizi P, Idris AS. (2015). Serine-rich protein is a novel positive regulator for silicon accumulation in mangrove. *Gene* 556: 170–181.

Said-Al Ahl HAH, Omer EA, Naguib NY. (2009). Effect of water stress and nitrogen fertilizer on herb and essential oil of oregano. *Int Agrophys* 23: 269–275.

Samuels AL, Glass ADM, Ehret DL, Menzies JG. (1991). Mobility and deposition of silicon in cucumber plants. *Plant Cell Environ* 14: 485–492.

Sangster AG, Hodson MJ, Tubb HJ. (2001). Silicon deposition in higher plants. In Datnoff LE, Snyder GH, Korndörfer GH (eds.), *Silicon in Agriculture*. Amsterdam, Elsevier, pp. 85–113.

Savant NK, Korndorfer GH, Datnoff LE, Snydre GH. (1999). Silicon nutrition and sugarcane production. *J Plant Nutr* 22: 1853–1903.

Savant NK, Snyder GH, Datnoff LE. (1997). Silicon management and sustainable rice production. *Adv Agron* 58: 151–190.

Schindler PW, Furst B, Dick R, Wolf PO. (1976). Ligand properties of surface silanol groups. I. Surface complex formation with Fe3, Cu2, Cd3, and Pb2. *J Colloid Interface Sci* 55: 469–475.

Schmidt RE, Zhang X, Chalmers DR. (1999). Response of photosynthesis and superoxide dismutase to silica applied to creeping bentgrass grown under two fertility levels. *J Plant Nutr* 22: 1763–1773.

Shen X, Lix X, Li Z, Duan L, Eneji AE. (2010). Growth, physiological attributes and antioxidant enzyme activities in soybean seedlings treated with or without silicon under UV-B radiation stress. *J Agron Crop Sci* 196: 431–439.

Shen X, Zhaohu L, Duan L, Eneji AE, Li J. (2014). Silicon mitigates ultraviolet-B radiation stress on soybean by enhancing chlorophyll and photosynthesis and reducing transpiration. *J Plant Nutr* 37: 837–849.

Shi QH, Bao ZY, Zhu ZJ, He Y, Qian QQ, Yu JQ. (2005). Silicon-mediated alleviation of Mn toxicity in *Cucumis sativus* in relation to activities of superoxide dismutase and ascorbate peroxidase. *Phytochemistry* 66: 1551–1559.

Singh M, Kumar J, Singh S, Singh VP, Prasad SM. (2015). Roles of osmoprotectants in improving salinity and drought tolerance in plants: A review. *Rev Environ Sci Biotechnol* 14: 407–426.

Singh VP, Kumar J, Singh S, Prasad SM. (2014). Dimethoate modifies enhanced UV-B effects on growth, photosynthesis and oxidative stress in mung bean (*Vigna radiata* L.) seedlings: Implication of salicylic acid. *Pest Biochem Physiol* 116: 13–23.

Singh VP, Tripathi DK, Kumar D, Chauhan DK. (2011). Influence of exogenous silicon addition on aluminium tolerance in rice seedlings. *Biol Trace Element Res* 144: 1260–1274.

Singha S, Choudhuri MA. (1990). Effect of salinity stress on H_2O_2 metabolism in *Vigna* and *Oryza* seedlings. *Biochem Physiol Pflanz* 186: 69–74.

Sommer M, Kaczorek D, Kuzyakov Y, Breuer J. (2006). Silicon pools and fluxes in soils and landscapes. *J Plant Nutr Soil Sci* 169: 310–329.

Sun W, Liang Y, Zhu YG, Christie P. (2007). Mechanisms of silicon mediated alleviation of abiotic stresses in higher plants: A review. *Environ Pollut* 147: 422–428.

Takahashi E. (1996). Uptake mode and physiological functions of silica. In *Science of the Rice Plant*, vol. 2: *Physiology*. Food and Agriculture Policy Research Center, Tokyo, pp. 420–433.

Takahashi E, Ma JF, Miyake Y. (1990). The possibility of silicon as an essential element for higher plants. *Comments Agric Food Chem* 2: 99–122.

Takahashi E, Miyake Y. (1982). The effect of silicon on the growth of cucumber plant. In Proceedings of the 9th International Plant Nutrition Colloquium, Warwick University, Coventry, UK, p. 669.

Tripathi DK, Chauhan DK, Kumar D, Tiwari SP. (2012a). Morphology, diversity and frequency based exploration of phytoliths in *Pennisetum typhoides*. *Rich Natl Acad Sci Lett* 35(4): 285–289.

Tripathi DK, Kumar R, Pathak AK, Chauhan DK, Rai AK. (2012b). Laser-induced breakdown spectroscopy and phytolith analysis: An approach to study the deposition and distribution pattern of silicon in different parts of wheat (*Triticum aestivum* L.) *Plant Agric Res* 1(4): 352–361.

Tripathi DK, Mishra S, Chauhan DK, Tiwari SP, Kumar C. (2013a). Typological and frequency based study of opaline silica (Phytolith) deposition in two common Indian *Sorghum* L. species. *Proc Natl Acad Sci India Sect B Biol Sci* 83(1): 97–104.

Tripathi DK, Pathak AK, Chauhan DK, Dubey NK, Rai AK, Prasad R. (2016). LIB spectroscopic and biochemical analysis to characterize lead toxicity alleviative nature of silicon in wheat (*Triticum aestivum* L.) seedlings. *Photochem Photobiol B* 154: 89–98.

Tripathi DK, Rajendra P, Chauhan DK. (2014). An overview of biogenic silica production pattern in the leaves of *Hordeum vulgare* L. *Ind J Plant Sci* 3(2): 167–177.

Tripathi DK, Singh S, Singh S, Chauhan DK, Dubey NK, Prasad R. (2015a). Micronutrients and their diverse role in agricultural crops: Advances and future prospective. *Acta Physiol Plant* 37(139): 1–14.

Tripathi DK, Singh VP, Gangwar S, Maurya JN, Chauhan DK, Prasad SM. (2013b). Role of silicon in enrichment of plant nutrients and protection from biotic and abiotic stresses. In *Improvement of Crops in the Era of Climate Change*. Vol. 1. Springer, Berlin, pp. 39–56.

Tripathi DK, Singh VP, Kumar D, Chauhan DK. (2012c). Impact of exogenous silicon addition on chromium uptake, growth, mineral elements, oxidative stress, antioxidant capacity, and leaf and root structures in rice seedlings exposed to hexavalent chromium. *Acta Physiol Plant* 34: 279–289.

Tripathi DK, Singh VP, Kumar D, Chauhan DK. (2012d). Rice seedlings under cadmium stress: Effect of silicon on growth, cadmium uptake, oxidative stress, antioxidant capacity and root and leaf structures. *Chem Ecol* 28: 281–291.

Tripathi DK, Singh VP, Prasad SM, Chauhan DK, Dubey NK, Rai AK. (2015b). Silicon-mediated alleviation of Cr(VI) toxicity in wheat seedlings as evidenced by chlorophyll florescence, laser induced breakdown spectroscopy and anatomical changes. *Ecotoxicol Environ Saf* 113: 133–144.

Tuna AL, Kaya C, Higg D, Murillo-Amador B, Aydemir AS, Girgin R. (2008). Silicon improves salinity tolerance in wheat plants. *Environ Exp Bot* 62: 10–16.

Vermeire ML, Kablan L, Dorel M, Delvaux B, Risède JM, Legrève A. (2011). Protective role of silicon in the banana—*Cylindrocladium spathiphylli* pathosystem. *Eur J Plant Pathol* 131: 621–630.

Vinocur B, Altman A. (2005). Recent advances in engineering plant tolerance to abiotic stress: Achievement and limitations. *Curr Opin Biotechnol* 16: 123–132.

Wang Y, Stass A, Horst WJ. (2004). Apoplastic binding of aluminum is involved in silicon-induced amelioration of aluminum toxicity in maize. *Plant Physiol* 136: 3762–3770.

Wu XC, Chen YK, Li QS, Fang CX, Xiong J, Lin WX. (2009). Effects of silicon nutrition on phenolic metabolization of rice (*Oryza sativa* L.) exposed to enhanced ultraviolet-B. *Chin Agric Sci Bull* 25: 225–230.

Xu CX, Ma YP, Liu YL. (2015). Effects of silicon (Si) on growth, quality and ionic homeostasis of aloe under salt stress. *S Afr J Bot* 98: 26–36.

Yamaji N, Ma JF. (2007). Spatial distribution and temporal variation of the rice silicon transporter *Lsi1*. *Plant Physiol* 143: 1306–1313.

Yamaji N, Mitatni N, Ma JF. (2008). A transporter regulating silicon distribution in rice shoots. *Plant Cell* 20: 1381–1389.

Yao X, Chu J, Cai K, Liu L, Shi J, Geng W. (2011). Silicon improves the tolerance of wheat seedlings to ultraviolet-B stress. *Biol Trace Elem Res* 143: 507–517.

Yoshida S, Forna DA, Cock JH, Gomez KA. (1976). *Physiological Studies of Rice*. International Rice Research Institute, Los Banos, Philippines, pp. 62–63.

Zancan S, Suglia I, La Rocca N, Ghisi R. (2008). Effects of UV-B radiation on antioxidant parameters of iron-deficient barley plants. *Environ Exp Bot* 63: 71–79.

Zeng F-R, Zhao F-S, Qiu B-Y, Ouyang Y-N, Wu F-B, Zhang G-P. (2011). Alleviation of chromium toxicity by silicon addition in rice plants. *Agric Sci China* 10: 1188–1196.

Zhang Q, Yan C, Liu J, Lu H, Duan H, Du J, Wang W. (2014). Silicon alleviation of cadmium toxicity in mangrove (*Avicennia marina*) in relation to cadmium compartmentation. *J Plant Growth Regul* 33: 233–242.

Zhu Z, Wei G, Li J, Qian Q, Yu J. (2004). Silicon alleviates salt stress and increases antioxidant enzymes activity in leaves of salt stressed cucumber (*Cucumis sativus* L.) *Plant Sci* 167: 527–533.

Zlatev ZS, Lidon FJC, Kaimakanova M. (2012). Plant physiological responses to UV-B radiation. *Emir J Food Agric* 24: 481–501.

12 Silicon and Chromium Toxicity in Plants
An Overview

CONTENTS

discuss the mechanisms involved in the alleviation of chromium toxicity by the application of silicon. Moreover, future perspectives in the field of soil and plants with respect to silicon are also discussed.

12.1   INTRODUCTION

Contamination of soil with heavy metals is one of the major environmental problems; it is mainly caused by anthropogenic activities such as mining, agriculture, and industry. In the periodic table, about 80% of the elements are metals, and the metals having a density of more than 5 g·cm^{-3} are the heavy metals, such as chromium (Cr), cadmium (Cd), lead (Pb), manganese (Mn), silver (Ag), copper (Cu), Zinc (Zn), and mercury (Hg). These contaminants come from natural as well as anthropogenic resources. Natural resources originate from rock weathering and atmospheric deposition, and anthropogenic resources include agriculture, industry, and mining (Nagajyoti et al., 2010). They are toxic to humans, animals, and plants due their high toxic potential and long persistency (Nascimento and Xing, 2006). Many studies declared that they are toxic to plants when their concentration exceeds the limits and results in decreased biomass, loss of productivity, high production of reactive oxygen species (ROS), low chlorophyll content, and other biochemical, physiological, and ultrastructural changes (Ehsan et al., 2014; Habiba et al., 2014; Tian et al., 2014; Gill et al., 2015a).

Chromium is one of those toxic heavy metals that can develop severe environment contamination in soil, sediments, and water resources (Bartlett, 1991; Dermou and Vayenas, 2008). It is considered the 7th most abundant element on earth and the 21st most abundant element in the earth's crust. It is extracted in the form of chromite, for example, iron chromite (FeCr$_2$O$_4$) (Barnhart, 1997; Cervantes and Campos-Garcia, 2007). Naturally, chromium is found in many oxidation states, ranging from –2 to +6; trivalent (Cr-III) and hexavalent (Cr-VI) are considered the most stable states, while the other oxidation states are unstable and do not persist long in the biological system (Cotton and Wilkinson, 1980; Barnhart, 1997; Avudainayagam et al., 2003). Both trivalent and hexavalent forms of chromium are used extensively in industries like leather, steel, and textile (Dixit et al., 2002). Cr-VI is more toxic than the trivalent form (Panda and Parta, 1997). These two valence states are different due to their individual properties of bioavailability, toxicity, and mobility (Panda and Choudhury, 2005). Cr-III is considered less toxic because it shows less mobility and generally bonds with organic matters in aquatic and soil environments, while Cr-VI is more soluble in water and has adverse effects on humans, animals, and plants and is usually found as chromate and dichromate (Cieślak-Golonka, 1996; Becquer et al., 2003). Cr-III seems to be essential for humans and a few other organisms in low concentrations (Katz and Salem, 1994), while Cr-VI is very harmful, as it causes different types of cancer, DNA damage, asthma, skin damage, and nasal epithelia in humans and other living organisms (Barceloux, 1999; OSHA, 2006; Linos et al., 2011). Cr-III and Cr-VI have the ability to pass through the biological membranes in plants (Gorbi et al., 2001).

Several studies have declared that chromium has many detrimental effects on plants, such as inhibiting plant growth and chlorophyll pigment, increasing the production of ROS, and decreasing the crop quality and yield production (Ali et al., 2015; Parmar and Patel, 2015; Qing et al., 2015).

In soil, silicon (Si) is the second most abundant element, after oxygen (Gong et al., 2006). A large portion of the earth's crust (more than 75%) is made up of silicon and oxygen (Sommer et al., 2006). However, the biosphere is comprised of only 0.03% silicon (Fauteux et al., 2005). Si is not considered essential for higher plants, but in many plants, it is a beneficial nutrient, as it is required for proper growth and development (Epstein, 1999; Epstein and Bloom, 2005). Many plants can accumulate Si up to 10% by mass, and this concentration is more than some macronutrients accumulated by plants (Epstein, 1994; Hodson et al., 2005). When a plant is under stress, Si improves its resistance by maintaining the rigidity of the cell wall (Hossain et al., 2007), increasing the net photosynthesis and light interception, and increasing the water use efficiency in different species of plants (Al-aghabary et al., 2004; Li et al., 2007). Many studies have shown that Si addition can improve the growth and yield of plants under biotic as well as abiotic stresses (Ma and Yamaji, 2006; Ali et al., 2013a; Tripathi et al., 2013). Silicon treatment shows positive behavior against chromium toxicity by reducing chromium uptake, improving the antioxidant defense, ameliorating the ultrastructural disorder, and so forth (Ali et al., 2013a; Tripathi et al., 2015). In this chapter, we discuss the beneficial role of silicon in protecting plants against chromium toxicity, the mechanisms involved, and the silicon efficacy to regulate nutrient uptake.

12.2 OCCURRENCE

12.2.1 CHROMIUM IN THE ENVIRONMENT

Chromium is the seventh most abundant element on earth (Cervantes et al., 2001). Naturally, Cr is found in many oxidation states, in which trivalent and hexavalent are considered the most stable (Barnhart, 1997), and its concentration in soil ranges from 0.01 to 0.05 $g \cdot kg^{-1}$ due to the parental material, while it can reach up to 125 $g \cdot kg^{-1}$ in ultramafic soils (Adriano, 1986).

It is used extensively in different industries, such as leather, textile, and steel (Dixit et al., 2002). In 2000, chromium production was estimated to be about 105.4 million tons per year (Han et al., 2004), in which 60%–70% is utilized in stainless steel and alloys, and 15% in electroplating, pigment, mainly tanning, wood preservation, and chemical industry processes (McGrath, 1995). Chromium's concentration ranges from 0.1 to 0.3 $mg \cdot kg^{-1}$ in the earth's crust. Just in India, a large amount of elemental chromium is released in the environment annually, which is estimated to be about 2000–3200 tons (Chandra and Sinha, 1997). In the United States, a hazardous level of Cr is reported in groundwater (14,800 $mg \cdot L^{-1}$) and soil (25,900 $mg \cdot kg^{-1}$) near United Chrome in Corvallis Oregon (Krishnamurthy and Wilkens, 1994). Generally, the Cr concentration found in freshwater ranges from 0.1 to 117 $mg \cdot L^{-1}$, and in seawater it ranges from 0.2 to 50 $mg \cdot L^{-1}$ (Nriagu, 1988).

12.2.2 SILICON IN THE ENVIRONMENT

In soil, silicon is the second most abundant element, after oxygen (Gong et al., 2006), and in combination with oxygen, it covers 75% of the earth's crust (Sommer et al., 2006). Si is found in soil at a concentration range of 50–400 $g \cdot kg^{-1}$ (Kovda, 1973), and it comprises only 0.3% of the biosphere (Fauteux et al., 2005). In combination with oxygen, silicon is found as silicon dioxide. Soil consists of 50%–70% by mass silicon dioxide, and in soil solution silicon is found as monosilicic acid (H_4SiO_4), with a concentration of 100–500 μM (Sommer et al., 2006). Many plants can take up silicon up to 10% (Hodson et al., 2005), which is more than for some of the macronutrients (Epstein, 1994).

12.3 CHROMIUM TOXICITY IN PLANTS

Chromium toxicity and its negative effects on plants were widely observed in many studies on different plants. The most stable valence states of chromium are Cr-VI and Cr-III, but Cr-VI is considered the most toxic because of its bioavailability, toxicity, and mobility. Both have deteriorative effects on plant growth and development. Some plants can tolerate stress below $3.8 \times 10^{-4} \mu M$ and show no significant effects (Huffman and Allaway, 1973a,b). Chromium has a high redox potential, which ranges from 1.33 to 1.38 eV, so it rapidly generates a high amount of ROS in plants (Shanker et al., 2004a,b). Cr shows physiological, biochemical, length, fresh weight, dry weight, and leaf area decrease. This was observed in many plants, such as wheat (Adrees et al., 2015; Nayak et al., 2015), *Brassica napus* (Gill et al., 2015a), and chickpea (Singh et al., 2014). Cr stress affects the photosynthetic pigment; for example, the total chlorophyll and carotenoid contents were significantly decreased (Adrees et al., 2015). Cr toxicity causes oxidative stress in plants; hence, the production of H_2O_2 and O^{2-} increases, as does the content of malondialdehyde (MDA) (Ali et al., 2013b). Under chromium stress, MDA is considered a sign of oxidative stress, as it is produced as a result of the peroxidation of membrane lipids (Malmir, 2011). Chromium stress also affects the performance of antioxidant enzymes and nonenzymes, such as the activities of catalase (CAT), ascorbate peroxidase (APX), superoxide dismutase (SOD), peroxidase (POD), and glutathione reductase (GR), and the contents of ascorbate (ASC) and glutathione (GSH).

As Cr is a nonessential element, no specific mechanism is associated with it in plants. However, uptake of Cr by a plant depends on the type of valence species. Cr uptake can be active or passive. Cr-III usually involves a passive mechanism, and no energy is required by plants for this phenomenon (Skeffington et al., 1976; Zayed and Terry, 2003). The uptake of Cr-VI involves an active mechanism, taken on by the carriers, that attaches to essential elements, for example, sulfates (Cervantes et al., 2001; Kim et al., 2014). There is a competition between Cr and other essential elements, such as phosphorus, sulfur, and iron, for binding with carriers (Shanker et al., 2005). Due to the similar structure of Cr and other important nutrient elements, it can negatively affect the mineral nutrition system of plants. Cr stress decreases the absorption of Cu in the roots of *Zea mays* (Mallick et al., 2010). Both macronutrients and micronutrients are affected by Cr toxicity: as Cr toxicity increases, the macronutrients (nitrogen, potassium, and phosphorus) and micronutrients (iron, magnesium, zinc, and copper) decrease (Liu et al., 2008; Sundaramoorthy et al., 2010). The highest Cr accumulation is reported in the root, followed by the leaves and then stem in wheat plants under Cr stress (Kumar et al., 2014). There are many plants that can hyperaccumulate Cr and transform it into a less toxic and immobilized form (Bluskov et al., 2005). Some hyperaccumulators of Cr can accumulate more than 1000 mg·kg^{-1} of dry weight of plant leaves. The tolerance of these plants toward Cr can be described as chelation with appropriate high-affinity ligands, reduction and biological transformation, and compartmentalization in the vacuole or cytoplasm, so the important mechanism of detoxification of Cr by plants can be accomplished by the vacuole of plant root cells (Pulford et al., 2001; Shanker et al., 2005).

12.4 PROTECTIVE ROLE OF SILICON

Silicon is not considered an essential element for the higher plants. Several studies have declared many beneficial effects on plants, particularly gramineous and cyperaceous plants. These effects can be observed on the growth and development, and can be direct or indirect (Epstein, 1994, 1999; Ma et al., 2001; Liang et al., 2007). Silicon's beneficial effects can be observed clearly when plants are grown under the abiotic and biotic stresses (Epstein, 1994; Ma et al., 2001). Silicon helps the plants to improve their tolerance against heavy metal stresses, such as Cr stress (Ali et al., 2013a). Many studies show the resistive behavior of Si against lodging and an increased plant strength, which maintains the turgidity of the plants. Scientists consider this phenomenon the result of attachment of Si to the cell walls of plants, which develop rigidity against stress in the cell walls (Jones and Handreck, 1967; Raven, 1983).

12.5 SILICON-MEDIATED ALLEVIATION OF CHROMIUM TOXICITY IN PLANTS

Chromium toxicity suppresses the growth, grain yield, and biomass production in all major crop plants (Nagarajan, 2014; Gill et al., 2015b; Hayyat et al., 2015). Many studies show the ameliorative effect of silicon against chromium toxicity in plants. This can be explained by different aspects, for example, Cr uptake, mineral nutrient uptake, photosynthetic pigment, and biomass production.

12.5.1 REDUCTION IN CHROMIUM UPTAKE

Silicon ameliorates chromium toxicity by decreasing the uptake of Cr in plants. Si also shows a similar behavior against the different heavy metal toxicities. Many studies show that Si develops the tolerance of plants against Cr toxicity by decreasing its uptake and translocation. Silicon application decreases the chromium concentration in different portions of plants. Ali et al. (2013a) reported that Si significantly reduces the concentration of Cr in different parts of plants, such as the leaves, stems, and roots in barley crop grown under chromium stress. Zhang et al. (2013b) did similar work on different crops. They concluded that under chromium stress, silicon decreases the uptake of chromium and significantly reduces the concentration of Cr in the leaves, stems, and roots of *Brassica chinensis* L. Reduction in Cr uptake due to application of Si can be explained in different ways. Si may enhance the production of root exudates, which can chelate the metal, for example, Cr and reduce the uptake of chromium in plants (Kidd et al., 2001). Si deposits lignin in the cell wall, which induces a metal ion bond to the cell wall. Due to this attachment of Cr to the cell wall, Cr uptake decreases and further translocation is inhibited (Ma and Yamaji, 2006). Furthermore, as silicon decreases, the metal concentration in apoplasm results in reeducation of the apoplasmic transport of metal, so Si can reduce chromium uptake in this way (Iwasaki et al., 2002; Rogalla and Römheld, 2002). Some physical barrier can also be responsible for the reeducation of Cr uptake in plants; for example, due to the accumulation of Si in the vicinity of the endoderm, a physical barrier can also be formed, which affects the inner root tissue by reducing the porosity of the cell wall, resulting in a reduction of the metal concentration in the xylem and the concentration of Cr in the shoots and leaves (Shi et al., 2005; da Cunha and do Nascimento, 2009; Keller et al., 2015). One of the major factors in the reduction of metal uptake is that Si accumulates in the roots' endodermis, which may block the way for Cr to go through and cause a reduction in the uptake by plants (Wang et al., 2004). Moreover, due to structural changes in the roots and shoots, chelation, coprecipitation, and compartmentation of metal in plants might cause a decrease in Cr uptake due to silicon application.

12.5.2 Development in Photosynthetic Pigment and Gas Exchange Attributes

Many studies have shown the beneficial effect of silicon on the enhancement of the photosynthetic pigment in different varieties of plants grown in various cultures, for example, soil, sand, and hydroponics under abiotic toxicity, for example, heavy metals. It was observed that chromium toxicity decreases the production of the photosynthetic pigment in plants. The chlorophyll contents and carotenoid production decrease significantly as Cr treatment increases (Singh et al., 2014; Adrees et al., 2015). Compared to other heavy metals, like cadmium, chromium toxicity shows a more distinct reduction in chlorophyll production (Gonzalez et al., 2015). There are many factors that may be involved in this influence. By altering the structure of the chloroplast, metals change the rate of photosynthesis, which finally results in the disturbance of the fatty acid composition and inhibits the enzymes' progress in the Calvin cycle and the photosynthetic pigment (Vazquez et al., 1987). Tripathi et al. (2015) observed the beneficial effect of silicon on wheat plant that was exposed to chromium toxicity by observing the light usage efficiency of the photosynthetic machinery by using chlorophyll fluorescence, which is considered a rapid detection tool. This study reported that a high concentration of Cr caused a decline in maximum fluorescence but increased the minimal fluorescence, which shows a reduction in the relative rations of minimal fluorescence, maximum fluorescence, and variable fluorescence. In the photosystem II (PS II) complex, minimal fluorescence is considered the peak calculation of the relative size of the antenna pigments (Huang et al., 2004) and damaging symptoms of this reaction center, which is caused by a reduction in light absorbance, and finally, unused extra light emits (Schnettger et al., 1994). It was also reported that this increase in minimal fluorescence was caused by a reduction in the efficiency to absorb light by the antenna chlorophyll a (Briantais et al., 1986), while the addition of silicon under chromium toxicity shows positive effects. Si remediates the harmful effects of chromium on photosynthetic parameters; for example, it improve the relative ratios between minimal fluorescence, maximum fluorescence, and variable fluorescence, with better growth of the plant, which enhances the production and strengthens the plant (Tripathi et al., 2015).

12.5.3 Nutrient Uptake, Plant Growth, and Biomass

Nutrients and minerals are important components of plants that develop the structure and provide support to the plant directly and indirectly by biochemical, physiological, and ultrastructural processes, and provide strength against biotic and abiotic stresses. They are the essential part of plant life. Many heavy metals are essential for plants, as they are required for proper growth, but an excess of any heavy metal is toxic to plants. Heavy metals may interfere in a plant's nutrient uptake when its concentration exceeds a specific limit, which affects the nutrient composition of a plant and significantly decreases the yield and biomass. Cr possesses high potential to disturb nutrient uptake by plants, which reduces the growth and development and finally decreases the yield. In this toxic situation, Si plays an important role in regulating the nutrient uptake by plants. Si enhances a plant's ability against the toxicity of Cr by reducing its uptake in the plant. It was observed that under the toxicity of Cr, Si promotes the uptake of micronutrients (e.g., Zn and Mn) and macronutrients (e.g., Mg, K, Ca, and P) by wheat plants (Tripathi et al., 2015). It is widely accepted that Si increases biomass and helps plant growth. When Si was supplied to a plant facing Cr toxicity, significant positive results were observed in the form of development in growth and biomass (Zeng et al., 2011). However, the mechanisms involved in these processes are still unclear. In the future, scientists may explore the mechanisms involved with Si, plants, and other mineral nutrients, but in the meantime, they have made suggestions regarding them. These beneficial effects can be indirect, such as the development of disease resistance in a plant system, or direct, for example, improvement in mineral and nutrient regulation. Another possibility is the improvement in the cell wall extensibility of the cell wall due to Si (Hossain et al., 2002). Si may improve plant tolerance against toxic metals by decreasing the uptake as well as translocation of toxic heavy metals in plants, which results in the promotion of growth and an increase in biomass (Shi et al., 2005; Nwugo and Huerta, 2008; Kaya et al., 2009).

12.6 MECHANISMS INVOLVED IN ALLEVIATION OF CHROMIUM TOXICITY

In Section 12.5, we discussed the beneficial effects of Si on plants under Cr stress; for example, Si reduces Cr uptake, enhances growth and biomass, improves photosynthetic pigment production, and regulates mineral and nutrient uptake. These beneficial effects may be the result of different mechanisms performed by Si at the plant and soil level. It is important to understand these mechanisms involved in the Si-mediated alleviation of Cr toxicity. These mechanisms have been widely studied in different plant species. At the soil level, this alleviation may be the result of immobilization of Cr by Si, which inhibits the uptake of Cr and reduces the toxic effects. At the plant level, there are several mechanisms, such as coprecipitation of chromium with silicon, structural alteration, promotion of the activity of antioxidants against oxidative stress, compartmentation of Cr, structural modification, chelation, and changes in gene expression. These mechanisms are discussed in detail below.

12.6.1 Immobilization

This phenomenon takes place at the soil level. Many studies have reported the efficacy of Si in reducing the availability of toxic metals in culture media (Liang et al., 2005; da Cunha et al., 2008; Naeem et al., 2014). Cr shows similar behavior in toxicity and the alleviation of its effect by Si amendment. It has been widely accepted that soil pH and the amount of organic matter in soil are very important factors that affect the bioavailability of Cr and other metals in plants. Many studies show that application of Si increases the pH (Chen et al., 2000; Barančíková et al., 2004; Gu et al., 2011), which decreases the bioavailability of Cr in the plant. Moreover, Si affects soil properties by increasing silicate-induced pH (Liang et al., 2005), which decreases the uptake of chromium by plants. It has also been reported that due to Fe^{2+} and organic matter, Cr-VI reduces to Cr-III (Zhang et al., 2013b), and as we discussed in Section 12.1, Cr-III has less mobility than Cr-VI due to its binding with organic matter. So with the conversion of Cr, mobility and bioavailability decrease and reduction in Cr uptake takes place, which enhances plant growth.

12.6.2 IMPROVEMENT IN ANTIOXIDANT DEFENSE SYSTEM

In plants, oxidative stress can be caused by many environmental stresses, which can be biotic or abiotic (Ali et al., 2015b; Noman et al., 2015). We have already discussed in detail that Cr stress causes oxidative stress by producing ROS such as H_2O_2, O^{2-}, and MDA (Ali et al., 2013b). On the other hand, plants have an antioxidant defense system against oxidative stress. Plants can tolerate oxidative stress to some degree, which is different for different plants and varieties. Once that limit is exceeded, plants undergo oxidative stress and the antioxidant defense system, which aids the plant in proper growth, weakens (Adrees et al., 2015). Silicon may improve the antioxidant defense system of plants under Cr stress by increasing the performance of antioxidant enzymes (Zeng et al., 2011) against oxidative stress. Ali et al. (2013a) reported that silicon decreased the production of H_2O_2, MDA, and electrolyte leakage in the roots and shoots of barley plants under Cr stress. The antioxidant defense system comprises enzymatic as well as nonenzymatic antioxidants in plants, and they develop many mechanisms that enhance the tolerance of plants (Suzuki et al., 2012) under Cr stress. In some studies, a decrease in the activity of antioxidants with the application of Si was also reported when plants were exposed to metal stress. For example, in a study on *Slanum nigrum* L., the performance of CAT, APX, and SOD decreased under Cd stress with the application Si, compared to without Si in cadmium stress (Liu et al., 2013). But in the case of Cr, many studies have declared the enhancement of antioxidant activities (Ali et al., 2013a; Zhang et al., 2013b). This change in behavior of Si may be due to the variance in plant species, experimental conditions, types of stress, and duration of treatment (Sharma and Dietz, 2009; Shi et al., 2010; Tang et al., 2015).

12.6.3 COMPARTMENTATION WITHIN PLANTS

Many scientists have suggested another mechanism of compartmentation of Cr in the shoots and roots to describe the mode of Si alleviation in plants. Si increases the concentration of metal in roots, compared to the concentration in the shoots of plants (Kuliková and Lux, 2010; Keller et al., 2015; Srivastava et al., 2015). Zhang et al. (2013b) observed that when Si was used under Cr stress, the concentration of Cr significantly decreased in the shoots and leaves, while the concentration in the roots of pakchoi plants increased. Most of the Cr was retained in roots. Silicon reduced the translocation of chromium from root to shoot. Si deposits lignin in the cell wall, which induces a metal ion bond to the cell wall. Due to this attachment of Cr to the cell wall, Cr uptake decreases and translocation is inhibited (Ma and Yamaji, 2006). Ali et al. (2013a) described the Si-mediated effect on barley crops under Cr stress in a hydroponic experiment. It was reported that with an increase of Cr concentration in the treatment of barley plants, the chromium concentration also increased in different parts, such as the root, shoot, and leaf. Moreover, Si reduces the concentration of Cr when Si is applied at concentrations of 1 and 2 mM. Cr accumulation in the roots was significantly higher than that in the stems and leaves, and the observed effects of Si were greater in the 2 mM application than in the 1 mM one of Si in barley plants.

12.6.4 COPRECIPITATION OF SILICON WITH CHROMIUM

Coprecipitation of silicon with chromium might be a cause of alleviation of Cr toxicity by the application of silicon in plants. Many scientists have reported the mechanism of coprecipitation of silicon with chromium and many other heavy metals. Zhang et al. (2013b) described that with the application of silicon, the uptake and translocation of chromium was significantly decreased because of the formation of a precipitation bond, which reduces the amount of exchangeable chromium. Another possibility is the "organic matter bond," which decreases the mobility of chromium and reduces its toxicity in contaminated soil. In this study, chromium was mixed with soil at concentrations ranging from 0 to 200 mg·kg⁻¹ Na₂SiO₃, whereas the Si-amended levels were from 0 to 200 g·kg⁻¹. Different forms of Cr were calculated in this treatment, such as exchangeable, precipitated, bound to organic matter, and residual, in the pakchoi plant with the application of silicon. It was observed that precipitated chromium increased significantly with an increase in silicon concentration for all levels of chromium treatment. Previous studies also reported the coprecipitation mechanism of silicon with other heavy metals, such as Zn (Gu et al., 2012), Cd (Zhang et al., 2008), Cu (Oliva et al., 2011), and Mn (Iwasaki et al., 2002). Zeng et al. (2011) also suggested a similar mechanism of coprecipitation of chromium with silicon in the rice plant, as silicon has the ability to develop complex forms with chromium and retain chromium in the cell wall, decreasing its mobility and reducing its the toxicity in the rice plant.

12.6.5 Structural and Ultrastructural Changes in Plants

Many studies reported that Si increases the amount of physical parameters, such as plant height, root length, leaf area, number of leaves, dry and fresh weight of root, stem, and leaves.

Tripathi et al. (2015) reported that chromium treatment decreased the shoot and root growth significantly, and root and shoot fresh weight was suppressed up to 40% and 37%, while dry weight decreased by 48% and 46%, respectively. Moreover, both plant height and root length declined by 40% compared to control. The leaf area was decreased by 47%. But when Si was applied under chromium stress, significant enhancements in the physical parameters were observed. Si enhanced parameters by reducing the effect on root fresh weight from 40% to 17%, shoot fresh weight from 37% to 28%, dry weight of root from 48% to 27% and of shoot from 11% to 28%, root length from 40% to 27%, and shoot length from 40% to 32% in wheat plants under chromium stress, compared to untreated plants.

Zeng et al. (2011) observed the toxicity of Cr in rice plant and alleviation by Si in a hydroponic experiment and reported that a 100 μM·L⁻¹ concentration of chromium reduced the plant growth by reducing the plant height and dry biomass significantly. With the addition of 1.25 and 2.5 mM silicon, an increase in plant growth was observed. Three levels of Si were used. As the Si concentration was increased from 1.25 to 2.5 mM, the seedling height and dry biomass increased, compared to rice plants that were under chromium stress.

There are several mechanisms involved in the reduction of the growth parameters of plants under chromium stress. Chromium can destroy the protein in the cell wall (Song et al., 2014; Wang et al., 2014) and cause an imbalance of nutrients, as discussed above. Si reduces the uptake of Cr and ameliorates its toxic effects through many mechanisms, such as developing complexes with metal, coprecipitation, and compartmentation. Cr reduces the performance antioxidant enzymes, which inhibit plant growth, such as plant height and root length. Silicon enhances the activities of antioxidants and increases the plant growth parameters in cucumber plant (Dragišić Maksimović et al., 2007).

Ali et al. (2013a) observed the ultrastructure of barley under chromium toxicity and its alleviation by Si application in a hydroponic experiment. Transmission electron micrographs showed the ultrastructural injuries due to Cr toxicity. In leaf, irregular swelling was observed in chloroplast, while the amount of plastoglobuli was increased under Cr stress. When Si + Cr were applied, they loosened and disintegrated the disclosed thylakoid membrane. In roots, electron-dense granule and crystal forms of Cr were observed in the vacuoles, whereas those found attached to the cell wall resulted in shrinkage of the cell membrane under chromium stress. Disruption of the nuclear membrane, disruption of the nucleus, and absence of the nucleus were observed in plants treated with Cr alone. With the addition of silicon under chromium stress, these affects were less than those of the plants treated with Cr alone. This might be the cause of less chromium uptake due to silicon's ameliorative effects and less accumulation and translocation in plants, which decreases ultrastructural damage.

12.7 CONCLUSION AND PERSPECTIVE

In this current era, the toxicity of heavy metals such as chromium is a huge problem for living organisms. It causes adverse effects on plants and inhibits their growth and reduces their yield significantly, which causes a big economy loss and food scarcity. We noted many effects of chromium on plants, for example, low fresh weight and dry weight of biomass, inhibition of growth parameter, and reduction in yield. In chromium-contaminated soil, its uptake in plants increases with an increase of chromium in the soil. It may enter our food chain and disturb our ecosystem. By consumption of these plants as food, its quantity may accumulate in the bodies of consumers and cause severe problems. Silicon has the ability to reduce the uptake of chromium and enhance the growth of plants by increasing tolerance in plants under stress. We reviewed many mechanisms involving the S-mediated alleviation of chromium stress. Si reduces the toxicity by reducing the mobility and activity of chromium ions with different mechanisms, such as complex formation, coprecipitation, enhancement of the performance of antioxidants, compartmentation, and structural and ultrastructural changes. We discussed these mechanisms to understand the alleviation action of Si under chromium toxicity.

We anticipate growing interest in the field of soil and plant sciences toward understanding the mechanisms of the alleviation of chromium and other heavy metal toxicity by the application of silicon in the near future. Current knowledge regarding the mechanisms in these cases is developing day by day. However, more studies are required for better understanding the mechanism involved in Si-mediated alleviation on a large scale. Moreover, the majority of experiments we reviewed are conducted in hydroponics performing in controlled conditions. So, more studies should be conduct in naturally contaminated fields, which will support step forwards for using silicon in farming practices in chromium-contaminated soil.

REFERENCES

Adrees, M., Ali, S., Rizwan, M., Zia-ur-Rehman, M., Ibrahim, M., Abbas, F. et al. 2015. Mechanisms of silicon-mediated alleviation of heavy metal toxicity in plants: A review. *Ecotoxicol. Environ. Saf.*, 119, 186–197.

Adriano, D.C. 1986. *Trace Elements in the Terrestrial Environment*. New York: Springer Verlag, pp. 105–123.

Al-aghabary, K., Zhu, Z., and Shi, Q. 2004. Influence of silicon supply on chlorophyll content, chlorophyll fluorescence, and antioxidative enzyme activities in tomato plants under salt stress. *J. Plant Nutr.*, 27, 2101–2115.

Ali, S., Bharwana, S.A., Rizwan, M., Farid, M., Kanwal, S., Ali, Q., Ibrahim, M., Gill, R.A., and Khan, M.D. 2015a. Fulvic acid mediates chromium (Cr) tolerance in wheat (*Triticum aestivum* L.) through lowering of Cr uptake and improved antioxidant defense system. *Environ. Sci. Pollut.*, 1–9.

Ali, S., Chaudhary, A., Rizwan, M., Anwar, H.T., Adrees, M., Farid, M., and Anjum, S.A. 2015b. Alleviation of chromium toxicity by glycinebetaine is related to elevated antioxidant enzymes and suppressed chromium uptake and oxidative stress in wheat (*Triticum aestivum* L.). *Environ. Sci. Pollut. Res.*, 22(14), 10669–10678.

Ali, S., Farooq, M.A., Jahangir, M.M., Abbas, F., Bharwana, S.A., and Zhang, G.P. 2013b. Effect of chromium and nitrogen form on photosynthesis and anti-oxidative system in barley. *Biol. Plant.*, 57(4), 758–763.

Ali, S., Farooq, M.A., Yasmeen, T., Hussain, S., Arif, M.S., Abbas, F., Bharwana, S.A., and Zhang, G.P. 2013a. The influence of silicon on barley growth, photosynthesis and ultra-structure under chromium stress. *Ecotoxicol. Environ. Saf.*, 89, 66–72.

Avudainayagam, S., Megharaj, M., Owens, G., Kookana, R.S., Chittleborough, D., and Naidu, R. 2003. Chemistry of chromium in soils with emphasis on tannery waste sites. *Rev. Environ. Contam. Toxicol.*, 178, 53–91.

Barančíková, G., Madams, M., and Rybàr, O. 2004. Crop contamination by selected trace elements. *J. Soils Sediments*, 4(1), 37–42.

Barceloux, D.G. 1999. Chromium. *Clin. Toxicol.*, 37, 173–194.

Barnhart, J. 1997. Occurrences, uses, and properties of chromium. *Regul. Toxicol. Pharmacol.*, 26, S3–S7.

Bartlett, R.J. 1991. Chromium cycling in soils and water: Links, gaps, and methods. *Environ. Health Perspect.*, 92, 17–24.

Becquer, T., Quantin, C., Sicot, M., and Boudot, J.P. 2003. Chromium availability in ultramafic soils from New Caledonia. *Sci. Total Environ.*, 301(1), 251–261.

Bluskov, S., Arocena, J.M., Omotoso, O.O., and Young, J.P. 2005. Uptake, distribution, and speciation of chromium in *Brassica juncea*. *Int. J. Phytoremediation*, 7(2), 153–165.

Briantais, J.M., Vernotte, C., Krause, G.H., and Weis, E. 1986. Chlorophyll fluorescence of higher plant: Chloroplasts and leaves. In Amesz, J., Govindjee, and Fork, D.C. (eds.), *Light Emission by Plants and Photosynthetic Bacteria*. New York: Academic Press, pp. 539–577.

Cervantes, C., and Campos-Garcia, J. 2007. Reduction and efflux of chromate by bacteria. In *Molecular Microbiology of Heavy Metals*. Heidelberg: Springer-Verlag, pp. 407–419.

Cervantes, C., Campos-Garcia, J., Devars, S., Gutiérrez-Corona, F., Loza-Tavera, H., Torres-Guzmàn, J.C., and Moreno-Sànchez, R. 2001. Interactions of chromium with microorganisms and plants. *FEMS Microbiol. Rev.*, 25, 335–347.

Chandra, P., Sinha, S., and Rai, U.N. 1997. Bioremediation of Cr from water and soil by vascular aquatic plants. In Kruger, E.L., Anderson, T.A., and Coats, J.R. (eds.), *Phytoremediation of Soil and Water Contaminants*. ACS Symposium Series, vol. 664. Washington, DC: American Chemical Society, pp. 274–282.

Chen, H.M., Zheng, C.R., Tu, C., and Shen, Z.G. 2000. Chemical methods and phytoremediation of soil contaminated with heavy metals. *Chemosphere*, 41(1), 229–234.

Cieślak-Golonka, M. 1996. Toxic and mutagenic effects of chromium (VI). A review. *Polyhedron*, 15(21), 3667–3689.

Cotton, F.A., and Wilkinson, G. 1980. *Advanced Inorganic Chemistry*. 4th ed. New York: Wiley.

da Cunha, K.P.V., and do Nascimento, C.W.A. 2009. Silicon effects on metal tolerance and structural changes in maize (*Zea mays* L.) grown on a cadmium and zinc enriched soil. *Water Air Soil Pollut.*, 197(1–4), 323–330.

da Cunha, K.P.V., do Nascimento, C.W.A., and José da Silva, A. 2008. Silicon alleviates the toxicity of cadmium and zinc for maize (*Zea mays* L.) grown on a contaminated soil. *J. Plant Nutr. Soil Sci.*, 171(6), 849–853.

Dermou, E., and Vayenas, D.V. 2008. Biological Cr(VI) reduction in a trickling filter under continuous operation with recirculation. *J. Chem. Technol. Biotechnol.*, 83, 871–877.

Dixit, V., Pandey, V., Shyam, R. 2002. Chromium ions inactivate electron transport and enhance superoxide generation in vivo in pea (*Pisum sativum* L.) root mitochondria. *Plant Cell Environ.*, 25, 687–693.

Dragišić Maksimović, J., Bogdanović, J., Maksimović, V., and Nikolic, M. 2007. Silicon modulates the metabolism and utilization of phenolic compounds in cucumber (*Cucumis sativus* L.) grown at excess manganese. *J. Plant Nutr. Soil Sci.*, 170(6), 739–744.

Ehsan, S., Ali, S., Noureen, S., Mahmood, K., Farid, M., Ishaque, W., Shakoor, M.B., Rizwan, M. 2014. Citric acid assisted phytoremediation of cadmium by *Brassica napus* L. *Ecotox. Environ. Saf.*, 106, 164–172.

Epstein, E. 1994. The anomaly of silicon in plant biology. *Proc. Natl. Acad. Sci. U.S.A.*, 91(1), 11–17.

Epstein, E. 1999. Silicon. *Plant Mol. Biol.*, 50, 641–664.

Epstein, E., and Bloom, A.J. 2005. *Mineral Nutrition of Plants: Principles and Perspectives*. 2nd ed. Sinauer Associates, Sunderland, MA.

Fauteux, F., Remus-Borel, W., Menzies, J.G., and Belanger, R.R. 2005. Silicon and plant disease resistance against pathogenic fungi: Mini review. *FEMS Microbiol Lett.*, 249, 1–6.

Gill, R.A., Ali, B., Islam, F., Farooq, M.A., Gill, M.B., Mwamba, T.M., and Zhou, W. 2015b. Physiological and molecular analyses of black and yellow seeded *Brassica napus* regulated by 5-aminolivulinic acid under chromium stress. *Plant Physiol. Biochem.*, 94, 130–143.

Gill, R.A., Zang, L., Ali, B., Farooq, M.A., Cui, P., Yang, S., Ali, S., Zhou, W. 2015a. Chromium-induced physio-chemical and ultrastructural changes in four cultivars of *Brassica napus* L. *Chemosphere*, 120, 154–164.

Gong, H.J., Randall, D.P., and Flowers, T.J. 2006. Silicon deposition in the root reduces sodium uptake in rice (*Oryza sativa* L.) seedlings by reducing bypass flow. *Plant Cell Environ.*, 29, 1970–1979.

Gonzalez, M.R., Pereyra, A.M., Zerbino, R., and Basaldella, E.I. 2015. Removal and cementitious immobilization of heavy metals: Chromium capture by zeolite-hybridized materials obtained from spent fluid cracking catalysts. *J. Cleaner Prod.*, 91, 187–190.

Gorbi, G., Corradi, M.G., Invidia, M., and Bassi, M. 2001. Light intensity influences chromium bioaccumulation and toxicity in *Scenedesmus acutus* (Chlorophyceae). *Ecotoxicol. Environ. Saf.*, 48, 36–42.

Gu, H.H., Qiu, H., Tian, T., Zhan, S.S., Chaney, R.L., Wang, S.Z. et al. 2011. Mitigation effects of silicon rich amendments on heavy metal accumulation in rice (*Oryza sativa* L.) planted on multi-metal contaminated acidic soil. *Chemosphere*, 83(9), 1234–1240.

Gu, H.H., Zhan, S.S., Wang, S.Z., Tang, Y.T., Chaney, R.L., Fang, X.H., and Qiu, R.L. 2012. Silicon-mediated amelioration of zinc toxicity in rice (*Oryza sativa* L.) seedlings. *Plant and Soil*, 350(1–2), 193–204.

Habiba, U., Ali, S., Farid, M., Shakoor, M.B., Rizwan, M., Ibrahim, M., Abbasi, G.H., Hayat, T., and Ali, B. 2014. EDTA enhanced plant growth, antioxidant defense system, and phytoextraction of copper by *Brassica napus* L. *Environ. Sci. Pollut. Res.*, 22, 1534–1544.

Han, F.X., Sridhar, B.B.M., Monts, D.L., and Su, Y. 2004. Phytoavailability and toxicity of trivalent and hexavalent chromium to *Brassica juncea*. *New Phytol.*, 162, 489–499.

Hayyat, M.U., Mahmood, R., Akram, S., Haider, Z., Rizwan, S.T., Siddiq, Z., and RabNawaz. 2015. Use of alleviators to reduce toxicity of chromium in germination of common agricultural crops. *Environ. Ecol. Res.*, 3(2), 44–49.

Hodson, M.J., White, P.J., Mead A., and Broadley M.R. 2005. Phylogenetic variation in the silicon composition of plants. *Ann. Bot.*, 96, 1027–1046.

Hossain, M.T., Mori, R., Soga, K., Wakabayashi, K., Kamisaka, S., Fujii, S., and Hoson, T. 2002. Growth promotion and an increase in cell wall extensibility by silicon in rice and some other Poaceae seedlings. *J. Plant Res.*, 115(1), 0023–0027.

Hossain, M.T., Soga, K., Wakabayashi, K., Kamisaka, S., Fujii, S., Yamamoto, R., and Hoson, T. 2007. Modification of chemical properties of cell walls by silicon and its role in regulation of the cell wall extensibility in oat leaves. *J. Plant Physiol.*, 164(4), 385–393.

Huang, Z.A., Jiang, D.A., Yang, Y., Sun, J.W., and Jin, S.H. 2004. Effects of nitrogen deficiency on gas exchange, chlorophyll fluorescence, and antioxidant enzymes in leaves of rice plants. *Photosynthetica*, 42, 357–364.

Huffman Jr., E.W.D., and Allaway, H.W. 1973a. Chromium in plants: Distribution in tissues, organelles, and extracts and availability of bean leaf Cr to animals. *J. Agric. Food Chem.*, 21, 982–986.

Huffman Jr., E.W.D., and Allaway, W.H. 1973b. Growth of plants in solution culture containing low levels of chromium. *Plant Physiol.*, 52, 72–75.

Iwasaki, K., Maier, P., Fecht, M., and Horst, W.J. 2002. Leaf apoplastic silicon enhances manganese tolerance of cowpea (*Vigna unguiculata*). *J. Plant Physiol.*, 159(2), 167–173.

Jones, L.H.P., and Handreck, K.A. 1967. Silica in soils, plants and animals. *Adv. Agron.*, 19(1), 107–149.

Katz, S.A., and Salem, H. 1994. *The Biological and Environmental Chemistry of Chromium*. New York: VCH Publishers.

Kaya, C., Tuna, A.L., Sonmez, O., Ince, F., and Higgs, D. 2009. Mitigation effects of silicon on maize plants grown at high zinc. *J. Plant Nutr.*, 32(10), 1788–1798.

Keller, C., Rizwan, M., Davidian, J.C., Pokrovsky, O.S., Bovet, N., Chaurand, P., and Meunier, J.D. 2015. Effect of silicon on wheat seedlings (*Triticum turgidum* L.) grown in hydroponics and exposed to 0 to 30 µM Cu. *Planta*, 241(4), 847–860.

Kidd, P.S., Llugany, M., Poschenrieder, C.H., Gunse, B., and Barcelo, J. 2001. The role of root exudates in aluminium resistance and silicon-induced amelioration of aluminium toxicity in three varieties of maize (*Zea mays* L.). *J. Exp. Bot.*, 52(359), 1339–1352.

Kim, Y.H., Khan, A.L., Kim, D.H., Lee, S.Y., Kim, K.M., Waqas, M. et al. 2014. Silicon mitigates heavy metal stress by regulating P-type heavy metal ATPases, *Oryza sativa* low silicon genes, and endogenous phytohormones. *BMC Plant Biol.*, 14(1), 13.

Kovda, V.A. 1973. *The Bases of Learning about Soils*. 2 vols. Moscow: Nayka.

Krishnamurthy, S., and Wilkens, M.M. 1994. Environmental chemistry of Cr. *Northeastern Geol.*, 16(1), 14–17.

Kulikova, Z.L., and Lux, A. 2010. Silicon influence on maize, *Zea mays* L., hybrids exposed to cadmium treatment. *Bull. Environ. Contam. Toxicol.*, 85(3), 243–250.

Kumar, D., Tripathi, D., and Chauhan, D. 2014. Phytoremediation potential and nutrient status of Gaerth. Tree seedlings grown under different chromium (CrVI) treatments. *Biol. Trace Element Res.*, 2(157), 164–174.

Li, G.J., Liu, Y.H., Wu, X.H., Zhu, Z.J., Wang, B.G., and Lu, Z.F. 2007. Effects of exogenous silicon on reactive oxygen species metabolism of asparagus bean seedlings under rust stress. *Acta Hortic. Sin.*, 34, 1207–1212.

Liang, Y., Wong, J.W.C., and Wei, L. 2005. Silicon-mediated enhancement of cadmium tolerance in maize (*Zea mays* L.) grown in cadmium contaminated soil. *Chemosphere*, 58(4), 475–483.

Liang, Y., Sun, W., Zhu, Y.G., and Christie, P. 2007. Mechanisms of silicon-mediated alleviation of abiotic stresses in higher plants: A review. *Environ. Pollut.*, 147(2), 422–428.

Linos, A., Petralias, A., Christophi, C.A., Christoforidou, E., Kouroutou, P., Stoltidis, M., Veloudaki, A., Tzala, E., Makris, K.C., Karagas, M.R. 2011. Oral ingestion of hexa-valent chromium through drinking water and cancer mortality in an industrial area of Greece—An ecological study. *Environ. Health Glob.*, 10, 50.

Liu, D., Zou, J., Wang, M., and Jiang, W. 2008. Hexavalent chromium uptake and its effects on mineral uptake, antioxidant defence system and photosynthesis in *Amaranthus viridis* L. *Bioresour. Technol.*, 99(7), 2628–2636.

Liu, J., Zhang, H., Zhang, Y., and Chai, T. 2013. Silicon attenuates cadmium toxicity in *Solanum nigrum* L. by reducing cadmium uptake and oxidative stress. *Plant Physiol. Biochem.*, 68, 1–7.

Ma, J.F., Miyake, Y., and Takahashi, E. 2001. Silicon as a beneficial element for crop plants. *Stud. Plant Sci.*, 8, 17–39.

Ma, J.F., and Yamaji, N. 2006. Silicon uptake and accumulation in higher plants. *Trends Plant Sci.*, 11, 392–397.

Mallick, S., Sinam, G., Mishra, R.K., and Sinha, S. 2010. Interactive effects of Cr and Fe treatments on plants growth, nutrition and oxidative status in *Zea mays* L. *Ecotoxicol. Environ. Saf.*, 73(5), 987–995.

Malmir, H.A. 2011. Comparison of antioxidant enzyme activities in leaves stems and roots of sorghum (*Sorghum bicolor* L) exposed to chromium (VI). *Afr. J. Plant Sci.*, 5(5), 436–444.

McGrath, S.P. 1995. Chromium and nickel. In Alloway, B.J. (ed.), *Heavy Metal in Soils*. 2nd ed. London: Chapman & Hall, pp. 152–178.

Naeem, A., Ghafoor, A., and Farooq, M. 2014. Suppression of cadmium concentration in wheat grains by silicon is related to its application rate and cadmium accumulating abilities of cultivars. *J. Sci. Food Agric.*, 95(12), 2467–2472.

Nagajyoti, P.C., Lee, K.D., and Sreekanth, T.V.M. 2010. Heavy metals, occurrence and toxicity for plants: A review. *Environ. Chem. Lett.*, 8, 199–216.

Nagarajan, M. 2014. Effect of chromium on growth, biochemicals and nutrient accumulation of paddy (*Oryza sativa* L.). *Int. Lett. Nat. Sci.*, 18.

Nascimento, C.W.A.D., and Xing, B. 2006. Phytoextraction: A review on enhanced metal availability and plant accumulation. *Sci. Agric.*, 63(3), 299–311.

Nayak, A.K., Jena, R.C., Jena, S., Bhol, R., and Patra, H.K. 2015. Phytoremediation of hexavalent chromium by *Triticum aestivum* L. *Scientia*, 9(1), 16–22.

Noman, A., Ali, S., Naheed, F., Ali, Q., Farid, M., Rizwan, M., and Irshad, M.K. 2015. Foliar application of ascorbate enhances the physiological and biochemical attributes of maize (*Zea mays* L.) cultivars under drought stress. *Arch. Agron. Soil Sci.*, 61(12), 1–14.

Nriagu, J.O. 1988. Production and uses of chromium. In *Chromium in Natural and Human Environment*. New York: John Wiley & Sons, pp. 81–105.

Nwugo, C.C., and Huerta, A.J. 2008. Effects of silicon nutrition on cadmium uptake, growth and photosynthesis of rice plants exposed to low-level cadmium. *Plant Soil*, 311(1–2), 73–86.

Occupational Safety and Health Administration (OSHA), Department of Labor. 2006. Occupational exposure to hexavalent chromium. Final rule. *Fed. Regist.* 71, 10099–10385.

Oliva, S.R., Mingorance, M.D., and Leidi, E.O. 2011. Effects of silicon on copper toxicity in *Erica andevalensis* Cabezudo and Rivera: A potential species to remediate contaminated soils. *J. Environ. Monitor.*, 13(3), 591–596.

Panda S.K., and Choudhury, S. 2005. Chromium stress in plants. *Braz. J. Plant Physiol.*, 17(1), 95–102.

Panda, S.K., and Parta, H.K. 1997. Physiology of chromium toxicity in plants—A review. *Plant Physiol. Biochem.*, 24(1), 10–17.

Parmar, J.K., and Patel, K.P. 2015. Remediation of phytotoxic effect of chromium by different amendments in rice-wheat sequence. *Nat. Environ. Pollut. Technol.*, 14(1), 77.

Pulford, I.D., Watson, C., and McGregor, S.D. 2001. Uptake of chromium by trees: Prospects for phytoremediation. *Environ. Geochem. Health*, 23(3), 307–311.

Qing, X., Zhao, X., Hu, C., Wang, P., Zhang, Y., Zhang, X., and Qu, C. 2015. Selenium alleviates chromium toxicity by preventing oxidative stress in cabbage (*Brassica campestris* L. ssp. Pekinensis) leaves. *Ecotoxicol. Environ. Saf.*, 114, 179–189.

Raven, J.A. 1983. The transport and function of silicon in plants. *Biol. Rev. Cambr. Phil. Soc.*, 58, 179–207.

Rogalla, H., and Römheld, V. 2002. Role of leaf apoplast in silicon-mediated manganese tolerance of *Cucumis sativus* L. *Plant Cell Environ.*, 25(4), 549–555.

Schnettger, B., Critchley, C., and Santore, U.J. 1994. Relationship between photo-inhibition of photosynthesis, D1 protein turnover and chloroplast structure: Effect of protein synthesis. *Plant Cell Environ.*, 17, 55–64.

Shanker, A.K., Cervantes, C., Loza-Tavera, H., and Avudainayagam, S. 2005. Chromium toxicity in plants. *Environ. Int.*, 31(5), 739–753.

Shanker, A.K., Djanaguiraman, M., Sudhagar, R., Chandrashekar, C.N., and Pathmanabhan, G. 2004a. Differential antioxidative response of ascorbat glutathione pathway enzymes and metabolites to chromium speciation stress in green gram (*Vigna radiata* (L) R Wilczek, cv CO 4) roots. *Plant Sci.*, 166, 1035–1043.

Shanker, A.K., Djanaguiraman, M., Sudhagar, R., Jayaram, R., and Pathmanabhan, G. 2004b. Expression of metallothionein 3 (MT3) like protein mRNA in *Sorghum* cultivars under chromium (VI) stress. *Curr. Sci.*, 86(7), 901–902.

Sharma, S.S., and Dietz, K.J. 2009. The relationship between metal toxicity and cellular redox imbalance. *Trends Plant Sci.*, 14(1), 43–50.

Shi, G., Cai, Q., Liu, C., and Wu, L. 2010. Silicon alleviates cadmium toxicity in peanut plants in relation to cadmium distribution and stimulation of antioxidative enzymes. *Plant Growth Regul.*, 61(1), 45–52.

Shi, Q., Bao, Z., Zhu, Z., He, Y., Qian, Q., and Yu, J. 2005. Silicon-mediated alleviation of Mn toxicity in *Cucumis sativus* in relation to activities of superoxide dismutase and ascorbate peroxidase. *Phytochemistry*, 66(13), 1551–1559.

Singh, N.K., Rai, U.N., Verma, D.K., and Rathore, G. 2014. *Kocuria flava* induced growth and chromium accumulation in *Cicer arietinum* L. *Int. J. Phytoremediation*, 16(1), 14–28.

Skeffington, R.A., Shewry, P.R., and Peterson, P.J. 1976. Chromium uptake and transport in barley seedlings (*Hordeum vulgare* L.). *Planta*, 132(3), 209–214.

Sommer, M., Kaczorek, D., Kuzyakov, Y., and Breuer, J. 2006. Silicon pools and fluxes in soils and landscapes—A review. *J. Plant Nutr. Soil Sci.*, 169(4), 582–582.

Song, A., Li, P., Fan, F., Li, Z., and Liang, Y. 2014. The effect of silicon on photosynthesis and expression of its relevant genes in rice (*Oryza sativa* L.) under high-zinc stress. *PLoS One*, 9(11), e113782.

Srivastava, R.K., Pandey, P., Rajpoot, R., Rani, A., Gautam, A., and Dubey, R.S. 2015. Exogenous application of calcium and silica alleviates cadmium toxicity by suppressing oxidative damage in rice seedlings. *Protoplasma*, 4(252), 959–975.

Sundaramoorthy, P., Chidambaram, A., Ganesh, K.S., Unnikannan, P., and Baskaran, L. 2010. Chromium stress in paddy: (i) Nutrient status of paddy under chromium stress; (ii) Phytoremediation of chromium by aquatic and terrestrial weeds. *Comp. Rend. Biol.*, 333(8), 597–607.

Suzuki, N., Koussevitzky, S., Mittler, R.O.N., and Miller, G.A.D. 2012. ROS and redox signalling in the response of plants to abiotic stress. *Plant Cell Environ.*, 35(2), 259–270.

Tang, H., Liu, Y., Gong, X., Zeng, G., Zheng, B., Wang, D. et al. 2015. Effects of selenium and silicon on enhancing antioxidative capacity in ramie (*Boehmeria nivea* (L.) Gaud.) under cadmium stress. *Environ. Sci. Pollut. Res.*, 1–10.

Tian, T., Ali, B., Qin, Y., Malik, Z., Gill, R.A., Ali, S., and Zhou, W. 2014. Alleviation of lead toxicity by 5-aminolevulinic acid is related to elevated growth, photosynthesis and suppressed ultrastructural damages in oilseed rape. *Biomed Res. Int.*, 2014, 1–11.

Tripathi, D.K., Singh, V.P., Prasad, S.M., Chauhan, D.K., Dubey, N.K., and Rai, A.K. 2015. Silicon mediated alleviation of Cr (VI) toxicity in wheat seedlings as evidenced by chlorophyll florescence, laser induced breakdown spectroscopy and anatomical changes. *Ecotoxicol. Environ. Saf.*, 113, 133–144.

Tripathi, P., Tripathi, R.D., Singh, R.P., Dwivedi, S., Goutam, D., Shri, M., and Chakrabarty, D. 2013. Silicon mediates arsenic tolerance in rice (*Oryza sativa* L.) through lowering of arsenic uptake and improved antioxidant defense system. *Ecol. Eng.*, 52, 96–103.

Vazquez, M.D., Poschenrieder, C.H., and Barcelo, J. 1987. Chromium VI induced structural and ultrastructural changes in bush bean plants (*Phaseolus vulgaris* L.). *Ann. Bot.*, 59(4), 427–438.

Wang, S., Wang, F., and Gao, S. 2014. Foliar application with nano-silicon alleviates Cd toxicity in rice seedlings. *Environ. Sci. Pollut. Res.*, 22(4), 2837–2845.

Wang, Y., Stass, A., and Horst, W.J. 2004. Apoplastic binding of aluminum is involved in silicon-induced amelioration of aluminum toxicity in maize. *Plant Physiol.*, 136(3), 3762–3770.

Zayed, A.M., and Terry, N. 2003. Chromium in the environment: Factors affecting biological remediation. *Plant Soil*, 249(1), 139–156.

Zeng, F.R., Zhao, F.S., Qiu, B.Y., Ouyang, Y.N., Wu, F.B., and Zhang, G.P. 2011. Alleviation of chromium toxicity by silicon addition in rice plants. *Agric. Sci. China*, 10(8), 1188–1196.

Zhang, C., Wang, L., Nie, Q., Zhang, W., and Zhang, F. 2008. Long-term effects of exogenous silicon on cadmium translocation and toxicity in rice (*Oryza sativa* L.). *Environ. Exp. Bot.*, 62(3), 300–307.

Zhang, Q., Yan, C., Liu, J., Lu, H., Wang, W., Du, J., and Duan, H. 2013a. Silicon alleviates cadmium toxicity in *Avicennia marina* (Forsk.) Vierh. seedlings in relation to root anatomy and radial oxygen loss. *Mar. Pollut. Bull.*, 76(1), 187–193.

Zhang, S., Li, S., Ding, X., Li, F., Liu, C., Liao, X., and Wang, R. 2013b. Silicon mediated the detoxification of Cr on pakchoi (*Brassica chinensis* L.) in Cr-contaminated soil. *Int. J. Food Agric. Environ.*, 11(2), 814–819.

13 Role of Silicon in Plants
Present Scenario and Future Prospects

Shweta, Swati Singh, Bishwajit Kumar Kushwaha,
Parvaiz Ahmad, Durgesh Kumar Tripathi,
Nawal Kishore Dubey, and Devendra Kumar Chauhan

CONTENTS

13.1 INTRODUCTION

The world's population increases day by day, especially in developing countries, and people are facing the problem of meeting nutritious food and health security requirements. It has become a major concern for scientists to improve the fertility of agricultural soil, so that the productivity, as well as the quality, increases and one can have a sufficient and healthy source of food. Soil contains many macronutrients (N, P, Mg, Ca, K, and S) and micronutrients (B, Mn, P, Cl, Cu, Fe, Mo, and Zn) for the growth and development of plants, but due to the increasing content of heavy metals by any means, like high atmospheric deposition, improper weathering of parental rocks, high mineralization, industrial waste inputs, and some other anthropogenic activities, the adequate concentration of beneficial elements decreases and that of heavy metals (macro- and micronutrients) increases excessively in the soil or agricultural field (Chen et al., 2010a,b; Elbaz et al., 2010; Singh and Prasad, 2011; Gill et al., 2013; Gangwar et al., 2014). This increased level of heavy metals may lead to an alteration in the properties and pH of soils and ultimately affect the uptake of nutrients in plants, hence lowering the quality and quantity of a plant's yield (Tripathi et al., 2014a,b, 2015a,b). Generally, heavy metals are known to cause deleterious effects in the morphology, as well as physiology, of plants, such as uptake of minerals, photosynthesis, and alteration in many genetic processes (Rodriguez et al., 2012; Ali et al., 2013; Li et al., 2013; Gangwar et al., 2014). Moreover, several methods have been used to overcome the adverse effects of heavy metals in plants, and the application of silicon is regarded as one of the most effective and beneficial elements, which increases the uptake of nutrients, and thus the growth of plants. However, it has been found that silicon plays remarkably many roles in plants, in which some are most beneficial to plants. Tripathi et al. (2012a) tested the effect of silicon on the growth of rice seedlings against cadmium and concluded that silicon successfully alleviated the toxicity of cadmium in plants and reduced the its accumulation. Similar beneficial effect of silicon was also demonstrated by Hussain et al. (2015) in *Triticum aesativum* under the cadmium stress. In addition to this, Malčovská et al. (2014) found that, in the presence of silicon, oxidative stresses were significantly reduced in *Zea mays* caused by cadmiun stress. Furthermore, it has been also observed that silicon can also alleviate aluminum, manganese, iron, and heavy metal toxicity as well as drought, water, salinity, stresses, and UV radiation damage (Cocker et al., 1998; Epstein, 1999; Britez et al., 2002; Liang et al., 2006; Shen et al., 2010; Chen et al., 2011, Singh et al., 2011; Tripathi et al., 2011, 2012b; De Souza et al., 2013; Mehrabanjoubani et al., 2015; Tripathi et al., 2016b). Silicon in the form of fertilizers is also used against the biotic and abiotic stresses and also in the case of other nutrient deprivation condition. Many scientists also observed that silica enhances the yield and quality, and make appropriate defense against fungal and other pathogenic attack (Lanning et al., 1966; Brwen et al., 1992; Korndorfer et al., 2001; Gillman et al., 2003; Rodrigues et al., 2004; Liang et al., 2006).

Silica also exists in nature in various forms, and its properties vary from substrate to substrate. In plants, silica accumulates in the form of silicic acid (monosilicic acid), having the formula $Si(OH)_4$. In nature, many water-soluble forms of silica (orthosilicates, metasilicates, disilicates, and trisilicates) are found, and they are present in surface water of about 1–100 mg/L (Liang et al., 2007). Silicon has proved to be more beneficial elements and has been reported as a good source in some plants like Conifers, Ferns, Mosses, and *Equisetum*; some families including Cyperaceae, Poaceae, etc. (Epstein, 1994, 1999; Liang, 1999; Ma et al., 2001; Liang et al., 2005; Chauhan et al., 2011a,b,c; Tripathi et al., 2012a,b, 2013, 2014a,b, 2016a; Ma and Yamaji, 2015). As it has been well reported, silicon is accumulated in the plants in the form of silicic acid and deposited in and between the cells, which is ultimately known as phytoliths (Chauhan et al., 2011a,b; Guntzer et al., 2012; Liang et al., 2007; Tripathi et al., 2011, 2012a,b, 2014a,b). Phytolith is involved in the biogeochemical cycle of silica by maintaining the amount of silica in soil (Gérard et al., 2008). Since silica is associated with carbon and hence involved in many other terrestrial geological cycles. It also shows an important role in atmospheric CO_2 regulation (Street-Perrott and Barker, 2008; Li et al., 2011; Song et al., 2012). Silicon is ubiquitously present in most of the plants because water always contains some amount of silica (Cheng, 1982; Guntzer et al., 2012; Tripathi et al., 2014a). Additionlly, in other countries, silicon has been used in combination with other macro- and micronutrients and thus significantly increased plant growth (Guntzer et al., 2012). The present chapter deals with the manifold roles of silicon and its efficacy against the stress condition; however, some recent studies have also been well discussed in the chapter.

13.2 DYNAMICS OF SILICON

After oxygen, silicon is considered the second most abundant element present in soil; it encompasses 25% of the total earth's crust (Sommer et al., 2006). It is considered as a nonessential element for most of the plant species, but the uptake of silicon assists in producing advanced effects in the majority of plants by enhancing their pathogenic resistance (Meyer and Keeping, 2001; Shiguro, 2001; Richmond and Sussman, 2003; Reynolds et al., 2009), decreasing the effect of drought (Lux et al., 2002; Richmond and Sussman, 2003), providing tolerance against heavy metals (Neumann and Zur Nieden, 2001; Richmond and Sussman, 2003), and improving the quality and quantity of the agricultural crop yields (Korndorfer and Lepsch, 2001; Richmond and Sussman, 2003). Beneficial or toxic responses of silicon differ from species; and their effects have also been noted during biotic and abiotic stresses (Epstein, 1999; Richmond and Sussman, 2003). Applications of silicon in animal cells are uncertain, while in plant life, it has profound applications. The reservoir of silicon in plants (phytolith) is also used in studies of evolutionary evidences, since it does not rot after the death of plants. In nature, silicon occurs in various forms because of its variable oxidation states; they vary from 4 to −4 (4, 3, 2, 1, −1, −2, −3, −4) (Eranna, 2014).

13.3   SILICON IN SOIL AND PLANTS

13.3.1   SILICA IN SOIL

Generally, silicon that present in the soil are rich in minerals, as evidenced by kaolin, smectite, vermiculite, and so forth, and some silicates are found in the form of crystals, like feldspars, orthoclase, and plagioclase; on the other hand, some are quartz and amorphous silica (Orlov, 1985; Sahebi et al., 2014; Tubaña and Heckman, 2015). However, the above forms are biogeochemically immobile (Sahebi et al., 2014). Monosilicic acid and polysilicic acid are two major forms of silicon that solubilize in soil more easily (Iler, 1979; Matychenkov and Snyder, 1996; Sahebi, 2015). The tendency of silicic acid to be absorbed by soil is very slow, and hence its mobilization in soil is lower (Matychenkov and Snyder 1996; Matichenkov et al., 1997; Sahebi, 2015). The uptake of phosphates directly depends on the concentration of monosilicic acid; as the concentration of monosilicic acid increases, the uptake of phosphates by the plant also increases (Matychenkov and Ammosova, 1996; Sahebi, 2015), but it starts decreasing after interaction with heavy metals like aluminum, iron, and manganese (Lumsdon and Farmer, 1995). The properties of monosilicic and polysilicic acids are different: polysilicic acid affects the physical properties of soil, while monosilicic acid forms colloidal atoms due to its chemically immobile nature (Exley and Birchall, 1993; Sahebi, 2015). It has been reported that polysilicic acid affects the soil's water holding capacity and structure formation (Matychenkov et al., 1995; Sahebi, 2015). The major drawback of silicon is its inability to form chemical bonds with other elements or compounds because of its small size—hence, it cannot take part much in plant metabolism.

13.3.2   SILICON IN PLANTS

Silicon is considered as one of the most essential elements because of its manifold beneficial role in several crop plants; as evidence, it is used to alleviate different kinds of biotic and abiotic stresses (Epstein, 1994, 1999; Marschner, 1995; Ma, 2004; Liang et al., 2006; Waraich et al., 2011) by generating various antioxidants, promoting growth, yield, metabolism, and thus quality and quantity of plant products (Gonzalo et al., 2013). The imbalance in soil nutrients and their poor fertilities lead to high troubles in agricultural foods or resources, which ultimately lead to decrease in global food production (Peng and Zhou, 2010; Moharana et al., 2012; Waraich et al., 2012; Tripathi et al., 2014).

The place of silicon deposition in plants is different and it depends on the type of species. Some plants of family Poaceae like *Triticum* spp., *Saccharum officinarum*, *Zea mays*, *Hordeum vulgare* L., *Lolium perenne*, and *Oryza sativa* accumulated more amount of silicon as compared to others (Barber and Shone, 1966; Jarvis, 1987; Casey et al., 2004; Snyder et al., 2006; Bakhat et al., 2009; Sahebi, 2015). Only a few forms of silicon can be taken up by plant roots, like monosilicic acid or orthosilicic acid (H_4SiO_4), which generally gathers in epidermal tissues and the cellulose membrane (only in the presence of pectin and calcium ions). In the sap of plants, silica is polymerized with water, and this is called the silica gel (Yamaji et al., 2008; Sahebi, 2015). Nanometer-sized biogenic silica is found in intercell structures of plants (Woesz et al., 2006; Sahebi, 2015).

In plants, silicon uptake is generally mediated by silicon channels or transporter genes, which includes, *OsLsi1*, *OsLsi2*, *HvLsi1*, and *HvLsi2*, found in *O. sativa* and *H. vulgare*, respectively (Figure 13.1). Transporter proteins are ubiquitously present in many plants including *Triticum aesativum*, *O. sativa*, *Cucurbita moschata*, *H. vulgare*, *Z. mays*, *Equisetum arvense* etc. (Grégoire et al., 2012). Out of two transporter proteins, Lsi1 belongs to the Nod26-like major intrinsic protein (NIP) subfamily of aquaporin-like proteins, which act as a Si-permeable channel (Ma et al., 2006). Their position in rice is at the distal side of the exo- and endodermis of roots, with the exception of *Cucurbita pepo*, *H. vulgare*, and *Z. mays*, in which they are located at the cortex and epidermis (Ma et al., 2006, Chiba et al., 2009; Mitani et al., 2009, 2011; Yamaji et al., 2012). Similarly, the Lsi2 genes act as efflux Si transporters and belong to the anion transporter family; they are also found in the endo- and exodermis, but a difference occurs in their polarity (Ma and

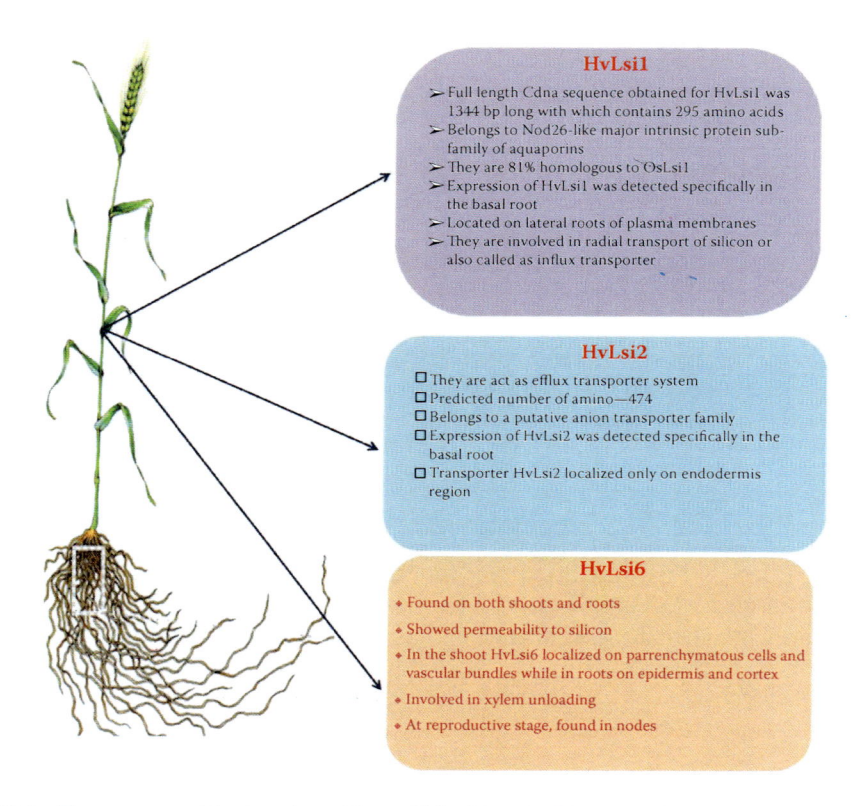

FIGURE 13.1 Transporters of barley plant. (From Chiba Y et al., *Plant J.* 57:810–818, 2009; Mitani N et al., *Plant Cell Physiol.* 50:5–12, 2009; Yamaji N et al., *Plant Physiol.* 160(3):1491–1497, 2012.)

Yamaji, 2006; Ma et al., 2006; Ma, 2010). Lsi2 genes are found at the proximal side and localized polarly in many plants, except *Z. mays* and *H. vulgare*, in which they are found at the root epidermis of the plants (Mitani et al., 2009b; Yamaji et al., 2012). The translocation of silicon aboveground occurs by Lsi6 (homologue of Lsi1), which is localized on xylem parenchymatous cells, especially on the adaxial side of leaf blades and leaf sheaths (Figure 13.1). Their absence results in decreased distribution of silicon in the panicle and alters the distribution of silicon in leaf cells. Similar functions of the transporter also occur in barley, having HvLsi6 transporter genes (Yamaji et al., 2012). Ma et al. (2004) reported two different silicon transporters of a mutant cultivar, SIT1 and SIT2; the first one is responsible for radial transport, while the second one transports silica up to the xylem from cortical cells. The process of transportation of silicon in plants, via transporter genes, occurs by energy consumption, which work opposite to the concentration gradient, but it is also reported that both active and passive processes of Si uptake can coexist in the same plants (Mitani and Ma, 2005; Henriet et al., 2006; Liang et al., 2007; Ding et al., 2008; Gérard et al., 2008; Guntzer et al., 2012). It is also reported that the uptake of silicon occurs only from the lateral roots, in spite of root hairs. After entry of silicon in the xylem, it penetrates the shoots of the plant, where transpiration plays a main role in the regulation of Si transportation and deposition (Guntzer et al., 2011; Raven, 2001). It is proved that transportation of silicon is primarily driven by transpiration and governed by the period of plant growth, hence the deposition of silicon in plants found with variable concentrations, i.e., they found more in older leaves while comparatively less in younger leaves (Henriet et al., 2006; Guntzer et al., 2012). Casey et al. (2004) reported that only mono- and disilicic acids forms of silicon are deposited in wheat sap at a ratio of 7:1. Sangster et al. (2001) demonstrated the effect of silicon on wheat after 8–10 days and found that silica was deposited in a solid form in the aerial parts, and then as precipitate in the form of amorphous silica. It is reported that deposition of silicon forms a link with carbohydrates and lignin, but no evidence has yet been found regarding the bond between Si and C (Perry and Keeling-Tucker, 1998; Guntzer et al., 2012). These amorphous silica particles deposited in plant cells are known as phytoliths, and during the isolation of phytoliths from the plants, it has been found that Ti, Mn, P, Cu, Al, Fe, N, and C were also present (Clarke, 2003; Cooke and Leishman, 2011; Guntzer et al., 2012; Tripathi et al., 2012a, 2014). Their location of deposition varies from species to species, as well as depends on the increasing age of the plant (Sangster et al., 2001; Ponzi and Pizzolongo, 2003; Guntzer et al., 2012). It has also been reported that phytoliths are also present in the root endodermis, leaf epidermis, and plasma membrane of vascular bundles. They are evenly deposited in whole plant cells (Sangster et al., 2001; Ponzi and Pizzolongo, 2003; Prychid et al., 2003; Guntzer et al., 2012). The preferential depositions of silicon in leaf occur in the abaxial epiderm, followed by both epidermal layers (Guntzer et al., 2012). The main storage of phytoliths is the silica cells, which are located on vascular bundles or silica bodies of fusoid or bulliform cells or prickle hairs of *O. sativa* (Ma and Yamaji, 2006; Guntzer et al., 2012), Bambuseae (Motomura et al., 2006; Guntzer et al., 2012), and *T. aesativum* (Dietrich et al., 2003; Guntzer et al., 2012).

It has been well demonstrated that both abiotic and biotic stresses may significantly reduce the growth and yield of mono- and dicotyledonous plant species; however, silicon effectively controlled the harmful impacts of these stresses (Belanger et al., 2003; Ma et al., 2004; Datnoff and Rodrigues, 2005; Fauteux et al., 2005; Van Bockhaven et al., 2013). Furthermore, it is well recognized that *O. sativa* is a typical silicon accumulator; it facilitates high and sustainable productivity of *O. sativa*, while its deficiency resulted in a sensitivity of plants toward diseases and pests, which leads to low fertility and ultimately decreased overall productivity (Tamai and Ma, 2008; Tripathi et al., 2014b). That is why, for the accurate production of rice, silicon is considered an agronomically essential element, which has been used in practice in the paddy fields of several countries. Silicon deposits in plants in the form of SiO_2, which act as a physical barrier, help to strengthen the plants and provide rigidity to the tissues; they also mechanically hamper the penetration of fungi and other pests (Tripathi et al., 2014b). Solubilized form of silicon plays very interesting and active role in plants by interacting with various types of stress signaling mechanisms of plants by stimulating and enhancing host resistance against diseases (Fawe et al., 1998; Rodrigues et al., 2004). Upon applying silicon to monocotyledons (especially *O. sativa* and *T. aesativum*) and dicotyledons (especially *Cucumis sativus*), they produce some secondary metabolites, like phenolics and phytoalexins, against fungal infections that cause rice blast and powdery mildew (Fawe et al., 1998; Rodrigues et al., 2004). Recently, it was discovered that silicon is involved in priming some defense responses, like the jasmonate-mediated antiherbivore defense mechanism (Ye et al., 2013). It has been observed that the integrity of plants is established by many elements, as evidenced by the presence of Cr toxicity; the level of Ca in plants is reduced, but the addition of silicon stimulates its accumulation in plants and helps in the development of the cell wall and its maintenance (Marschner, 1999; Waraich et al., 2011).

13.4 SILICON UNDER DEFICIENCY AND TOXICITY OF NUTRIENT UPTAKE AND OTHER STRESSES IN PLANTS

Due to anthropogenic activities, accumulation of an excess of metals, and widespread edaphic conditions, soil contaminations occur. Contaminations lead to major physiological disturbances in nutrient uptake or photosynthesis inhibition. Tripathi et al. (2012a,b, 2014a,b) demonstrated that heavy metal stress leads to a reduction in macro- and micronutrients in plants, but the exogenous application of silicon overcomes this type of nutrient-deficient trouble. As already discussed, silicon plays a significant role in the growth and development of plants; it has been also revealed that silicon can alleviate the deficiency symptoms of iron and alter their distribution in plants (Pavlovic et al., 2013). It has been experimentally proved that an exogenously supply of silicon plays a vital role against the toxicity of heavy metals in plants (Pankovic et al., 2000; Hassan et al., 2005; Sarwar et al., 2010; Tripathi et al., 2014a,b). Until the present day, only a few studies have proved the possible role of silicon in nutrient management and its contribution to stress physiology. The C and K contents in maize plants decreased in water stress conditions, but Si improves the status of nutrients in the same plant (Kaya et al., 2006).

Studies of silicon were performed by Brenchley and Maskell (1927) and Fisher (1929) on barley crops to observe its effect on the uptake of phosphorus, and it was found that silicon proved to be very effective in case of P limitation. They concluded that the main reason for the enhancement of yield is the silicon fertilizer. A correlation between the phosphorus uptake and silicon was also found. However, in a phosphorus deficiency, silicon can improve phosphorus utilization by increasing phosphorylation (Cheong and Chan, 1973; Guntzer et al., 2012) by decreasing the concentration of Mn (Ma and Takahashi, 1990a,b; Guntzer et al., 2012). In contrast, in the case of an excess supply of phosphorus, silicon limits the phosphorus uptake and manifestation of chlorosis, and reduces the transpiration rate (Ma et al., 2001b; Guntzer et al., 2012).

Mali and Aery (2008a) reported that the uptake of potassium in the presence of silicon is enhanced in hydroponic medium and soil through H-ATPase activation. They also observed that in cowpea and wheat, the incorporation of N and Ca is enhanced at higher doses of sodium metasilicate (50–800 mg Si·kg^{-1}); in addition, better nodulation is also seen with N_2 fixation (Mali and Aery, 2008a,b). In the presence of an excessive concentration of N in nutrient solution, the erectness of the rice leaf decreases, while it can be mitigated with the supply of silicon (Guntzer et al., 2012).

Silicon enhances the capacity of oxidation in the roots of *O. sativa*; it changes ferrous ion into ferric ion, hence limiting its toxicity (Ma and Takahashi, 2002; Guntzer et al., 2012). It is also observed that silicon supplementation regulates the uptake of iron and release of OH^- from roots (Wallace, 1993; Guntzer et al., 2012). However, some recent studies clearly proved that supplementation of silicon in iron-deficient plants may drastically maintain the deficiency of iron and repair morphophysiological, biochemical, and anatomical injuries caused by iron deficiency in plants (Chalmardi and Zadeh, 2013; Chalmardi et al., 2014).

When silicon interacts with aluminum, silicon reduced the phytotoxicity of aluminum found in soil solutions by formation of subcolloidal and inert aluminosilicate (Li et al., 1996; Liang et al., 2007; Guntzer et al., 2012). Silicon can also chelate and reduce Al absorption around the roots by stimulating phenolic exudation (Kidd et al., 2001; Guntzer et al., 2012). Detoxification of aluminum in the shoots occurs through sequestration in the phytoliths (Hodson and Sangster, 1993, 2002; Guntzer et al., 2012) and the formation of hydroxyl-aluminum silicates in the apoplasts (Ryder et al., 2003; Wang et al., 2004; Guntzer et al., 2012).

The toxicity of manganese, cadmium, copper, zinc, and arsenic can be reduced by adding supplementation of silicon. In the case of Mn toxicity, Si enhances the binding of Mn to the cell wall, which leads to form a semipermeable hindrance to the cytoplasmic concentrations (Rogalla and Romheld, 2002; Liang et al., 2007; Guntzer et al., 2012). In the leaves of cowpea, it was observed that silicon modifies the cation binding capacity of cell walls and facilitates the homogenous allocation of nutrients (Guntzer et al., 2012).

The concentration of arsenic in rice shoots can be decreased by the addition of silicic acid in a hydroponic medium. It also alleviates metal stress in the same plant (Guntzer et al., 2012; Sanglard et al., 2014).

13.5 SILICON AND OTHER ENVIRONMENTAL STRESSES IN PLANTS

When plants are exposed to various environmental stresses, they are considered as affected by oxidative stress. This may occur by biotic or abiotic stresses (Tripathi et al., 2014b; Adrees et al., 2015; Ali et al., 2015a,b; Habiba et al., 2015). Silica is known to suppress both chemical stresses (as evidenced by nutrient imbalances, metal toxicity, and salinity) and physical stresses like freezing, UV, high temperature, and drought (Ma et al., 2002, 2004; Tuna et al., 2008). The concerns of drought stress are increasing worldwide due to its high impact on the production of crops. The effects of silicon are studied on the yield of biomass with shortage of irrigation (Eneji et al., 2008); it is observed that Si leads to an enhanced biomass or grain yield in large sets of crops (e.g., Eneji et al., 2008; Pei et al., 2010; Shen et al., 2010). When silicon-treated plants are subjected to drought stress, Si manages their relative water content and higher stomatal conductance and water potential than untreated plants. Silicon leads to an increase in the size and thickness of leaves, which limits the loss of water through the transpiration (Gong et al., 2003a; Hattori et al., 2005), and hence the consumption of water also reduces (Eneji et al., 2005). Rigidity of the silicon-treated rice plant is gained by silification of the shoots, which enhances the plant's resistance toward typhoons (Ma et al., 2001b). Fertilization of silicon also plays a critical role in the formation of endodermis, especially secondary and tertiary cells, thus facilitating faster growth and enhanced resistance of roots (Hattori et al., 2003, 2005). Si also enhances the antioxidant defense mechanism by maintaining several physiological processes, like photosynthesis. Salt stress is another major abiotic stress in plants, which affects the production of salt worldwide, but it can be mitigated through silicon. It was experimentally reported that translocation of Na^+ decreases in shoots upon supply of silicate in the same plant, while it augments the formation of dry matter in rice (Zhua et al., 2004). It has been reported that silicon stimulates the growth of salt-treated barley and enhances the chlorophyll contents and activity of photosynthesis in barley.

Silicon has been found to be very effective against pathogens in a wide variety of plants. In the case of *T. aesativum*, silicon proves its effective role against fungal attacks and other diseases (Rodgers-Gray and Shaw, 2004; Guntzer et al., 2012), eyespot (*Oculimacula yallundae*), powdery mildew (*Blumeria graminis*), and septoria (*Phaeosphaeria nodorum* and *Mycosphaerella graminicola*). Similarly, in *O. sativa*, the consequences of silicon have been found in *Fusarium* wilt (*Fusarium* sp.), rice blast (*Magnaporthe grisea*), stalk rot (*Leptosphaeria salvinii*), tan spot (*Cochliobolus miyabeanus*), melting seedlings (*Thanatephorus cucumeris*), and leaf spots (*Monographella albescens*) (Ma and Takahashi, 2002; Tripathi et al., 2014b). It was experimentally shown that silica acts as a mechanical barrier by forming amorphous silica precipitation in plants, and thus helps in boosting other defense mechanisms, with the deposition of phenolic compounds, lignin, and phytoalexins (Epstein, 1999; Fawe et al., 2001; Ma and Yamaji, 2006; Guntzer et al., 2012). When a plant is infected with powdery mildew after being fertilized with silicon, the development of infection is minimal. Various experimental studies have shown that the plant defense mechanism stimulates the assembly of callose or methylaconitate (phytoalexin) and phenolic compounds (Belanger et al., 2003; Ghanmi et al., 2004; Rémus-Borel et al., 2005, 2009; Guntzer et al., 2012). Guevel et al. (2007) observed the application of Si in foliar leaves to reduce powdery mildew, but the proper mechanism is still unknown. A prophylactic effect for different pathogens is lost when silicon fertilization is limited, instead of amorphous silica bein deposited (Fawe et al., 2001; Fauteux et al., 2005; Guntzer et al., 2012). Silicon is not only regulates biological stresses, but also controls damage caused by insects and animals that are harmful for plants. Cotterill et al. (2007) and Hunt et al. (2008) reported that grasses that are fertilized by Si are less grazed by animals like locusts and wild rabbits.

13.6 SILICON AND ANTIOXIDANT ACTIVITIES

Generally, environmental stress stunts the photosynthesis and growth of plants, which leads to an imbalance between the reactive oxygen species (ROS) and antioxidant defense; this ultimately allows the accumulation of ROS and provokes oxidative stress to lipids, proteins, and other cellular components (Fu and Huang, 2001; Monakhova and Chernyad'ev, 2002). In the plants, two types of antioxidant defense systems exist: first one is enzymatic, while the other is nonenzymatic, which includes superoxide dismutase (SOD), ascorbate peroxidase (APX), guaiacol peroxidase (POD), glutathione reductase, catalase, as well as reduced glutathione (GSH), cysteine (Cys), and ascorbic acid (Asc) (Gill and Tuteja et al., 2010; Rico et al., 2015). However, formation of superoxide free radicals is associated with hydrogen peroxide (H_2O_2) removed by SOD that can be detoxified by POD and CAT (Sudhakar et al., 2001; Gong et al., 2005). In the case of the ascorbate–glutathione cycle, however, H_2O_2 is reduced by APX via ascorbate as an electron donor, and then the oxidized ascorbate further undergoes reduction by GSH formed from glutathione disulfide (GSSG) and is catalyzed by GR in the presence of NADPH (Lin and Kao, 2000; Gong et al., 2005). Since the roles of silicon in plants are poorly known, it is still not considered a fundamental element for plants (Lin and Kao, 2000; Liang et al., 2003; Zhu et al., 2004). Generation of ROS, O_2^-, OH·, $^1O_2$, and H_2O_2 is the major risk for living organisms; they are generated via a number of pathways (Becana et al., 2000; Kanazawa et al., 2000), like photorespiration, photosynthesis (Foyer and Noctor, 2000), oxidation of fatty acid, and senescence (Vitória et al., 2001). Zhu et al. (2004) reported that supplementation of silicon in the plant medium leads to a decreased level of elastin-like polypeptides (ELP), lipid peroxidation (LPO), and H_2O_2 concentrations, while SOD, GPX, APX, DR, and GR activities seem to be increased in salt-stressed cucumber leaves. The function of silicon in salt-stressed cucumber leaves is to decrease plasma membrane permeability and peroxidation of the lipid membrane, without changing the integrity of plants. Gong et al. (2005) reported that in the case of an optimum supply of water, silicon can not influence the activities of APX and SOD, whereas in wheat plants, CAT and POD activities were found to be reduced (Gong et al., 2003a,b). However, with the addition of silicon in salt-stressed barley, SOD activities were enhanced in the leaves, while SOD, POD, CAT, and GR activities were enhanced in the roots.

13.7 SILICON AND ANATOMICAL STRUCTURES

Anatomical studies are one of the most valuable and analytical parts of plant science due to their wide and significant role in explaining the positive or negative changes in plants under various environmental stresses. It has been well discussed that silicon helps the plants from overcoming various defects like improper photosynthesis and attacks of pests, fungus, etc., along with the proper maintenance of injured anatomical structures (Ma and Yamaji, 2006; Tripathi et al., 2012a,b, 2013, 2014a,b, 2015, 2016b; Vaculík et al., 2012). Tripathi et al. (2011) reported that plants treated with Si + Cr show better results in leaf anatomy than those treated with Cr alone. It protects leaf cells by preventing the leaf epidermal cells from decreasing due to Cr stress. Similarly, Hossain et al. (2002) observed that the addition of silicon enhances the length and growth of the leaf, preventing a decline in the stomatal frequency against Cr stress or any other type of stresses. Cr alone is capable of increasing the root hairs; however, Cr + Si together show a pronounced effect, and they adapt their strategy to absorb more elements for seedlings, thus increasing the number of root hairs as compared to the control and Cr-treated plants (Tripathi et al., 2012c). Singh et al. (2011) reported that aluminum-treated seedlings show smaller vascular bundles, larger and prominent air cavities, prominent sieve elements in phloem (but less phloem parenchyma), and chlorosis in mesophyll cells. In contrast, plants treated with Al + Si show a large number of xylem and phloem cells having sieve elements and mesophyll cells, and the length of epidermal cells also increases in plants treated with both Al and Si. They have also seen a decrease in the frequency and length of root hairs. Similarly, in *Oryza sativa*, Cd and Si combined showed a more pronounced effect, as evidenced by the comparatively increased root length, leaf thickness, and chloroplast disorganization (Vaculík et al., 2015).

13.8 CO_2 AND SILICA UPTAKE IN PLANTS

Day by day, the human population continues to expand, consequently it resulted in increasing requirements for food and energy, for which some additional resources, like fossil fuels, are needed. Fossil fuel is the major cause of atmospheric carbon dioxide production. One of the major function of plants is to participate in the global silicon cycle (Struyf et al., 2009). CO_2 biological pumping in the soil from the atmosphere occurs through plant respiration, translocation of carbon in roots, and the soil biota of dead and decaying plant materials (Lucas, 2001). Pearson and Palmer (2000) observed that the earth's biosphere is exposed to atmospheric carbon dioxide. At the time of the industrial revolution, the concentration of CO_2 increased by 40% (comparatively from the normal days). According to the Intergovernmental Panel on Climate Change (IPCC) 2013, this occurrence will continue to predict the future and impact the enhanced CO_2 level. From over the last several decades, terrestrial atmospheric biosphere concentrations are the major concern of scientists. Plants respond to CO_2 with regional and global climate changes, including increased temperature and enhanced N availability (Norby and Luo, 2004; Ainsworth and Long, 2005; Bernhardt et al., 2006; Finzi et al., 2006). Fulweiler et al. (2015) demonstrated that the level of CO_2 was increased with decreased silicon content, compared to the control; however, it was not proved to be significant statistically. The net primary production (NPP) increased up to 28% by elevation of CO_2, N, and combined N and CO_2. It was hypothesized that the foliar Si content may decrease in the presence of elevated CO_2; however, Fulweiler et al. (2014) found no significant effect on providing Si treatments to plants. Further, it has been concluded that elevated CO_2 results in increased NPP, and thus the silicon uptake rates are also enhanced (Fulweiler et al., 2014). These changes occur mainly from anthropogenic activities; they elevate CO_2 and cause various changes, like increased silicon pumping and an increased magnitude of silicon pumping.

13.9 ASSESSMENT OF CURRENT STATUS AND FUTURE CONJECTURE OF SILICON RESEARCH

As it has already been propounded the manifold roles of silicon in plants, we are still far from formulating the interminable measures of silicon in several other attractive fields of the modern era. We have a significant body of knowledge regarding the magnificent role of silicon in soil health. The essential role of silicon in enhancing crop yield and productivity could easily be manifested by studies from the rice industry in Japan to the sugarcane industry in North America.

Beyond this, silicon is also endowed with the potential to enhance the efficacy of delivering other elements in broader fertilization programs. Besides behaving as a soil conditioner to lock up the toxic elements and enhance the uptake of nutrients by soil, such as N, P, and K, the role of silicon can also be manifested in the robust field of nanotechnology. Nanotechnology has been hailed as the upcoming great thing for decades, in which the role of silicon as a nanoparticle is truly immense and remarkable, enabling researchers to more easily and effectively envisage the impact of various means of silicon in nutrient programs, as well as consider its cost-effectiveness (Tripathi et al., 2015b, 2016b,c). It is now considered one of the most imperative tools of modern agriculture, and agrifood nanotechnology are predicted to be the driving economic force in the near future to satisfy the needs of a burgeoning world population. Agrifood themes concern the sustainability and protection of agricultural crops, comprising crops for animal feeding and human consumption. The role of silicon in facilitates new delivery methods and agrochemical agents for the advancement of crop productivity and the sustainability of agricultural crops.

Likewise, Siddiqui and Al-Whaibi (2014) demonstrated the notable role of nano-SiO_2 in enhancing the seed germination potential of *Lycopersicum esculentum*. The extent of the n-SiO_2 holds a promising role in protection of plants and their nutrition because of the higher surface-to-volume ratio, size-dependent qualities, and unique optical properties. Several diverse opportunities have opened up new avenues for the potential use of silicon in several other fields, as its benefits are enormous. Better monitoring and targeted action of silicon in plants is still desirable, with a view to achieving sustainable agriculture in the near future.

REFERENCES

Adrees M, Ali S, Rizwan M, Zia-ur-Rehman M, Ibrahim M, Abbas F et al. (2015). Mechanisms of silicon-mediated alleviation of heavy metal toxicity in plants: A review. *Ecotoxicol. Environ. Saf.* 119:186–197.
Ainsworth EA, Long SP. (2005). What have we learned from 15 years of free-air CO2 enrichment (FACE)? Ameta—Analytic review of the responses of photosynthesis, canopy properties and plant production to rising CO_2. *New Phytol.* 165:351–371.

Ali S, Farooq MA, Yasmeen T, Hussain S, Arif MS, Abbas F et al. (2013). The influence of silicon on barley growth, photosynthesis and ultra-structure under chromium stress. *Ecotoxicol. Environ. Saf.* 89:66–72.

Bakhat HF, Hanstein S, Schubert S. (2009). Optimal level of silicon for maize (*Zea mays* L. cv AMADEO) growth in nutrient solution under controlled conditions. Presented at Proceedings of the 16th International Plant Nutrition Colloquium, Davis, CA.

Barber DA, Shone MGT. (1966). The absorption of silica from aqueous solutions by plants. *J. Exp. Bot.* 17(3):569–578.

Becana M, Dalton DA, Moran JF, Iturbe OI, Matamoros MA, Rubio MC. (2000). Reactive oxygen species and antioxidants in legume nodules. *Physiol. Plant.* 109:372–381.

Belanger RR, Benhamou N, Menzies JG. (2003). Cytological evidence of an active role of silicon in wheat resistance to powdery mildew (*Blumeria graminis* f. sp. *tritici*). *Phytopathology* 93:402–412.

Berner EU, Berner RA. (1996). *Global Environment. Water, Air and Geochemical Cycles*. Uppersaddle River, NJ, USA: Prentice Hall.

Bernhardt ES, Barber JJ, Pippen JS, Taneva L, Andrews JA, Schlesinger WH. (2006). Long-term effects of free air CO_2 enrichment (FACE) on soil respiration. *Biogeochemistry* 77:91–116.

Brenchley WE, Maskell EJ. (1927). The inter-relation between silicon and other elements in plant nutrition. *Ann. Appl. Biol.* 14:45–82.

Britez RM, Watanabe T, Jansen S, Reissmann CB, Osaki M. (2002). The relationship between aluminium and silicon accumulation in leaves of *Faramea merginata* (Rubiaceae). *New Phytol.* 156:437–444.

Brwen P, Menzies J, Ehert D, Samuels LADM. (1992). Glass soluble silicon sprays inhibit powdery mildew development on grape leaves. *J. Am. Soc. Hortic. Sci.* 117:906–912.

Casey WH, Kinrade SD, Knight CTG, Rains DW, Epstein E. (2004). Aqueous silicate complexes in wheat, *Triticum aestivum* L. *Plant Cell Environ.* 27(1):51–54.

Chalmardi ZK, Abdolzadeh A, Sadeghipour HR. (2014). Silicon nutrition potentiates the antioxidant metabolism of rice plants under iron toxicity. *Acta Physiol. Plant.* 36(2):493–502.

Chalmardi ZK, Zadeh AA. (2013). Role of silicon in alleviation of iron deficiency and toxicity in hydroponically-grown rice (*Oryza sativa* L.) plants. *J. Sci. Technol. Greenhouse Cult.* 3(12):Pe79–Pe88.

Chauhan DK, Tripathi DK, Kumar D, Kumar Y. (2011a). Diversity, distribution and frequency based attributes of phytolith in *Arundo donax* L. *Int. J. Innov. Biol. Chem. Sci.* 1:22–27.

Chauhan DK, Tripathi DK, Rai NK, Rai AK. (2011b). Detection of biogenic silica in leaf blade, leaf sheath, and stem of Bermuda grass (*Cynodon dactylon*) using LIBS and phytolith analysis. *Food Biophys.* 6:416–423.

Chauhan DK, Tripathi DK, Pathak AK, Rai S, Rai AK. (2011c). Detection of electrolytically active elements in (tulsi) using Libs *Ocimum sanctum* L. *Int. J. Eng. Sci. Manag.* 1:66–70.

Chen F, Wang F, Wu F, Mao W, Zhang G, Zhou M. (2010a). Modulation of exogenous glutathione in antioxidant defense system against Cd stress in the two barley genotypes differing in Cd tolerance. *Plant Physiol. Biochem.* 48:663–672.

Chen W, Yao X, Cai K, Chen J. (2011). Silicon alleviates drought stress of rice plants by improving plant water status, photosynthesis and mineral nutrient absorption. *Biol. Trace Elem. Res.* 142:67–76.

Chen WW, Yang JL, Qin C, Jin CW, Mo JH, Ye T et al. (2010b). Nitric oxide acts downstream of auxin to trigger root ferric-chelate reductase activity in response to iron deficiency in *Arabidopsis*. *Plant Physiol.* 154:810–819.

Cheng BT. (1982). Some significant functions of silicon to higher plants. *J. Plant Nutr.* 5:1345–1353.

Cheong YWY, Chan PY. (1973). Incorporation of P32 in phosphate esters of the sugarcane plant and the effect of Si and Al on the distribution of these esters. *Plant Soil* 38:113–123.

Chiba Y, Mitani N, Yamaji N, Ma JF. (2009). HvLsi1 is a silicon influx transporter in barley. *Plant J.* 57:810–818.

Clarke J. (2003). The occurrence and significance of biogenic opal in the regolith. *Earth-Sci. Rev.* 60(3):175–94.

Cocker KM, Evans DE, Hodson MJ. (1998). The amelioration of aluminum toxicity by silicon in wheat (*Triticum aestivum* L.) malate exudation as evidence for an in planta mechanism. *Planta* 204:318–323.

Cooke J, Leishman MR. (2011). Is plant ecology more siliceous than we realise? *Trends Plant Sci.* 16(2):61–68.

Cotterill JV, Watkins RW, Brennon CB, Cowan DP. (2007). Boosting silica levels in wheat leaves reduces grazing by rabbits. *Pest Manag. Sci.* 63:247–253.

Datnoff LE, Rodrigues FA. (2005). The role of silicon in suppressing rice diseases. APSnet Features. February. http://dx.doi.org/10.1094/APSnetFeature-2005-0205.

De Souza PF, Santos D, de Carvalho GGA, Nunes LC, da Silva Gomes M, Guerra MBB, Krug FJ. (2013). Determination of silicon in plant materials by laser-induced breakdown spectroscopy. *Spectrochim. Acta B Atomic Spectrosc.* 83:61–65.

Dietrich D, Hinke S, Baumann W, Fehlhaber R, Baucher E, Ruhle G, Wienhaus O, Marx G. (2003). Silica accumulation in *Triticum aestivum* L. and *Dactylis glomerata* L. *Anal. Bioanal. Chem.* 376:399–404.

Ding TP, Zhou JX, Wan DF, Chen ZY, Wang CY, Zhang F. (2008). Silicon isotope fractionation in bamboo and its significance to the biogeochemical cycle of silicon. *Geochim. Cosmochim. Acta* 72: 1381–1395.

Doncheva SN, Poschenrieder C, Stoyanova Z, Georgieva K, Velichkova M, Barceló J. (2009). Silicon amelioration of manganese toxicity in Mn-sensitive and Mn-tolerant maize varieties. *Environ. Exp. Bot.* 65(2):189–197.

Elbaz A, Wei YY, Meng Q, Zheng Q, Yang ZM. (2010). Mercury-induced oxidative stress and impact on antioxidant enzymes in *Chlamydomonas reinhardtii*. *Ecotoxicology* 19:1285–1293.

Eneji AE, Inanaga S, Muranaka S, Li J, Hattori T, An P, Tsuji W. (2008). Growth and nutrient use in four grasses under drought stress as mediated by silicon fertilisers. *J. Plant Nutr.* 31:355–365.

Eneji E, Inanaga S, Muranaka S, Li J, An P, Hattori T, Tsuji W. (2005). Effect of calcium silicate on growth and dry matter yield of *Chloris gayana* and *Sorghum sudanense* under two soil water regimes. *Grass Forage Sci.* 60:393–398.

Epstein E. (1994). The anomaly of silicon in plant biology. *Proc. Natl. Acad. Sci. U.S.A.* 91:11e17.

Epstein E. (1999). Silicon. *Annu. Rev. Plant Physiol. Plant Mol. Biol.* 50:641–664.

Eranna G. (2014). *Crystal Growth and Evaluation of Silicon for VLSI and ULSI*. CRC Press, Boca Raton, FL, p. 7.

Exley C, Birchall JD. (1993). A mechanism of hydroxyaluminosilicate formation. *Polyhedron* 12(9):1007–1017.

Fauteux F, Rémus-Borel W, Menzies JG, Bélanger RR. (2005). Silicon and plant disease resistance against pathogenic fungi. *FEMS Microbiol. Lett.* 249:1–6.

Fawe A, Abou-Zaid M, Menzies JG, Bélanger RR. (1998). Silicon-mediated accumulation of flavonoid phytoalexins in cucumber. *Phytopathology* 88:396–401.

Fawe A, Menzies JG, Cherif M, Bélanger RR. (2001). Silicon and disease resistance in dicotyledons. In Datnoff LE, Snyder GH, Korndorfer GH (eds.), *Silicon in Agriculture*, vol. 8: *Studies in Plant Science*. Elsevier, Amsterdam, pp. 159–169.

Feng J, Shi Q, Wang X, Wei M, Yang F, Xu H. (2010). Silicon supplementation ameliorated the inhibition of photosynthesis and nitrate metabolism by cadmium (Cd) toxicity in *Cucumis sativus* L. *Sci. Hortic.* 123:521–530.

Finzi AC, Allen AS, Delucia EH, Ellsworth DS, Schlesinger WH. (2001). Forest litter production, chemistry, and decomposition following two years of free-air CO_2 enrichment. *Ecology* 82:470–484.

Finzi AC, Moore DJP, Delucia EH, Lichter J, Hofmockel KS, Jackson RB et al. (2006). Progressive N limitation of ecosystem processes under elevated CO2 in a warm-temperate forest. *Ecology* 87:15–25.

Fisher RA. (1929). A preliminary note on the effect of sodium silicate in increasing the yield of barley. *J. Agric. Sci.* 19:132–139.

Foyer CH, Noctor G. (2000). Oxygen processing in photosynthesis: Regulation and signaling. *New Phytol.* 146:359–388.

Fu J, Huang B. (2001). Involvement of antioxidants and lipid peroxidation in the adaptation of two cool-season grasses to localized drought stress. *Environ. Exp. Bot.* 45:105–114.

Fulweiler RW, Maguire TJ, Carey JC, Finzi AC. (2014). Does elevated CO_2 alter silica uptake in trees? *Front. Plant Sci.* 5:793.

Fulweiler RW, Maguire TJ, Carey JC, Finzi AC. (2015). Does elevated CO_2 alter silica uptake in trees? *Front. Plant Sci.* 13(5):793.

Gangwar S, Singh VP, Tripathi DK, Chauhan DK, Prasad SM, Maurya JN. (2014). Plant responses to metal stress: The emerging role of plant growth hormones in toxicity alleviation. In Ahmad P (ed.), *Emerging Technologies and Management of Crop Stress Tolerance*. Vol. 2, pp. 215–248. Academic Press.

Gérard F, Mayer KU, Hodson MJ. (2008). Modelling the biogeochemical cycle of silicon in soils: Application to a temperate forest ecosystem. *Geochim. Cosmochim. Acta* 72:A304, 741–758.

Ghanmi D, McNally DJ, Benhamou N, Menzies JG, Belanger RR. (2004). Powdery mildew of Arabidopsis thaliana: A pathosystem for exploring the role of silicon in plant–microbe interactions. *Physiol. Mol. Plant Pathol.* 64(4):189–199.

Gill SS, Tuteja N. (2010). Reactive oxygen species and antioxidant machinery in abiotic stress tolerance in crop plants. *Plant Physiol. Bioch.* 48(12):909–930.

Gill SS, Hasanuzzaman M, Nahar K, Macovei A, Tuteja N. (2013). Importance of nitric oxide in cadmium stress tolerance in crop plants. *Plant Physiol. Biochem.* 63:254–261.

Gillman JH, Zlesak DC, Smith JA. (2003). Applications of potassium silicate decrease black spot infection in *Rosa hybrid* Meipelta (Fuschia Meidland). *Hortic. Sci.* 38:1144–1147.

Gong H, Chen K, Chen G, Wang S, Zhang C. (2003a). Effects of silicon on the growth of wheat and its antioxidative enzymatic system. *Chin. J. Soil Sci.* 34:55–57.

Gong H, Zhu X, Chen K, Wang S, Zhang C. (2005). Silicon alleviates oxidative damage of wheat plants in pots under drought. *Plant Sci.* 169(2):313–321.

Gong HJ, Chen KM, Chen GC, Wang SM, Zhang CL. (2003b). Effects of silicon on growth of wheat under drought. *J. Plant Nutr.* 26:1055–1063.

Gonzalo MJ, Lucena JJ, Hernández-Apaolaza L. (2013). Effect of silicon addition on soybean (*Glycine max*) and cucumber (*Cucumis sativus*) plants grown under iron deficiency. *Plant Physiol. Biochem.* 70:455–461.

Grégoire C, Rémus-Borel W, Vivancos J, Labbé C, Belzile F, Bélanger RR. (2012). Discovery of a multigene family of aquaporin silicon transporters in the primitive plant *Equisetum arvense*. *Plant J.* 72(2):320–330.

Guevel MH, Menzies JG, Bélanger RR. (2007). Effect of root and foliar applications of soluble silicon on powdery mildew control and growth of wheat plants. *Eur. J. Plant Pathol.* 119:429–436.

Guntzer F, Keller C, Meunier JD. (2012). Benefits of plant silicon for crops: A review. *Agron. Sustain. Dev.* 1(32):201–213.

Habiba U, Ali S, Farid M, Shakoor MB, Rizwan M, Ibrahim M et al. (2015). EDTA enhanced plant growth, antioxidant defence system and phytoextraction of copper by *Brassica napus* L. *Environ. Sci. Pollut. Res.* 22:1534–1544.

Hassan MJ, Zhang G, Wu F, Wei K, Chen Z. (2005). Zinc alleviates growth inhibition and oxidative stress caused by cadmium in rice. *J. Plant Nutr. Soil Sci.* 168:255–261.

Hattori T, Inanaga H, Araki H, An P, Morita S, Luxova M, Lux A. (2005). Application of silicon enhanced drought tolerance in *Sorghum bicolor*. *Physiol. Plant* 123:459–466.

Hattori T, Inanaga S, Tanimoto E, Lux A, Luxova M, Sugimoto Y. (2003). Silicon-induced changes in viscoelastic properties of sorghum root cell walls. *Plant Cell Physiol.* 44:743–749.

Henriet C, Draye X, Oppitz I, Swennen R, Delvaux B. (2006). Effects, distribution and uptake of silicon in banana (*Musa* spp.) under controlled conditions. *Plant Soil* 287:359–374.

Hodson MJ, Sangster AG. (1993). The interaction between silicon and aluminum in *Sorghum bicolor* (L.) Moench—Growth analysis and x-ray microanalysis. *Ann Bot* 72:389–400.

Hodson MJ, Sangster AG. (2002). X-ray microanalytical studies of mineral localization in the needles of white pine (*Pinus strobes* L.). *Ann. Bot.* 89:367–374.

Hossain MT, Mori R, Wakabayashi KSK, Fujii SKS, Yamamoto R, Hoson T. (2002). Growth promotion and an increase in cell wall extensibility by silicon in rice and some other Poaceae seedlings. *J. Plant Res.* 115:23–27.

Hunt JW, Dean AP, Webster RE, Johnson GN, Ennos AR. (2008). A novel mechanism by which silica defends grasses against herbivory. *Ann. Bot.* 102:653–656.

Hussain I, Ashraf MA, Rasheed R, Asghar A, Sajid MA, Iqbal M. (2015). Exogenous application of silicon at the boot stage decreases accumulation of cadmium in wheat (*Triticum aestivum* L.) grains. *Braz. J. Bot.* 38(2):223–234.

IPCC. (2013). "Summary for policymakers," in *Climate Change 2013: The Physical Science Basis. Contribution of Working Group I to the Fifth Assessment Report of the Intergovernmental Panel on Climate Change*, TF Stocker, D Qin, G-K Plattner, M Tignor, SK Allen, J Boschung et al., eds. Cambridge: Cambridge University Press.

Iler RK. (1979). *The Chemistry of Silica*. New York: Wiley, p. 621.

Jarvis SC. (1987). The uptake and transport of silicon by perennial ryegrass and wheat. *Plant Soil* 97(3):429–437.

Kanazawa S, Sano S, Koshiba T, Ushimaru T. (2000). Changes in antioxidative in cucumber cotyledons during natural senescence: Comparison with those during dark-induced senescence. *Physiol. Plant.* 109:211–216.

Kaya C, Tuna L, Higgs D. (2006). Effect of silicon on plant growth and mineral nutrition of maize. *J. Plant Nutr.* 29:1469–1480.

Kidd PS, Lluugan M, Poschhenrieder C, Gunsě B, Barceló J. (2001). The role of root exudates in aluminum resistance and silicon induced amelioration of aluminum toxicity in three varieties of maize (*Zea mays* L.). *Exp. Bot.* 52(359):1339–1352.

Korndorfer GH, Lepsch I. (2001). Effect of silicon on plant growth and crop yield. In Datnoff LE, Snyder GH, Korndorfer GH (eds.), *Silicon in Agriculture*, vol. 8: *Studies in Plant Science*. Elsevier, Amsterdam, pp. 133–147.

Lanning CF. (1966). Relation of silica in wheat to disease and pest resistance. *J. Agric. Food Chem.* 14:350–352.

Li DD, Lerman A, Mackenzie FT. (2011). Human perturbations on the global biogeochemical cycles of coupled Si–C and responses of terrestrial processes and the coastal ocean. *Appl. Geochem.* 26:S289–S291.

Li X, Yang Y, Jia L, Chen H, Wei X. (2013). Zinc-induced oxidative damage, antioxidant enzyme response and proline metabolism in roots and leaves of wheat plants. *Ecotoxicol. Environ. Saf.* 89:150–157.

Li YC, Summer ME, Miller WP, Alva AK. (1996). Mechanism of silicon induced alleviation of aluminum phytotoxicity. *J. Plant Nutr.* 19:1075–1087.

Liang Y. (1999). Effects of silicon on enzyme activity and sodium, potassium and calcium concentration in barley under salt stress. *Plant Soil* Feb 1;209(2):217–24.

Liang Y, Chen Q, Liu Q, Zhang W, Ding R. (2003). Exogenous silicon (Si) increases antioxidant enzyme activity and reduces lipid peroxidation in roots of salt-stressed barley (*Hordeum vulgare* L.). *J. Plant Physiol.* 160:1157–1164.

Liang Y, Hua H, Zhu YG, Zhang J, Cheng C, Römheld V. (2006). Importance of plant species and external silicon concentration to active silicon uptake and transport. *New Phytol.* 172(1):63–72.

Liang Y, Sun W, Zhu YG, Christie P. (2006). Mechanisms of silicon-mediated alleviation of abiotic stresses in higher plants: A review. *Environ. Pollut.* 20:1–7.

Liang Y, Sun W, Zhu YG, Christie P. (2007). Mechanisms of silicon-mediated alleviation of abiotic stresses in higher plants: A review. *Environ. Pollut.* 147(2):422–428.

Lin CC, Kao CH. (2000). Effect of NaCl stress on H2O2 metabolism in rice leaves. *Plant Growth Regul.* 30:151–155.

Lucas Y. (2001). The role of plants in controlling rates and products of weathering: Importance of biological pumping. *Annu. Rev. Earth Planet. Sci.* 29:135–163.

Lumsdon DG, Farmer VC. (1995). Solubility characteristics of proto-imogolite sols: How silicic acid can de-toxify aluminium solutions. *Eur. J. Soil Sci.* 46(2):179–186.

Lux A, Luxova M, Hattori T, Inanaga S, Sugimoto Y. (2002). Silicification in sorghum (*Sorghum bicolor*) cultivars with different drought tolerance. *Physiol. Plant* 115:87–92.

Ma J, Takahashi E. (1990a). Effect of silicon on the growth and phosphorus uptake of rice. *Plant Soil* 126(1):115–119.

Ma J, Takahashi E. (1990b). The effect of silicic acid on rice in a P-deficient soil. *Plant Soil* 126(1):121–125.

Ma J, Takahashi E. (2002). *Soil, Fertilizer and Plant Silicon Research in Japan.* Elsevier, Amsterdam.

Ma JF. (2004). Role of silicon in enhancing the resistance of plants to biotic and abiotic stresses. *Soil Sci. Plant Nutr.* 50:11–18.

Ma JF, Goto S, Tamai K, Ichii M. (2001a). Role of root hairs and lateral roots in silicon uptake by rice. *Plant Physiol.* 127:1773–1780.

Ma JF, Mitani N, Nagao S, Konishi S, Tamai K, Iwashita T, Yano M. (2004). Characterization of the silicon uptake system and molecular mapping of the silicon transporter gene in rice. *Plant Physiol.* 136:3284–3289.

Ma JF, Miyake Y, Takahashi E. (2001b). Silicon as a beneficial element for crop plants. In Datnoff LE, Snyder GH, Korndorfer GH (eds.), *Silicon in Agriculture*, vol. 8: *Studies in Plant Science.* Elsevier, Amsterdam, pp. 17–39.

Ma JF, Tamai K, Yamaji N, Mitani N, Konishi S, Katsuhara M, Ishiguro M, Murata Y, Yano M. (2006). A silicon transporter in rice. *Nature* 440:688–691.

Ma JF, Yamaji N. (2006). Silicon uptake and accumulation in higher plants. *Trends Plant Sci.* 11:392–397.

Ma JF, Yamaji N. (2015). A cooperated system of silicon transport in plants. *Trends Plant Sci.* 20(7):435–442.

Ma JF. (2010). Silicon transporters in higher plants. *In MIPs and Their Role in the Exchange of Metalloids.* New York: Springer, pp. 99–109.

Malčovská SM, Dučaiová Z, Maslaňáková I, Bačkor M. (2014). Effect of silicon on growth, photosynthesis, oxidative status and phenolic compounds of maize (*Zea mays* L.) grown in cadmium excess. *Water Air Soil Pollut.* 225:1–11.

Mali M, Aery NC. (2008a). Influence of silicon on growth, relative water contents and uptake of silicon, calcium and potassium in wheat grown in nutrient solution. *J. Plant Nutr.* 31:1867–1876.

Mali M, Aery NC. (2008b). Silicon effects on nodule growth, dry matter production, and mineral nutrition of cowpea (*Vigna unguiculata*). *J. Plant Nutr. Soil Sci.* 171:835–840.

Marafon AC, Endres L. (2013). Silicon: Fertilization and nutrition in higher plants. *Amaz. J. Agric. Environ. Sci.* 56(4):380–388.

Marschner H. (1999). *Mineral Nutrition of Higher Plants.* London: Academic, p. 889.

Marschner H. (1995). Beneficial mineral elements. In *Mineral Nutrition of Higher Plants.* Academic, San Diego.

Matichenkov VV, Ammosova YM, Bocharnikova EA. (1997). The method for determination of plant available silica in soil. *Agrochemistry* 1:76–84.

Matychenkov VV, Snyder GS. (1996). Mobile silicon-bound compounds in some soils of southern Florida. *Eurasian Soil Sci.* 12:1165–1173.

Matychenkov VV, Pinskiy DL, Bocharnikova YA. (1995). Influence of mechanical compaction of soils on the state and form of available silicon. *Eurasian Soil Sci.* 27(120):58–67.

Matychenkov VV, Ammosova YM. (1996). Effect of amorphous silica on some properties of a sod-podzolic acid. *Eurasian Soil Sci.* 28(10):87–99.

Mehrabanjoubani P, Abdolzadeh A, Sadeghipour HR, Aghdasi M. (2015). Silicon affects transcellular and apoplastic uptake of some nutrients in plants. *Pedosphere* 25(2):192–201.

Meyer JH, Keeping MG. Past, present and future research of the role of silicon for sugarcane in southern Africa. In Datnoff LE, Snyder GH, Korndorfer GH (eds.), *Silicon in Agriculture*, vol. 8: *Studies in Plant Science*. Elsevier, Amsterdam, pp. 257–275.

Mitani N, Ma JF. (2005). Uptake system of silicon in different plant species. *J. Exp. Bot.* 56:1255–1261.

Mitani N, Yamaji N, Ago Y, Iwasaki K, Ma JF. (2011). Isolation and functional characterization of an influx silicon transporter in two pumpkin cultivars contrasting in silicon accumulation. *Plant J.* 66: 231–232.

Mitani N, Yamaji N, Ma JF. (2009). Identification of maize silicon influx transporters. *Plant Cell Physiol.* 50:5–12.

Moharana PC, Sharma BM, Biswas DR, Dwivedi BS, Singh RV. (2012). Long-term effect of nutrient management on soil fertility and soil organic carbon pools under a 6-year old pearl millet wheat cropping system in an Inceptisol of subtropical India. *Field Crop Res.* 136:32–41.

Monakhova OF, Chernyad'ev II. (2002). Protective role of kartolin-4 in wheat plants exposed to soil drought. *Appl. Biochem. Microbiol.* 38:373–380.

Motomura H, Fujii T, Suzuki M. (2006). Silica deposition in abaxial epidermis before the opening of leaf blades of *Pleioblastus chino* (Poaceae, Bambusoideae). *Annals Bot.* 2006 Apr 1;97(4):513–519.

Neumann D, Zur Nieden U. (2001). Silicon and heavy metal tolerance of higher plants. *Phytochemistry* 56:685–692.

Norby RJ, Luo YQ. (2004). Evaluating ecosystem responses to rising atmospheric CO2 and global warming in a multi-factor world. *New Phytol.* 162:281–293.

Nwugo CC, Huerta AJ. (2008). Effects of silicon nutrition on cadmium uptake: Growth and photosynthesis of rice plants exposed to low level cadmium. *Plant Soil* 311:73–86.

O'Mara, William C. (1990). *Handbook of Semiconductor Silicon Technology*. William Andrew, Norwich, NY, pp. 349–352.

Orlov DS. (1985). *Humus Acids of Soils*. Oxonian Press, Oxford.

Pankovic D, Plesnicar M, Maksimoviic IA, Petrovic N, Sakac Z, Kastori R. (2000). Effects of nitrogen nutrition on photosynthesis in Cd-treated sunflower plants. *Ann. Bot. Lond.* 86(4):841–847.

Pavlovic J, Samardzic J, Maksimović V, Timotijevic G, Stevic N, Laursen KH et al. (2013). Silicon alleviates iron deficiency in cucumber by promoting mobilization of iron in the root apoplast. *New Phytol.* 198(4):1096–107.

Pearson PN, Palmer MR. (2000). Atmospheric carbon dioxide concentrations over the past 60 million years. *Nature* 406:695–699.

Pei ZF, Ming DF, Liu D, Wan GL, Geng XX, Gong HJ, Zhou WJ. (2010). Silicon improves the tolerance of water-deficit stress induced by polyethylene glycol in wheat (*Triticum aestivum* L.) seedlings. *J. Plant Growth Regul.* 29:106–115.

Peng Q, Zhou Q. (2010). Effects of enhanced UV-B radiation on the distribution of mineral elements in soybean (*Glycine max*) seedlings. *Chemosphere* 78:859–863.

Perry CC, Keeling-Tucker T. (1998). Aspects of the bioinorganic chemistry of silicon in conjunction with the biometals calcium, iron and aluminium. *J. Inorg. Biochem.* 69:181–191.

Ponzi R, Pizzolongo P. (2003). Morphology and distribution of epidermal phytoliths in *Triticum aestivum* L. *Plant Biosyst.* 137:3–10.

Prychid CJ, Rudall PJ, Gregory M. (2003). Systematics and biology of silica bodies in monocotyledons. *Bot. Rev.* 69:377–440.

Raven JA. (2001). Silicon transport at the cell and tissue level. In Datnoff LE, Snyder GH, Korndorfer GH (eds.), *Silicon in Agriculture*, vol. 8: *Studies in Plant Science*. Elsevier, Amsterdam, pp. 41–55.

Rémus-Borel W, Menzies JG, Bélanger RR. (2005). Silicon induces antifungal compounds in powdery mildew-infected wheat. *Physiol. Mol. Plant Pathol.* 66:108–115.

Rémus-Borel W, Menzies JG, Bélanger RR. (2009). Aconitate and methyl aconitate are modulated by silicon in powdery mildew-infected wheat plants. *J. Plant Physiol.* 166:1413–1422.

Reynolds OL, Keeping MG, Meyer JH. (2009). Silicon-augmented resistance of plants to herbivorous insects: A review. *Ann. Appl. Biol.* 155(2):171–86.

Richmond KE, Sussman M. (2003). Got silicon? The non-essential beneficial plant nutrient. *Curr. Opin. Plant Biol.* 6(3):268–272.

Rico CM, Peralta-Videa JR, Gardea-Torresdey JL. (2015). Chemistry, biochemistry of nanoparticles, and their role in antioxidant defense system in plants. In *Nanotechnology and Plant Sciences.* Springer International Publishing, pp. 1–17.

Rizwan M, Meunier JD, Miche H, Keller C. (2012). Effect of silicon on reducing cadmium toxicity in durum wheat (*Triticum turgidum* L. cv. Claudio W.) grown in a soil with aged contamination. *J. Hazard. Mater.* 209–210:326–334.

Rodgers-Gray BS, Shaw MW. (2004). Effects of straw and silicon soil amendments on some foliar and stem-base diseases in pot-grown winter wheat. *Plant Pathol.* 53(6):733–740.

Rodrigues FA, McNally DJ, Datnoff LE, Jones JB, Labbe C, Benhamou N, Menzies JG, Bélanger RR. (2004). Silicon enhances the accumulation of diterpenoid phytoalexins in rice: A potential mechanism for blast resistance. *Phytopathology* 94:177–183.

Rodriguez E, Santos C, Azevedo R, Moutinho-Pereira J, Correia C, Dias MC. (2012). Chromium (VI) induces toxicity at different photosynthetic levels in pea. *Plant Physiol. Biochem.* 53:94–100.

Rogalla H, Romheld V. (2002). Role of leaf apoplast in silicon mediated manganese tolerance of *Cucumis sativus* L. *Plant Cell Environ.* 25:549–555.

Ryder M, Gérard F, Evans DE, Hodson MJ. (2003). The use of root growth and modelling data to investigate amelioration of aluminium toxicity by silicon in *Picea abies* seedlings. *J. Inorg. Biochem.* 97:52–58.

Sahebi M, Hanafi MM, Rafii MY, Azizi P, Tengoua FF, Shabanimofrad M. (2014). Importance of silicon and mechanisms of biosilica formation in plants. *Biomed Res. Int.* 2015:396010.

Sanglard LMVP, Martins SCV, Detmann KC, Silva PEM, Lavinsky AO, Silva MM, Detmann E, Araújo WL, DaMatta FM. (2014). Silicon nutrition alleviates the negative impacts of arsenic on the photosynthetic apparatus of rice leaves: An analysis of the key limitations of photosynthesis. *Physiol. Plant.* 152(2):355–366.

Sangster AG, Hodson MJ, Tubb HJ. (2001). Silicon deposition in higher plants. In Datnoff LE, Snyder GH, Korndorfer GH (eds.), *Silicon in Agriculture*, vol. 8: *Studies in Plant Science.* Elsevier, Amsterdam, pp. 85–113.

Sarwar N, Saifullah SSM, Munir HZ, Asif N, Sadia B, Ghulam F. (2010). Role of mineral nutrition in minimizing cadmium accumulation by plants. *J. Sci. Food Agric.* 90:925–937.

Shen XF, Zhou YY, Duan LS, Li ZH, Eneji AE, Li JM. (2010). Silicon effects on photosynthesis and antioxidant parameters of soybean seedlings under drought and ultraviolet-B radiation. *J. Plant Physiol.* 167:1248–1252.

Shiguro K. (2001). Review of research in Japan on the roles of silicon in conferring resistance against rice blast. In Datonoff L, Korndorfer G, Synder G (eds.), *Silicon in Agriculture*, vol. 8: *Studies in Plant Science.* Elsevier, Amsterdam, pp. 8, 277–291.

Siddiqui MH, Al-Whaibi MH. (2014). Role of nano-SiO_2 in germination of tomato (*Lycopersicum esculentum* seeds Mill.). *Saudi J. Biol. Sci.* 21(1):13–17.

Singh A, Prasad SM. (2011). Reduction of heavy metal load in food chain: Technology assessment. *Rev. Environ. Sci. Biotechnol.* 10:199–214.

Singh VP, Tripathi DK, Kumar D, Chauhan DK. (2011). Influence of exogenous silicon addition on aluminium tolerance in rice seedlings. *Biol. Trace Elem. Res.* 144(1–3):1260–1274.

Snyder GH, Matichenkov VV, Datnoff LE. (2006). Silicon. In *Plant Nutrition.* Taylor & Francis, Belle Glade, FL, pp. 551–562.

Sommer M, Kaczorek D, Kuzyakov Y, Breuer J. (2006). Silicon pools and fluxes in soils and landscapes—A review, *J. Plant Nutr. Soil Sci.* 169:310–329.

Song Z, Wang H, Strong, PJ, Li Z, Jiang P. (2012). Plant impact on the coupled terrestrial biogeochemical cycles of silicon and carbon: Implications for biogeochemical carbon sequestration. *Earth Sci. Rev.* 115(4):319–331.

Street-Perrott FA, Barker PA. (2008). Biogenic silica: A neglected component of the coupled global continental biogeochemical cycles of carbon and silicon. *Earth Surf. Processes Landforms* 33:1436–1457.

Struyf E, Smis A, Van Damme S, Meire P, Conley DJ. (2009). The global biogeochemical silicon cycle. *Silicon.* 1(4):207–13.

Sudhakar C, Lakshmi A, Giridarakumar S. (2001). Changes in the antioxidant enzyme efficacy in two high yielding genotypes of mulberry (*Morus alba* L.) under NaCl salinity. *Plant Sci.* 161:613–619.

Tamai K, Ma JF. (2008). Reexamination of silicon effects on rice growth and production under field conditions using a low silicon mutant. *Plant Soil* 307:21–27.

Tripathi DK, Kumar R, Chauhan DK, Rai AK, Bicanic D. (2011). Laser-induced breakdown spectroscopy for the study of the pattern of silicon deposition in leaves of *Saccharum* species. *Instrum. Sci. Technol.* 39:510–521.

Tripathi DK, Chauhan DK, Kumar D, Tiwari SP. (2012a). Morphology, diversity and frequency based exploration of phytoliths in *Pennisetum typhoides* Rich. *Natl. Acad. Sci. Lett.* 35(4):285–289.

Tripathi DK, Kumar R, Pathak AK, Chauhan DK and Rai AK. (2012b). Laser-induced breakdown spectroscopy and phytolith analysis: An approach to study the deposition and distribution pattern of silicon in different parts of wheat (*Triticum aestivum* L.) plant. *Agric. Res.* 1(4):352–361.

Tripathi DK, Singh VP, Kumar D and Chauhan DK. (2012c). Impact of exogenous silicon addition on chromium uptake, growth, mineral elements, oxidative stress, antioxidant capacity, and leaf and root structures in rice seedlings exposed to hexavalent chromium. *Acta. Physiol. Plant.* 34(1):279–289.

Tripathi DK, Mishra S, Chauhan DK, Tiwari SP, Kumar C. (2013). Typological and frequency based study of opaline silica (Phytolith) deposition in two common Indian *Sorghum* L. species. *Proc. Natl. Acad. Sci.*, India, *Sect. B Biol. Sci.* 83(1):97–104.

Tripathi DK, Singh VP, Gangwar S, Prasad SM, Maurya JN and Chauhan DK. (2014a). Role of silicon in enrichment of plant nutrients and protection from biotic and abiotic stresses. In *Improvement of Crops in the Era of Climatic Changes*. New York: Springer, pp. 39–56.

Tripathi DK, Prasad R, Chauhan DK. (2014b). An overview of biogenic silica production pattern in the leaves of *Hordeum vulgare* L. *Indian J. Plant Sci.* 3(2):167–177.

Tripathi DK, Singh VP, Prasad SM, Chauhan DK, Dubey NK and Rai AK. (2015a). Silicon-mediated alleviation of Cr (VI) toxicity in wheat seedlings as evidenced by chlorophyll florescence, laser induced breakdown spectroscopy and anatomical changes. *Ecotox. Environ. Safe.* 113:133–144.

Tripathi DK, Singh S, Singh S, Mishra S, Chauhan DK, Dubey NK. (2015b). Micronutrients and their diverse role in agricultural crops: Advances and future prospective. *Acta. Physiol. Plant.* 37(7):1–14.

Tripathi DK, Singh VP, Prasad SM, Chauhan DK, Dubey NK. (2015c). Silicon nanoparticles (SiNp) alleviate chromium (VI) phytotoxicity in *Pisum sativum* (L.) seedlings. *Plant Physiol. Bioch.* 96:189–198.

Tripathi DK, Pathak AK, Chauhan DK, Dubey NK, Rai AK, Prasad R. (2016a). LIB spectroscopic and biochemical analysis to characterize lead toxicity alleviative nature of silicon in wheat (*Triticum aestivum* L.) seedlings. *Photochem. Photobiol. B* 154:89–98.

Tripathi DK, Singh VP, Singh S, Prasad SM, Chauhan DK, Dubey NK. (2016b). Silicon nanoparticles more efficiently alleviate arsenate toxicity than silicon in maize cultivar and hybrid differing in arsenate tolerance. *Front. Env. Sci.* doi:10.3389/fenvs.2016.00046.

Tripathi DK, Singh S, Singh VP, Prasad SM, Dubey NK, Chauhan DK. (2016c). Silicon nanoparticles more effectively alleviated UV-B stress than silicon in wheat (*Triticum aestivum*) seedlings. *Plant Physiol. Bioch.* doi:10.1016/j.plaphy.2016.06.026.

Tubaña BS, Heckman R. (2015). Silicon in Soils and Plants. In *Silicon and Plant Diseases*. Springer International Publishing, pp. 7–51.

Tuna AL, Kaya C, Higg D, Murillo-Amador B, Aydemir AS, Girgin R. (2008). Silicon improves salinity tolerance in wheat plants. *Environ. Exp. Bot.* 62:10–16.

Vaculík M, Landberg T, Greger M, Luxová M, Stoláriková M, Lux A. (2012). Silicon modifies root anatomy, and uptake and subcellular distribution of cadmium in young maize plants. *Ann. Bot.* 110(2):433–43.

Vaculík M, Pavlovič A, Lux A. (2015). Silicon alleviates cadmium toxicity by enhanced photosynthetic rate and modified bundle sheath's cell chloroplasts ultrastructure in maize. *Ecotoxicol. Environ. Saf.* 120:66–73.

Van Bockhaven J, De Vleesschauwer D, Höfte M. (2013). Towards establishing broad-spectrum disease resistance in plants: Silicon leads the way. *J. Exp. Bot.* 64(5):1281–93.

Vitória AP, Lea PJ, Azevedo RA. (2001). Antioxidant enzymes responses to cadmium in radish tissues. *Phytochemistry* 57:710.

Wallace A. (1993). Participation of silicon in cation–anion balance as a possible mechanism for aluminum and iron tolerance in some Gramineae. *J. Plant Nutr.* 16:547–553.

Wang Y, Stass A, Horst WJ. (2004). Apoplastic binding of aluminum is involved in silicon-induced amelioration of aluminum toxicity in maize. *Plant Physiol.* 136:3762–3770.

Waraich EA, Ahmad R, Ashraf S, Ehsanullah MY. (2011). Role of mineral nutrition in alleviation of drought stress in plants. *Aust. J. Crop. Sci.* 5(6):764–777.

Waraich EA, Ahmad R, Halim A, Aziz T. (2012). Alleviation of temperature stress by nutrient management in crop plants: A review. *J. Soil Sci. Plant Nutr.* 12(2):221–244.

Woesz A, Weaver JC, Kazanci M, Dauphin Y, Aizenberg J, Morse DE, Fratzl P. (2006). Micromechanical properties of biological silica in skeletons of deep-sea sponges. *J. Mater. Res.* 21(8):2068–2078.

Yamaji N, Chiba Y, Mitani-Ueno N, Ma JF. (2012). Functional characterization of a silicon transporter gene implicated in silicon distribution in barley. *Plant Physiol.* 160(3):1491–1497.

Yamaji N, Mitatni N, Jian FM. (2008). A transporter regulating silicon distribution in rice shoots. *Plant Cell* 20(5):1381–1389.

Yang YH, Chen SM, Chen Z, Zhang HY, Shen HG, Hua ZC, Li N. (1999). Silicon effects on aluminum toxicity to mungbean seedling growth. *J. Plant Nutr.* 22:693–700.

Ye M, Song Y, Long J, Wang R, Baerson SR, Pan Z et al. (2013). Priming of jasmonate-mediated antiherbivore defense responses in rice by silicon. *Proc. Natl. Acad. Sci. U.S.A.* 110:E3631–E3639.

You-Qiang FU, Shen H, Dao-Ming WU, Kun-Zheng CAI. (2012). Silicon-mediated amelioration of Fe 2+ toxicity in rice (*Oryza sativa* L.) roots. *Pedosphere* 22(6):795–802.

Zhang Q, Liu J, Lu H, Zhao S, Wang W, Du J, Yan C. (2015). Effects of silicon on growth, root anatomy, radial oxygen loss (ROL) and Fe/Mn plaque of *Aegiceras corniculatum* (L.) Blanco seedlings exposed to cadmium. *Environ. Nanotechnol. Monit. Manag.* 4:6–11.

Zhu Z, Wei G, Li J, Qian Q, Yu J. (2004). Silicon alleviates salt stress and increases antioxidant enzymes activity in leaves of salt-stressed cucumber (*Cucumis sativus* L.). *Plant Sci.* 167(3):527–533.

14 Silicon and Cadmium Toxicity in Plants
An Overview

Fakhir Hannan, Shafaqat Ali, Rehan Ahmad,
Muhammad Rizwan, Muhammad Iqbal,
Hina Rizvi, and Muhammad Zia-ur-Rehman

CONTENTS

ABSTRACT

In the present scenario, cadmium toxicity is one of the major problems in the agricultural field. A higher concentration of Cd is toxic to plants and causes a reduction in plant growth and biomass. Cadmium toxicity can reduce photosynthetic pigments, respiration, and water relations and cause oxidative damage and lipid peroxidation in plants. In the literature, several methods have been described to mitigate the effect of cadmium toxicity in plants. Silicon is present prominently in the environment, but it is not considered an essential element for plant growth and development. However, many studies have shown the positive effects of Si on the growth and physiology of plants, especially under biotic and abiotic stresses, including Cd stress. Silicon mitigates the toxicity of cadmium by immobilizing the cadmium into the growth media and also by reducing the Cd uptake and translocation. Under Cd stress, Si application enhances the activities of antioxidant enzymes and reduces the oxidative stress by decreasing the malondialdehyde (MDA) and electrolyte leakage (EL) and hydrogen peroxide (H_2O_2) production. Silicon could reduce Cd stress in plants by complexation and coprecipitation and deposition of Cd with Si in different plant parts. Silicon also reduces Cd toxicity by structural alterations in plants and by regulating gene expression. However, these mechanisms may vary with plant species, growth conditions, and duration of Cd stress applied.

14.1   SOURCES AND CONTENT OF Cd IN SOIL

Cadmium (Cd) is located in 2B, period 5 of the periodic table and is a transition metal. The atomic mass of Cd is 112.411 u ± 0.008 with atomic number 48 and having a specific density of 8.65 g cm^{-3}. It melts at 320.9°C and has a boiling point of 765°C (Wuana and Okieimen, 2011). The wide application of cadmium is in manufacturing nickel (Ni) and Cd batteries. Moreover, it is produced as a by-product in lead (Pb) and zinc (Zn) refining, in the manufacturing of different alloys; it is also used as a pigmentation agent and stabilizer for plastics and other electronic compounds (Di Toppi and Gabbrielli, 1999). However, Cd is nonessential and produces detrimental effects to animals and plant life. The ingestion rate of cadmium in humans adds up to the great amount of cadmium uptake (EC, 2000). For instance, in Japan the higher concentrations of Cd in rice were the cause of its entrance into the food chain, which disrupted the internal structure of humans. In the 1950s and 1960s, the itai-itai disease in Japan was considered to be a result of the larger Cd concentrations in rice (Yamagata and Shigematsu, 1970). Presently in Japan, it is acknowledged that rice is the fundamental source of Cd entry into humans (Watanabe et al., 2000), which is a risk to human well-being (Ueno et al., 2010). A mean tolerable weekly intake (TWI) of 2.5 µg kg^{-1} of Cd by body weight was established for the European population. Nonetheless, a twofold increase in uptake of Cd is observed in people living in highly Cd-toxic areas (EFSA, 2009). Several types of alteration such as structural, biochemical, and physiochemical have been reported in different plant species (Das et al., 1997; Di Toppi and Gabrielli, 1999; Benavides et al., 2005). Cd has been reported to disrupt the plant's anatomy and the photosynthetic rates carried out by plants (Sandalio et al., 2001). The significant observable symptom of Cd in plants is the significant reduction in root length (Guo and Marschner, 1995). Cd is also responsible for affecting the rates of translocation and absorption of micronutrients (zinc, copper, manganese) and macronutrients (N, P, K, Ca, Mg) by plants (Jalil et al., 1994; Sarwar et al., 2010). The heavy metal enters plants by a pathway through roots, which come into direct contact with soil solution containing heavy metals (Lux et al., 2011). Presently, many studies demonstrated in situ and ex situ remediation techniques to remediate soils containing Cd like phytoextraction, immobilization, phytostabilization, extraction, physical separation, etc. (Wuana and Okieimen, 2011). The widely used technique among these is phytoremediation, which is used for cleaning up metals and other toxic compounds from soils. This technique consists of using green plants for the removal of toxic soils containing metals and other harmful substances. There are a number of phytoremediation techniques: phytoextraction, phytovolatilization, and phytostabilization. *Phytostabilization* checks the mobility of heavy metal in soil. Metal immobilization can be done by declining the windblown dust, erosion of soil, and contaminant bioavailability to the food chain (Adriano, 2001; Wuana and Okieimen, 2011). Toxicants are absorbed and accrued in the roots or precipitated in the rhizosphere, so in this way, their solubility and mobility are constrained (Bes and Mench, 2008). This technique is useful for quick stabilization of heavy metals including Cd and is beneficial since disposal of hazardous material is not required (Mench et al., 2006).

14.2   NATURAL CADMIUM LEVELS IN SOIL

Cadmium is an extremely scattered and naturally present (0.1–0.2 mg kg^{-1}) element in the environment. Naturally, Cd can be found in the bedrocks below the soil and is present in complex forms in ores comprising copper, lead, and zinc (UNEP, 2010). Cadmium in the earths's crust is mostly present from 0.1 to 25 ppm, but different factors can affect its concentration rates, either increasing or decreasing its levels. The levels of Cd present in sedimentary rocks range from 0.1 to 25 ppm, and the levels of Cd in metamorphic rocks range from 0.02 to 0.2 ppm (Cook and Morrow, 1995). When associated with Cu, Zn, and Pb ores, the Cd concentrations are higher. The levels of Cd present in the phosphate fertilizers range from 2 to 200 mg kg^{-1}. Marine phosphates hold about 15 mg kg^{-1} levels of Cd (EC, 2000). Phosphate rocks (PRs) of igneous origin typically incorporate less than 15 mg Cd/kg P_2O_5 (phosphate fertilizer), in contrast to 20–245 mg Cd kg^{-1} in sedimentary counterparts (Çotuk et al., 2010). The normal range of Cd present in the earth is reported to be 0.02–6.2 mg Cd kg^{-1}, and the soils having 5–20 mg Cd kg^{-1} probably need remedial actions, as these levels are considered toxic to the surrounding area (Adriano, 2001). In agricultural or horticultural soil, Cd concentrations approximately range from 0.2 to 1.0 mg kg^{-1} in rural areas and 0.5 to 1.5 mg kg^{-1} in urban areas (EC, 2000). On the other hand, anthropogenic or natural activities can augment the Cd levels in the soil (He et al., 2005).

14.3 SOURCES OF CADMIUM CONTAMINATION OF AGRICULTURAL SOILS

The main causes of cadmium contamination in agricultural soil are pedogenic physical processes, aerial depositions, agricultural use of pesticides and phosphate fertilizers, sewage sludge, and waste disposal from industries (Grant et al., 1998; Wuana and Okieimen, 2011). In France, the Cd concentration is augmented to more than 100 mg Cd kg^{-1} of soil (Baize et al., 1999). In accordance with the French ASPITET program, the Cd in agricultural soils can range from 0.02 to 6.9 mg kg^{-1} (Baize, 1997; Mench et al., 1997). However, the concentrations of Cd in soil solution are less and vary from 0.2 to 6.0 μg L^{-1}. In addition, greater values ranging up to 300 μg L^{-1} of cadmium have been observed, which is above the critical level of cadmium tolerance (Itoh and Yumura, 1979). Furthermore, when compared with our heavy metals, the levels of Cd in the domestic sewage sludge are higher, which includes the Cd from car tires and cigarette butts flushed down toilets having Cd in them. Additionally, composted sludge can have greater concentrations of Cd. For instance, the composted sludge from Topeka, Kansas, applied to cropland, possesses up to 4.2 mg Cd kg^{-1} compost (Liphadzi and Kirkham, 2006). Phosphate fertilizers, manures, pesticides, and industrial wastes are the main source of cadmium pollution (Nagajyoti et al., 2010; Wuana and Okieimen, 2011). The application of such agricultural proportions of substances having Cd augments the total Cd in soils. Nonetheless, Cd concentrations of less than 3 mg kg^{-1} available in dry soils are usually advised for agricultural soils in order to lessen its levels in edible parts (Lux et al., 2011).

14.4 CADMIUM UPTAKE AND ACCUMULATION IN PLANTS

Plants primarily absorb Cd because of the pathway of roots by the soil solution. Cd uptake via roots seems to occur by means of different transporters like Mn^{2+}, in addition to Zn^{2+}, Ca^{2+}, and Fe^{2+} (Clemens, 2006). The absorption of cadmium mainly depends on the species and variety of crop plants (Grant et al., 1998, 2008; Liu et al., 2003). Cd absorption in plant tissues generally occurs through three transport systems: (1) root uptake associated with Cd, (2) xylem loading, and (3) retranslocation to help seeds. Roots may be the main pathway where water, nutrients, and pollutants, just like heavy metals, penetrate into the plant body. Cadmium accommodates in plant tissues through the root from the contaminated soil. Cd compartmentation present in the soil solution is usually likewise adsorbed toward the surface connected with plant roots. Cd uptake via roots is increased by the increasing exposure intervals and Cd concentrations (Hentz et al., 2012). Accumulation of Cd in roots can also be affected simply by root morphology, root apices, and the root surface area (Kubo et al., 2011). The root endodermis, in addition to the exodermis, plays a vital role in Cd accumulation in maize plants and behaves as a barrier toward the solute circulation (Redjala et al., 2011), and also, the presence of high Cd accelerates maturation with the maize underlying the endodermis (Lux et al., 2011). During the CD uptake by plants, Cd ions can easily compete for the same transmembrane, like the ones that are helpful in taking up plant nutrients (Benavides et al., 2005). This presence of other aspects can slow down the Cd uptake, like Zn^{2+} from the nutrient solution inhibits the taking up of Cd within wheat crop (Welch et al., 1999; Hart et al., 2002). Molecular systems involving Cd that simply uptake by their roots are nevertheless poorly comprehended. After adsorption on the underlying surface area of roots, Cd enters from the underlying roots as Cd^{2+} via ZIP transporters including Zn-regulated transporter- or even Fe-regulated transporter-like protein (Lux et al., 2011). As soon as Cd continues to be adopted by means of the root system, the area of the Cd amasses inside roots, and an additional element is actually translocated to the shoots (Kabata-Pendias and Pendias, 2001; Gill et al., 2011). Normally, Cd ions tend to be retained inside roots, and a modest percentage is also actually transferred to the shoots; nonetheless, it depends on the plant type (Abe et al., 2008). Following absorption by means of roots, Cd can move toward the xylem through apoplastic or symplastic trails (Salt et al., 1995; Benavides et al., 2005). In the beginning, Cd may also be complexed using several ligands, including organic and natural acids or phytochelatins (PCs), and it is generally focused within vacuoles as well as nuclei (Hart et al., 2006; Lux et al., 2011). Xylem packing is a crucial method regarding long-distance transportation associated with Cd (Clemens et al., 2002). Cd transportation in the central cylinder can be managed by means of the Casparian strip, as well as the plasmalemma of the endodermis (Seregin et al., 2004). After absorption within the roots, Cd is actually carried by means of the xylem, as well as the phloem to the aerial regions of the plant (Tudoreanu and Phillips, 2004). Cadmium transportation within plant tissues depends on the species and genotype of plants (Dunbar et al., 2003). Translocation associated with Cd via roots to shoots is an unaggressive method motivated by means of transpiration (Salt et al., 1995; Hart et al., 2006), or it is even translocated positively through diverse transporters like Fe (Nakanishi et al., 2006). It has been proposed that the citrate might engage in a large function within Cd transportation inside the xylem vessels (Zorrig et al., 2010). Likewise, Van der Vliet et al. (2007) noted that much of the Cd translocation through the roots to the shoots within durum grain crop occurs by way of the symplastic path. Research on Cd translocation confirmed that inside the edible regions of plants, it depends on the genotypes (Meyer et al., 1982; Cakmak et al., 2000). Cd piling up within grains additionally depends on Cd attention within shoots, as well as the whole leaf (Greger and Löfstedt, 2004). Within grains, Cd piling up may possibly happen through phloem-mediated Cd transportation through the leaves, as well as stalks, for the maturation of grains (Hart et al., 1998; Harris and Taylor, 2001; Liu et al., 2007; Yoneyama et al., 2010). Occurrence associated with different ions may possibly prevent or even enhance phloem packing associated with Cd (Welch et al., 1999; Cakmak et al., 2000). Cadmium does not seem to be an important source of nourishment regarding plant growth or development, given that it does not have a known biological purpose (Marschner, 1995). Among the poisonous heavy metals, Cd is actually associated with a lot more problem than others due to its higher toxicity at very low concentrations as well as higher solubility within water (Benavides et al., 2005). Greater Cd piling up in plants induces a few pressure variables in them. The main toxic outcomes (direct or indirect) with this heavy metal on plant species growth, as well as physiological procedures, are usually (1) enzyme inactivation, (2) generation of ROS, (3) damage to lipids, (4) cell death, (5) reduction in plant biomass, (6) decrease in water status, (7) pigment contents reduction, (8) affect photosynthetic rates, (9) decrease in protein contents, (10) root respiration, (11) root elongation, (12) mineral uptake, (13) root biomass, and (14) seed germination.

14.5 FACTORS INVOLVING CADMIUM UPTAKE AND ACCUMULATION IN PLANTS

The soil and plant factors and microorganisms are mainly responsible for the uptake of Cd in crop plants.

14.5.1 SOIL FACTORS

The behavior of Cd in soil can be correlated with major contamination problems. Cd is present in soils in different forms, such as an exchangeable free ion of Cd^{2+} and other metal ions present in soil having Cd bonding, adsorbed with organic matter present in soil, clay particles, and iron oxides. Cd is also present in soils in the form of amino acids complexes, COOH- complexes, Cl^-, and SO_4^{2-} (Sammut et al., 2010; Vega et al., 2010). Cd behavior in soil depends on these diverse forms and is primarily ascertained by complex interactions (Adriano et al., 2004). Less important is the total Cd levels present in soil solution, compared to the bioavailable Cd for plants. This is important because plants can mainly take up bioavailable Cd (Ok et al., 2004; Kirkham, 2006). Soil factors such as available Cd and total Cd levels largely affect the amount of bioavailablity of Cd. Other factors can include type of plant species taking up Cd, depending on pH and type of organic matter present in soil (Sauve et al., 2000a,b; Kirkham, 2006; Jung, 2008), cation exchange capacity (CEC) speciation (Lehoczky et al., 2000; Sammut et al., 2010; Vega et al., 2010), and chloride (Grant et al., 1998; Degryse et al., 2004; Weggler et al., 2004; Kirkham, 2006). The dissolved organic carbon that can react with metals such as Cd, Cu, and Zn can affect the rates of Cd bioavailability, solubility, and the rates of plant uptake in the soil (Antoniadis and Alloway, 2002; Zhao et al., 2007). Cd bioavailability can be lowered with time (Kirkham, 2006). Nitrogen fertilizers, genotypes, types of managing of soils, and organized agricultural practices can affect Cd bioavailability (Zhang et al., 2009; Perilli et al., 2010; Gao et al., 2010, 2011). The amount of nutrient concentrations in plants also affects Cd bioavailability (Kirkham, 2006; Sarwar et al., 2010). The amount of Cd soil solution is affected by the metals adsorbed by the dry soil particles that control the presence of metals in the soil. Furthermore, the levels of Cd in soil solution are affected by the sorption and complexation capabilities of soil particles. The greater the rates of adsorption and sorption, the more likely that the seed in soil solution will decrease (Grant et al., 1998). The most important factor is the pH of the soil, which influences the Cd bioavailability (McBride et al., 1997; Grant et al., 1998; Jung, 2008). The uptake of Cd is higher in low-pH soils and low in high-pH alkaline soils. The main reason is that Cd is more mobile at lower pH values, ranging from 4.5 to 5.5, and less mobile or immobile in alkaline soils (Kirkham, 2006). The pH affects the rates of adsorption, sorption, desorption, and complexation, which in turn affects the levels of bioavailability of cadmium (Naidu et al., 1994, 1997; Bolan et al., 1999). In addition to this, generally cadmium controls the processing of soil, especially the root developing zone of soil.

14.5.2 PLANT FACTORS

The type of plants species also affects the rates of Cd bioavailability (Mench et al., 1989). Cd in soil is present in the form of chelates, but in the soil, Cd is predominantly present in the form of Cd^{2+} (Tudoreanu and Phillips, 2004). Unique chelating materials are also introduced by means of some specific plants called phytosiderophores, as well as underlying root exudates (Mench and Martin, 1991). Normally, phytosiderophores are manufactured by graminaceous facilities like barley, wheat, or grain, as well as almond under Fe deficiency, which usually mobilizes Fe by infrequently soluble forms (Marschner et al., 1986; Reichard et al., 2005). Most of these phytosiderophores can complex metals, including Cd and Cu, and affect their bioavailability. Root exudates strongly affect the bioavailability of soil by impacting the actual characteristics regarding the rhizosphere (Hill et al., 2002; Dong et al., 2007). Root exudates influence their own Cd bioavailability and also toxicity via modifying the rhizospheric pH, in addition to chelation/complexion, redox potential (Eh), and deposition of Cd ions. Additionally, numerous plant species will certainly exude organic acids that lead to the formation of metal complexes in the rhizosphere. Organic acids additionally increase uptake associated with Cd by solubilizing particulate-bound Cd directly into the soil solution (Cieslinski et al., 1998). Similarly in the plants of *Zea mays*, the presence of organic acids inclined Cd mobilization, plant availability, and accumulation (Nigam et al., 2001; Han et al., 2006). Such factors alter Cd behavior in the soil, in addition to affecting the Cd uptake through the plants and altering the current community, along with activities connected with the microbes present within the rhizosphere (Shenker et al., 2001; Dong et al., 2007).

14.5.3 Microorganisms

It is the characteristic of microorganisms to release specific chelating compounds known as siderophores (Neubauer et al., 2000 and references therein) that will solubilize Cd (Dimkpa et al., 2009). A bacterium can make numerous kinds of siderophores, such as hydroxomates and carboxylic acids (Klumpp et al., 2005). These siderophores are able to desorb Cd from a moderate pH limit (Hepinstall et al., 2005). Microorganisms may furthermore decrease Cd solubility because of the formation of insoluble metal sulfides and also sequestration of the toxic metal via the cell walls or by proteins and extracellular polymers, and so forth (Francis, 1990; Dong et al., 2007).

14.5.4 Climatic Factors

Metal mobility and bioavailability of cadmium are also influenced by the change in climate. Primarily, the concentration of organic matter (which affects those Cd that are available in plants) is strongly affected by the climate. Lower soil organic matter and humid climate, which is a characteristic of arid climates, contain large amounts of organic matter. Organic matter may bind Cd to exchange complexes within tropical climate conditions. Its presence associated with iron, manganese, and aluminum oxide minerals with soil profiles can limit mobility, along with the bioavailability involving Cd. Temperature exerts a great click effect in metal speciation, because just about all chemical reaction rates are highly sensitive to temperature changes (Elder, 1989). The increase related to 10°C may supply biochemical reaction rates and enhance the tendency of a system to reach equilibrium. Temperature will certainly furthermore affect the amounts connected to metal uptake via an organism (Prosi, 1989). Acid rain also affects Cd bioavailablity by inducing the release of metals due to cation exchange with Mg^{2+}, Ca^{2+}, and H^+, etc. (Probst et al., 2000; Hernandez et al., 2003).

14.6 TOXIC EFFECTS OF Cd IN PLANTS

14.6.1 Effects on Plant Growth and Biomass

The dangerous results of Cd on plant expansion, along with biomass, tend to be the most studied. Reduction in root length is just about the majority of visible indicators involving Cd toxicity in plants (Guo and Marschner, 1995; Lux et al., 2011; Haouari et al., 2012). Increased Cd levels in the root base multiplied the development involving the endodermis of the roots and furthermore transformed the relative size, along with proportion, of the root tissue (Seregin et al., 2004; Lux et al., 2011). Cd is well known to induce negative effects on plant expansion and growth, along with the biomass. One example is the pea plant grown hydroponically together with 50 µM Cd, which after 31 days appreciably reduces plant expansion and growth, along with the biomass (Sandalio et al., 2001). Recently, Haouari et al. (2012) noted that the root, along with the shoot's fresh and dried out biomass in tomato plants grown hydroponically together with raised Cd concentrations, has been negatively impacted. Wójcik and Tukiendorf (2005) demonstrated the cadmium toxicity in *Thlaspi caerulescens* in the form of reduced weight. Similarly, higher Cd (100 mg Cd kg⁻¹ involving soil) perturbs the expansion and growth involving garden cress (*Lepidium sativum* D.) plants (Gill et al., 2012). With cowpea plants, raising Cd (10, 30, and 50 mg Cd kg⁻¹ soil) concentrations furthermore reduced the development, along with nutrient uptake (Vijayaragavan et al., 2011). Quite a few preceding studies have noted that the Cd source in durum grain cultivars decreases the shoot along with root dried out matter, and the root length along with leaf area (Jalil et al., 1994). Increased Cd attention in facilities triggers leaf chlorosis, wilting along with leaf abscission in plants (Bavi et al., 2011). Similarly, Shi et al. (2010a) showed that direct exposure to Cd depresses peanut plant expansion. This kind of lower ingrowth, along with biomass beneath Cd strain, is related to declined photosynthetic rate, inhibition of metabolic digestive enzyme production, and a decrease in the roots, along with translocation involving macro- and micronutrient leaves in plants (Sandalio et al., 2001; Gonçalves et al., 2009; Feng et al., 2010). On the other hand, hazardous dangerous results involving Cd on growth of plants and their development along with biomass depend on the dose amount and the period of exposure to Cd toxicity (Das et al., 1997; Di Toppi and Gabrielli, 1999). Similarly, chlorosis indicators with necrotic spots have shown up in tomato leaves (*Lycopersicon esculentum*) from 10 to 100 μM Cd in nutrient solution (López-Millán et al., 2009). Additionally, the toxic results involving Cd vary, together with the plant species (Das et al., 1997; Di Toppi and Gabrielli 1999).

14.6.2 SIDE EFFECTS ON NUTRIENTS

Cd absorption with nutrient solution tends to be an open matter mainly because the decrease in healthy eating of nutrients through Cd is actually directly connected with crop yield rates. Quite a bit of research regarding Cd results on nutrient solution has shown contradictory final results. Harmful results involving Cd on mineral nutrients rely on the accumulation of Cd in plants and happen to be commonly reported (Ouzounidou et al., 1997; Sandalio et al., 2001; Wang et al., 2008; Gonçalves et al., 2009). Harmful results involving Cd on mineral nutrients depend on the intensity of toxicity and how long the Cd stress is imposed (Hernandez et al., 1998; Street et al., 2010). Harmful results involving Cd on mineral nutrient uptake and translocation fluctuate, together with the plants and types of nutrients (Hernandez et al., 1998; Dong et al., 2006). The results connected with Cd on Zn uptake and also deposition within crops are not reliable. One example is 0.2 μM Cd concentrations, which diminished Zn uptake within durum wheat or grain when Zn levels were reduced (1.0 μM) in the source of nourishment remedy, although with higher Zn (10 and 19 μM) levels, this specific result was synergistic (Welch et al., 1999). Conversely, in spring wheat with 20 μM Cd in the nutrient solution decreased Zn concentration and increased Cd concentrations in roots and shoots when Zn was lower (1 ~ 200 μM) in the nutrient medium while at higher (>200 μM) Zn levels Cd concentration significantly reduced in seedlings while Zn concentration increased indicating antagonistic effects on each other (Zhao et al., 2005). In tomato crops, excess Cd (50 and 100 μM) in the source of nourishment remedy diminished the particular uptake connected with K^+, Ca^{2+}, and Mg^{2+} ions through sources and consequently diminished within shoots (Haouari et al., 2012). Extra Cd in the solution altered not only the particular uptake connected with nutrients, but also the particular deposition and translocation. Nonetheless, once again, this specific craze varies according to seed species and Cd stress induced (Yang et al., 1998; Zhang et al., 2002). One example is within new soybean plants, where Cd levels diminished Cu, Zn, and Mn levels within sources, although there was no impact within shoots (Drazic et al., 2004). In comparison, Cd in the growth and development medium raised the K, P, and Mn levels within the wheat or grain sources and also inhibited the translocation to be able to reach shoots (Zhang et al., 2002). In the same way, yet another example connected interactions with essential elements and Cd, and they affect Cd when there is element deficiency in plants. For example, Ca deficiency increases Cd toxicity in rice seedlings (Cho et al., 2012).

14.6.3 EFFECTS ON PHOTOSYNTHETIC PIGMENTS

Cd is remarkably poisonous to the photosynthetic machines of crops. A decrease in the rates of photosynthesis is often a well-known warning sign connected with Cd toxicity within crops. This specific lowering of photosynthesis occurs through many operations, such as loss of net photosynthetic price, chlorophyll, and carotenoid articles (Sandalio et al., 2001; Mobin and Khan, 2007; Vijayaragavan et al., 2011; Haouari et al., 2012). Additionally, in cucumber (*Cucumis sativus* L.) plant, Cd treatment decreased photosynthesis which was related to damage or functional loss of the photosynthetic machinery and enzyme inhibition of nitrate metabolism (Feng et al.,

2010). In the same way, Cd request within new wheat or grain plants breaks the particular leaf photosystem (PS II), and also the composition connected with chloroplasts leads to reduction in chlorophyllous content and also inhibition associated with photosynthesis (Ouzounidou et al., 1997; Ci et al., 2009). In peanut (*Arachis hypogaea*) crops, Cd therapies net photosynthetic rate which is often caused by a reduction connected to stomatal conductance and photosynthetic colors, as well as amendments within leaf composition (Shi and Cai, 2008). More recently, Gill et al. (2012) reported that a greater amount of Cd diminished particular photosynthesis and the nitrogen metabolic process within garden cress (*Lepidium sativum* L.). In the same way, in tomato (*Lycopersicon esculentum*) plants, low Cd (10 µM) levels failed to impact photosynthesis, but greater Cd (100 µM) levels within the nutrient solution decreased the photosynthetic rates and photosynthetic pigment concentrations (López-Millán et al., 2009). The toxic results of Cd on photosynthetic machinery rely not only on the quantity of Cd, but also on the coverage time. By way of example, Chugh and Sawhney (1999), inside one-month-old pea plants, documented distinct Cd remedies over 6 nights, with a far more obvious effect on the experience regarding PS II, but upon prolongation of the Cd coverage time (12 nights), the functioning of PS I had been equally afflicted.

14.6.4 CD-INDUCED OXIDATIVE STRESS IN PLANTS

Extra Cd yields totally free radicals in addition to the reactive oxygen species (ROS) variety, bringing about oxidative pressure inside crops (Khan et al., 2007; Mobin and Khan, 2007). ROS react with proteins and pigments and cause lipid peroxidation, inactivation of enzymes and membrane damage (Khan et al., 2007). Extra Cd inside crops also triggered the customization regarding antioxidant minerals, in addition to sulfur compression activity (Shi et al., 2010b; Gill et al., 2012). Ascorbic acidity attentiveness decreased within the shoots, in addition to sources in durum wheat or grain cultivars, by supplying 75–150 µM Cd within the nutrient medium (Ozturk et al., 2003). Extra Cd also decreased the glutamine synthetase (GS) enzyme in several seed varieties (Balestrasse et al., 2006). Extra Cd increased the output of H_2O_2 inside pea crops, and therefore boosted activity of superoxide dismutase (SOD) in addition to the small activity regarding peroxidase (POD), catalase (CAT), and ascorbate peroxidase (APX) (Pandey and Singh, 2012). Even so, physiological effects on seedlings as a result of Cd pressure are dependent on the variety and cultivars (Shi et al., 2010a).

14.6.5 MISCELLANEOUS TOXIC EFFECTS

Cd toxicity afflicts typical seedling increases, as well as advancement, through several means. Even so, there are still greater numbers of toxic results of Cd on crops. Extra Cd lowered the percentage of H_2O content in roots and shoots inside pea crops (Lozano-Rodriguez et al., 1997). The effect of Cd on plant–water associations is usually relevant to the lowering of the absorption surfaces by inhibiting the creation of root hairs (Pál et al., 2006). Cd application inside new wheat or grain plants decreases the total soluble sugar concentrations, as well as increases free amino acid levels inside both roots and shoots (Ci et al., 2009). Additionally, Cd affects stomatal conductance inside safflower (*Carthamus tinctorius* T.) crops (Shi et al., 2010b).

14.7 SILICON IN SOIL

Silicon is that element that ranks second in terms of abundance, and in the soil, they range from 50 to 400 g Si kg^{-1} of soil (Kovda, 1973). In the soil, silicon is found in the form of SiO_2 (primarily) and also in various forms of alumina-silicate. Silicon dioxide comprises about 50%–70% of the soil mass (Ma and Yamaji, 2006). Externally Si can be obtained from magnesium and calcium silicates, dolomite, rock phosphate, silicate slag, and diatomite (Savant et al., 1997; Guntzer et al.,

2012; Rizwan et al., 2012). Silicon compounds exist in both the liquid and phase in soil. In the solid phase, Si compounds can be majorly categorized into poorly crystalline, amorphous, and crystalline forms. Amorphous silica (ASi) is comprised from plants, including inorganic forms and in the form of phytoliths (Sauer et al., 2006), and it is considered to be the first pool of available Si for plants (Alexandre et al., 1997). The amount of biogenic silica in the form of phytoliths in soils ranges from 0.03 to 0.06 wt% (Desplanques et al., 2006). Phytoliths of grassland soils possess up to only 1%–3% of the total Si pool (Blecker et al., 2006). The major composition of phytoliths contains about 92 wt% silica plus 6 wt% water, with little amounts of carbon and traces of Al and Fe (Meunier et al., 1999). In the liquid phase, Si is present as mono- and polysilicic acids, complexed with organic and inorganic compounds (Cornelis et al., 2011). In natural soil solution, Si is mainly present in the form of uncharged orthosilicic acid, H_4SiO_4, ranging from 0.1 to 0.6 mM (Epstein, 1994; Sommer et al., 2006), and it is thought to be the only form that is taken up by plants (Epstein, 1994, 1999; Ding et al., 2005).

14.8 SILICON UPTAKE AND ACCUMULATION IN PLANTS

All plants comprise the 0.1% to 10% dry weight of Si in tissues (Epstein, 1994, 1999; Ma and Yamaji, 2008). Silicon absorption in plants depends on species, and cultivators like gramineous versus leguminous plants are divided into high, intermediate, and non-Si accumulators (Takahashi et al., 1990; Ma et al., 2001). However, Si uptake mechanisms vary among plant species and depend on transport mechanism and the genotype of plants (Epstein, 1994; Casey et al., 2003; Ma and Yamaji, 2006; Ma and Yamaji, 2008; Ding et al., 2008). The process of silicon uptake in plant roots is still not known, but it occurs mainly in higher plants by the active, passive, and rejective mechanism (Ma et al., 2004; Mitani and Ma, 2005). Passive uptake of Si takes part along the transpiration stream, and it is shown that most dicotyledonous plants absorb Si passively (Ma et al., 2001). Active Si uptake generally occurs in rice, maize, sugarcane, and wheat (Casey et al., 2003; Rains et al., 2006).

Transporters responsible for Si uptake by roots have been discovered in various plants such as rice, barley, maize, wheat, and cucumber (Ma et al., 2007, 2011; Chiba et al., 2009; Mitani et al., 2009). Once Si is absorbed by roots, it travels to the shoots by transpirational water flow via the xylem (Ma et al., 2006). Silicon mainly flows in the xylem sap in the form of monosilicic and disilicic (Casey et al., 2003; Mitani and Ma, 2005). Recently, intervascular transport has been recognized and also that Si transporters are helpful in the xylem unloading in leaves (Lsi6) identified (Yamaji et al., 2008, 2012). After transportation to shoot, Si is deposited in transpiration sites in the immobile and stable form of amorphous silica, $SiO_2 \cdot nH_2O$ (Hodson and Sangster, 1990; Ma and Takahashi, 2002). Deposition of silica occurs mainly in different plant parts such as the epidermis of roots and shoots (Lux et al., 2003; Keller et al., 2015). On the other hand, the phytolith deposition in different plant families depends on type, composition, localization, and different environmental circumstances of plant species (Guntzer et al., 2012; Li et al., 2014). Ameliorative effects of Si on the toxicity of certain metals such as Zn, Mn, Al, Cu, and Cd in many plant species have been well reported. Si concentration also affects the metal, mineral absorption, photosynthetic pigments, and plant growth and weight.

14.9 SILICON AND Cd TOXICITY IN PLANTS

On comparing with zinc and copper, the absorption coefficient proved to be higher for cadmium (Llamas et al., 2000). Cadmium is the main cause of soil contamination in the environment from several sources such as phosphate fertilizer and sewage sludge. Excess concentration of cadmium induces toxicity in plants, and the adsorption of cadmium depends on the cadmium concentration present in soil as well as ions available for adsorption in plant species (Marrs and Walbot, 1997). Cadmium availability to plant tissues is affected mainly by CEC, pH, fertilizer, and redox state of cadmium (Yin-Ming et al., 1994; Hegedüs et al., 2001). Cadmium toxicity appears in the form of chlorosis, necrosis, and red and dark red appearance of the leaf margin and root tissues (Prasad,

1997; Fediuk and Erdei, 2002). Cadmium also stimulates the oxidative burst in plant tissues, but the exact pathways and signaling of cadmium toxicity is still unknown (Haag-Kerwer et al., 1999). It can inhibit the activity of some antioxidant enzymes or induce their activity. The effects of cadmium on activities of some antioxidant enzymes, like peroxidase, catalase, ascorbate peroxidase, and glutathione reductase, have been studied (Marrs and Walbot, 1997). Lipid peroxidation is the major index of the increase in active free radicals, and MDA is the main by-product of the lipid peroxidation process. Some research studies also reveal the role of anthocyanin in the quenching of free radicals, and the reaction depends on the redox state of metal. This reaction is reversible. Cd^{2+} increases the antioxidant enzyme of *Zea mays* (Yin-Ming et al., 1994). Cadmium also induces the expression of the *Bronze2* gene, which is responsible for the synthesis of glutathione-S-transferase (GST) enzyme. The GST enzyme takes an active role in the biosynthesis of anthocyanins and its localization into the vacuoles (Balestrasse et al., 2001). The MDA content is an index of stress conditions in different cadmium concentrations. The detoxification of cadmium by flavonoid and anthocyanin has been studied in different crops species. Silicon, being the second most abundant element, is not an essential element for the growth of plants (Schickler and Caspi, 1999; Vega et al., 2010). However, there is increasing evidence that it has a number of beneficial impacts on plant growth in different stress conditions (Marrs and Walbot, 1997). Regarding abiotic stresses, Si shows mitigating effect on Cd toxicity in different plant species (Haag-Kerwer et al., 1999).

14.10 CHANGES IN MINERAL NUTRIENT UPTAKE AND PLANT GROWTH AND BIOMASS DUE TO Si

Mineral nutrients are the major constituents of plant structural components and provide mechanical, biochemical, and physical strengths to the plants. Mineral nutrients also play a major role in the growth and development of plants, and heavy metals may interfere with essential nutrient uptake and transport, thereby disturbing the mineral nutrition composition. Under stress circumstance, the qualitative and quantitative analysis provides the information about mineral uptake and transport with silicon and the detoxification pathways of silicon in retort to different stress conditions. Silicon takes part in the absorption of nutrients by plants under metal stress. Silicon treatment in the hydroponic medium augmented Zn and Mn, and macronutrients, Ca, Mg, P, and K, by wheat plants under Cr, Cd, and Cu (Rizwan et al., 2012; Keller et al., 2015; Tripathi et al., 2015). Wang et al. (2014) have stated that foliar treatment with nano-Si amplified Mg, Fe, and Zn nutrition in rice seedlings under Cd toxicity. Silicon enhances the Mg and Zn concentrations in shoots and root tissues of *Oryza sativa* in Al stress, while with only Si application, there is no change in the Mn concentration (Singh et al., 2011). In copper stress, silicon persuades the mineral nutrient in Zinnia (*Zinnia elegans*) and *Erica andevalensis* plants (Frantz et al., 2011; Oliva et al., 2011). Under Zn toxicity, Si and Zn application increases K and Fe in maize and rice seedlings compared to Zn stress alone (Kaya et al., 2009; Mehrabanjoubani et al., 2014). Silicon treatment elevates the mineral content in various heavy metal concentrations, whereas Bokor et al. (2014b) illustrated the diminished mineral (Mn, Fe, Ca, P, Mg, Ni, Co, and K) concentration in Si + Zn treated seedlings. These experiments show that the homeostatic system of mineral elements is interrupted and caused depressing alteration in mineral uptake. Silicon reduces the toxicity in plants by interacting with nutrient elements, but the mechanism and pathway of these interactions are still unknown. Furthermore, it remains for future studies why this positive or negative correlation occurred between Si and mineral nutrients under different metal toxicity and study expression level of other transporter genes. The effect of silicon in the plant growth and biomass at different stress has been broadly accepted. Si application increases the biomass and diminishes metal toxicity by the dilution effect, but the mechanism of these pathways is still not well stated. Silicon also ameliorates the dry weight of plant tissues under chromium, cadmium, and aluminum stress in barley, rice, and mung bean (Liang et al., 2005; da Cunha and do Nascimento, 2009; Vaculik et al., 2009), wheat (Rizwan et al., 2012; Naeem et al., 2014), strawberry

(Treder and Cieslinski, 2005), and rice (Corrales et al., 1997; Hara et al., 1999; Yang et al., 1999; Morikawa and Saigusa, 2002; Nwugo and Huerta, 2008; Zhang et al., 2008; Gu et al., 2011; Singh et al., 2011; Zeng et al., 2011; Tripathi et al., 2012a,b, 2015; Ali et al., 2013; Zhang et al., 2013a,b; Srivastava et al., in press). In contrast, 5 mM Si and 150 mM Zn in hydroponics decrease shoots and root weight of sorghum compared to Zn treatment within 20 days (Masarovic et al., 2012). Similarly, 5 mM Si and 800 mM Zn addition declined the shoot and root biomass of maize plants grown in hydroponics for 10 days, compared to the same Zn treatment without Si (Bokor et al., 2014a,b). In general, increases in plant growth and biomass might be due to decreases in toxic metal uptake by plants or increases in nutrient uptake by plants, increases in photosynthetic pigments, and maintenance of the structure of photosynthetic machinery. Overall, Si is beneficial to plants under metal stress and also improves the plant biomass.

14.11 ENHANCEMENT IN GAS EXCHANGE ATTRIBUTES AND PHOTOSYNTHETIC PIGMENTS

Positive impacts of Si on chlorophyll biosynthesis and photosynthetic mechanisms under HM-induced toxicity have been widely described. For instance, Si and Cd application elevated the level of chlorophyll a, chlorophyll b, and carotenoid contents in leaves compared to application of only Cd in maize (Malčovská et al., 2014), wheat (Rizwan et al., 2012; Hussain et al., 2015), cucumber (Feng et al., 2010), and rice plant (Nwugo and Huerta, 2008). Silicon application increased the chlorophyll levels under Al, Cr, and As stress in rice, mung bean, wheat, and barley (Yang et al., 1999; Singh et al., 2011; Ali et al., 2013; Sanglard et al., 2014; Tripathi et al., 2015). Hydroponically applied silicon improved the gas exchange activities (NPR, SC, TR, and water transportation efficiency), chlorophyll, and carotenoid contents in Cd-stressed plants of cotton, rice, and cucumber (Nwugo and Huerta, 2008; Feng et al., 2010; Farooq et al., 2013). Similarly, silicon increases the gas exchange process under Al and Cr toxicity in peanut and barley (Ali et al., 2013; Shen et al., 2014), under Pb toxicity in cotton (Bharwana et al., 2013), and under Zn toxicity in rice (Song et al., 2014). Similarly, Hu et al. (2013) reported that silicon delivery enhanced the net photosynthetic rate and decreased stomatal constraint in rice under arsenic (As) stress. Furthermore, Sanglard et al. (2014) carefully studied thorough gas exchange capacity with chlorophyll fluorescence analysis to examine the impacts of Si nutrition on photosynthetic performance in rice plants under As (arsenite) toxicity. Several authors have reported that Si considerably relapsed As-mediated destruction in carbon fixation and leaf conductance at the stomata and mesophyll levels, and the response occurred in a time- and genotype-dependent manner. Furthermore, Si enhanced photosynthetic activity and chlorophyll fluorescence in different plant species under HM toxicity, such as barley and wheat under Cr toxicity (Ali et al., 2013; Tripathi et al., 2015). Moreover, Nwugo and Huerta (2010) reported that in rice leaves, Si elevated the Cd tolerance by increasing the water use efficiency, carboxylation efficiency of ribulose-1,5-bisphosphate carboxylase oxygenase (RuBisCO), and light use efficiency. In general, the positive effect of Si on photosynthetic machinery might be due to lower uptake of metals by the plants, which reduces the damage to photosynthetic machinery or due to activation of the plant defense system and morphological, anatomical, and ultrastructual alterations in the photosynthetic apparatus.

14.12 MODIFICATION OF GENE EXPRESSION

To clarify the roles of Si in mitigating heavy toxicity at the gene level, Kim et al. (2014) investigated that in rice plants, Si application activated *OsHMA3* and under Cu/Cd stress, compared to metal treatments alone, which might ameliorate the negative effects of the metal stress in plants. Similarly, Li et al. (2008) demonstrated that 1.5 mM Si treatment under Cu toxicity induced the expression of genes responsible for the construction of metallothioneins (MTs) in *A. thaliana* that take part in chelating toxic heavy metals. *COPT1* and *HMA5* transporter genes were induced by Cu, and Si and Cu supply reduced these genes. In addition, Si may decrease the expression of phenylalanine ammonia lyase *(PAL)* genes under Cu stress. However, their role in decreasing Cu toxicity is not well understood (Khandekar and Leisner, 2011). Bokor et al. (2014b) stated that maize plants treated with Zn and also with Zn and Si showed downregulation of *ZmLsi1* and *ZmLsi2* genes in roots and an upregulated expression level of ZmLsi6 in the first leaf and decreased expression of ZmLsi6 in the second leaf. Nwugo and Huerta (2010) investigated the effect of Si on the leaf proteome of *Oryza sativa* under Cd stress and identified 60 protein spots that were regulated due to Cd and Si application. More recently, Ma et al. (2015) reported that in rice plants, the expression of the Lsi1 gene was upregulated in the presence of Si in the medium, whereas the expression of the natural resistance-associated macrophage protein 5 *(Nramp5)* gene, involved in the transport of Cd, was downregulated. In addition, Song et al. (2014) reported that Si application enhanced the expression of genes (Os08g02630 [PsbY], Os05g48630 [PsaH], Os07g37030 [PetC], Os03g57120 [PetH], Os09g26810, and Os04g38410) associated with photosynthesis in rice under Zn toxicity. So far, mechanisms of Si-mediated alleviation of heavy metal toxicity are poorly understood at the molecular and genetic levels, and more genetic experiments are required to resolve linkage relationships between Si and metal stress to investigate the gene expression related to absorbance and mobility of metals and Si in different plant species.

14.13   CONCLUSION AND PERSPECTIVES

Cadmium level in agricultural field is increasing continuously due to numerous natural and anthropogenic activities. Cd contamination into agricultural soil occurs as a result of natural activities such as mineral dissociation, weathering of parent material, and atmospheric deposition, as well as anthropogenic activities related to mining, industrial emissions, industrial wastes, application of sewage sludge to agricultural soils, and use of fertilizer and pesticides. Uptake and augmentation of cadmium in plants take place due to composite contact among soil, plant, and environmental factors that take part in influencing cadmium phytoavailability. Cadmium behavior in soil depends on soil and biological factors. The soil factors can be described as the parent material, organic matter, pH, CEC, redox setting, opposing ions, and the quantity of organic and inorganic ligands, and biological factors as the plants species, root morphology, and microbial environment. Root is the primary corridor through which cadmium enters into the plant. When cadmium sequesters in the root, it further moves to the other parts by different pathways, depending on the plant species. High level of cadmium augmentation in plants leads to several toxicities; for instance, it reduces seed germination and decreases root elongation and root–shoot biomass; furthermore, it also causes chlorosis, hinders photosynthesis, and decreases the concentration of essential elements in different parts of the plant. Cadmium on interaction with plant stimulated oxidative stress by generating ROS that resulted in the production of antioxidant enzymes. However, the strength of these effects diversified and relied on the amount of metal dosage, period of exposure, plant species, and plant phase. Worldwide HM toxicity leads to vast fatalities in agriculture productivity. It is generally accepted that silicon can alleviate numerous stresses such as biotic and abiotic, particularly the HM stress in plants. Si has the potential to diminish HM accumulation in different plant species and improved the growth and biomass of plants. The key mechanisms suggest sinking of active HM ions in growth media, chelation and production of antioxidant enzymes in plants, intrication and coprecipitation of toxic HMs, partition and structural modification, and the regulation of expression of metal transport genes. However, these mechanisms might be associated with plant species, genotypes, metal elements, growth conditions, duration of the stress imposed, and so on. Therefore, on taking a broad review of silicon, mediated alleviation of metal toxicity should be made with caution. Furthermore, this review in this chapter should contribute to a better understanding of the mechanisms of Si-mediated enhanced metal tolerance of plants, as well as increasing the productivity of crops under stress conditions. However, more work needs to be done to explore fully the mechanisms involved.

REFERENCES

Abe, T., Fukami, M., Ogasawara, M. 2008. Cadmium accumulation in the shoots and roots of 93 weed species. *Soil Sci. Plant Nutr.* 54, 566–573.

Adriano, D.C. 2001. *Trace Elements in Terrestrial Environments: Biogeochemistry, Bioavailability, and Risks of Metals*, 2nd ed. Springer, New York.

Adriano, D.C., Wenzel, W.W., Vangronsvel, J., Bolan, N.S. 2004. Role of assisted natural remediation in environmental cleanup. *Geoderma* 122, 121–142.

Alexandre, A., Meunier, J.D., Lezine, A.M., Vincens, A., Schwartz, D. 1997. Phytoliths: Indicators of grass land dynamics during the late Holocene in intertropical Africa. Palaeogeogr. *Palaeoclimatol. Palaeoecol.* 136, 213–229.

Ali, S., Farooq, M.A., Yasmeen, T., Hussain, S., Arif, M.S., Abbas, F., Bharwana, S.A., Zhang, G.P. 2013. The influence of silicon on barley growth, photosynthesis and ultra-structure under chromium stress. *Ecotoxicol. Environ. Saf.* 89, 66–72.

Antoniadis, V., Alloway, B.J. 2002. The role of dissolved organic carbon in the mobility of Cd, Ni and Zn in sewage sludge-amended soils. *Environ. Pollut.* 117, 515–521.

Baize, D. 1997. *Teneurs Totales en Eléments Traces Métalliques dans les Sols Français. Références et Stratégies d'Interprétation.* INRA Editions, Paris.

Baize, D., Deslais, W., Gaiffe, M. 1999. Anomalies naturelles en cadmium dans les sols de France. *Étude Gestion Sols* 6, 85–104.

Balestrasse, K.B., Gallego, S.M., Tomaro, M.L. 2006. Oxidation of the enzymes involved in nitrogen assimilation plays an important role in the cadmium induced toxicity in soybean plants. *Plant Soil* 284, 187–194.

Balestrasse, K.B., Garbey, L., Gallego, S.M., and Tomaro, M.L. 2001. Response of antioxidant defense system in soybean nodules and root subjected to cadmium stress. *Aust. J. Plant Physiol.* 28, 49.

Bavi, K., Kholdebarin, B., Moradshahi, A. 2011. Effect of cadmium on growth, protein content and peroxidase activity in pea plants. *Pak. J. Bot.* 43, 1467–1470.

Benavides, M.P., Gallego, S.M., Tomaro, M.L. 2005. Cadmium toxicity in plants. *Braz. J. Plant Physiol.* 17, 21–34.

Bes, C., Mench, M. 2008. Remediation of copper-contaminated topsoils from a wood treatment facility using in situ stabilisation. *Environ. Pollut.* 156(3), 1128–1138.

Bharwana, S.A., Ali, S., Farooq, M.A., Iqbal, N., Abbas, F., Ahmad, M.S.A. 2013. Alleviation of lead toxicity by silicon is related to elevated photosynthesis, antioxidant enzymes suppressed lead uptake and oxidative stress in cotton. *J. Bioremediat. Biodegrad.* 4, 1–11.

Blecker, S.W., McCulley, R.L., Chadwick, O.A., Kelly, E.F. 2006. Biologic cycling of silica across a grass land bioclimosequence. *Glob. Biogeochem. Cycles* 20. doi:10.1029/2006GB002690.

Bokor, B., Bokorová, S., Ondoš, S., Švubová, R., Lukačová, Z., Hýblová, M., Szemes, T., Lux, A. 2014a. Ionome and expression level of Si transporter genes (Lsi1, Lsi2, and Lsi6) affected by Zn and Si interaction in maize. *Environ. Sci. Pollut. Res.* http://dx.doi.org/10.1007/s11356-014-3876-6.

Bokor, B., Vaculik, M., Slováková, L., Masarovič, D., Lux, A. 2014b. Silicon does not always mitigate zinc toxicity in maize. *Acta Physiol. Plant.* 36, 733–743.

Bolan, N.S., Naidu, R., Syers, J.K., Tillman, R.W. 1999. Surface charge and solute interactions in soils. *Adv. Agron.* 67, 88–141.

Cakmak, I., Welch, R.M., Hart, J., Norvell, W.A., Ozturk, L., Kochian, L.V. 2000. Uptake and retranslocation of leaf-applied cadmium (Cd109) in diploid, tetraploid and hexaploid wheat. *J. Exp. Bot.* 51, 221–226.

Casey, W.H., Kinrade, S.D., Knight, C.T.G., Rains, D.W., Epstein, E. 2003. Aqueous silicate complexes in wheat, *Triticum aestivum* L. *Plant Cell Environ.* 27, 51–54.

Cho, S.C., Chao, Y.Y., Kao, C.H. 2012. Calcium deficiency increases Cd toxicity and Ca is required for heat-shock induced Cd tolerance in rice seedlings. *J. Plant Physiol.* 169, 892–898.

Chugh, L.K., Sawhney, S.K. 1999. Photosynthetic activities of *Pisum sativum* seedlings grown in presence of cadmium. *Plant Physiol. Biochem.* 37, 297–303.

Ci, D., Jiang, D., Dai, T., Jing, Q., Cao, W. 2009. Effects of cadmium on plant growth and physiological traits in contrast wheat recombinant inbred lines differing in cadmium tolerance. *Chemosphere* 77, 1620–1625.

Cieslinski, G., Van Rees, K.C.J., Szmigielska, A.M., Krishnamurti, G.S.R., Huang, P.M. 1998. Low-molecular-weight organic acids in rhizosphere soils of durum wheat and their effect on cadmium bioaccumulation. *Plant Soil* 203, 109–117.

Clemens, S. 2006. Toxic metal accumulation, responses to exposure and mechanisms of tolerance in plants. *Biochemistry* 88, 1707–1719.

Clemens, S., Palmgren, M.G., Kramer, U. 2002. A long way ahead: Understanding and engineering plant metal accumulation. *Trends Plant Sci.* 7, 309–315.

Cook, M.E., Morrow, H. 1995. Anthropogenic sources of cadmium in Canada. Presented at National Workshop on Cadmium Transport into Plants, Ottawa, ON, June 20–21.

Cornelis, J.T., Titeux, H., Ranger, J., Delvaux, B. 2011. Identification and distribution of the readily soluble silicon pool in a temperate forest soil below three distinct tree species. *PlantSoil* 342, 369–378.

Çotuk, Y., Belivermis, M., Kiliç, O. 2010. Environmental biology and pathophysiology of cadmium. *IUFS J. Biol.* 69, 1–5.

da Cunha, K.P.V., do Nascimento, C.W.A. 2009. Silicon effects on metal tolerance and structural changes in maize (*Zea mays* L.) grown on a cadmium and zinc enriched soil. *Water Air Soil Pollut.* 197, 323–330.

Das, P., Samantaray, S., Rout, G.R. 1997. Studies on cadmium toxicity in plants: A review. *Environ. Pollut.* 98, 29–36.

Degryse, F., Buekers, J., Smolders, E. 2004. Radiolabile cadmium and zinc in soil as affected by pH and source of contamination. *Eur. J. Soil Sci.* 55, 113–121.

Desplanques, V., Cary, L., Mouret, J.C., Trolard, F., Bourrié, G., Grauby, O., Meunier, J.D. 2006. Silicon transfers in a rice field in Camargue (France). *J. Geochem. Explor.* 88, 190–193.

Di Toppi, S.L., Gabrielli, R. 1999. Response to cadmium in higher plants. *Environ. Exp. Bot.* 41, 105–130.

Dimkpa, C.O., Merten, D., Svatos, A., Buchel, G., Kothe, E. 2009. Siderophores mediate reduced and increased uptake of cadmium by *Streptomyces tendae* F4 and sunflower (*Helianthus annuus*), respectively. *J. Appl. Microbiol.* 107, 1687–1696.

Ding, T.P., Ma, G.R., Shui, M.X., Wan, D.F., Li, R.H. 2005. Silicon isotope study on rice plants from the Zhejiang province, China. *Chem. Geol.* 218, 41–50.

Ding, T.P., Zhou, J.X., Wan, D.F., Chen, Z.Y., Zhang, F. 2008. Silicon isotope fractionation in bamboo and its significance to the biogeochemical cycle of silicon. *Geochim. Cosmochim. Acta* 72, 1381–1395.

Dong, J., Mao, W.H., Zhang, G.P., Wu, F.B., Cai, Y. 2007. Root excretion and plant tolerance to cadmium toxicity—A review. *Plant Soil Environ.* 53, 193–200.

Dong, J., Wu, F., Zhang, G. 2006. Influence of cadmium on antioxidant capacity and four microelement concentrations in tomato seedlings (*Lycopersicon esculentum*). *Chemosphere* 64, 1659–1666.

Drazic, G., Mihailovic, N., Stojanovic, Z. 2004. Cadmium toxicity: The effect on macro- and micronutrient contents in soybean seedlings. *Biol. Plant* 4, 605–607.

Dunbar, K.R., McLaughlin, M.J., Reid, R.J. 2003. The uptake and partitioning of cadmium in two cultivars of potato (*Solanum tuberosum* L.). *J. Exp. Bot.* 54, 349–354.

EC. 2000. Ambient air pollution by As, Cd and Ni compounds. Position paper.

European Commission, Directorate—General Environment. October. http://ec.europa.eu/environment/air /pdf/pp_as_cd_ni.pdf.

EFSA (European Food Safety Authority). 2009. Cadmium in food: Scientific opinion of the Panel on Contaminants in the Food Chain. *ESFA J.* 980, 1–139.

Elder, J.F. 1989. Metal biogeochemistry in surface-water systems—A review of principles and concepts. *U.S. Geological Survey Circular 1013*.

Farooq, M.A., Ali, S., Hameed, A., Ishaque, W., Mahmood, K., Iqbal, Z. 2013. Alleviation of cadmium toxicity by silicon is related to elevated photosynthesis, antioxidant enzymes; suppressed cadmium uptake and oxidative stress in cotton. *Ecotoxicol. Environ. Saf.* 96, 242–249.

Fediuk, E., Erdei, L. 2002. Physiological and biochemical aspects of cadmium toxicity and protective mechanisms induced in *Phragmites australia* and *Typha latifolia*. *J. Plant Physiol.* 159, 265–271.

Feng, J., Shi, Q., Wang, X., Wei, M., Yang, F., Xu, H. 2010. Silicon supplementation ameliorated the inhibition of photosynthesis and nitrate metabolism by cadmium (Cd) toxicity in *Cucumis sativus* L. *Sci. Hortic.* 123, 521–530.

Francis, A.J. 1990. Microbial dissolution and stabilization of toxic metals and radionuclides in mixed wastes. *Experientia* 46, 840–851.

Frantz, J.M., Khandekar, S., Leisner, S. 2011. Silicon differentially influences copper toxicity response in silicon-accumulator and non-accumulator species. *J. Am. Soc. Hortic. Sci.* 136, 329–338.

Gao, X., Brown, K.R., Racz, G.J., Grant, C.A. 2010. Concentration of cadmium in durum wheat as affected by time, source and placement of nitrogen fertilization under reduced and conventional-tillage management. *Plant Soil* 337, 341–354.

Gao, X., Mohr, R.M., McLaren, D.L., Grant, C.A. 2011. Grain cadmium and zinc concentrations in wheat as affected by genotypic variation and potassium chloride fertilization. *Field Crops Res.* 122, 95–103.

Gill, S.S., Khan, N.A., Tuteja, N. 2011. Differential cadmium stress tolerance in five Indian mustard (*Brassica juncea* L.) cultivars. An evaluation of the role of antioxidant machinery. *Plant Signal. Behav.* 6, 293–300.

Gill, S.S., Khan, N.A., Tuteja, N. 2012. Cadmium at high dose perturbs growth, photosynthesis and nitrogen metabolism while at low dose it upregulates sulfur assimilation and antioxidant machinery in garden cress (*Lepidium sativum* L.). *Plant Sci.* 182, 112–120.

Gonçalves, J.F., Antes, F.G., Maldaner, J., Pereira, L.B., Tabaldi, L.A., Rauber, R., Rossato, L.V., Bisognin, D.A., Dressler, V.L., de Moraes Flores, E.M., Nicoloso, F.T. 2009. Cadmium and mineral nutrient accumulation in potato plantlets grown under cadmium stress in two different experimental culture conditions. *J. Plant Physiol. Biochem.* 47, 814–821.

Grant, C.A., Buckley, W.T., Bailey, L.D., Selles, F. 1998. Cadmium accumulation in crops. *Can. J. Plant Sci.* 78, 1–17.

Grant, C.A., Clarke, J.M., Duguid, S., Chaney, R.L. 2008. Selection and breeding of plant cultivars to minimize cadmium accumulation. *Sci. Total Environ.* 390, 301–310.

Greger, M., Löfstedt, M. 2004. Comparison of uptake and distribution of cadmium in different cultivars of bread and durum wheat. *Crop Sci.* 44, 501–507.

Gu, H.H., Qiu, H., Tian, T., Zhan, S.S., Deng, T.H.B., Chaney, R.L., Wang, S.Z., Tang, Y.T., Morel, J.L., Qiu, R.L. 2011. Mitigation effects of silicon rich amendments on heavy metal accumulation in rice (*Oryza sativa* L.) planted on multi-metal contaminated acidic soil. *Chemosphere* 83, 1234–1240.

Gu, H.H., Zhan, S., Wang, S.Z., Tang, Y.T., Chaney, R.L., Fang, X.H., Cai, X.D., Qiu, R.L. 2012. Silicon-mediated amelioration of zinc toxicity in rice (*Oryza sativa* L.) seedlings. *Plant Soil* 350, 193–204.

Guntzer, F., Keller, C., Meunier, J.D. 2012. Benefits of plant silicon for crops: A review. *Agron. Sustain. Dev.* 32, 201–213.

Guo, Y., Marschner, H. 1995. Uptake, distribution and binding of cadmium and nickel in different plant species. *J. Plant Nutr.* 18, 2691–2706.

Haag-Kerwer, A., Schäfer, H.J., Heiss, S., Walter, C., and Rausch, T. 1999. Cadmium exposure in *Brassica juncea* causes a decline in transpiration rate and leaf expansion without effect on photosynthesis. *J. Exp. Bot.* 341(50), 1827–1835.

Han, F., Shan, X., Zhang, S., Wen, B., Owens, G. 2006. Enhanced cadmium accumulation in maize roots— The impact of organic acids. *Plant Soil* 289, 355–368.

Haouari, C.C., Nasraoui, A.H., Bouthour, D., Houda, M.D., Daieb, C.B., Mnai, J., Gouia, H. 2012. Response of tomato (*Solanum lycopersicon*) to cadmium toxicity: Growth, element uptake, chlorophyll content and photosynthesis rate. *Afr. J. Plant Sci.* 6, 1–7.

Hara, T., Gu, M.H., Koyama, H. 1999. Ameliorative effect of silicon on aluminum injury in the rice plant. *Soil Sci. Plant Nutr.* 45, 929–936.

Harris, N.S., Taylor, G.J. 2001. Remobilization of cadmium in maturing shoots of near isogenic lines of durum wheat that differ in grain cadmium accumulation. *J. Exp. Bot.* 52, 1473–1481.

Hart, J.J., Welch, R.M., Norvell, W.A., Kochian, L.V. 2002. Transport interactions between cadmium and zinc in roots of bread and durum wheat seedlings. *Physiol. Plant.* 116, 73–78.

Hart, J.J., Welch, R.M., Norvell, W.A., Kochian, L.V. 2006. Characterization of cadmium uptake, translocation and storage in near-isogenic lines of durum wheat that differ in grain cadmium concentration. *New Phytol.* 172, 261–271.

Hart, J.J., Welch, R.M., Norvell, W.A., Sullivan, L.A., Kochian, L.V. 1998. Characterization of cadmium binding, uptake and translocation in intact seedlings of bread and durum wheat cultivars. *Plant Physiol.* 116, 1413–1420.

Hegedüs, A., Erdei, S., and Horváth, G. 2001. Comparative studies of H_2O_2 detoxifying enzymes in green and greening barley seedlings under cadmium stress. *Plant Sci.* 160(6), 1085–1093.

Hentz, S., McComb, J., Miller, G., Begonia, M., Begonia, G. 2012. Cadmium uptake, growth and phytochelatin contents of *Triticum aestivum* in response to various concentrations of cadmium. *World Environ.* 2, 44–50.

Hepinstall, S.E., Turner, B.F., Maurice, P.A. 2005. Effects of siderophores on Pb and Cd adsorption to kaolinite. *Clays Clay Miner.* 53, 557–563.

Hepler, P.K., Wayne, R.O. 1985. Calcium and plant development. *Ann. Rev. Plant Physiol.* 36, 391–439.

Hernandez, L., Probst, A., Probst, J.L., Ulrich, E. 2003. Heavy metal distribution in some French forest soils: Evidence for atmospheric contamination. *Sci. Total Environ.* 312, 195–219.

Hernandez, L.E., Lozano, E., Garate, A., Carpena, R. 1998. Influence of cadmium on the uptake, tissue accumulation and subcellular distribution of manganese in pea seedlings. *Plant Sci.* 132, 139–151.

Hill, O.A., Lion, L.W., Ahner, B.A. 2002. Reduced Cd accumulation in *Zea mays*: A protective role for phytosiderophores? *Environ. Sci. Technol.* 36, 5363–5368.

Hodson, M.J., Sangster, A.G. 1990. Techniques for the microanalysis of higher plants, with particular reference to silicon in cryofixed wheat tissues. *Scanning Microsc.* 4, 407–418.

Hu, H., Zhang, J., Wang, H., Li, R., Pan, F., Wu, J., Liu, Q. 2013. Effect of silicate supplementation on the alleviation of arsenite toxicity in 93-11 (*Oryza sativa* L. *indica*). *Environ. Sci. Pollut. Res.* 20, 8579–8589.

Hussain, I., Ashraf, M.A., Rasheed, R., Asghar, A., Sajid, M.A., Iqbal, M. 2015. Exogenous application of silicon at the boot stage decreases accumulation of cadmium in wheat (*Triticum aestivum* L.) grains. *Braz. J. Bot.* http://dx.doi.org/10.1007/s40415-014-0126-6.

Itoh, S., Yumura, Y. 1979. Studies on the contamination of vegetables crops by excessive absorption of heavy metals. *Bull. Veg. Ornam. Crops Res. Stn.* 6a 123 (Ja).

Jalil, A., Selles, F., Clark, J.M. 1994. Effect of cadmium on growth and uptake of cadmium and other elements by durum wheat. *J. Plant Nutr.* 17, 1839–1858.

Jung, M.C. 2008. Heavy metal concentrations in soils and factors affecting metal uptake by plants in the vicinity of a Korean Cu-W mine. *Sensors* 8, 2413–2423.

Kabata-Pendias, A., Pendias, H. 2001. *Trace Elements in Soils and Plants*, 3rd ed. CRC Press, Boca Raton, FL.

Kaya, C., Tuna, A.L., Sonmez, O., Ince, F., Higgs, D. 2009. Mitigation effects of silicon on maize plants grown at high zinc. *J. Plant Nutr.* 32, 1788–1798.

Keller, C., Rizwan, M., Davidian, J.C., Pokrovsky, O.S., Bovet, N., Chaurand, P., Meunier, J.D. 2015. Effect of silicon on wheat seedlings (*Triticum turgidum* L.) grown in hydroponics and exposed to 0 to 30 m M Cu. *Planta* 241, 847–860.

Khan, N.A., Samiullah, A., Singh, S., Nazar, R. 2007. Activities of antioxidative enzymes, sulphur assimilation, photosynthetic activity and growth of wheat (*Triticum aestivum*) cultivars differing in yield potential under cadmium stress. *J. Agron. Crop Sci.* 193, 435–444.

Khandekar, S., Leisner, S. 2011. Soluble silicon modulates expression of *Arabidopsis thaliana* genes involved in copper stress. *J. Plant Physiol.* 168, 699–705.

Kim, Y.H., Khan, A.L., Kim, D.H., Lee, S.Y., Kim, K.M., Waqas, M., Lee, I.J. 2014. Silicon mitigates heavy metal stress by regulating P-type heavy metal ATPases, *Oryza sativa* low silicon genes, and endogenous phytohormones. *BMC Plant Boil.* 14, 13. http://www.biomedcentral.com/1471-2229/14/13.

Kirkham, M.B. 2006. Cadmium in plants on polluted soils: Effects of soil factors, hyperaccumulation, and amendments. *Geoderma* 137, 19–32.

Klumpp, C., Burger, A., Mislin, G.L., Abdallah, M.A. 2005. From a total synthesis of cepabactin and its 3:1 ferric complex to the isolation of a 1:1:1 mixed complex between iron (III), cepabactin and pyochelin. *Bioorg. Med. Chem. Lett.* 15, 1721–1724.

Kovda, V.A. 1973. *The Bases of Learning About Soils*, vol. 2. Nayka, Moscow.

Kubo, K., Watanabe, Y., Matsunaka, H., Seki, M., Fujita, M., Kawada, N., Hatta, K., Nakajima, T. 2011. Differences in cadmium accumulation and root morphology in seedlings of Japanese wheat varieties with distinctive grain cadmium concentration. *Plant Prod. Sci.* 14, 148–155.

Lehoczky, E., Marth, P., Szabados, I., Lukacs, P. 2000. Influence of soil factors on the accumulation of cadmium by lettuce. *Commun. Soil Sci. Plant Anal.* 31, 2425–2431.

Li, J., Leisner, M., Frantz, J. 2008. Alleviation of copper toxicity in *Arabidopsis thaliana* by silicon addition to hydroponic solutions. *J. Am. Soc. Hortic. Sci.* 133, 670–677.

Li, Z., Song, Z., Cornelis, J.T. 2014. Impact of rice cultivar and organ on elemental composition of phytoliths and the release of bio-available silicon. *Front Plant Sci.* 5. http://dx.doi.org/10.3389/fpls.2014.00529.

Liang, Y.C., Wong, J.W.C., Long, W. 2005. Silicon-mediated enhancement of cadmium tolerance in maize (*Zea mays* L.) grown in cadmium contaminated soil. *Chemosphere* 58, 475–483.

Liphadzi, M.S., Kirkham, M.B. 2006. Availability and plant uptake of heavy metals in EDTA-assisted phytoremediation of soil and composted biosolids. *S. Afr. J. Bot.* 72, 391–397.

Liu, J.G., Li, K., Xu, J.K., Liang, J.S., Lu, X.L., Yang, J.C., Zhu, Q.S. 2003. Interaction of Cd and five mineral nutrients for uptake and accumulation in different rice cultivars and genotypes. *Field Crop Res.* 83, 271–281.

Liu, J.G., Qian, M., Cai, G.L., Yang, J.C., Zhu, Q.S. 2007. Uptake and translocation of Cd in different rice cultivars and the relation with Cd accumulation in rice grain. *J. Hazard. Mater.* 143, 443–447.

Llamas, A., Cornelia, I.U., and Sanz, A. 2000. Cadmium effects on transmembrane electrical potential difference, respiration and membrane permeability of rice (*Oryza sativa*) roots. *Plant Soil* 219, 21–28.

López-Millán, A.F., Sagardoy, R., Solanas, M., Abadia, A., Abadia, J. 2009. Cadmium toxicity in tomato (*Lycopersicon esculentum*) plants grown in hydroponics. *Environ. Exp. Bot.* 65, 376–385.

Lozano-Rodriguez, E., Hernandez, L.E., Bonay, P., Carpena-Ruiz, R.O. 1997. Distribution of cadmium in shoot and root tissues of maize and pea plants: Physiological disturbances. *J. Exp. Bot.* 48, 123–128.

Lux, A., Luxová, M., Abe, J., Tanimoto, E., Hattoriand, T., Inanaga, S. 2003. The dynamics of silicon deposition in the sorghum root endodermis. *New Phytol.* 158, 437–441.

Lux, A., Martinka, M., Vaculik, M., White, P.J. 2011. Root responses to cadmium in the rhizosphere: A review. *J. Exp. Bot.* 62, 21–37.

Ma, J.F., Takahashi, E. 2002. *Soil, Fertiliser, and Plant Silicon Research in Japan.* Elsevier.

Ma, J.F., Yamaji, N. 2008. Functions and transport of silicon in plants. *Cell Mol. Life Sci.* 65, 3049–3057.

Ma, J.F., Yamaji, N. 2006. Silicon up take and accumulation in higher plants. *Trends Plant Sci.* 11, 392–397.

Ma, J.F., Miyake, Y., Takahashi, E. 2001. Silicon as a beneficial element for crop plants. In: Datnoff, L.E., Snyder, G.H., Korndorfer, G.H. (Eds.), *Silicon in Agriculture.* Elsevier Science Publishing, Amsterdam, pp. 17–39.

Ma, J.F., Mitani, N., Nagao, S., Konishi, S., Tamai, K., Iwashita, T., Yano, M. 2004. Characterization of the silicon uptake system and molecular mapping of the silicon transporter gene in rice. *Plant Physiol.* 136, 3284–3289.

Ma, J.F., Tamai, K., Yamaji, N., Mitani, N., Konishi, S., Katsuhara, M., Ishiguro, M., Murata, Y., Yano, M. 2006. A silicon transporter in rice. *Nature* 440, 688–691.

Ma, J.F., Yamaji, N., Mitani, N., Tamai, K., Konishi, S., Fujiwara, T., Katsuhara, M., Yano, M. 2007. An efflux transporter of silicon in rice. *Nature* 448, 209–212.

Ma, J.F., Yamaji, N., Mitani-Ueno, N. 2011. Transport of silicon from roots to panicles in plants. *Proc. Jpn. Acad. Ser. B: Phys. Biol. Sci.* 87, 377.

Ma, J., Cai, H., He, C., Zhang, W., Wang, L. 2015. A hemicellulose-bound form of silicon inhibits cadmium ion uptake in rice (*Oryza sativa*) cells. *New Phytol.* http://dx.doi.org/10.1111/nph.13276.

Malčovská, S.M., Dučaiová, Z., Maslaňáková, I., Bačkor, M. 2014. Effect of silicon on growth, photosynthesis, oxidative status and phenolic compounds of maize (*Zea mays* L.) grown in cadmium excess. *Water Air Soil Pollut.* 225, 1–11.

Marrs, K.A., Walbot, V. 1997. Expression and RNA splicing of the maiz glutathione S-transferase Bronze2 gene is regulated by cadmium and other stresses. *Plant Physiol.* 113(1), 93–102.

Marschner, H. 1995. *Mineral Nutrition of Higher Plants,* 2nd ed. Academic Press, San Diego.

Marschner, H., Romheld, V., Kissel, M. 1986. Different strategies in higher-plants in mobilization and uptake of iron. *J. Plant Nutr.* 9, 695–713.

Masarovic, D., Slovakova, L., Bokor, B., Bujdos, M., Lux, A. 2012. Effect of silicon application on Sorghum bicolor exposed to toxic concentration of zinc. *Biologia* 67, 706–712.

McBride, M., Sauve, S., Hendershot, W. 1997. Solubility control of Cu, Zn, Cd and Pb in contaminated soils. *Eur. J. Soil Sci.* 48, 337–346.

Mehrabanjoubani, P., Abdolzadeh, A., Sadeghipour, H.R., Aghdasi, M. 2014. Impacts of silicon nutrition on growth and nutrient status of rice plants grown under varying zinc regimes. *Theor. Exp. Plant Physiol.* http://dx.doi.org/10.1007/s40626-014-0028-9.

Mench, M., Baize, D., Mocquot, B. 1997. Cadmium availability to wheat in five soil series from the Yonne district, Burgundy, France. *Environ. Pollut.* 95, 93–103.

Mench, M., Martin, E. 1991. Mobilization of cadmium and other metals from two soils by root exudates of *Zea mays* L., *Nicotiana tabacum* L. and *Nicotiana rustica* L. *Plant Soil* 132, 187–196.

Mench, M., Renella, G., Gelsomino, A., Landi, L., Nannipieri, P. 2006. Biochemical parameters and bacterial species richness in soils contaminated by sludge-borne metals and remediated with inorganic soil amendments. *Environ. Pollut.* 144(1), 24–31.

Mench, M., Tancogne, J., Gomez, A., Juste, C. 1989. Cadmium bioavailability to *Nicotiana tabacum* L., *Nicotiana rustica* L., and *Zea mays* L. grown in soil amended or not amended with cadmium nitrate. *Biol. Fertil. Soils* 8, 48–53.

Meunier, J.D., Colin, F., Alarcon, C. 1999. Biogenic silica storage in soils. *Geology* 27, 835–838.

Meyer, M.W., Fricke, F.L., Holmgren, G.G.S., Kubota, J., Chaney, R.L. 1982. Cadmium and lead in wheat grain and associated surface soils of major wheat production areas of the United States. In *Agronomy Abstracts.* American Society of Agronomy, Madison, WI, p. 34.

Mitani, N., Yamaji, N., Ma, J.F. 2009. Identification of maize silicon influx transporters. *Plant Cell Physiol.* 50, 5–12.

Mobin, M., Khan, N.A. 2007. Photosynthetic activity, pigment composition and antioxidative response of two mustard (*Brassica juncea*) cultivars differing in photosynthetic capacity subjected to cadmium stress. *J. Plant Physiol.* 164, 601–610.

Morikawa, C.K., Saigusa, M. 2002. Si amelioration of Al toxicity in barley (*Hordeum vulgare* L.) growing in two Andosols. *Plant Soil* 240, 161–168.

Naeem, A., Ghafoor, A., Farooq, M. 2014. Suppression of cadmium concentration in wheat grains by silicon is related to its application rate and cadmium accumulating abilities of cultivars. *J. Sci. Food Agric.* http:// dx.doi.org/10.1002/ jsfa.6976.

Nagajyoti, P.C., Lee, K.D., Sreekanth, T.V.M. 2010. Heavy metals, occurrence and toxicity for plants: A review. *Environ. Chem. Lett.* 8, 199–216.

Naidu, R., Bolan, N.S., Kookana, R.S., Tiller, K.G. 1994. Ionic strength and pH effects on the sorption of cadmium and the surface charge of soils. *Eur. J. Soil Sci.* 45, 419–429.

Naidu, R., Kookana, R.S., Sumner, M.E., Harter, R.D., Tiller, K.G. 1997. Cadmium sorption and transport in variable charge soils: A review. *J. Environ. Qual.* 26, 602–617.

Nakanishi, H., Ogawa, I., Ishimaru, Y., Mori, S., Nishizawa, N.K. 2006. Iron deficiency enhances cadmium uptake and translocation mediated by the Fe2+ transporters OsIRT1 and OsIRT2 in rice. *Soil Sci. Plant Nutr.* 52, 464–469.

Neubauer, U., Nowack, B., Furrer, G., Schulin, R. 2000. Heavy metal sorption on clay minerals affected by the siderophore desferrioxamine B. *Environ. Sci. Technol.* 34, 2749–2755.

Nigam, R., Srivastava, S., Prakash, S., Srivastava, M.M. 2001. Cadmium mobilisation and plant availability— The impact of organic acids commonly exuded from roots. *Plant Soil* 230, 107–113.

Nwugo, C.C., Huerta, A.J. 2008. Effects of silicon nutrition on cadmium uptake: Growth and photosynthesis of rice plants exposed to low level cadmium. *Plant Soil* 311, 73–86.

Ok, Y.S., Lee, H., Jung, J., Song, H., Chung, N., Lim, S., Kim, J.G. 2004. Chemical characterization and bio-availability of cadmium in artificially and naturally contaminated soils. *Agric. Chem. Biotechnol.* 47, 143–146.

Oliva, S.R., Mingorance, M.D., Leidi, E.O. 2011. Effects of silicon on copper toxicity in *Erica andevalensis* Cabezudo and Rivera: A potential species to remediate contaminated soils. *J. Environ. Monit.* 13, 591–596.

Ouzounidou, G., Moustakas, M., Eleftheriou, E.P. 1997. Physiological and ultrastructural effects of cadmium on wheat (*Triticum aestivum* L.) leaves. *Arch. Environ. Contam. Toxicol.* 32, 154–160.

Ozturk, L., Eker, S., Ozkutlu, F. 2003. Effect of cadmium on growth and concentrations of cadmium. Ascorbic acid and sulphydryl groups in durum wheat cultivars. *Turk. J. Agric For.* 27, 161–168.

Pál, M., Horvath, E., Janda, T., Paldi, E., Szalai, G. 2006. Physiological changes and defense mechanisms induced by cadmium stress in maize. *J. Plant Nutr. Soil Sci.* 169, 239–246.

Pandey, N., Singh, G.K. 2012. Studies on antioxidative enzymes induced by cadmium in pea plants (*Pisum sativum* L.). *J. Environ. Biol.* 33, 201–206.

Perilli, P., Mitchell, L.G., Grant, C.A., Pisante, M. 2010. Cadmium concentration in durum wheat grain (*Triticum turgidum*) as influenced by nitrogen rate, seeding date and soil type. *J. Sci. Food Agric.* 90, 813–822.

Prasad, M.N.V. 1997. *Plant Ecophysiology*. John Wiley & Sons, New York.

Probst, A., El Ghmari, A., Aubert, D., Fritz, B., McNutt, R. 2000. Strontium as a tracer of weathering processes in a silicate catchment polluted by acid atmospheric inputs, Strengbach, France. *Chem. Geol.* 170, 203–219.

Prosi, F. 1989. Factors controlling biological availability and toxic effects of lead in aquatic organisms. *Sci. Total Environ.* 79, 157–169.

Rains, D.W., Epstein, E., Zasoski, R.J., Aslam, M. 2006. Active silicon uptake by wheat. *Plant Soil* 280, 223–228.

Redjala, T., Zelkoa, I., Sterckemana, T., Leguec, V., Lux, A. 2011. Relationship between root structure and root cadmium uptake in maize. *Environ. Exp. Bot.* 71, 241–248.

Reichard, P.U., Kraemer, S.M., Frazier, S.W., Kretzschmar, R. 2005. Goethite dissolution in the presence of phytosiderophores: Rates, mechanisms, and the synergistic effect of oxalate. *Plant Soil* 276, 115–132.

Rizwan, M., Meunier, J.D., Miche, H., Keller, C. 2012. Effect of silicon on reducing cadmium toxicity in durum wheat (*Triticum turgidum* L. cv. Claudio W.) grown in a soil with aged contamination. *J. Hazard. Mater.* 209–210, 326–334.

Salt, D.E., Prince, R.C., Pickering, I.J., Raskin, I. 1995. Mechanisms of cadmium mobility and accumulation in Indian mustard. *Plant Physiol.* 109, 1427–1433.

Sammut, M.L., Noack, Y., Rose, J., Hazemann, J.L., Proux, O., Depoux, M., Ziebel, A., Fiani, E. 2010. Speciation of Cd and Pb in dust emitted from sinter plant. *Chemosphere* 78, 445–450.

Sandalio, L.M., Dalurzo, H.C., Gomez, M., Romero-Puertas, M.C., Del Rio, L.A. 2001. Cadmium-induced changes in the growth and oxidative metabolism of pea plants. *J. Exp. Bot.* 52, 2115–2126.

Sanglard, L.M.V.P., Martins, S.,C.V., Detmann, K.C., Silva, P.E.M., Lavinsky, A.O., Silva, M.M., Detmann, E., Araújo, W.L., DaMatta, F.M. 2014. Silicon nutrition alleviates the negative impacts of arsenic on the photosynthetic apparatus of rice leaves: An analysis of the key limitations of photosynthesis. *Physiol. Plant.* http://dx. doi.org/10.1111/ppl.12178.

Sarwar, N., Saifullah, Malhi, S.S., Zia, M.H., Naeem, A., Bibi, S., Farid, G. 2010. Role of mineral nutrition in minimizing cadmium accumulation by plants. *J. Sci. Food Agric.* 90, 925–937.

Sauve, S., Norvell, W.A., McBride, M., Hendershot, W. 2000a. Speciation and complexation of cadmium in extracted soil solutions. *Environ. Sci. Technol.* 34, 291–296.

Sauve, S., Hendershot, W., Allen, H.E. 2000b. Solid–solution partitioning of metals in contaminated soils: Dependence on pH, total metal burden, and organic matter. *Environ. Sci. Technol.* 34(7), 1125–1131.

Schickler, H., Caspi, H. 1999. Response of antioxidative enzymes to nickel and cadmium stress in hyperaccumulator plants of the genus *Alyssum*. *Physiol. Plant.* 105, 39–44.

Seregin, I.V., Shpigun, L.K., Ivanov, V.B. 2004. Distribution and toxic effects of cadmium and lead on maize roots. *Russ. J. Plant Physiol.* 51, 525–533.

Shen, X., Xiao, X., Dong, Z., Chen, Y. 2014. Silicon effects on antioxidative enzymes and lipid peroxidation in leaves and roots of peanut under aluminum stress. *Acta Physiol. Plant.* 36, 3063–3069.

Shenker, M., Fan, T.W.M., Crowley, D.E. 2001. Phytosiderophores influence on cadmium mobilization and uptake by wheat and barley plants. *J. Environ. Qual.* 30, 2091–2098.

Shi, G., Cai, Q., Liu, C. 2010a. Silicon alleviates cadmium toxicity in peanut plants in relation to cadmium distribution and stimulation of antioxidative enzymes. *Plant Growth Regul.* 61, 45–52.

Shi, G., Liu, C., Cai, Q., Liu, Q.Q., Hou, C. 2010b. Cadmium accumulation and tolerance of two safflower cultivars in relation to photosynthesis and antioxidantive enzymes. *Bull. Environ. Contam. Toxicol.* 85, 256–263.

Shi, G.R., Cai, Q.S. 2008. Photosynthetic and anatomic responses of peanut leaves to cadmium stress. *Photosynthetica* 46, 627–630.

Singh, V.P., Tripathi, D.K., Kumar, D., Chauhan, D.K. 2011. Influence of exogenous silicon addition on aluminium tolerance in rice seedlings. *Biol. Trace Elem. Res.* 144, 1260–1274.

Sommer, M., Kaczorek, D., Kuzyakov, Y., Breuer, J. 2006. Silicon pools and fluxes in soils and landscapes: A review. *J. Plant Nutr. Soil Sci.* 169,310–329.

Song, A., Li, P., Li, P., Fan, F., Nikolic, M., Liang, Y. 2011. The alleviation of zinc toxicity by silicon is related to zinc transport and antioxidative reaction in rice. *Plant Soil* 344, 319–333.

Song, A., Li, P., Fan, F., Li, Z., Liang, Y. 2014. The effect of silicon on photosynthesis and expression of its relevant genes in rice (*Oryza sativa* L.) under high-zinc stress. *PLoS One* 9(11). http://dx.doi.org/10.1371/journal.pone.0113782.

Srivastava, R.K., Pandey, P., Rajpoot, R., Rani, A., Gautam, A., Dubey, R.S. 2014. Exogenous application of calcium and silica alleviates cadmium toxicity by suppressing oxidative damage in rice seedlings. http://dx.doi.org/10.1007/ s00709-014-0731-z.

Street, R.A., Kulkarni, M.G., Stirka, W.A., Southway, C., Staden, J.V. 2010. Effect of cadmium on growth and micronutrient distribution in wild garlic (*Tulbaghia violacea*). *S. Afr. J. Bot.* 76, 332–336.

Takahashi, E., Ma, J.F., Miyake, Y. 1990. The possibility of silicon as an essential element for higher plants. *Comm. Agric. Food Chem.* 2, 99–122.

Treder, W., Cieslinski, G. 2005. Effect of silicon application on cadmium uptake and distribution in strawberry plants grown on contaminated soils. *J. Plant Nutr.* 28, 917–929.

Tripathi, D.K., Singh, V.P., Kumar, D., Chauhan, D.K. 2012a. Impact of exogenous silicon addition on chromium uptake, growth, mineral elements, oxidative stress, antioxidant capacity, and leaf and root structures in rice seedlings exposed to hexavalent chromium. *Acta Physiol. Plant.* 34, 279–289.

Tripathi, D.K., Singh, V.P., Kumar, D., Chauhan, D.K. 2012b. Rice seedlings under cadmium stress: Effect of silicon on growth, cadmium uptake, oxidative stress, antioxidant capacity and root and leaf structures. *Chem. Ecol.* 28, 281–291.

Tripathi, D.K., Singh, V.P., Prasad, S.M., Chauhan, D.K., Dubey, N.K., Rai, A.K. 2015. Silicon-mediated alleviation of Cr (VI) toxicity in wheat seedlings as evidenced by chlorophyll fluorescence, laser induced breakdown spectroscopy and anatomical changes. *Ecotoxicol. Environ. Saf.* 113, 133–144.

Tudoreanu, L., Phillips, C.J.C. 2004. Modeling cadmium uptake and accumulation in plants. *Adv. Agron.* 84, 121–157.

Ueno, D., Yamaji, N., Kono, I., Huang, C.F., Ando, T., Yano, M., Ma, J.F. 2010. Gene limiting cadmium accumulation in rice. *Proc. Natl. Acad. Sci. U.S.A.* 107, 16500–16505.

UNEP (United Nations Environment Program). 2010. Final review of scientific information on cadmium. December. UNEP, Geneva.

Vaculik, M., Lux, A., Luxova, M., Tanimoto, L. 2009. Silicon mitigates cadmium inhibitory effects in young maize plants. *Environ. Exp. Bot.* 67, 52–58.

Van der Vliet, L., Peterson, C., Hale, B. 2007. Cd accumulation in roots and shoots of durum wheat: The roles of transpiration rate and apoplastic bypass. *J. Exp. Bot.* 58, 2939–2947.

Vega, F., Andrade, M., Covelo, E. 2010. Influence of soil properties on the sorption and retention of cadmium, copper and Pb, separately and together, by 20 soil horizons: Comparison of linear regression and tree regression analyses. *J. Hazard. Mater.* 174(1–3), 522–533.

Vijayaragavan, M., Prabhahar, C., Sureshkumar, J., Natarajan, A., Vijayarengan, P., Sharavanan, S. 2011. Toxic effect of cadmium on seed germination, growth and biochemical contents of cowpea (*Vigna unguiculata* L.) plants. *Int. Multidis. Res. J.* 1(5), 1–6.

Wang, S., Wang, F., Gao, S. 2014. Foliar application with nano-silicon alleviates Cd toxicity in rice seedlings. *Environ. Sci. Pollut. Res.*, 1–9.

Wang, L., Zhou, Q., Ding, L., Sun, Y. 2008. Effect of cadmium toxicity on nitrogen metabolism in leaves of *Solanum nigrum* L. as a newly found cadmium hyperaccumulator. *J. Hazard. Mater.* 154, 818–825.

Watanabe, T., Moon, C.S., Zhang, Z.W., Shimbo, S., Nakatsuka, H., Matsuda-Inoguchi, N., Higashikawa, K., Ikeda, M. 2000. Cadmium exposure of women in general populations in Japan during 1991–1997 compared with 1977–1991. *Int. Arch. Occup. Environ. Health* 73, 26–34.

Weggler, K., McLaughlin, M.J., Graham, R.D. 2004. Effect of chloride in soil solution on the plant availability of biosolid-borne cadmium. *J. Environ. Qual.* 33, 496–504.

Welch, R.M., Hart, J.J., Norvell, W.A., Sullivan, L.A., Kochian, L.V. 1999. Effects of nutrient solution zinc activity on net uptake, translocation, and root export of cadmium and zinc by separated sections of intact durum wheat (*Triticum turgidum* L. var *durum*) seedling roots. *Plant Soil* 208, 243–250.

Wójcik, M., Tukiendorf, A. 2005. Cadmium uptake, localization and detoxification in *Zea mays*. *Biol. Plant* 49, 237–245.

Wuana, R.A., Okieimen, F.E. 2011. Heavy metals in contaminated soils: A review of sources, chemistry, risks and best available strategies for remediation. *ISRN Ecol.* 2011, 402647. http://dx.doi.org/10.5402/2011/402647.

Yamagata, N., Shigematsu I. 1970. Cadmium pollution in perspective. *Bull. Inst. Publ. Health* 19, 1–27.

Yamaji, N., Chiba, Y., Mitani-Ueno, N., Ma, J.F. 2012. Functional characterization of a silicon transporter gene implicated in silicon distribution in barley. *Plant Physiol.* 160, 1491–1497.

Yamaji, N., Mitatni, N., Ma, J.F. 2008. A transporter regulating silicon distribution in rice shoots. *Plant Cell* 20, 1381–1389.

Yang, Y.H., Chen, S.M., Chen, Z., Zhang, H.Y., Shen, H.G., Hua, Z.C., Li, N. 1999. Silicon effects on aluminum toxicity to mungbean seedling growth. *J. Plant Nutr.* 22, 693–700.

Yang, M.G., Lin, X.Y., Yang, X.E. 1998. Impact of Cd on growth and nutrient accumulation of different plant species. *Chin. J. Appl. Ecol.* 9, 89–94.

Yin-Ming, L., Chaney, R.L., and Schyciter, A.A. 1994. Effect of soil chloride on cadmium concentration in sunflower kernels. *Plant Soil* 167, 275–280.

Yoneyama, T., Gosho, T., Kato, M., Goto, S., Hayashi, H. 2010. Xylem and phloem transport of Cd, Zn and Fe into the grains of rice plants (*Oryza sativa* L.) grown in continuously flooded Cd-contaminated soil. *Soil Sci. Plant Nutr.* 56, 445–453.

Zeng, F.R., Zhao, F.S., Qiu, B.Y., Ouyang, Y.N., Wu, F.B., Zhang, G.P. 2011. Alleviation of chromium toxicity by silicon addition in rice plants. *Agric. Sci. China* 10, 1188–1196.

Zhang, C., Wang, L., Nie, Q., Zhang, W., Zhang, F. 2008. Long-term effects of exogenous silicon on cadmium translocation and toxicity in rice (*Oryza sativa* L.). *Environ. Exp. Bot.* 62, 300–307.

Zhang, G.P., Fukami, M., Sekimoto, H. 2002. Influence of cadmium on mineral concentration and yield components in wheat genotypes differing in cadmium tolerance at seedling stage. *Field Crop Res.* 79, 1–7.

Zhang, H., Dang, Z., Zheng, L.C., Yi, X.Y. 2009. Remediation of soil co-contaminated with pyrene and cadmium by growing maize (*Zea mays* L.). *Int. J. Environ. Sci. Technol.* 6, 249–258.

Zhang, Q., Yan, C., Liu, J., Lu, H., Wang, W., Du, J., Duan, H. 2013a. Silicon alleviates cadmium toxicity in *Avicennia marina* (Forsk.) Vierh. seedlings in relation to root anatomy and radial oxygen loss. *Mar. Pollut. Bull.* 76, 187–193.

Zhang, S., Li, S., Ding, X., Li, F., Liu, C., Liao, X., Wang, R. 2013b. Silicon mediated the detoxification of Cr on pakchoi (*Brassica chinensis* L.) in Cr-contaminated soil. *J. Food Agric. Environ.* 11, 814–819.

Zhao, L.Y.L., Schulin, R., Weng, L., Nowack, B. 2007. Coupled mobilization of dissolved organic matter and metals (Cu and Zn) in soil columns. *Geochim. Cosmochimi. Acta* 71, 3407–3418.

Zhao, M.T., Wang, J., Lu, B., Lu, H. 2005. Certification of the cadmium content in certified reference materials for Cd rice flour. *Rapid Commun. Mass Spectrom.* 19, 910–914.

Zorrig, W., Rouached, A., Shahzad, Z., Abdelly, C., Davidian, J.C., Berthomieu, P. 2010. Identification of three relationships linking cadmium accumulation to cadmium tolerance and zinc and citrate accumulation in lettuce. *J. Plant Physiol.* 167, 1239–1247.

15 Roles of Silicon in Improving Drought Tolerance in Plants

Hemmat Khattab

CONTENTS

ABSTRACT

Nowadays, the water resources required for irrigation have declined and intensify the problem of water scarcity. Crop plants are mainly affected due to water scarcity. Water deficit may result in drought stress and thereby affect a plant's water status. Drought stress is a major environmental issue that reduces the relative water content in crop plants and thereby decreases plant growth and yield. Hence, it is necessary to provide plants with some nutrients that can sustain water status in drought-stressed plants. Silicon is considered one of the salient elements that show hope, particularly under abiotic stress conditions such as drought stress. Fertilization with silicon has shown a positive role in the improvement of the growth and productivity of plants, particularly those belonging to the family Poacea. Numerous positive potential effects of silicon (Si) on plant growth and development have been documented, particularly under water deficit conditions. There are several strategies showing that silicon improves plant drought tolerance, through formation of a mechanical barrier, as Si is deposited on cell walls, lumens, and intercellular spaces, as well as xylem vessels. In addition, silicon modulates plant metabolism and alters physiological activities, thus enhancing drought stress tolerance. Likewise, a variety of genes are overexpressed by silicon in many drought-stressed plants. The products of these genes participate in the regulation of some drought-responsive gene expressions and signal transduction pathways under drought stress conditions. Transcription factors (TFs) are among the stress-induced genes and play vital roles in drought stress tolerance via transcriptional regulation of up- and downstream genes. Furthermore, the TFs contribute to the regulation of some drought-inducible gene expressions in an abscisic acid (ABA)–independent or ABA-dependent manner in response to water deficit. The overexpression of such genes encodes proteins, which have a role in the production and accumulation of some osmoregulators and antioxidants, the protection of macromolecules, and particularly the increase of stomatal closure in silicon-treated plants experiencing drought stress. The mechanism of silicon-enhanced drought stress tolerance is discussed in this chapter at the cellular and molecular levels.

15.1 INTRODUCTION

Water deficiency is one of the major problems that affect the growth and productivity of crops worldwide. Drought stress resulting from a water deficit causes about a 20% reduction in crop yield worldwide (Schiermeier, 2008; Shehab et al., 2010; Hussain et al., 2011; Emam et al., 2014). Likewise, the greatest reduction in water availability required for farming has also been stimulated by global climatic changes (Bates et al., 2008). However, water is necessary for all plant growth phases (Athar and Ashraf, 2005); thus, the deficiency in water uptake adversely affects the growth and yield of many crops (Wang et al., 2001). Meanwhile, improving and increasing productivity under different stress conditions is of prime importance in ensuring future food security. Thus, it is necessary to control and manage the available water supply and search for suitable and safe alleviators to achieve both water and food security. Elements such as silicon (Si) display a positive role in the growth and productivity of many crop plants, particularly under stress conditions. Si protects plants from abiotic stresses such as drought stress. Si is the most plentiful element in soil. Moreover, Si is one of the major constituents of many plants, specifically plants belonging to the family Poacea, which possess an equal amount of macronutrients (Epstein, 1999). The roles of silicon in plant defense against drought stress have been poorly understood (Naeem et al., 2010; Khattab et al., 2014). Si has been shown to be able to restore the growth and development of many stressed plants (Hattori et al., 2005; Gong et al., 2008; Emam et al., 2012a). The beneficial roles of Si in stressed plants are attributed to the formation of mechanical barriers as Si is deposited on cell walls, lumens, and intercellular spaces (Sacala, 2009), and thus increases cell wall thickness, providing mechanical resistance (Yoshida et al., 1962). Si has a key role in improving plant tolerance to stress adversities by acting as an "antitranspirant and natural antistress" mechanism that provides higher yields with better quality. It was reported that Si mitigates the adverse effects of drought stress in cucumber (Zhu et al., 2004), tomato (Al-Aghabary et al., 2004), wheat (Pei et al., 2010), and rice (Emam et al., 2014; Khattab et al., 2014). However, the mechanisms for such effects are still unclear. Silicon increased the antioxidant capacity and enhanced the accumulation of starch and sugars (Xue et al., 2001; Hartikainen, 2005; Emam et al., 2014). Moreover, silicon increases the productivity in terms of number of spikelets or panicles (Deren et al., 1994; Emam et al., 2014), spikelet fertility (Takahashi, 1995), and total number of grains or panicles (Balastra et al., 1989; Emam et al., 2014).

15.2 SILICON FERTILIZER

In the soil, silicon is generally found in the nondissociable form of silicic or monosilicic acid (H_4SiO_4), which is easily absorbed by plant tissues. Si releases in the soil solution from various sources such as decomposition of plant residue, dissociation of monosilicic acid, release of silicon from the iron, and aluminum oxides and hydroxide. The dissolution of noncrystalline and crystalline minerals and silicon fertilizers is also a major source of soil silica (Korndörfer et al., 2004). Silicon is the most abundant element in the earth's crust. Generally, sandy soils exhibit a low level of Si in the upper layers (Marschner, 1995). However, clay soils display higher concentrations of phyllosilicates and a higher concentration of Si than sandy soils. Sand fraction contains mainly complex quartz (SiO_2) mineral, which makes sandy soil more responsible for the silicate application in contrast to clay soil (Demattê et al., 2011). Due to desilication process, tropical and subtropical soils have low silicon availability (Epstein, 1999). Hence, these soils respond to Si fertilization. Si fertilizers are mostly neutral or slightly alkaline. It was recorded that Si has a neutralizing effect on soil acidity that occurs through the following reactions of SiO_3^{-2} anions with H^+ protons in the soil solution: $CaSiO_3 + H_2O \leftrightarrow Ca^{2+} + SiO_3^{-2} + H_2O$ $SiO_3^{-2} + 2H- \leftrightarrow H_2SiO_3$ $H_2SiO_3 + H_2O \leftrightarrow H_4SiO_4$ (Savant et al., 1999). Moreover, silicates promote soil chemical reactions that increase the pH, precipitate toxic Al and Mn, increase the exchangeable Ca and Mg levels and saturation of bases, and reduce the saturation of Al^{3+}, with the additional benefit of increasing the Si levels in soils (Epstein, 1999; Savant et al., 1999). Slag, wollastonite and by-products of elemental phosphorous production, calcium, sodium, potassium and magnesium silicates (serpentinite), cement, fused phosphate, and diatomite are used as sources of Si by plants (Lima Filho et al., 1999; Abdalla, 2011). Silicates are supplemented in soil as fertilizers or plants as foliar spray. The ideal features of a good silicate are its high concentration of available Si, good physical properties, simplicity of mechanical application, ready availability to plants, high amounts of Ca and Mg, low concentration of heavy metals, and low cost (Prado et al., 2001). Slag is rich in Si as well as Ca and Mg oxides, which makes it an excellent source of nutrients at low cost, and it acts as a soil acidity neutralizer because of its high pH, especially in sandy and low-fertility soils (Datnoff et al., 2001). In addition, the action of silicates is remarkably similar to that of limestone. However, the calcium silicate is more effective and faster in pH neutralization in semiperennial crops in contrast to limestone (Alcarde, 1992). Additionally, Si applications have been shown to break down a compacted clay soil to improve aeration and water penetration (Crooks and Prentice, 2011). It was also reported that Si application increases water retention via improving the soil structure, thereby increasing the moisture retention capacity (Crooks and Prentice, 2011). Furthermore, Si fertilizers have many potential benefits and promote the growth and productivity of plants (Crooks and Prentice, 2011). Silicon fertilizers also enhance biotic and abiotic stress tolerance.

15.3 SILICON IN PLANTS

Silicon is a significant component of plant constituent. The accumulation of Si in some plants seems to be similar or higher than that of the other macronutrients (Savant et al., 1997; Epstein, 1999). The silicon content of diverse plants is incredibly variable, running from around 0.1% to 10% (w/w) Si on a dry weight basis (Takahashi et al., 1990; Epstein, 1994). Furthermore, the silicon content was positively related to the plant capabilities for Si uptake (Takahashi et al., 1990), which consecutively are dependent on the plant species (Takahashi et al., 1990; Ma et al., 2001). Different plant species differ in both their contents of Si and their uptake capability, and thereby accumulation of Si. Plants take up supplemented Si in the form of silica or silicates (Ma et al., 1989) as silicic acid (Chen et al., 2000) once the pH of the soil reaches around 9 (Ma et al., 2004). The silicic acid absorbed via roots is then transported to shoots through the xylem and constantly deposited as amorphous silica in the plant tissues (Ranganathan et al., 2006). The distribution of silicon in shoots is controlled by transpiration. It was recorded that the accumulation of silicon was greater in the older tissues. Therefore, excess accumulation of silicon does not damage plants due to its undissociation and polymerization properties (Ma et al., 2001). Silicon is deposited beneath the cuticle layer; thus, silicon can minimize the stress-adverse effects and be used as a stress alleviator (Ma and Yamaji, 2006). The potential encouraging properties of silicon on plant growth under stress conditions have been reported by many investigators. It was indicated that Si application nullifies the negative adverse effects of water stress on plant growth (Gong et al., 2005; Hattori et al., 2005; Emam et al., 2012a; Khattab et al., 2014).

15.4 Si-MEDIATED DROUGHT STRESS TOLERANCE

Plants regularly face adverse effects of drought by modulation of plant metabolism and gene expression and, consequently, the physiological and morphological criteria. Silicon application has been described to improve plant tolerance against drought stress (Ma, 2004; Hattori et al., 2008; Khattab et al., 2014) and increase water uptake, thereby helping the plant to produce a higher yield. Plants cope with abiotic stress by different recognized strategies; however, those underlying Si-mediated mitigation of drought stress harms are poorly understood (Khattab et al., 2014).

15.4.1 MECHANISMS OF SI-MEDIATED MITIGATION OF DROUGHT STRESS DAMAGES

15.4.1.1 Roles of Si on the Morphological Adaptation and Water Relations

Water stress is a serious problem worldwide, which usually causes disturbance in metabolism, particularly photosynthesis (Ben Ahmed et al., 2009), and stimulates the production of reactive oxygen species (ROS) in mitochondria and chloroplasts (Asada, 1999). It was recorded that water stress reduced the relative water content in both drought-sensitive and drought-tolerant maize plants, kochtia, and rice leaves (Valentovic et al., 2006; Masoumi et al., 2010; Emam et al., 2012a). Therefore, it is necessary to provide plants with some nutrients that can uphold water status in the stressed plants. Silicon is considered to be one of the beneficial elements that can improve drought stress tolerance. Silicon is considered an antitranspirant, which enhances drought tolerance in plants by reducing the transpiration rate, thus upholding leaf water potential and the CO_2 assimilation rate (Hattori et al., 2005). It has also been experimentally proved that, silicon promotes plant growth and production, preclude the lodging and support the leaves to get the sufficient light and enhance plant tolerance to biotic and abiotic stresses (Coors, 1987; Epstein, 2001; Fawe et al., 2001; Voogt and Sonneveld, 2001; Liang et al., 2005; Hamayun et al., 2010; Lee et al., 2010), and influences the composition of elements (nitrogen, phosphorus, etc.) in plant tissue (Epstein and Bloom, 2005).

15.4.1.2 Role of Silicon on Growth

Drought stress imposes negative effects on plant growth and development. It was recorded that drought stress stimulates the inhibition of shoot growth in several plant species (Nonami and Boyer, 1990; Chazen and Neumann, 1994; Helal, 2013). Under drought stress, sometimes the root system continuously elongates while the shoot growth is completely inhibited (Spollen et al., 1993). The increase in drought-stressed root length has been attributed to the higher osmoregulation capacity of drought-tolerant genotypes (Rauf and Sadaqat, 2008). The higher root/shoot ratio was measured in drought-stressed tomato and rice seedlings compared to those of control plants (Mingo et al., 2004; Khattab et al., 2014). The differences in growth of roots and shoots under water deficit conditions are considered to be among the morphological adaptations of plants to drought stress, since continued root elongation facilitates water uptake from the soil (Djibril et al., 2005).

On the other hand, application of Si reduced the stress-induced adverse effects. The positive effects of silicon were displayed by rice (Emam et al., 2012a,b) and wheat (Karmollachaab et al., 2013) plants in terms of morphological and physiological characteristics under stress conditions. Silicon fertilizer results in increased growth rate of roots and, consequentially, increased total and adsorbing surfaces (Matichenkov, 1996; Khattab et al., 2014; Saud et al., 2014). Furthermore, an increment in root growth due to application of Si in water-stressed rice plants has been reached by Khattab et al. (2014). Previously, some authors related such an increase in the root/shoot ratio to the abscisic acid (ABA) content of roots and shoots under drought stress conditions (Sharp and Lenoble, 2002; Manivannan et al., 2007). Silicon application improved the growth of several plant species, including wheat (Gong et al., 2005), soybean (Shen et al., 2010), and rice (Chen et al., 2011; Khattab et al., 2014) plants grown under water scarcity stressful conditions. The positive effects of silicon on the healthy growth and development have been confirmed in many plant species, particularly Gramineae (Broadley et al., 2012). Some previous studies revealed that Si fertilizer increases the weight of citrus (Matichenkov et al., 2000) and anbahia grass (*Paspalum notatum* Flügge) (Savant et al., 1997) roots more than that of shoots. It has also been reported that Si fertilization stimulates the silicification of leaves (Cooke and Leishman, 2011) and so decreases the cuticular transpiration concomitant with a reduction in the stomatal pores' diameter (Snyder and Matichenkov, 2007). Hence, such effects can ultimately enhance the drought tolerance of Si-treated plants. In addition, Si promotes root elongation and protects the stele via hardening of the stele and endodermal cell walls. Moreover, Si stimulates the cell elongation but not cell division, perhaps as a result of the positive effects of Si on the cell wall extensibility (Hossain et al., 2002). It was recorded that Si improved the cell wall extensibility of the growing zone while diminishing cell wall extensibility in the basal zone of sorghum roots (Hattori et al., 2003; Lux et al., 2003), thus increasing root growth and enhancing water uptake, which contributes to drought resistance.

In addition, Gholami and Falah (2013) observed that silicone fertilizers significantly increase tiller number, leaf dry weight, and yield components of the rice plant. Many other studies also observed the positive effects of silicon on plant growth, obviously by increasing dry mass and yield, enhancing pollination, and increasing stress resistance. However, Ahmed et al. (2011) recorded that application of silicon increases chlorophyll content, shoot and root dry weights associated with remarkable decreases in leaf water potential and shoot/root ratio in sorghum.

Additionally, it was reported that Si application significantly increased the dry weights of drought-stressed sorghum and soybean plants (Shen et al., 2010; Sonobe et al., 2010). Likewise, application of silicon improves the drought tolerance of plants as a result of the increments of water uptake capability and enhancement of water use efficiency under stress conditions (Hattori et al., 2005; Gunes et al., 2007; Liang et al., 2007; Cooke and Leishman, 2011; Ahmed et al., 2011).

Furthermore, applications of silicon fertilizers interact positively with many other applied elements and thereby improve their agronomic performance and efficiency. Maintenance of greater leaf water content under stress conditions is one of the remarkable effects of silicon supplementation on stressed plants (Lux et al., 2002). Furthermore, silicon fertilizers can help reduce water loss and transpiration, increase P availability, reduce lodging, and improve leaf and branch erectness (Narayanan et al., 2008). Si treatment basically improves the dry mass, chlorophyll content, relative water content, electrolyte leakage, and root/shoot ratios of water-stressed maize and rice plants (Kaya et al., 2006; Helal, 2013). Application of silicon decreases the decline in fresh and dry weights of drought-stressed maize and wheat plants (Kaya et al., 2006; Karmollachaab et al., 2013). Drought stress and silicon have a significant interaction on leaf electrolyte leakage.

15.4.1.3 Roles of Silicon Deposition in Drought Tolerance Mechanism

Distinctive plant species are skilled to absorb and aggregate diverse measures of silicon, for instance, grasses assemble the most extreme amount of Si (Ma et al., 2006). The silicon is absorbed and transported in the form of uncharged silicic acid from roots to shoots through xylem. Then, silicic acid is irreversibly deposited into the plant tissues in the form of amorphous silica (Ranganathan et al., 2006). The transpiration rate is regulated by silicon accumulation. A greater level of silicon accumulation has basically been observed in plant tissues experiencing high transpiration. Nevertheless, it was confirmed that silicon accumulation in or on plant roots may achieve drought tolerance. Lux et al. (2002) reported that the high root endodermal silicification may serve as major aspect of the sorghum drought tolerance strategy. Furthermore, deposition of silica in the plant cell walls enhanced the rigidity and strength of plant cell walls and accordingly increased the plant architecture and leaf erectness (Ma and Takahashi, 2002). The accumulation of a greater level of Si in leaves stimulates the formation of a double layer of cuticle silica, which consequently reduces the transpiration rate, decreasing the opening of stoma and limiting the loss of water in plants. Such an effect can be significant, particularly for Gramineae (Nolla et al., 2012). Likewise, Si deposited in the interfibrillar spaces reduces the movement of water through the cell wall, thus conserving water by decreasing the transpiration rate (Savant et al., 1999). Moreover, the deposition of silicon in trichomes increases the boundary layer between the leaf and the atmosphere, thus protecting the plant from excessive water loss and decreasing transpiration (Emadian and Newton, 1989). Silicon is also involved in reducing the plant transpiration rate, by controlling the stomatal opening and closing mechanism, which provides further tolerance to water scarcity (Korndörfer et al., 2002). The silicification of epidermal cell membranes also protects plants against excessive loss of water via transpiration (Waterkey et al., 1982; Efimova and Dokynchan, 1986; Gao et al., 2006) and results in the reduction in diameter of the stomatal opening (Barcelo et al., 1993). Romero-Arnada et al. (2006) reported that the precipitation of silicate in the epidermal cells forms a barrier that reduces cuticular transpiration and improves plant tissue water content. Silica deposition checks the compression of xylem tissue due to high transpiration rate under drought stress. Hence, Si induces stem cell wall rigidity, thereby reducing lodging (Savant et al., 1999) and improving leaf positioning of rice plants (Savant et al., 1999; Ando et al., 2002). Silicon increases the stress tolerance by enhancing the cell wall extensibility, inhibiting the transpiration and increasing the efficiency of water utilization (Hossain et al., 2002; Savant et al., 1999).

It was evident that the deposition of silicon is under temporal and metabolic controls. It was reported that the ultrastructural shape of silicon changed from sheetlike to a globular shape during the transition of the primary cell wall of grasses to the secondary cell wall. Such changes are brought by the cell wall metabolites, which interact with silicic acid, resulting in the deposition of a greater amount of silicon in the cell wall (Perry et al., 1987).

15.4.1.4 Role of Si on Leaf Rolling under Drought Stress

Additionally, plants protect themselves against drought stress by induction of leaf rolling (Figure 15.1). In physiological studies of cereal species, relatively little attention has been given to leaf forms or shapes under drought conditions. Leaf rolling and leaf wilting occur as a result of water deficits. Moreover, visual leaf rolling is correlated with leave water potential (O'Toole and Moya, 1978; Emam et al., 2012a) in rice. The reduction in leaf water content results in a decline

(a) (b) (c)

(d) (e) (f)

FIGURE 15.1 Growth and morphological changes under pretreatment with Si on two rice cultivars in drought stress. (a) and (d) represent normal well water control, (b) and (e) represent drought control, and (c) and (f) represent drought + Si, for Giza 177 and IET 1444 rice cultivars, respectively.

in the turgor pressure, and thereafter the leaves wilt and roll. Consequently, the stimulation of leaf rolling by drought or other osmotic stress reduces the photosynthetic activities (Terzi et al., 2009; Saruhan et al., 2010). Several researchers have proved that rolling of leaf causes inadequate light interception and absorption, reduction of photosynthetic activities, and injury to the membrane system (Li-feng et al., 2012). Pei et al. (2010) reported that addition of 1.0 mM Si could partially improve leaf rolling, which is helpful in keeping the leaf erect, as well as improve light reception and water transpiration (Chen et al., 2004). It was also indicated that Si treatments could improve the water status of water-stressed plants and increase leaf erectness, thus decreasing the leaf rolling score (Emam et al., 2012a; Saud et al., 2014).

Meanwhile, it was reported that application of Si improves water use efficiency and stimulates the enzymatic and nonenzymatic antioxidant defense systems (Liang et al., 2003, 2007; Hattori et al., 2005; Gunes et al., 2007; Cooke and Leishman, 2011), thereby enhancing drought tolerance.

15.4.1.5 Roles of Si in Restoring Cellular Structure and Functions under Drought Stress

It was investigated that Si modulates the harsh effects induced by drought stress on the growth, relative water content, chlorophyll (Chl) content, and properties, as evident from fluorescence parameters, metabolism and gas exchange patterns, and the antioxidants. In addition, Si application boosts the antioxidant defense systems and buffers the ROS, consequently mitigating the photooxidative damage and maintaining the cell membranes' integrity, and hence enhancing the plant drought tolerance (Waraich et al., 2011).

15.4.1.5.1 Roles of Si in the Repair of Photosynthesis Machineries

The photosynthetic pigments play a vital role in harvesting the light energy and production of chemical energy required for all cell activities. Drought stress causes a reduction in chlorophyll content and damages the photosynthetic machineries, and subsequently inhibits the photosynthesis (Nikoleava et al., 2010). On the other hand, carotenoids play fundamental roles and protect the plants from the adverse effects of drought (Jaleel et al., 2009). Meanwhile, it was recorded that drought stress modified the ratio of chlorophyll ($a + b$) and carotenoids (Anjum et al., 2003; Farooq et al., 2009). Drought stress usually causes disturbance in metabolism, especially in photosynthesis (Ben Ahmed et al., 2009), reduction in dry weight, and formation of ROS in mitochondria and chloroplasts (Asada, 1999). Drought stress stimulates stomatal closure and thus limits gas exchange (Jaleel et al., 2007; Singh and Reddy, 2011), subsequently decreasing the rate of photosynthesis (Jaleel et al., 2009; Liu et al., 2012) in several plant species. Severe drought stress also modifies the photosynthesis of plants by reducing the chlorophyll content and damaging the photosynthetic systems (Nikolaeva et al., 2010). Therefore, to know the physiological state of stressed plants, the evaluation of pigment content and ratio of (chl a/b) is mandatory. In normal plants, the ratio of chl a/b is 2.6–3.1, while under stress conditions, it declines to 2.1 (Lichtenthaler and Rinderle, 1988). The reduction in chlorophyll content was determined in drought-stressed cotton (Massacci et al., 2008); *Catharanthus roseus* (Jaleel et al., 2008), sunflower (Kiani et al., 2008), maize (Liu et al., 2012) and rice (Emam et al., 2012b). Chlorophyll a is the main photosynthetic pigment in higher plants that is responsible for light absorption, which participates in the photosynthesis process (Taiz and Zeiger, 2002). Light absorbed by chlorophylls and carotenoids of leaves is mainly used for photosynthetic quantum conversion, whereas small proportions are reemitted as heat and red chlorophyll fluorescence (Schweiger et al., 1996; Hura et al., 2006). Chlorophyll a naturally absorbs blue light and emits fluorescence red light. At room temperature, the emitted red fluorescence has been shown in the PSII and only traces from PSI (Papageorgiou, 1975; Kyle et al., 1984). Under optimum photosynthetic conditions, fluorescence emission of chlorophyll is low. Environmental stress inhibits photosynthesis parallel with the increments in both dissipation processes, including heat and fluorescence emission (Lichtenthaler, 1987). It was reported that the intensity and form of the fluorescence emission spectra are influenced by environmental conditions (Theisen, 1988). The changes in the fluorescence parameters concomitant with the reduction in chlorophyll level, particularly at the plant's early growth stages, measure the stress intensity (Larsson et al., 1998). Retention of photosynthetic pigments in leaves increases crop productivity (Fukai and Cooper, 1995).

Additionally, the changes in chlorophyll fluorescence emissions offer knowledge about the photochemistry of PSII and demonstrate the modifications in the photosynthetic activity of stressed leaves (Maxwell and Johnson, 2000). It was shown that drought stress reduces the maximum PSII efficiency (Fv/Fm) while increasing the nonphotochemical quenching (qN); hence, these parameters seem to be signs for severe damage in PSII (Maxwell and Johnson, 2000). Consequently, photoinhibition leads to the destruction of the photosynthetic systems in drought-stressed plants (Cronic, 1994). Furthermore, stomatal closure limits CO_2 availability assimilated in dark reactions and contributes to stimulating drought-stressed photoinhibition (Reddy et al., 2004). In addition, drought stress causes metabolic impairment, including reduction of Ribulose-1, 5-biphophate carboxylase/oxygenase (Rubisco) activity and ATP synthesis (Reddy et al., 2004). However, silicon treatment increases the net photosynthetic rate (Hattori et al., 2005; Gong et al., 2005; Shen et al., 2010; Ahmed et al., 2011) in several stressed plant species, concomitant with an increment in the activity of some photosynthetic enzymes, such as RuBP carboxylase (Adatia and Besford, 1986).

Si treatment stimulates stomatal conductance in plant under water scarcity condition. The elevated stomatal conductance, particularly in drought-stressed plants, may be one of the mechanisms for current dry matter production in drought-stressed plants. In contrast, Si treatment is also involved in controlling the stomatal opening and closing mechanism (Korndörfer et al., 2002), thereby regulating the optimum CO_2 concentration required for carbon fixation, which may prevent photoinhibition, as reflected by the increment in Fv/Fm value in drought-stressed plants. Xie et al. (2014) indicated that the proper concentration of silicon can be beneficial in increasing the photosynthetic capacity of maize, and thereby the grain yield and growth of maize, not only by increasing the values of the total chlorophyll contents, photosynthetic rate, and stomatal conductance, but also by decreasing the values of the transpiration rate and intercellular CO_2 concentration. Furthermore, Si treatment stimulated the accumulation of the leaf anthocyanin and soluble phenols, which participate in the protection of chloroplasts from the negative adverse effects of drought stress (Emam et al., 2012a). The effects of Si treatment on photochemical parameters are limited to the species belonging to the family Gramineae (Gao et al., 2011). Silicon treatment significantly increased plant dry weight and relative water content under drought stress. Silicon mitigates the effects of drought stress by increasing the maximum quantum yield and also alleviates the diminution in net assimilatory rate, as well as increased the stomatal conductance and thus enhanced plant biomass.

Likewise, silicon treatment resulted in higher catalase and superoxide dismutase activities, and thereby lowered lipid peroxidation of drought-stressed plants. The mitigation of the negative effects of drought stress upon silicon application may be through the enhancement of photochemical efficiency and photosynthetic gas exchange, as well as an activation of the antioxidant defense systems of water-stressed pistachio and rice plants (Habibi and Hajiboland, 2013; Helal, 2013).

Moreover, silicon has positive effects on the physiology and metabolism of different crops. Application of silicate stimulates plant growth in crops such as gerbera (Savvas et al., 2002) and rice (Yeo et al., 1999; Khattab et al., 2014). It was recorded that Si application enhances chlorophyll contents of cucumber and tomato (Miyake, 1992), barley (Liang, 1999), and rice (Emam et al., 2012b) under drought stress condition. Thus, the photosynthetic activity is depending on the chlorophyll content. It is greatly affected by stress conditions and is also considered an important indicator for the assessment of the effects and the extent of drought stress on plants. Several stressed plants, such as wheat (Ahmad et al., 1992), cucumber (Li and Ma, 2002), and rice (Emam et al., 2012b), maintain an adequate chlorophyll content by silicon application. It was also reported that the chlorophyll content was markedly increased in several plant species in response to Si application (Shen et al., 2010; Nasseri et al., 2012; Sayed and Gadallah, 2014).

Moreover, the increments in ratio of chlorophyll a/b demonstrate the positive effects of silicon application (Liang et al., 1996; Emam et al., 2012b). The increase in the chlorophyll a/b ratio in the

Si-treated plants is directly proportional to the benefits of chlorophylls *a* and *b* in increments of light absorption and utilization. It was proposed that Si confers stress tolerance by inhibiting stress-induced premature senescence and cell death as a result of impairment of the photosynthesis systems (van Bockhaven, 2014). Hence, Si application improves the photosynthetic rate, which might be concomitant with the increase in the activities of photosynthetic enzymes, and in the amount of total soluble sugars in drought stressed plants (Shen et al., 2010). Consequently, silicon application enhances the tolerance of drought-stressed plants (Sayed and Gadallah, 2014). Similarly, silicon also increases the net photosynthetic rates, and the biomass and total protein content in drought-stressed rice plants (Emam et al., 2012).

Moreover, the sustainable increase in the photosynthesis concomitant with the increases in biomass and water content of stressed cucumber leaves treated with silicon may be attributed to the reduction of the stomatal conductance, enhancement of the water holding capacity, and maintaining the steady rate of transpiration during drought stress (Li et al., 2007). On the other hand, the increments in water content of silicon-stressed plants due to the enhancement of water uptake ability was attributed to the effect of Si on a stabilization of cellular membranes (Ahmed et al., 2011).

Silicon application increases the soluble sugar concentration which regulates the photosynthetic rate (Ding et al., 2007). Therefore, Si treatment improves the total dry weight of maize plants grown under water deficit conditions (50% of field capacity). The increments in the dry weights of Si-treated plants may be attributed to the increment in the chlorophyll contents, which improved the photosynthetic efficiency to utilize more light energy. Moreover, the constructive role of Si in sustaining chlorophyll contents established that the accumulated polymerized Si inside leaves is able to protect chloroplasts. It can be assumed that the deposition of a Si layer beneath the cuticle acted as a shelter and thus reduced the pigment photooxidation to maintain the chlorophyll pigment content under drought stress conditions. It was also reported that the reduction in chlorophyll degradation in Si-stressed plants was attributed to the reduction in ROS generation and, subsequently, lessening of photooxidative damage and conservation of chloroplast membrane integrity. Thus, enhancement of plant drought tolerance was ascribed to lessening of ROS and, along these lines, a decrease of photooxidative harm, supporting chloroplast film respectability and hence improving plant dry spell resistance (Waraich et al., 2011).

15.4.1.5.2 Role of Silicon in Osmoregulation

Survival under drought stress conditions and, accordingly, the tolerance mechanism were also associated with osmotic adjustment and turgor maintenance. This process involves accumulation of soluble osmoregulators such as sugars, proline protein alcohols, glycine and organic acids, as well as some ions like Ca and K. These compounds reduce the osmotic potential and enhance the turgidity of leaf (Mahajan and Tuteja, 2005; Trovato et al., 2008), buffer ROS, protect membranes, and stabilize protein (Ashraf and Foolad, 2007; Liu et al., 2011). The accumulation of osmregulators is one of the most important mechanisms involved in plant drought stress tolerance (Serraj and Sinclair, 2002). Water scarcity stress stimulates an alteration in the endogenous plant water content and lowers the water potential of the stressed cells (Babu et al., 1999). Consequently, some osmolytes are vigorously accumulated within the stressed cells in order to adjust their water potential (Blum, 1998), facilitate water uptake, and sustain cell turgor, gas exchange, and growth under severe stress conditions (White et al., 2000; Chaves et al., 2003). Furthermore, osmotic adjustment regulates the cell activities and helps plants to improve growth and photosynthesis and assimilate partitioning, thereby increasing the yield (Subbarao et al., 2000: Emam et al., 2014). A discrepancy in osmotic adjustment among drought-stressed plants belonging to different species has been observed, and depends on the extent of osmotic stress (Moinuddin and Khanna-Chopra, 2004; Khattab et al., 2014). Compatible solutes are involved in osmotic regulation, buffering ROS, maintaining membrane integrity, protecting macromolecules such as DNA, enzymes, and proteins, and consequently enhancing stress resistance (Ashraf and Foolad, 2007).

The morphology of the plant body and biochemical characters of photosynthetic tissues has been involved in photon capture optimization and consumption of these photons in CO_2 fixation (Madhava Rao et al., 2006) to produce carbohydrates, which are the primary energy storage compounds, and other organic substances (Lindhorst, 2000). Carbohydrates play roles in osmotic adjustment, protection of membranes from damage, and stabilization of proteins and enzymes (Villadsen et al., 2005; Lee et al., 2008; Ben Ahmed et al., 2009; Hessini et al., 2009). Moreover, carbohydrates are essential for the synthesis of numerous compounds, energy products, and signaling molecules (Gibson, 2005). Drought stress largely increases the accumulation of total soluble carbohydrate rice leaves (Mostajeran and Rahimi-Eichi, 2009; Emam et al., 2012b), *Arabidopsis thaliana* leaves (Moustakas et al., 2011), and wheat (Qayyum et al., 2011; Loutfy et al., 2012). Accommodation of soluble sugars is one of the most significant osmoregulators in stressed plants (Chaves et al., 2003; Ben Ahmed et al., 2009; Hessini et al., 2009). In addition, sugar alcohol and monosaccharide are also involved in the osmotic adjustment that plays a part in the plant stress tolerance mechanism (Wang and Stutte, 1992).

Many studies have also indicated that water stress causes disturbances in the nitrogen metabolism of many plants, which results in the accumulation of high levels of soluble proteins and total free amino acids in rice plants grown under drought stress (Babita et al., 2010; Shehab et al., 2010; Helal, 2013). When water deficit was experienced, the accumulation of free amino acids increased in stressed plants. Such an increment resulted from the higher activity of protease enzymes, forming free amino acid that contributes to osmoregulation (Costa et al., 2006; Delfini et al., 2010; Lobato et al., 2013; Helal, 2013). It was suggested that the increase in the accumulation of amino acids serves as a nitrogen source, mainly for the synthesis of enzymes, and contributes to an increase in osmotic adjustment, thus enhancing the resistance of the plant to drought stress (Navari-Izzo et al., 1990). In drought-stressed plants, the accumulation of free amino acids and the reduction in protein level has been noticed due to improper protein synthesis and increase cleavage of proteins (Navari-Izzo et al., 1990; Gong et al., 2005; Zhang et al., 2013). The adverse effects of drought stress hinge on species, tissue, and age, as well as the nature, duration, and extent of the stress. The amino acid pools may be influenced by nitrogen assimilation into amino acids, amino acid interconversions, protein biosynthesis and degradation, amino acid transport and degradation, or utilization of amino acids (Noctor et al., 2002). Amino acid accumulation, particularly proline, has been widely recommended to participate in the improvement of plant drought stress tolerance (Hare and Cress, 1997; Rhodes et al., 1999; Martinelli et al., 2007; Khattab et al., 2014). Many scientists have reported the higher proline content in plants that are grown in water-deficient condition (Pei et al., 2010; Anjum et al., 2012; Cha-um et al., 2013; Khattab et al., 2014). The accumulation of greater amounts of proline was correlated with an increase in the membrane stability and improves drought tolerance in barley (Bandurska, 2000). In fact, proline and quaternary ammonium compounds, such as glycine betaine (GB) and choline, are key osmolytes participating in osmotic adjustment (Kavikishore et al., 2005; Moussa and Abdel-Aziz, 2008; Wani et al., 2013; Khattab et al., 2014). It was reported that a greater amount of GB accumulated in plants experiencing dehydration (Venkatesan and Chellappan, 1998; Yang et al., 2003; Moussa and Abdel-Aziz, 2008; Chaitanya et al., 2009). Moreover, the accumulation of glycine betaine (GB), particularly in the chloroplast, plays a vigorous part in the stabilization and protection of chloroplast membranes, thus achieving a sustainable photosynthetic efficiency (Genard et al., 1991). The defensive role of betaine in stressed condition occurs via the signal transduction pathways (Subbarao et al., 2000). GB accumulation was observed in many stressed crop plants, including sugar beet, spinach, barley, wheat, and sorghum (Yang et al., 2003; Chaitanya et al., 2009). The accumulation of GB under stress conditions was greater in the tolerant plant genotypes than in the sensitive ones (Ashraf and Foolad, 2007). GB is synthesized by either oxidation of choline or N-methylation of glycine (Chen and Murata, 2002). In a plant cell, choline is transformed into betaine aldehyde by the enzyme choline monooxygenase and finally is converted into glycine betaine via the NAD^+ dependent aldehyde dehydrogenase enzyme. These enzymes generally synthesized in chloroplast stroma and are activated under stress conditions (Giri, 2011).

In some plants, the increase in citrulline level has been seen in contrast to proline and GB. Citrulline is an amino acid that is not encoded via a nuclear gene; however, it is found in several proteins (Kawasaki et al., 2000). Citrulline accumulation occurs in wild watermelon in contrast to proline and GB (Kawasaki et al., 2000). Citrulline is the most effective scavenger of OH⁻ radicals, and it is involved in the protection of DNA and enzymes against oxidative stress harms (Akashi et al., 2001; Bektasoglu et al., 2006).

Likewise, the nonprotein amino acid γ-aminobutyric acid accumulation was detected in some stressed plant tissues. γ-Aminobutyric acid acts as a zwitterion, occurs in free highly water-soluble form (Shelp et al., 1999), and can be found in a cyclic structure that is similar to that of proline. γ-Aminobutyric acid may serve as a signaling molecule in higher plants exposed to abiotic stress conditions (Serraj et al., 1998). γ-Aminobutyric acid is involved in osmotic regulation of drought-stressed tissues, and hence it may play a role in stress tolerance (Shelp et al., 1999), scavenging of ROS, formation of proline from putrescine, and intracellular signal transduction (Kinnersley and Turano, 2000). It was recorded that the level of proline was more than 50% after imposition of drought stress; however, the γ-aminobutyric acid level reached about 27% at the end of the recovery period (Simon-Sarkadi et al., 2006).

On the other hand, silicon treatments increased the accumulation of proline in drought-stressed plants. These secondary impacts are produced by Si on transpiration (Pereira et al., 2013). The increase in proline in Si-treated plants causes a higher affinity for water (H_2O) under water deficit, thereby increasing water retention in tissue, which consequently increases plant tolerance to water deficit (Lobato et al., 2009; Pereira et al., 2013). The application of Si stimulates a massive accumulation of proline in drought-stressed pepper and rice plants (Pereira et al., 2013; Khattab et al., 2014). The accumulation of free proline depends on the synthesis, catabolism, and transportation between the adjacent cells and cellular organelles (Szabados and Savoure, 2010).

Likewise, GB is an inert molecule in plant cells and protects the oxygen-evolving PSII complex and ATP synthesis. Similarly, GB protects the enzymes of onion and rice cell membranes (Mansour, 1998; Rahman et al., 2002). Such effects of GB participate in the improvement of stress tolerance and productivity (Shirasawa et al., 2006). GB formed from choline and glycine (Meneses et al., 2006) is extensively accumulated in some stressed plants (Ashraf and Foolad, 2007). The silicon prominently increased the accumulation of glycine betaine in stressed plants (Pereira et al., 2013; Khattab et al., 2014). The increments in GB accumulation could be attributed to the acceleration of its biosynthesis, as verified by the overexpression of *OsCMO* encoding the CMO oxidase enzyme involved in GB biosynthesis (Khattab et al., 2014).

15.4.1.5.3 Role of Silicon on Nutrient Uptake

The osmotically active molecules and ions, including calcium, potassium, and chloride ions, beside soluble sugars, sugar alcohols, proline, glycine betaine, and organic acids, are accumulated under stress conditions. The decline in water absorptions under drought stress conditions usually limits the total nutrient uptake in plants. Moreover, the deficiency in nutrient absorption under stressed condition could also be attributed to a decline in the available energy used for assimilation under drought conditions (Farooq et al., 2009). The low transpiration rate and mineral absorption has been noticed in drought-stressed plants (Saud et al., 2014). Therefore, silicon fertilizer could improve nutrient uptake under nutrient-deficient conditions. Si is involved in the improvement of nutrient use efficiency under nutrient-deficient conditions. It significantly enhances the accumulation of K in soybean plants grown in K-deficient medium, thus nullifying the membrane lipid peroxidation and oxidative stress damages through alteration of the antioxidant enzyme activities (Miao et al., 2010). Silicon increases the efficiency of N and P fertilization (Ma and Takahashi, 1989; Savant et al., 1999) and affects the contents of N, P, K, Ca, and Mg in drought-stressed rice plants (Helal, 2013). It was also recorded that the C/N ratio in Kentucky bluegrass leaves increases by exposure to drought, which might be attributed to a reduction in leaf N content (Saud et al., 2014). Similarly, Si nullifies the reduction in nutrient levels, including Ca and K ions, in water-stressed maize plants

(Kaya et al., 2006). The efficiency of silicon in resisting drought stress was associated with increasing H^+-ATPase activity in the root membrane and H^+-PPase in the tonoplast of salt-stressed barley (Liang et al., 2007). Moreover, an increase in K ion absorption increased its accumulation and concentration inside the cell, as well as the uptake of water, thus affecting activities of some enzymes and physiological processes (Liang et al., 2007). Thus, application of Si may be one of the policies that can be piloted to improve the growth and productivity of crops in arid or semiarid areas and increase crop's water deficiency tolerance (Kaya et al., 2006). Application of silicon in drought-stressed rice leads to significant increases in the Si and K concentrations in shoots and roots compared to control. Such increases may be due to increments in the H^+-ATPase activity in plant roots (Chen et al., 2011). Consequently, Si application overcomes chlorosis, improves chloroplast constructions, and significantly increases pigment contents, which may be positively related to the photosynthetic activity, plant growth, and productivity. Silicon supplementation increases the capability of photosynthetic pigments, leaf water potential, and growth rate, as well as water use efficiency of drought-stressed Kentucky bluegrass (Liu et al., 2008). Such effects might be attributed to the stimulative effects of Si on water uptake, root growth, and leaf erectness induced by Si application. Under water stress condition, the application of Na_2SiO_3 increases the net assimilation rate of plants (Hattori et al., 2005; http://www.hindawi.com/journals/tswj/2014/368694/; Gong et al., 2005; Gao et al., 2006). It was also recorded that silicon application also enhances the water use efficiency in maize plants (Gao et al., 2005).

Actually, plants can tolerate drought stress by sustaining tissue water content mainly by osmoregulation, improving the antioxidant defense system, and thereby protection of the cell membranes. Some compounds, such as plant growth regulators, polyamines, γ-aminobutyric acid, free amino acids, and soluble carbohydrates, contribute to improving drought tolerance by scavenging the ROS, regulation of stomatal movement, protection of the essential cellular macromolecules, and preservation of the cell water status.

15.4.1.5.4 Roles of Si in Regulation of ROS Production and Detoxification

Drought stress induces the formation of ROS that impair the cellular macromolecules, including proteins, membranes, and DNA, as well as other cellular components. Imbalance in the neutralization and generation of ROS results in oxidative stress damage (Smirnoff, 1993). Plants detoxify these ROS by enhancing the antioxidant defense system (Ashraf, 2010; Gill and Tuteja, 2010; Helal, 2013), which includes both enzymatic antioxidants (Alscher et al., 2002) and nonenzymatic antioxidants (Peltzer et al., 2002). In addition to the well-known antioxidant system, flavonoids and phenolic compounds are effective antioxidants (Michalak, 2006). Superoxide dismutase catalase (SOD), peroxidases (POX, APX), and glutathione reductase (GR) are the antioxidant defense enzyme system intended to buffer free radicals (Peltzer et al., 2002). Drought stress stimulates some antioxidant enzymes, such as POX, GR, and reduced CAT, in the leaves of hyacinth bean seedlings (Myrene et al., 2011). The nonenzymatic antioxidant compounds, such as ascorbic acid (ASA) and reduced glutathione (GSH), are responsible for buffering H_2O_2 via the Hallwell–Asada pathway (Horemans et al., 2000). Alteration of antioxidant metabolism, especially at early growth stages, is considered an adaptive strategy that may impact drought tolerance (Da Costa and Huang, 2007; Shehab et al., 2010).

Silicon (Si) belongs to metalloids and is considered a signaling molecule that has been involved in the activation of plant defense systems. Silicon enhances the antioxidant enzyme activities, including SOD, which is involved in the production of H_2O_2 in plants, thereby mitigating the negative effects of stress and increasing plant stress tolerance (Datnoff et al., 2001). Furthermore, Si enhances SOD, CAT, and peroxidases and hence maintains membrane integrity and nullifies membrane leakage in drought-stressed leaves of cucumber and rice (Li et al., 2007; Emam et al., 2012a). Feng et al. (2009) stated that silicon increases the enzyme activities involved in the ascorbate–glutathione cycle, including peroxidase and dehydroascorbate reductase in cucumber chloroplasts. It was also recorded that application of silicon under drought stress conditions enhances the activities of some antioxidant enzymes, including SOD, APX, GR, guaiacol peroxidase (GPX), and CAT, concomitant with the stimulation of fatty acid unsaturation of lipids (Gong et al., 2005; Balakhnina and Borkowska, 2013). Similarly, silicon application improves the drought tolerance of maize plants by increasing the oxidative defense abilities through enhancement of ASA formation, decreasing the concentration of H_2O_2 and polyphenol intensity, reducing membrane injury, and improving the water status (Sayed and Gadallah, 2014).

15.4.1.5.5 Effect of Si on Restoration of Cell Membranes and Cell Integrity

The foremost target of stress condition is the cell membrane. So, the stability and integrity of the cell membrane is a significant character for the stress tolerance. It was reported that stress induced leakage of organic compounds (De and Mukherjee, 1996), amino acids (Shcherbakova and Kacperska, 1983), and sugars (Palta et al., 1977). The electrolyte leakage from the cells is a good indicator for the estimation of cell membrane damage extent resulting from lipid peroxidation induced by water stress. The electrolyte leakage is a common process in the oxidative burst (Raza et al., 2007; Noreen et al., 2010). The electrolyte leakage of the sensitive maize and rice cultivars was much higher than that of tolerant cultivars (Valentovic et al., 2006; Helal, 2013). It has been confirmed that the extent of electrolyte leakage may be controlled by some physiological and biochemical factors that participate in the plants' tolerance against environmental stresses. Such factors include production of antioxidant enzymes (Liu and Huang, 2000; Sreenivasulu et al., 2000), membrane acyl lipid content (Lauriano et al., 2000), efficiency of water utilization (Costa Franca et al., 2000; Saelim and Zwiazek, 2000), stomatal movement, osmotic pressure, and leaf rolling index (Premachandra et al., 1989). Hence, electrolyte leakage has been used for estimating the extent of stress tolerance in several crop plants (Leopold et al., 1981; Stevanovic et al., 1997). It was observed that Si treatment can diminish the electrolyte leakage from rice leaves, increasing the photosynthetic activity in drought-stressed plants (Agarie et al., 1998; Helal, 2013).

Aside from these, silicon treatment nullified membrane impairment in stressed plants by reducing the H_2O_2 level. It was investigated that silicon application reduces membrane leakage and lipid peroxidation of stressed plants (Kaya et al., 2006; Gunes et al., 2008; Shen et al., 2010; Helal, 2013). Silicate nullifies the stress injuries by decreasing the permeability of the membrane, and thereby maintaining membrane integrity (Liang, 1998). Such an effect could be due to its deposition in the cell membrane and silicification and hardening (Liang et al., 2007). Zhu et al. (2004) reported that application of Si significantly decreased the electrolytic leakage percentage due to a decline in H_2O_2 production, and thus lipid peroxidation and thiobarbituric acid reactive substances in cucumber plants. The addition of silicon reduced membrane injury through its role in the reduction of the plasma membrane permeability and membrane lipid peroxidation; thus, it could enhance the stability of lipids and maintain integrity and function in cell membranes of stressed plants (Shen et al., 2010).

Furthermore, ion leakage from the cell not only resulted from lipid peroxidation of the membranes, but also may be ascribed to the inhibition of membrane-bound enzymes that are involved in maintaining chemical potential gradients in the plant cells (Reynolds et al., 2001). It was also stated that in the drought condition, *Arabdiopsis* plants have a resistant membrane due to polar lipid content and more soluble membrane components (Gigon et al., 2004).

The roles of Si application on the stimulation of the morphological, physiological, and biochemical traits (transpiration, photosynthesis, respiration, antioxidant compounds, and enzymes) of drought-stressed plants concomitant with the accumulation of osmoregulators were greatly controlled by the optimization of hormonal balance (Abdalla, 2011).

15.4.2 Silicon–Hormone Interactions

The balances in the endogenous hormone are another mechanism by which silicon mediates stress tolerance. It has been proved that silicon can interact with plant hormone signaling. Plant hormones are involved in the regulation of plant growth and development through signal transduction pathways that are persuaded by changes in the environmental conditions. Phytohormones are rapidly synthesized in response to various abiotic fluctuations, such as drought and heat, that displayed significant roles in the signaling pathways (Miyazono et al., 2009; Sheard and Zheng, 2009). Recently, it was suggested that ABA-pretreated plants such as maize (*Zea mays*) show much higher osmolyte accumulation and antioxidant enzyme activity (Jiang and Zhang, 2002; Costa et al., 2011). Plant growth regulators can improve plant growth, absorption and mobility of ions, and nutrients in different plant species under stress conditions. They contribute in the regulation of growth in terms of cell elongation and division as well as flowering, and prevent chlorophyll cleavage and diminshed ROS levels that lead to cell death. They stabilize microtubules in plant organs against depolymerization (Wen et al., 2010; Janda et al., 2013; Bose et al., 2014). It has been proposed that hormones can control a signal transduction pathway that is associated with hormonal responses (Davies, 1995).

Silicon can mediate the endogenous gibberellins in plants, but the mechanism is still unknown. Meanwhile, it has been proved that N and Si increase the level precursor molecule GA_{20} and thus GA_1 content. It was also deduced that the endogenous bioactive GA_1 and GA_4 contents in soybean increased with elevated Si, while they decreased with the imposition of NaCl and PEG compared to the control (Hamayun et al., 2010). It was reported that gibberellins regulate all plant growth characteristics (Ritchie and Gilroy, 1998).

It was reported that water stress induced a regular decline in all growth-promoting hormone levels (auxins, gibberellins, and cytokinins). Meanwhile, stresses reversibly raised the growth inhibitors, including the ABA level (Zhang et al., 2006; Abdalla and El-Khoshiban, 2007). On the other hand, Si application mitigated the adverse effects of stress on endogenous plant hormone levels via increasing growth promoter levels, including auxins, gibberellins, and cytokinins, in parallel with increasing the growth inhibitor contents, particularly ABA (Hanafy Ahmed et al., 2008; Abdalla, 2011). In contrast, it was investigated that the increment in the ABA level in silicon-treated cowpea plants was concomitant with a reduction in the zeatin content under stress conditions (Dakora and Nelwamondo, 1996). The accumulation of ABA in Si-treated leaves experiencing drought stress stimulated the partial closure of stomata and consequently maintained the plant water relations under stress conditions (Zhang et al., 2006), thereby improving plant drought tolerance.

In addition, the endogenous jasmonic acid (JA) contents significantly increased with abiotic stress application, while decreasing after addition of Si to stressed plants (Wang et al., 2001; Hamayun et al., 2010).

Van Bockhaven et al. (2012) postulated that silicon does not cause a continuous change in phytohormone homeostasis, yet it simulates hormone biosynthesis and signaling pathways that induce the defense responses of the plant to stress.

15.4.3 MOLECULAR MECHANISM OF SILICON-MEDIATED DROUGHT TOLERANCE

15.4.3.1 Silicon-Mediated Expression of Transcription Factors and Some Associated Drought-Responsive Genes

The stress stimuli are perceived by plants and trigger a cascade of signal transduction pathways that upregulate gene expression and consequently lead to metabolic changes that achieve plant stress tolerance (Agarwal et al., 2007). Stress responses initiate a complex gene regulatory network that may be regulated and dependent on or independent of the phytohormone ABA. ABA persuades plants to deal with abiotic stresses such as water scarcity and finally achieve stress tolerance in plants via ABA-dependent signaling systems (Cutler et al., 2010; Kim et al., 2010). The ABA-dependent signaling schemes intercede in stress tolerance by stimulation of at least two separate regulons (a group of genes controlled by a certain transcription factor [TF], such as the *AREB/ABF* (ABA-responsive element-binding protein/ABA-binding factor) regulon and the *MYC* (myelocytomatosis oncogene)/*MYB* (myeloblastosis oncogene) regulon (Abe et al., 1997; Busk and Pagès, 1998; Saibo et al., 2009), whereas the ABA-independent regulons include the *CBF/DREB* regulon, the *NAC* (*NAM, ATAF,* and *CUC*) regulon, and the *ZF-HD* (zinc finger homeodomain) regulon (Nakashima et al., 2009; Saibo et al., 2009). Previous studies have recognized the existence of both ABA-dependent and -independent pathways of stress mechanism that act through *AP2/EREBP* (*ERF*) family members (Yamaguchi-Shinozaki and Shinozaki, 1994; Kizis and Pagès, 2002). Moreover, a large number of other TFs also participate in the abiotic stress responses; thus, they are mainly involved in the stimulation of plant stress tolerance.

It was suggested that the mechanism involved in the interaction of plants to drought stress is very complex. Some genes are overexpressed in response to abiotic stresses, and thus cause direct protection of the plant cells, while others are involved in various signaling pathways (Blumwald et al., 2004). Furthermore, a number of TFs that can alter various stress-inducible genes contribute to the control of the plant response to water shortage conditions (Bartels and Sunkar, 2005). Drought stress induces the expression of some genes at the transcriptional level. The products of these drought-inducible genes are believed to be involved in drought tolerance (Kavar et al., 2007). The overexpression of drought-inducible genes may be provoked directly by the drought stress conditions or promoted by secondary stresses or injury responses (Agarwal et al., 2007; Cattivelli et al., 2008). Likewise, the overexpression of some TFs can control a wide range of signaling pathways that are involved in stress tolerance attainment (Umezawa et al., 2006). Additionally, abiotic stresses persuade the synthesis of some vital metabolic proteins, including those required for the formation of osmolytes and regulatory proteins involved in signal transduction pathways, such as kinases or the TFs (Chaves and Oliveira, 2004). TFs such as *CBF1/DREB1B*, *CBF2/DREB1C*, and *CBF3/DREB1A* are involved in the transcriptional responses to abiotic stresses (Shinozaki and Yamaguchi-Shinozaki, 2007; Thomashow, 2010). It was reported that overexpression of *CBF1/DREB1B*, *CBF2/DREB1C*, and *CBF3/DREB1A* TFs in *Arabidopsis* enhanced the resistance to abiotic stresses (Liu et al., 1998; Kasuga et al., 1999; Gilmour et al., 2004). It was also investigated that the overexpression of *CBF/DREB1* in plants was concomitant with the accumulation of a greater amount of proline and soluble carbohydrates under normal growth conditions and during cold adaptaion (Cook et al., 2004; Achard et al., 2008). The expression of *DREB2A* (dehydration-responsive element-binding protein 2A) is activated by osmotic stress, and thus improves dehydration tolerance (Sakuma et al., 2006). It was observed that the TF *OsDREB2A* was weakly expressed in control plants and stimulated by dehydration (Chen et al., 2008; Khattab et al., 2014). The pretreatment with Si upregulated the expression of *OsDREB2A* in two drought-stressed rice cultivars compared to the untreated and drought-stressed ones (Khattab et al., 2014). Such results might be attributed to the enhancement of phosphorylation of the *DREB2A* protein to their active forms in Si-pretreated shoots. Overexpression of *OsDREB2A* in rice could protect cells during drought stress (Matsukura et al., 2010).

In addition, heat shock transcription factors (HSFs) are encoded in plants by a large gene family with different expression forms and functions (von Koskull-Döring et al., 2007). The overexpression of *HsFA2* induced the accumulation of galactinol and raffinose, and thereby enhanced the resistance of *Arabidopsis* plants to different environmental stresses (Nishizawa et al., 2006, 2008; Ogawa et al., 2007). Similarly, the overexpression of the *NAC* domain family of plant-specific transcriptional regulators contributed to the developing processes, as well as to the hormonal mechanism and stress protection (Olsen et al., 2005). Furthermore, osmotic stress and ABA treatment induced the overexpression of *OsNAC5*. Overexpression of *OsNAC5* boosted stress tolerance and accumulation of proline and soluble sugar, thus enhancing abiotic stress tolerance (Takasaki et al., 2010; Song et al., 2011). Moreover, the overexpression of the rice *Myb* TF, *OsMyb4*, under normal and stress conditions was concomitant with the accumulation of soluble sugars (glucose, fructose, and sucrose) and proline, as well as glycine betaine, therefore enhancing drought stress endurance (Mattana et al., 2005). Similarly, expression of the TF *OsNAC5* was detected in water-stressed rice plants that were left untreated or pretreated with Si (Khattab et al., 2014). Notably, the level of expression was much higher in Si-treated and drought-stressed rice plants. Thus, *OsNAC5* improved the stress tolerance of different rice cultivars, particularly the drought-resistant ones (Figure 15.2) (Khattab et

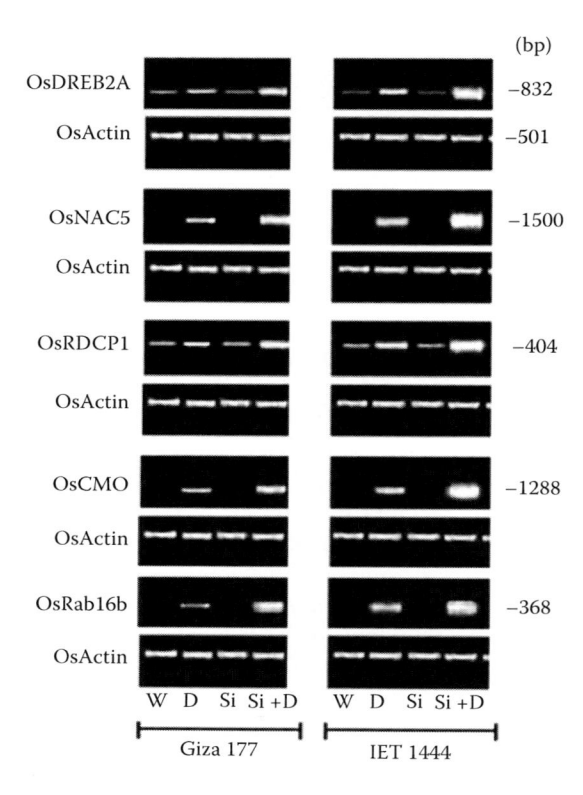

FIGURE 15.2 Polymerase chain reaction analyses of two TFs (OsDREB2A and OsNAC5) and two downstream genes responsible for plant tolerance to drought (OsRDCP1 and OsCMO) in the shoots of two rice cultivars, Giza 177 and IET 1444. OsRab16b was used as a positive control for drought-induced genes and actin was used as a loading control. Lanes: W, well-watered control; D, drought; Si, control pretreated with Si; Si + D, drought-stressed shoots pretreated with Si. (Adapted from Khattab HI et al., *Biol. Plant.*, 58(2): 265–273, 2014. With permission.)

al., 2014). Overexpression of *OsNAC5* leads to stress resistance via upregulation of the expression of some stress-responsive rice genes, such as *LEA3* genes that encode LEA proteins (Takasaki et al., 2010). Furthermore, *OsRDCP1, OsCMO*, and *OsRAB16b* are stress-inducible genes that were upregulated under drought stress. Likewise, Si pretreatment stimulated the overexpression of these genes in drought-stressed rice shoots compared to the untreated drought-stressed ones (Khattab et al., 2014). A background expression level of *OsRDCP1* was seen in the normal control shoots, while *OsCMO* was not expressed. The overexpression of *OsRDCP1*, under drought conditions, may be involved in the inactivation or degradation of water stress-related proteins, which would result in high stimulation of defense protein synthesis in rice (Park et al., 2011). Recently, it was reported that the expression of these previous genes was higher in rice cv. IET 1444 (drought tolerant) than in cv. Giza 177 (drought sensitive). The highest level of expression of such genes was observed in the Si-pretreated shoots. These results might be due to a difference in their Si uptake. It was deduced that increasing the resistance of plants to various stresses depends on the levels of Si uptake and accumulation in plant tissues. Plants differ widely in the amount of Si uptake (Mitani and Ma, 2005). Furthermore, a highly sustainable rice production requires a greater amount of Si (Savant et al., 1997).

In higher plants, the choline monooxygenase (CMO) gene encodes the important ferredoxin-dependent enzyme CMO, which catalyzes the synthesis of glycine betaine (Burnet et al., 1995). It was reported that *OsCMO* plays a vital role in the responses of rice plants to environmental stresses

(Lijuan et al., 2007). It was also postulated that the *OsCMO* genes are regulated by abiotic stresses under the control of *DREB* TFs (Shinozaki et al., 2003). Furthermore, it was postulated that overexpression of *OsCMO* under drought stress conditions was concomitant with the accumulation of glycine betaine in rice-stressed shoots (Khattab et al., 2014). Rice is considered a typical nonglycine betaine accumulation species under normal growth conditions, due to the presence of inactive gene for *CMO*, which becomes active under stress conditions (Rathinasabapathi et al., 1993; Shirasawa et al., 2006). LEA genes, particularly *OsRAB16b*, encode LEA proteins, which display a role in the acclimatization of plants to environmental stresses. The expression of several groups of LEA genes under stress conditions and their roles in stress tolerance have been recently reported in several species (Lenka et al., 2011). In addition, *OsRAB16b* genes are stimulated in vegetative and reproductive tissues exposed to either drought, salinity, or extreme temperatures or by exogenous ABA (Tunnacliffe and Wise, 2007; Bies-Ethève et al., 2008).

Recently, it was demonstrated that silicon pretreatments of rice cultivars could mitigate the adverse effects of drought stress. Interestingly, silicon enhanced the expression of *OsCMO* and *OsRDCP1* via the activation of the *DREB2A* and *NAC* TF pathways (Figure 15.2). This activation leads to the production of important osmoprotectant molecules (proline and glycine betaine) and important enzymes (CMO and *RING* E3 Ub ligases), which improve the oxidative defense system and consequently enhance rice tolerance to drought stress (Khattab et al., 2014). Furthermore, Song et al. (2014) reported that the addition of Si could, in leaves, nullify damages of the chloroplast and stimulate overexpression of *Os08g02630* (*PsbY*), *Os05g48630* (*PsaH*), *Os07g37030* (*PetC*), *Os03g57120* (*PetH*), *Os09g26810*, and *Os04g38410* genes, which link with protection of the photosynthetic pigments and consequently improve the photosynthesis, thereby improving the tolerance of rice plants to Zn stressful conditions.

15.4.3.2 Role of Si on Aquaporins

The movement of water between cells can be greatly regulated by the activity of aquaporins, which are rapidly and reversibly affected by the environmental conditions (Vandeleur et al., 2009). Aquaporins belong to a highly conserved family of major intrinsic membrane proteins (Tyerman et al., 2002). Aquaporins are located in the plasma membrane and vacuolar membrane of the plant cells, and they mediate membrane water transport. Aquaporins control the membranes' hydraulic conductivity and water permeability (Maurel and Chrispeels, 2001). The greater increments in the aquaporins' activity greatly improve plant water relations, and thereby improve stress tolerance. Hence, silicon application enhances the expression of several root aquaporin genes at the transcription levels under osmotic stress conditions (Liu et al., 2014). Application of silicon increases the expression of several *SbPIP* genes in sorghum plants exposed to osmotic stress in parallel with the transpiration rate (Liu et al., 2015). Silicon improves rice, maize, and cucumber drought tolerance via decreasing the transpiration rate. Likewise, aquaporin activity is affected by a number of stimuli, including ABA, ethylene, Ca^{2+}, and ROS (Azad et al., 2004; Boursiac et al., 2008; Parent et al., 2009; Hu et al., 2012). It has been described that salt stress and H_2O_2 can reduce the activity of aquaporin by direct oxidant gating, regulating the phosphorylation status, and stimulating the relocalization of aquaporin and consequently reducing water uptake (Boursiac et al., 2008). The enhancement of aquaporin activity may be attributed to the upregulation of the aquaporin genes at the transcriptional level or a decrease in the H_2O_2 level (Liu et al., 2015). It was also investigated that ABA can motivate the TFs, which contribute to the regulation of the expression of *PIP* (for plasma membrane intrinsic protein) aquaporins (Kaldenhoff et al., 1996; Shinozaki et al., 1998). Likewise, ABA treatment alters a larger number of *PIP* isoforms than drought stress (Jang et al., 2004), which may suggest the independence between ABA and drought signal transduction cascades (Mariaux et al., 1998). However, the overexpression of *PIP* mRNA is often temporary and ABA concentration dependent (Zhu et al., 2005; Beaudette et al., 2007), as well as not necessarily associated with an increment in *PIP* protein content (Morillon and Chrispeels, 2001; Aroca et al., 2006).

15.4.3.3   Effect of Silicon on Stress Proteins

Production of stress proteins is another protective mechanism implicated in drought stress tolerance (Taiz and Zeiger, 2006). Most of the stress proteins are water soluble and participate in the enhancement of water stress tolerance via the hydration of cellular structures (Wahid et al., 2007). A variety of gene products (proteins) are implicated in plant stress tolerance, including LEAs and *HSPs*. It was observed that the expression levels of late embryogenesis abundant/dehydrin-type genes and molecular chaperones are greatly influenced by drought stress and contribute to the protection of the cellular proteins from denaturation (Mahajan and Tuteja, 2005). Heat shock proteins (HSPs) are among the molecular chaperones that are usually produced only under stressful conditions (Coca et al., 1994; Wahid et al., 2007) and play an essential role in plant stress tolerance through stabilization of other protein structures. It was reported that *HSPs* participate in adenosine triphosphate–dependent protein unfolding or assembly or disassembly reactions and prevent protein denaturation under stress conditions (Gorantla et al., 2007).

On the other hand, application of silicon enhanced drought stress tolerance at both the cellular and genetic levels. Si confers drought tolerance not only by formation of mechanical barriers, but also by affecting some genes' expression, and thereby the genes products (proteins). It was observed that application of silicon increases the overexpression of several stress proteins, including membrane-stabilizing proteins and late embryogenic abundant proteins, which can confer drought tolerance for rice plants, as apparent from the overexpression of LEA genes (Khattab et al., 2014). These proteins enhance the capacity of water binding by the formation of defensive environment for protein or structure and known as dehydrins. Dehydrins are formed in stressed plants and also known as late embryogenesis abundant proteins (Close, 1997). Dehydrins are involved in hydrophobic interactions, inducing stabilization of macromolecules (Svensson et al., 2002). Likewise, application of Si significantly regulates several stressed proteins, including LEA proteins associated with redox homeostasis, photosynthesis, regulation of protein synthesis, and chaperone activity. Remarkably, it was detected that Si treatment persuaded the upregulation of a class III peroxidase and a thaumatin-like protein in Cd-stressed rice plants (Nwugo and Huerta, 2011).

15.5   CONCLUSION

Application of Si ameliorates drought stress tolerance of plants. Si enhanced the reduced growth rate induced by drought stress via several strategies, including the formation of mechanical barriers and thereby the reduction of transpiration regulated by ABA or better water retention. In addition, Si improved drought tolerance through other molecular strategies involving the regulation of several drought-responsive genes at the transcription level. Recently, it was documented that Si significantly enhances the expression of TFs such as *DREB2A* and *NAC5*, as well as the expression of the *OsRDCP1* (ring domain-containing protein) gene and some drought-specific genes, such as *OsCMO* (rice choline monooxygenase) and *OsRAB16b* (dehydrins; late embryogenesis abundant protein). Expression of such TFs and the associated studied genes was markedly enhanced in Si-stressed shoots of rice plants, which favors drought tolerance through the activation of antioxidant defense systems and overproduction of osmoregulators, hence reducing membrane lipid peroxidation and improving membrane stability and integrity, as well as cell metabolism (Figure 15.3).

15.6   FUTURE PROSPECTS

Nowadays, water scarcity causes major trouble for plants worldwide. Plants constantly experience several stresses that Si may alleviate. Accordingly, plants belonging to the family Poacea are able to accumulate greater quantities of Si. However, most dicots are incapable of accumulating Si in sufficient beneficial amounts, so genetic manipulation of the Si uptake ability of the plants might improve the accumulation capacity of Si, and hence enhance the tolerance against both biotic and

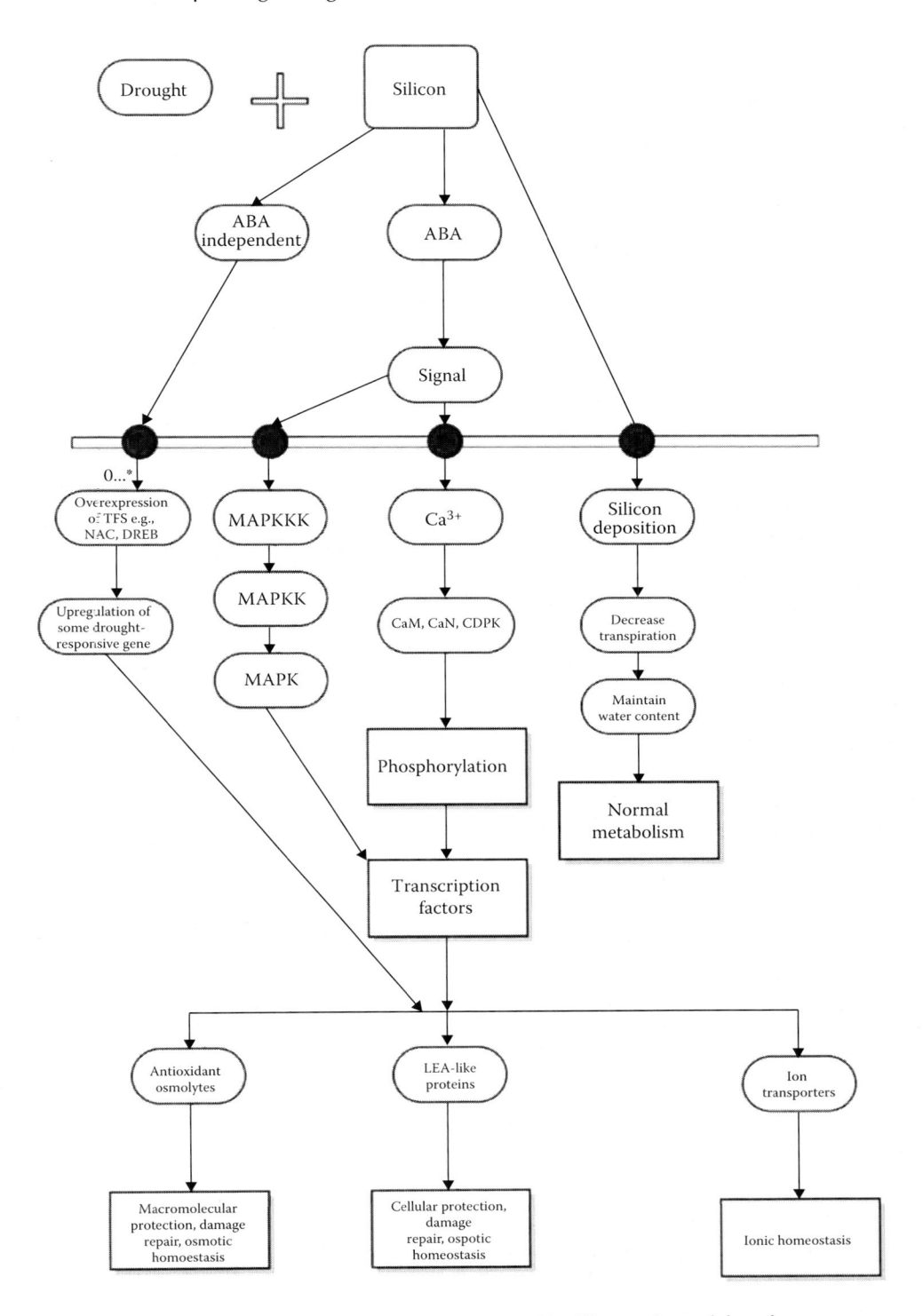

FIGURE 15.3 Proposed signal transduction pathways involved in silicon-enhanced drought stress responses. The stress signal is perceived through membrane receptors and transduced to the intracellular targets. Application of Si may activate the mitogen-activated protein kinase (MAPK) cascade, which is initiated by drought stress, and then MAPK induces the overexpression of transcription factors thereafter, inducing a cellular response.

abiotic stresses. Thus, further work is required to illustrate the roles of various chemical signals, such as free radicals, calcium, and plant hormones, which might mediate the defense systems in Si-stressed plants by acting through transduction cascades and activating genomic reprograming. The role of Si in the mitogen-activated protein (MAP) kinases, which are important mediators in signal transduction, connecting the perception of external stimuli and silicon to cellular responses, needs further work. Future investigation on the characterization of more specific Si-responsive genes, particularly heat shock genes, which are associated with drought-stress tolerance, is required; hence, these can be manipulated in stress-sensitive plants to enhance their tolerance and improve plant growth and productivity (Figure 15.3).

REFERENCES

Abdalla MM. (2011). Beneficial effects of diatomite on the growth, the biochemical contents and polymorphic DNA in *Lupinus albus* plants grown under water stress. *Agric. Biol. J. N. Am.*, 2(2): 207–220.

Abdalla MM, El-Khoshiban NH. (2007). The influence of water stress on growth, relative water content, photosynthetic pigments, some metabolic and hormonal contents of two *Triticum aestivum* cultivars. *J. Appl. Sci. Res.*, 3(12): 2062–2074.

Abe H, Yamaguchi-Shinozaki K, Urao T, Iwasaki T, Hosokawa D, Shinozaki K. (1997). Role of *Arabidopsis MYC* and *MYB* homologs in drought- and abscisic acid–regulated gene expression. *Plant Cell*, 9: 1859–1868.

Achard P, Gong F, Cheminant S, Alioua M, Hedden P, Genschik P. (2008). The cold-inducible CBF1 factor-dependent signaling pathway modulates the accumulation of the growth repressing DELLA proteins via its effect on gibberellin metabolism. *Plant Cell*, 20(8): 2117–2129.

Agarie S, Hanaoka N, Ueno O, Miyazaki A, Kubota F, Agata WD, Kaufman PB. (1998). Effects of silicon on tolerance to water deficit and heat stress in rice plants (*Oryza sativa* L.), monitored by electrolyte leakage. *Plant Prod. Sci.*, 1: 96–103.

Agarwal P, Agarwal PK, Nair S, Sopory SK, Reddy MK. (2007). Stress inducible DREB2A transcription factor from *Pennisetum glaucum* is a phosphoprotein and its phosphorylation negatively regulates its DNA binding activity. *Mol. Genet. Genom.*, 277: 189–198.

Ahmad R, Zaheer SH, Ismail S. (1992). Role of silicon in salt tolerance of wheat (*Triticum aestivum* L.). *Plant Sci.*, 85: 43–50.

Ahmed M, Fayyaz-ul-Hassen, Qadeer UM, Aslam A. (2011). Silicon application and drought tolerance mechanism of sorghum. *Afr. J. Agric. Res.*, 6: 594–607.

Akashi K, Miyake C, Yokota A. (2001). Citrulline, a novel compatible solute in drought-tolerant wild watermelon leaves, is an efficient hydroxyl radical scavenger. *FEBS Lett.*, 508: 438–442.

Al-Aghabary K, Zhu Z, Shi Q. (2004). Influence of silicon supply on chlorophyll content, chlorophyll fluorescence, and antioxidative enzyme activities in tomato plants under salt stress. *J. Plant Nutr.*, 12: 2101–2115.

Alcarde JC. (1992). Corretivos da acidez do solo: Características e interpretações técnicas. Boletim Técnico 6. Associação Nacional para Difusão de Adubos e Corretivos (ANDA), São Paulo—SP.

Alscher RG, Erturk N, Heath LS. (2002). Role of superoxide dismutase (SODs) in controlling oxidative stress in plants. *Exp. Bot.*, 53: 133–141.

Ando H, Kakuda K, Fujii H, Suzuki K, Ajiki T. (2002). Growth and canopy structure of rice plants grown under field conditions as affected by Si application. *Soil Sci. Plant Nutr.*, 48(3): 429–432.

Anjum F, Yaseen M, Rasul E, Wahid A, Anjum S. (2003). Water stress in barley (*Hordeum vulgare* L.). II. Effect on chemical composition and chlorophyll contents. *Pakistan J. Agric. Sci.*, 40: 45–49.

Anjum SA, Saleem MF, Wang LC, Bilal FM, Saeed A. (2012). Protective role of glycine betaine in maize against drought-induced lipid peroxidation by enhancing capacity of antioxidative system. *Aust. J. Crop Sci.*, 6: 576–583.

Aroca R, Ferrante A, Vernieri P, Chrispeels MJ. (2006). Drought, abscisic acid, and transpiration rate effects on the regulation of PIP aquaporin gene expression and abundance in *Phaseolus vulgaris* plants. *Ann. Bot.*, 98: 1301–1310.

Asada K. (1999). The water-water cycle in chloroplasts: Scavenging of active oxygen and dissipation of excess photons. *Annu. Rev. Plant Physiol. Plant Mol. Biol.*, 50: 601–639.

Ashraf M, Foolad MR. (2007). Roles of glycine betaine and proline in improving plant abiotic stress resistance. *Environ. Exp. Bot.*, 59: 206–216.

Adatia MH, Besford RT. (1986). The effects of silicon on cucumber plants grown in recirculating nutrient solution. *Ann. Bot.*, 58: 343–351.

Ashraf M. (2010). Inducing drought tolerance in plants: Some recent advances. *Biot. Adv.*, 28: 169–183.

Athar HUR, Ashraf M. (2005). Photosynthesis under drought stress. In Pessarakli M (ed.), *Hand Book Photosynthesis*. 2nd ed. CRC Press, New York, pp. 795–810.

Azad AK, Sawa Y, Ishikawa T, Shibata H. (2004). Phosphorylation of plasma membrane aquaporin regulates temperature-dependent opening of tulip petals. *Plant Cell Physiol.*, 45: 608–617.

Babita M, Maheswari M, Rao LM, Shanker AK, Rao DG. (2010). Osmotic adjustment, drought tolerance and yield in caster (*Ricinuscommunis* L.) hybrids. *Environ. Exp. Bot.*, 69: 243–249.

Babu RC, Pathan MS, Blum A, Nguyen HT. (1999). Comparison of measurement methods of osmotic adjustment in rice cultivars. *Crop Sci.*, 39: 150–158.

Balakhnina T, Borkowska A. (2013). Effects of silicon on plant resistance to environmental stresses: Review. *Int. Agrophys.*, 27: 225–232.

Balastra ML, Juliano CM, Villreal P. (1989). Effect of silica level on some proprieties *Oryza sativa* straw and Hult. *Can. J. Bot.*, 67: 2356–2363.

Bandurska H. (2000). Does proline accumulated in leaves of water deficit stressed barley plants confine cell membrane injury? I. Free proline accumulation and membrane injury index in drought and osmotically stressed plants. *Acta Physiol. Plant.*, 22: 409–415.

Barcelo J, Guevara P, Poschenrieder C. (1993). Silicon amelioration of aluminium toxicity in teosinte (*Zea mays* L. ssp. mexicana). *Plant Soil*, 154: 249–255.

Bartels D, Sunkar R. (2005). Drought and salt tolerance in plants. *Crit. Rev. Plant Sci.*, 24: 23–58.

Bates BC, Kundzewicz ZW, Wu S, Palutikof JP. (2008). Climate change and water. Technical Paper of the Intergovernmental Panel on Climate Change. IPCC Secretariat, Geneva.

Beaudette PC, Chlup M, Yee J, Emery RJN. (2007). Relationships of root conductivity and aquaporin gene expression in *Pisum sativum*: Diurnal patterns and the response to HgCl2 and ABA. *J. Exp. Bot.*, 58: 1291–1300.

Bektasoglu B, Esin CS, Ozyürek Mustafa O, Kubilay G, Resat A. (2006). Novel hydroxyl radical scavenging antioxidant activity assay for water-soluble antioxidants using a modified CUPRAC method. *Biochem. Biophys. Res. Commun.*, 345: 1194–2000.

Ben Ahmed C, Ben Rouina B, Sensoy S, Boukhris M, Ben Abdallah F. (2009). Changes in gas exchange, praline accumulation and antioxidative enzyme activities in three olive cultivars under contrasting water availability regimes. *Environ. Exp. Bot.*, 67: 345–352.

Bies-Ethèva N, Gaubier-Comella P, Debures A, Lasserre E, Jobet E, Raynal M, Cooke R, Delseny M. (2008). Inventory, evolution and expression profiling diversity of the LEA (late embryogenesis abundant) protein gene family in *Arabidopsis thaliana*. *Plant Mol. Biol.*, 67: 107–124.

Blum A. (1998). Improving wheat grain filling under stress by stem reserve mobilization. *Euphytica*, 100: 77–83.

Blumwald E, Grover A, Good AG. (2004). Breeding for abiotic stress resistance: Challenges and opportunities. In *New Directions for a Diverse Planet, Proceedings of the 4th International Crop Science Congress*, Brisbane, Australia, September 26–October 1. http://www.cropscience.org.au.

Bose J, Rodrigo-Moreno A, Shabala S. (2014). ROS homeostasis in halophytes in the context of salinity stress tolerance. *J. Exp. Bot.*, 65(5): 1241–1257.

Boursiac Y, Boudet J, Postaire O, Luu DT, Tournaire-Roux C, Maurel C. (2008). Stimulus-induced downregulation of root water transport involves reactive oxygen species-activated cell signalling and plasma membrane intrinsic protein internalization. *Plant J.*, 56: 207–218.

Broadley M, Brown P, Cakmak I, Ma JF, Rengel Z, Zhao F. (2012). Beneficial elements. In Marschner P (ed.), *Marschner's Mineral Nutrition of Higher Plants*. 3rd ed. Academic Press, London.

Burnet M, Lafontaine PJ, Hanson AD. (1995). Assay, purification, and partial characterization of choline monooxygenase from spinach. *Plant Physiol.*, 108: 581–588.

Busk PK, Pagès M. (1998). Regulation of abscisic acid-induced transcription. *Plant Mol. Biol.*, 37: 425–435.

Cattivelli L, Rizza F, Badeck F-W, Mazzucotelli E, Mastrangelo AM, Francia E, Marè C, Tondelli A, Stanca AM. (2008). Drought tolerance improvement in crop plants: An integrated view from breeding to genomics. *Field Crop Res.*, 115: 1–14.

Chaitanya KV, Rasineni GK, Reddy AR. (2009). Biochemical responses to drought stress in mulberry (*Morus alba* L.): Evaluation of proline, glycine betaine and abscisic acid accumulation in five cultivars. *Acta Physiol. Plant.*, 31: 437–443.

Cha-um S, Samphumphuang T, Kirdmanee C. (2013). Glycine betaine alleviates water deficit stress in *indica* rice using proline accumulation, photosynthetic efficiencies, growth performances and yield attributes. *Aust. J. Crop. Sci.*, 7: 213–218.

Chaves MM, Maroco JP, Pereira JS. (2003). Understanding plant responses to drought from genes to the whole plant. *Funct. Plant Biol.*, 30: 239–264.

Chaves MM, Oliveira MM. (2004). Mechanisms underlying plant resilience to water deficits: Prospects for water-saving agriculture. *J. Exp. Bot.*, 55: 2365–2384.

Chazen O, Neumann PM. (1994). Hydraulic signals from the roots and rapid cell-wall hardening in growing maize (*Zea mays* L.) leaves are primary responses to polyethylene glycol-induced water deficits. *Plant Physiol.*, 104: 1385–1392.

Chen JQ, Meng QP, Zhang Y, Xia M, Wang XP. (2008). Over-expression of OsDREB genes lead to enhanced drought tolerance in rice. *Biotechnol. Lett.*, 30: 2191–2198.

Chen TH, Murata N. (2002). Enhancement of tolerance of abiotic stress by metabolic engineering of betaines and other compatible solutes. *Curr. Opin. Plant Biol.*, 5: 250–257.

Chen W, Yao X, Cai K, Chen J. (2011). Silicon alleviates drought stress of rice plants by improving plant water status, photosynthesis and mineral nutrient absorption. *Biol. Trace Elem. Res.*, 142: 67–76.

Chen Y, Cao XD, Lu Y, Wang XR. (2000). Effects of rare earth metal ions and their EDTA complexes on anti-oxidant enzymes of fish liver. *Bull. Environ. Contam. Toxicol.*, 65: 357–365.

Chen ZX, Hu J, Chen G, Pan XB. (2004). Effects of rolled leaf gene RI (t) on economic traits of hybrid rice. *Acta Agron. Sin.*, 30: 465–469.

Close TJ. (1997). Dehydrins: A commonality in the response of plants to dehydration and low temperature. *Physiol. Plant.*, 100(2): 291–296.

Coca MA, Almoguera C, Jordano J. (1994). Expression of sunflower low-molecular-weight heat-shock proteins during embryogenesis and persistence after germination: Localization and possible functional implications. *Plant Mol. Biol.*, 25: 479–492.

Cook D, Fowler S, Fiehn O, Thomashow MF. (2004). A prominent role for the CBF cold response pathway in configuring the low temperature metabolome of *Arabidopsis*. *Proc. Natl. Acad. Sci. U.S.A.*, 101: 15243–15248.

Cooke J, Leishman MR. (2011). Is plant ecology more siliceous than we realise? *Trends Plant Sci.*, 16(2): 61–68.

Coors JG. (1987). Resistance to the European corn borer, *Ostrinia nubilalis* (Hubner), in maize, *Zea mays* L., as affected by soil silica, plant silica, structural carbohydrates, and lignin. In Gabelman HW, Laughman B (eds.), *Genetic Aspects of Plant Mineral Nutrition*. Martinus Nijhof, Dordrecht, pp. 445–456.

Cronic G. (1994). Drought stress and high light effects on leaf photosynthesis. In Baker NR, Bowyer JR (eds.), *Photoinhibition of Photosynthesis*. BIOS Scientific Publishers, Oxford, pp. 297–329.

Costa D, Lobato S, Silvera D, Laughinghouse HD. (2011). ABA-mediated proline synthesis in cowpea leaves exposed to water deficiency and rehydration. *Turk. J. Agric. For.*, 35: 309–317.

Costa RCL, Lobato AKS, Oliveira Neto CF. (2006). Variation in content of total soluble amino acids in leaves of cowpea under water stress. In *Congresso Nacional de feijão caupi*, Teresina, PI, Maio, pp. 1–19.

Costa Franca MG, Pham Thi AT, Pimentel C, Pereyra Rossiello RO, Zuily-Fodil, Y, Laffray D. (2000). Differences in growth and water relations among *Phaseolus vulgaris* cultivars in response to induced drought stress. *Environ. Exp. Bot.*, 43: 227–237.

Crooks R, Prentice P. (2011). The benefit of silicon fertilizer for sustainably increasing crop productivity. Presented at International Conference on Silicon Agriculture, Beijing.

Cutler SR, Rodriguez PL, Finkelstein RR, Abrams SR. (2010). Abscisic acid: Emergence of a core signaling network. *Annu. Rev. Plant Biol.*, 61: 651–679.

Da Costa M, Huang B. (2007). Changes in antioxidant enzyme activities and lipid peroxidation for bent grass species in responses to drought stress. *J. Am. Soc. Hortic. Sci.*, 132: 319–326.

Dakora FD, Nelwamondo A. (1996). Silicon nutrition promotes root growth and tissue mechanical strength in symbiotic cow pea. *Funct. Plant Biol.*, 30(9): 947–953.

Datnoff LE, Snyder GH, Korndörfer GH. (2001). *Silicon in Agriculture*. Elsevier, New York.

Davies PJ. (1995). *Plant Hormones: Physiology, Biochemistry and Molecular Biology*. Kluwer, Dordercht.

De B, Mukherjee AK. (1996). Mercuric chloride induced membrane damage in tomato cultured cells. *Biol. Plant.*, 38: 469–473.

Delfini R, Belgoff C, Fernández E, Fabra A, Castro S. (2010). Symbiotic nitrogen fixation and nitrate reduction in two peanut cultivars with different growth habit and branching pattern structures. *Plant Growth Reg.*, 6: 153–159.

Deren CW, Datnoff LE, Snyder GH, Martin FG. (1994). Silicon concentration, disease response, and yield components of rice genotypes grown on flooded organic histosols. *Crop Sci.*, 34: 733–737.

Demattê JLI, Paggiaro CM, Beltrame JA, Ribeiro SS. (2011). Uso de silicatos em cana-de-açúcar. *Inform. Agron.*, 133: 7–12.

Ding YF, Liang YC, Zhu J, Li ZJ. (2007). Effects of silicon on plant growth, photosynthetic parameters, and soluble sugar content in leaves of wheat under drought stress. *Plant Nutr. Fert. Sci.*, 13: 471–478.

Djibril S, Mohamed OK, Diaga D, Diégane D, Abaye BF, Maurice S, Alain B. (2005). Growth and development of date palm (*Phoenix dactylifera* L.) seedlings under drought and salinity stresses. *Afr. J. Biotechnol.*, 4: 968–972.

Efimova GV, Dokynchan SA. (1986). Anatomo-morphological construction of epidermical tissue of rice leaves and increasing of its protection function under silicon effect. *Agric. Biol.*, 3: 57–61.

Emadian SF and Newton RJ. (1989) Growth Enhancement of Loblolly Pine (*Pinus taeda* L.) Seedlings by Silicon. *Journal of Plant Physiology*, 134(1): 98–103.

Emam MM, Khattab HI, Helal NM. (2012a). Effect of selenium or silicon on photosynthetic apparatus and antioxidant capacity of rice plant grown under drought condition. *Egypt. J. Exp. Bot.*, 8(2): 271–283.

Emam MM, Khattab HI, Helal NM, Deraz AE. (2012b). Improving drought tolerance of two rice cultivars (*Oryza sativa* L.) by using selenium and silicon. *Assiut Univ. J. Bot.*, 62: 45–45.

Emam MM, Khattab HI, Helal NM, Deraz AE. (2014). Effect of selenium and silicon on yield quality of rice plant grown under drought stress. *Aust. J. Crop Sci.*, 8(4): 596–605.

Epstein E. (1994). The anomaly of silicon in plant biology. *Proc. Natl. Acad. Sci. U.S.A.*, 91: 11–17.

Epstein E. (1999). Silicon. *Annu. Rev. Plant Physiol. Plant Mol. Biol.*, 50: 641–664.

Epstein E. (2001) Silicon in plants: Facts vs. concepts In: Silicon in Agriculture (Datnoff, L.E., Snyder, G.H. and Korndo¨fer, G.H., Eds.), pp. 1–16. Elsevier, Amsterdam.

Epstein E, Bloom AJ. (2005). *Mineral Nutrition of Plants: Principles and Perspectives*. 2nd ed. Sinauer Associates, Sunderland, MA.

Farooq M, Wahid A, Kobayashi N, Basra SMA. (2009). Plant drought stress: Effects, mechanisms, and management. *Agron. Sustain. Dev.*, 29: 185–212.

Fawe A, Menzies AJG, Chérif M, Bélanger RB. (2001). Silicon and disease resistance in dicotyledons. In Datnoff LE, Snyder GH, Korndörfer GH (eds.), *Silicon in Agriculture*. Elsevier, New York, pp. 1–15.

Feng JP, Shi QH, Wang XF. (2009). Effects of exogenous silicon on photosynthetic capacity and antioxidant enzyme activities in chloroplast of cucumber seedlings under excess manganese. *Agric. Sci. China*, 8: 40–50.

Fukai S, Cooper M. (1995). Development of drought-resistant cultivars using physiological traits in rice. *Field Crops Res.*, 40: 67–86.

Gao D, Cai K, Chen J, Luo S, Zeng R, Yang J, Zhu X. (2011). Silicon enhances photochemical efficiency and adjusts mineral nutrient absorption in *Magnaporthe oryzae* infected rice plants. *Acta Physiol. Plant.*, 33: 675–682.

Gao X, Zou Ch, Wang L, Zhang F. (2005) Silicon Improves Water Use Efficiency in Maize Plants. HYPERLINK "http://www.tandfonline.com/toc/lpla20/27/8" *Journal of Plant Nutrition*, 27 (8): 1457–1470.

Gao X, Zou C, Wang L, Zhang F. (2006). Silicon decreases transpiration rate and conductance from stomata of maize plants. *J. Plant Nutr.*, 29: 1637–1647.

Genard H, Saos JL, Billard JP, Tremolieres A, Boucaud J. (1991). Effect of salinity on lipid composition, glycine betaine content and photosynthetic activity in chloroplasts *Suaeaa maritima*. *Plant Physiol. Biochem.*, 29: 421–427.

Gholami Y, Falah A. (2013). Effects of two different sources of silicon on dry matter production, yield and yield components of rice, Tarom Hashemi variety and 843 lines. *Int. J. Agric. Crop Sci.*, 5: 227–231.

Gibson SI. (2005). Control of plant development and gene expression by sugar signaling. *Curr. Opin. Plant Biol.*, 8: 93–102.

Gigon A, Matos A, Laffray D, Zuily-fodil Y, Pham-Thi A. (2004). Effect of drought stress on lipid metabolism in the leaves of *Arabidopsis thaliana* (ecotype Columbia). *Ann. Bot.*, 94: 345–351.

Gill SS, Tuteja N. (2010). Reactive oxygen species and antioxidant machinery in abiotic stress tolerance in crop plants. *Plant Physiol. Biochem.*, 48: 909–930.

Gilmour SJ, Fowler SG, Thomashow MF. (2004). Arabidopsis transcriptional activators CBF1, CBF2, and CBF3 have matching functional activities. *Plant Mol. Biol.*, 54(5): 767–781.

Giri J. (2011). Glycine betaine and abiotic stress tolerance in plants. *Plant Signal. Behav.*, 6: 1746–1751.

Gong HJ. Chen KM, Zhao ZG, Chen GC, Zhou WJ. (2008). Effects of silicon on defense of wheat against oxidative stress under drought at different developmental stages. *Biol. Plant.*, 52: 592–596.

Gong HJ, Zhu X, Chen K, Wang S, Zhang C. (2005). Silicon alleviates oxidative damage of wheat plants in pots under drought. *Plant Sci.*, 169: 313–321.

Gorantla M, Babu PR, Lachagari VBR, Reddy AMM, Wusirika R, Bennetzen JL, Reddy AR. (2007). Identification of stress-responsive genes in an indica rice (*Oryza sativa* L.) using ESTs generated from drought-stressed seedlings. *J. Exp. Bot.*, 58(2): 253–265.

Gunes A, Inal A, Bagci EG, Pilbeam DJ. (2007). Silicon-mediated changes of some physiological and enzymatic parameters symptomatic for oxidative stress in spinach and tomato grown in sodic-B toxic soil. *Plant Soil*, 290: 103–114.

Gunes A, Pilbeam DJ, Inal A, Coban S. (2008). Influence of silicon on sunflower cultivars under drought stress. I. Growth, antioxidant mechanisms, and lipid peroxidation. *Commun. Soil Sci. Plant Anal.*, 39: 1885–1903.

Habibi G, Hajiboland R. (2013). Alleviation of drought stress by silicon supplementation in pistachio (*Pistacia vera* L.) plants. *Folia Hortic.*, 25(1): 21–29.

Hamayun M, Sohn EY, Khan SA, Shinwari ZK, Khan AL, Lee IJ. (2010). Silicon alleviates the adverse effects of salinity and drought stress on growth and endogenous plant growth hormones of soybean (*Glycine max* L.). *Pak J. Bot.*, 42: 1713–1722.

Hanafy Ahmed AH, Harb EM, Higazy MA and Morgan ShH. (2008) Effect of Silicon and Boron Foliar Applications on Wheat Plants Grown under Saline Soil Conditions. *International Journal of Agricultural Research*, 3: 1–26.

Hare PD, Cress WA. (1997). Metabolic implications of stress induced proline accumulation in plants. *Plant Growth Regul.*, 21: 79–102.

Hartikainen H. (2005). Biogeochemistry of selenium and its impact on food chain quality and human health. *J. Trace Elem. Med. Biol.*, 18: 309–318.

Hattori T, Inanaga S, Araki H, An P, Morita S, Luxová M, Lux A. (2005). Application of silicon enhanced drought tolerance in *Sorghum bicolour*. *Physiol. Plant.*, 123: 459–466.

Hattori T, Inanaga S, Tanimoto E, Lux A, Luxová M, Sugimoto Y. (2003). Silicon-induced changes in viscoelastic properties of sorghum root cell walls. *Plant Cell Physiol.*, 44(7): 743–749.

Hattori T, Sonobe K, Inanaga S, An P, Morita S, Morita S. (2008). Effects of silicon on photosynthesis of young cucumber seedlings under osmotic stress. *J. Plant Nutr.*, 31: 1046–1058.

Helal NM. (2013). Enhancement of drought tolerance of different rice cultivars. PhD thesis, Faculty of Science, Ain Shams University, Egypt.

Hessini K, Martinez JP, Gandour M, Albouchi A, Soltani A, Abdelly C. (2009). Effect of water stress on growth, osmotic adjustment, cell wall elasticity and water-use efficiency in *Spartina alterniflora*. *Environ. Exp. Bot.*, 67: 312–319.

Horemans N, Foyer CH, Potters G, Asard H. (2000). Ascorbate function and associated transport systems in plants. *Plant Physiol. Biochem.*, 38: 531–540.

Hossain MT, Mori R, Soga K, Wakabayashi K, Kamisaka S, Fujii S, Yamamoto R, Hoson T. (2002). Growth promotion and an increase in cell wall extensibility by silicon in rice and some other Poaceae seedlings. *J. Plant Res.*, 115: 23–27.

Hu W, Yuan Q, Wang Y, Cai R et al. (2012). Overexpression of a wheat aquaporin gene, *TaAQP8*, enhances salt stress tolerance in transgenic tobacco. *Plant Cell Physiol.* 53: 2127–2141.

Hura T, Grzesiak S, Hura K, Grzesiak MT, Rzepka A. (2006). Differences in the physiological state between triticale and maize plants during drought stress and followed rehydration expressed by the leaf gas exchange and spectrofluorimetric methods. *Acta Physiol. Plant.*, 28: 433–443.

Hussain SS, Kayani MA, Amjad M. (2011). Transcription factors as tools to engineer enhanced drought tolerance in plants. *Biotechnol. Prog.*, 27: 297–306.

Jaleel CA, Gopi R, Sankar B, Gomathinavagam M, Panneerselvam R. (2008). Alterations in morphological parameters and photosynthetic pigment responses of *Catharanthus roseus* under soil water deficit. *Colloids Surf. B Biointerfaces*, 61: 298–303.

Jaleel CA, Manivannan P, Sankar B, Kishorekumar A, Gopi R, Somasundaram R, Panneerselvam R. (2007). Induction of drought stress tolerance by ketoconazole in *Catharanthus roseus* is mediated by enhanced antioxidant potentials and secondary metabolite accumulation. *Colloids Surf. B Biointerfaces*, 60: 201–206.

Jaleel CA, Manivannan P, Wahid A, Farooq M, Al Juburi HJ, Somasundaram R, Panneerselvam R. (2009). Drought stress in plants: A review on morphological characteristics and pigments composition. *Int. J. Agric. Biol.*, 11: 100–105.

Janda M, Planchais S, Djafi N, Martinec J, Burketova L, Valentova O, Zachowski A, Ruelland E. (2013). Phosphoglycerolipids are master players in plant hormone signal transduction. *Plant Cell Rep.*, 32(6): 839–851.

Jang JK, Kim DG, Kim YO, Kim JS, Kang HS. (2004) An expression analysis of a gene family encoding plasma membrane aquaporins in response to abiotic stresses in *Arabidopsis thaliana*. *Plant Molecular Biology*, 54: 713–725.

Jiang M, Zhang J. (2002). Water stress-induced abscisic acid accumulation triggers the increased generation of reactive oxygen species and upregulates the activities of antioxidant enzymes in maize leaves. *J Exp Bot.*, 53: 2401–2410.

Kaldenhoff R, Kolling A, Richter G. (1996). Regulation of the *Arabidopsis thaliana* aquaporin gene AthH2 (PIP1b). *J. Photochem. Photobiol. B*, 36: 351–354.

Karmollachaab A, Bakhshandeh A, Gharineh MH, Telavat MRM, Fathi G. (2013). Effect of silicon application on physiological characteristics and grain yield of wheat under drought stress condition. *Int. J. Agron. Plant Prod.*, 4: 30–37.

Kasuga M, Liu Q, Miura S, Yamaguchi-Shinozaki K, Shinozaki K. (1999) Improving plant drought, salt, and freezing tolerance by gene transfer of a single stress-inducible transcription factor. *Nat Biotechnol.*;17(3): 287–91.

Kavar T, Maras M, Kidric M, Sustar-Vozlic J, Meglic V. (2007). Identification of genes involved in the response of leaves of *Phaseolus vulgaris* to drought stress. *Mol. Breed.*, 21: 159–172.

Kavikishore PB, Sangam S, Amrutha RN, Srilaxmi P, Naidu KR, Rao KRSS, Rao S, Reddy KJ, Theriappan P, Sreenivasulu N. (2005). Regulation of proline biosynthesis, degradation, uptake and transport in higher plants: Its implications in plant growth and abiotic stress tolerance. *Curr. Sci.*, 88: 424–438.

Kawasak S, Miyake C, Kohchi T, Fujii S, Uchida M, Yokota A. (2000). Responses of wild watermelon to drought stress: Accumulation of an ArgE homologue and citrulline in leaves during water deficits. *Plant Cell Physiol.*, 41: 864–873.

Kaya C, Tuna L, Higgs D. (2006). Effect of silicon on plant growth and mineral nutrition of maize grown under water-stress conditions. *J. Plant Nutr.*, 29: 1469–1480.

Khattab HI, Emam MA, Emam MM, Helal NM, Mohamed MR. (2014). Effect of selenium and silicon on transcription factors *NAC5* and *DREB2A* involved in drought-responsive gene expression in rice. *Biol. Plant.*, 58(2): 265–273.

Kiani SP, Maury P, Sarrafi A, Grieu P. (2008). QTL analysis of chlorophyll fluorescence parameters in sunflower (*Helianthusannuus* L.) under well-watered and water-stressed conditions. *Plant Sci.*, 175: 565–573.

Kinnersley AM, Turano FJ. (2000). Gamma aminobutyric acid (GABA) and plant responses to stress. *Crit. Rev. Plant Sci.*, 19: 479–509.

Kim T-H, Böhmer M, Hu H, Nishimura N, Schroeder JI. (2010). Guard cell signal transduction network: Advances in understanding abscisic acid, CO_2, and Ca^{2+} signaling. *Annu. Rev. Plant Biol.*, 61: 561–591.

Kizis D, Pagès M. (2002). Maize DRE-binding proteins DBF1 and DBF2 are involved in rab17 regulation through the drought-responsive element in an ABA-dependent pathway. *Plant J.*, 30: 679–689.

Korndörfer GH, Pereira HS, Camargo MS. (2002). Papel do silício na produção de cana-de-açúcar. *Revista STAB*, 21(1): 6–9.

Korndörfer GH, Pereira HS, Nolla A. (2004). Análise de silício: Solo, planta e fertilizante. Boletim Técnico n. 2. GPSi/ICIAG/UFU, Uberlândia.

Kyle DJ, Ohad I, Arntzen CJ. (1984). Membrane protein damage and repair: Selective loss of a quinoneprotein function in chloroplast membranes. *Proc. Natl. Acad. Sci. U.S.A.*, 81: 4070–4074.

Larsson EH, Bornman JF, Asp H. (1998). Influence of UV-B radiation and Cd^{2+} on chlorophyll fluorescence, growth and nutrient content in *Brassica napus*. *J. Exp. Bot.*, 323: 1031–1039.

Lauriano JA, Lidon FC, Carvalho CA, Campos PS, Matos MD. (2000). Drought effects on membrane lipids and photosynthetic activity in different peanut cultivars. *Photosynthetica*, 38: 7–12.

Lee BR, Jin YL, Jung WJ, Avice JC, Morvan-Bertrand A, Ourry A, Park CW, Kim TH. (2008). Water-deficit accumulates sugars by starch degradation—not by de novo synthesis—in white clover leaves (*Trifolium repens*). *Physiol. Plant.*, 134: 403–411.

Lee SK, Sohn EY, Hamayun M, Yoon JY, Lee IJ. (2010) Effect of silicon on growth and salinity stress of soybean plant grown under hydroponic system. *Agroforest. Syst.* 80:333–340.

Leopold AC, Musgrave ME, Williams KM. (1981). Solute leakage resulting from leaf desiccation. *Plant Physiol.*, 68: 1222–1225.

Lenka SK, Katiyar A, Chinnusamy V, Bansal KC. (2011). Comparative analysis of drought-responsive transcriptome in Indica rice genotypes with contrasting drought tolerance. *Plant Biotechnol. J.*, 9: 315–327.

Li M, Liu H, Li L, Li L, Yi X, Zhu X. (2007). Carbon isotope composition of plants along an altitudinal gradient and its relationship to environmental factors on the Qinghal-Tibet Plateau. *Polish J. Ecol.*, 55: 67–78.

Li QF, Ma CC. (2002). Effect of available silicon in soil on cucumber seed germination and seedling growth metabolism. *Acta Hortic. Sin.*, 29: 433–437.

Liang Y, Chen Q, Liu Q, Zhang W, Ding R. (2003). Exogenous silicon (Si) increases antioxidant enzyme activity and reduces lipid peroxidation in roots of salt-stressed barley (*Hordeumvulgare* L.). *J. Plant Physiol.*, 160: 157–164.

Liang YC. (1998). Effects of Si on leaf ultrastructure, chlorophyll content, and photosynthetic activity in barley under salt stress. *Pedosphere*, 8: 289–296.

Liang YC. (1999). Effects of silicon on enzyme activity, and sodium, potassium and calcium concentration in barley under salt stress. *Plant Soil*, 209: 217–224.

Liang YC, Shen QR, Shen ZG, Ma TS. (1996). Effects of silicon on salinity tolerance of two barley cultivars. *J. Plant Nutr.*, 19: 173–183.

Liang YC, Sun WC, Zhu YG, Christie P. (2007). Mechanisms of silicon-mediated alleviation of abiotic stresses in higher plants: A review. *Environ. Pollut.*, 147: 422–428.

Liang YC, Wong JWC, Wei L. (2005). Silicon-mediated enhancement of cadmium tolerance in maize (*Zea mays* L.) grown in cadmium contaminated soil. *Chemosphere*, 58: 475–483.

Lichtenthaler HK. (1987). Chlorophylls and carotenoids, the pigments of photosynthetic biomembranes. *Methods Enzymol.*, 148: 350–382.

Lichtenthaler HK, Rinderle U. (1988). The role of chlorophyll fluorescence in the detection of stress conditions in plants. *CRC Crit. Rev. Ann. Chem.*, 19: 529–585.

Li-feng W, Hao F, Yun-He J. (2012). Photosynthetic characterization of a rolled leaf mutant of rice (*Oryza sativa* L.). *Afr. J. Biotechnol.*, 11(26): 6839–6846.

Lijuan P, Ji H, Zhoufei W, Hongsheng Z. (2007). Molecular cloning and expression of *OsCMO* encoding a putative choline monooxygenase in rice (*Oryza sativa* L.). *Mol. Plant Breed.*, 5: 8–14.

Lima Filho OF, Lima MTG, Tsai SMO. (1999). Silício na agricultura. *Inform. Agron.*, 87: 1–7.

Lindhorst TK. (2000). *Essentials in Carbohydrate Chemistry and Biochemistry.* Wiley-VCH, Weinheim, pp. 3–31, 195–208.

Liu C, Liu Y, Guo K, Fan D, Li G, Zheng Y. (2011). Effect of drought on pigments, osmotic adjustment and antioxidant enzymes in six woody plant species in karst habitats of southwestern China. *Environ. Exp. Bot.*, 71: 174–183.

Liu C, Mao B, Ou S, Wang W, Liu L, Wu Y, Chu C, Wang X. (2014). OsbZIP71, a bZIP transcription factor, confers salinity, and drought tolerance in rice. *Plant Mol. Biol.*, 84: 19–36.

Liu K, Ye Y, Tang C, Wang Z, Yang J. (2008). Responses of ethylene and ACC in rice grains to soil moisture and their relations to grain filling. *Front. Agric. China*, 2: 172–180.

Liu M, Qi H, Zhang Z, Song Z, Kou TG, Zhang WJ, Yu JL. (2012). Response of photosynthesis and chlorophyll fluorescence to drought stress in two maize cultivars. *Afr. J. Agric. Res.*, 7: 4751–4760.

Liu Q, Kasuga M, Sakuma Y, Habe H, Miura S, Yamaguchi-Shinozaki KY, Shinozaki K. (1998). Two transcription factors, DREB1 and DREB2, with an EREBP/AP2 DNA binding domain separate two cellular signal transduction pathways in drought- and low-temperature responsive gene expression, respectively, in *Arabidopsis*. *Plant Cell*, 10: 1391–1406.

Liu SC, Yaoa MZ, Maa CL, Jina JQ, Maa JQ, Li CF, Chen L. (2015). Physiological changes and differential gene expression of tea plant under dehydration and rehydration conditions. *Sci. Hortic.*, 184: 129–141.

Liu XZ, Huang BR. (2000). Heat stress injury in relation to lipid peroxidation in creeping bentgrass. *Crop Sci.*, 40: 503–510.

Lobato AKS, Luz LM, Costa RCL, Santos Felho BG, Meirelles ACS, Neto CFO. (2009). Silicon exercises influence on nitrogen components in pepper subjected to water deficit. *Res. J. Biol. Sci.*, 4: 1048–1055.

Lobato AKS, Guedes EMS, Marques DJ, Neto CF de O. (2013). Silicon: A benefic element to improve tolerance in plants exposed to water deficit. In *Responses of Organisms to Water Stress*. INTECH, pp. 95–113. http://dx.doi.org/10.5772/53765.

Loutfy N, El-Tayeb MA, Hassanen AM, Moustafa MFM, Sakuma Y, Inouhe M. (2012). Changes in the water status and osmotic solute contents in response to drought and salicylic acid treatments in four different cultivars of wheat (*Triticum aestivum*). *J. Plant Res.*, 125: 73–184.

Lux A, Luxová M, Abe, Morita S, Inanaga S. (2003). Silicification of bamboo (*Phyllostachys heterocycla* Mitf.) root and leaf. *Plant Soil*, 255(1): 85–91.

Lux A, Luxova M, Hattori T, Inanaga S, Sugimoto Y. (2002). Silicification in sorghum (*Sorghum bicolor*) cultivars with different drought tolerance. *Physiol. Plant.*, 115: 87–92.

Ma JF. (2004). Role of silicon in enhancing the resistance of plants to biotic and abiotic stresses. *Soil Sci. Plant Nutr.*, 50: 11–18.

Ma JF, Goto S, Tamai K, Ichii M. (2001). Role of root hairs and lateral roots in silicon uptake by rice. *Plant Physiol.*, 127: 1773–1780.

Ma JF, Mitani N, Nagao S, Konishi S, Tamai K, Iwashita T, Yano M. (2004). Characterization of the silicon uptake and molecular mapping of the silicon transporter gene in rice. *Plant Physiol.*, 136: 3284–3289.

Ma JF, Nishimura K, Takahashi E. (1989). Effect of silicon on the growth of rice plant at different growth stages. *Soil Sci. Plant Nutr.*, 35: 347–356.

Ma JF, Takahashi E. (2002). *Soil, Fertilizer, and Plant Silicon Research in Japan.* Elsevier Science, Amsterdam.

Ma JF, Takahashi E. (1989). Effect of silicic acid on phosphorus uptake by rice plant. *Soil Sci. Plant Nutr.*, 35: 663–667.

Ma JF, Tamai K, Yamaji N, Mitani N, Konishi S, Katsuhara M, Ishiguro M, Murata Y, Yano M. (2006). A silicon transporter in rice. *Nature*, 440: 688–691.

Ma JF, Yamaji N. (2006). Silicon uptake and accumulation in higher plants. *Trends Plant Sci.*, 11: 392–397.

Madhava Rao KV, Raghavendr AS, Reddy J. (2006). *Physiology and Molecular Biology of Stress Tolerance in Plants.* Springer, Berlin, p. 83.

Mahajan S, Tuteja N. (2005). Cold, salinity and drought stresses: An overview. *Arch. Biochem. Biophys.*, 444: 139–158.

Manivannan P, Jaleel CA, Kishorekumar A, Sankar B, Somasundaram R, Sridharan R, Panneerselvam R. (2007). Changes in antioxidant metabolism of *Vigna unguiculata* L. Walp. by propiconazole under water deficit stress. *Colloids Surf. B Biointerfaces*, 57: 69–74.

Mansour MM. (1998). Protection of plasma membrane of onion epidermal cells by glycinebetaine and proline against NaCl stress. *Plant Physiol. Biochem.*, 35: 767–772.

Mariaux JB, Bockel C, Salamini F, Bartels D. (1998). Desiccation- and abscisic acid-responsive genes encoding major intrinsic proteins (MIPs) from the resurrection plant *Craterostigma plantagineum*. *Plant Mol. Biol.*, 38: 1089–1099.

Marschner H. (1995). *Mineral Nutrition of Higher Plants.* 2nd ed. Academic Press, London.

Martinelli T, Whittaker A, Bochicchio A, Vazzana C, Suzuki A, Masclaux-Daubresse C. (2007). Amino acid pattern and glutamate metabolism during dehydration stress in the resurrection plant *Sporobolus stapfianus.* A comparison between desiccation-tolerant and dessication-sensitive leaves. *J. Exp. Bot.*, 58: 3037–3046.

Masoumi A. Kafi M, Khazaei H, Davari K. (2010). Effect of drought stress on water status, electrolyte leakage and enzymatic antioxidants of kochia (*Kochiascoparia*) under saline condition. *Pak. J. Bot.*, 42: 3517–3524.

Massacci A, Nabiev SM, Pietrosanti L, Nematov SK, Chernikova TN, Thor K, Leipner J. (2008). Response of the photosynthetic apparatus of cotton (*Gossypium hirsutum*) to the onset of drought stress under field conditions studied by gas-exchange analysis and chlorophyll fluorescence imaging. *Plant Physiol. Biochem.*, 46: 189–195.

Matichenkov VV. (1996). The silicon fertilizer effect of root cell growth of barley. In *Abstract in the 5th Symposium of the International Society of Root Research*, Clemson, SC, p. 110.

Matichenkov VV, Calvert DV, Snyder GH. (2000). Prospective silicon fertilization for citrus in Florida. *Soil Crop Sci. Proc.*, 59: 137–141.

Matsukura S, Mizoi J, Yoshida T, Todaka D, Ito Y, Maruyama K, Shinozaki K, Yamaguchi-Shinozaki K. (2010). Comprehensive analysis of rice DREB2-type genes that encode transcription factors involved in the expression of abiotic stress responsive genes. *Mol. Genet. Genom.*, 283: 185–196.

Mattana M, Biazzi E, Consonni R, Locatelli F, Vannini C, Provera S, Coraggio I (2005) Overexpression of Osmyb4 enhances compatible solute accumulation and increases stress tolerance of *Arabidopsis thaliana*. *Physiol Plant* 125:212–223.

Maurel C, Chrispeels MJ. (2001). Aquaporins. A molecular entry into plant water relations. *Plant Physiol.*, 125: 135–138.

Maxwell K, Johnson GN. (2000). Chlorophyll fluorescence—A practical guide. *J. Exp. Bot.*, 51: 659–668.

Meneses CHSG, Lima LHGM, Lima MMA, Vidal MS. (2006). Genetic and molecular aspects of plants submitted at water stress. *Rev. Bras. Ol. Fibros*, 10: 1039–1072.

Michalak A. (2006). Phenolic compounds and their antioxidant activity in plants growing under heavy metal stress. *Polish J. Environ. Stud.*, 15(4), 523–530.

Miao B-H, Han X-G, Zhang W-H. (2010). The ameliorative effect of silicon on soybean seedlings grown in potassium-deficient medium. *Ann. Bot.*, 105: 967–973.

Mingo DM, Theobald JC, Bacon MA, Davies WJ, Dodd IC. (2004). Biomass allocation in tomato (*Lycopersicon esculentum*) plants grown under partial rootzone drying: Enhancement of root growth. *Funct. Plant Biol.*, 31: 971–978.

Mitani N, Ma JF. (2005). Uptake system of silicon in different plant species. *J. Exp. Bot.*, 56: 1255–1261.

Miyake Y. (1992) The effect of silicon on the growth of the different groups of rice (*Oryza sativa*) plants. The Agriculture, Forestry and Fisheries Research Information Technology Center. Htt://www.affrc.go.jp /en/10903.

Miyazono K, Miyakawa T, Sawano Y. (2009). Structural basis of abscisic acid signalling. *Nature* 462: 609–614.

Moinuddin, Khanna-Chopra R. (2004). Osmotic adjustment in chickpea in relation to seed yield and yield parameters. *Crop Sci.*, 44: 449–455.

Morillon R, Chrispeels MJ. (2001). The role of ABA and the transpirationstream in the regulation of the osmotic water permeability of leaf cells. *Proc. Natl. Acad. Sci. U.S.A.*, 98: 14138–14143.

Mostajeran A, Rahimi-Eichi V. (2009). Effects of drought stress on growth and yield of rice (*Oryza sativa* L.) cultivars and accumulation of proline and soluble sugars in sheath and blades of their different ages leaves. *Am. Eurasian J. Agric. Environ. Sci.*, 5: 264–272.

Moussa HR, Abdel-Aziz SM. (2008). Comparative response of drought tolerant and drought sensitive maize genotypes to water stress. *Aust. J. Crop Sci.*, 1: 31–36.

Moustakas M, Sperdouli I, Kouna T, Antonopoulou CI and Therios I (2011) Exogenous proline induces soluble sugar accumulation and alleviates drought stress effects on photosystem II functioning of Arabidopsis thaliana leaves. *Plant Growth Regul.* 65:315–325.

Myrene R, Souza D, Devaraj VR. (2011). Specific and non-specific responses of Hyacinth bean (*Dolichoslablab*) to drought stress. *Ind. J. Biotechnol.*, 10: 130–139.

Naeem MS, Jin L, Wan GL, Liu D, Liu HB, Yoneyama K, Zhou WJ. (2010). 5-Aminolevulinic acid improves photosynthetic gas exchange capacity and ion uptake under salinity stress in oilseed rape (*Brassicanapus* L.). *Plant Soil*, 332: 405–415.

Nakashima K, Ito Y, Yamaguchi-Shinozaki K. (2009). Transcriptional regulatory networks in response to abiotic stresses in *Arabidopsis* and grasses. *Plant Physiol.*, 149: 88–95.

Narayanan S, Sandy A, Shu D, Sprung M, Preissner C, Sullivan J. (2008). Design and performance of an ultrahigh-vacuum-compatible artificial channel-cut monochromator. *J. Synchrotron Rad.*, 15: 12–18.

Nasseri M, Arouiee K, Kafi M, Neamati H. (2012). Effect of silicon on growth and physiological parameters in fenugreek (*Trigonella foenum-graceum* L.) under salt stress. *Int. J Agric. Crop Sci.*, 21: 1554–1558.

Navari-Izzo F, Quartacci MF, Izzo R. (1990). Water-stress induced changes in protein and free amino acids in field-grown maize and sunflower. *Plant Physiol. Biochem.*, 28: 531–537.

Nikolaeva MK, Maevskaya SN, Shugaev AG, Bukhov NG. (2010). Effect of drought on chlorophyll content and antioxidant enzyme activities in leaves of three wheat cultivars varying in productivity. *Russ. J. Plant Physiol.*, 57(1): 57–87.

Nishizawa A, Yabuta Y, Shigeoka S. (2008). Galactinol and raffinose constitute a novel function to protect plants from oxidative damage. *Plant Physiol.*, 147: 1251–1263.

Nishizawa A, Yabuta Y, Yoshida E, Maruta T, Yoshimura K, Shigeoka S. (2006). *Arabidopsis* heat shock transcription factor A2 as a key regulator in response to several types of environmental stress. *Plant J.*, 48: 535–547.

Noctor G, Veljovic-Jovanovic S, Driscoll S, Novitskaya L, Foyer CH. (2002). Drought and oxidative load in wheat leaves: A predominant role for photorespiration? *Ann. Bot.*, 89: 841–850.

Nolla A, de Faria RJ, Korndörfer GH, da Silva TRB. (2012). Effect of silicon on drought tolerance of upland rice. *J. Food Agric. Environ.*, 10: 269–272.

Nonami H, Boyer JS. (1990). Primary events regulating stem growth at low water potentials. *Plant Physiol.*, 94: 1601–1609.

Noreen Z, Ashraf M, Akram NA. (2010). Salt-induced regulation of some key antioxidant enzymes and physiobiochemical phenomena in five diverse cultivars of turnip (*Brassicarapa* L.). *J. Agron. Crop Sci.*, 196: 273–285.

Nwugo CC, Huerta AJ. (2011). The effect of silicon on the leaf proteome of rice (*Oryza sativa* L.) plants under cadmium-stress. *J. Proteome Res.*, 10(2): 518–528.

Ogawa D, Yamaguchi K, Nishiuchi T. (2007). High-level overexpression of the *Arabidopsis HsfA2* gene confers not only increased themotolerance but also salt/osmotic stress tolerance and enhanced callus growth. *J. Exp. Bot.*, 58: 3373–3383.

Olsen AN, Ernst HA, Leggio LL, Skriver K. (2005). NAC transcription factors: Structurally distinct, functionally diverse. *Trends Plant Sci.*, 10(2): 79–87.

O'Toole JC, Moya TB. (1978). Genotypic variation in maintenance of leaf water potential in rice. *Crop Sci.*, 18: 873–876.

Palta J, Levitt J, Stadelmann EJ. (1977). Freezing injury in onion bulb cells. I. Evaluation of the conductivity method and analysis of ion and sugar efflux. *Plant Physiol.*, 60: 393–397.

Papageorgiou GC. (1975). Chlorophyll fluorescence: An intrinsic probe of photosynthesis. In Govindjee (ed.), *Bioenergetics of Photosynthesis*. Academic Press, New York, pp. 319–371.

Parent B, Hachez C, Redondo E, Simonneau T, Chaumont F, Tardieu F. (2009). Drought and abscisic acid effects on aquaporin content translate into changes in hydraulic conductivity and leaf growth rate: A transscale approach. *Plant Physiol.*, 149: 2000–2012.

Park JJ, Yi J, Yoon J, Cho LH, Ping J, Jeong HJ, Cho SK, Kim WT. (2011). OsPUB15, an E3 ubiquitin ligase, functions to reduce cellular oxidative stress during seedling establishment. *Plant J.*, 65: 194–205.

Pei ZF, Ming DF, Liu D, Wan GL. (2010). Silicon improves the tolerance to water-deficit stress induced by polyethylene glycol in wheat (*Triticum aestivum* L.) seedlings. *J. Plant Growth Regul.*, 29: 106–115.

Peltzer D, Dreyer E, Polle A. (2002). Differential temperature dependencies of antioxidative enzymes in two contrasting species: *Fagussylvatica* and *Coleusblumei*. *Plant Physiol. Biochem.*, 40: 141–150.

Pereira TS, Lobato AKS, Tan DKY, da Costa DV, Uchôa EB, Ferreira R do N, Pereira E dos S, Ávila FW, Marques DJ, Guedes EMS. (2013). Positive interference of silicon on water relations, nitrogen metabolism, and osmotic adjustment in two pepper (*Capsicum annuum*) cultivars under water deficit. *Aust. J. Crop Sci.*, 7(8): 1064–1071.

Perry CC, Williams RJP, Fry SC. (1987). Cell wall biosynthesis during silicification of grass hairs. *J. Plant Physiol.*, 126: 437–448.

Prado RM, Fernandes FM, Natale W. (2001). Uso agrícola da escória de siderurgia no Brasil: Estudos na cultura da cana-de-açúcar. FUNEP, Jaboticabal.

Premachandra GS, Saneoka H, Ogata S. (1989) Nutrio-physiological evaluation of polyethylene glycol test of cell membrane stability in maize. *Crop Sci.* 29: 1287–1292.

Qayyum A, Razzaq A, Ahmad M, Jenks MA. (2011). Water stress causes differential effects on germination indices, total soluble sugar and proline content in wheat (*Triticum aestivum* L.) genotypes. *Afr. J. Biotechnol.*, 10: 14038–14045.

Rahman MS, Miyake H, Takeoka Y. (2002). Effects of exogenous glycinebetaine on growth and ultrastructure of salt-stressed rice seedlings (*Oryza sativa* L.). *Plant Prod. Sci.*, 5: 33–44.

Ranganathan S, Suvarchala V, Rajesh YBRD, Prasad MS, Padmakumari AP, Voleti SR. (2006). Effects of silicon sources on its deposition, chlorophyll content, and disease and pest resistance in rice. *Biol. Plant.*, 50: 713–716.

Rathinasabapathi B, Gage DA, Mackill DJ, Hanson AD. (1993). Cultivated and wild rices do not accumulate glycine betaine due to deficiencies in two biosynthetic steps. *Crop Sci.*, 33: 534–538.

Rauf S, Sadaqat HA. (2008). Effect of osmotic adjustment on root length and dry matter partitioning in sunflower (*Helianthus annuus* L.) under drought stress. *Acta Agric. Scand. B Soil Plant Sci.*, 58: 252–260.

Raza SH, Athar HR, Ashraf M, Hameed A. (2007). Glycinebetaine-induced modulation of antioxidant enzymes activities and ion accumulation in two wheat cultivars differing in salt tolerance. *Environ. Exp. Bot.*, 60(3): 368–376.

Reddy AR, Chaitanya KV, Jutur PP, Sumithra K. (2004). Differential antioxidative responses to water stress among five mulberry (*Morusalba* L.) cultivars. *Environ. Exp. Bot.*, 52: 33–42.

Reynolds M.P., Oritz-Monasterio J.I., Mc Nab A. (2001) Application of physiology in wheat breeding, CIMMYT, Mexico.

Rhodes D, Verslues PE, Sharp RE. (1999). Role of amino acids in abiotic stress resistance. In Singh BK (ed.), *Plant Amino Acids: Biochemistry and Biotechnology*. Marcel Dekker, New York, pp. 319–356.

Ritchie S, Gilroy ST. (1998). Gibberellins: Regulating genes and germination. *New Phytol.*, 140: 363–383.

Rodrigues FA, Verslues PE, Sharp RE. (1999). Role of amino acids in abiotic stress resistance. In Singh BK (ed.), *Plant Amino Acids: Biochemistry and Biotechnology*. Marcel Dekker, New York, pp. 319–356.

Romero-Arnada MR, Jourado O, Cuartero J. (2006). Silicon alleviates the deleterious salt effects on tomato plant growth by improving plant water status. *J. Plant Physiol.*, 163: 847–855.

Sacala E. (2009). Role of silicon in plant resistance to water stress. *J. Elementol.*, 14: 619–630.

Saelim S, Zwiazek JJ. (2000). Preservation of thermal stability of cell membranes and gas exchange in high temperature-acclimated *Xylia xylocarpa* seedlings. *J. Plant Physiol.*, 156: 380–385.

Saibo NJM, Lourenço T, Oliveira MM. (2009). Transcription factors and regulation of photosynthetic and related metabolism under environmental stresses. *Ann. Bot.*, 103: 609–623.

Sakuma Y, Maruyama K, Qin F, Osakabe Y, Shinozaki K, Yamaguchi-Shinozaki K. (2006). Dual function of an *Arabidopsis* transcription factor DREB2A in water-stress-responsive and heat-stress-responsive gene expression. *Proc. Natl. Acad. Sci. U.S.A.*, 103: 18822–18827.

Saruhan N, Terzi R, Saglam A, Nar H, Kadioglu A. (2010). Scavenging of reactive oxygen species in apoplastic and symplastic areas of rolled leaves in *Ctenanthe setosa* under drought stress. *Acta Biol. Hung.*, 61(3): 282–298.

Saud S, Li X, Chen Y, Zhang L, Fahad S, Hussain S, Sadiq A, Chen Y. (2014). Silicon application increases drought tolerance of Kentucky bluegrass by improving plant water relations and morphophysiological functions. *Sci. World J.*, 2014: 1–10.

Savant NK, Korndörfer GH, Datnoff LE, Snyder GH. (1999). Silicon nutrition and sugarcane production: A review. *J. Plant Nutr.*, 22: 1853–1903.

Savant NK, Snyder GH, Datnoff LE. (1997). Silicon management and sustainable rice production. *Adv. Agron.*, 58: 151–199.

Savvas D, Manos G, Kotsiras A, Souvaliotis S. (2002). Effects of silicon and nutrient-induced salinity on yield, flower quality and nutrient uptake of gerbera grown in a closed hydroponic system. *J. Appl. Bot.*, 76: 153–158.

Sayed SA, Gadallah MAA. (2014). Effect of silicon on *Zea mays* exposed to water deficit and oxygen deficiency. *Russ. J. Plant Physiol.*, 61(4): 460–466.

Schiermeier Q. (2008). A long dry summer. *Nature*, 452: 270–273.

Schweiger J, Lang M, Lichtenthaler HK. (1996). Differences in fluorescence excitation spectra of leaves between stressed and non-stressed plants. *J. Plant Physiol.*, 148: 536–547.

Serraj R, Sinclair TR. (2002). Osmolyte accumulation: Can it really help increase crop yield under drought conditions? *Plant Cell Environ.*, 25: 333–341.

Serraj R, Vasquez-Diaz H, Drevon JJ. (1998). Effects of salt stress on nitrogen fixation, oxygen diffusion, and ion distribution in soybean, common bean, and alfalfa. *J. Plant Nutr.*, 21: 475–488.

Sharp RE, Lenoble ME. (2002). ABA, ethylene and the control of shoot and root growth under water stress. *J. Exp. Bot.*, 53(366): 33–37.

Sheard LB, Zheng N. (2009). Plant biology: Signal advance for abscisic acid. *Nature*, 462: 575–576.

Shehab GG, Ahmed OK, El-Beltagi HS. (2010). Effects of various chemical agents for alleviation of drought stress in rice plants (*Oryzasativa* L.). *Not. Bot. Hortic. Agrobot. Cluj.*, 38: 139–148.

Shelp BL, Bown AW, McLean MD. (1999). Metabolism and functions of gamma-aminobutyric acid. *Trends Plant Sci.*, 11: 446–452.

Shen X, Zhou Y, Duan L, Li Z, Eneji AE, Li J. (2010). Silicon effects on photosynthesis and antioxidant parameters of soybean seedlings under drought and ultraviolet-B radiation. *J. Plant Physiol.*, 167: 1248–1252.

Shcherbakova A. and Kacperska A. (1983) Water stress injuries and tolerance as related to potassium efflux from winter rape hypocotyls. *Physiol. Plant.* 57: 296–300.

Shinozaki K, Yamaguchi-Shinozaki K. (2007). Gene networks involved in drought stress response and tolerance. *J. Exp. Bot.*, 58(2): 221–227.

Shinozaki K, Yamaguchi-Shinozaki K, Mizoguchi T et al. (1998). Molecular responses to water stress in *Arabidopsis thaliana*. *J. Plant Res.*, 111: 345–351.

Shinozaki K, Yamaguchi-Shinozaki K, Seki M. (2003). Regulatory network of gene expression in the drought and cold stress responses. *Curr. Opin. Plant Biol.*, 6: 410–417.

Shirasawa K, Takabe T, Takabe T, Kishitani K. (2006). Accumulation of glycinebetaine in rice plants that overexpress choline monooxygenase from spinach and evaluation of their tolerance to abiotic stress. *Ann. Bot.*, 98: 565–571.

Simon-Sarkadi L, Kocsy G, Várhegyi A, Galiba G, De Ronde JA. (2006). Effect of drought stress at supra-optimal temperature on polyamine concentrations in transgenic soybean with increased proline levels. *Z. Naturforsch.*, 61c: 833D839.

Singh SK, Reddy KR. (2011). Regulation of photosynthesis, fluorescence, stomatal conductance, and water-use efficiency of cowpea under drought. *J. Photochem. Photobiol.*, 105: 40–50.

Smirnoff N. (1993). The role of active oxygen in the response of plants to water deficit and desiccation. *New Phytol.*, 125: 27–58.

Snyder GH, Matichenkov VV. (2007). Silicon. In Barker AV, Pilbeam DJ (eds.), *Handbook of Plant Nutrition*. Taylor & Francis Group, CRC Press, Boca Raton, FL, pp. 551–568.

Song A, Li P, Fan F, Li Z, Liang Y. (2014). The effect of silicon on photosynthesis and expression of its relevant genes in rice (*Oryza sativa* L.) under high-zinc stress. *PLoS One*, 9(11): e113782.

Song SY, Chen Y, Chen J, Dai XY, Zhang WH. (2011). Physiological mechanisms underlying OsNAC5-dependent tolerance of rice plants to abiotic stress. *Planta*, 234: 331–345.

Sonobe K, Hattorri T, An P, Tsuji W, Eneji AE, Kobayashi S, Kawamura Y, Tanaka K, Inanaga S. (2010). Effect of silicon application on sorghum root responses to water stress. *J. Plant Nutr.*, 34: 71–82.

Spollen WG, Sharp RE, Saab IN, Wu Y. (1993). Regulation of cell expansion in roots and shoots at low water potentials. In Smith JAC, Griffiths H (eds.), *Water Deficits: Plant Responses from Cell to Community*. BIOS Scientific Publishers, Oxford, pp. 37–52.

Sreenivasulu N, Grinm B, Wobus U, Weschke W. (2000). Differential response of antioxidant compounds to salinity stress in salt-tolerant and salt-sensitive seedlings of foxtail millet (*Setaria italica*). *Physiol. Planta*, 109: 435–442.

Stevanovic B, Sinzar J, Glisic O. (1997). Electrolyte leakage differences between poikilohydrous and homoio-hydrous species of Gesneriaceae. *Biol. Plant.*, 40: 299–303.

Subbarao GV, Chauhan YS, Johansen C. (2000). Patterns of osmotic adjustment in pigeonpea—Its importance as a mechanism of drought resistance. *Eur. J. Agron.*, 12: 239–249.

Svensson J, Ismail AM, Palva ET, Close TJ. (2002). Dehydrins. In Storey KB, Storey JM (eds.), *Sensing, Signalling and Cell Adaptation*. Elsevier Science, Amsterdam, pp. 155–71.

Szabados L, Savoure A. (2010). Proline: A multifunctional amino acid. *Trends Plant Sci.*, 15: 89–97.

Taiz L, Zeiger E. (2002). *Plant Physiology*. 3rd ed. Sinauer Associates, Sunderland, MA.

Taiz L, Zeiger E. (2006). *Plant Physiology*. 4th ed. Sinauer Associates, Sunderland, MA.

Takahashi E. (1995). Uptake mode and physiological functions of silica. In Matusuo T, Kumazawa K, Ishii R, Ishihara K, Hirata H (eds.), *Science of Rice Plant Physiology*. Vol. 2. Nobunkyo, Tokyo, pp. 420–433.

Takahashi E, Ma JF, Miyake Y. (1990). The possibility of silicon as an essential element for higher plants. *Comments Agric. Food Chem.*, 2: 99–112.

Takasaki H, Maruyama K, Kidokoro S, Ito Y, Fujita Y, Shinozaki K. (2010). The abiotic stress-responsive NAC-type transcription factor OsNAC5 regulates stress-inducible genes and stress tolerance in rice. *Mol. Genet. Genom.*, 284: 173–183.

Terzi R, Saruhan N, Saglam A, Nar H, Kadioglu A. (2009). Photosystem II functionality and antioxidant system changes during leaf rolling in post-stress emerging *Ctenanthe setosa* exposed to drought. *Acta. Biol. Hung.*, 60: 417–431.

Theisen AF. (1988). Fluorescence changes of a drying maple leaf observed in the visible and near-infrared. In Lichtenthaler HK (ed.), *Applications of Chlorophyll Fluorescence in Photosynthesis Research, Stress Physiology, Hydrobiology and Remote Sensing*. Kluwer Academic, Dordrecht, pp. 197–201.

Thomashow MF. (2010). Molecular basis of plant cold acclimation: Insights gained from studying the CBF cold response pathway. *Plant Physiol.*, 154: 571–577.

Trovato M, Matioli R, Costantino P. (2008). Multiple roles of proline in plant stress tolerance and development. *Rendiconti Lincei*, 19: 325–346.

Tunnacliffe A, Wise MJ. (2007). The continuing conundrum of the LEA proteins. *Naturwissenschaften*, 94: 791–812.

Tyerman SD, Niemietz CM, Bramley H. (2002). Plant aquaporins: Multifunctional water and solute channels with expanding roles. *Plant Cell Environ.*, 25: 173–194.

Umezawa T, Fujita M, Fujita Y, Yamaguchi-Shinozaki K, Shinozaki K. (2006). Engineering drought tolerance in plants: Discovering and tailoring genes unlock the future. *Curr. Opin. Biotechnol.*, 17: 113–122.

Valentovic P, Luxova M, Kolarovic L, Gasparikova O. (2006). Effect of osmotic stress on compatible solutes content, membrane stability and water relations in maize cultivars. *Plant Soil Environ.*, 52: 186–191.

Van Bockhaven J. (2014). Silicon-induced resistance in rice (*Oryza sativa* L.) against the brown spot pathogen *Cochliobolus miyabeanus*. PhD thesis, Ghent University, Belgium.

Van Bockhaven J, de Vleesschauwer D, Hofte M. (2012). Towards establishing broad-spectrum disease resistance in plants: Silicon leads the way. *J. Exp. Bot.*, 64: 1281–1293.

Vandeleur RK, Mayo G, Shelden MC, Gilliham M, Kaiser BN, Tyerman SD. (2009). The role of plasma membrane intrinsic protein aquaporins in water transport through roots: Diurnal and drought stress responses reveal different strategies between isohydric and anisohydric cultivars of grapevine. *Plant Physiol.*, 149: 445–460.

Venkatesan A, Chellappan KP. (1998). Accumulation of proline and glycine betaine in *Ipomoea pes-caprae* induced by NaCl. *Biol. Plant.*, 41: 271–276.

Villadsen D, Rung JH, Nielsen TH. (2005). Osmotic stress changes carbohydrate partitioning and fructose-2,6-bisphosphate metabolism in barley leaves. *Funct. Plant Biol.*, 32: 1033–1043.

von Koskull-Döring P, Scharf KD, Nover L. (2007). The diversity of plant heat stress transcription factors. *Trends Plant Sci.*, 12(10): 452–457.

Voogt W. Sonneveld C. (2001). Silicon in horticultural crops in soilless culture. In Datnoff LE, Snyder GH, Korndörfer GH (eds.), *Silicon in Agriculture*. Elsevier, New York, pp. 115–131.

Wahid A, Gelani S, Ashraf M, Foolad MR. (2007). Heat tolerance in plants: An overview. *Environ. Exp. Bot.*, 61 (3): 199–223.

Wang XQ, Ullah H, Jones AM, Assmann SM. (2001). G protein regulation of ion channels and abscisic acid signaling in *Arabidopsis* guard cells. *Science*, 292: 2070–2072.

Wang Z, Stutte GW. (1992). The role of carbohydrates in active osmotic adjustment in apple under water stress. *J. Am. Soc. Hortic. Sci.*, 117: 816–823.

Wani SH, Singh NB, Haribhushan A, Mir JI. (2013). Compatible solute engineering in plants for abiotic stress tolerance—Role of glycine betaine. *Curr. Genom.*, 14: 157–165.

Waraich EA, Ahmad R, Ashraf MY, Saifullah Ahmad M. (2011). Improving agricultural water use efficiency by nutrient management in crop plants. *Acta Agric. Scand. B Plant Soil Sci.*, 61(4): 291–304.

Waterkey L, Bientait A, Peeters A. (1982). Callose et silice epidermiques rapports avec la transpiration cuticulaire. *La Cellule*, 73: 263–287.

Wen FP, Zhang ZH, Bai T, Xu Q, Pan YH. (2010). Proteomics reveals the effects of gibberellic acid (GA3) on salt-stressed rice (*Oryza sativa* L.) shoots. *Plant Sci.*, 178: 170–175.

White DA, Turner NC, Galbraith JH. (2000). Leaf water relations and stomatal behavior of four allopatric *Eucalyptus* species planted in Mediterranean south Western Australia. *Tree Physiol.*, 20: 1157–1165.

Xie Z, Song F, Xu H, Shao H, Song R. (2014). Effects of silicon on photosynthetic characteristics of maize (*Zea mays* L.) on alluvial soil. *Sci. World J.*, 2014: 718716. http://dx.doi.org/10.1155/2014/718716.

Xue T, Hartikainen H, Piironen V. (2001). Antioxidative and growth-promoting effect of selenium in senescing lettuce. *Plant Soil*, 237: 55–61.

Yamaguchi-Shinozaki K, Shinozaki K. (1994). A novel *cis*-acting element in an *Arabidopsis* gene is involved in responsiveness to drought, low-temperature, or high-salt stress. *Plant Cell*, 6(2): 251–264.

Yang WJ, Rich PJ, Axtell JD, Wood KV, Bonham CC, Ejeta G, Mickelbart MV, Rhodes D. (2003). Genotypic variation for glycine betaine in sorghum. *Crop Sci.*, 43: 162–169.

Yeo AR, Flowers SA, Rao G, Welfare K, Senanayake N, Flowers TJ. (1999). Silicon reduces sodium uptake in rice (*Oryza sativa* L.) in saline conditions and this is accounted for by a reduction in the transpirational bypass flow. *Plant Cell Environ.*, 22: 559–565.

Yoshida S, Ohnishi Y, Kitagishi K. (1962). Chemical forms, mobility and deposition of silicon in rice plant. *Soil Sci. Plant Nutr.*, 8: 15–21.

Zhang T, Liu Y, Yang T, Zhang L, Xu S, Xue L, An L. (2006). Diverse signals converge at MAPK cascades in plant. *Plant Physiol. Biochem.*, 44: 274–283.

Zhang W, Tian Z, Pan X, Zhao X, Wang F. (2013). Oxidative stress and non-enzymatic antioxidants in leaves of three edible canna cultivars under drought stress. *Hortic. Environ. Biotechnol.*, 54: 1–8.

Zhu J, Verslues PE, Zheng X et al. (2005). HOS10 encodes an R2R3-type MYB transcription factor essential for cold acclimation in plants. *Proc. Natl. Acad. Sci. U.S.A.*, 102: 9966–9971.

Zhu Z, Wei G, Li J, Qian Q, Yu J. (2004). Silicon alleviates salt stress and increases antioxidant enzymes activity in leaves of salt-stressed cucumber (*Cucumissativus* L.). *Plant Sci.*, 167: 527–533.

16 Silicon Nutrition and Crop Improvement
Recent Advances and Future Perspective

Arkadiusz Artyszak and Katarzyna Kucińska

CONTENTS

ABSTRACT

Nowadays, there are new challenges for agriculture, like safety, sustainability, and environmental care, all while increasing the yield and quality of crops at the same time. Therefore, new, more effective solutions are necessary. The role and beneficial effects of silicon (Si) in plant biology start to play an important role. It has a favorable effect on the growth and development of many plant species, like cereals, root crops, vegetables, and fruits. Silicon plays a very important role in the mitigation of plants' vulnerability to biotic and abiotic environmental stresses and increases plants' resistance to pathogens and pests. One of the most important beneficial effects of silicon on plant growth is increased resistance to water stress conditions like drought or salt stress. Silicon improves the uptake of some important nutrients, like potassium, calcium, and some important microelements. It also protects or mitigates the effects of the contents of some other microelements, like heavy metals, that are too high. What is very important is that it is safe to use for the environment as well. Some fertilizers with silicon are authorized for use in organic farming; that is, silicon application is a novel idea of fertilization. Due to all its beneficial effects, it seems that silicon should be included as one of the important nutrients in modern crops' fertilization in the near future.

16.1   INTRODUCTION

Fertilizers with macronutrients like N, P, K, Mg, Ca, and S, as well as micronutrients like B, Cu, Cl, Fe, Mo, Mn, Ni, and Zn, are commonly used in crop fertilization. However, there is another group of plant nutrients, the "beneficial mineral elements." Mineral nutrients that are not very important or are essential only for certain plant species, or under specific conditions, are defined as beneficial elements (Marschner, 2002). They are cobalt, sodium, and silicon. We know that the majority of the crops produced worldwide are silicon accumulators.

Silicon fertilizers are quite commonly used in orchard and vegetable crops. Moreover, interest in the use of such fertilizers in field crops has been growing recently in Europe. Additionally, they are also becoming more and more popular in sustainable crop production technologies, such as the rapidly developing organic agriculture system. In this chapter, the authors detail the scientific field and laboratory research on the use of different soils and foliar silicon fertilizers worldwide during the last decade (mostly since 2004). The main focus of the presented results concerns yields and some key quality parameters of crops.

16.2   FIELD CROPS

16.2.1   Root Crops

16.2.1.1   Sugar Beet

Sugar beet (*Beta vulgaris* L.) is the only source of plant raw material for sucrose production in Europe. Currently, the process to maximize the yield of technological sugar is the main challenge in sugar beet cultivation. The technological sugar is the biological sugar yield minus the content of molassigenic components like amino-alpha-nitrogen, potassium, and sodium ions. The most important and decisive role for biological sugar is increasing the roots' yield. The sugar (sucrose) content in roots is considerably less important. Therefore, in scientific research on sugar beet production, increasing the roots' yield is most important. For this reason, in 2010–2012, Artyszak et. al. (2016) conducted scientific field research on the effect of Herbagreen Basic fertilizer on the yield and technological quality parameters of roots in sugar beet (Britannia variety) cultivation. Herbagreen Basic (also know as Megagreen) is one of the foliar environmentally friendly fertilizers, and it is recommended for improving the yield of sugar beets *inter alia*. This is a natural product, as it is marine-micronized calcite (Figure 16.1). For this reason, it is safe for the environment and even approved for use in organic agriculture practices.

According to information from the producer, the activity of this fertilizer is based on the role of calcium carbonate ($CaCO_3$), which is decomposed into calcium oxide (CaO) and carbon dioxide (CO_2) in the leaves' stomata, and this carbon dioxide increases the intensity of photosynthesis. However, Herbagreen Basic contains a significant amount of silicon. The content of Herbagreen Basic fertilizer is as follows: Ca, 262 g/kg; Si, 80 g/kg; Fe, 23.8 g/kg; Mg, 14.5 g/kg; K, 4.2 g/kg; Na, 3.7 g/kg; Ti, 3.0 g/kg; P, 2.2 g/kg; S, 1.6 g/kg; and Mn, 0.8 g/kg. There are also trace amounts of boron, cobalt, cuprum, and zinc.

FIGURE 16.1 Herbagreen Basic—containing mainly Ca and Si. (Courtesy of A. Artyszak.)

In the mentioned field experiment, two variants of foliar fertilization were used. Each variant contained two terms of application, as follows:

1. First variant: First application of 262 g/ha Ca and 80 g/ha Si in four to six leaves' stage of sugar beet, and 21 days later, second application of 524 g/ha Ca and 160 g/ha Si.
2. Second variant: First application of 524 g/ha Ca and 160 g/ha Si in four to six leaves' stage of sugar beet, and 21 days later, second application of 524 g/ha Ca and 160 g/ha Si.

In total, the amounts of calcium and silicon in the first variant were 786 and 240 g/ha respectively, and in the second variant, 1048 and 320 g/ha, respectively. The effects of the applied fertilizer were compared with the control (variant 0, without foliar fertilization). In every spraying, 250 dm^3/ha water was used.

Averaged over three years of research, foliar fertilization with silicon and calcite resulted in significant increase in the yield of roots and leaves, and the biological and technological sugar yield, in comparison with the control variant. A 21.7%–21.8% increase in root yield and 21.7%–35.8% increase in leaf yield, compared with the variant, were observed. Similarly, a 23.9%–24.8% increase in biological sugar yield and a 24.3%–25.2% increase in technological sugar yield were obtained. What is interesting is that the foliar fertilizer dose had a significant effect on the leaf yield only; it had no effect on the root yield increase. Additionally, what is extremely important is that foliar fertilization with silicon and calcite effected a significant root yield increase, but in spite of this, it did not reduce the technological quality of roots.

In 2011–2012, the same authors (Artyszak et al., 2014) conducted a similar field experiment on the effect of silicon and calcium foliar fertilization on sugar beet. They used another variety, named Danuśka KWS. Calcium and silicon foliar fertilization were applied in the same two variants, as in the case of the Britannia variety, and resulted in increases of (1) the roots' yield (average for both treatments about 13.1%), (2) the leaves' yield (about 21.0%), (3) the biological sugar yield (about 15.5%), and (4) the technological yield of sugar (about 17.7%), compared with the control treatment. Additionally, at the same time, some beneficial effects on the roots' technological quality were found. There was a significant reduction of the alpha-amino-nitrogen content, and the content of potassium and sodium ions also tended to decrease. Moreover, in both experiments (Artyszak et al., 2014, 2016), foliar silicon and calcium application resulted in a few days' delay of infection by cercospora blight (*Cercospora beticola* Sacc.). It is the most dangerous fungal pathogen of sugar beet in Poland.

Differentiated silicon and calcium foliar fertilization contained in the Herbagreen Basic fertilizer and silicon fertilization contained in the Optysil fertilizer were factors for Artyszak et al. (2015) in another experiment (Figure 16.2).

Both fertilizers were applied one, two, or three times during the vegetation period. A single dose for Herbagreen Basic was 1.5 kg/ha, and for Optysil it was 0.5 dm³/ha. The term of the first application was four to six sugar beet leaves in the growth stage. The second application was applied one week later, and the third two weeks after the first application. In every spraying, 250 dm³/ha water was used. The concentration of Herbagreen Basic was 0.6%, and of Optysil, 0.2%. The content of Optysil is 94.1 g Si and 25 g Fe per 1 dm³. Similar to Herbagreen Basic, Optysil is approved for use in organic farming in Poland and can be recommended for use in the fertilization of organic sugar beet. In comparison with the control variant, foliar fertilization with calcium and silicon resulted in a 12.6% increase of root yield, on average, and varied from 10.4% to 16.2%, relatively, of the applied dose. At the same time, foliar fertilization with silicon (without calcium) resulted in a 14.5% increase of root yield, on average, and varied from 13.7% to 15.9%, relatively, of the applied dose (Figure 16.3). The leaves' yield was similar in every variant (36.2–41.7 t/ha) (Figure 16.4).

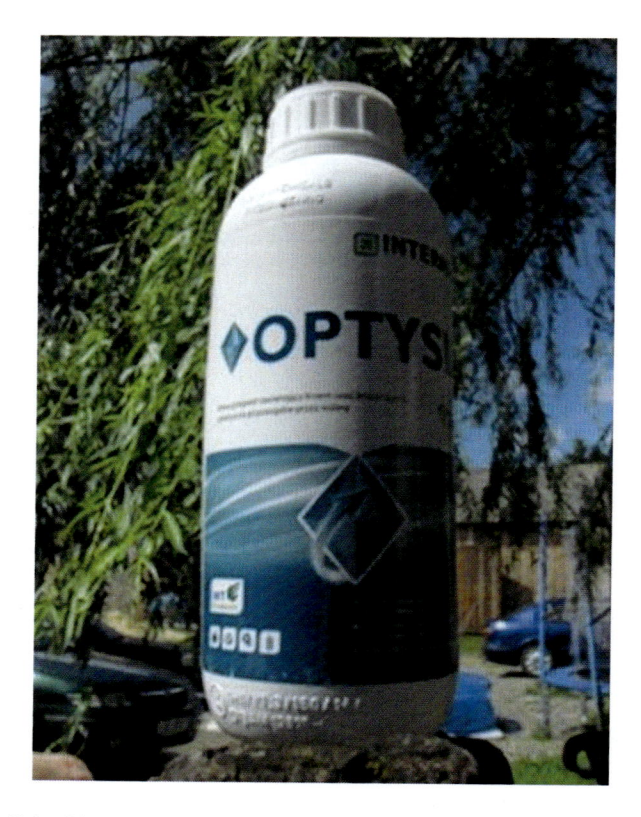

FIGURE 16.2 Optysil fertilizer contains 94.1 g Si and 24 g Fe per 1 dm³. (Courtesy of A. Artyszak.)

FIGURE 16.3 Sugar beet foliar fertilization with silicon (July 2013). (Courtesy of A. Artyszak.)

FIGURE 16.4 Sugar beet harvest from field research on sugar beet foliar fertilization with silicon and calcite and with silicon only (October 2014). (Courtesy of A. Artyszak.)

Foliar fertilization in both silicon with and silicon without calcium resulted in an increase of the biological and technological sugar yields. In the case of the biological sugar yield, an increase of 14.3% was obtained, on average, for silicon with calcium fertilization and was in the range of 11.4%–18.1%, relatively, of the applied variant. Silicon without calcium resulted in a 14.5% increase and was in the range of 12.7%–15.7%. Similarly, in the case of the technological sugar yield, an increase of 13.5% was obtained, on average, for silicon with calcium fertilization and was in the range of 12.2%–17.7%, relatively, of the applied variant. Silicon without calcium resulted in a 14.0% increase and was in the range of 12.2%–15.6%.

The technological root quality was estimated by such traits as sucrose, alpha-amino-nitrogen, potassium, and refined sugar content, which were not modified significantly by foliar fertilization variants in comparison with the control. However, a tendency for a sodium content decrease was observed.

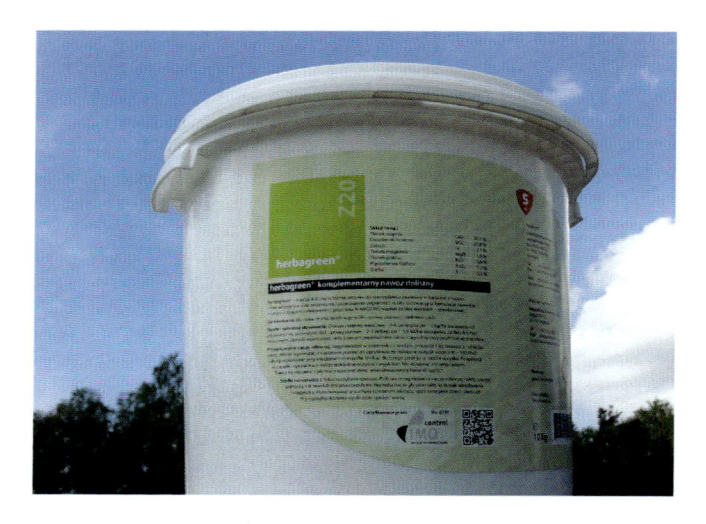

FIGURE 16.5   Fertilizers containing Ca and Si are constantly used in Europe. (Courtesy of A. Artyszak.)

TABLE 16.1

Effect of Foliar Fertilization with Silicon and Calcite on Yield and Technological Quality of Sugar Beet Roots (Jagusia Variety), Average Results from Two Locations (2014)

Parameter of Yield	Control	Ca + Si[a]	LSD$_{0.05}$
Roots yield, t/ha	66	66	ns
Biological sugar yield, t/ha	11.2	11.5	ns
Technological sugar yield, t/ha	9.60	10.07	ns
Sucrose content, %	16.9	17.4	0.4*
Content of amino-alpha-nitrogen, μM/kg of root pulp	25.8	20.6	3.1*
Content of potassium, μM/kg of root pulp	50.7	44.7	5.1*
Content of sodium, μM/kg of root pulp	4.45	3.75	ns

Source: Artyszak A, Study on the effect of Herbagreen Z20 application on yielding, and technological quality of sugar beet roots, 2014, unpublished.

[a]   4 × 220 g/ha Ca and 130 g/ha Si.

In 2014, Artyszak (unpublished) conducted field research with a new type of Herbagreen fertilizer, named Herbagreen Z20, in two locations in Poland. Herbagreen Z20's content is as follows: Ca, 220 g/kg; Si, 130 g/kg; Fe, 21 g/kg; Mg, 9.6 g/kg; K, 5.0 g/kg; P, 0.9 g/kg; and S, 1 g/kg (Figure 16.5).

This fertilizer was used in four applications: the first application was at four to six leaves of the sugar beet growth stage, and the next three treatments occurred every seven days. Each dose per one application of fertilizer was 2 kg/ha. A significant improvement of the technological root quality parameters, like an increase in the sucrose content and a reduction of the molassigenic components content, was observed (Table 16.1). However, there was no effect on root yield. Therefore, differences in biological and technological yield of sugar were not significant.

16.2.1.2   Potato

Nowadays, the quality of potato tubers is becoming more and more important in potato (*Solanum tuberosum* L.) cultivation in Europe. In 2009–2011, Wróbel (2012) conducted research on the productive and quality effects of foliar fertilization on potatoes (Jelly variety). He applied YaraVita Potato and Actisil fertilizers. The content of the YaraVita fertilizer is as follows: P, 192 g/L; K, 62 g/L; MgO, 40 g/L; Mn, 10 g/L; and Zn, 5 g/L. Actisil contains silicon only at a concentration of 0.6%. The author did not observe any effects of either fertilizers on tuber yield. However, he found a significant effect from Actisil's application, an approximately 50% reduction of the smallest tuber amount (diameter less than 30 mm) in the total yield, as well as a significant (23%) increase of the largest tubers' share (more than 60 mm). In variant where Actisil was applied together with YaraVita Potato, he obtained only a 10% increase.

During the growing season, any significant differences in the rate of late blight of tomato (*Phytophtora infestans*) spread between plants with foliar fertilization and the control variant were not observed. Only in 2011, which was very favorable for the development of this disease, was there a slightly slower potato blight spread among foliar fertilized plants than among the control variant plant. In particular, Actisil foliar application conducted with standard chemical protection significantly inhibited the development of potato late blight on leaves during the growing season. Evaluation of tuber disease infestation after harvest was not clear in each year of the study due to significant differences in the severity of particular diseases' infection pressure during the vegetation period. Therefore, there is a lack of scientific information on the effect of potato foliar silicon fertilization on the spread of *P. infestans*. Such research with silicon would be very useful for organic agriculture to decrease tomato late blight infective pressure. Use of Actisil fertilizer significantly decreased the amount of tubers with symptoms of wet rot (*Erwinia carotovora* var. *carotovora*), while it increased the number of tubers with sclerotia (spores of *Rhizoctonia solani*). No significant effects of applied silicon foliar fertilization on tuber physiological disorders, like hollow heart and rust internal spot, and tuber diseases, like potato common scab (*Streptomyces scabies*) and potato dry rot (*Fusarium* spp.), were noticed. In 2011, another study was conducted on the effect of YaraVita Potato and Actisil foliar fertilizers on potato tubers' quality parameters, like darkening of raw tubers and the contents of phosphorus, potassium, magnesium, and calcium. No significant differences were observed. However, applied fertilizer resulted in a significant reduction of harmful nitrate accumulation in tubers. There was an almost 60% reduction of nitrate content in the fertilization variant where Actisil was applied together with YaraVita Potato fertilizer.

Silicon fertilization had no significant effect on natural losses in potato tubers after six months of storage. In 2009 and 2011, applied silicon fertilizers effected a trend of the reduction of tuber diseases like wet rot (*Erwinia carotovora* var. *carotovora)* and dry rot (*Fusarium* spp.) infestation. The differentiation observed by authors was not statistically significant. The smallest bulbs' infestation of wet rot was found for the variant where Actisil only was used. After storage in the respective year, 0.2%–0.5% of the tuber mass was infected by wet rot only, while 1.6%–5.7% of bulbs from the control variant, without foliar fertilization, were infected.

In Crusciol et al.'s (2009) research on potato cultivation under greenhouse conditions, silicon foliar fertilization effected a higher tuber yield from each plant. This beneficial effect was stronger under no drought stress conditions.

In 2011–2012, Trawczyński (2013) conducted field research on the effects of Herbagreen Basic fertilizer use (the content of this fertilizer is indicated in Section 16.2.1.1) on tuber yield and quality parameters of the Finezja variety potato. This variety matures semiearly and is recommended for the production of French fries. For foliar applications, Herbagreen Basic with differentiated ammonium nitrate doses (50, 75, and 100 kg N/ha) was used. Herbagreen Basic was applied three times during the growing season:

1. First application—at the beginning of June, but no later than June the 10th
2. Second application—two weeks after the first application
3. Third application—in the relative year, two to three weeks after the second application, at the end of flowering time

In each application, fertilizer was used at 2.5 kg/ha (655 g CaO and 200 g Si) with 500 L/ha water. In 2011, weather (water and thermal) conditions were unfavorable for potato cultivation. In May and June, the beginning of potato vegetation in this part of Europe, it was dry and moderately warm. Then in July, during the crop accumulation time, there was heavy rainfall and low temperatures, followed by a drought in August and September. Under such unfavorable weather conditions, the foliar use of Herbagreen Basic fertilizer resulted in a significantly higher growth and tuber yield than in 2012, when these stressors had limited impact. In 2011, the author noticed that Herbagreen foliar application effected an average 5.6 t/ha (14.8%) higher total tuber yields compared with the control variant. Similarly, use of Herbagreen resulted in a 5.1 t/ha (16.4%) average increase of commercial yield of tubers compared with the control. However, in 2012 the beneficial effect of such foliar fertilization with silicon and calcium on potato tuber yield was significantly weaker. Total and commercial tuber yields were 3.7 t/ha (5.5%) and 4.1 t/ha (7.7%), respectively, higher than in the variant without foliar fertilization. On average, in 2011–2012 foliar fertilization with Herbagreen resulted in an approximately 5 t/ha higher yield of potato tubers: 9% and 11% of total and commercial tuber yields, respectively. Therefore, based on the obtained results, the author suggested the possibility of soil fertilization with mineral nitrogen reduction. The beneficial effect of nitrogen use at a dose of 75 kg/ha and foliar fertilization with Herbagreen on tuber yield was similar to that from nitrogen use at a dose of 100 kg/ha, but without foliar fertilization. Consequently, under good weather conditions, a 25% reduction of soil nitrogen dose was possible, but under unfavorable weather conditions, even 50% was obtained.

Foliar fertilization with silicon and calcium (Herbagreen Basic) had a beneficial effect on tuber size in the tuber yield structure as well. The shares of tubers with diameters of 50–60 mm and above 60 mm (big tubers) were approximately 2% and 4% higher than those of the control variant.

Additionally, foliar application of silicon with calcium resulted in a lower share of total external defects in the tubers' yield, like deformation, greening, common scab (*Streptomyces scabies*), and infestation. It was mainly the effect of smaller greened tubers' weight percentage in yield. Foliar fertilization had no effect on the contents of starch, nitrates, and dry matter in tubers.

16.2.2 Oil Crops

16.2.2.1 Oilseed Rape

In the central and north part of Europe, winter oilseed rape (*Brasssica napus* L.) is the main oil crop. Yields of rapeseed are mainly determined by overwintering. In the 2010–2011 and 2011–2012 growing seasons, Artyszak (2012, unpublished) conducted field research on the effect of differentiated foliar fertilization with silicon and calcium (Herbagreen Basic fertilizer—chemical content is indicated in Section 16.2.1.1) on rape's overwintering and yield. During fall season, at the growing stage of four to six rosette leaves, foliar fertilization with 393 g/ha Ca and 120 g/ha Si improved the winter hardiness of the plants and effected a significant reduction of plant winter losses compared with the variant without foliar fertilization applied before the end of the autumnal vegetation time (Table 16.2). Among the plants from variants with autumnal silicon and calcite (Herbagreen Basic) foliar fertilization, winter losses of 5.3%–5.7% were noted, while at the same time, among plants from variants without autumnal silicon foliar fertilization, 14.7%–16.0% losses were observed.

Good wintering of winter rape is the effect of several important morphological features of plants, such as the number of leaves in the rosette before winter, apical bud's seat height, root collar's diameter, and tap root's length (Figures 16.6 and 16.7). In autumn, in the presented experiment, from variant 1, where foliar fertilization with 393 g Ca and 120 g/ha Si at the four to six leaves' growing stage of rape was applied, plant morphological features were compared with those of the control variant—without foliar fertilization. Foliar fertilization resulted in a significant seat height reduction of the apical bud (Table 16.3), which could have a beneficial impact on better plants' wintering. However, no other beneficial effect of foliar fertilization on plants' morphological features was observed.

TABLE 16.2

Effect of Autumnal Foliar Fertilization with Silicon and Calcium on Number of Winter Oilseed Rape (Visby Variety) Plants per m² in 2010–2011 and 2011–2012 Growing Seasons

Fertilization Variant	No. of Plants in Autumn, pcs./m²	No. of Plants in Spring, pcs./m²	No. of Plants during Harvest, pcs./m²	Plant Winter Losses	
				pcs./m²	%
2010–2011					
Control	47.8	38.5	38.2	9.25	19.63
1	45.5	43.0	42.8	2.50	5.63
2	54.0	52.0	51.2	2.00	3.79
3	46.0	37.2	37.0	8.75	18.41
4	53.5	43.0	42.8	10.50	19.62
$LSD_{0.05}$	ns	10.6*	11.0*	3.92*	6.37*
2011–2012					
Control	28.2	25.5	25.5	2.75	9.75
1	24.8	23.5	23.5	1.25	5.04
2	26.0	24.0	24.0	2.00	7.69
3	27.0	23.5	23.5	3.50	12.96
4	30.2	26.5	26.5	3.75	12.42
$LSD_{0.05}$	ns	ns	ns	1.35*	4.54*
2010–2011 and 2011–2012					
Control	38.0	32.0	31.9	6.00	14.69
1	35.2	33.3	33.2	1.88	5.34
2	40.0	38.0	37.6	2.00	5.74
3	36.5	30.4	30.3	6.13	15.69
4	41.9	34.8	34.7	7.13	16.02
$LSD_{0.05}$	ns	ns	ns	2.56*	–

Source: Artyszak A, Study on the effect of Herbagreen Basic application on overwintering, yielding, and technological quality of winter oilseed rape, 2012, unpublished.

Note: 1 = autumn: 393 g Ca and 120 g Si/ha at 4–6 rape leaves' growing stage; 2 = autumn: 393 g Ca and 120 g Si/ha at 4–6 rape leaves' growing stage, spring—start of vegetation: 393 g Ca and 120 g Si/ha; 3 = spring—start of vegetation: 393 g Ca and 120 g Si/ha; 4 = spring—start of vegetation: 393 g Ca and 120 g Si/ha, 2 weeks later 393 g Ca and 120 g Si/ha. LSD, least significant difference; ns, no significant differences.

* Significant differences at the level of $\alpha = 0.05$.

The effects of foliar fertilization with silicon and calcite on seed yield, content, and yield of oil were differentiated, respectively, by the growing season (Table 16.4). In 2011, the highest seed yield with the highest oil content and, consequently, the highest yield of oil was observed in the variant with one autumnal application of 393 g Ca and 120 g Si/ha at the four to six rape leaves' growing stage. In 2012, the most beneficial effect for the yield of seeds and oil was on the variant with one spring application of 393 g Ca and 120 g Si/ha. The oil contents in seeds were similar in every variant that year. Considering two years' average results, the differences in seed yield, oil content, and yield decreased.

The yield of winter oilseed rape is significantly modified by characteristics such as number of plants during harvest, number of siliques per plant, number of seeds per silique, and mass of 1000 seeds. Two years of research on foliar fertilization with silicon and calcite resulted, on average, in a significant effect on the plants' height before harvest (Table 16.5). However, other effects were differentiated each year.

FIGURE 16.6 Autumnal foliar fertilization with silicon and calcite of winter oilseed rape (November 2011). (Courtesy of A. Artyszak.)

FIGURE 16.7 Winter oilseed rape collar diameter measurement (November 2011). (Courtesy of A. Artyszak.)

In the 2011–2012 and 2012–2013 growing seasons, Ciecierski and Kardasz (2014) conducted field research in four locations of experimental stations of the Research Center of Cultivar Testing in Poland. They used foliar fertilizer Optysil (94.1 g Si and 25 g Fe per 1 dm^3) in winter oilseed rape (*Brasssica napus* L.) cultivation. They noticed a 1.7%–17% seed yield increase, depending on the respective rape variety and location. Additionally, the effect of a 1.4%–19% increase of 1000-seed weight was observed.

16.2.2.2 Sunflower

South African research on silicon's protective role in sunflower was conducted by Elungi et al. (2009). The authors observed silicon to be of significant help in the reduction of plant infection by important sunflower diseases that are caused by pathogens such as *Sclerotinia sclerotiorum* and *Rhizoctonia solani*. Use of silicon fertilizers effected a higher number of leaves and a head dry

TABLE 16.3
Effect of Foliar Silicon Fertilization on Morphological Features of Winter Oilseed Rape Plants before Winter (Average Results from 2010–2011)

	Fertilization Variant		
Morphological Feature	Control	1	LSD$_{0.05}$
Tap root's length, cm	16.6	17.3	ns
Apical bud's seat height, cm	6.00	4.88	0.73*
Root collar's diameter, cm	1.06	1.07	ns
Number of leaves in the rosette, pcs.	6.25	6.20	ns
Dry mass of rosette, g	13.1	11.5	ns
Dry mass of root, g	4.56	4.25	ns
Dry mass of plant, g	17.8	15.5	ns

Source: Artyszak A, Study on the effect of Herbagreen Basic application on overwintering, yielding, and technological quality of winter oilseed rape, 2012, unpublished.

Note: 1—autumn: 393 g Ca and 120 g Si/ha at 4–6 rape leaves' growing stage. LSD, least significant difference; ns, no significant differences.

* Significant differences at the level of $\alpha = 0.05$.

TABLE 16.4
Effect of Foliar Fertilization with Silicon and Calcium on Oilseed Yield, Oil Content, and Oil Yield in 2010–2011 and 2011–2012 Growing Seasons

Fertilization Variant	Seed Yield, kg/ha	Oil Content, %	Oil Yield, kg/ha
	2010–2011		
Control	3163	43.6	1367
1	4285	46.5	1996
2	2995	43.6	1311
3	3196	45.2	1447
4	2616	44.3	1156
LSD$_{0.05}$	1043*	2.79*	474*
	2011–2012		
Control	4027	41.6	1642
1	3893	41.4	1581
2	4170	42.8	1756
3	5281	42.6	2208
4	3914	42,0	1625
LSD$_{0.05}$	1376*	ns	592*
	2010–2011 and 2011–2012		
Control	3595	42.6	1505
1	4089	44.0	1789
2	3583	43.2	1534
3	4239	43.9	1828
4	3265	43.2	1391
LSD$_{0.05}$	ns	ns	ns

Source: Artyszak A, Study on the effect of Herbagreen Basic application on overwintering, yielding, and technological quality of winter oilseed rape, 2012, unpublished.

Note: Variant descriptions are the same as in Table 16.2.

TABLE 16.5

Effect of Foliar Silicon and Calcium Fertilization on Rape Plants' Morphological Features during Harvest in the 2010–2011 and 2011–2012 Growing Seasons

Fertilization Variants	Plant Height during Harvest, cm	No. of Siliques, pcs.		No. of Seeds per One Silique, pcs.			Mass of 1000 Seeds, g
		On the Main Stem	On Lateral Shoots	Lower Part of Plant	Central Part of Plant	Upper Part of Plant	
2010–2011							
Control	157	58.6	113	17.5	21.6	19.8	4.70
1	156	55.2	106	15.8	23.2	25.8	4.82
2	155	50.7	101	17.3	20.2	21.9	4.52
3	152	59.6	105	16.5	23.6	20.8	4.52
4	141	50,5	72.0	21.5	18.1	18.3	4.52
$LSD_{0.05}$	8.58*	6.56*	32.4*	3.26*	4.14*	6.29*	ns
2011–2012							
Control	167	40.7	173	19.3	22.7	19.5	5.13
1	160	35.6	152	18.3	22.6	17.8	4.74
2	152	40.4	160	19.4	22.2	19.0	4.94
3	155	36.8	156	24.2	24.8	23.1	4.96
4	158	43.8	186	19.3	26.1	20.4	4.83
$LSD_{0.05}$	ns	8.07*	ns	ns	ns	3.9*	ns
2010–2011 and 2011–2012							
Control	162	50.0	143	18.4	22.2	19.7	4.92
1	154	45.4	129	17.1	22.9	21.8	4.78
2	154	45.6	131	18.4	21.2	20.5	4.73
3	154	48.2	131	20.4	24.2	22.0	4.74
4	150	47.2	129	20.4	22.1	19.4	4.68
$LSD_{0.05}$	8.61*	ns	ns	ns	ns	ns	ns

Source: Artyszak A, Study on the effect of Herbagreen Basic application on overwintering, yielding, and technological quality of winter oilseed rape, 2012, unpublished.

Note: Variant descriptions are the same as in Table 16.2.

weight increase from 47% to 146%, depending on the respective fertilization variant, compared with the *Sclerotinia*-infested control.

16.2.2.3 Oil Palm

Researchers from Colombia (Cristancho and Restrepo, 2014) observed the effect of silicon fertilization in a nursery and young oil palm stages. The results showed that an acidulated magnesium silicate application increased P, Mg, B, and Si uptake. The treatments improved the leaf area index and petiole cross section. The treated plants accelerated plant recovery from bud rot disease attack. Treatments with acidulated magnesium silicate showed healthier growth and a higher yield. The authors confirmed that silicon applications had significant potential to improve plant nutrient efficiencies, improve plant disease recovery, and increase yields and business sustainability.

16.2.3 CEREALS

For higher plants, silicon is not an essential element, but in general, the beneficial effect of this element has been demonstrated for many plant species, especially from the Poaceae family. In addition to the importance of silicon in plant nutrition, optimal growth, and development, it plays a significant role in the alleviation of various symptoms of biotic and abiotic stresses (Vaculík et al., 2014). Especially grasses accumulate silicon in higher concentrations, and we should expect a significant impact of silicon fertilization on grain yield and quality for this reason. Although there are many publications on the role of silicon in different kinds of biotic and abiotic stress mitigation, there is a lack of information on the role of silicon in yield and crop quality increases.

16.2.3.1 Rice

Rice (*Oryza sativa*) is one of the most important foods for more than half of the world's human population. Although it is one of the largest silicon-accumulating crops, silicon cycling in rice paddies is still understudied (Marxen et al., 2014).

One of the older and well-known studies on the effect of silicon on rice yields was conducted by Seebold et al. (2000). The described experiment was carried out at two locations in Colombia, where rice was sown under upland conditions. The silicone fertilizer was applied in two doses. The authors observed the effect of silicon fertilization on complementation of inherent rice disease resistance and reduction of leaf and neck blast severity of sensitive rice cultivars. The effects of silicon fertilization were similar to those of comparable blast-resistant cultivars not treated with silicon fertilization. The effects of silicon application on grain yield, grain quality, and silicon content in plant tissue were also evaluated. In that experiment, silicon was applied in the form of calcium inosilicate mineral (wollastonite [$CaSiO_3$]) and was used according to rate. There were three variants of silicon doses in both locations: (1) control with no fertilization, (2) 500 kg Si/ha, and (3) 1000 kg Si/ha.

A significant effect of silicon rate on the mean grain yield across all cultivars was not observed at any location. At the same time, interactions between silicon fertilization variants and rice cultivars were not observed. The tendencies of higher mean grain yields as an effect of silicon fertilization were only observed across all cultivars at both locations. However, use of silicon at rates of 500 and 1000 kg/ha effected a significant increase in the rice grain yield, compared with the same rice cultivars without silicon fertilization: 19% and 24% ($p \leq 0.05$), respectively, to the variant, at Santa Rosa, and 40% and 72% at La Libertad. At Santa Rosa, silicon fertilization at the lower rate (500 kg/ha) had no significant effect on rice grain quality features like milling quality from cultivars sown. However, the higher silicon dose (1000 kg/ha) increased the head rice (whole grains) by 20%. At the same time, a 15% reduction of broken grains was observed. Both beneficial effects were noticed across all planted rice cultivars in comparison with the same cultivars, but without silicon fertilization. Additionally, averaging for all tested varieties, silicon fertilization at the higher dose (1000 kg/ha) effected a 26% reduction of grain discoloration. At La Libertad, the significant beneficial effects of silicon fertilization on quality features of grain across all varieties were observed at the lower and higher dose. There were 16% and 26% increases, respectively, of head rice, in comparison with the control variants, and a 68% reduction of grain discoloration at both silicon rates, compared with the control.

According to Ghanbari-Malidarreh and Dastan (2014), silicon and phosphorus are two important elements in rice production, especially in areas where rice is frequently cultivated. The aim of these authors' research was to determine the uptake and content of these nutrients and estimate the silicon and phosphorus effect on the grain yield of rice. Rice was grown in a field without silicon (control variant) and with two silicon rates (500 and 1000 kg/ha) and with three phosphorus levels (P low, 0 kg/ha; P medium, 50 kg/ha; and P high, 100 kg/ha). The obtained results showed that the maximum plant height, panicle stage, panicle per square meter, and grain yield were produced with 1000 kg Si/ha; in this treatment, compare with the control, the grain yield increased about 23%. Also, the highest silicon content and the highest silicon uptake by whole aboveground parts of plants (grain and straw) and the highest phosphorus content and uptake by whole aboveground parts of plants (grain and straw) were obtained at a higher dose of silicon (1000 kg Si/ha). Medium phosphorus fertilization resulted in an increase in the panicle length, number of spikelets per panicle, 1000-grain weight, and grain yield. Also, this fertilization variant increased the phosphorus content of grain. Infected leaf blast percentage, leaf blast diameter, panicle-infected blast, and grain-infected blast decreased with a 1000 kg/ha silicon dose, but only panicle-infected blast and grain-infected blast decreased with phosphorus application, compared with the control treatment. The highest grain yield and biological yield were obtained by a fertilization variant where 1000 kg/ha silicon and 50 kg/ha phosphorus doses were used. Generally, silicon application increased phosphorus uptake, as this element has a synergism effect that causes Si and P to increase uptake and grain yield in rice, while also decreasing disease parameters in rice.

Another new study on silicon fertilization needs and the effect on rice yields was conducted by Mahendran et al. (2014) in the region of India where rice has been cultivated in an intensive mono-culture for many years under Periyar Vaigai Irrigation System conditions. This is one of the oldest and second largest irrigation systems in India (since 1898), and rice is grown as a monocrop. The authors found that high accumulation of silicon in rice is necessary for healthy plant growth and high stable yields. Depletion of silicon available for plants in traditional rice soils from the continu-ous monoculture of intensive high-yield cultivars could be one of the most important limiting fac-tors, causing a decline or stagnation in grain yields in many countries where rice is grown. Although there is ample research on macro- and micronutrients based on needs for sustaining rice productiv-ity, work related to silicon is very much limited in India, in spite of its heavy feeding by rice crops. Hence, the investigation set out to assess the status of available silicon and fix the critical limit for silicon fertilization in the intensive rice-growing soils of the Periyar Vaigai Command (PVC) area. In total, 100 surface soil samples were collected and analyzed for plant-available silicon. In order to assess the response of rice to applied silicon for arriving at the optimum silicon level, and also to fix the critical limit of silicon, a pot culture experiment was conducted, making use of bulk soil samples collected from 20 locations (out of 100) of rice-growing tracts in the PVC area, with five levels of Si (0, 75,150, 225, and 300 ppm) using rice variety ADT-45 as the test crop. The effects of silicon on plant growth and development (number of productive tillers, number of filled grains per panicle, and percent chaffiness) and rice grain yield were recorded. This investigation revealed that available silicon in the rice soils ranged from 4 to 250 ppm. The authors observed that the various yield attributes and grain yield of rice were significantly increased with silicon application up to the 225 ppm level. However, the optimum physical and economic doses of silicon were found to be 368 and 310 ppm, respectively, in the rice-growing soils of the PVC area. The soil-available silicon levels were categorized into low, medium, and high based on the percent relative yield. Accordingly, soil-available Si up to 122 ppm (<75% relative yield) was considered low, 122–181 ppm (75 to 95% relative yield) was medium, and more than 181 ppm (>95% relative yield) was high in Si availability. Per this categorization, out of the 100 soil samples tested in the PVC area, 64% of the soil samples were deficient in silicon supply, warning of the seriousness of silicon deficiency. Hence, the soils shown to have less than 122 ppm of available silicon need balanced fertilization, including silicon to avoid its deficiency.

16.2.3.2 Wheat

Wheat (*Triticum aestivum* L.) is one of the most important cereal crops in the world, and it plays a very important role in human and animal nutrition. Wheat is a source of food ingredients, and for this reason, it is the most important crop from a food security perspective. In contrast to rice, wheat is grown around the world. Often, field conditions are unsuitable for wheat cultivation, especially in areas where soil salinity and drought occur frequently during the growing season.

Hanafy Ahmed et al. (2008) conducted two pot experiments under differentiated salinity levels of the soil (0, 2000, 4000, and 6000 ppm of salinity) and a field experiment under a saline soil condition. The aim of such experiments was to observe the effect of silicone and boron fertilization on wheat (*Triticum aestivum* L. Seets 1 cultivar) growth, yield, and chemical composition. Under the pot experiments, the total doses of silicon fertilizers (0, 250, and 1000 ppm SiO_2) and boron fertilizers (0 and 25 ppm B) were sprayed in two divided rates (40 and 70 days after sowing). In the field experiment, the total doses of silicon (0, 250, 500, and 1000 ppm SiO_2) and boron (0, 25, and 50 ppm B) were applied in three rates (at 35, 60, and 85 days after sowing). Regardless of the silicone dose and use of boron additive, silicon fertilization significantly increased characteristics of plant growth, such as the shoot height, leaf area, grain yield per plant, and 1000-seed weight. In the field experiment, the lowest silicon rate significantly increased all growth features. But all silicon rates significantly increased the number of spikes and grains in spikes, as well as the yield of grain, in comparison with the control fertilization variants. The authors underlined the superior effect of the lower silicon rate.

Similar to soil salinity, drought is considered to be one of the most important agricultural abiotic stresses, limiting wheat production in warmer parts of the world. Silicon is also known as a nutrient that can mitigate such kinds of stress; however, the mechanism of this mitigation remains unclear. In Passala et al.'s (2014) research conducted in India, the authors observed the effect of Silixol's efficacy in alleviating drought stress in wheat. The yield in terms of seed weight was increased with increasing Silixol concentration treatment, and the increase was 5%–10% over that of the control under severe stress conditions. Therefore, the authors recommend using Silixol as a fertilizer because of its strong impact on alleviating drought stress and minimizing yield losses in wheat under drought conditions.

In 2009–2011, during two growing seasons, Hellal et al. (2012) conducted a field experiment to determine the effect of organic and inorganic nitrogen applied together with silicon fertilizer (in the form of diatomite) on the chemical composition of wheat plants and grain yield of wheat. The diatomite chemical component was as follows: 33.7% silicon, 19.0% calcium, 5.1% sodium, 3.4% magnesium, and 2.9% iron, and many other trace nutrients, such as boron, copper, manganese, titanium, and zirconium. For the organic nitrogen source, the authors used organic material (farmyard manure [FYM]), and for the inorganic nitrogen source, ammonium sulfate (AS) was applied. The total rate of nitrogen applied as FYM was 100 kg N/fed. The organic manure was mixed with 0–30 cm of the surface soil layer before wheat sowing. Silicon fertilizer was applied as foliar spray. Total doses of this nutrient were used at four rates of 0.0%, 0.2%, 0.4%, and 0.6% and applied in two sprays at four-week intervals. The first doses of fertilizers were applied after 60 days of cultivation. In total, the experiments included nine combinations of fertilization:

1. T1: Control (without fertilizers)
2. T2: AS
3. T3: FYM
4. T4: AS + 0.2% Si
5. T5: AS + 0.4% Si
6. T6: AS + 0.6% Si
7. T7: FYM + 0.2% Si
8. T8: FYM + 0.4% Si
9. T9: FYM + 0.6% Si

Silicon foliar fertilization at a rate of 0.4% and FYM or AS application resulted in the highest grain yields and values of wheat straw. Based on the obtained results, the authors strongly recommended silicon foliar fertilization combined with application of FYM or AS for the best wheat growth and yields, especially under conditions of clay soil.

Another investigation of silicon's effect on wheat yield was conducted in 2013–2014 by White et al. (2014) in Louisiana. These authors checked the effect of silicon on excessive nitrogen application in wheat. On average, higher grain yields were achieved at 101 kg N/ha, compared with 145 kg N/ha in 2013, while in 2014, higher yields were observed at the 145 kg N/ha rate. In 2013, the highest yield was obtained from plots that received 2 Mt $CaSiO_3$ slag/ha and 101 kg N/ha. At the 145 N kg/ha rate, the grain yield was significantly raised from that in 1987 to 2123 kg/ha with the addition of 4.5 Mt $CaSiO_3$ slag/ha ($p < 0.05$). The initial results of this study showed that the wheat yield level was improved at certain combinations of N and Si rates. The authors pointed out that the outcomes of this research would help establish links among silicon fertilization rates, raise the level of soil silicon and plant-essential nutrients, increase resistance to lodging and diseases, and improve grain yield.

Similarly, Ciecierski and Kardasz (2014) conducted laboratory and field research on the effect of the application of liquid silicon–based fertilizer Optysil (Si) on drought stress reduction in wheat (*Triticum aestivum* L.). The trials were conducted in the Department of Plant Physiology of the Polish Academy of Science. The yield quality and quantity were checked in experimental stations of the Research Center of Cultivar Testing, located in four different parts of Poland. The liquid silicon application on drought stress reduction resulted in very promising effects. Laboratory tests on wheat in stimulated drought conditions showed lower electrolyte leakage by 41%, and at the same time, protein production increased by 40%. Field trials showed a yield increase from 8% to 20.5% on winter wheat, depending on the location and variety. At the same time, the 1000-grain weight increased from 1.6% to 7%.

16.2.3.3 Maize

Some research on silicon was conducted with maize (*Zea mays* L.) also. This plant species, similarly to rice and wheat, is one of the most important crops, and its cultivation is popular worldwide. Most recent publications refer to the role of silicon in mitigating the salinity stress and interactions between silicon and some heavy metals like zinc or antimony (Bokor et al., 2014; Fialová et al., 2014; Khan et al., 2014).

In every mentioned publication, the authors confirmed a significant role of silicon in helping to better maize germination and growth parameters, like the total length of the primary seminal roots, fresh and dry weights of biomass, and some physiological parameters and biochemical activity, or even toxicity. While there is a lack of information about the effect of silicon on maize grain yield, the authors (Vaculíková et al., 2014) observed no positive or negative effect of silicon on plant biomass.

Zhiming et al. (2014) conducted field research in China, under conditions of alluvial soils, on the effects of silicon fertilization on maize photosynthetic characteristics like total chlorophyll content, photosynthetic rate, stomatal conductance, intercellular CO_2 concentration, and transpiration rate. Five rates of silicon (0, 45, 90, 150, and 225 kg Si/ha) were factors of the experiment. The silicon fertilization effected an increase in the values of total chlorophyll content and maize photosynthetic efficiency. The authors recommended 150 kg/ha as the optimal silicon dose for use under conditions of alluvial soils. Additionally, the obtained results showed that silicon applied at proper rates can be very helpful in increasing the photosynthetic ability of maize. Such beneficial effects can significantly increase maize growth and yields.

16.2.4 Legumes

Other crops that are important worldwide are legumes, among which the most important species in the world is soybeans.

Crusciol et al. (2013) conducted some research on the effect of foliar fertilization with silicon on legume seed crops. He noticed that foliar applications of stabilized silicic acid in soybean (*Glycine max* L. *Merr.*) effected a 14% increase of seed yields, as well as an increase of pod numbers. Similarly, such silicon fertilization resulted in a 15% increase of common bean (*Phaseolus vulgaris* L.) yield and a 9.6% increase of peanut (*Arachis hypogaea*) yield.

Some research with silicon fertilization was conducted in South Africa by Visser et al. (2009). Yields of soybean seeds from plants after silicon application with rates of 1000 (×3) and 2000 (×3) mg Si/L were higher (267.5 and 283.2 g, respectively) than those from the control plants (221.5 g). A good correlation between the application of potassium silicate to soybeans and a reduction of soybean rust was observed at the same time. In conclusion, the authors observed that silicon consistently reduced the levels of soybean rust and increased yields.

Ciecierski and Kardasz (2014) obtained interesting results in their two-year study on the effect of silicon foliar application on soybean yield and quality under drought stress. They use the liquid silicon–based fertilizer Optysil. Small field trials were conducted in the Department of Plant Physiology of the Polish Academy of Science. Similar to the results obtained on winter wheat and winter oilseed rape, the effect of liquid silicon application on drought stress reduction was very promising. The number of pods on soya increased by 18%, and the average seed weight per plant increased by 21%. In their conclusion, the authors underline that their results confirmed the positive effect of silicon-based fertilizers on yield. However, the experiments also showed that the more severe the stress, the higher the yield increase that can be observed.

Alfalfa (*Medicago sativa*) is another plant species from the legumes, and it plays an important role in ruminant feeding in some regions. It is mostly is used as hay content. Fuping et al. (2005) observed a significant increase of such yield parameters as leaf fresh weight (16.4%), leaf dry weight (24.1%), and inflorescence weight as the effects of silicon fertilization. However, with no effect on stem weight, silicon additions at the mature stage increased the stem dry weight and root number. No effect on seed yield was seen in year 1, but the 1000-seed weight increased by 29%.

Guo et al. (2006) observed better growth of alfalfa (*Medicago sativa* L.) shoots and roots as the effect of silicon fertilization. They used a silicon-deficient topsoil in pots with silicon addition rates of 0, 0.025, 0.05, 0.10, 0.20, and 0.30 g/kg for growing alfalfa. The silicon content of the shoots and roots increased as the rates of the applied silicon increased. Silicon effected an increase in the plants' height, leaf area, shoots per plant, and forage yields, during the growing season. Additionally, silicon fertilization increased root biomass and their volume, as well as the number of secondary roots.

16.2.5 VEGETABLES

16.2.5.1 Lettuce

According to Ugrinović et al. (2011), foliar application, or watering with liquid or soluble fertilizers, is one of the methods recommended for solving problems with the excessive use of mineral fertilizers in vegetable crop production. Foliar fertilization facilitates the efficient use and faster absorption of nutrients. For this reason, such fertilization should be recommended as a preventive or curative way to reduce nutrient deficiency. Foliar fertilization can play a significant role in the intensive production of vegetables. The authors conducted experiments under green house conditions on the effect of differentiated foliar fertilization on lettuce (*Lactuca sativa* L. cv. Neva) yields. The effects of five liquid or soluble fertilizers on yield characteristics such as weight of the single plant (g), number of leaves per single plant, weight of leaf rosette (g), weight of single head (g), and head index were observed. The head index was calculated as the ratio of head weight to leaf rosette weight. The authors used six different foliar fertilizers. One of them (Herbagreen) contained 4.2% silicon. The application of five kinds of fertilizers resulted in a significant increase of lettuce yield compared with the control. Additionally, use of fertilizer with silicon resulted in a significantly better yield compared with other fertilization variants, as well as Drin + Foliacon or Bioactive. Interesting results were observed in quality features of lettuce. On the one hand, the use of silicon resulted in the highest number of leaves from all kinds of fertilization, but on the other hand, the head index was the lowest compared to those obtained in other fertilization variants. According to the authors' recommendation, the lettuce head index should be as high as possible because lettuce with highly developed rosettes is not preferred by consumers.

On the other hand, Bacchus (2010), in a field experiment, noticed the positive effect of lettuce fertilization with a volcanic sandy loam in combination with organic composts. This beneficial effect was significantly better than that observed with fertilization with a diammonium, phosphate, and calcium ammonium nitrate mixture. In both combinations of fertilizers, the rates of nitrogen and phosphorus were similar. Furthermore, when the compost and plants were sprayed with a biodynamic silica preparation, the beneficial effect on lettuce yields was even better. However, the variation in lettuce fresh yield was very high (coefficient of variation = 28%), and for this reason, the effects of the fertilization variants were not statistically significant. Regardless of the fertilizer type (compost or soluble fertilizer), silica application significantly affected the higher dry matter (DM) content of lettuce. However, it did not significantly affect the fresh head yield of lettuce, nitrogen uptake, amino acid content, or plant sap nitrate concentrations.

The newest Polish study of Kleiber et al. (2015), with lettuce, was conducted on the effect of differentiated silicon fertilization in combination with other nutrient solutions on such character-istics as the content of macro- and micronutrients (including silicon and manganese) in the leaves and yields of lettuce. The lettuce plants were planted under strong manganese stress conditions (19.2 mg/dm^3 nutrient solution). Experimental fertilizers were used as foliar applications. The two cycles of the experiment were carried out in 2014 in a chamber with a controlled environment, at the experimental station of the Departments of the Faculty of Horticulture and Landscape Architecture, Poznań University of Life Sciences. Foliar silicon fertilization applied at increasing rates effected significantly higher lettuce yields. Generally, silica sol fertilization improved the growth and yield of plants. The average head weight of lettuce and the number of leaves per single plant were higher. The increasing silicon fertilization significantly increased the relative water content (RWC) and, at the same time, reduced the dry matter content of leaves. Furthermore, foliar silicon fertilization significantly changed the nutrient content of plants. A significant increase of the silicon content in leaves was observed. The total uptake of such elements as N, P, K, Ca, Mg, Na, and Fe was higher, and the uptake of Zn and Cu was lower after silicone application, compared with the control (with-out silicon).

16.2.5.2 Cucumber and Other Cucurbits

Cucumber (*Cucumis sativus* L.) is another popular vegetable all over the world. In a South African study on the effect of silicon application on cucumber conducted by Matichenkov and Bocharnikova (2008), a 46% increase of cucumber yields was observed as a result of foliar silicon application at a 10 kg/ha rate at the third leaf stage and subsequent biweekly applications during the season. Soil silicon applications at a 40 kg/ha rate also increased yields. Additional benefits of silicon included stimulation of fruit formation and accelerated fruit maturation.

At the same time, in 2009–2010, Polish experiments on the effect of silicon fertilization on cucumber yields were conducted in a greenhouse of the Department of Cultivation and Fertilization of Horticular Plants, University of Life Sciences, Lublin (Jarosz, 2013). Cucumber plants were watered with differentiated fertilizations. The chemical compositions of fertilizing combinations were as follows (in mg/dm^3): 0.02 N-NH$_4$, 5.0 N-NO$_3$, 4.0 P-PO$_4$, 1.4 K, 120 Ca, 13.8 Mg, 32.0 S-SO$_4$, 9.5 Cl, 6.7 Si, 2.7 Na, 0.24 Fe, 0.018 B, 0.026 Mn, 0.038 Zn, and 0.001 Cu. Additionally, every 8–10 days, cucumber plants were fertilized with a liquid solution of Yara fertilizer and calcium nitrate, potassium nitrate, and magnesium sulfate (13% Mg). The amounts of nutrients were the same in all treatments, adjusted according to the growing medium and the cucumber nutritional requirements. In the experiment, four silicone rates were applied (0, 250, 500, and 750 mg Si/dm^3). Silicon was applied to the roots in the form of a liquid solution of silica sol. The total dose was divided and applied at five equal doses during the fertilization of the plants. In this study, the author observed a significant increase of cucumber fruit yield in the treat-ments where silicon was applied at the rates of 500 and 750 mg Si/dm^3, compared with the control (without silicon). This increase was 8.60% and 5.89%, respectively, to the doses. The highest rate of silicon also resulted in a significant content increase of dry matter and total soluble solid, in comparison with the control variant.

In Toresano-Sánchez et al. (2010), research applications of a 2.5% monosilicic acid solution sup-plied through drip irrigation to triploid watermelon (*Citrullus lanatus Thunb.* cv. Queen of Hearts) grafted onto a squash hybrid (*Cucurbita maxima* × *Cucurbita moschata* cv. *Strongtosa*) increased fruit quality by 10% in pulp consistency and rind width, but did not affect overall yields in such parameters as harvest weight and fruit number.

Silicon nutrition with other elements as a solution can play an important role for plant disease control, growth, and yield of crops growing under hydroponic conditions. For some vegetables, such cultivation and fertilization is crucial. Silicon fertilization of zucchini planted under hydroponic conditions was another subject of the newest study. In Tesfagiorgis and Laing (2009), research with zucchini cultivated under a hydroponic system using differentiated solutions of silicone affected plant resistance and powdery mildew infestation. The higher the concentration of silicon in the solution, the smaller the infection of powdery mildew that was observed. Furthermore, the silicon contents in roots and leaves were higher, without affecting silicon distribution to other plant parts. Moreover, silicone nutrition improved plant growth and total uptake of nutrients like phosphorus, calcium, and magnesium, as well as potassium accumulations, in the shoots. However, the increased silicon concentration in the solution effected reduction of the calcium accumulation in plants and their growth and yield. Based on the obtained results, the authors recommend a silicon rate of 50–100 mg/L as the optimal concentration that should be applied for the best disease control and yield of zucchini.

16.2.6 Berries

16.2.6.1 Strawberry

Some research on the effect of silicon fertilization on strawberry (*Fragaria* × *ananassa* Duchesne) production was conducted in 2005 and 2006. However, Polish (Mikiciuk and Mikiciuk, 2008, 2009) and British (Jin et al., 2014) investigations were focused on the protective role of silicon against the most frequent diseases of strawberry. The obtained British results showed that silicon fertilizers significantly reduced powdery mildew. In the Polish results, the severity of gray mold and leather rot on unprotected plants of strawberry was 10% and 14%, respectively. A liquid silicon preparation (Optysil) used at 0.5 and 1.0 L/ha four times from the beginning of flowering (BBCH 61) reduced the severity of both diseases. Its efficacy ranged from 40% to 80%, depending on the dose and term of assessment, and it was lower than (first assessment) or similar to (second assessment) that obtained with standard fungicides like cyprodinil + fludioxonil, or piraclostrobin and boskalid and a biological product containing *Pythium oligandrum*. The yield from plots treated with liquid silicon was significantly higher than that from the control plots. The preliminary results indicate the possibility of using silicon in an Integrated Pest Management (IPM) program against fungal diseases. Wang et al. (2009) wrote about a similar yield increase as a result of silicon application at the same time. In Chinese research, field experiments and lab analyses were combined to study the effect of silicon application on photosynthetic characteristics and the yield of open cultivated strawberry. These studies provided results that suggest that silicon supplied to strawberry can increase the leaf net photosynthetic rate, stomata conductance, and intercellular CO_2. Silicon also decreased the transpiration rate and promoted the water use efficiency of strawberry. By adding silicon to strawberry, it can advance the individual fruit number and the mean fruit weight, debase the number of malformed fruit and diseased fruit, and obviously enhance the strawberry yield. The yield increasing range of these different treatments was 10.3%–29.2% more than that of the control. When the amount of silicon application reached 2.93 kg/667 m^2, the peak output was obtained.

16.2.7 Pomiculture

16.2.7.1 Grapevine

In the study of Kara and Sabir (2010), the effects of 100% natural fertilizer (Herbagreen—chemical content is indicated in Section 16.2.1.1) containing silicon on young grapevine cultivation was observed. The experiments were conducted under grapevine nursery conditions where Herbagreen was pulverized. Silicon application affected the formation of longer shoots, compared with the control variant. The average shoot lengths of plants after silicon application were longer than 20 cm, compared with the shoot lengths of the control plants. Besides, Herbagreen application affected some important vegetative features, like the level of shoot development or number of lateral shoots. Additionally, silicon fertilization with Herbagreen effected better leaf development and their thickening. Furthermore, the leaves were greener as a result of improved photosynthesis. Based on the obtained results, the authors suggested that Herbagreen would be helpful in the production of robust saplings. They underline that silicon plant nutrition as an environmentally friendly fertilization could improve vegetative development without environmental hazard. For this reason, silicon application can play an important role as a modern agent for plant growth promotion, aiding in healthy and vigorous vine development, and should be recommended for sustainable viticulture.

According to Lynch (2008), silicates have been used in Australian viticulture for more than 20 years with outstanding results in yield, fruit quality, and soil health. These results have been seen under multiple climatic conditions, irrespective of soil type, texture, or pH. Superior grapes (both table and wine) have resulted from silicon supplements, with superior skin quality, higher Brix, a more uniform bunch size, and the near absence of fungal diseases. Sustainability is increased, with increases of beneficial insects apparent.

16.3   PERSPECTIVES FOR SILICON FERTILIZATION

By far, a significant majority of the presented results from scientific research on silicon foliar fertilization clearly show the beneficial effects of such fertilization. In most cases, silicon provided a potential and environmentally friendly method for the control of numerous plant diseases and also improves the development and growth of plants by significantly increasing their yields and improving the crop's raw material. While silicon fertilization has so far been mainly used in rice, sugarcane in warmer parts of the world, and horticulture crops in Europe, we expected the rapid dissemination of silicon fertilizer use for foliar fertilization of agricultural crops worldwide in the near future. The increasing role of silicon in plant foliar fertilization is due to several reasons:

- Climate change and increasingly intensified droughts repeated more often
- Soil salinity
- Immunization of pathogens responsible for plant diseases and pests on the available plant protection products and the associated increasing problems with plant protection
- Need to reduce the intensity of fertilization with mineral nitrogen (food safety, environmental protection, and economic issues)
- Increasing demands of customers and consumers about the quality of plant raw materials
- Health considerations (silicon is essential to the proper functioning of the human body)
- Increasing need for environment protection
- Economic issues—the necessity of production cost reduction while raising yields at the same time

In the European market, we can meet the increasing offer of silicon foliar fertilizers. Nowadays, we have many results showing that liquid foliar fertilizers can play another important role—as a biostimulant. When applied to the plants, the fertilizers can modify physiological processes in such a way that provides potential benefits to the growth, development, and abiotic and biotic stress responses. Some of them induce in almost any crop increased growth, more biomass, higher yield, higher quality, and longer shelf life. For this reason, the process of permanently improving silicon fertilizers continues, in terms of application convenience. Still, the main issues in this case are the reduction of problems with dissolution and spray liquid preparation and clogging of filters and tips of sprayers. Also, silicon is very hard for plants to take up from soil fertilizers. Therefore, based on the latest results and growing human knowledge of plant metabolism and the biological mechanism of silicon uptake and transport by plants, we suspect that soil silicon fertilizers are likely to play an important role in the near future.

Additionally, we have to realize that according to the European Regulations, some silicon fertilizers, even foliar, are authorized for use in organic agriculture systems, where their protective and yield-promoting roles are crucial.

However, the problems with silicon fertilization require us to carry out several more studies to find answers to the following questions:

- What should be the optimal doses of silicon in the particular plant species cultivated?
- What is the optimal time for the silicon application in various crops?
- How do we differentiate silicon doses and the application terms under different environmental conditions?

In conclusion, we suspect that silicon fertilization will soon become as important as the now commonly used macro- and micronutrients. Silicon foliar fertilization is a very promising technology for the twenty-first century.

REFERENCES

Artyszak A. 2012. Study on the effect of Herbagreen Basic application on overwintering, yielding, and technological quality of winter oilseed rape. Unpublished.

Artyszak A. 2014. Study on the effect of Herbagreen Z20 application on yielding, and technological quality of sugar beet roots. Unpublished.

Artyszak A, Gozdowski D, Kucińska K. 2014. The effect of foliar fertilization with marine calcite in sugar beet. *Plant Soil Environ* 60(9), 413–417.

Artyszak A, Gozdowski D, Kucińska K. 2015. The effect of silicon foliar fertilization in sugar beet— *Beta vulgaris* (L.) ssp. *vulgaris* conv. *crassa* (Alef.) prov. *Altissima* (Döll). *Turk J Field Crops* 20(1), 115–119.

Artyszak A, Gozdowski D, Kucińska K. 2016. The effect of calcium and silicon foliar fertilization in sugar beet. *Sugar Tech* 18(1), 109–114.

Bacchus GL. 2010. An evaluation of the influence of biodynamic practices including foliar-applied silica spray on nutrient quality of organic and conventionally fertilised lettuce (*Lactuca sativa* l.). *J Org Syst* 5(1), 4–13.

Bokor B, Bokorová S, Ondoš S, Švubová R, Lukačová Z, Szemes T, Lux A. 2014. Ionome and Si transporters affected by zinc and silicon interaction in maize. In *6th International Conference on Silicon in Agriculture*, Stockholm, Sweden, August 26–30, p. 44.

Ciecierski W, Kardasz H. 2014. Impact of silicon based fertilizer Optysil on abiotic stress reduction and yield improvement in field crops. In *6th International Conference on Silicon in Agriculture*, Stockholm, Sweden, August 26–30, pp. 54–55.

Cristancho RJA, Restrepo F. 2014. Silicon in agriculture new developments in Latin America In *6th International Conference on Silicon in Agriculture*, Stockholm, Sweden, August 26–30, p. 60.

Crusciol CAC, Pulz AL, Lemos LB, Soratto RP, Lima GPP. 2009. Effects of silicon and drought stress on tuber yield and leaf biochemical characteristics in potato. *Crop Sci* 49, 949–954.

Crusciol CAC, Soratto RP, Castro GSA, Costa CHMD, Ferrari Neto J. 2013. Foliar application of stabilized silicic acid on soybean, common bean, and peanut. *Rev Ciênc Agron* 44(2), 404–410.

Elungi K, Caldwell KP, Laing MD, Yobo KS. 2009. The effect of Eco-T (*Trichoderma harzianum*) and silicon for the control of *Rhizoctonia solani* and *Sclerotinia sclerotiorum* on selected sunflower cultivars. In *Silicone in Agriculture: 1st Annual Research Day*, Durban, South Africa, August 4, p. 12.

Fialová I, Šimková L, Vaculíková M, Sedláková B, Luxová M. 2014. The effect of silicon on the antioxidative response of maize young plants to salinity and zinc stress. In *6th International Conference on Silicon in Agriculture*, Stockholm, Sweden, August 26–30, p. 70.

Fuping T, Zihe Z, Zixuan C, Suomin W. 2005. Effect of Si fertilizer on yield of alfalfa. *Gansu Nongye Daxue Xuebao* 40, 42–47.

Ghanbari-Malidarreh A, Dastan S. 2014. Effect of silicon and phosphorus rates on Si and P content and uptake of rice (*Oryza sativa* L. cv. Tarom) and disease. In *6th International Conference on Silicon in Agriculture*, Stockholm, Sweden, August 26–30, p. 76.

Guo ZG, Liu HX, Tian FP, Zhang ZH, Wang SM. 2006. Effect of silicon on the morphology of shoots and roots of alfalfa (*Medicago sativa*). *Aust J Exp Agric* 46(9), 1161–1166.

Hanafy Ahmed AH, Harb EM, Higazy MA, Morgan ShH. 2008. Effect of silicon and boron foliar applications on wheat plants grown under saline soil conditions. *Int J Agric Res* 3, 1–6.

Hellal FA, Zeweny RM, Yassen AA. 2012. Evaluation of nitrogen and silicon application for enhancing yield production and nutrient uptake by wheat in clay soil. *J Appl Sci Res* 8(2), 686–692.

Jarosz Z. 2013. The effect of silicon application and type of substrate on yield and chemical composition of leaves and fruit of cucumber. *J Elem* 18(3), 403–414.

Jin X, Hall AM, Huang Y, Fitt BDL. 2014. The role of Si and potassium bicarbonate in controlling epidemics of *Podospheara aphanis* on strawberry in the field. In *6th International Conference on Silicon in Agriculture*, Stockholm, Sweden, August 26–30, p. 96.

Kara Z, Sabir A. 2010. Effects of Herbagreen application on vegetative developments of some grapevine rootstocks during nursery propagation in glasshouse. In *2nd International Symposium on Sustainable Development*, Sarajevo, June 8–9, pp. 127–132.

Khan W, Aziz T, Maqsood MA, Khalid M. 2014. Screening and categorization of maize cultivars for silicon acquisition under salinity stress. In *6th International Conference on Silicon in Agriculture*, Stockholm, Sweden, August 26–30, p. 102.

Kleiber T, Krzesiński W, Przygocka-Cyna K, Spiżewski T. 2015. The response of hydroponically grown lettuce under Mn stress to differentiated application of silica sol. *J Elem* 20(3), 609–619.

Lynch M. 2008. Silicates in contemporary Australian farming: A 20-year review. In *Silicon in Agriculture Conference South Africa 2008: 4th International Conference Abstracts*, KwaZulu-Natal, South Africa, p. 68.

Mahendran PP, Sreya UP, Balasubramaniam P. 2014. Fixing critical limit for silicon fertilization in the intensively rice growing soils of Periyar Vaigai Command area, Tamil Nadu, India. In *6th International Conference on Silicon in Agriculture*, Stockholm, Sweden, August 26–30, p. 118.

Marschner H. 2002. *Mineral Nutrition of Higher Plants*. 2nd ed. Academic Press, Amsterdam.

Marxen A, Klotzbücher T, Vetterlein D, Jahn R. 2014. Controls on Si uptake by rice in Southeast-Asian paddy soils. In *6th International Conference on Silicon in Agriculture*, Stockholm, Sweden, August 26–30, p. 122.

Matichenkov VV, Bocharnikova EA. 2008. New generation of silicon fertilizers. In *Silicon in Agriculture Conference South Africa 2008: 4th International Conference Abstracts*, KwaZulu-Natal, South Africa, p. 71.

Mikiciuk G, Mikuciuk M. 2008. A physiological response of strawberry (*Fragaria annanassa* Duch.) variety Senga Sengana to foliar application of potassium and silicon fertilizer. *Ann UMCS Sec E* LXIII(2), 81–85.

Mikiciuk G, Mikuciuk M. 2009. The influence of foliar application of potassium and silicone fertilizer on some physiological features of strawberry (*Fragaria ananassa* Duch.) variety Elvira. *Ann UMCS Sec E* LXIV(2), 20–27.

Passala R, Jain N, Deokate PP, Rao V, Minhas PS. 2014. Assessment of Silixol (OSA) efficacy on wheat physiology: Growth and nutrient content under drought conditions. In *6th International Conference on Silicon in Agriculture*, Stockholm, Sweden, August 26–30, p. 138.

Seebold KW, Datnoff LE, Correa-Victoria FJ, Kucharek TA, Snyder GH. 2000. Effect of silicon rate and host resistance on blast, scald, and yield of upland rice. *Plant Dis* 8, 871–876.

Tesfagiorgis HB, Laing MD. 2009. Effects of silicon concentrations on management of powdery mildew and growth of zucchini and zinnia. In *Silicone in Agriculture: 1st Annual Research Day*, Durban, South Africa, August 4, p. 3.

Toresano-Sánchez F, Díaz-Pérez M, Diánez-Martínez F, Camacho-Ferre F. 2010. Effect of the application of monosilicic acid on the production and quality of triploid watermelon. *J Plant Nutr* 33(10), 411–1421.

Trawczyński C. 2013. The effect of foliar fertilization with Herbagreen on potato yielding. *Ziemniak Polski* 2, 29–33.

Ugrinović M, Oljača S, Brdar-Jokanović M, Zdravković J, Girek Z, Zdravković M. 2011. The effect of liquid and soluble fertilizers on lettuce yield. Contemporary agriculture. *Serbian J Agric Sci* 60(1–2), 110–115.

Vaculík M, Vaculíková M, Lux A, Frossard E, Schulin R. 2014. Influence of silicon on zinc toxicity in wheat. In *6th International Conference on Silicon in Agriculture*, Stockholm, Sweden, August 26–30, p. 170.

Vaculíková M, Vaculík M, Šimková L, Fialová I, Luxová M, Tandy S. 2014. Growth and antioxidative response of maize roots exposed to antimony and silicon. In *6th International Conference on Silicon in Agriculture*, Stockholm, Sweden, August 26–30, p. 172.

Visser DD, Caldwell PM, Laing MD. 2009. The use of silicon (potassium silicate) to control *Phakopsora pachyrhizi* (soybean rust) on soybean. In *Silicone in Agriculture: 1st Annual Research Day*, Durban, South Africa, August 4, p. 10.

Wang Y, Liu M, Li D. 2009. Effects of silicon enrichment on photosynthetic characteristics and yield of strawberry. Northern Horticulture. http://en.cnki.com.cn/Article_en/CJFDTOTAL-BFYY200912034.htm.

White EB, Tubana SB, Dupree P, Yzenas J, Datnoff L, Mascagni H Jr. 2014. Effect of enhanced soil silicon on grain yield and agronomic parameters of wheat under sufficient and high nitrogen application rates. In *6th International Conference on Silicon in Agriculture*, Stockholm, Sweden, August 26–30, p. 178.

Wróbel S. 2012. Effects of fertilization of potato cultivar Jelly with foliar fertilizers YaraVita. *Biul. IHAR* 266, 295–305.

Zhiming X, Fengbin S, Hongwen X, Hongbo S, Ri S. 2014. Effects of silicon on photosynthetic characteristics of maize (*Zea mays* L.) on alluvial soil. *Sci. World J. The Scientific World Journal Volume* 2014, Article ID 718716, 6 pp.

17 Silicon and Antioxidant Defense System against Abiotic Stresses in Plants

An Overview

Muhammad Zia-ur-Rehman, Hinnan Khalid,
Fatima Akmal, Maqsooda Waqar, Muhammad Rizwan,
Farooq Qayyum, and Muhammad Nadeem

CONTENTS

ABSTRACT

The world's population is on an ever-increasing track, and to feed the huge community, demand for food is increasing with the same rate. Urbanization and industrialization are the major components that reduce plant growth and development by imposing different stresses like drought stress, salinity stress, and heavy metal stress. Less growth and development ultimately leads toward reduced quantity and poor quality of food. Present agricultural yield seems to fail in providing food to a common person on a sustainable basis. This situation leads toward numerous socioeconomic problems. Keeping in view this scenario, it is the need of hour to use strategies that can enhance the yield by supporting plant growth under such awful stresses. Silicon is one of the most important elements that play a beneficial role in plant growth and development. It helps the plants and crops attain their maximum potential of yield. Along with high yield, Si also has majestic effects in stress conditions for plants. Silicon enhances the resistance of plants against pathogens and diseases. It also protects plants against the harms and fluctuations of temperature, metal, salinity, and water deficit stress. Silicon also improves the mechanism of enzymes under stress conditions. The role of Si for supporting plants to combat abiotic stresses needs further exploration as our present knowledge is not up to the mark. This chapter is very effective in solving this critical situation and gives us a complete detail of silicon's supportive mechanism against various abiotic stresses.

17.1 DROUGHT IS AN ABIOTIC STRESS

Drought is a serious problem for agriculture in arid and semiarid regions of the globe. Drought poses drastic effects on the growth and yield of agricultural products and is a major and foremost door-knocking threat in the near future for crop production (Ranjan et al. 1996; Wang and Jiang 2007). According to Boyer (1982), "Drought is an abiotic stress caused when precipitation is below the minimum requirement of the plant productivity." Due to such water deficiency, crops cannot reach their maximum expected yield because this is one of the most limiting factors in crop productivity (Martin and Mitra 2001).

17.2 DROUGHT LIMITS PLANT GROWTH

Drought has a direct effect on the photosynthetic activity of plants and causes harm to photosynthetic pigments and their components (Iturbe-Ormaetxe et al. 1998). Monakhova and Chernyad'ev (2002) reported that drought lowers the carboxylation activity of Rubisco, the key enzyme involved in the photosynthetic mechanisms, and thus decreases plant growth. Drought causes severe damage to plant growth and development through different ways; the main consequences are

- Accumulation of reactive oxygen species (ROS) in the plant, which disrupts the cell membrane, protein, nucleic acid, and lipids of plant cells, leading to cell death (Iturbe-Ormaetxe et al. 1998; Monakhova and Chernyad'ev 2002)
- Diminution in stomatal conductance and escalation in the foliar temperature (Mohammadian et al. 2005)

- Lowering of the photosynthetic rate (Razzaghi et al. 2011) as a result of the denaturing of different enzymes
- Decreases in the plant growth rate, osmotic potential, and leaf water potential (Kusvuran 2011)
- Reduced seed yield and seed quality, for example, in tomato (Pervez et al. 2009) and soybean (Heatherly 1993)

17.3 DEFENSE STRATEGIES AGAINST DROUGHT

Plants respond differently to drought stress depending on the plant species, its developmental stage and duration, and intensity of drought. Drought-tolerant crops like wheat and barley can easily withstand the drought, whereas drought-moderate crops (corn, sunflower, etc.) can prevent their cells and tissues from severe water stress (Heszky 2007). Various plant defense mechanisms against drought are

- Maintaining the water potential in plants. The dehydrated roots increase the release of abscisic acid, which further accumulates in the leaves of plants and causes stomatal closure (Parry et al. 2002).
- Reduction in the tissue water loss by regulating the stomatal behavior (Gomes et al. 2004), changing the morphology (Du et al. 2009), and osmotic adjustment in which amino acids, sugars, and quaternary ammonia accumulate to draw water into the cell.
- Drought tolerance by effecting biochemical and molecular changes to maintain osmotic adjustment (Streeter et al. 2001; El-Tayeb 2006).
- Antioxidative defense machinery of plant constitutes an antioxidative defense enzymes and molecules, higher efficacy of these machinery could alleviates the damage imposed by oxidative stress (Talaat and Shawky 2013, 2014).
- Adaptive responses like increased root activity, biomass, proliferation, and root size (Shen et al. 2010).

17.4 ROLE OF Si UNDER DROUGHT ON PLANT GROWTH

Silicon has not yet been declared an essential element for plants, but it still has beneficial effects on their life, especially under stress conditions (Ma 2004). Silicon enhances the activities of the antioxidants, including superoxide dismutase (SOD), catalase (CAT), peroxidase (POD), ascorbate peroxidase (APX), and glutathione reductase, as well as nonenzymatic antioxidants such as cystein, glutathione, and ascorbic acid and ameliorates toxicity imposed by drought stress in plants (Ashraf and Harris 2004; Gunes et al. 2007b; Ashraf 2009). SOD detoxifies the ROS by converting them into hydrogen peroxide (H_2O_2), and this H_2O_2 can be further reduced by the action of antioxidant enzymes like CAT and POD (Sudhakar et al. 2001).

It has been widely reported that Si improves the growth of different plants under drought stress by activating the antioxidant defense mechanism, but much work still needs to be done (Agarie et al. 1998; Hattori et al. 2001; Lux et al. 2002). It was concluded by different studies that Si improves water availability to plants and increases the dry matter contents under drought conditions (Gong et al. 2003b). Different experiments have shown that Si enhances the resistance of plants against drought stress and plays a beneficial role in the physiological activities, as well as in the metabolic activities, of the plants by activating the antioxidant mechanism (Gong et al. 2005).

Drought stress causes a restriction in the uptake of the water; silicon addition maintains the water uptake by plants compared with Si-deficient plants (Hattori et al. 2001; Gong et al. 2003b). Silicon increases the wheat yield under drought stress. It has been reported that wheat crop had a high amount of the H_2O_2 under drought conditions when grown without treatment with Si (Gong et al. 2008). Silicon increases the activities of SOD and POD in the wheat plant by the detoxifying mechanism (Gong et al. 2003a). Furthermore, application of the proper amount of Si enhances the stability of wheat under drought stress (Tuna et al. 2008).

Xiong and Zhu (2002) reported that an increased proline concentration in plants under drought is controlled by Si addition. Accumulation of solutes in the plants increase the water availability in the plants as well as it regulates the damage imposed by reactive oxygen species that eventually enhance the tolerance of plants towards drought stress. Under drought, membrane permeability, lipid peroxidation (malondialdehyde [MDA]), and proline concentration increase in sunflower, while application of Si improves the tolerance strategy of sunflower by reducing the level of proline and increasing the antioxidant concentrations, like glutathione reductase, SOD, CAT, and POD (Gunes et al. 2008). Various studies have revealed the improvement in the tolerance of sunflower against drought stress by stimulation of the antioxidants up to a beneficial level (Sairam and Saxena 2000; Sairam et al. 2005; Tan et al. 2006; Zlatev et al. 2006).

Silicon application reduces the harm of drought in the chickpea plant by improving its resistance against stress (Anbessa and Bejiga 2002), as well as enhances the tolerance of the cucumber plant against this stress by reducing the activity of ROS and improving the net photosynthetic rate (Ma et al. 2004). Application of Si to soybean seedling enhances the rate of seed germination and growth potential of seedling by enhancing their physiological activities (Li et al. 2009).

17.5 MECHANISMS OF Si-MEDIATED ALLEVIATION OF DROUGHT STRESS IN PLANTS

As mentioned earlier, Si ameliorates drought stress and improves plant growth and development. This promising role of Si might be due to different mechanisms that affect plants positively, either directly or indirectly.

17.5.1 INCREASED UPTAKE OF ESSENTIAL NUTRIENTS BY PLANTS

For proper growth and development, all plants need essential nutrients in adequate amounts to provide mechanical, biochemical, and physical strength to the plants. It has been observed that nutrients in plants become imbalanced under abiotic stresses (Gunes et al. 2008; Chen et al. 2011; Emam et al. 2014). Silicon plays a very important role in the uptake of nutrients that are restricted due to drought. Its application increases the content of phosphorus in wheat under drought (Gong and Chen 2012). Similarly, it has been reported that Si application increases the contents of K and Ca in wheat leaf and Ca in roots under drought stress (Kaya et al. 2006). Emam et al. (2014) reported that under drought, enhanced P and K contents in rice straw are observed.

In a sunflower drought-stressed cultivar, Si application improved the uptake of various nutrients, including P, K, Ca, Fe, and Mn (Gunes et al. 2008). In contrast, Chen et al. (2011) reported that in rice plants, Si application decreased the mineral elements of K, Ca, Mg, and Fe, which were increased by drought. Thus, several studies need to be done to clearly understand this positive or negative correlative role of Si in nutrient uptake.

17.5.2 REGULATION OF GAS EXCHANGE ATTRIBUTES

Under drought stress, stomata regulate the water level inside the plant and conserve water by controlling the transpiration rate. The positive effect of Si in gas exchange attributes has been widely studied under drought stress (Gong et al. 2005; Lu et al. 2009). Abdalla (2011) reported that application of diatomite in soil enhances the stomatal conductance and photosynthetic rate while regulating the transpiration rate in plants. Similarly, Hattori et al. (2005) reported that supply of Si increases the net stomatal conductance in sorghum plants (Boyer 1982). An increase of Si accumulation in plants improves the tolerance against drought by increasing the CO_2 accumulation, lowering the transpiration rate, and maintaining the leaf water potential by leaf adjustment (Hattori et al. 2005).

Furthermore, it is observed that Si affects the stomata of plants, thus improving the water use efficiency of drought-stressed maize crops by lowering the transpiration rate of the plant leaf (Xu et al. 2005). Similar results were revealed by Amin et al. (2014), who found that in drought-stressed maize, addition of Si improved the photosynthetic rate of plants while reducing the transpiration rate. It has also been reported that Si application affects the diurnal changes in the gas exchange characteristics of plants, which further increases the photosynthetic rate, transpiration rate, and stomatal conductance of plants, compared with drought-stressed maize crop without Si application (Gong and Chen 2012). In short, application of Si enhances the gas exchange capacity of plants, which in turn increases the photosynthetic rate, water use efficiency, and growth of plants under drought stress.

17.5.3 OSMOTIC ADJUSTMENT

Plants can maintain their optimum water potential under drought stress by osmotic adjustment. Many researchers concluded that Si application could enhance the water contents in plants under drought conditions. It has been reported that silicon application increases the water potential in many plant species, such as in wheat (Gong et al. 2003a; Kaya et al. 2006; Tale Ahmad and Haddad 2011; Gong and Chen 2012; Ali et al. 2013), rice (Ming et al. 2012), chickpea (Gunes et al. 2007a; Kurdali et al. 2013), sunflower (Gunes et al. 2008b), cucumber (Ma 2004), and sorghum (Hattori et al. 2005). Application of 2.11 mM sodium silicate increased the water potential in the leaf up to ~0.2 MPa in drought-stressed wheat crop (Gong et al. 2003b).

In Kentucky bluegrass, application of Si increased the leaf water content up to 33% in drought, compared to without Si application (Saud et al. 2014). Si application lowers the osmotic potential in the roots of rice while showing no effects on the water content in sorghum plant under drought conditions (Sonobe et al. 2009). Gong et al. (2003) revealed that under drought stress, when Si is applied, the water potential inside the leaf increases in wheat plant, which might be due to an increase in the thickness of the wheat leaf. It might be possible that Si deposition inside the leaf of the plant does not allow the water molecules to leave the leaf surface, and thus maintains the water contents under drought conditions (Peters et al. 2007).

17.5.4 REDUCTION IN OXIDATIVE STRESS

Silicon application enhances the activity of antioxidant defense when plants are subjected to oxidative stress under drought by decreasing the over expression of ROS. Reactive oxygen species produced during drought stress are the superoxide radical (O^{2-}), hydroxyl radical (OH^-), application and H_2O_2 (Mittler et al. 2004).

Many researchers have revealed that Si application decreases the oxidative damage to plants under drought stress (Ma 2004; Gong et al. 2005). Si application decreases the oxidative stress, lipid peroxidation, and H_2O_2 in chickpea (Gunes et al. 2007b), sunflower (Gunes et al. 2008), and wheat (Gong et al. 2003b, 2008).

It has been reported that Si reduces the contents of MDA, electrolyte leakage (EL), and hydrogen peroxide in wheat crop under drought stress (Pei et al. 2010). Karmollachaab et al. (2013) discussed the role of Si as it increased the activity of SOD while reducing the EL when Si was applied at a rate of 30 kg ha^{-1} in the field. It has been reported that Si enhances the activity of antioxidant enzymes, including SOD, POD, and CAT, in wheat (Gong et al. 2005), chickpea (Gunes et al. 2007b), tomato (Shi et al. 2014), and sunflower (Ma 2004). It has also been noted that some studies explain that Si addition under drought stress can reduce these antioxidant enzymes; for example, Si decreased SOD and POD activity in cucumber (Ma et al. 2004) and soybean (Shen et al. 2010). This variation of the effect of Si on antioxidant enzymes is clearly illustrated by Gong et al. (2008), who showed that such variations are dependent not only on plant species, but also on the different growth stages at which Si is applied.

Like antioxidant enzymes, nonenzymatic antioxidants also play a vital role in mitigating oxidative stress. Glutathione reductase (GSH) activity in wheat increases with Si application compared to Si-deficient plant (Gong et al. 2005). It is also notable that the ascorbic acid (AA) concentration increases in wheat plants under drought stress (Pei et al. 2010).

17.5.5   MODIFICATION IN PLANT STRUCTURE

Modification in certain attributes of plants, like regulating leaf size, increasing root length, or rolling of leaves, helps to maintain the osmotic adjustment to combat water deficiency. Several evidences are available that Si application alleviates the drought stress by altering the plant structure. Si application of Ca and Mg silicates in tomato plant increased the tomato plant's height, caused a higher yield, and decreased the stem lodging under drought conditions (Pulz et al. 2008). Lux et al. (2002) reported that Si addition improved the water potential of plants under drought stress, basically due to the formation of a silica–cuticle double layer on the leaf epidermal cells.

Chen et al. (2010) demonstrated that exogenous application of 1.5 mM Si resulted in an increase in the root length, root activity, and surface area in drought-stressed plants. Furthermore, supply of Si at the rates of 100 and 200 mg L^{-1} mitigated salt and drought stress by increasing the plant shoot length and shoot and root mass (Hamayun et al. 2010). Similarly, addition of 3 mM Na_2SiO_3 in hydroponic solution reduced the leaf area (Bakhat et al. 2009), thus contributing largely to the reduction of the transpiration rate.

17.6   HEAVY METALS ABIOTICALLY STRESS PLANT GROWTH

Heavy metals are continuously entering into the soil, water (Guo et al. 2006; Liao and Xie 2007; Nehnevajova et al. 2007), and plants due to anthropogenic activities, as well as naturally by mineral dissolution (Wu and Cederbaum 2003; Sharpe-Young 2005). These metals eventually arrive in the food chain through plants and affect human and animal health (Zhang et al. 2007). Accumulation of nonessential trace elements (Cd, Pb, Cr, and Hg) various plant parts would adversely affects the uptake and transport of the essential nutrients, which ultimately results in the growth reduction of different plants (Wang et al. 2007a). Seedling growth of cotton and rice (Zhang et al. 2007), spinach (Maksymiec et al. 2005), and *Brasenia schreberi* L. (Yang et al. 1999) was adversely inhibited by different heavy metal concentrations.

Heavy metal stress reduces root vitality, and high uptake of heavy metals (Zn, Cd, Cu, and Pb) or absorption of inorganic nutrients is prevented, which leads to evident chlorosis. Toxicity imposed by heavy metals at gene levels would influence the overall synthesis and duplication of DNA and chromosomes, inducing chromosomal aberration (Zhang et al. 2007). Heavy metals affect the ultra-structure of plant cells. The grana cascade of chloroplast mitochondria decreases, disappears, or swells (Feng et al. 2007), and polypeptide compositions of the thylakoid membrane are degraded under heavy metal stress. The enzyme system is damaged by Cd, and penetration of the cell membrane is increased (Chen and Cutright 2001); as there is a significant relationship between penetration of the cell membrane and the Cd concentration, increasing membrane penetration will lead to increased Cd toxicity (Sun et al. 2007).

17.7   PLANT RESISTANCE AGAINST HEAVY METAL STRESS

Plants naturally detoxify heavy metal stress to varying degrees in a number of ways, as follows:

- Reduction in the absorption of heavy metals and their contents in plants, therefore decreasing their impact on plants and the ecosystem as a whole (Hall 2002)
- Accumulation of heavy metals in trichomes of the epidermis in order to prevent the direct effect of heavy metals on the mesophyll (Sarret et al. 2013)
- Precipitation or chelation of heavy metals in special sites in the plant for detoxification
- Increase in antioxidizing enzyme activities and removal of free radicals to prevent damage by the free radicals to the plant's production of different proteins in response to the heavy metals to combat metal stress (Zhang et al. 2007)

17.8 ROLE OF Si IN HEAVY METAL-CONTAMINATED SOIL

Silicon potentially combats with heavy metals, supporting plants in resisting their toxicity. Silicon also promotes the growth and yield of plants, even under metal stress.

17.8.1 Silicon's Role as a Growth Promoter under Metal Stress

Application of Si improves the growth and yield potential of several plant species, even though it is not considered an essential nutrient for plants (Epstein 2001). Many studies have revealed that Si enhances the growth of wheat, rice, cucumber, maize, and bamboo under metal stress (Feng et al. 2007; Zhang et al. 2007; da Cunha and do Nascimento 2009). Besides this, application of Si also provides protection to plants from several other abiotic stresses like radiation, drought and heat stress (Epstein 2001; Baudrimont et al. 2003).

17.9 SILICON ALLEVIATES METAL TOXICITY IN PLANTS

Elevated concentrations of heavy metals negatively correlate with biomass production and the grain yield of major field-grown crops (Nagajyoti et al. 2010; Adrees et al. 2015). Many studies support the ameliorative effects of Si on metal toxicity in many plant species. The common mechanisms are discussed below.

17.9.1 Decrease in Metal Uptake by Plants

Silicon application enhances tolerance to toxic metals in many plant species by reducing the uptake and translocation of metals within the plant body (Liang et al. 2007). Effect of Si on acquisition and translocation of metal differs with plant species and genotypes (Gulli et al. 2005; Martín et al. 2006; Verbruggen et al. 2009).

Application of Si limits the acquisition and agglomeration of Cd in maize (Liang et al. 2005), rice (Shi et al. 2005b), wheat (Rizwan et al. 2012), cotton (Farooq et al. 2013) and peanut (Shi et al. 2010). The concentration of Al is decreased in roots, stems, and leaves of peanut and rice seedlings by Si application (Shen et al. 2014). A Si-mediated decrease in Cu uptake was observed in wheat (Nowakowski and Nowakowski 1997; Keller et al. 2015). Si also decreases the Cr concentration in the shoots and roots of wheat, rice, and barley (Zeng et al. 2011; Ali et al. 2013; Tripathi et al. 2015). Exogenous application of Si significantly reduces the agglomeration of Zn in several parts of plant, for example, roots and leaves, of maize (Bokor et al. 2014).

This decrease in metal uptake in the presence of Si can be explained in many ways; for example, Si can enhance the uptake of Ca ions and reduce the uptake of Al by plants (Liang et al. 2007). Application of Si may activate the generation of root exudates, which can chelate the metals and reduce their uptake by roots (Nwugo and Huerta 2008). Besides this, application of Si also delimits the uptake of metal in plants by reducing the appoplasmic transport of metal which in turns reduces the concentration of free metals in the appoplasm (Iwasaki et al. 2002). Furthermore, the physical barrier produced by the deposition of Si in the vicinity of the endoderm may decreased the cell wall porosity of inner root tissues, thus decreases the metal concentration in the xylem (Shi et al. 2005; Gunes et al. 2007a).

17.10 ENHANCEMENT IN GAS EXCHANGE ATTRIBUTES AND PHOTOSYNTHETIC PIGMENT

It has been well documented that Si has significant effect on the biosynthesis of chlorophyll and photosynthetic machinery of plant under heavy metal toxicity. Application of silicon and Cd increases the chlorophyll a, chlorophyll b, and carotenoid contents in the leaves of maize (Malcovska et al. 2014), rice (Nwugo and Huerta 2008), wheat (Rizwan et al. 2012; Hussain et al. 2015), and cucumber (Feng et al. 2007). Application of Si application improves the rate of net photosynthetic activity, stomatal conductance, transpiration rate, and water use efficiency (Nwugo and Huerta 2008; Feng et al. 2009; Farooq et al. 2013).

17.11 CHANGES IN MINERAL NUTRIENT UPTAKE AND PLANT GROWTH AND BIOMASS

Significant role of Si in plant growth and biomass has been widely accepted. Si enhances the biomass of plants under heavy metal stress by reducing the metal toxicity occurs due to the dilution effect. Si displays a phenomenous role in the acquisition of nutrient by plants subjected under metal toxicity. In hydroponic solution, Si application increases the contents and accumulation of macronutrients (Ca, Mg, P, and K) and micronutrients (Mn and Zn) by wheat under Cr, Cd, and Cu toxicity (Rizwan et al. 2012; Keller et al. 2015; Tripathi et al. 2015). It has been well documented that foliar application with nano-Si enhances the concentration of Mg, Zn, and Fe in rice under Cd stress growth conditions (Wang et al. 2014). The combined application of Zn and Si increases Fe, P, and K in maize and rice seedlings under Zn toxicity (Kaya et al. 2009).

17.12 MECHANISM OF Si-MEDIATED METAL DETOXIFICATION IN PLANTS

Si-mediated metal detoxification mechanisms have been widely studied in many plant species. At the soil level, these may include immobilization of metals in the soil.

17.12.1 IMMOBILIZATION OF TOXIC METALS IN GROWTH MEDIA

In the culture media, Si-mediated amelioration of heavy metal toxicity in plants might occurs due to an external plant effect by reducing the availability of phytotoxic metals (Liang et al. 2005; Wan et al. 2011). Application of Si affects the several properties of soil, which in return control the heavy metal availability for plants (Liang et al. 2007; Rizwan et al. 2012). Soil organic matter and pH have been recognized as two of the most important parameters that regulates the availability of metals since Si reduces the metal bioavailability by accelerating soil pH (Morikawa and Saigusa 2004; Gu et al. 2011).

Application of Si also alters the speciation in soil solutions by forming of silicate complexes (Street 2012). Hara et al. (2012) explained that Si changes the chemical forms of Al in solution. Aluminum forms hydroxyl aluminosilicate (HAS) species in growth media by the addition of Si, which contributes to Al detoxification in plants (Hodson and Evans 1995).

A credible number of studies support that application of Si immobilizes Cd, Cu, and Zn in multimetal contaminated acidic soils by enhancing soil pH (Gu et al. 2011). In a study, it was reported that silica application in contaminated soil decreased the metal mobility in the soil, and x-ray diffraction data showed the generation of insoluble lead silicate in soil. Taken together, the application of Si could immobilize toxic metals in soil by enhancing soil pH and altering metal speciation in soil.

At plant levels, these mechanisms may include

1. **Enhancement of the antioxidant defense system**. Occurrence of oxidative stress in plant might be considered as one of the most unfortunate consequence of varied environmental stress, including heavy metal stress (Habiba et al. 2015). Si decreases oxidative stress in plant by decreasing ROS and enhancing the activities of antioxidant enzymes. For instance it was reported that application of Si in metal-stressed plant reduces the contents of MDA, hydrogen peroxide, and EL in shoots and roots of cotton plants under Pb, Cd, and Zn stress (Bharwana et al. 2013; Farooq et al. 2013).

   Application of Si reduces MDA concentration in rice, peanut, and pak choi under Cd stress (Wang et al. 2009; Kim et al. 2014) and in peanut under Al stress (Shen et al. 2014). Likewise, Liu et al. (2013) propounded that application of Si decreases EL in *Solanum nigrum* seedlings under Cd stress. Silicon reduces the thiobarbituric acid reactive substance contents in leaves and roots of rice under Cr stress (Zeng et al. 2011).

To combat the toxicity posed by heavy metal stress, plants have evolved a complex antioxidative defense system, including both enzymatic and nonenzymatic antioxidants (Solanki and Dhankhar 2011). Stimulation of antioxidants under metal appears to be a mechanism that accounts for the increased metal tolerance in plants. However, antioxidant enzyme activities decrease under higher metal stress (Ehsan et al. 2014; Adrees et al. 2015). Application of Si may increase antioxidant defense activity of plant (e.g., SOD, CAT, POD, and APX) under metal stress to help plants cope. Si-mediated activation of antioxidants has been recognized under Cd stress in pak choi (Song et al. 2009), peanut (Shi et al. 2010), cotton (Farooq et al. 2013), maize (Lukacova et al. 2013), and wheat (Hussain et al. 2015). Si also activates the antioxidant defense enzymes under Mn toxicity in cucumber (Shi et al. 2005b) and Mn tolerant cultivar of rice (Li et al. 2012), under Zn toxicity in rice roots (Song et al. 2011) and under Cu toxicity in *A. thaliana* (Khandekar and Leisner 2011).

2. **Compartmentation within plants**. Several studies have observed the compartmentation of metals in plant shoots and roots to understand the Si detoxification mechanisms in plants. In 1957, Williams and Vlamis demonstrated that application of Si did not affect the concentration of Mn in barley leaves, but Mn was evenly distributed in leaves, instead of discrete necrotic spots. So in this way, Si decreases the effect of Mn toxicity. Si decreases the metal concentration in roots compared to shoots (Peters et al. 2007). Naeem et al. (2014) reported that in wheat plants, Si application increases Cd in roots and decreases translocation to shoots and grains.

   Sequestration of heavy metals in metabolically less active cell compartments, such as the cell walls, might be an important mechanism for Si-mediated heavy metal tolerance in plants. The effect of Si on metal distribution varies not only among plants, but also among genotypes. Ye et al. (2012) reported that in the *Kandelia obovata* plant, Si enhances binding of Cd to the cell walls and restricts the apoplasmic transport of Cd in root tips. The effects of Si on metal compartmentation in the leaves and roots might be the crucial mechanism in metal uptake and detoxification by plants.

3. **Coprecipitation of Si with metals**. The coprecipitation of Si with metals in plants might be a mechanism that reduces metal toxicity in plants. Neuman et al. (2001) demonstrated that in *Minuartia verna*, a Si-accumulating dicotyledon, Zn was co-deposited as Zn silicates in the cell walls of the leaf epidermis. In addition, Gu et al. (2011) demonstrated that lower Cd concentrations in rice leaves might be due to co-deposition of Si with heavy metals in the stem. Iwasaki et al. (2002) suggested the coprecipitation of Si and Mn in the leaf apoplast of cowpea plants.

4. **Chelation**. The mechanisms by which Si mitigates heavy metal stress in plants might be related to the chelation of flavonoid-phenolic organic acids with metals. It has been reported that phenol exudation increases in maize plants treated with Si + Al, and phenolics like catechin and quercetin have strong Al-chelating capability, which might be considered as a factor leading to the Si-enhanced Al resistance in the root tip apoplast of maize (Barcela and Poschenrieder 2004; Zhang et al. 2007). Studies have shown that the Si-mediated release of chelates that can complex metals might be a mechanism, which could reduce the toxic effects of metal in plants.

5. **Structural alteration in plants**. Exogenous application of Si enhances the height of plants, root length, number of leaves per plant, and leaf area of cotton plants under Cd, Pb, and Zn stress (Bharwana et al. 2013; Farooq et al. 2013; Anwaar et al. 2015). Keller et al. (2015) reported that Si increases the root length of wheat seedlings under Cu stress. Si decreases leaf damage and necrosis under Cu stress in rice, and also metal toxic effects on the root architecture, and increases root length, compared with plants treated with metals only (Kim et al. 2014). Si alleviates the ultrastructural disorders imposed by chromium stress in

many plant species (Ali et al. 2013). Si plays a sumptuous role in defending the photosynthetic machinery from damage under metal stress. Si protects the chloroplast ultrastructure from disorientation due to Cd, Zn, and Cr stress in cucumber, maize, rice, and barley plants (Doncheva et al. 2009; Kaya et al. 2009; Feng et al. 2009; Ali et al. 2013).

17.13   SOIL SALINITY

A large concentration of salts in the soil that affect plant growth and production is called salt-affected land. The most common soluble ions that are responsible for salinization are sodium, calcium, magnesium, potassium cations, and carbonate, bicarbonate, chloride, sulfate, and nitrate anions (Shi et al. 2005; Wang and Han 2007; Afzal et al. 2014). It is feared that by 2050, more than 50% of cultivated land will be converted into salt-affected land (Ashraf and Harris 2004).

17.13.1   CAUSES OF SOIL SALINITY

Soil salinity can be due to natural conditions, which include the parent material having a high amount of soluble salts. It can also occur as a result of human activities, like unsuitable methods of irrigation, poor and imperfect cropping patterns and crop rotation, improper drainage systems, and contamination caused by different chemical industries and pesticide companies (Gomes 2009).

17.13.2   HARM OF SALINITY ON PLANTS

Salinity causes a lot of problems for plants, which result in low yield, poor economy returns, and low productivity (Hu et al. 2007). Specifically, salinity has a severe impact on seed germination, the plant's ability to take up water and nutrients, and plant growth, by causing changes in plants at multiple levels (morphological, physiological and biochemical processes) (Akbarimoghaddam et al. 2011). Soil salinity causes nutrient deficiency, such as nitrogen (N), calcium (Ca), potassium (K), phosphorus (P), zinc (Zn), and iron (Fe). Salinity is also the major cause of less phosphorous availability, because phosphate ions become precipitated with calcium ions (Bano and Fatima 2009). Thus, salinity causes a nutritional imbalance, osmotic effect, and specific ion toxicity effects, and damages plants (Ashraf 2004; Lauchli and Grattan 2007). Cell death of the plant occurs due to the high accumulation of sodium ions in the plant cells (Munns and Tester 2008). The photosynthetic rate decreases due to the smaller leaf area, less chlorophyll content, and poor stomatal conductance (Netondo et al. 2004; Moradi and Ismail 2007). With all these effects, the growth and yield of plants decrease up to the critical level (Ashraf and Harris 2004).

17.13.3   PHYSIOLOGICAL AND BIOCHEMICAL PROCESSES OF PLANTS AFFECTED BY SALINITY

It is well documented that high salt concentrations reduce the bioavailability of water for plants by creating osmotic stress and lowering the water potential (Kausar and Shahbaz 2013). Under the salinity stress, Na^+ and Cl^- are available in excess amounts, and their high availability can reduce the availability of other ions. This is characterized as specific ion toxicity and ion imbalance, which are the basic factors that reduce the growth of plants (Nasim et al. 2007; Gogile et al. 2013). High availability of the Na^+ reduces the availability of K^+, and thus restricts the activity of certain enzymes (de Azevedo Neto et al. 2006; Gorai and Neffati 2011; Doayan 2013). Salt stress causes oxidative stress, along with osmotic stress and specific ion toxicity (Hoque et al. 2007). Oxidative stress stimulates the agglomeration of ROS, such as superoxide, hydrogen peroxide, hydroxyl radical, and singlet oxygen (Chinnusamy et al. 2004). These ROS severely affect the nucleic acid, as well as proteins and lipids of the plants (Menezes-Benavente et al. 2004). This damage to the nucleic acid, protein, and lipids raises the permeability of the cell membrane (Lokhande et al. 2010).

17.13.4 SILICON AND ABIOTIC STRESSES, ALONG WITH ANTIOXIDANT DEFENSE SYSTEM

Much work has been done on the ability of Si to either lessen the effects of abiotic stress or produce resistance against it. There are several physical stresses, like temperature, ultraviolet radiation, water deficiency, and freezing. Chemical stresses are also present in huge volumes in the environment, such as salt stress, nutrient imbalance, and toxicity caused by metals. Silicon helps the plants cope with these stresses by creating and promoting the antioxidant defense mechanism. Silicon supports plants in producing such antioxidants that cause resistance in the plants against abiotic stresses.

17.13.5 SILICON'S ROLE IN COMBATING SALINITY STRESS

Silicon can reduce the precarious effects of salinity and thus improve the plant's growth and productivity (Liang et al. 2003). It has been stated that there are several methods by which we can reduce the effect of salinity on the growth of plants (Liang et al. 2007). Silicon enhances the water level of plants (Ghanem et al. 2008). It also activates the photosynthetic rate of plants (Shu and Liu 2001). With the application of silicon to plants, the ROS scavenging mechanism is activated (Zhu et al. 2004). Sodium availability to plants is reduced due to the presence of silicon (Liang et al. 2007). It has been reported that silicon forms complexes with toxic sodium, and thus reduces its availability to plants (Bybordi and Tabatabaei 2009). Due to the complexation of sodium with silicon, sodium uptake is lower and potassium uptake is enhanced (Tahir et al. 2010; Shi et al. 2013). This greater uptake of potassium, compared with sodium, occurs due to the high availability of potassium (Hasegawa et al. 2000). Silicon also activates the antioxidant mechanism and reduces salinity stress. Thus, with all these mechanisms, silicon combats salinity stress in different plants and helps them to grow in a better way.

17.13.6 SILICON AND WHEAT CROP

In wheat, growth is reduced by unavailability of the proper nutrients due to salinity stress. When saline conditions are available to the wheat crop, salts gather at a high concentration in different parts, especially the leaves (Gong et al. 2006). Any element that can reduce the uptake of sodium is helpful in combating the harmful effects of salinity. It has been reported that silicon reduces the uptake of sodium in saline environments (Raza et al. 2006). Silicon resists the upward movement of sodium by forming a complex with it (Ahmad et al. 1992; Tuna et al. 2008). Less sodium uptake also causes less uptake of the singlet oxygen (Lee et al. 2001). Fewer uptakes of radicals and oxygen singlet reduce the damages to the nucleic acid, proteins and cell membranes in wheat (Menezes-Benavente et al. 2004). Salinity causes harm to the wheat crop by making the enzymatic processes slower and also by decreasing the formation of proteins (Tester and Davenport 2003). Through an experiment it was concluded that the wheat crop shows reduced activity of antioxidants in salt stress conditions due to the higher uptake of sodium, but silicon decreases its effect (Ali et al. 2013). In this way, silicon also helps the wheat crop to activate the antioxidant system in salt stress conditions. Silicon reduces the harmful effects of salinity by raising the activity of CAT and SOD, thus improving the antioxidant system (Al-aghabary et al. 2005).

17.13.7 SILICON AND RICE CROP

Salinity causes hazards to rice by making water less available, causing the toxicity of some ions, like Na^+, Cl^-, and SO_4^{2-}, and a nutritional imbalance (Munns and Tester 2008). An excess quantity of salts raises the cytosolic Ca^+ concentration and stimulates the plasma membrane anion channel (Lee et al. 2010). ROS change the metabolism of the cell, and thus are harmful for the proteins, lipid membranes, and plasma membranes (Imlay 2003). To overcome this damage caused by ROS, rice plants have modified their antioxidant enzyme mechanism (Guo et al. 2007). Hazards of ROS can be reduced by antioxidants like CAT, POD, and polyphenol oxidase. In salt stress conditions, plants become unable to alleviate the effects of ROS. Lipid peroxidation causes oxidative stress and activates the root H⁺-ATPase in the membranes of the plants. It is said that rice accumulates silicon in its different parts, and therefore it is required to obtain the proper target yield (Ma et al. 2007). Thus, salinity has created resistance in rice plants against salinity stress (Lee et al. 2010). When silicon is present in higher concentrations in rice plants, it provides strength to the cell membrane by making silicified cells (Ma and Yamaji 2008). It has also been stated that with all these effects, silicon increases the ROS in the rice plants (Zhang et al. 2006; Rouhier and Jacquot 2008; Kim et al. 2014). Actually, ROS are needed to stimulate the signaling mechanism, growth, and development of the rice crop, but an excess concentration is dangerous for plant metabolism (Rouhier and Jacquot 2008; Choudhury et al. 2011). These ROS are eliminated by CAT and POD by a direct mechanism, as well as indirectly by ascorbate and glutathione, which form complex molecules to eliminate the ROS (Imlay 2003; Mittler et al. 2004). Different experiments also show that silicon accumulates in rice plants with the passage of time and reduces the effects of salinity (Kim et al. 2014).

17.13.8 Silicon and Maize Crop

A high concentration of salt damages the nitrogen metabolism of maize crop (Kant et al. 2007). Salinity also hinders the assimilation of ammonia in the plant, and thus affects its proper growth (Sahu et al. 2001). Various metabolic pathways, including photorespiration, phenylpropanoid metabolism, use of the nitrogen transport compounds, and amino acid catabolism from symbiotically fixed nitrogen, produce ammonia within the plant (Hirel and Lea 2001). Because ammonia is very damaging for the plant, it must be converted to less toxic compounds to save the plant from its damaging effects (Kronzucker et al. 2001). Salinity stress alters the ammonia assimilation mechanism within the plant (Wang et al. 2007b; Wang et al. 2009). The results of an experiment showed that silicon improves the root length and cell strength of maize crop. It also reduces salinity stress by lowering the high accumulation of salts like NaCl and improves the antioxidant mechanism of maize plant. Roots of the maize crop show lower growth under salinity conditions when not treated with silicon (Cicek and Cakirlar 2002; Zhang et al. 2007). It has also been reported that ammonia produced as a result of salinity stress was detoxified by the presence of glutamate dehydrogenase (GDH), thus helping in the control of damage caused by salinity (Kochanova et al. 2014).

17.13.9 Silicon and Cucumber

Salinity is a major hindrance in the growth and development of cucumber. During the different metabolic processes, ROS are released, such as superoxide radical (O^{2-}), hydroxyl radical (OH'), singlet oxygen ($^1O_2$), and hydrogen peroxide (H_2O_2), and all these are very harmful in excess concentrations in cucumber (Becana et al. 2000; Kmlazawa et al. 2000). ROS have serious effects that harm the membranes of the plants, nucleic acids, lipids, photosynthetic pigments, and proteins (Gill and Tuteja 2010). As these ROS are harmful for cucumber plants, the plants have modified their antioxidant mechanism to lessen their damaging effects. When silicon was applied to cucumber plants, they showed much better growth in salinity conditions, compared with the cucumber plants grown in salinity conditions but not treated with silicon (Zhu et al. 2004).

17.13.10 Silicon and Alfalfa Plant

Salinity promotes the activity and also produces active oxygen species (AOS), including hydrogen peroxide, superoxide, and hydroxyl radicals, in high concentrations in alfalfa. When alfalfa faces stress caused by salt, it starts showing a high concentration of MDA. This compound is produced when the decomposition of polyunsaturated fatty acids of membranes occurs occur (Bless 2016).

But this is the compound that causes serious oxidant damage to plants. This damage reduces the yield of plants and is a big factor in causing economic loss in countries. To control this damage, the maize plant modifies the antioxidant enzyme system.

Different species and cultivars of alfalfa, however, show different effects from salt stress (Wang and Han 2007; Wang et al. 2009). Silicon shows better results in salt stress conditions. When silicon is provided to alfalfa, it activates the antioxidant enzyme mechanism. This causes resistance in alfalfa under salt stress conditions.

17.13.11 SILICON AND GRAPES

Salt stress and boron toxicity severely reduce the productivity and growth of grapes (Alpaslan and Gunes 2001). Salinity has damaging effects on grapes due to its high osmotic potential, which causes resistance in the availability of water to the plant. An excess amount of salt is also the source of a nutrient imbalance, due to the presence of some ions in high concentrations, compared to others (Munns 2005; Yamaguchi and Blumwald 2005). Under salt stress conditions, silicon helps grapes to activate the antioxidant defense mechanism. These antioxidants convert damaging compounds like ROS into less damaging compounds like hydrogen peroxide and water (Sairam et al. 2005; Zhu and Gong 2011). Antioxidants also stabilize the macromolecule concentration in the plant by reducing the damaging effect of ROS (Xiong and Zhu 2002). Thus, although grapevines show poor growth under salinity stress (Gunes et al. 2006), silicon reduces the damage that occurs (Gunes et al. 2006, 2007a). The results of different experiments have also proved the supportive role of silicon for grapevines under salinity stress (Soylemezoglu et al. 2009).

17.13.12 SILICON AND TOMATO

In salinity stress, tomato has to overcome the effect of the ROS, and for this purpose, tomato activates the antioxidant enzyme mechanism (Al-aghabary et al. 2005). SOD is supposed to be the basic enzyme that reduces the damaging effect of the ROS and forms the hydrogen peroxide as the end product. CAT converts to the hydrogen peroxide to a less toxic form by reducing the electrons from oxygen (Romero-Aranda et al. 2006). CAT usually will not be present in the chlorophyll, but APX will be the component of the chlorophyll, and thus detoxification of the hydrogen peroxide will occur there (Gunes et al. 2007a). POD is another antioxidant enzyme, and it also helps in reducing the damaging effect of the salinity stress and provides resistance to the tomato against the salt stress conditions by converting the hydrogen peroxide to water. Silicon improves the dry matter content, and also, growth of the tomato will be enhanced by use of the silicon (Al-aghabary et al. 2005). Antioxidant enzyme activity was inhibited under salt stress conditions in the different cultivars of tomatoes. SOD reduced its activity drastically in the tomatoes in the saline environment. But addition of silicon improved the activity of SOD in tomatoes, and thus tolerance against the salinity stress was produced in them.

17.13.13 SILICON AND OKRA

Salinity stress causes damage to okra in different ways. It may be affected by the low availability of water as a result of a high concentration of salt or by a nutrient imbalance as a result of a high concentration of salt reducing the availability of the other nutrients (Munns 2005; Yamaguchi and Blumwald 2005). In response to these damages, okra activates the certain defense mechanisms, including the stimulation of the antioxidant enzyme system. It has been revealed by the experiments that proline induces resistance against salinity stress in the plant (Ronde et al. 2000). Glycine betaine is a compound that encourages the plant to tolerate abiotic stresses like salinity by detoxifying the ROS. Glycine betaine can easily dissolve in water and will reach to the plastids and chloroplasts, thus helping the plant to overcome salinity stress. Salinity stress also causes harm to the protein content of the plant (Rahnama and Ebrahimzadeh 2004). Okra plant stimulates different antioxidants, like SOD,

POD, CAT, and APX, in salinity stress. But studies have revealed that the activity of the antioxidants will be reduced due to the presence of a high concentration of salts, like NaCl. In such situations, silicon helps the plant to raise the activities of the antioxidants. Experiments have shown that silicon improves the growth of okra in salt stress conditions. In salt stress conditions, silicon also provides a supportive role to the plants by reducing the availability of salts. Silicon has no role in the catalytic activities during the metabolism of the plants (Richmond and Sussman 2003). It has been stated that in stress conditions like salinity stress, silicon has an effect on the different activities of the plants, including physiological activities and structural activities (Shen et al. 2010). Results of different experiments have shown that silicon improves the growth and productivity of field crops, including okra (Alvarez and Datnoff 2001). From an experiment that was conducted on okra, it was revealed that silicon also enhances the concentration of proline in salt stress conditions, which provides tolerance against salinity stress (Abbas et al. 2012). Silicon also improves the osmotic potential of the okra plant in salt stress conditions by stimulating some enzymes, as well as improving the antioxidant enzyme mechanism and reducing the effect of salinity stress by lowering the concentration of the MDA, which causes damage to the plant in salinity conditions (Zhu et al. 2004). It has also been reported that silicon has a remarkable ability to enhance the activity of antioxidant enzymes (Kusvuran 2012).

17.13.14 SILICON AND SUNFLOWER

Silicon improves the growth of sunflower by activating the antioxidant defense mechanism (Abdolzadeh et al. 2014). It has been stated that salinity lowers the activities of different antioxidants, such as SOD, CAT, and POX, in sunflower, and thus its growth is affected. Sunflower also accumulates ROS in its different plant parts, which damages normal growth of the sunflower plant. CAT was considered the most important antioxidant enzyme that decreases the damage of salinity by detoxifying the ROS. Sunflower also shows a high concentration of SOD and POX in its plant parts in the presence of silicon, which reduces the hazards of salinity (Azooz et al. 2015) and improves growth of the sunflower. Tolerance in the sunflower is due to a reduction of the availability of NaCl, and thus less oxidation of the membrane lipids. Glutathione (GSH) and ascorbic acid (AA), both of which are nonenzymatic antioxidants, also showed a high concentration in the sunflower due to abiotic stress such as salinity (Ruiz and Blumwald 2002). It is said that GSH and AA both have the ability to significantly control salinity stress in sunflower (Tausz et al. 2004). Silicon improves the antioxidant mechanism in sunflower, and an enhanced yield of sunflower was also observed.

17.13.15 SILICON AND SORGHUM

Sorghum can be grown in salinized areas, as it is somewhat resistant to salinization and overcomes the hazards of salinity stress to some extent (Hashemi et al. 2010). In severe salinization growth, the productivity of the sorghum will also be reduced. Mostly, damage to the plants starts in the cell membrane (Ashraf 2009). It has been stated that the silicon improves the membrane stability index (MSI) and supports the cells against salinity stress (Hayat et al. 2008). Silicon enhances the activity of the antioxidants. High activity of the antioxidants will provide resistance to the sorghum plant against the damage and harm of salinity stress. Silicon has also been shown to have a role in lowering the oxidation of the lipids of the membrane (Kafi et al. 2011). In this way, salinity stress can be overcome in sorghum by the application of silicon.

REFERENCES

Abbas W, Ashraf M, Akram NA. (2012). Alleviation of salt-induced adverse effects in eggplant (*Solanum melongena* L.) by glycinebetaine and sugarbeet extracts. *Scientia Horticulturae* 125: 188–195.
Abdalla MM. (2011). Beneficial effects of diatomite on growth, the biochemical contents and polymorphic DNA in *Lupinus albus* plants grown under water stress. *Agric Biol. North Am* 2(2): 207–220.

Abdolzadeh A, Chalmardi ZK, Sadeghipour HR. (2014). Silicon nutrition potentiates the antioxidant metabolism of rice plants under iron toxicity. *Acta Physiologiae Plantarum* 36: 493–502.

Adrees M, Ali S, Rizwan M, Zia-ur-Rehman, M, Ibrahim M, Abbas F, Farid M, Qayyum MF, Irshad MK. (2015). Mechanisms of silicon-mediated alleviation of heavy metal toxicity in plants: A review. *Ecotoxicol Environ Saf* 119: 186–197.

Afzal M, Ahmad A, Alderfasi AA, Ghoneim A, Saqib M. (2014). Physiological tolerance and cation accumulation of different genotypes of *Capsicum annum* under varying salinity stress. *Proceedings of the International Academy of Ecology and Environmental Sciences* 4: 39–49.

Agarie S, Hanaoka N, Ueno O, Miyazaki A, Kubota F, Agata W, Kaufman PB. (1998). Effects of silicon on tolerance to water deficit and heat stress in rice plants (*Oryza sativa* L.), monitored by electrolyte leakage. *Plant Production Science* 1: 96–103.

Ahmad R, Zaheer SH, Ismail S. (1992). Role of silicon in salt tolerance of wheat (*Triticum aestivum* L.). *Plant Science* 85: 43–50.

Akbarimoghaddam H, Galavi M, Ghanbari A, Panjehkeh N. (2011). Salinity effects on seed germination and seedling growth of wheat bread cultivars. *Trakia J Sci* 9: 43–50.

Al-aghabary K, Zhu Z, Shi Q. (2005). Influence of silicon supply on chlorophyll content, chlorophyll fluorescence, and antioxidative enzyme activities in tomato plants under salt stress. *Journal of Plant Nutrition* 27: 2101–2115.

Ali S, Farooq MA, Yasmeen T, Hussain S, Arif MS, Abbas F, Bharwana SA, Zhang G. (2013). The influence of silicon on barley growth, photosynthesis and ultra-structure under chromium stress. *Ecotoxicology and Environmental Safety* 89: 66–72.

Alpaslan M, Gunes A. (2001). Interactive effects of boron and salinity stress on the growth, membrane permeability and mineral composition of tomato and cucumber plants. *Plant and Soil* 236: 123–128.

Alvarez J, Datnoff LE. (2001). The economic potential of silicon for integrated management and sustainable rice production. *Crop Protection* 20: 43–48.

Amin M, Ahmad R, Basra SM, Murtaza G. (2014). Silicon induced improvement in morpho-physiological traits of maize (*Zea mays* L.) under water deficit. *Pak J Agric Sci* 51: 187–196.

Anbessa Y, Bejiga G. (2002). Evaluation of Ethiopian chickpea landraces for tolerance to drought. *Genetic Resources and Crop Evolution* 49: 557–564.

Anwaar SA, Ali S, Ali S, Ishaque W, Farid M, Farooq MA, Najeeb U, Abbas F, Sharif M. (2015). Silicon (Si) alleviates cotton (*Gossypium hirsutum* L.) from zinc (Zn) toxicity stress by limiting Zn uptake and oxidative damage. *Environ Sci Pollut Res* 22: 3441–3450.

Ashraf M. (2004). Some important physiological selection criteria for salt tolerance in plants. *Flora-Morphology, Distribution, Functional Ecology of Plants* 199: 361–376.

Ashraf M. (2009). Biotechnological approach of improving plant salt tolerance using antioxidants as markers. *Biotechnology Advances* 27: 84–93.

Ashraf M, Harris PJC. (2004). Potential biochemical indicators of salinity tolerance in plants. *Plant Science* 166: 3–16.

Azooz MM, Metwally A, Abou-Elhamd MF. (2015). Jasmonate-induced tolerance of *Hassawi okra* seedlings to salinity in brackish water. *Acta Physiologiae Plantarum* 37: 1–13.

Bakhat HF, Hanstein S and Schubert S. (2009). Optimal level of silicon for maize (*Zea mays* L. c.v. Amadeo) growth in nutrient solution under controlled conditions. Proceedings of the International Plant Nutrition Colloquium XVI, Davis, CA.

Bano A, Fatima M. (2009). Salt tolerance in *Zea mays* (L). following inoculation with *Rhizobium* and *Pseudomonas*. *Biology and Fertility of Soils* 45: 405–413.

Barcela J, Poschenrieder C. (2004). Structural and ultrastructural changes in heavy metal exposed plants. In Prasad MNV (ed.), *Heavy Metal Stress in Plants*, pp. 223–248. Springer, Berlin.

Baudrimont M, Andres S, Durrieu G, Boudou A. (2003). The key role of metallothioneins in the bivalve *Corbicula fluminea* during the depuration phase, after in situ exposure to Cd and Zn. *Aquatic Toxicology* 63: 89–102.

Becana M, Dalton DA, Moran JF, Iturbe-Ormaetxe I, Matamoros MA, Rubio CM. (2000). Reactive oxygen species and antioxidants in legume nodules. *Physiologia Plantarum* 109: 372–381.

Bharwana SA, Ali S, Farooq MA, Iqbal N, Abbas F, Ahmad MSA. (2013). Alleviation of lead toxicity by silicon is related to elevated photosynthesis, antioxidant enzymes suppressed lead uptake and oxidative stress in cotton. *J. of Biorem. Biodegrad.* 4: 187.

Bokor B, Vaculík M, Slováková L, Masarovič D, Lux A. (2014). Silicon does not always mitigate zinc toxicity in maize. *Acta Physiol. Plant.* 36: 733–743.

Boyer JS. (1982). Plant productivity and environment. *Science* 218: 443–448.

Bybordi A, Tabatabaei J. (2009). Effect of salinity stress on germination and seedling properties in canola cultivars (*Brassica napus* L.). *Notulae Botanicae Horti Agrobotanici Cluj-Napoca* 37: 71.

Chen H, Cutright T. (2001). EDTA and HEDTA effects on Cd, Cr, and Ni uptake by *Helianthus annuus*. *Chemosphere* 45: 21–28.

Chen W, Yao X, Cai K, Chen J. (2010). Silicon alleviates drought stress of rice plants by improving plant water status, photosynthesis and mineral nutrient absorption. *Biol Trace Elem Res*. Available at http://www.springerlink.com/content/nv2685m42700841p (accessed June 6, 2011).

Chen W, Yao X, Cai K, Chen J. (2011). Silicon alleviates drought stress of rice plants by improving plant water status, photosynthesis and mineral nutrient absorption. *Biol Trace Elem Res* 142(1):67–76.

Chinnusamy V, Schumaker K, Zhu JK. (2004). Molecular genetic perspectives on cross-talk and specificity in abiotic stress signalling in plants. *Journal of Experimental Botany* 55: 225–236.

Choudhury S, Panda P, Sahoo L, Panda SK. (2011). Reactive oxygen species signaling in plants under abiotic stress. *Plant Signaling & Behavior* 8: e23681.

Cicek N, Cakirlar H. (2002). The effect of salinity on some physiological parameters in two maize cultivars. *Bulgarian Journal of Plant Physiology* 28: 66–74.

da Cunha KPV, do Nascimento CWA. (2009). Silicon effects on metal tolerance and structural changes in maize (*Zea mays* L.) grown on a cadmium and zinc enriched soil. *Water, Air, and Soil Pollution* 197: 323–330.

de Azevedo Neto AD, Prisco JT, Enéas-Filho J, de Abreu CEB, Gomes-Filho EA. (2006). Effect of salt stress on antioxidative enzymes and lipid peroxidation in leaves and roots of salt-tolerant and salt-sensitive maize genotypes. *Environmental and Experimental Botany* 56: 87–94.

Doayan M. (2013). Antioxidative and proline potentials as a protective mechanism in soybean plants under salinity stress. *African Journal of Biotechnology* 10: 5972–5978.

Doncheva SN, Poschenrieder C, Stoyanova Z, Georgieva K, Velichkova M, Barceló J. (2009). Silicon amelioration of manganese toxicity in Mn-sensitive and Mn-tolerant maize varieties. *Environ Exp Bot* 65: 189–197.

Du W, Wang M, Fu S, Yu D. (2009). Mapping QTLs for seed yield and drought susceptibility index in soybean (*Glycine max* L.) across different environments. *Journal of Genetics and Genomics* 36: 721–731.

Ehsan S, Ali S, Noureen S, Mahmood K, Farid M, Ishaque W, Shakoor MB, Rizwan M. (2014). Citric acid assisted phytoremediation of cadmium by *Brassica napus* L. *Ecotoxicol Environ Saf* 106: 164–172.

El-Tayeb MA. (2006). Differential response of two *Vicia faba* cultivars to drought: Growth, pigments, lipid peroxidation, organic solutes, catalase and peroxidase activity. *Acta Agronomica Hungarica* 54: 25–37.

Emam MM, Khattab HE, Helal NM, Deraz AE. (2014). Effect of selenium and silicon on yield quality of rice plant grown under drought stress. *Aust J Crop Sci* 8(4): 596.

Epstein E. (2001). Silicon in plants: Facts vs. concepts. *Studies in Plant Science* 8: 1–15.

Farooq MA, Ali S, Hameed A, Ishaque W, Mahmood K, Iqbal Z. (2013). Alleviation of cadmium toxicity by silicon is related to elevated photosynthesis, antioxidant enzymes; suppressed cadmium uptake and oxidative stress in cotton. *Ecotoxicol Environ Saf* 96: 242–249.

Feng S, Wang X, Wei G, Peng P, Yang Y, Cao Z. (2007). Leachates of municipal solid waste incineration bottom ash from Macao: Heavy metal concentrations and genotoxicity. *Chemosphere* 67: 1133–1137.

Feng J, Shi Q, Wang X, Wei M, Yang F, Xu H. (2010). Silicon supplementation ameliorated the inhibition of photosynthesis and nitrate metabolism by cadmium (Cd) toxicity in *Cucumis sativus* L. *Sci Hort* 123: 521–530.

Ghanem ME, Albacete A, Martínez-Andújar C, Acosta M, Romero-Aranda R, Dodd IC, Lutts S, Pérez-Alfocea F. (2008). Hormonal changes during salinity-induced leaf senescence in tomato (*Solanum lycopersicum* L.). *Journal of Experimental Botany* 59: 3039–3050.

Gogile A, Andargie M, Muthuswamy M. (2013). The response of some cowpea (*Vigna unguiculata* L. Walp.) genotypes for salt stress during germination and seedling stage. *Journal of Stress Physiology & Biochemistry* 9: 73–84.

Gomes CP. (2009). Computational sustainability: Computational methods for a sustainable environment, economy, and society. *Bridge* 39: 5–13.

Gomes MdMdA, Lagôa AMMA, Medina CL, Machado EC, Machado MA. (2004). Interactions between leaf water potential, stomatal conductance and abscisic acid content of orange trees submitted to drought stress. *Brazilian Journal of Plant Physiology* 16: 155–161.

Gong H, Chen K. (2012). The regulatory role of silicon on water relations, photosynthetic gas exchange, and carboxylation activities of wheat leaves in field drought conditions. *Acta Physiol Plant* 34(4): 1589–1594.

Gong H, Zhu X, Chen K, Wang S, Zhang C. (2005). Silicon alleviates oxidative damage of wheat plants in pots under drought. *Plant Science* 169: 313–321.

Gong HJ, Chen KM, Chen GC, Wang SM, Zhang C. (2003a). Effects of silicon on the growth of wheat and its antioxidative enzymatic system. *Chinese Journal of Soil Science* 34: 55–57.

Gong HJ, Chen KM, Chen GC, Wang SM, Zhang CL. (2003b). Effects of silicon on growth of wheat under drought. *Journal of Plant Nutrition* 26: 1055–1063.

Gong HJ, Chen KM, Zhao ZG, Chen GC, Zhou WJ. (2008). Effects of silicon on defense of wheat against oxidative stress under drought at different developmental stages. *Biologia Plantarum* 52: 592–596.

Gong HJ, Randall DP, Flowers TJ. (2006). Silicon deposition in the root reduces sodium uptake in rice (*Oryza sativa* L.) seedlings by reducing bypass flow. *Plant, Cell & Environment* 29: 1970–1979.

Gorai M, Neffati M. (2011). Osmotic adjustment, water relations and growth attributes of the xero-halophyte *Reaumuria vermiculata* L. (Tamaricaceae) in response to salt stress. *Acta Physiologiae Plantarum* 33: 1425–1433.

Gu H-H, Qiu H, Tian T, Zhan S-S, Chaney RL, Wang S-Z, Tang Y-T, Morel J-L, Qiu R-L. (2011). Mitigation effects of silicon rich amendments on heavy metal accumulation in rice (*Oryza sativa* L.) planted on multi-metal contaminated acidic soil. *Chemosphere* 83: 1234–1240.

Gulli M, Rampino P, Lupotto E, Marmiroli N, Perrotta C. (2005). The effect of heat stress and cadmium ions on the expression of a small hsp gene in barley and maize. *Journal of Cereal Science* 42: 25–31.

Gunes A, Inal A, Bagci EG, Pilbeam DJ. (2007a). Silicon-mediated changes of some physiological and enzymatic parameters symptomatic for oxidative stress in spinach and tomato grown in sodic-B toxic soil. *Plant and Soil* 290: 103–114.

Gunes A, Pilbeam DJ, Inal A, Bagci EG, Coban S. (2007b). Influence of silicon on antioxidant mechanisms and lipid peroxidation in chickpea (*Cicer arietinum* L.) cultivars under drought stress. *Journal of Plant Interactions* 2: 105–113.

Gunes A, Pilbeam DJ, Inal A, Coban S. (2008). Influence of silicon on sunflower cultivars under drought stress. I. Growth, antioxidant mechanisms, and lipid peroxidation. *Communications in Soil Science and Plant Analysis* 39: 1885–1903.

Gunes A, Soylemezoglu G, Inal A, Bagci EG, Coban S, Sahin O. (2006). Antioxidant and stomatal responses of grapevine (*Vitis vinifera* L.) to boron toxicity. *Scientia Horticulturae* 110: 279–284.

Guo G, Zhou Q, Ma LQ. (2006). Availability and assessment of fixing additives for the in situ remediation of heavy metal contaminated soils: A review. *Environmental Monitoring and Assessment* 116: 513–528.

Guo W, Zhu YG, Liu WJ, Liang YC, Geng CN, Wang SG. (2007). Is the effect of silicon on rice uptake of arsenate (As V) related to internal silicon concentrations, iron plaque and phosphate nutrition? *Environmental Pollution* 148: 251–257.

Habiba U, Ali S, Farid M, Shakoor MB, Rizwan M, Ibrahim M, Abbasi GH, Hayat T, Ali B. (2015). EDTA enhanced plant growth, antioxidant defense system, and phytoextraction of copper by *Brassica napus* L. *Environmental Science and Pollution Research* 22: 1534–1544.

Hara T, Ichikuni N, Shimazu S. (2012). A novel preparation method of Ni–Sn alloy catalysts supported on aluminium hydroxide: Application to chemoselective hydrogenation of unsaturated carbonyl compounds. *Chem. Lett.* 41: 769–771.

Hall JL. (2002). Cellular mechanisms for heavy metal detoxification and tolerance. *J Exp Bot* 53(366): 1–11.

Hamayun M, Sohn EY, Khan SA, Shinwari ZK, Khan AL, Lee IJ. (2010). Silicon alleviates the adverse effects of salinity and drought stress on growth and endogenous plant growth hormones of soybean (*Glycine max* L.). *Pak J Bot* 42: 1713–1722.

Hasegawa PM, Bressan RA, Zhu J-K, Bohnert HJ. (2000). Plant cellular and molecular responses to high salinity. *Annual Review of Plant Biology* 51: 463–499.

Hashemi A, Abdolzadeh A, Sadeghipour HR. (2010). Beneficial effects of silicon nutrition in alleviating salinity stress in hydroponically grown canola, *Brassica napus* L., plants. *Soil Science and Plant Nutrition* 56: 244–253.

Hattori T, Inanaga S, Araki H, An P, Morita S, Luxová M, Lux A. (2005). Application of silicon enhanced drought tolerance in *Sorghum bicolor*. *Physiologia Plantarum* 123: 459–466.

Hattori T, Lux A, Tanimoto E, Luxova M, Sugimoto Y, Inanaga S. (2001). The effect of silicon on the growth of sorghum under drought. Presented at Proceedings of the 6th Symposium of the International Society of Root Research, Nagoya, Japan.

Heatherly LG. (1993). Drought stress and irrigation effects on germination of harvested soybean seed. *Crop Sci* 33(4): 777–781.

Heszky L. (2007). Szárazság és a növény kapcsolata. *Agrofórum* 18. (11/M): 37–41 (in Hungarian).

Hirel B, Lea PJ. (2001). Ammonia assimilation. In Lea PJ, Morot-Gaudry J-F (eds.), *Plant Nitrogen*, pp. 79–99. Springer, Berlin.

Hodson MJ, Evans DE. (1995). Aluminium/silicon interactions in higher plants. *J. Exp. Bot.* 46: 161–171.

Hoque MA, Okuma E, Banu MNA, Nakamura Y, Shimoishi Y, Murata Y. (2007). Exogenous proline mitigates the detrimental effects of salt stress more than exogenous betaine by increasing antioxidant enzyme activities. *Journal of Plant Physiology* 164: 553–561.

Hu Y, Burucs Z, von Tucher S, Schmidhalter U. (2007). Short-term effects of drought and salinity on mineral nutrient distribution along growing leaves of maize seedlings. *Environmental and Experimental Botany* 60: 268–275.

Imlay JA. (2003). Pathways of oxidative damage. *Annual Reviews in Microbiology* 57: 395–418.

Iturbe-Ormaetxe I, Escuredo PR, Arrese-Igor C, Becana M. (1998). Oxidative damage in pea plants exposed to water deficit or paraquat. *Plant Physiology* 116: 173–181.

Iwasaki K, Maier P, Fecht M, Horst WJ. (2002). Effects of silicon supply on apoplastic manganese concentrations in leaves and their relation to manganese tolerance in cowpea (*Vigna unguiculata* (L.) Walp.). *Plant and Soil* 238: 281–288.

Kafi M, Nabati J, Zare Mehrjerdi M. (2011). Effect of salinity and silicon application on oxidative damage of sorghum [*Sorghum bicolor* (L.) Moench]. *Pakistan Journal of Botany* 43: 2457–2462.

Kant S, Kant P, Lips H, Barak S. (2007). Partial substitution of NO_3^- by NH_4^+ fertilization increases ammonium assimilating enzyme activities and reduces the deleterious effects of salinity on the growth of barley. *Journal of Plant Physiology* 164: 303–311.

Karmollachaab A, Bakhshandeh A, Gharineh MH, Moradi Telavat MR, Fathi G. (2013). Effect of silicon application on physiological characteristics and grain yield of wheat under drought stress condition. *Int J Agron Plant Prod* 4: 30–37.

Kausar F, Shahbaz M. (2013). Interactive effect of foliar application of nitric oxide (NO) and salinity on wheat (*Triticum aestivum* L.). *Pakistan Journal of Botany* 45: 67–73.

Kaya MD, Okçu G, Atak M, Çıkılı Y, Kolsarıcı O. (2006). Seed treatments to overcome salt and drought stress during germination in sunflower (*Helianthus annuus* L.). *European Journal of Agronomy* 24: 291–295.

Keller C, Rizwan M, Davidian JC, Pokrovsky OS, Bovet N, Chaurand P, Meunierm JD. (2015). Effect of silicon on wheat seedlings (*Triticum turgidum* L.) grown in hydroponics and exposed to 0 to 30 μM Cu. *Planta* 241: 847–860.

Khandekar S, Leisner S. (2011). Soluble silicon modulates expression of Arabidopsis thaliana genes involved in copper stress. *J. Plant Physiol* 168: 699–705.

Kim YH, Khan AL, Waqas M, Shim JK, Kim DH, Lee KY, Lee IJ. (2014). Silicon application to rice root zone influenced the phytohormonal and antioxidant responses under salinity stress. *Journal of Plant Growth Regulation* 33: 137–149.

Kmlazawa S, Sano S, Koshiba T. (2000). Changes in antioxidative in cucumber cotyledons during natural senescence: Comparison with those during dark-induced senescence. *Physiologia Plantarum* 109: 211–216.

Kochanova Z, Jaikoi K, Sedlaikovai B, Luxovai M. (2014). Silicon improves salinity tolerance and affects ammonia assimilation in maize roots. *Biologia* 69: 1164–1171.

Kronzucker HJ, Britto DT, Davenport RJ, Tester M. (2001). Ammonium toxicity and the real cost of transport. *Trends in Plant Science* 6: 335–337.

Kurdali F, Mohammad AC, Ahmad M. (2013). Growth and nitrogen fixation in silicon and/or potassium fed chickpeas grown under drought and well watered conditions. *J Stress Physiol Biochem* 9: 385–406.

Kusvuran S. (2012). Influence of drought stress on growth, ion accumulation and antioxidative enzymes in okra genotypes. *International Journal of Agriculture and Biology* 14: 401–406.

Läuchli A, Grattan SR. (2007). Plant growth and development under salinity stress. In *Advances in molecular breeding toward drought and salt tolerant crops* (pp. 1–32). Springer Netherlands.

Lee DH, Kim YS, Lee CB. (2001). The inductive responses of the antioxidant enzymes by salt stress in the rice (*Oryza sativa* L.). *Journal of Plant Physiology* 158: 737–745.

Lee SK, Sohn EY, Hamayun M, Yoon JY, Lee IJ. (2010). Effect of silicon on growth and salinity stress of soybean plant grown under hydroponic system. *Agroforestry Systems* 80: 333–340.

Li Y, Zhou W, Wang H, Xie L, Liang Y, Wei F, Idrobo JC, Pennycook SJ, Dai H. (2012). An oxygen reduction electrocatalyst based on carbon nanotube-graphene complexes. *Nat Nanotechnol* 7: 394–400.

Li RY, Stroud JL, Ma JF, McGrath SP, Zhao FJ. (2009). Mitigation of arsenic accumulation in rice with water management and silicon fertilization. *Environ Sci Technol* 43(10): 3778–3783.

Liang Y, Chen Q, Liu Q, Zhang W, Ding R. (2003). Exogenous silicon (Si) increases antioxidant enzyme activity and reduces lipid peroxidation in roots of salt-stressed barley (*Hordeum vulgare* L.). *Journal of Plant Physiology* 160: 1157–1164.

Liang Y, Sun W, Zhu Y-G, Christie P. (2007). Mechanisms of silicon-mediated alleviation of abiotic stresses in higher plants: A review. *Environmental Pollution* 147: 422–428.

Liang Y, Zhang W, Chen Q, Ding R. (2005). Effects of silicon on H+-ATPase and H+-PPase activity, fatty acid composition and fluidity of tonoplast vesicles from roots of salt-stressed barley (*Hordeum vulgare* L.). *Environmental and Experimental Botany* 53: 29–37.

Liao M, Xie XM. (2007). Effect of heavy metals on substrate utilization pattern, biomass, and activity of microbial communities in a reclaimed mining wasteland of red soil area. *Ecotoxicology and Environmental Safety* 66: 217–223.

Liu J, Zhang H, Zhang Y, Chai T. (2013). Silicon attenuates cadmium toxicity in *Solanum nigrum* L. by reducing cadmium uptake and oxidative stress. *Plant Physiol Biochem* 68: 1–7.

Lukačová Z, Švubová R, Kohanová J, Lux A. (2013). Silicon mitigates the Cd toxicity in maize in relation to cadmium translocation, cell distribution, antioxidant enzymes stimulation and enhanced endodermal apoplasmic barrier development. *Plant Growth Regul* 70: 89–103.

Lokhande VH, Nikam TD, Penna S. (2010). Biochemical, physiological and growth changes in response to salinity in callus cultures of *Sesuvium portulacastrum* L. *Plant Cell, Tissue and Organ Culture (PCTOC)* 102: 17–25.

Lu G, Gao C, Zheng X, Han B. (2009). Identification of OsbZIP72 as a positive regulator of ABA response and drought tolerance in rice. *Planta* 229: 605–615.

Lux A, Luxová M, Hattori T, Inanaga S, Sugimoto Y. (2002). Silicification in sorghum (*Sorghum bicolor*) cultivars with different drought tolerance. *Physiologia Plantarum* 115: 87–92.

Ma CC, Li QF, Gao YB, Xin TR. (2004). Effects of silicon application on drought resistance of cucumber plants. *Soil Science and Plant Nutrition* 50: 623–632.

Ma JF. (2004). Role of silicon in enhancing the resistance of plants to biotic and abiotic stresses. *Soil Science and Plant Nutrition* 50: 11–18.

Ma JF, Yamaji N. (2008). Functions and transport of silicon in plants. *Cellular and Molecular Life Sciences* 65: 3049–3057.

Ma JF, Yamaji N, Mitani N, Tamai K, Konishi S, Fujiwara T, Katsuhara M, Yano M. (2007). An efflux transporter of silicon in rice. *Nature* 448: 209–212.

Maksymiec W, Wianowska D, Dawidowicz AL, Radkiewicz Sa, Mardarowicz M, Krupa Z. (2005). The level of jasmonic acid in *Arabidopsis thaliana* and *Phaseolus coccineus* plants under heavy metal stress. *Journal of Plant Physiology* 162: 1338–1346.

Malčovská SM, Dučaiová Z, Maslaňáková I, Bačkor M. (2014). Effect of silicon on growth, photosynthesis, oxidative status and phenolic compounds of maize (*Zea mays* L.) grown in cadmium excess. *Water Air Soil Pollut.* 225: 1–11.

Martín JAR, Arias ML, Corbí JMG. (2006). Heavy metals contents in agricultural topsoils in the Ebro basin (Spain). Application of the multivariate geoestatistical methods to study spatial variations. *Environmental Pollution* 144: 1001–1012.

Martin W, Mitra D. (2001). Productivity growth and convergence in agriculture versus manufacturing. *Economic Development and Cultural Change* 49: 403–422.

Menezes-Benavente L, Kernodle SP, Margis-Pinheiro MR, Scandalios JG. (2004). Salt-induced antioxidant metabolism defenses in maize (*Zea mays* L.) seedlings. *Redox Report* 9: 29–36.

Ming DF, Pei ZF, Naeem MS, Gong HJ, Zhou WJ. (2012). Silicon alleviates PEG-induced water-deficit stress in upland rice seedlings by enhancing osmotic adjustment. *J Agron Crop Sci* 198: 14–26.

Mittler R, Vanderauwera S, Gollery M, Van Breusegem F. (2004). Reactive oxygen gene network of plants. *Trends in Plant Science* 9: 490–498.

Mohammadian R, Moghaddam M, Rahimian H, Sadeghian SY. (2005). Effect of early season drought stress on growth characteristics of sugar beet genotypes. *Turkish Journal of Agriculture and Forestry* 29: 357.

Monakhova OF, Chernyad'ev II. (2002). Protective role of kartolin-4 in wheat plants exposed to soil draught. *Applied Biochemistry and Microbiology* 38: 373–380.

Moradi F, Ismail AM. (2007). Responses of photosynthesis, chlorophyll fluorescence and ROS-scavenging systems to salt stress during seedling and reproductive stages in rice. *Annals of Botany* 99: 1161–1173.

Morikawa CK, Saigusa M. (2004). Mineral composition and accumulation of silicon in tissues of blueberry (*Vaccinum corymbosus* cv. Bluecrop) cuttings. *Plant and Soil* 258: 1–8.

Munns R. (2005). Genes and salt tolerance: Bringing them together. *New Phytologist* 167: 645–663.

Munns R, Tester M. (2008). Mechanisms of salinity tolerance. *Annual Review of Plant Biology* 59: 651–681.

Nagajyoti PC, Lee KD, Sreekanth TVM. (2010). Heavy metals, occurrence and toxicity for plants: A review. *Environ Chem Lett* 8: 199–216.

Nasim M, Qureshi RH, Aziz T, Saqib M, Nawaz S, Sahi ST, Pervaiz S. (2007). Screening trees for salinity tolerance: A case-study with ten *Eucalyptus* species. *Pakistan Journal of Agricultural Sciences* 44: 3.

Nehnevajova E, Herzig R, Federer G, Erismann K-H, Schwitzguébel J-P. (2007). Chemical mutagenesis—A promising technique to increase metal concentration and extraction in sunflowers. *International Journal of Phytoremediation* 9: 149–165.

Netondo GW, Onyango JC, Beck E. (2004). Sorghum and salinity. *Crop Science* 44: 797–805.

Neumann D, zur Nieden U, Schwieger W, Leopold I, Lichtenberger O. (1997). Heavy metal tolerance of Minuartia verna. *J Plant Physiol* 151: 101–108.

Nowakowski W, Nowakowska J. (1997). Silicon and copper interaction in the growth of spring wheat seedlings. *Biol Plant* 39: 463–466.

Nwugo CC, Huerta AJ. (2008). Effects of silicon nutrition on cadmium uptake, growth and photosynthesis of rice plants exposed to low-level cadmium. *Plant and Soil* 311: 73–86.

Parry MAJ, Andralojc PJ, Khan S, Lea PJ, Keys AJ. (2002). Rubisco activity: Effects of drought stress. *Annals of Botany* 89: 833–839.

Pei ZF, Ming DF, Liu D, Wan GL, Geng XX, Gong HJ, Zhou WJ. (2010). Silicon improves the tolerance to water-deficit stress induced by polyethylene glycol in wheat (*Triticum aestivum* L.) seedlings. *J Plant Growth Regul* 29: 106–115.

Pervez MA, Ayub CM, Khan HA, Shahid MA, Ashraf I. (2009). Effect of drought stress on growth, yield and seed quality of tomato (*Lycopersicon esculentum* L.). *Pak J Agric Sci* 46: 174–178.

Peters S, Mundree SG, Thomson JA, Farrant JM, Keller F. (2007). Protection mechanisms in the resurrection plant *Xerophyta viscosa* (Baker): Both sucrose and raffinose family oligosaccharides (RFOs) accumulate in leaves in response to water deficit. *Journal of Experimental Botany* 58: 1947–1956.

Pulz AL, Crusciol CAC, Lemos LB, Soratto RP. (2008). Influência de silicato e calcário na nutrição, produtividade e qualidade da batata sob deficiência hídrica. *Rev Bras Ciênc Solo* 32(4): 1651–1659.

Rahnama H, Ebrahimzadeh H. (2004). The effect of NaCl on proline accumulation in potato seedlings and calli. *Acta Physiologiae Plantarum* 26: 263–270.

Ranjan G, Vasan RM, Charan HD. (1996). Probabilistic analysis of randomly distributed fiber-reinforced soil. *Journal of Geotechnical Engineering* 122: 419–426.

Raza SH, Athar H-U-R, Ashraf M. (2006). Influence of exogenously applied glycinebetaine on the photosynthetic capacity of two differently adapted wheat cultivars under salt stress. *Pakistan Journal of Botany* 38: 341–351.

Razzaghi F, Ahmadi SH, Adolf VI, Jensen CR, Jacobsen SE, Andersen MN. (2011). Water relations and transpiration of quinoa (*Chenopodium quinoa* Willd.) under salinity and soil drying. *J Agron Crop Sci* 197(5): 348–360.

Richmond KE, Sussman M. (2003). Got silicon? The non-essential beneficial plant nutrient. *Current Opinion in Plant Biology* 6: 268–272.

Rizwan M, Ali S, Ibrahim M, Farid M, Adrees M, Bharwana SA, Zia-ur-Rehman M, Qayyum MF, Abbas F. (2012). Mechanisms of silicon-mediated alleviation of drought and salt stress in plants: A review. *Environmental Science and Pollution Research* 22: 15416–15431.

Ronde JAD, Spreeth MH, Cress WA. (2000). Effect of antisence 1-Δ-pyrroline-5-5 carboxylate reductase transgenic soyabean plants subjected to osmotic and drought stress. *Plant Growth Regul* 32: 13–26.

Romero-Aranda MR, Jurado O, Cuartero JS. (2006). Silicon alleviates the deleterious salt effect on tomato plant growth by improving plant water status. *Journal of Plant Physiology* 163: 847–855.

Rouhier N, Jacquot JP. (2008). Getting sick may help plants overcome abiotic stress. *New Phytologist* 180: 738–741.

Ruiz J, Blumwald E. (2002). Salinity-induced glutathione synthesis in *Brassica napus*. *Planta* 214: 965–969.

Sahu AC, Sahoo SK, Sahoo N. (2001). NaCl-stress induced alteration in glutamine synthetase activity in excised senescing leaves of a salt-sensitive and a salt-tolerant rice cultivar in light and darkness. *Plant Growth Regulation* 34: 287–292.

Sairam RK, Saxena DC. (2000). Oxidative stress and antioxidants in wheat genotypes: Possible mechanism of water stress tolerance. *Journal of Agronomy and Crop Science* 184: 55–61.

Sairam RK, Srivastava GC, Agarwal S, Meena RC. (2005). Differences in antioxidant activity in response to salinity stress in tolerant and susceptible wheat genotypes. *Biologia Plantarum* 49: 85–91.

Sarret G, Pilon Smits EAH, Castillo Michel H, Isaure MP, Zhao FJ, Tappero R. (2013). Use of synchrotron-based techniques to elucidate metal uptake and metabolism in plants. *Adv Agron* 119: 1–82.

Saud S, Li X, Chen Y, Zhang L, Fahad S, Hussain S, Chen Y. (2014). Silicon application increases drought tolerance of Kentucky bluegrass by improving plant water relations and morphophysiological functions. *Sci World J.*

Sharpe-Young G. (2005). *New Wave of American Heavy Metal*. Zonda Books, New Plymouth, New Zealand.

Shen X, Zhou Y, Duan L, Li Z, Eneji AE, Li J. (2010). Silicon effects on photosynthesis and antioxidant parameters of soybean seedlings under drought and ultraviolet-B radiation. *Journal of Plant Physiology* 167: 1248–1252.

Shen MY, Li BR, Li YK. 2014. Silicon nanowire field-effect-transistor based biosensors: From sensitive to ultra-sensitive. *Biosens Bioelectron* 60: 101–111.

Shi G, Cai Q, Liu C, Wu L. (2010). Silicon alleviates cadmium toxicity in peanut plants in relation to cadmium distribution and stimulation of antioxidative enzymes. *Plant Growth Regul* 61: 45–52.

Shi X, Zhang C, Wang H, Zhang F. (2005). Effect of Si on the distribution of Cd in rice seedlings. *Plant and Soil* 272: 53–60.

Shi Y, Wang Y, Flowers TJ, Gong H. (2013). Silicon decreases chloride transport in rice (*Oryza sativa* L.) in saline conditions. *Journal of Plant Physiology* 170: 847–853.

Shi Y, Zhang Y, Yao H, Wu J, Sun H, Gong H. (2014). Silicon improves seed germination and alleviates oxidative stress of bud seedlings in tomato under water deficit stress. *Plant Physiol Biochem* 78: 27–36.

Shu LZ, Liu YH. (2001). Effects of silicon on growth of maize seedlings under salt stress. *Agro-environ Prot* 20: 38–40.

Solanki R, Dhankhar R. (2011). Biochemical changes and adaptive strategies of plants under heavy metal stress. *Biologia* 66: 195–204.

Song A, Li Z, Zhang J, Xue G, Fan F, Liang Y. (2009). Silicon-enhanced resistance to cadmium toxicity in *Brassica chinensis* L. is attributed to Si-suppressed cadmium uptake and transport and Si-enhanced antioxidant defense capacity. *J Hazard Mater* 172: 74–83.

Sonobe K, Hattori T, An P, Tsuji W, Eneji E, Tanaka K, Inanaga S. (2009). Diurnal variations in photosynthesis, stomatal conductance and leaf water relation in sorghum grown with or without silicon under water stress. *Journal of Plant Nutrition* 32: 433–442.

Soylemezoglu G, Demir K, Inal A, Gunes A. (2009). Effect of silicon on antioxidant and stomatal response of two grapevine (*Vitis vinifera* L.) rootstocks grown in boron toxic, saline and boron toxic-saline soil. *Scientia Horticulturae* 123: 240–246.

Street RA. (2012). Heavy metals in medicinal plant products—An African perspective. *South African Journal of Botany* 82: 67–74.

Streeter JG, Lohnes DG, Fioritto RJ. (2001). Patterns of pinitol accumulation in soybean plants and relationships to drought tolerance. *Plant, Cell & Environment* 24: 429–438.

Sudhakar C, Lakshmi A, Giridarakumar S. (2001). Changes in the antioxidant enzyme efficacy in two high yielding genotypes of mulberry (*Morus alba* L.) under NaCl salinity. *Plant Science* 161: 613–619.

Sun Q, Ye ZH, Wang XR, Wong MH. (2007). Cadmium hyperaccumulation leads to an increase of glutathione rather than phytochelatins in the cadmium hyperaccumulator *Sedum alfredii*. *Journal of Plant Physiology* 164: 1489–1498.

Tahir MA, Rahmatullah T, Aziz T, Ashraf M. (2010). Wheat genotypes differed significantly in their response to silicon nutrition under salinity stress. *Journal of Plant Nutrition* 33: 1658–1671.

Talaat NB, Shawky BT. (2013). 24-Epibrassinolide alleviates salt-induced inhibition of productivity by increasing nutrients and compatible solutes accumulation and enhancing antioxidant system in wheat (*Triticum aestivum* L.). *Acta Physiol Plant* 35(3): 729–740.

Talaat NB, Shawky BT. (2014). Modulation of the ROS-scavenging system in salt-stressed wheat plants inoculated with arbuscular mycorrhizal fungi. *J Plant Nutr Soil Sci* 177(2): 199–207.

Tale Ahmad S, Haddad R. (2011). Study of silicon effects on antioxidant enzyme activities and osmotic adjustment of wheat under drought stress. *Czech J Genet Plant Breed* 47: 17–27.

Tan Y, Liang Z, Shao H, Du F. (2006). Effect of water deficits on the activity of anti-oxidative enzymes and osmoregulation among three different genotypes of Radix astragali at seeding stage. *Colloids and Surfaces B: Biointerfaces* 49: 60–65.

Tausz M, Sircelj H, Grill D. (2004). The glutathione system as a stress marker in plant ecophysiology: Is a stress-response concept valid? *Journal of Experimental Botany* 55: 1955–1962.

Tester M, Davenport R. (2003). Na$^+$ tolerance and Na$^+$ transport in higher plants. *Annals of Botany* 91: 503–527.

Tripathi DK, Singh VP, Prasad SM, Chauhan DK, Dubey NK, Rai AK. (2015). Silicon-mediated alleviation of Cr (VI) toxicity in wheat seedlings as evidenced by chlorophyll florescence, laser induced breakdown spectroscopy and anatomical changes. *Ecotoxicol Environ Saf* 113: 133–144.

Tuna AL, Kaya C, Higgs D, Murillo-Amador B, Aydemir S, Girgin AR. (2008). Silicon improves salinity tolerance in wheat plants. *Environmental and Experimental Botany* 62: 10–16.

Verbruggen N, Hermans C, Schat H. (2009). Mechanisms to cope with arsenic or cadmium excess in plants. *Current Opinion in Plant Biology* 12: 364–372.

Wan G, Najeeb U, Jilani G, Naeem MS, Zhou W. (2011). Calcium invigorates the cadmium-stressed *Brassica napus* L. plants by strengthening their photosynthetic system. *Environmental Science and Pollution Research* 18: 1478–1486.

Wang CF, Mäkilä EM, Kaasalainen MH, Liu D, Sarparanta MP, Airaksinen AJ, Salonen JJ, Hirvonen JT, Santos HA. (2014). Copper-free azide–alkyne cycloaddition of targeting peptides to porous silicon nanoparticles for intracellular drug uptake. *Biomaterials* 35: 1257–1266.

Wang S, Jiang L. (2007). Definition of superhydrophobic states. *Advanced Materials* 19: 3423–3424.

Wang W-B, Kim Y-H, Lee H-S, Kim K-Y, Deng X-P, Kwak S-S. (2009). Analysis of antioxidant enzyme activity during germination of alfalfa under salt and drought stresses. *Plant Physiology and Biochemistry* 47: 570–577.

Wang XS, Han JG. (2007). Effects of NaCl and silicon on ion distribution in the roots, shoots and leaves of two alfalfa cultivars with different salt tolerance. *Soil Science and Plant Nutrition* 53: 278–285.

Wang Y-P, Shi J-Y, Qi LIN, Chen X-C, Chen Y-X. (2007a). Heavy metal availability and impact on activity of soil microorganisms along a Cu/Zn contamination gradient. *Journal of Environmental Sciences* 19: 848–853.

Wang Z-Q, Yuan Y-Z, Ou J-Q, Lin Q-H, Zhang C-F. (2007b). Glutamine synthetase and glutamate dehydrogenase contribute differentially to proline accumulation in leaves of wheat (*Triticum aestivum*) seedlings exposed to different salinity. *Journal of Plant Physiology* 164: 695–701.

Williams DE, Vlamis J. (1957). The effect of silicon on yield and manganese-54 uptake and distribution in the leaves of barley plants grown in culture solutions. *Plant Physiol.* 32: 404.

Wu D, Cederbaum AI. (2003). Alcohol, oxidative stress, and free radical damage. *Alcohol Research and Health* 27: 277–284.

Xiong L, Zhu JK. (2002). Molecular and genetic aspects of plant responses to osmotic stress. *Plant, Cell & Environment* 25: 131–139.

Xu JL, Lafitte HR, Gao YM, Fu BY, Torres R, Li ZK. (2005). QTLs for drought escape and tolerance identified in a set of random introgression lines of rice. *Theoretical and Applied Genetics* 111: 1642–1650.

Yamaguchi T, Blumwald E. (2005). Developing salt-tolerant crop plants: Challenges and opportunities. *Trends in Plant Science* 10: 615–620.

Yang YH, Chen SM, Chen Z, Zhang HY, Shen HG, Hua ZC, Li N. (1999). Silicon effects on aluminum toxicity to mungbean seedling growth. *J Plant Nutr* 22: 693–700.

Ye J, Yan C, Liu J, Lu H, Liu T, Song Z. (2012). Effects of silicon on the distribution of cadmium compartmentation in root tips of Kandelia obovata (S., L.) Yong. *Environ Pollut* 162: 369–373.

Zeng FR, Zhao FS, Qiu BY, Ouyang YN, Wu FB, Zhang GP. (2011). Alleviation of chromium toxicity by silicon addition in rice plants. *Agric Sci China* 10: 1188–1196.

Zhang F-Q, Wang Y-S, Lou Z-P, Dong J-D. (2007). Effect of heavy metal stress on antioxidative enzymes and lipid peroxidation in leaves and roots of two mangrove plant seedlings (*Kandelia candel* and *Bruguiera gymnorrhiza*). *Chemosphere* 67: 44–50.

Zhang J-H, Liu Y-P, Pan Q-H, Zhan J-C, Wang X-Q, Huang W-D. (2006). Changes in membrane-associated H⁺-ATPase activities and amounts in young grape plants during the cross adaptation to temperature stresses. *Plant Science* 170: 768–777.

Zhu Y, Gong H. (2011). Beneficial effects of silicon on salt and drought tolerance in plants. *Agronomy for Sustainable Development* 34: 455–472.

Zhu Z, Wei G, Li J, Qian Q, Yu J. (2004). Silicon alleviates salt stress and increases antioxidant enzymes activity in leaves of salt-stressed cucumber (*Cucumis sativus* L.). *Plant Science* 167: 527–533.

Zlatev ZS, Lidon FC, Ramalho JC, Yordanov IT. (2006). Comparison of resistance to drought of three bean cultivars. *Biologia Plantarum* 50: 389–394.

18 Silicon Nutrition and Rice Crop Improvement in Iran
Recent Advances and Future Perspectives

Allahyar Fallah

CONTENTS

ABSTRACT

Rice (*Oryza sativa* L.) is the main crop after wheat in Iran, and these two are the main suppliers of calories for the Iranian people. Overusing fertilizers and toxins of chemicals has caused environmental pollution. One of the applied methods for reducing this status is using silicon fertilizer in the paddy field. The process of exchange and absorption of nutrient elements is used in the rhizosphere environment. Silicon absorption was independent of the water absorption rate in rice plants. The state of the dissolved silicon differs depending on the pH of the solution and nutritional status. Addition of calcium silicate increased the levels of silicon, phosphorus, calcium, and copper, and reduced the levels of nitrogen, potassium, manganese, and zinc in the leaf of rice. It has been held that an abundant portion of silicon is needed for rice to grow, indicating the predominant role of silicon in rice production. Photosynthesis produces 90% of dry matter and is influenced by diverse factors, including light, CO_2, water, and mineral nutrients. Leaf area increases are augmented by silicon; furthermore, silicon causes the leaves to be erect, positively influencing crop photosynthesis. The addition of silicon does not play a role in the photosynthesis rate, stomatal conductance, and transpiration. However, it has been observed that the amounts of chlorophyll a and b and total chlorophyll are augmented in flowering stage. These amendments could be correlated with an enhancement in the level of sodium and potassium silicate fertilizers. Silicon at the cellular scale increased the cuticle thickness and cell diameter of the parenchyma, xylem, and vessel phloem. Silicon decreased the protein, Fe, Mn, and Zn of leaves of the Shiroodi variety. Application of 250–500 kg·ha^{-1} silicon fertilizer also decreased the amount of stem borer pollution and blast, and as a result, yield improved about 5% in the field and 24% in pot experiment. Therefore, it is recommended that silicon fertilizers like granule, ligules, and nanosilicon be used in the paddy fields of north Iran to improve the growth and yield and reduce the biotic and abiotic stresses in paddy fields.

18.1 INTRODUCTION

In Iran, rice (*Oryza sativa* L.) is regarded as a pivotal cultivation crop, second only after wheat; thus, people abundantly consume rice. Wheat and rice crops are the main suppliers of calories for the Iranian people. The rice cultivation area of Iran ranges from 550,0000 to 620,000 ha. The average yield of rice is 4783 kg·ha^{-1}, which is lower than that of Egypt but higher than the average yield worldwide. About 75% of rice is produced in north Iran (RRII 2014), where it plays a dominant role in the economy there. Rice is produced primarily by small farmers who cultivated less than 1 ha. Between the years 2000 and 2013, Iran required an additional 1 million tons (t) of rice to maintain self-sufficiency. The lack of supply of rice motivated Iranians to design a plan to improve the variety of rice via plant breeding methods against abiotic stresses, such as drought, salinity, and cold and heat stress, and to optimize fertilizers used in paddy fields. The target average yield of rice is 8 t·ha^{-1}, with a plan to reach that goal in 5 years (RRII 2014). Hence, productivity will have to grow.

The crop production of rice in the world is complex for several reasons. Overusing fertilizers and toxins of chemicals has culminated in environmental pollution. One of the applied ways to reduce this status is to use silicon fertilizer in paddy fields (Fallah 2010). The utilization of silicon fertilizer is not common in Iran; however, it has been suggested that by extension of the technical knowledge of rice farmers in the future, promotion of information about the utilization of silicon fertilizer would result in a sustainable agro-ecosystem of paddy fields in north Iran (Fallah 2010).

Due to the deficiency of water in Iran, the plan is to have rice production based on water use per hectare. Although this situation may decrease the rice cultivation of Iran, using silicon fertilizer will improve plant nutrition. Therefore, rice plant nutrition is more important than rice cultivation. As a result, rice cultivation will be optimized using water and fertilizer, simultaneously decreasing production costs and increasing the income of farmers. This chapter discusses the effects of silicon on the sustainability of rice production in Iran.

18.2 PADDY SOILS

Paddy soils are categorized as flooded water soils and land preparation special method soils (Malakouti and Kavousi 2004). The first plowing of the land is done without water, but the second and third plowings are done with water, with flattening and boundaries to keep the water in the plot. Paddy fields in north Iran have about 3–5 cm of water for the duration of the rice plant's growth; however, there is no water in field at maximum tillering, about five to seven days and two to three weeks before harvesting of rice crops (Solimani and AmiriLarijani 2004).

Flooding of paddy soil causes chemical and electrochemical changes in the soil and, as a result, affects the soil fertility of the paddy field (Malakouti and Kavousi 2004). This situation increases the pH of acidic soil, which is common in the paddy fields of north Iran. The electrical conductivity of the paddy soil will increase one or two weeks after flooding and then stabilize. The maximum electrical conductivity of the soil after flooding is 2–4 ds·m^{-1}, but it decreases to 0.7–1.9 ds·m^{-1} (Malakouti and Kavousi 2004). The complete submergence of soil causes nutrient elements, such as nitrogen, phosphorus, sulfur, and silicon, to be more available (Malakouti and Kavousi 2004).

18.3 RICE RHIZOSPHERE

The rhizosphere is the part of the soil around rice roots that it is affected by root activities (Hoshikawa 1989). The diameter of the rhizosphere ranges from a few millimeters to a centimeter, depending on the variety of rice and environment (Hoshikawa 1989). The process of exchange and absorption of nutrient elements takes place in the rhizosphere. Therefore, plant growth depends on the root growth and diameter of rhizosphere environment, because the absorption of nutrient elements and water depends on the condition of the rhizosphere (Hoshikawa 1989). Malakouti and Kavousi (2004) reported that the pH of soil in the rice rhizosphere is less than 2 units. Hoshikawa (1989) obtained the same result. So, these changes improved absorption of nutrient elements by rice plant and their growth as well (Malakouti and Kavousi 2004). Patrick and Reddy (1978) reported the total loss of 143 mg $N \cdot m^{-2} \cdot d^{-1}$ recorded in the level of nitrogen in the rhizosphere of rice plant by the usual process of nitrification and denitrification. There is no information about the losses of silicon in the paddy fields of north Iran.

Three main processes modify soil conditions near rice roots (rhizosphere) growing in anaerobic soil: release of O_2 from roots, causing ferrous iron (formed due to soil reduction); release of acidity, $4Fe(OH)_3 + 8H^+$; and release of H^+ ions from the roots to balance the cation–anion intake (i.e., maintain electrical neutrality across the root–soil interface), with nitrogen being taken up chiefly as a cation, $NH4^+$.

Since under the higher strain of CO_2 generated in anaerobic soil, roots may either liberate or accumulate CO_2 from the soil, with corresponding change in the pH (Malakouti and Kavousi 2004).

18.4 SILICON IN PADDY SOIL

The silicon chemistry in flooded is still not well inferred. Silicon is not involved in oxidation–reduction, but flooding influences its behavior in rice soils. Silicon concentration tends to increase in rice paddy soils after submergence due to the release of absorbed and occluded Fe(III) oxyhydroxides (Patrick and Reddy 1978). The silicon behavior in flooded soil is affected by pH, kind and amount of clay minerals, and their transformation in different conditions, such as electrolyte levels, and competitive reaction with many other ligands. Such factors could be firmly correlated or coordinated with one another (Kundu 1987). In soils of temperate regions, like in north Iran, subject to little desilication, silicon is concentrated in stable silicate minerals. The silicon concentration in soil solution of paddy fields in the Caspian Sea area is 80–190 ppm (Fallah 2010). Parent material, vegetation, climate, texture, pedogenesis, and intensity of weathering are among the factors that affect the quantity of silicon, as well as the way silicon in paddy soils is distributed (Kundu 1987).

18.5 SILICON IN RICE PLANTS

In 1962, Yoshida et al. utilized a hydrofluoric acid etching method to determine the process of localization of silicon within diverse tissues of rice plants. In the husk (6.9% Si), the heaviest deposition took place in the interspace between the cuticle and epidermal cell. In the leaf blade (5.5% Si), the epidermis, vascular bundles, bundle sheath and sclernchyma are among the most important and notorious parts in which the process of localization of silicon takes place. Silicon is made up of the leaf sheath (4.6%) and epidermis (2.3%). This silicon is located in the sclernchyma, bundle sheath, and parenchyma along the cell wall. Furthermore, in the leaf, epidermal cell silicon and cellulose are concomitted (Yoshida et al. 1962).

More recent studies with microprobe analysis and scanning electron microscopy have shown a great deal of structural detail of silicon accumulation in the entire epidermal system, except in the guard cells along the cell walls, in the silicon cells, and in the trichomes of the leaf blade and sheath and internodes. Lower amounts of silicon occur in the stomatal apparatus of the internodes. No detectable silicon was present in the stomatal apparatus of the leaf blades and sheath (Soni et al. 1972).

Rice is a known silicon accumulator plant, and it benefits from silicon nutrition (Takahashi 1995). Deposition of silicon in rice roots occurred only in the endodermis, and the heaviest accumulation was associated with the inner tangential wall of the endodermal cells. However, there are reports that silicon is distributed in all tissues of rice roots (Takahashi 1995).

18.6   SILICON UPTAKE

Acquisition of Silicon does not rely or depends on the rate of water absorptions by rice plants. Okuda and Takahashi (1965) found that some elements, including sodium cyanide, sodium fluoride, 2,4-D-iodoacetate, 2,4-dinitrophenol, and D-glucosamine, but not phloridzin or sodium malanate, prevented silicon uptake by the rice plant. So, they came to the conclusion that the uptake of silicon by the rice plant requires energy arising from aerobic respiration, perhaps in the form of higher-energy phosphorus. Okuda and Takahashi (1965) also suggested that there was a relationship between silicon uptake and aerobic glycolysis, suggesting that silicon uptake was affected by the excised top and not decreased by metabolic inhibitors, such as sodium cyanide, dinitrophenol, and iodoacetate. Observing these phenomena, Okuda and Takahashi (1965) emphasized the potential of rice roots for metabolic absorption and concentration. Moreover, three Si transporters (Lsi_1, Lsi_2, and Lsi_6) have been acknowledged as well-known species that can accumulate and play an important role in silicon uptake and distribution (Ma et al. 2007). Lsi_1 and Lsi_2 are primarily expressed in the roots, and Lsi_6 is observed in the leaf sheath and blades. The localization of Lsi_1 and Lsi_2 is observed in the exodemis and endodermis, with the former on the distal side and the latter on the proximal side (Ma et al. 2007). They came to the conclusion that Lsi_1 and Lsi_2 are two important elements that play a crucial role in Si uptake, and the former element has a crucial role in transporting Si out of the xylem. Silicon uptake occurs in an area of about 1–3 cm of rice root tip. There is the same Si transporter for maize (*Zea mays* L.).

18.7   FACTORS AFFECTING SILICON UPTAKE

The factors that affect silicon uptake are the soil's pH and nutritional status.

The state of the dissolved silicon differs, depending on the pH of the solution. When the pH is below 8, silicon is dissolved as nondissociated silicic acid ($H_4SiO_4)_n$, where $n = 2$ or 3 (Takahashi and Hino 1987). Therefore, the silicon absorbed by rice plants in paddy fields is considered to take a molecular form, which seems to be superior to the ionic form. When a silicate anion was applied to solution at a given pH, the absorption declined. Furthermore, coexisting anions, such as phosphate and nitrate, mutually inhibit the absorption of silicate anions (Takahashi 1995).

The silicic acid which rice usually absorbs has no electrical charge, so the effect of coexisting ions is directly generated. However, it has been observed that excessive application of nitrogen affects the physiological condition of rice plants and decreases the silicon concentration or number of silicified cells. Ammonium ions among cations substantially decreased the amount of silicon uptake (Takahashi 1995).

Application of nitrogen tends to decrease silicon uptake in rice, as well as fertilizer containing ammonium nitrogen ($NH4^+$-N) (Wallace 1989). At the earlier growth stage, a high amount of $NH4^+$-N in soil from high nitrogen rates may limit silicon uptake by the rice plants (Sumida 1992). Takahashi and Nishi (1982) have concluded that without potassium, magnesium, and sulfur treatments, the silicon absorption slightly declined. However, without nitrogen treatment, the silicon absorption clearly increased. In the phosphorus treatment, silicon absorption is promoted. The addition of calcium silicate increased the levels of silicon, phosphorus, calcium, and copper, and reduced the levels of nitrogen, potassium, manganese, and zinc in the leaf of rice. The concentration of iron and manganese in rice shoots was decreased by silicon at all phosphorus levels (Ma et al. 1989).

18.8   RECENT ADVANCES OF SILICON IN IMPROVING
THE GROWTH AND YIELD OF RICE IN IRAN

Silicon is an essential element for the growth of rice plants. It can be regarded as a useful element, playing a defining role in rice production (Ma 2004; Fallah 2012). Generally speaking, plant growth could be affected by nutrient supply through influencing dry matter production. The photosynthesis process could produce 90% of the dry matter, influencing diverse sources, including light, CO_2, water, and mineral nutrients. The greater the supply of these inputs from the deficiency range, the greater the plant growth rate (Marschner 1995). The necessity of silicon for rice plants has not been verified in experiments. The exclusion of silicon from nutrient experiments is not particularly a viable task due to the widespread presence of silicon in glass, nutrient salts, water, and atmospheric dust (Fallah 2000). Still, the positive influence of silicon on rice crops has been shown through a large number of studies (Yoshida 1981; Marschner 1995; Takahashi 1995; Fallah 2000). Agarie et al. (1993) showed that through utilizing silicon, we could efficiently increase dry matter production in rice crops. As a whole, silicon could increase the leaf area and also cause leaves to be erect, concomitantly improving crop photosynthesis. Whenever the supply of nitrogen is suboptimal, or there is a low rate of net photosynthesis or an inadequate expansion of cells, these factors could restrict the rate of leaf growth. In the cells of green leaves, chloroplasts, being primarily an enzyme protein, are the basic locations of more than 75% of whole organic nitrogen; however, photosynthetic efficiency is decreased by nitrogen deficiency, and is directly involved in protein synthesis or chloroplast pigment or electron transfer (Takahashi 1995).

In 2008, the greenhouse of the Deputy of Rice Research Institute of Iran (Amol) was the main location for administering a hydroponics culture experiment. The treatment arrangements were in a factorial randomized complete design; that is, at three levels (0, 50, and 100 ppm), a silicon factor and a nitrogen fertilizer as another factor with two levels (30 and 60 ppm) with four replicates were administered. The variety for testing was Nemat, an Iranian improved and high-yielding variety. At panicle initiation as well as the flowering stages, the concentration of silicon in the stems and green leaves was augmented. Silicon treatment did not influence the chlorophyll meter reading. Addition of Silicon did not influence the rate of photosynthesis, transpiration and stomatal conductance in plant however, the net rate of photosynthesis could be significantly enhanced by nitrogen (Fallah 2012). Based on the study performed by Takahashi et al. (1966), it was found that assimilation of CO2 in the leaf blades and their significant transport in the panicle could be conspicuously elevated by Silicon. According to Kaufman et al. (1979), in the epidermal system of sugarcane, silica cells might be regarded as a "window"; that is, silica cells are the tools whereby light can be transmitted to photosynthetic mesophyll tissue below the epidermal layer. Agarie et al. (1996) claimed that in this window hypothesis, leaves in which the density of silica bodies is high may exhibit greater photosynthetic production, concluding that silica bodies are treated as a window in the epidermal systems of rice leaf. In our research, an effect of silicon on the net photosynthetic rate at the reproductive stage was not observed; hence, the efficacy of silicon for rice plants was not proved.

The specific leaf weight (SLW) or leaf thickness is one of many indicators of the internal anatomy in which Hayashi (1968) obtained a close positive correlation between the net assimilation rate and SLW. Although increasing the supply of silicon could increase the leaf thickness, increasing the supply of nitrogen could decrease it. While an increase of the nitrogen supply increased the leaf length, width, and area, the impact of silicon on these parameters was not significantly observed (Yoshida et al. 1969). Leaf thickness is genetically a fixed character for variety, but it was affected by mineral nutrition status.

Garami et al. (2012) hinted that the increase in levels of sodium silicate resulted in the enrichment of the tiller number and leaf area of the Tarom Mahali (Local Tarom) variety; an increasing trend in the height of the plant and total dry weight was also observed. Moreover, when the levels of the potassium silicate in the stem and leaf tissue increased, it had a positive impact on the morphological characters in the tillering stage. Besides this, in the flowering stage, it was also found that the increment in the level of Sodium and potassium silicate fertilizer could consequentially enhance the total count of chlorophyll a and b. Hosini Motlagh et al. (2012) claimed that at the flowering stage, when the levels of nitrogen and silicon were high, soluble sugars and chlorophyll a, b, and (a + b) were at their maximum states.

Takahashi (1995) recommended that silicon causes the plant to grow, hence increasing the amount of dry production. An increase in silicon levels is concomitant with an increase in the height and number of the tiller. According to Yoshida (1981), silicon could have an effect on the vertical position of the leaves. Snyder et al. (1986) hinted that the height of the plant is increased by silicon. Ravent et al. (1983) proposed that implication of materials made up of peculiarly silicon would have remarkable effect on the height of plant. The efficiency of silicon by setting over the cell walls, together with lignin, over the dry weight has been verified. Agarie et al. (1993) have assessed 10 samples of diverse rice cultivations, recommending that when the amount of silicon fertilizers increases, the dry weight of the plant and its height also increase. Furthermore, it has been indicated that this has the greatest impact on the "suween" because when silicon levels advance, the leaf area and number of tiller increase. Agarie et al. (1993) proposed that if the silicon amount is short, the chlorophyll and photosynthesis amounts will decrease as well. In Quanzhi and Erming's (1998) viewpoint, silicon could culminate in leaf broadcasting, hence assisting in the improvement of photosynthesis. The investigations indicated that a silicon decrease could result in the decrease of absorption of carbon dioxide. So, a silicon decrease could restrain the hatches (Garami et al. 2012).

Soils and waters contaminated with pesticide in rice fields cause environmental problems (Fallah 2010). To analyze the impact of silicon on the prevalence of pest and disease, and the growth and yield of rice (Tarom Hashemi variety), pot and field experiments as random block designs were conducted, with three replicates in a greenhouse and a field, by the Iran (Amol) Deputy of Rice Research Institute between 2009 and 2010. To do so, three levels of N fertilizer (69, 92, and 115 kg N·ha^{-1}), as well as three levels of silicon fertilizer (0, 250, and 500 kg·ha^{-1}) as a factorial, were conducted. The silicon content, rates of blast, and sheath blight were measured at the flowering stage based on the International Rice Research Institute (IRRI) standard (Yoshida et al. 1976). The results indicated that when the amount of silicon as a fertilizer increases in the field and pot, the total dry weight of plants also increases. With increased use of Si fertilizer percent, white head and blast disease decreased. In a study of sheath blight spots, they decreased with an increasing use of silica fertilizer. Application of 250–500 kg·ha^{-1} silicon fertilizer also decreased the amount of stem borer pollution, and as a result, the yield improved about 5% in the field and 24% in the pot experiment (Fallah 2010). There is evidence indicating that mechanical stabilization (rigidity) of the tissue could be due to silicon deposition in the cell wall as a physical process; in fact, silicon can be regarded as a mechanical hindrance to pathogens (Marschner 1995). According to Ma (2004), different plant species' resistance to diseases caused by both fungi and bacteria, such as powdery mildew, rice blast, and sheath blight, could be enhanced through the use of silicon. Rice plants with a high level of silicon have normal growth compared to those with a low level of silicon, and the rice plant is associated with decreasing grain discoloration levels. In the wetlands, in rice that lacks silicon, the reduction of grain production and vegetative growth is prevalent, and it would not be uncommon to observe deficiency symptoms, including the necrosis of mature leave and wilting of plants (Dobermann and Fairhurst 2000). Thus, it could be suggested, but not verified, that silicon be treated as a necessary element, playing a crucial role in rice growth because it assist plants in overcoming a large number of biotic and abiotic stresses.

There are several kinds of silicon fertilizer used in crop production. In order to demonstrate the influence of two sources of Silicon at three level, a study was performed based on the amount of Silicon acquisitions, dry biomass generation, yield productivity in the 843 line and Tarom Hashemi variety. Based on a randomized complete block design in three replications, it could be concluded that the conducted research was factorial. There were three factors in this experiment: varieties (843 line and Tarom Hashemi variety), silicon sources (73% SiO_2 and 22% of Savadkoh mine), and silicon rates (0, 250, and 500 kg·ha^{-1}). The results of the study demonstrates that the implication of silicon could eventually enhanced the level of silicon in the stem and leaves, tiller number, dry weight of leaf, 1000 grain weight, and yield. The results also indicate that two silicon sources were not significantly different in all characters, and the 843 line is better than the Tarom Hashemi variety. Hence, the 250 kg ha–1 treatement was recorded as the ever level of Silicon consumption marked for farms for the reduction of production chargés (Gholami and Fallah 2013). The characteristic differences among rice plants could be observed with respect to their potential to take up silicon. We suggest using the remains of the Savadkoh mine (22% SiO_2) as a silicon fertilizer for paddy fields. Silicon improves the nutrient imbalance in rice plants.

Aghajani Mir (2013) conducted a field experiment in which he studied how liquid silicon fertilizers could have an effect on the yield and growth of the Anbarboo variety, in Ghaemiheh village of Bahnamier in the province of Mazandaran. Treatment included liquid silicon fertilizer, nanosilicon, and potassium silicate with 2, 4, and 6 $ml \cdot L^{-1}$ concentration and check (no using silicon); there were a total of 10 treatments. Silicon fertilizer spray was done at 25 and 50 days after transplanting. The results showed that silicon spray increased the plant height and total dry weight at the flowering and physiological maturity stages, but had no affect at the midtillering stage. Utilizing silicon spray at the midtillering and flowering stages could increase leaf area. The treatment of 6 $ml \cdot L^{-1}$ of silicon fertilizer caused an increase in the growth and yield of the Anbarboo rice variety of about 10–15% compared to check (Aghajani Mir 2013). We suggest using silicon fertilizer as spray in the flowering stages.

In 2012, a pot experiment was conducted to analyze the impacts of silicon spray and the environment on the vegetative growth stages of varieties of rice at a pilot site. A combined analysis and factorial layout was used with three replicates at the Deputy of Rice Research Institute of Iran (Amol), with five varieties of rice (Tarom Mahali, Shiroodi, Fajr, Keshvari, and Kohsar) in outdoor and greenhouse environments. The results showed that the environment had an important role in increasing the percent of SiO_2 in the stems, leaves, and roots. The Soil Plant Analysis Development (SPAD) or chlorophyll meter value was increased the indoor environment, as well as in the outdoor one. This increase was significant, about 23% in the greenhouse environment for the silica content of the stems and leaves. From this study it can be concluded that the foliar silica is not significant in two environments (Nori 2014). That is, silicon does not improve rice growth under cold stress.

Paddy husk or rice hull has 20% silica. To analyze whether silicate fertilizers could have an influence on the growth of rice, a study was done with eight different compound treatments of silicon, husk, and husk ash based on randomized complete design in greenhouse. Pot has 10 kg of soil with clay–loam tissue. Agronomic characteristics were measured at the stage in which the physiological status of the plant reached maturity. The results of the research indicated that these treatments cause a significant increase in plant height, leaf area, leaf, stem, panicle dry weight, and total dry weight, and an increase in the harvest index (HI) and yield compared with the control. The greatest increases in these traits were observed in treatments with compounds of 500 g of husk and 210 g of husk ash or 250 g of husk and 105 g of husk ash. The final results obtained indicate that husk and husk ash can be used as a renewable organic substance instead of chemical silicate fertilizer (Jalali 2011).

Silicon increases the height and fresh weight of the plant. As a result, silicon increases the bending moment, too. Culm thickness was increased at 100 ppm silicon treatment. It was significantly different from that without silicon. The concentration of silicon in the culture solution increases the break resistance. The index of lodging was lower with the addition of silicon than without it (Fallah 2008). The stems and leaves of cereals are believed to stiffen as a result of silicon deposits in the form of silica gel. The treatments that included silicon, compared to those without, imparted more strength to the stem, making it capable of resisting breakage, and thus increasing the number of silicate cells, as well as the silicon concentration in the stalk, even at higher levels of nitrogen (Sadanandan and Varghese 1968). In sum, the contribution of silica deposition in the rice plant to advancing the mechanical strength of the culm was highlighted. Moreover, it has been acknowledged that silica deposition reduces lodging.

In order to study the impact of utilizing a granule as well as a spray form of silicon fertilizer on morphological characters of the Shiroodi variety in the vegetative growth stage, a pot experiment was carried out in 2012 at the Deputy of Rice Research Institute with a completely random design composed of four replicates. The treatments included check (without silicon fertilizer) and using silicon fertilizer (73% SiO_2) in the granule form, with three levels (1.5, 3, and 4.5 g) in each pot, with 10 kg loamy soils. The spray form of silicon fertilizer (having 32% SiO_2) was used in 2, 4, and 6 ml·L^{-1} concentrations in the shoot of the Shiroodi variety. Silicon fertilizer was given two times over 20 days after transplant in pots. The results showed that with the use of silicon fertilizer, compared to without it, there were increases of 5%, 31%, 22%, 34%, and 21% of plant height, leaf area, stems, leaves, and total dry weight, respectively. Use of silicon fertilizer spray was more efficient than use of silicon fertilizer granule, producing greater vegetative growth (Roodi 2012). Foliar application of silicon fertilizer by means of spray offers a method of supplying silicon fertilizer to rice plants more rapidly than methods involving root application, particularly for leaves with thick cuticles.

Percent of the silica concentration (SiO_2) of leaves and stems was increased by using silicon fertilizer as a granule or spray; however, the manganese and iron content of the leaves and stems decreased (Roodi 2012).

Roodi (2012) concluded that the silica in rice grains was between 2% and 4%, and there was no significant difference in treatment between the granule and spray concentrations of rice grains, but the concentrations of Mn, Fe, and Zn decreased by increasing the silicon levels. Dobermann and Fairhust (2000) cited that the availability of P in soil and root oxidation was increased by Si. According to them, Si alleviates the toxicity of Fe and Mn by decreasing the uptake of these elements.

Roodi (2012) mentioned that the grains had 0.5%–0.7% lipids and 8%–9.7% proteins. There was a significant difference in the effect of silicon on proteins.

She (2012) concluded that the thickness of the stem's silica layer, the diameter of the leaf's parenchyma cell, and the xylem and phloem of the stem at the panicle initiation and flowering stages were increased by using silica as a spray or granule fertilizer. On a cellular scale, silicon increased the cuticle thickness and cell diameter of the parenchyma, xylem, and vessel phloem (Roody 2012). Manganese is contained in crops having normal growth at a density of 50–100 ppm in general. In paddy rice, it is reported that the deficiency is accompanied by the incidence of *Helminthosporium* leaf spot.

After yield, rice quality is the most important subject to study. Amylose is a noticeable determinant of rice quality and is regarded as one of starch components and have positive linear. In 2011, an experiment was conducted at the Rice Research Institute of Iran using a factorial and completely randomized block design with three replications and nine combined treatments. This experiment was administered to assess whether the process of spraying potassium silicate and nitrogen could affect the physicochemical properties of Tarom Mahali ratoon. The treatments used in the research included spraying potassium silicate, as well as nitrogen, at zero level (control), the amount of which was three and six ml per 1000 ml. These elements were used in two steps 20 and 40 days after harvest of the main crop. Variance analysis indicated that there was a statistically significant impact of potassium silicate and nitrogen treatments or their interactions at any level. However, a significant difference between the control and other treatment combinations was observed with respect to a means comparison of the treatments indicated. Thus, the measurement was conducted for the increased rate of 2%–3% for all attributes of quality properties and conversion efficiency (MosavaiAzandahi et al. 2013). Several researchers do not consider the amount of rice grain protein, unlike wheat, to be a qualitatively effective factor. However, it is known that rice protein is nutritionally superior to that of wheat, and a predominant percentage of human energy, especially in Asian countries, is due to rice consumption (MosavaiAzandahi et al. 2013). One strategy by which we could increase the rice harvest is through reuptake (known as ratoon). According to Buck et al. (2008), less fertilizer, inexpensiveness, easy usage, and high quality (very low pollution) are among the many benefits of foliar spraying. Babaeian and Tavasoli (1999) indicated that the gelatinization temperature of the ratoon crop was more than that of the original crop. However, there were no significant differences between the gel and amylose contents of the ratoon crop compared with those of the main crop. In an experiment analyzing the effects of nitrogen on rice quality factors, it has been indicated that nitrogen application could increase grain protein (Bahmanyar and Ranjbar 2007).

Barnyardgrass (*Echinochloa crus-galli* L.) is an annual plant belonging to the Gramineae family. It is one of the most important weeds in rice paddies in Iran and worldwide. An experiment was administered to assess the impacts of aqueous extracts obtained from the roots and shoots of barnyardgrass at 50% and 100% concentrations of silica bodies on the leaf and stem of rice plants. The results demonstrates that aqueous extract of barnyardgrass root had significant influence on the number and total size of silica grains and knob of the leaf and stem (EsmaeilyKenary 2012), which resulted in more resistant and stronger rice stems, which led to a greater production yield of rice. Zarinkamar (2008) demonstrated that silica bodies are horizontally elongated with round ends and sinuous walls. Silica bodies are narrow and tall, especially over the veins. The shape and size of silica bodies are different according to the genus in the Gramineae family (Zarinkamar 2008).

Across the stem of a check plant, a large vascular bundle was usually arranged in lower position of small vascular bundle and some of them were arranged between them. Rice plants were treated by 50% aqueous extracts of barnyardgrass root, and the size and number of knob silica increased. The diameter of the large vascular bundle increased as well. Rice plants were treated by 100% aqueous extracts of barnyardgrass roots, and the number of knob silica increased in the overall under epidermis layer (EsmaeilyKenary 2012).

Silicon content was higher in grains and straw than the nitrogen content, nitrogen content of grains more than straw. The methods of irrigation were significant only for the nitrogen content of straw. With an increasing of nitrogen level, the silicon content of grains and straw decreased (Ghanbari Malidareh et al. 2011); however, the nitrogen content of grains and straw increased. There was no difference between irrigation methods on the yield of rice, but the nutrients of nitrogen and silicon affected the yield. The effect of nitrogen was more than that of the silicon fertilizer on the rice crop's yield. Therefore, rice crops need more silicon than nitrogen; however, it does not have the same importance for crop performance as nitrogen.

Silicon use could increase cellulose up to 7.6%. The greatest amount of cellulose noted (46.7%) was for 600 kg·ha^{-1} of silicon usage, and hemicellulose was higher as well (15.2%). Silicon application increased lignin by 26.3%, because lignin was 12.5% in 600 kg silicon rate (Dastan et al. 2012). Lignification protects aerenchyma by providing mechanical toughness to roots; consequently, lignification protects the root from flooding injury (Takahashi 1995).

18.9 FUTURE PERSPECTIVES

During the industrial age, we added carbon dioxide to the atmosphere by burning oil, gas, and coal, and we have destroyed forests of trees, which remove carbon dioxide by photosynthesis. The concentration of carbon dioxide is increasing in the atmosphere, and the average temperature is also increasing. This global warming could cause a mean temperature increase of 2°C to 3°C in the future (Peng et al. 1995). This seems like to little to worry about changing of the last freeze in spring planting time and increasing of temperature in summer, as a result, deficiency of water for rice crop production.

The growth response of rice to carbon dioxide is not linear. Growth increases only with carbon dioxide concentrations up to 500 µl·L^{-1}. The sink limitations, acclimation of Rubisco (Ribulose – 1, 5 – bisphosphate carboxylase- oxygenase), plant nutrition, developmental responses, and carbon allocation processes must be investigated for their possible involvement in this phenomenon (Peng et al. 1995).

Elevated atmospheric CO_2 concentrations promote the growth and yield of rice by way of enhancement of the net photosynthetic rate; however the level of endorsement is influenced by associated factors such as light intensity, temperature, and mineral nutrition (Peng et al. 1995). There is an interactive effect between elevated CO_2 and temperature on rice growth and development. The optimum temperature for rice growth is 25°C–35°C. High temperatures can reduce crop productivity because rice plants are most susceptible to damage at the flowering stage. Therefore, grain quality will be improved or deteriorate in response to elevated CO_2. Such changes should be corrected by cultural practices, that is, improved fertilization or by breeding cultivars with grain quality characters that are insensitive to high temperatures (Peng et al. 1995).

A great diversity in agroclimatic factors affects both yield and cooking quality characteristics of rice grains. It seems that using silicon fertilizer may improve rice production under these circumstances. Therefore, it is recommended that silicon fertilizers like granule, ligules, and nanosilicons be used in the paddy fields of northern Iran to improve the growth and yield and reduce the biotic and abiotic stresses. The foliar application of silicon fertilizer by means of spray offers a method of supplying silicon fertilizer to rice plants more rapidly than methods involving root application. We are interested in using nanosilicon fertilizer as a spray because this novel technology could expand the producing area and fruitfulness of the paddy fields, making use of silicon fertilizer in rice crops a constructive alternative strategy. We are interested in conducting research at the cell level by electron microscope, studying the enzyme and cytokine activity, which may be affected by silicon. Strategies for silicon management in Iran in the future follow:

Natural inputs: In some parts of Iran, outstanding input of Si from irrigation water may happen, especially if we are going to irrigate using groundwater from landscapes with volcanic geology.

Management of straw: Over a long period of time, through not eliminating straw from the field after harvest, Si shortage can be hindered. Recycle the straw of rice (5–6% Si), as well as the husks (10%).

Management of fertilizer: It is highly recommended that extra amounts of N fertilizer not be used, as they will not only increase the yield and total uptake of N and Si, but also decrease the concentration of Si in straw because of excessive biomass growth.

Postharvest measure: If available, recycle rice hulls or rich hull ash to reproduce Si.

REFERENCES

Agarie S, Agata W, Uchida H, Kubota F, Kaufman PB. (1996). Function of silica bodies in the epidermal system of rice (*Oryza sativa* L.): Testing the window hypothesis. *J Exp Bot* 47(299): 655–660.

Agarie S, Uchida H, Agata W, Kubota F, Kaufman PB. (1993). Effect of silicon on growth, dry matter production and photosynthesis in rice plant (*Oryza sativa* L.). CPITA 225-234. Korean-Society-for-Cognitive-Science (KSCS), South Korea.

Aghajani Mir SH. (2013). The effect of different levels of liquid silicon fertilizer on growth and yield of Anbarboo rice variety in Bahnamirarea [in Persian]. MSc thesis, Islamic Azade University of Shareh Ghods Branch.

Babaeian N, Tavasoli F. (1999). Evaluation of quality and quantity of ratoon crop in Iranian rice varieties [in Persian]. In Proceeding 6th Congress of Agronomy and Plant Breeding, Babolsar, Iran, June 16–19, p. 498.

Bahmanyar MA, Ranjbar GA. (2007). Response of rice (*Oryza sativa* L.) cooking quality properties to nitrogen and potassium application [in Persian]. *Pak J Biol Sci* 10(11): 1880–1884.

Buck GB, Korndorfer GH, Nolan A, Coelho L. (2008). Potassium silicate as foliar spray and rice blast control. *J Plant Nutr* 31(2): 231–237.

Dastan S, Siavoshi M, Zakavi D, Ghanbari Malidareh A, Yadi R, GhorbanniaDelavar E, Nasiri AR. (2012). Application of nitrogen and silicon rates on morphological and chemical lodging related characteristics in rice (*Oryza sativa* L.). *J Agric Sci* 4(6): 12–18.

Dobermann A, Fairhurst T. (2000). Mineral deficiencies. In *Rice Nutrient Disorders and Nutrient Management*. International Rice Research Institute, Manila, Philippines, pp. 28–104.

EsmaeilyKenary S. (2012). Allelopathic effects of *Oryza sativa* against some weeds [in Persian]. MSc thesis, Biology Science College, Alzahra University.

Fallah A. (2000). Effects of silicon and nitrogen on growth, lodging and spikelet filling in rice (*Oryza sativa* L.). PhD dissertation, University of the Philippines Los Baños, Laguna.

Fallah A. (2008). Studies effect of silicon on lodging parameters in rice plant under hydroponics culture in a greenhouse experiment. Presented at 4th Silicon in Agriculture Conference, Wild Coast Sun, South Africa, October 26–31.

Fallah A. (2010). Study of silicate fertilizer on the rate of important disease and pests on rice [in Persian]. Deputy of Rice Research Institute of Iran, Amol.

Fallah A. (2012). Study of silicon and nitrogen effects on some physiological characters of rice. Available at www.ijagcs.com.IJACS/2012/4-5/238-241.

Garami M, Fallah A, Khatami Moghadam MR. (2012). Study of potassium and sodium silicate on the morphological and chlorophyll content on the rice plant in pot experiment (*Oryza sativa* L.). Available at www.ijagcs.com.IJACS/2012/4-10/658-661.

Ghanbari Malidareh A, Kashani A, Nourmohamadi G, Mobasser HR, Alavi SV. (2011). Evaluation of silicon application and nitrogen rate on yield, yield components in rice (*Oryza sativa* L.) in two irrigation systems. *Am Eurasian J Agric Environ Sci* 10(4): 532–543.

Gholami Y, Fallah A. (2013). Effects of two different sources of silicon on dry matter production, yield and yield components of rice, Tarom Hashemi variety and 843 lines. Available at www.ijagcs.com.IJACS/2013/5-3/227-231.

Hayashi K. (1968). Responses of net assimilation rate of differing intensity of sunlight in rice varieties. *Proc Crop Sci Jap* 37: 528–533.

Hoshikawa K. (1989). Root development in relation to environmental conditions. In *The Rice Growing Plant: An Anatomical Monograph*. Nobunkyo, Tokyo, pp. 199–205.

Hosini Motlagh SP, Fallah A, Rahdari P. (2012). Interactive effect of silicon and nitrogen spray on some physiological characteristics of ratton Tarom Mahali (*Oryza sativa* L.) in pot experiment [in Persian]. Presented at the 17th National and 5th International Iranian Bilology Conference, Kerman.

Jalali M. (2011). The effect of silicon fertilizer and ash hull of rice paddy on growth and yield of Shiroodi variety in pot experiment [in Persian]. MSc thesis, Islamic Azade University of Qaemshar Branch.

Kaufman PB, Takeoka Y, Carlson TJ, Bigelow WC, Jones JD, Moor PH, Ghoshed NS. (1979). Studies on silica deposition in sugarcane, using scanning electron microscopy, energy dispersive x-ray analysis, neutron activation analysis, and light microscopy. *Phytomorphology* 29: 185–193.

Kundu DK. (1987). Chemical kinetics of aerobic soils and rice growth. PhD thesis, Indian Agricultural Research Institute, New Delhi.

Ma JF. (2004). Role of silicon in enhancing the resistance of plants to biotic and a biotic stresses. *Soil Sci Plant Nutr* 50: 11–18.

Ma JF, Nishimura K, Takahashi E. (1989). Effect of silicon on growth of rice plant at different growth stages. *Soil Sci Plant Nutr* 35: 347–356.

Ma JF, Yamaji N, Mitani N, Tamai K, Konishi S, Fujiwara T, Katsuhara M, Yano M. (2007). An efflux transporter of silicon in rice. *Nature* 448: 209–212.

Malakouti MJ, Kavousi M. (2004). Characteristics of paddy soil. In *Balanced Nutrition of Rice* [in Persian]. Ministry of Jihad-e-Agriculture Deputy, Agronomy Affairs, Tehran, pp. 3–44.

Marschner H. (1995). Beneficial mineral elements. In *Mineral Nutrition of Higher Plants*. 2nd ed. Academic Press Harcourt Brace, London, pp. 405–435.

MosavaiAzandahi SM, Fallah A, Taghizadeh R. (2013). Effect of spraying potassium silicate and nitrogen on the physicochemical characteristics of ratoon rice (Tarom's local variety). *Int J Agron Plant Prod* 4(10): 2748–2750.

Nori S. (2014). Effects of silicon spray and environment on vegetative growth stages of rice varieties in pot experiment [in Persian]. MSc thesis, Islamic Azade University of Ayatollah, Amoli Branch.

Okuda A, Takahashi E. (1965). The role of silicon. In *Mineral Nutrition of the Rice Plant, Proceedings of the International Rice Research Institute*, February 1964, pp. 123–146.

Patrick WH, Reddy CN. (1978). Chemical changes in rice soils. In *Soils and Rice*. International Rice Research Institute, Los Banos, Philippines, pp. 361–379.

Peng S, Ingram KT, Neue HU, Ziska LH. (1995). *Climate Change and Rice*. International Rice Research Institute, Manila, Philippines, p. 374.

Quanzhi Z, Erming G. (1998). Effect of silicon application on rice in a rice area along the yellow river. No. 32. Department of Agronomy, pp. 308–313.

Ravent JA. (1983). The transport and function of silicon implants. *Rev Cambridge Philos* 58: 179–207.

Rice Research Institute of Iran (RRII). (2014). National program on self-sufficiency in rice production [in Persian]. RRII, Rasht, Iran.

Roodi S. (2012). Study of silicon effect on number of traits anatomy and physiology in *Oryza sativa* (L.) Shiroodi [in Persian]. MSc thesis, Bu-AliSina University.

Sadanandan AK, Varghese EJ. (1968). Studies on the silicate nutrition of rice in the laterite soil of Kerala. I. Effect on growth and yield. *Madras Agric J* 11: 261–264.

Snyder GH, Jones DB, Gascho GJ. (1986). Silicon fertilization of rice on everglades histosols. *Soil Sci Soc Am J* 50: 1259–1263.

Solimani A, AmiriLarijani B. (2004). *Principles of Rice Production* [in Persian]. Arevij, Tehran, Iran, p. 303.

Soni SL, Kaufman PB, Jones RA. (1972). Electron microprobe analysis of the distribution of silicon and other elements in rice leaf epidermis. *Bot Gaz* 133(1): 6–72.

Sumida H. (1992). Effect of nitrogen on silica uptake by rice plant. *Jpn J Soil Sci Plant Nutr* 63: 633–639.

Takahashi E. (1995). Uptake mode and physiological functions of silica. In Matuso T, Kumazawa K, Ishii R, Ishihara K, Hirata J (eds.), *Science of the Rice Plant*. Food and Agriculture Policy Research Center, Tokyo, pp. 420–433.

Takahashi E, Arai K, Kasiad Y. (1966). Studies on physiological role of silicon in crop plants (Part 14). Effect of silicon on CO_2 assimilation and translocation of assimilate to panicle. *J Sci Soil Manure Jap* 37: 597–598.

Takahashi E, Hino K. (1978). The silica uptake by rice plants with special reference to the form of dissolved silica. *J Sci Soil Manure Jap* 49: 537–540.

Takahashi E, Nishi T. (1982). Comparative studies on the silica nutrition in plants. 21. Effect of nutritional condition on the silicon uptake by rice plants. *J Sci Soil Manure Jap* 53: 395–401.

Wallace A. (1989). Relationships among nitrogen, silicon and heavy metal uptake by plants. *Soil Sci* 147: 457–460.

Yoshida S. (1981). *Fundamentals of Rice Crop Science*. International Rice Research Institute, Los Baños, Laguna, Philippines.

Yoshida S, Fornod A, Cock JH, Gomez KA. (1976). *Laboratory Manual for Physiological Studies of Rice*. 3rd ed. International Rice Research Institute, Los Baños, Laguna, Philippines.

Yoshida S, Navasero SA, Ramirez EA. (1969). Effect of silica and nitrogen supply on some leaf characters of the rice plant. *Plant Soil* 31: 48–56.

Yoshida S, Ohnishi Y, Kitagishi K. (1962). Histochemistry of silicon in the rice plant. 2. Localization of silicon within rice tissues. *Soil Sci Plant Nutr* 8(1): 36–41.

Zarinkamar F. (2008). Gramineae family. In *Iranian Atlas of Foliar Anatomy* [in Persian]. Noorbakhsh Press, Tehran, Iran, pp. 32–52.

19 Efficacy of Silicon against Aluminum Toxicity in Plants

An Overview

Durgesh Kumar Tripathi, Gausiya Bashri, Shweta,
Swati Singh, Parvaiz Ahmad, Vijay Pratap Singh,
Sheo Mohan Prasad, Nawal Kishore Dubey,
and Devendra Kumar Chauhan

CONTENTS

ABSTRACT

The bulk of potentially arable land (30%–40%) in tropical and subtropical regions of the world is acidic, and one of the major menaces particularly evident in acidic soil is the phytotoxicity of aluminum (Al) that limits the growth and productivity of agricultural crops. At neutral pH of the soil, Al resides in stable form, but when the pH of soil drops below 5.5, solubilization of the toxic form of Al occurs, making an organism specifically vulnerable to the precarious effects of the ion. Plants under Al stress incur several incurable toxicity symptoms such as rapid and transient overgeneration of reactive oxygen species $\left(\text{ROS}: O_2^{\cdot-}, H_2O_2, {}^{\cdot}OH\right)$ that resulted in oxidative burst; further, toxic symptoms involve subsequent reduction in the acquisition of water and mineral uptake by the plant that consequently results in significant reduction of crop yield. Quick and rapid advancement in understanding the tolerance mechanism of Al has been achieved in the last decades because of the intensified research on the specific area of Al biochemistry and molecular biology. Besides the exclusion and detoxification mechanism employed by plants, it has been found after exhaustive research that the macronutrient silicon (Si) also plays a central role in the amelioration of Al toxicity. Substantial progress has been made to develop our understanding of the physiology of Si-induced amelioration of Al toxicity in plants.

19.1 INTRODUCTION

Environmental degradation due to severe pervasiveness of acid precipitation in tropical and subtropical zones has turned out to be a major societal issue in recent years, since it serves as the core ground behind the steep run-up in the global food crisis and spreading hunger in millions of people worldwide. In accordance with peer-reviewed studies and available data, it has been shown that approximately 30%–40% of the total world's usable land, and about 70% of the potentially arable land, has been covered by acidic soil (Kochian et al., 2004; Chen et al., 2010; Bian et al., 2013). A considerably large number of developing and developed countries (e.g., Australia, Europe, Africa, Asia, North and South America, East Asia, and Southeast Pacific) are located where the majority of acid soil resides. Such distribution of acid soils on the global level hampers crop productivity worldwide. A micromolar concentration of trivalent aluminum (Al^{3+}) ions is enough to alter the overall physiology and genetic complexity of living organisms (Silva et al., 2001; Ryan et al., 2011; Inostroza-Blancheteau et al., 2012). Imposition of aluminum leads to intense phytotoxicity in acidic soil enriched with mineral content.

Aluminum (Al) is a ubiquitously found element with atomic number 13, and it ranks third (after silicon and oxygen) among the most distributed metallic elements on the earth's crust. In neutral and weakly acidic soil, Al resides in the form of a stable complex with oxygen and silicate and pervades any interaction with the biological processes of living organisms. However, when the pH drops below 5.5, Al is solubilized in toxic forms, such as $Al(OH)^{2+}$, $Al(OH)^{3+}$, and $Al(H_2O)^{3+}$ (Krstic et al., 2012). The biochemistry of aluminum is extremely complex and is not fully understood, the cause behind such intricate trouble is the extremely large spectrum of polynuclear inorganic and organometallic complexes of varying stability that occur in natural soils and waters (Krstic et al., 2012).

Aluminum is a highly reactive metal and it has several potential binding sites, such as the cell wall, plasma membrane, cytoskeleton, and nucleus (Panda et al., 2009). Over- and rapid accumulation of aluminum in the apoplast and symplasm of sensitive plants instigates an array of altered physiological and cellular processes. In this chapter, an attempt is made to briefly localize the particular and intense phototoxicity, mediated by aluminum interference with several biological processes of plants.

Marked damaging impacts incurred by aluminum toxicity include decreased root hair elongation and development, lipid peroxidation (LPO) of the plasma membrane, altered membrane surface potential (zeta potential), overexpression of highly toxic reactive oxygen species (ROS), reduced transport and acquisition of mineral elements, excess accumulation of callose in the plasmodesmata, leading to an impediment in cell-to-cell trafficking, disrupted cytoskeleton and apoplastic processes (Matsumoto, 2000; Yamamoto et al., 2002; Sivaguru et al., 2003; Kochian et al., 2004; Horst et al., 2010), increased rigidity of the DNA double helix by reduced DNA replication (Rout et al., 2001), cross-linked pectins, increased cell wall rigidity (Rout et al., 2001), diminished activity of enzymes employed in sugar phosphorylation processes (Panda et al., 2009), and disruption of cytoplasmic Ca^{2+} homeostasis (Panda et al., 2009). Besides this, Al toxicity relentlessly alters the respiratory function and redox status of mitochondria. In addition, the activity of a key signaling enzyme named phospholipase C (PLC), which is employed in signal transduction, has also been masked under Al stress, and thereby interferes with the phosphoinositide signaling pathway (Panda et al., 2009).

Upon Al exposure, an array of genes has been identified that are induced or repressed (Ryan et al., 2011). Besides plants, the precarious impacts of aluminum can also be noted in animals, as it is responsible for severe nervous, kidney, and lung disorders. Scientists from many laboratories around the world now indulged in studying the physiology, morphology, and molecular biology of Al tolerance in plants.

To curb the toxicity imposed by Al, crop plants have fortuitously evolved several tolerance mechanisms that enable them to tolerate intense toxicity of aluminum in acidic soils. Over the decades, overwhelming research has been conducted with a view to establishing the mechanisms inherent to aluminum toxicity and tolerance, ranging from the global to the molecular level. Tolerance to Al can be achieved via either exclusion of Al (apoplastic mechanism) from the root apex or intracellular or internal detoxification tolerance (symplastic mechanism) by the chelation and sequestration (Al-OA complexes in the vacuole) of Al that enters the plant's symplasm (i.e., true tolerance) (Panda et al., 2009). A deeper understanding of aluminum's resistance mechanism in plants and the role of mitochondria in stress signaling is still needed so that breeders will be able to grow tolerant varieties of their interest in acidic soil.

Intensified research over the past couple of decades has elucidated the intricate role of silicon in metal detoxification (Singh et al., 2011; Tripathi et al., 2012a,b). To date, the mechanism lying behind such silicon-mediated amelioration of metal toxicity is still lacking. But on the basis of previous studies, it can be inferred that the mechanism contributing to Si-mediated detoxification of metal includes reduced uptake and transport of metals and maintenance of redox homeostasis via favorable anatomical amendments of the root and leaf (Singh et al., 2011; Tripathi et al., 2013). Over the past decades, workers from many laboratories around the world have focused their work on elucidating the immense significance of silicon in eliminating metal toxicity. Silicon is one of the most beneficial elements absorbed by plants, in the form of silicic acid ($Si(OH)_4$), when the pH of soil is below 9. It ranks second among the most abundant elements in the earth's crust, after oxygen. Because silicon is abundant in soil solution, the roots of almost all plants contain a significant fraction of it in their tissues. Silicon plays a beneficial role in plant growth and development, but until the twentieth century, its significance was overlooked or denied by a class of workers (Tripathi et al., 2014).

From last few decades silicon is not only regarded as an essential element, but also used as an ameliorative substance against various abiotic and biotic stress (Tripathi et al., 2014). In a wide variety of crop plants, like rice, sorghum, maize, and barley, it has been found that silicon ameliorates the intense abnormalities imposed by aluminum toxicity. But the descriptions associated with the mechanism of Si-induced amelioration of aluminum toxicity are still controversial. Some workers suggest that silicon alleviates aluminum toxicity by increasing the pH of the soil solution, or by reducing the hyperaccumulation of aluminum in the plant, or through the formation of complex aluminosilicate species, or via an internal *in planta* detoxification mechanism. The present status of our knowledge concerning the Si-induced alleviation of Al toxicity and tolerance in plants has been covered in numerous reviews, but still, exhaustive research is required in this field to pinpoint new avenues that may contribute significantly to the much differentiated view of the mechanism underlying Si-induced amelioration of aluminum toxicity in targeted tissues. In this respect, colossal publications should also be published consistently in the near future, particularly evidencing the intricate troubles of aluminum in acidic soil and the relevant response of silicon in its detoxification (Table 19.1).

19.2 SILICON DETOXIFICATION IN MORPHOPHYSIOLOGY OF PLANTS CAUSED BY ALUMINUM TOXICITY

Generally, the impacts of any abiotic stresses first appear on the morphology and physiology of plants, which may be directly observed and assumed at the toxicity level. As has been well documented, in the acid soils, aluminum toxicity is regarded as one of the key constraints for plant growth and development. An excess level of aluminum significantly influences the morphophysiological parameters of plants (Horst et al., 2010). Furthermore, Mossor-Pietraszewska (2001) reported that in the plant system, the principal target sites of Al toxicity are generally the root cells and plasma membrane, particularly of the root apex. However, it has been shown that the trivalent (Al^{3+}) form of aluminum is highly toxic, and its interactions with oxygen donor ligands like proteins, nucleic acids, and polysaccharides have resulted in the inhibition of cell division, cell extension, and transport (Mossor-Pietraszewska, 2001). Moreover, it has also been well documented that aluminum hampers

TABLE 19.1

Summary of Si-Mediated Amelioration of Aluminum Toxicity in Certain Crop Plants

No.	Plant Species	Treated with Aluminum (Al)	Treatment with Silicon (Si)	Observation on Growth Parameters	References
1.	Rice (*Oryza sativa*) seedling	50 μM 0–800 μM	10 μM 0–2000 μM	Exogenous application of silicon ameliorates Al toxicity by reducing the Al accumulation or maintaining the levels of other essential mineral nutrients.	Singh et al., 2011; Nhan and Hai, 2013
2.	Sorghum (*Sorghum bicolor* L.)	296 μM	0–3600 μM	Silicon does not stimulate shoot growth, but shoot/root ratio is restored.	Galvez et al., 1987; Galvez and Clark, 1991; Baylis et al., 1994
3.	Sorghum (*Sorghum bicolor* L.)	0–100 μM	0–2800 μM	Silicon ameliorates the toxic effects of Al on root growth and the root/shoot ratio.	Hodson and Sangster, 1993
4.	Wheat (*Triticum aestivum*)	0–100 μM	0–2800 μM	Amelioration of aluminum by silicon occurs not because of the reduced level of aluminum, but because of the formation of aluminosilicate; it has an *in planta* component.	Cocker et al., 1997; Hodson and Sangster, unpublished
5.	Barley (*Hordeum vulgare*)	0–100 μM	0–2800 μM	Treatment of silicon significantly ameliorates aluminum toxicity by reducing the level of aluminum uptake by plants.	Hammond et al., 1995; Liang et al., 2001
6.	Teosinte (*Zea mays* L. ssp. *mexicana*)	0–120 μM	0–4 μM	Silicon ameliorates Al toxicity by adding a significantly less amount of it in the nutrient solution.	Barcelo et al., 1993
7.	Cotton (*Gossypium hirsutum*)	0–100 μM	0–2800 μM	A slight reduction has been observed in the nutrient solution of aluminum by adding silicon.	Li et al., 1989
8.	Pea (*Pisum sativum*)	0–100 μM	0–2800 μM	Silicon application confers little amelioration of aluminum toxicity; mechanisms considered for amelioration are solution effects and the codeposition of Al and Si within the plant.	Hammond and Hodson, unpublished
9.	Soyabean (*Glycine max*)	0–100 μM	0–2800 μM	Exogenous application of silicon improves aluminum toxicity in plants, but this was dependent on the pH; interaction of Si ameliorated the toxicity imposed by aluminum.	Baylis et al., 1994; Hiradate et al., 1998
10.	Maize (*Zea mays*) seedling	50 μM	1000 μM	The flavonoid-type phenolics were not considered, but they appear to play a significant role in the mechanisms of Si-induced amelioration of Al toxicity.	Ma et al., 1997; Kidd et al., 2001
11.	Peanut (*Arachis hypogaea*)	80 mg/L	160 mg/L	Silicon ameliorates the toxicity imposed by aluminum by forming a complex.	Shen et al., 2014

various physiological processes, like photosynthesis, respiration, transpiration, stomatal frequency, and root hair frequency (Silva et al., 2008).

Silicon has been reported as one of the most beneficial element for plants and enhanced the growth by supporting the plant metabolic system; however, in some cases, it only provides a physical and mechanical barrier for plants (Tripathi et al., 2011, 2012a through d, 2013, 2014, 2015a through c, 2016). Barbosa et al. (2015) explained that silicon is commonly connected with the physiological, morphological, nutritional, and molecular aspects in plants. Their study proposed that the exogenous treatment of silicon does not make any considerable alterations in hydrogen peroxide, glutathione (GSH), electrolyte leakage, and malondialdehyde (MDA) in the maize plant. Besides this, silicon significantly increases the total content of chlorophyll, particularly chlorophyll a and chlorophyll b. Thus, it may be concluded that silicon plays a beneficial role in the stability of photosynthetic pigments and in the maintenance of cell membrane damage due to oxidant compounds in young *Zea mays* plants.

Similarly, Shahnaz et al. (2011) investigated the role of silicon application against aluminum stress in plants and found that Si successfully alleviates aluminum toxicity in *Borago officinalis* plants. The results of Shahnaz et al. (2011) further demonstrated that the treatment of Al^{3+} significantly increased the level of MDA, aldehydes, proline content, and proteins, and caused a drastic decline in phenolic compounds. On the other hand, silicon treatments against aluminum toxicity increased the tolerance of *Borago officinalis* to Al stress by enhancing phenolic compounds and proline contents and decreasing the level of lipid peroxidation.

Additionally, Baylis et al. (1994) reported the positive effect of silicon against the toxicity of aluminum in soybean roots; this study provides evidence that the effective rate of silicon is dependent on the pH value. At lower pH levels, aluminum is more toxic; thus, to reduce the toxicity level of aluminum, greater concentrations of Si is required. The author suggested that in plants, the formation of subcolloidal inert hydroxyaluminosilicate species in response to aluminum exposure is the prime reason behind Si-induced alleviation of aluminum toxicity.

The interaction of aluminum and silicon in plants is a great matter of interest between the global scientific communities. Hodson and Evans (1995) investigated eight plant species to test whether silicon is capable of alleviating aluminum toxicity in all tested plants, and successfully reported that silicon is not too effective in reducing the toxicity of aluminum in all plants. However, in the case of rice, wheat, cotton, and pea, Si showed little or no amelioration of Al toxicity. Consequently, Hodson and Sangster (1993) further reported a highly significant Al and Si relationship in sorghum root growth, while at the same time, no major impact could be seen in the case of total plant dry weight and the shoot/root ratio. Hammond et al. (1995) have also reported that silicon is capable of reducing aluminum toxicity by about 80%, related to all growth parameters in barley seedlings. Additionally, Barcelo et al. (1993) noticed a remarkable ameliorative impact of silicon against the aluminum toxicity of teosinte (*Zea mays* L. ssp. *mexicana*) grown under full nutrient media.

Liang et al. (2001) suggested that silicon reduces the toxicity symptoms of Al when plants are exposed to Al. Their study also revealed that at a higher level of aluminum treatment (100 µmol L^{-1}), it caused a significant imbalance in the uptake and transport of N, P, and K by plants, and thus this nutrient discrepancy caused by Al toxicity could not be alleviated by adding Si. Conversely, this study also suggested that a lower dose of Al (50 µmol L^{-1}) may be successfully controlled by silicon addition, and damages related to the growth and nutrient imbalances are partially overcome by added Si.

Wang et al. (2004) extended the hypothesis of a silicon-induced alleviative response to aluminum toxicity in plants. Their study suggested that the mechanisms of aluminum alleviation by silicon are based on *ex* or *in planta* mechanisms.

Furthermore, the interaction of aluminum and silicon has been investigated in some crop plants, like wheat (*Triticum aestivum* L.), durum wheat (*T. durum* Desf.), triticale (× *Triticosecale* Wittmack), and rye (*Secale cereale* L.), where a significant reduction in root and shoot growth of plants was found when different aluminum treatments were applied. In contrast, the application of silicon successfully reduced the toxic impacts of aluminum and enhanced the growth of plants

(Vashegyi et al., 2002). Singh et al. (2011) examined a lower dose of silicon against aluminum stress in rice seedlings and reported that aluminum decreased the growth of the rice seedlings, which was correlated with an increase in Al accumulation. Application of silicon significantly alleviated the lethal property of aluminum and led to a decline in Al accumulation. Further, Al exposure reduced the frequency of stomata and root hairs, length of root hairs, and leaf epidermal cells; in contrast, silicon drastically repaired these abnormalities (Singh et al., 2011). There are several other studies that have clearly demonstrated the positive role of silicon in plants against aluminum toxicity and proven it to be one of the most beneficial elements and effective alleviators against aluminum toxicity (Galvez et al., 1987; Hammond et al., 1995; Vázquez et al., 1999).

19.3   SILICON EFFECTIVELY REPAIRS THE ANATOMICAL STRUCTURES OF PLANTS INJURED UNDER ALUMINUM TOXICITY

Due to the changing scenario of environmental conditions, the volume of different stresses has been drastically increased, and may be successfully studied at the physiological, morphological, cellular, biochemical, anatomical, or even molecular level (Bohnert et al., 1995; Sairam and Tyagi, 2004; De Micco and Aronne, 2012; Hasanuzzaman et al., 2013). Over the past decades, anatomical studies have received significant importance in understanding the attributes of plants from the very beginning; however, less attention has been paid to deducing anatomical adaptation with respect to modifying environmental conditions. Under Al stress, plasticity varies in the terms of morphological and anatomical adaptation, which contributes significantly to stress tolerance (Deng et al., 2009; Tripathi et al., 2012a,b; Vaculík et al., 2012). Plants under Al stress incurred reduced frequency of stomata, impediments in the elongation and development of root hairs, disrupted structure and integrity of mesophyll cells and phloem, and so forth; however, the application of silicon significantly reduces such abnormalities. Singh et al. (2011) find out in their study that the addition of Si reduces the Al toxicity in rice seedling by modifying the key structures of leaves and roots. Si alleviates the overall abnormalities imposed by Al such as elongated root hair and an increased number of protoxylem elements. Anatomical parameters of plants get influenced by the toxicity imposed by metals, and plants have developed adaptive strategies to cope with the toxicity; these strategies were further enhanced by exogenous application of Si. For instance, the anatomical parameter of plants treated with Al was affected, and implication of Si repaired the damage successfully since it is responsible for curing the overall abnormalities imposed by any stress.

19.4   SILICON AND MINERAL REGULATION IN PLANTS UNDER ALUMINUM TOXICITY

Aluminum is one of the major constraints of plants in acid soils. It occupies about 30% of the land in the world and primarily occurs in the humid tropics and boreal region. These soils (about 67%) are mainly covered by forests and woodland (von Uexküll and Mutert, 1995). Mostly, Al is present as aluminosilicates in soils and other precipitated forms, which are nontoxic to plants (Brunner and Sperisen, 2013). But under acidic conditions, these minerals are solubilized and converted into the toxic ion Al^{3+} in soil (Kinraide, 1997). The characteristics of acid soils are that, they have greater levels of Al^{3+} and small levels of base cations (i.e., Ca^{2+}, Mg^{2+}, and K^+), which are key plant nutrients. It was previously reported that Al^{3+} and base cations interact at the plasma membrane surface, which determines the plant's response to Al^{3+} exposure (Sverdrup and Warfvinge, 1993; Cronan and Grigal, 1995; Kinraide, 2003). The ratio of base cations/Al^{3+} in the soil solution is a commonly used indicator of Al^{3+} stress; if this ratio is lower than 1, that shows an adverse effect on nutrient imbalance and ultimately on plant growth (Brunner et al., 2004; Richter et al., 2007; Vanguelova et al., 2007). Excess Al^{3+} in soil is toxic to plants, causing stunted roots and reduced availability of phosphorus (P), sulfur (S), and other nutrient cations through competitive interaction. Furthermore, an exogenous supply of Si has been reported to balance the mineral uptake and mobility in plant tissues under different abiotic stresses, particularly when plants are subjected to high concentrations of toxic elements (Hernandez-Apaolaza, 2014; Zhu and Gong, 2014). Beneficial effects of silicon on uptake and *in planta* mobility have been reported for both macro- and micronutrient in plants (Epstein, 1994). Si also has tremendous quality to control the soil properties and can decrease the mobility of toxic metals, like aluminum (Al), iron (Fe), and manganese (Mn), into plants (Maichenkov, 2002; Sadgrove, 2006). Hodson and Sangster (1999) reported that Al toxicity resulted in the deficiency of calcium (Ca) or magnesium (Mg) in conifers, which was reversed by the Si fertilization, alleviating the deficiency of these nutrients in conifers (*Picea glauca, Pinus strobus, Larix laricina*, and *Abies balsamea*) under Al stress. Similar results were reported in *Sorghum* plants, where Al decreased the contents of Mg, Ca, Cu, and Zn, while Si application enhanced the uptake of these nutrients and maintained the greater root mass (root length) under Al stress in sorghum plants (Galvez and Clark, 1991). Recently, Singh et al. (2011) reported the decrease in Mg and Zn contents under Al stress, while Si supplementation alleviated the decrease in Mg and Zn contents. These results may be explained on the basis of Si and Al interaction in soil that ultimately leads to the formation of subcolloidal and inert aluminosilicates, which finally declined the Al levels from the soil (Guntzer et al., 2012; Wu et al., 2013).

19.5 SILICON PROMOTES THE ANTIOXIDANT DEFENSE SYSTEM OF PLANTS UNDER ALUMINUM TOXICITY

A negative effect of stress on plants is the excess generation of ROS, which disturbs several metabolic processes and ultimately causes damage to proteins, lipids, carbohydrates, and DNA. This state in plants is generally termed as *oxidative damage* (Gill and Tuteja, 2010). The production of ROS under Al stress has also been reported in many plants, which is a major growth-limiting factor (Ryan et al., 1993; Wang et al., 2004; Kováčik et al., 2008). The ROS include singlet oxygen ($_1O^2$), superoxide radical $\left(O_2^{\cdot-}\right)$, hydroxyl radical ($^{\cdot}OH$), and toxic hydrogen peroxide (H_2O_2) molecules. To combat these, plants have developed a number of nonenzymatic and enzymatic defense systems to limit the oxidative damage (Parida et al., 2004; Kováčik et al., 2008; Triantaphylidès and Havaux, 2009; Gill and Tuteja, 2010). The antioxidant defense system constitutes nonenzymatic antioxidants, that is, GSH, ascorbic acid (AsA), a-tocopherol, proline, and carotenoids, and enzymatic antioxidants, that is, superoxide dismutase (SOD, EC 1.15.1.1), catalase (CAT, EC 1.11.1.6), and peroxidase (POD, EC 1.11.1.7) (Farooq et al., 2013). SOD catalyses the toxic $O_2^{\cdot-}$ radicals and results in formation of H_2O_2, which is subsequently converted to H_2O by POD and CAT (Triantaphylidès and Havaux, 2009). So, the antioxidant systems play an important role in protecting plants from oxidative damage (Shi et al., 2005). Al^{3+} can also bind with lipid in the plasma membrane, causing rigidification of the membrane (Akeson et al., 1989). Aluminum also enhances lipid peroxidation (Yamamoto et al., 2001) and the activities of antioxidant enzymes (SOD, CAT, POD, and GR) (Kuo and Kao, 2003). A number of studies have demonstrated that Si improved the antioxidant defense system to scavenge ROS under abiotic stress conditions and thus attenuates the oxidative stress (Liang et al., 2003; Song et al., 2011; Khoshgoftarmanesh et al., 2014; Shi et al., 2014). In addition to enzymatic antioxidants, Si has also been reported to increase the levels of non-enzymatic antioxidants like reduced glutathione, polyamins, and phenolic compounds (chlorogenic and caffeic acids), which also protected the plants under abiotic stress (Liu and Xu, 2007; Maksimović et al., 2007; Saqib et al., 2008; Pei et al., 2010). Furthermore, Si supplementation under abiotic stress was correlated with decreased levels of ROS, causing oxidative damage (i.e., H_2O_2 and metabolites resulting from lipid peroxidation [MDA]) (Zhu et al., 2004; Zhu and Gong, 2014). Shahnaz et al. (2011) reported that Si induced an increase in phenolic compounds and proline contents and a decline in MDA accumulation, giving tolerance to *Borago officinalis* under Al stress. Similar results was observed in *Arachis hypogaea* L. upon Si addition, which decreases the MDA content and membrane damage under Al stress by increasing the activities of CAT, SOD, and POD (Shen et al., 2014). The above-mentioned observation of Si signaling–mediated enhancement in the antioxidant defense system is shown in Figure 19.1.

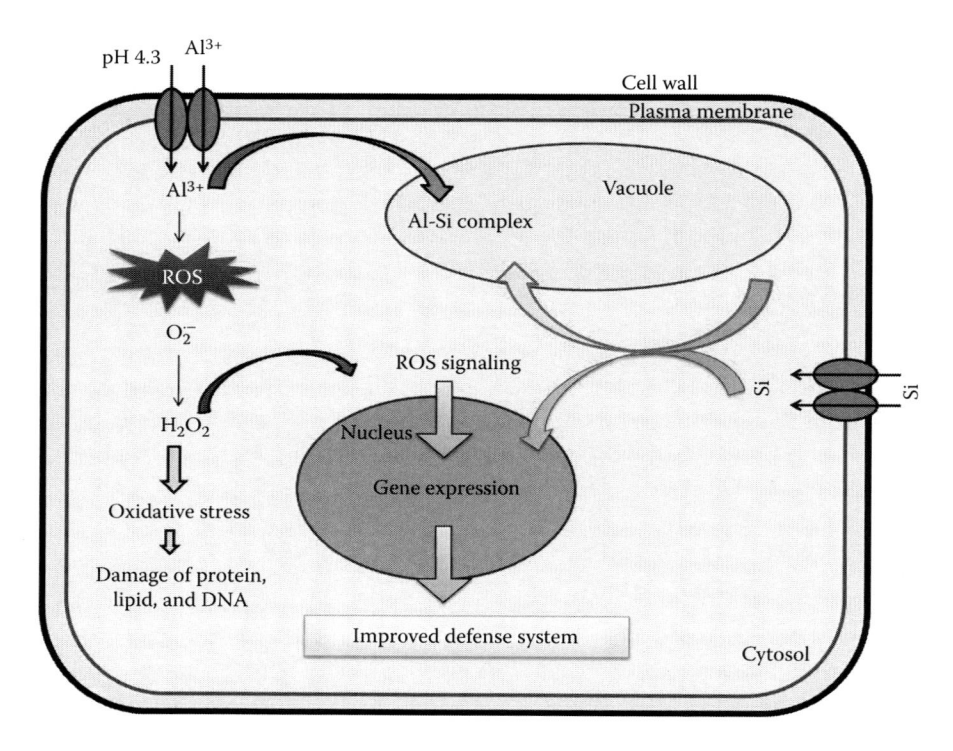

FIGURE 19.1 The schematic representation of Al³⁺-induced oxidative damage that ultimately damages the protein, lipids, and DNA in plants. Si-mediated signaling improved the defense system to protect the plants.

19.6 ROLE OF SILICON IN GENE EXPRESSION UNDER ALUMINUM TOXICITY

Aluminum has the capability to bind with DNA, and therefore it can alter the expressions of genes in plants (Matsumoto et al., 1996; Vitorello et al., 2005). Plants are equipped with an array of cellular mechanisms to mitigate the Al toxicity. Many plants also have a genetically complex trait, where different tolerance mechanisms for Al toxicity coexist. Several studies have reported a number of genes responsible for Al tolerance. Simoes et al. (2012) reported many membrane transporter genes, i.e., ALMT, MATE, and ABC families, and functional homologs of these genes were varied from species to species. For example, transcription factor 1, *ART1*, is a rice homologue of *AtSTOP1*, which controlled the expression of *STAR1* and *STAR2* (Yamaji et al., 2009), *Nrat1* (Xia et al., 2010), *OsFRDL4* (Yokosho et al., 2010), and *OsALS1* (Huang et al., 2012) genes related to Al tolerance in rice plants. Similarly, Fujii et al. (2012) observed the role of *HvAACT1* gene that enhanced citrate exudation, which accordingly improved Al tolerance in barley plants. Recently, Arenhart et al. (2013) said that rice is considered the most tolerant to aluminum among cereal crops. They suggested that the ASR5 protein acts as a transcription factor to regulate the expression of different genes that collectively protect rice cells from Al-induced stress responses. In recent years, silicon has been considered an important element in the signaling and regulation of gene expression related to increases in stress tolerance. There are many studies that prove this: Si application results in enhanced SOD activity under Cu toxicity and is linked with the altered gene expression in *Arabidopsis* plants (Li et al., 2008; Khandekar and Leisner, 2011). But there is no evidence that proved that these effects are Si mediated. So a study in this area is needed to find the relation between Al/Si interaction and their gene expression.

19.7 CONCLUSION AND FUTURE PERSPECTIVE

Al toxicity is a major factor limiting crop production in acid soils. Al^{3+} inhibits root growth and nutrient uptake and can generate ROS that cause oxidative damage to plants. Supplementation of Si has been reported to ameliorate Al by enhancing the nutrient uptake and antioxidative defense system of plants. However, other mechanisms for an alleviative effect of Si have also been proposed, including codeposition of Al with Si in the plant and action in the cytoplasm. The alleviative effect of Si on Al toxicity varies with plant species, probably due to the difference in Al tolerance or differences in the mechanisms involved. In addition to this, there is a big gap in Si-induced alleviation underlying molecular mechanism in plants under abiotic stress especially to Al. So there is a need for studies in this area to find the relationship between Al/Si interaction and their gene expression.

REFERENCES

Akeson M, Munns DN, Burau RG. (1989). Adsorption of Al^{3+} to phosphotidylcholine vesicles. *Biochem Biophys Acta* 986: 33–40.

Arenhart RA, De Lima JC, Pedron M, Carvalho FEL, Da Silveira JAG, Rosa SB, Caverzan A, Andrade CMB, Schünemann M, Margis R, Margis-Pinheiro M. (2013). Involvement of ASR genes in aluminium tolerance mechanisms in rice. *Plant Cell Environ* 36: 52–67.

Barbosa MAM, da Silva MHL, Viana GDM, Ferreira TR, de Carvalho Souza CLF, Lobato EMSG, da Silva Lobato AK. (2015). Beneficial repercussion of silicon (Si) application on photosynthetic pigments in maize plants. *Aust J Crop Sci* 9(11): 1113.

Barcelo J, Guevara P, Poschenrieder C. (1993). Silicon amelioration of aluminum toxicity in teosinte (*Zea mays* L. ssp. *mexicana*). *Plant Soil* 154:249–255.

Baylis AD, Gragopoulou C, Davidson KJ, Birchall JD. (1994). Effects of silicon on the toxicity of aluminium to soybean. *Commun Soil Sci Plant Anal* 25(5–6): 537–546.

Bian R, Chen D, Liu X et al. (2013). Biochar soil amendment as a solution to prevent Cd-tainted rice from China: Results from a cross-site field experiment. *Ecol Eng* 58, 378–383.

Bohnert HJ, Nelson DE, Jensen RG. (1995). Adaptations to environmental stresses. *Plant Cell* 7 (7), 1099.

Brunner I, Sperisen C. (2013). Aluminum exclusion and aluminum tolerance in woody plants. *Front Plant Sci* 4: 172.

Brunner I, Zimmermann S, Zingg A, Blaser P. (2004). Wood-ash recycling affects forest soil and tree fine-root chemistry and reverses soil acidification. *Plant Soil* 267: 61–71.

Chen ZC, Zhao XQ, Shen RF. (2010). The alleviating effect of ammonium on aluminum toxicity in *Lespedeza bicolor* results in decreased aluminum-induced malate secretion from roots compared with nitrate. *Plant Soil* 337(1–2): 389–398.

Cocker KM, Hodson MJ, Evans DE, Sangster AG. (1997). Interaction between silicon and aluminium in Triticum aestivum L. (cv. Celtic). *Isr J Plant Sci* 45, 285–292.

Cronan CS, Grigal DF. (1995). Use of calcium/aluminum ratios as indicators of stress in forest ecosystems. *J Environ Qual* 24: 209–226.

De Micco V, Aronne G. (2012). Morpho-anatomical traits for plant adaptation to drought. In *Plant Responses to Drought Stress*. Springer, Berlin, pp. 37–61.

Deng H, Ye ZH, Wong MH. (2009). Lead, zinc and iron (Fe^{2+}) tolerances in wetland plants and relation to root anatomy and spatial pattern of ROL. *Environ Exp Bot* 65: 353–362.

Epstein E. (1994). The anomaly of silicon in plant biology. *Proc Natl Acad Sci USA* 91: 11–17.

Farooq MA, Ali S, Hameed A, Ishaque W, Mahmood K, Iqbal Z. (2013). Alleviation of cadmium toxicity by silicon is related to elevated photosynthesis, antioxidant enzymes, suppressed cadmium uptake and oxidative stress in cotton. *Ecotoxicol Environ Saf* 96: 242–249.

Fujii M, Yokosho K, Yamaji N, Saisho D, Yamane M, Takahashi H et al. (2012). Acquisition of aluminium tolerance by modification of a single gene in barley. *Nat Commun* 3: 713.

Galvez L, Clark RB. (1991). Effects of silicon on growth and mineral composition of sorghum (*Sorghum bicolor*) grown with toxic levels of aluminium. In Wright RJ, Baligar PC, Murrmann RP (eds.), *Plant-Soil Interactions at Low pH*. Kluwer Academic, Dordrecht, The Netherlands, pp. 815–823.

Galvez L, Clark RB, Gourley LM, Maranville JW. (1987). Silicon interactions with manganese and aluminum toxicity in sorghum. *J Plant Nutr* 10(9–16): 1139–1147.

Gill SS, Tuteja N. (2010). Reactive oxygen species and antioxidant machinery in abiotic stress tolerance in crop plants. *Plant Physiol Biochem* 48: 909–930.

Guntzer F, Keller C, Meunier J-D. (2012). Benefits of plant silicon for crops: A review. *Agron Sustain Dev* 32: 201–213.

Hammond KE, Evans DE, Hodson MJ. (1995). Aluminium/silicon interactions in barley (*Hordeum vulgare* L.) seedlings. *Plant Soil* 173(1): 89–95.

Hasanuzzaman M, Nahar K, Fujita M. (2013). Plant response to salt stress and role of exogenous protectants to mitigate salt-induced damages. In: Ahmad, P., Azooz, M.M., Prasad, M.N.V. (Eds.), *Ecophysiology and Responses of Plants under Salt Stress*, Springer-Verlag, New York, pp. 25–87.

Hernandez-Apaolaza L. (2014). Can silicon partially alleviate micronutrient deficiency in plants? A review. *Planta* 240: 447–458.

Hiradate S, Taniguchi S, Sakurai K. (1998). Aluminum speciation in aluminum-silica solutions and potassium chloride extracts of acidic soils. *Soil Sci Soc Am J* 62: 630–636.

Hodson MJ, Evans DE. (1995). Aluminium/silicon interactions in higher plants. *J Exp Bot* 46(2): 161–171.

Hodson MJ, Sangster AG. (1993). The interaction between silicon and aluminum in *Sorghum bicolor* L. Moench: Growth analysis and X-ray microanalysis. *Ann Bot* 72: 389–400.

Hodson MJ, Sangster AG. (1999). Aluminium/silicon interactions in conifers. *J Inorg Biochem* 76: 89–98.

Horst WJ, Wang Y, Eticha D. (2010). The role of the root apoplast in aluminium-induced inhibition of root elongation and in aluminium resistance of plants: A review. *Ann Bot* 106(1): 185–197.

Huang CF, Yamaji N, Chen Z, Ma JF. (2012). A tonoplast-localized half-size ABC transporter is required for internal detoxification of aluminum in rice. *Plant J* 69(5): 857–867.

Inostroza-Blancheteau C, Rengel Z, Alberdi M, de la Luz Mora M, Aquea F, Arce-Johnson P, Reyes-Díaz M. (2012). Molecular and physiological strategies to increase aluminum resistance in plants. *Mol Biol Rep* 39(3): 2069–2079.

Khandekar S, Leisner S. (2011). Soluble silicon modulates expression of *Arabidopsis thaliana* genes involved in copper stress. *J Plant Physiol* 168: 699–705.

Khoshgoftarmanesh AH, Khodarahmi S, Haghighi M. (2014). Effect of silicon nutrition on lipid peroxidation and antioxidant response of cucumber plants exposed to salinity stress. *Arch Agron Soil Sci* 60: 639–653.

Kidd PS, Llugany M, Poschenrieder C, Gunse B, Barcelo J. (2001). The role of root exudates in aluminium resistance and silicon-induced amelioration of aluminium toxicity in three varieties of maize (*Zea mays* L.). *J Exp Bot* 52: 1339–1352.

Kinraide TB. (1997). Reconsidering the rhizotoxicity of hydroxyl, sulphate, and fluoride complexes of aluminium. *J Exp Bot* 48: 1115–1124.

Kinraide TB. (2003). Toxicity factors in acidic forest soils: Attempts to evaluate separately the toxic effects of excessive Al^{3+} and H^+ and insufficient Ca^{2+} and Mg^{2+} upon root elongation. *Eur J Soil Sci* 54: 323–333.

Kochian LV, Hoekenga OA, Piñeros MA. (2004). How do crop plants tolerate acid soils? Mechanisms of aluminum tolerance and phosphorous efficiency. *Annu Rev Plant Biol* 55: 459–493.

Kováčik J, Grúz J, Bačkor M, Tomko J, Strnad M, Repčák M. (2008). Phenolic compounds composition and physiological attributes of *Matricaria chamomilla* grown in copper excess. *Environ Exp Bot* 62: 145–152.

Krstic D, Djalovic I, Nikezic D, Bjelic D. (2012). Aluminium in acid soils: Chemistry, toxicity and impact on maize plants. *Food Prod Approaches Challenges Tasks* 231–242.

Kuo MC, Kao CH. (2003). Aluminium effects on lipid peroxidation and antioxidative enzyme activities in rice leaves. *Biol Plant* 46: 149–152.

Li J, Frantz J, Leisner S. (2008). Alleviation of copper toxicity in *Arabidopsis thaliana* by silicon addition to hydroponic solutions. *J Am Soc Hortic Sci* 133: 670–677.

Li YC, Alva AK, Sumner ME, (1989). Response of cotton cultivars to aluminum in solutions with varying silicon concentrations. *J. Plant Nutr* 12: 881–892.

Liang Y, Yang C, Shi H. (2001). Effects of silicon on growth and mineral composition of barley grown under toxic levels of aluminum. *J Plant Nutr* 24(2): 229–243.

Liang YC, Chen Q, Liu Q, Zhang WH, Ding RX. (2003). Exogenous silicon (Si) increases antioxidant enzyme activity and reduces lipid peroxidation in roots of salt-stressed barley (*Hordeum vulgare* L.). *J Plant Physiol* 160: 1157–1164.

Liu YX, Xu XZ. (2007). Effects of silicon on polyamine types and forms in leaf of *Zizyphus jujuba* cv. Jinsi-xiaozao under salt stress. *J Nanjing For Univ* 31: 27–32.

Ma JF, Sasaki M, Matsumoto H. (1997). Al-induced inhibition of root elongation in corn, *Zea mays* L. is overcome by Si addition. *Plant Soil* 188: 171–176.

Maksimovíc I, Kastori R, Krstic L, Lukovic J. (2007). Steady presence of cadmium and nickel affects root anatomy, accumulation and distribution of essential ions in maize seedlings. *Biol Plant* 51: 589–592.

Matichenkov VV, Calvert DV. (2002). Silicon as a beneficial element for sugarcane. *J Am Soc Sugarcane Technol* 22: 21–29.

Matsumoto H. (2000). Cell biology of aluminum toxicity and tolerance in higher plants. *Int Rev Cytol* 200: 1–46.

Matsumoto H, Senoo Y, Kasai M, Maeshima M. (1996). Response of plant root to aluminium stress: Analysis of the inhibition of the root elongation and changes in membrane function. *J Plant Res* 109: 99–105.

Mossor-Pietraszewska T. (2001). Effect of aluminium on plant growth and metabolism. *Acta Biochim Pol Engl Ed* 48(3): 673–686.

Nhan PP, Hai NT. (2013). Amelioration of aluminum toxicity on OM4900 rice seedlings by sodium silicate. *Afr J Plant Sci* 30:208–212.

Panda SK, Baluska F, Matsumoto H. (2009). Al stress signaling in plants. *Plant Signal Behav* 7: 592–597.

Parida AK, Das AB, Mohanty P. (2004). Defense potentials to NaCl in a mangrove, *Bruguiera parviflora*: Differential changes of isoforms of some antioxidative enzymes. *J Plant Physiol* 161: 531–542.

Pei ZF, Ming DF, Liu D, Wan GL, Geng XX, Gong HJ et al. (2010). Silicon improves the tolerance to water deficit stress induced by polyethylene glycol in wheat (*Triticum aestivum* L.) seedlings. *J Plant Growth Regul* 29: 106–115.

Richter AK, Walthert L, Frossard E, Brunner I. (2007). Does low soil base saturation affect the vitality of fine roots of European beech? *Plant Soil* 298: 69–79.

Rout GR, Samantaray S, Das P. (2001). Aluminium toxicity in plants: A review. *Agronomie* 21(1): 3–21.

Ryan PR, DiTomaso JM, Kochian LV. (1993). Aluminium toxicity in roots: An investigation of spatial sensitivity and the role of the root cap. *J Exp Bot* 44: 437–446.

Ryan PR, Tyerman SD, Sasaki T, Furuichi T, Yamamoto Y, Zhang WH, Delhaize E. (2011). The identification of aluminium-resistance genes provides opportunities for enhancing crop production on acid soils. *J Exp Bot* 62(1): 9–20.

Sadgrove N. (2006). Nutrient and moisture economics in diatomaceous earth amended growth media. Southern Cross University.

Sairam RK, Tyagi A. (2004). Physiology and molecular biology of salinity stress tolerance in plants. *Curr Sci* 86:407–421.

Saqib M, Zörb C, Schubert S. (2008). Silicon-mediated improvement in the salt resistance of wheat (*Triticum aestivum*) results from increased sodium exclusion and resistance to oxidative stress. *Funct Plant Biol* 35: 633–639.

Shahnaz G, Shekoofeh E, Kourosh D, Moohamadbagher B. (2011). Interactive effects of silicon and aluminum on the malondialdehyde (MDA), proline, protein and phenolic compounds in *Borago officinalis* L. *J Med Plants Res* 5: 5818–5827.

Shen X, Xiao X, Dong Z, Chen Y. (2014). Silicon effects on antioxidative enzymes and lipid peroxidation in leaves and roots of peanut under aluminum stress. *Acta Physiol Plant* 36: 3063–3069.

Shi XH, Zhang CC, Wang H, Zhang FS. (2005). Effect of Si on the distribution of Cd in rice seedlings. *Plant Soil* 272: 53–60.

Shi Y, Zhang Y, Yao H, Wu J, Sun H, Gong H. (2014). Silicon improves seed germination and alleviates oxidative stress of bud seedlings in tomato under water deficit stress. *Plant Physiol Biochem* 78: 27–36.

Silva C, Martínez V, Carvajal M. (2008). Osmotic versus toxic effects of NaCl on pepper plants. *Biol Plant* 52: 72–79.

Silva IR, Smytha TJ, Rapera CD, Carterb TE, Rufty TW. (2001). Differential aluminum tolerance in soybean: An evaluation of the role of organic acids. *Physiol Plant* 112:200–210.

Simões CC, Melo JO, Magalhaes JV, Guimarães CT. (2012). Genetic and molecular mechanisms of aluminum tolerance in plants. *Genet Mol Res* 11: 1949–1957.

Singh VP, Tripathi DK, Kumar D, Chauhan DK. (2011). Influence of exogenous silicon addition on aluminium tolerance in rice seedlings. *Biol Trace Elem Res* 144(1–3): 1260–1274.

Sivaguru M, Ezaki B, He ZH, Tong H, Osawa H, Baluška F et al. (2003). Aluminum-induced gene expression and protein localization of a cell wall-associated receptor kinase in *Arabidopsis*. *Plant Physiol* 132(4): 2256–2266.

Song AL, Li P, Li ZJ, Fan FL, Nikolic M, Liang YC. (2011). The alleviation of zinc toxicity by silicon is related to zinc transport and antioxidative reactions in rice. *Plant Soil* 344: 319–333.

Sverdrup H, Warfvinge P. (1993). *The Effect of Soil Acidification on the Growth of Trees, Grass, and Herbs as Expressed by the (Ca + Mg + K)/Al Ratio*. Vol. 2, Report in Ecology and Environmental Engineering. Lund University, Lund.

Triantaphylidès C, Havaux M. (2009). Singlet oxygen in plants: Production, detoxification and signaling. *Trends Plant Sci* 14: 219–228.

Tripathi DK, Chauhan DK, Kumar D, Tiwari SP. (2012a). Morphology, diversity and frequency based exploration of phytoliths in *Pennisetum typhoides* Rich. *Natl Acad Sci Lett* 35(4): 285–289.

Tripathi DK, Kumar R, Chauhan DK, Rai AK, Bicanic D. (2011). Laser-induced breakdown spectroscopy for the study of the pattern of silicon deposition in leaves of *Saccharum* species. *Instrum Sci Technol* 39: 510–521.

Tripathi DK, Kumar R, Pathak AK, Chauhan DK, Rai AK. (2012b). Laser-induced breakdown spectroscopy and phytolith analysis: An approach to study the deposition and distribution pattern of silicon in different parts of wheat (*Triticum aestivum* L.) plant. *Agric Res* 1(4): 352–361.

Tripathi DK, Mishra S, Chauhan DK, Tiwari SP, Kumar C. (2013). Typological and frequency based study of opaline silica (phytolith) deposition in two common Indian *Sorghum* L. species. *Proc Natl Acad Sci India Sect B Biol Sci* 83(1): 97–104.

Tripathi DK, Pathak AK, Chauhan DK, Dubey NK, Rai AK, Prasad R. (2016). LIB spectroscopic and biochemical analysis to characterize lead toxicity alleviative nature of silicon in wheat (*Triticum aestivum* L.) seedlings. *Photochem Photobiol B* 154: 89–98.

Tripathi DK, Singh S, Singh S, Chauhan DK, Dubey NK, Prasad R. (2015a). Micronutrients and their diverse role in agricultural crops: Advances and future prospective. *Acta Physiol Plant* 37(139): 1–14.

Tripathi DK, Singh VP, Gangwar S, Prasad SM, Maurya JN, Chauhan DK. (2014). Role of silicon in enrichment of plant nutrients and protection from biotic and abiotic stresses. In *Improvement of Crops in the Era of Climatic Changes*. Springer, New York, pp. 39–56.

Tripathi DK, Singh VP, Kumar D, Chauhan DK. (2012c). Impact of exogenous silicon addition on chromium uptake, growth, mineral elements, oxidative stress, antioxidant capacity, and leaf and root structures in rice seedlings exposed to hexavalent chromium. *Acta Physiol Plant* 34(1): 279–289.

Tripathi DK, Singh VP, Kumar D, Chauhan DK. (2012d). Rice seedlings under cadmium stress: Effect of silicon on growth, cadmium uptake, oxidative stress, antioxidant capacity, and root and leaf structures. *Chem Ecol* 28: 281–291.

Tripathi DK, Singh VP, Prasad SM, Chauhan DK, Dubey NK. (2015b). Silicon nanoparticles (SiNp) alleviate chromium (VI) phytotoxicity in *Pisum sativum* (L.) seedlings. *Plant Physiol Biochem* 96: 189–198.

Tripathi DK, Singh VP, Prasad SM, Chauhan DK, Dubey NK, Rai AK. (2015c). Silicon-mediated alleviation of Cr (VI) toxicity in wheat seedlings as evidenced by chlorophyll florescence, laser induced breakdown spectroscopy and anatomical changes. *Ecotoxicol Environ Saf* 113: 133–144.

Vaculík M, Landberg T, Greger M, Luxová M, Stoláriková M, Lux A. (2012). Silicon modifies root anatomy, and uptake and subcellular distribution of cadmium in young maize plants. *Ann Bot* 110: 433–443.

Vanguelova EI, Hirano Y, Eldhuset TD, Sas-Paszt L, Bakker MR, Püttsepp Ü, Brunner I, Lõhmus K, Godbold D. (2007). Tree fine root Ca/Al molar ratio—Indicator of Al and acidity stress. *Plant Biosyst* 141: 460–480.

Vashegyi A, Zsoldos F, Pécsváradi A, Bona L. (2002). Aluminium/silicon interactions in cereal seedlings. *Acta Biol Szegediensis* 46(3–4): 129–130.

Vázquez MD, Poschenrieder C, Corrales I, Barceló J. (1999). Change in apoplastic aluminum during the initial growth response to aluminum by roots of a tolerant maize variety. *Plant Physiol* 119(2): 435–444.

Vitorello VA, Capald FR, Stefanuto VA. (2005). Recent advances in aluminium toxicity and resistance in higher plants. *Braz J Plant Physiol* 17: 129–143.

von Uexküll HR, Mutert E. (1995). Global extent, development and economic impact of acid soils. *Plant Soil* 171: 1–15.

Wang YX, Stass A, Horst WJ. (2004). Apoplastic binding of aluminum is involved in silicon-induced amelioration of aluminum. *Plant Physiol* 136: 3762–3770.

Wu JW, Shi Y, Zhu YX, Wang YC, Gong HJ. (2013). Mechanisms of enhanced heavy metal tolerance in plants by silicon: A review. *Pedosphere* 23: 815–825.

Xia J, Yamaji N, Kasai T, Ma JF. (2010). Plasma membrane-localized transporter for aluminum in rice. *Proc Natl Acad Sci USA* 107: 18381–18385.

Yamaji N, Huang CF, Nagao S, Yano M, Sato Y, Nagamura Y, Ma JF. (2009). A zinc finger transcription factor ART1 regulates multiple genes implicated in aluminum tolerance in rice. *Plant Cell* 21: 3339–3349.

Yamamoto Y, Kobayashi Y, Devi SR, Rikiishi S, Matsumoto H. (2002). Aluminum toxicity is associated with mitochondrial dysfunction and the production of reactive oxygen species in plant cells. *Plant Physiol* 128(1): 63–72.

Yamamoto Y, Kobayashi Y, Matsumoto H. (2001). Lipid peroxidation is an early symptom triggered by aluminum, but not the primary cause of elongation inhibition in pea roots. *Plant Physiol* 125: 199–208.

Yokosho K, Yamaji N, Ma JF. (2010). Isolation and characterization of two MATE genes in rye. *Funct Plant Biol* 37: 296–303.

Zhu Y, Gong H. (2014). Beneficial effects of silicon on salt and drought tolerance in plants. *Agron Sustain Dev* 34: 455–472.

Zhu Z, Wei G, Li J, Qian Q, Yu J. (2004). Silicon alleviates salt stress and increases antioxidant enzymes activity in leaves of salt-stressed cucumber (*Cucumis sativus* L.). *Plant Sci* 167: 527–533.

Index

Page numbers followed by f and t indicate figures and tables, respectively.